CONTINUOUS ADVANCES IN QCD 2002

ARKADYFEST

Honoring the 60th Birthday of Arkady Vainshtein

Proceedings of the Conference on

CONTINUOUS ADVANCES IN QCD 2002

ARKADYFEST
Honoring the 60th Birthday of Arkady Vainshtein

William I. Fine Theoretical Physics Institute
University of Minnesota, Minneapolis, USA
17 – 23 May 2002

Editors

K. A. Olive, M. A. Shifman and M. B. Voloshin

University of Minnesota, USA

World Scientific
New Jersey • London • Singapore • Hong Kong

Published by

World Scientific Publishing Co. Pte. Ltd.

5 Toh Tuck Link, Singapore 596224

USA office: Suite 202, 1060 Main Street, River Edge, NJ 07661

UK office: 57 Shelton Street, Covent Garden, London WC2H 9HE

British Library Cataloguing-in-Publication Data
A catalogue record for this book is available from the British Library.

ISBN 981-238-215-1

This book is printed on acid-free paper.

Printed in Singapore by Uto-Print

FOREWORD

The Theoretical Physics Institute hosted a Symposium and Workshop "Continuous Advances in QCD 2002/Arkadyfest," to honor the 60th birthday of Arkady Vainshtein from Friday, May 17 through Thursday, May 23, 2002.

This event was an occasion to exchange the latest ideas in QCD and gauge theories at strong coupling at large. We also looked back on the history of the subjects in which Arkady played such a central role: applications of PCAC, penguins, invisible axions, QCD sum rules, exact beta functions, condensates in supersymmetry, powerful heavy quark expansions, and new anomalies in 2D SUSY theories. The current status of these subjects was summarized in several excellent presentations that also outlined a historical perspective.

The Symposium (May 17-18) consisted exclusively of review talks presenting the big picture regarding QCD and related areas, with an emphasis on topics in which Arkady's contributions are most plentiful. The format of the Workshop (May 20-23) was chosen to create a stimulating atmosphere for breakthrough discussions and collaborations. The morning sessions consisted of two review talks presenting the most promising research topics of today. The afternoon sessions were reserved for shorter original presentations and discussions. The cosmological constant problem in extra-dimension theories, supersymmetric monopoles, solitons and confinement, AdS/CFT correspondence, and high density QCD were just a few of the hot topics at the focus of discussions.

Many of you may know that the creation of the Theoretical Physics Institute during the 1980s is directly connected to the life-long interest in the physical sciences of Bill Fine. We are all deeply saddened by his recent death which ocurred on May, 18, 2002 during the symposium. Without his gracious financial donation the Institute would not have become a reality. His generosity is never far from our minds and it is with forethought that we have chosen to honor Mr. Fine by officially changing our name. As of January 9, 2002, we are known as the William I. Fine Theoretical Physics Institute.

The organizers would like to take this opportunity to thank Sally Menefee and Catharine Grahm for their dedicated work on numerous aspects of organization before, during, and after the workshop.

K.A. Olive

M.A. Shifman

M.B. Voloshin

SYMPOSIUM AND WORKSHOP
"CONTINUOUS ADVANCES IN QCD 2002/ARKADYFEST"
honoring the 60th birthday of Prof. Arkady Vainshtein

CONTENTS

SECTION 1.
PERTURBATIVE AND
NONPERTURBATIVE QCD

PENGUINS 2002: PENGUINS IN $K \rightarrow \pi\pi$ DECAYS

JOHAN BIJNENS

Department of Theoretical Physics 2, Lund University,
Sölvegatan 14A, S22362 Lund, Sweden
E-mail: bijnens@thep.lu.se

This talk contains a short overview of the history of the interplay of the weak and the strong interaction and CP-violation. It describes the phenomenology and the basic physics mechanisms involved in the Standard Model calculations of $K \rightarrow \pi\pi$ decays with an emphasis on the evaluation of Penguin operator matrix-elements.

1. Introduction

In this conference in honour of Arkady Vainshtein's 60th birthday, a discussion of the present state of the art in analytical calculations of relevance for Kaon decays is very appropriate given Arkady's large contributions to the field. He has summarised his own contributions on the occasion of accepting the Sakurai Prize.[1] This also contains the story of how Penguins, at least the diagram variety, got their name. In Fig. 1 I show what a real (Linux) Penguin looks like and the diagram in a "Penguinized" version. This talk could easily have had other titles, examples are "QCD and Weak Interactions of Light Quarks" or "Penguins and Other Graphs." In fact I have left out many manifestations of Penguin diagrams. In particular I do not cover the importance in B decays where Penguins were first experimentally verified via $B \rightarrow K^{(*)}\gamma$, but are at present more considered a nuisance and often referred to as "Penguin Pollution." Penguins also play a major role in other Kaon decays, reviews of rare decays where pointers to the literature can be found are Refs [2,3,4].

In Sect. 2 I give a very short historical overview. Sects 3 and 4 discuss the main physics issues and present the relevant phenomenology of the $\Delta I = 1/2$ rule and the CP-violating quantities ε and ε'/ε. The underlying Standard Model diagrams responsible for CP-violation are shown there as well. The more challenging part is to actually evaluate these diagrams in the presence of the strong interaction. We can distinguish several regimes of momenta which have to be treated using different methods. An overview

Figure 1. The comparison between a (Linux) Penguin and the Penguin diagram. (Linux Penguin from Neal Tucker, (http://www.isc.tamu.edu/~lewing/linux/).)

is given in Sect. 5 where also the short-distance part is discussed. The more difficult long-distance part has a long history and some approaches are mentioned, but only my favourite method, the X-boson or fictitious Higgs exchange, is described in more detail in Sect. 6 where I also present results for the main quantities. For two particular matrix-elements, those of the electroweak Penguins Q_7 and Q_8, a dispersive analysis allows to evaluate these in the chiral limit from experimental data. This is discussed in Sect. 7. We summarise our results in the conclusions and compare with the original hopes from Arkady and his collaborators. This talk is to a large extent a shorter version of the review [5].

2. A short historical overview

The weak interaction was discovered in 1896 by Becquerel when he discovered spontaneous radioactivity. The next step towards a more fundamental study of the weak interaction was taken in the 1930s when the neutron was discovered and its β-decay studied in detail. The fact that the proton and electron energies did not add up to the total energy corresponding to the mass of the neutron, made Pauli suggest the neutrino as a solution. Fermi then incorporated it in the first full fledged theory of the weak interaction, the famous Fermi four-fermion [6] interaction.

$$\mathcal{L}_{\text{Fermi}} = \frac{G_F}{\sqrt{2}} \left[\bar{p}\gamma_\mu \left(1 - \gamma_5\right) n \right] \left[\bar{e}\gamma^\mu \left(1 - \gamma_5\right) \nu \right]. \tag{1}$$

The first fully nonhadronic weak interaction came after world-war two with the muon discovery and the study of its β-decay. The analogous Lagrangian to Eq. (1) was soon written down. At that point T.D. Lee and C.N. Yang [7] realized that there was no evidence that parity was conserved in the weak interaction. This quickly led to a search for parity violation both in nuclear decays [8] and in the decay chain $\pi^+ \to \mu^+ \nu_\mu \to e^+ \nu_e \bar{\nu}_\mu \nu_\mu$.[9] Parity violation was duly observed in both cases. These experiments and others led to the final form of the Fermi Lagrangian given in Eq. (1).[10,11]

During the 1950s steadily more particles were discovered providing many puzzles. These were solved by the introduction of strangeness,[12,13] of what is now known as the K_L and the K_S [14] and the "eightfold way" of classifying the hadrons into symmetry-multiplets.[15]

Subsequently Cabibbo realized that the weak interactions of the strange particles were very similar to those of the nonstrange particles.[16] He proposed that the weak interactions of hadrons occurred through a current which was a mixture of the strange and non-strange currents with a mixing angle now universally known as the Cabibbo angle. The hadron symmetry group led to the introduction of quarks [17] as a means of organising which $SU(3)_V$ multiplets were present in the spectrum.

In the same time period the Kaons provided another surprise. Measurements at Brookhaven [18] indicated that the long-lived state, the K_L, did occasionally decay to two pions in the final state as well, showing that CP was violated. Since the CP-violation was small, explanations could be sought at many scales, an early phenomenological analysis can be found in Ref. [19], but as the socalled superweak model [20] showed, the scale of the interaction involved in CP-violation could be much higher.

The standard model for the weak and electromagnetic interactions of leptons was introduced in the same period. The Fermi theory is nonrenormalisable. Alternatives based on Yang-Mills [21] theories had been proposed by Glashow [22] but struggled with the problem of having massless gauge bosons. This was solved by the introduction of the Higgs mechanism by Weinberg and Salam. The model could be extended to include the weak interactions of hadrons by adding quarks in doublets, similar to the way the leptons were included. One problem this produced was that loop-diagrams provided a much too high probability for the decay $K_L \to \mu^+ \mu^-$ compared to the experimental limits. These socalled flavour changing neutral currents (FCNC) needed to be suppressed. The solution was found in the Glashow-Iliopoulos-Maiani mechanism.[23] A fourth quark, the charm quark, was introduced beyond the up, down and strange quarks. If all the quark masses were equal, the dangerous loop contributions to FCNC processes cancel, the socalled GIM mechanism. This allowed a prediction of the charm quark mass, [24] soon confirmed with the discovery of the J/ψ.

In the mean time, QCD was formulated.[25] The property of asymptotic freedom [26] was established which explained why quarks at short distances could behave as free particles and at the same time at large distances be confined inside hadrons.

The study of Kaon decays still went on, and an already old problem, the $\Delta I = 1/2$ rule saw the first signs of a solution. It was shown [27,28] that the short-distance QCD part of the nonleptonic weak decays provided already an enhancement of the $\Delta I = 1/2$ weak $\Delta S = 1$ transition over the $\Delta I = 3/2$ one. The ITEP group extended first the Gaillard-Lee analysis for the charm mass,[29] but then realized that in addition to the effects that were included in Refs [27,28], there was a new class of diagrams that only contributed to the $\Delta I = 1/2$ transition.[30,31] While, as we will discuss in more detail later, the general class of these contributions, the socalled Penguin-diagrams, is the most likely main cause of the $\Delta I = 1/2$ rule, the short-distance part of them provide only a small enhancement contrary to the original hope. A description of the early history of Penguin diagrams, including the origin of the name, can be found in the 1999 Sakurai Prize lecture of Vainshtein.[1]

Penguin diagrams at short distances provide nevertheless a large amount of physics. The origin of CP-violation was (and partly is) still a mystery. The superweak model explained it, but introduced new physics that had no other predictions. Kobayashi and Maskawa [32] realized that the framework established by Ref. [23] could be extended to three generations. The really new aspect this brings in is that CP-violation could easily be produced at the weak scale and not at the much higher superweak scale. In this Cabibbo-Kobayashi-Maskawa (CKM) scenario, CP-violation comes from the mixed quark-Higgs sector, the Yukawa sector, and is linked with the masses and mixings of the quarks. Other mechanisms at the weak scale also exist, as e.g. an extended Higgs sector.[33]

The inclusion of the CKM mechanism into the calculations for weak decays was done by Gilman and Wise [34,35] which provided the prediction that ε'/ε should be nonzero and of the order of 10^{-3}. Guberina and Peccei [36] confirmed this. This prediction spurred on the experimentalists and after two generations of major experiments, NA48 at CERN and KTeV at Fermilab have now determined this quantity and the qualitative prediction that CP-violation at the weak scale exists is now confirmed. Much stricter tests of this picture will happen at other Kaon experiments as well as in B meson studies.

The K^0-$\overline{K^0}$ mixing has QCD corrections and CP-violating contributions as well. The calculations of these required a proper treatment of box diagrams and inclusions of the effects of the $\Delta S = 1$ interaction squared. This was accomplished at one-loop by Gilman and Wise a few years later.[37,38]

That Penguins had more surprises in store was shown some years later when it was realized that the enhancement originally expected on chiral grounds for the Penguin diagrams [30,31] was present, not for the Penguin diagrams with gluonic intermediate states, but for those with a photon.[39] This contribution was also enhanced in its effects by the $\Delta I = 1/2$ rule. This lowered the expectation for ε'/ε, but it became significant after it was found that the top quark had a very large mass. Flynn and Randall [40] reanalysed the electromagnetic Penguin with a large top quark mass and included also Z^0 exchange. The final effect was that the now rebaptized electroweak Penguins could have a very large contribution that could even cancel the contribution to ε'/ε from gluonic Penguins. This story still continues at present and the cancellation, though not complete, is one of the major impediments to accurate theoretical predictions of ε'/ε.

The first calculation of two-loop effects in the short-distance part was done in Rome [41] in 1981. The value of Λ_{QCD} has risen from values of about 100 MeV to more than 300 MeV. A full calculation of all operators at two loops thus became necessary, taking into account all complexities of higher order QCD. This program was finally accomplished by two independent groups. One in Munich around A. Buras and one in Rome around G. Martinelli.

3. $K \to \pi\pi$ and the $\Delta I = 1/2$ rule

The underlying qualitative difference we want to understand is the $\Delta I = 1/2$ rule. We can try to calculate $K \to \pi\pi$ decays by simple W^+ exchange. For $K^+ \to \pi^+\pi^0$ we can draw the two Feynman diagrams of Fig. 2(a). The W^+-hadron couplings are known from semi-leptonic decays. This approximation agrees with the measured decay within a factor of two.

A much worse result appears when we try the same for $K^0 \to \pi^0\pi^0$. As shown in Fig. 2(b) there is no possibility to draw diagrams similar to those in Fig. 2(a). The needed vertices always violate charge-conservation. So we expect that the neutral decay should be small compared with the ones with charged pions. Well, if we look at the experimental results, we see

$$\Gamma(K^0 \longrightarrow \pi^0\pi^0) = \frac{1}{2}\Gamma(K_S \longrightarrow \pi^0\pi^0) = 2.3 \cdot 10^{-12} \text{ MeV}$$

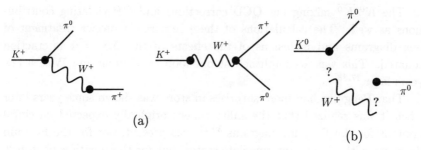

Figure 2. (a) The two naive W^+-exchange diagrams for $K^+ \longrightarrow \pi^+\pi^0$. (b) No simple W^+-exchange diagram is possible for $K^0 \longrightarrow \pi^0\pi^0$.

$$\Gamma(K^+ \longrightarrow \pi^+\pi^0) = 1.1 \cdot 10^{-14} \text{ MeV} \tag{2}$$

So the expected zero one is by far the largest !!!

The same conundrum can be expressed in terms of the isospin amplitudes: [a]

$$A[K^0 \to \pi^0\pi^0] \equiv \sqrt{1/3}A_0 - \sqrt{2/3}\,A_2$$
$$A[K^0 \to \pi^+\pi^-] \equiv \sqrt{1/3}A_0 + \sqrt{1/6}\,A_2$$
$$A[K^+ \to \pi^+\pi^0] \equiv (\sqrt{3}/2)A_2 . \tag{3}$$

The above quoted experimental results can now be rewritten as

$$|A_0/A_2|_{\exp} = 22 \tag{4}$$

while the naive W^+-exchange discussed would give

$$|A_0/A_2|_{\text{naive}} = \sqrt{2} . \tag{5}$$

This discrepancy is known as the problem of the $\Delta I = 1/2$ rule.

Some enhancement comes from final state $\pi\pi$-rescattering. Removing these and higher order effects in the light quark masses one obtains [42,43]

$$|A_0/A_2|_\chi = 17.8 . \tag{6}$$

This changes the discrepancy somewhat but is still different by an order of magnitude from the naive result (5). The difference will have to be explained by pure strong interaction effects and it is a *qualitative* change, not just a quantitative one.

We also use amplitudes without the final state interaction phase:

$$A_I = -ia_I e^{i\delta_I} \tag{7}$$

---

[a]The sign convention is the one used in the work by J. Prades and myself.

for $I = 0, 2$. δ_I is the angular momentum zero, isospin I scattering phase at the Kaon mass.

4. $K \to \pi\pi$, ε, ε'/ε

The K^0, $\overline{K^0}$ states have $\bar{s}d$, $\bar{d}s$ quark content. CP acts on these states as

$$CP|K^0\rangle = -|\overline{K^0}\rangle . \tag{8}$$

We can construct eigenstates with a definite CP transformation:

$$K^0_{1(2)} = \frac{1}{\sqrt{2}} \left(K^0 - (+)\overline{K^0} \right), \qquad CP|K_{1(2)}\rangle = +(-)|K_{1(2)}\rangle . \tag{9}$$

The main decay mode of K^0-like states is $\pi\pi$. A two pion state with charge zero in spin zero is always CP even. Therefore the decay $K_1 \to \pi\pi$ is possible but $K_2 \to \pi\pi$ is *impossible*; $K_2 \to \pi\pi\pi$ is possible. Phase-space for the $\pi\pi$ decay is much larger than for the three-pion final state. Therefore if we start out with a pure K^0 or $\overline{K^0}$ state, the K_2 component in its wavefunction lives much longer than the K_1 component such that after a long time only the K_2 component survives.

In the early sixties, as you see it pays off to do precise experiments, one actually measured [18]

$$\frac{\Gamma(K_L \to \pi^+\pi^-)}{\Gamma(K_L \to \text{all})} = (2 \pm 0.4) \cdot 10^{-3} \neq 0 , \tag{10}$$

showing that *CP is violated*. This leaves us with the questions:

??? Does K_1 turn in to K_2 (mixing or indirect CP violation)?
??? Does K_2 decay directly into $\pi\pi$ (direct CP violation)?

In fact, the answer to both is *YES* and is major qualitative test of the standard model Higgs-fermion sector and the CKM-picture of CP-violation.

The presence of CP-violation means that K_1 and K_2 are not the mass eigenstates, these are

$$K_{S(L)} = \frac{1}{\sqrt{1 + |\tilde{\varepsilon}|^2}} \left(K_{1(2)} + \tilde{\varepsilon}K_{2(1)} \right) . \tag{11}$$

They are not orthogonal since the Hamiltonian is not hermitian.

We define the observables

$$\varepsilon \equiv \frac{A(K_L \to (\pi\pi)_{I=0})}{A(K_S \to (\pi\pi)_{I=0})}$$

$$\varepsilon' = \frac{1}{\sqrt{2}} \left(\frac{A(K_L \to (\pi\pi)_{I=2})}{A(K_S \to (\pi\pi)_{I=0})} - \varepsilon\frac{A(K_S \to (\pi\pi)_{I=2})}{A(K_S \to (\pi\pi)_{I=0})} \right) . \tag{12}$$

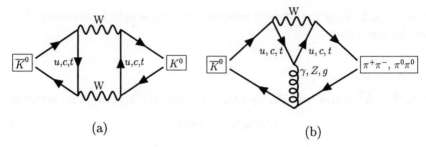

Figure 3. (a) The box diagram contribution to $K^0\overline{K^0}$ mixing. Crossed versions and diagrams with extra gluons etc. are not shown. (b) The Penguin diagram contribution to $K \to \pi\pi$. Extra gluons and crossed versions etc. are not shown.

The latter has been specifically constructed to remove the K^0-$\overline{K^0}$ transition. $|\varepsilon|$ is a directly measurable as ratios of decay rates.

We now make a series of experimentally valid approximations,

$$|\mathrm{Im}a_0|, |\mathrm{Im}a_2| << |\mathrm{Re}a_2| << |\mathrm{Re}a_0|, \quad |\varepsilon|, |\bar{\varepsilon}| << 1, \quad |\varepsilon'| << |\varepsilon|, \quad (13)$$

to obtain the usually quoted expression

$$\varepsilon' = \frac{i}{\sqrt{2}} e^{i(\delta_2 - \delta_0)} \frac{\mathrm{Re}a_2}{\mathrm{Re}a_0} \left(\frac{\mathrm{Im}a_2}{\mathrm{Re}a_2} - \frac{\mathrm{Im}a_0}{\mathrm{Re}a_0} \right). \quad (14)$$

Experimentally,[44]

$$|\varepsilon| = (2.271 \pm 0.017) \cdot 10^{-3}. \quad (15)$$

The set of diagrams, depicted schematically in Fig. 3(a), responsible for $K^0\overline{K^0}$ mixing are known as box diagrams. It is the presence of the virtual intermediate quark lines of up, charm and top quarks that produces the CP-violation.

The experimental situation on ε'/ε was unclear for a long time. Two large experiments, NA31 at CERN and E731 at FNAL, obtained conflicting results in the mid 1980's. Both groups have since gone on and build improved versions of their detectors, NA48 at CERN and KTeV at FNAL. ε'/ε is measured via the double ratio

$$\mathrm{Re}\left(\frac{\varepsilon'}{\varepsilon} \right) = \frac{1}{6} \left\{ 1 - \frac{\Gamma(K_L \to \pi^+\pi^-)/\Gamma(K_S \to \pi^+\pi^-)}{\Gamma(K_L \to \pi^0\pi^0)/\Gamma(K_S \to \pi^0\pi^0)} \right\}. \quad (16)$$

The two main experiments follow a somewhat different strategy in measuring this double ratio, mainly in the way the relative normalisation of K_L and K_S components is treated. After some initial disagreement with the first results, KTeV has reanalysed their systematic errors and the situation

Table 1.   Recent results on ε'/ε. The years refer to the data sets.

NA31	$(23.0 \pm 6.5) \times 10^{-4}$
E731	$(7.4 \pm 5.9) \times 10^{-4}$
KTeV 96	$(23.2 \pm 4.4) \times 10^{-4}$
KTeV 97	$(19.8 \pm 2.9) \times 10^{-4}$
NA48 97	$(18.5 \pm 7.3) \times 10^{-4}$
NA48 98+99	$(15.0 \pm 2.7) \times 10^{-4}$
ALL	$(17.2 \pm 1.8) \times 10^{-4}$

for ε'/ε is now quite clear. We show the recent results in Table 1. The data are taken from Ref. [45] and the recent reviews in the Lepton-Photon conference.[46,47]

The Penguin diagram shown in Fig. 3(b) contributes to the direct CP-violation as given by ε'. Again, W-couplings to all three generations show up so CP-violation is possible in $K \to \pi\pi$. This is a qualitative prediction of the standard model and borne out by experiment. The main problem is now to embed these diagrams and the simple W-exchange in the full strong interaction. The $\Delta I = 1/2$ rule shows that there will have to be large corrections to the naive picture.

5. From Quarks to Mesons: a Chain of Effective Field Theories

The full calculation in the presence of the strong interaction is quite difficult. Even at short distances, due to the presence of logarithms of large ratios of scales, a simple one-loop calculation gives very large effects. These need to be resummed which fortunately can be done using renormalisation group methods.

The three steps of the full calculation are depicted in Fig. 4. First we integrated out the heaviest particles step by step using Operator Product Expansion methods. The steps OPE we describe in the next subsections while step ??? we will split up in more subparts later.

5.1. Step I: from SM to OPE

The first step concerns the standard model diagrams of Fig. 5(a). We replace their effect with a contribution of an effective Hamiltonian given by

$$\mathcal{H}_{\text{eff}} = \sum_i C_i(\mu)Q_i(\mu) = \frac{G_F}{\sqrt{2}}V_{ud}V_{us}^* \sum_i \left(z_i - y_i\frac{V_{td}V_{ts}^*}{V_{ud}V_{us}^*}\right)Q_i. \quad (17)$$

In the last part we have real coefficients z_i and y_i and the CKM-matrix-elements occurring are shown explicitly. The four-quark operators Q_i can

ENERGY SCALE	FIELDS	Effective Theory

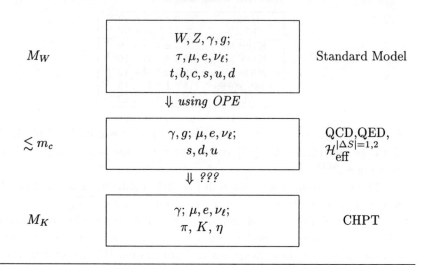

M_W	$W, Z, \gamma, g;$ $\tau, \mu, e, \nu_\ell;$ t, b, c, s, u, d	Standard Model

$\Downarrow$ *using OPE*

$\lesssim m_c$	$\gamma, g; \mu, e, \nu_\ell;$ s, d, u	QCD,QED, $\mathcal{H}_{\text{eff}}^{|\Delta S|=1,2}$

$\Downarrow$ *???*

M_K	$\gamma; \mu, e, \nu_\ell;$ π, K, η	CHPT

Figure 4. A schematic exposition of the various steps in the calculation of nonleptonic matrix-elements.

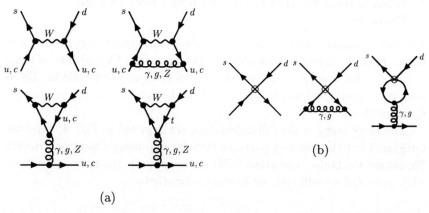

(a)

(b)

Figure 5. (a) The standard model diagrams to be calculated at a high scale. (b) The diagrams needed for the matrix-elements calculated at a scale $\mu_H \approx m_W$ using the effective Hamiltonian.

be found in e.g. Ref. [48].

We calculate now matrix-elements between quarks and gluons in the standard model using the diagrams of Fig. 5(a) and equate those to the same matrix-elements calculated using the effective Hamiltonian of Eq. (17)

Table 2. The Wilson coefficients and their main source at the scale $\mu_H = m_W$ in the NDR-scheme.

z_1	0.053	g, γ-box	y_6	-0.0019	g-Penguin
z_2	0.981	W^+-exchange g, γ-box	y_7	0.0009	γ, Z-Penguin
y_3	0.0014	g, Z-Penguin WW-box	y_8	0.	
y_4	-0.0019	g-Penguin	y_9	-0.0074	γ, Z-Penguin WW-box
y_5	0.0006	g-Penguin	y_{10}	0.	

and the diagrams of Fig. 5(b). This determines the value of the z_i and y_i. The top quark and the W and Z bosons are integrated out all at the same time. There should be no large logarithms present due to that. The scale $\mu = \mu_H$ in the diagrams of Fig. 5(b) of the OPE expansion diagrams should be chosen of the order of the W mass. The scale μ_W in the Standard Model diagrams of Fig. 5(a) should be chosen of the same order.

Notes:

• In the Penguin diagrams CP-violation shows up since all 3 generations are present.

• The equivalence is done by calculating matrix-elements between *Quarks and Gluons*

• The SM part is μ_W-independent to $\alpha_S^2(\mu_W)$.

• OPE part: The μ_H dependence of $C_i(\mu_H)$ cancels the μ_H dependence of the diagrams to order $\alpha_S^2(\mu_H)$.

This procedure gives at $\mu_W = \mu_H = M_W$ in the NDR-scheme [b] the numerical values given in Table 2. In the same table I have given the main source of these numbers. Pure tree-level W-exchange would have only given $z_2 = 1$ and all others zero. Note that the coefficients from γ, Z exchange are similar to the gluon exchange ones since α_S at this scale is not very big.

5.2. Step II

Now comes the main advantage of the OPE formalism. Using the renormalisation group equations we can calculate the change with μ of the C_i, thus resumming the log $\left(m_W^2/\mu^2\right)$ effects. The renormalisation group equations (RGEs) for the strong coupling and the Wilson coefficients are

$$\mu\frac{d}{d\mu}g_S(\mu) = \beta(g_S(\mu)), \quad \mu\frac{d}{d\mu}C_i(\mu) = \gamma_{ji}(g_S(\mu), \alpha)C_j(\mu). \tag{18}$$

β is the QCD beta function for the running coupling. The coefficients γ_{ij} are the elements of the anomalous dimension matrix $\hat{\gamma}$. They can be derived

---

[b]The precise definition of the four-quark operators Q_i comes in here as well. See the lectures by Buras [49] for a more extensive description of that.

Table 3. The Wilson coefficients z_i and y_i at a scale $\mu_{OPE} = 900$ MeV in the NDR scheme and in the X-boson scheme at $\mu_X = 900$ MeV.

i	z_i	y_i	z_i	y_i
	$\mu_{OPE} = 0.9$ GeV		$\mu_X = 0.9$ GeV	
z_1	-0.490	$0.$	-0.788	$0.$
z_2	1.266	$0.$	1.457	$0.$
z_3	0.0092	0.0287	0.0086	0.0399
z_4	-0.0265	-0.0532	-0.0101	-0.0572
z_5	0.0065	0.0018	0.0029	0.0112
z_6	-0.0270	-0.0995	-0.0149	-0.1223
z_7	$2.6\ 10^{-5}$	$-0.9\ 10^{-5}$	0.0002	-0.00016
z_8	$5.3\ 10^{-5}$	0.0013	$6.8\ 10^{-5}$	0.0018
z_9	$5.3\ 10^{-5}$	-0.0105	0.0003	-0.0121
z_{10}	$-3.6\ 10^{-5}$	0.0041	$-8.7\ 10^{-5}$	0.0065

from the infinite parts of loop diagrams and this has been done to one [50] and two loops.[51] The series in α and α_S is known to

$$\hat{\gamma} = \hat{\gamma}_S^0 \frac{\alpha_S}{4\pi} + \hat{\gamma}_S^1 \left(\frac{\alpha_S}{4\pi}\right)^2 + \hat{\gamma}_e \frac{\alpha}{4\pi} + \hat{\gamma}_{se} \frac{\alpha_S}{4\pi} \frac{\alpha}{4\pi} + \cdots \qquad (19)$$

Many subtleties are involved in this calculation.[49,51] They all are related to the fact that everything at higher loop orders need to be specified correctly, and many things which are equal at tree level are no longer so in $d \neq 4$ and at higher loops, see the lectures [49] or the review [52]. The numbers below are obtained by numerically integrating Eq. (18).[53,54]

We perform the following steps to get down to a scale μ_{OPE} around 1 GeV. Starting from the z_i and y_i at the scale μ_H:

(1) solve Eqs. (18); run from μ_H to μ_b.

(2) At $\mu_b \approx m_b$ remove b-quark and match to the theory without b by calculating matrix-elements of the effective Hamiltonian in the five and in the four-quark picture and putting them equal.

(3) Run from μ_b to $\mu_c \approx m_c$.

(4) At μ_c remove the c-quark and match to the theory without c.

(5) Run from μ_c to μ_{OPE}.

Then *all* large logarithms including m_W, m_Z, m_t, m_b and m_c, are summed.

With the inputs $m_t(m_t) = 166$ GeV, $\alpha = 1/137.0$, $\alpha_S(m_Z) = 0.1186$ which led to the initial conditions shown in Table 2, we can perform the above procedure down to μ_{OPE}. Results for 900 MeV are shown in columns two and three of Table 3. z_1 and z_2 have changed much from 0 and 1. This is the short-distance contribution to the $\Delta I = 1/2$ rule. We also see a large enhancement of y_6 and y_8, which will lead to our value of ε'.

5.3. *Step III: Matrix-elements*

Now remember that the C_i depend on μ_{OPE} (scale dependence) and on the definition of the Q_i (scheme dependence) and the numerical change in the coefficients due to the various choices for the Q_i possible is not negligible. It is therefore important both from the phenomenological and fundamental point of view that this dependence is correctly accounted for in the evaluation of the matrix-elements. We can solve this in various ways.

• **Stay in QCD** $\Rightarrow$ Lattice calculations.[55]

• **ITEP Sum Rules** or QCD sum rules. [56]

• **Give up** $\Rightarrow$ Naive factorisation.

• **Improved factorisation**

• X-**boson method** (or fictitious Higgs method)

• **Large** N_c (in combination with something like the X-boson method.) Here the difference is mainly in the treatment of the low-energy hadronic physics. Three main approaches exist of increasing sophistication.[c]

⊛ CHPT: As originally proposed by Bardeen-Buras-Gérard [57] and now pursued mainly by Hambye and collaborators.[58]

⊛ ENJL (or extended Nambu-Jona-Lasinio model [59]): As mainly done by myself and J. Prades.[60,48,61,53,54]

⊛ LMD or lowest meson dominance approach.[62] These papers stay with dimensional regularisation throughout. The X-boson corrections discussed below, show up here as part of the QCD corrections.

• **Dispersive methods** Some matrix-elements can in principle be deduced from experimental spectral functions.

Notice that there other approaches as well, e.g. the chiral quark model.[63] These have no underlying arguments why the μ-dependence should cancel, but the importance of several effects was first discussed in this context. I will also not treat the calculations done using bag models and potential models which similarly do not address the μ-dependence issue.

6. The X-boson Method and Results using ENJL for the Long Distance

We want to have a consistent calculational scheme that takes the scale and scheme dependence into account correctly. Let us therefore have a closer look at how we calculate the matrix-elements using naive factorisation. We start from the four-quark operator:

---

[c]Which of course means that calculations exist only for simpler matrix-elements for the more sophisticated approaches.

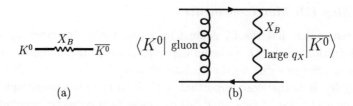

Figure 6. (a) The leading in $1/N_c$ contribution from X_B exchange. (b) The large momentum part of the X_B exchange matrix-element.

$\Rightarrow$ See it as a product of currents or densities.

$\Rightarrow$ Evaluate current matrix-elements in low energy theory or model or from experiment.

$\Rightarrow$ Neglect extra momentum transfer between the current matrix elements. The main lesson here is that *currents and densities are easier to deal with.* We also need to go beyond the approximation in the last step. To obtain well defined currents, we replace the four-quark operators by exchanges of fictitious massive X-bosons coupling to two-quark currents or densities.

$$\mathcal{H} = \sum_i C_i(\mu_{OPE})Q_i \quad \Longrightarrow \quad \sum_i g_i X_i J_i. \tag{20}$$

• This is a well defined scheme of nonlocal operators.

• The matching to obtain the coupling constants g_i from the C_i is done with matrix-elements of *quarks and gluons.*

A simple example is the one needed for the B_K parameter. The four-quark operator is replaced by the exchange of one X-boson X_B:

$$C(\mu)\bar{d}\gamma_\mu(1-\gamma_5)s\,\bar{d}\gamma_\mu(1-\gamma_5)s \quad \Longrightarrow \quad g_B X_B^\mu \bar{d}\gamma_\mu(1-\gamma_5)s. \tag{21}$$

Taking a matrix-element between quarks at next-to-leading order in α_S gives

$$C(\mu_{OPE})\left(1+r\alpha_S(\mu_{OPE})\right) = (g_B^2/M_{X_B}^2)\left(1+r'\alpha_S+r''\log\left(M_{X_B}^2/\mu^2\right)\right). \tag{22}$$

The coefficients r and r' take care of the scheme dependence. The l.h.s. is scale independent to the required order in α_S. The effect of these coefficients surprisingly always went in the direction to improve agreement with experiment [48,54] as can be seen from columns 4 and 5 in Table 3.

The final step is the matrix-element of X_B-boson exchange. For this we split the integral over the X_B momentum q_X in two parts

$$\int dq_X^2 \quad \Longrightarrow \quad \int_0^{\mu^2} dq_X^2 + \int_{\mu^2}^\infty dq_X^2. \tag{23}$$

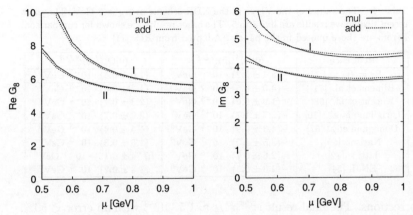

Figure 7. Left: the results for the real part of the octet $\mathrm{Re}G_8$ as a function of μ. Right: the same for the imaginary part.

The leading in N_c contribution is depicted in Fig. 6(a) and corresponds to the large N_c factorisation. The large momentum regime is evaluated by the diagram in Fig. 6(b), since the large momentum must flow back through quarks and gluons. Hadronic exchanges are power suppressed because of the form factors involved. The α_S present already suppresses by N_c so the matrix-element of this part can be evaluated using factorisation. This part cancels the $r'' \log(M_{X_B}^2/\mu^2)$ present in (22). The final part with small q_X momentum in the integral then needs to be evaluated nonperturbatively. Here one can use Chiral Perturbation Theory, the ENJL model or meson exchange approximations with various short-distance constraints.

Let me now show some results from Refs [48,54]. The chiral limit coupling G_8 responsible for the octet contribution, it is 1 in the naive approximation and about 6 when fitted to experiment,[42,43] is shown in Fig. 7. As can be seen, the matching between the short and long distance is reasonable both for the real and imaginary parts. The value of $\mathrm{Re}G_8$ is dominated by the matrix-element of Q_1 and Q_2 but about 30-60% comes from the long distance Penguin part of Q_2. The result for the $\mathrm{Im}G_8$ corresponds to a value of the Q_6 matrix-element much larger than usually assumed. We obtained $B_6 \approx$ 2-2.5 while it is usually *assumed* to be less than 1.5.

Putting our results in (14) we obtain a chiral limit value for $(\varepsilon'/\varepsilon)_\chi$ of about $6 \cdot 10^{-3}$.

We now add the main isospin breaking component Ω [64] and the effect of final state interaction (FSI).[65] The latter in our case has mainly effect on the forefactor $\mathrm{Re}a_2/\mathrm{Re}a_0$ in (14) since the ratios of imaginary parts have been evaluated to the same order in p^2 in CHPT and thus receive no FSI

Table 4. The values of the VEVs in the NDR scheme at $\mu_R = 2$ GeV. The most recent dispersive results are line 3 to 5. The other results are shown for comparison. Errors are those quoted in the papers. Adapted from Ref. [71].

Reference	$\langle 0|O_6^{(1)}|0\rangle_\chi^{NDR}$	$\langle 0|O_6^{(2)}|0\rangle_\chi^{NDR}$
$B_7 = B_8 = 1$	$-(5.4 \pm 2.2) \cdot 10^{-5}$ GeV^6	$(1.0 \pm 0.4) \cdot 10^{-3}$ GeV^6
Bijnens et al. [71]	$-(4.0 \pm 0.5) \cdot 10^{-5}$ GeV^6	$(1.2 \pm 0.5) \cdot 10^{-3}$ GeV^6
Knecht et al. [68]	$-(1.9 \pm 0.6) \cdot 10^{-5}$ GeV^6	$(2.3 \pm 0.7) \cdot 10^{-3}$ GeV^6
Cirigliano et al. [70]	$-(2.7 \pm 1.7) \cdot 10^{-5}$ GeV^6	$(2.2 \pm 0.7) \cdot 10^{-3}$ GeV^6
Donoghue et al.[67]	$-(4.3 \pm 0.9) \cdot 10^{-5}$ GeV^6	$(1.5 \pm 0.4) \cdot 10^{-3}$ GeV^6
Narison [69]	$-(3.5 \pm 1.0) \cdot 10^{-5}$ GeV^6	$(1.5 \pm 0.3) \cdot 10^{-3}$ GeV^6
lattice [72]	$-(2.6 \pm 0.7) \cdot 10^{-5}$ GeV^6	$(0.74 \pm 0.15) \cdot 10^{-3}$ GeV^6
ENJL [54]	$-(4.3 \pm 0.5) \cdot 10^{-5}$ GeV^6	$(1.3 \pm 0.2) \cdot 10^{-3}$ GeV^6

corrections. The final result is [54] $\varepsilon'/\varepsilon \approx 1.5 \cdot 10^{-3}$ with an error $\gtrsim 50\%$.

7. Dispersive Estimates for $\langle Q_7 \rangle$ and $\langle Q_8 \rangle$

Some of the matrix-elements we want can be extracted from experimental information in a different way. The canonical example is the mass difference between the charged and the neutral pion in the chiral limit which can be extracted from a dispersive integral over the difference of the vector and axial vector spectral functions.[66]

This idea has been pursued in the context of weak decay in a series of papers by Donoghue, Golowich and collaborators.[67] The matrix-element of Q_7 could be extracted directly from these data. To get at the matrix-element of Q_8 is somewhat more difficult. Ref. [67] extracted it first by requiring μ-independence, this corresponds to extracting the matrix element of Q_8 from the spectral functions via the coefficient of the dimension 6 term in the operator product expansion of the underlying Green's function. The most recent papers using this method are Refs. [68,69,70] and [71]. In the last two papers also some QCD corrections were included which had a substantial impact on the numerical results.

The results are given in Table. 4. The operator $O_6^{(1)}$ is related by a chiral transformation to Q_7 and $O_6^{(2)}$ to Q_8. The numbers are valid in the chiral limit. The various results for the matrix-element of $O_6^{(1)}$ are in reasonable agreement with each other. The underlying spectral integral, evaluated directly from data in Refs. [69],[70] and [71], or via the minimal hadronic ansatz [68] are in better agreement. The largest source of the differences is the way the different results for the underlying evaluation of $O_6^{(2)}$ come back into $O_6^{(1)}$.

The results for $O_6^{(2)}$ are also in reasonable agreement. Ref. [71] uses two approaches. First, the matrix-element for $O_6^{(2)}$ can be extracted via a similar dispersive integral over the scalar and pseudoscalar spectral functions. The requirements of short-distance matching for this spectral function combined with a saturation with a few states imposes that the nonfactorisable part is suppressed and the number and error quoted follows from this. Extracting the coefficient of the dimension 6 operator in the expansion of the vector and axial-vector spectral functions yields a result comparable but with a larger error of about 0.9. Ref. [68] uses a derivation based on a single resonance plus continuum ansatz for the spectral functions and assumes a typical large N_c error of 30%. This ansatz worked well for lower moments of the spectral functions which can be tested experimentally. Adding more resonances allows for a broader range of results.[71] Ref. [70] chose to enforce all the known constraints on the vector and axial-vector spectral functions to obtain a result. This resulted in rather large cancellations between the various contributions making an error analysis more difficult. A reasonable estimate lead to the value quoted.

The reason why the central value based on the same data can be so different is that the quantity in question is sensitive to the energy regime above 1.3 GeV where the accuracy of the data is rather low.

8. Conclusions

Penguins are alive and well, they provide a sizable part of the $\Delta I = 1/2$ enhancement though mainly through long distance Penguin like topologies in the evaluation of the matrix-element of Q_2. They have found a much richer use in the CP violation phenomenology. For the electroweak Penguins, calculations are in qualitative agreement but more work is still needed to get the errors down. For the strong Penguins, the work I have presented here shows a strong enhancement over factorisation with B_6 significantly larger than one. The latter conclusion is similar to the one derived from the older more phenomenological arguments where the coefficients were taken at a low scale and the matrix-elements for Q_6 taken from the the value of the $\Delta I = 1/2$ rule. This also indicated a rather large enhancement of the matrix-element of Q_6 over the naive factorisation.

Acknowledgements

This work has been partially supported by the Swedish Research Council and by the European Union TMR Network EURODAPHNE (Contract No. ERBFMX-CT98-0169). I thank the organisers for a nice and well organised meeting and Arkady for many discussions and providing a good reason to organise a meeting.

References

1. A. I. Vainshtein, the 1999 Sakurai Prize Lecture, *Int. J. Mod. Phys.* **A14**, 4705 (1999) [hep-ph/9906263].
2. G. Isidori, hep-ph/9908399, talk KAON99; G. Isidori, hep-ph/0110255; G. Buchalla, hep-ph/0110313.
3. L. Littenberg and G. Valencia, *Ann. Rev. Nucl. Part. Sci.* **43**, 729 (1993) [hep-ph/9303225].
4. A. J. Buras, "CP violation and rare decays of K and B mesons," hep-ph/9905437, Lake Louise lectures.
5. J. Bijnens, hep-ph/0204068, to be published in 'At the Frontier of Particle Physics/ Handbook of QCD', edited by M. Shifman, Volume 4.
6. E. Fermi, *Nuovo Cim.* **11**, 1 (1934); *Z. Phys.* **88**, 161 (1934).
7. T. D. Lee and C. N. Yang, *Phys. Rev.* **104**, 254 (1956).
8. C. S. Wu *et al.*, *Phys. Rev.* **105**, 1413 (1957).
9. J. I. Friedman and V. L. Telegdi, *Phys. Rev.* **105**, 1681 (1957).
10. E. C. Sudarshan and R. E. Marshak, *Phys. Rev.* **109**, 1860 (1958).
11. R. P. Feynman and M. Gell-Mann, *Phys. Rev.* **109**, 193 (1958).
12. A. Pais,*Phys. Rev.* **86**, 663 (1952).
13. M. Gell-Mann, *Phys. Rev.* **92**, 833 (1953).
14. M. Gell-Mann and A. Pais,*Phys. Rev.* **97**, 1387 (1955).
15. Y. Ne'eman, *Nucl. Phys.* **26**, 222 (1961); M. Gell-Mann, *Phys. Rev.* **125**, 1067 (1962).
16. N. Cabibbo, *Phys. Rev. Lett.* **10**, 531 (1963).
17. M. Gell-Mann, *Phys. Lett.* **8**, 214 (1964); G. Zweig, CERN report, unpublished.
18. J. H. Christenson *et al*, *Phys. Rev. Lett.* **13**, 138 (1964).
19. T. T. Wu and C. N. Yang, *Phys. Rev. Lett.* **13**, 380 (1964).
20. L. Wolfenstein, *Phys. Rev. Lett.* **13**, 562 (1964).
21. C. N. Yang and R. L. Mills, *Phys. Rev.* **96**, 191 (1954).
22. S. L. Glashow, *Nucl. Phys.* **22**, 579 (1961).
23. S. L. Glashow, J. Iliopoulos and L. Maiani, *Phys. Rev.* **D2**, 1285 (1970).
24. M. K. Gaillard and B. W. Lee, *Phys. Rev.* **D10**, 897 (1974).
25. H. Fritzsch, M. Gell-Mann and H. Leutwyler, *Phys. Lett.* **B47**, 365 (1973).
26. D. J. Gross and F. Wilczek, *Phys. Rev. Lett.* **30**, 1343 (1973); H. D. Politzer, *Phys. Rev. Lett.* **30**, 1346 (1973).
27. M. K. Gaillard and B. W. Lee, *Phys. Rev. Lett.* **33**, 108 (1974).
28. G. Altarelli and L. Maiani, *Phys. Lett.* **B52**, 351 (1974).
29. A. I. Vainshtein, V. I. Zakharov, V. A. Novikov and M. A. Shifman, *Sov. J. Nucl. Phys.* **23**, 540 (1976) [*Yad. Fiz.* **23**, 1024 (1976)].
30. A. I. Vainshtein, V. I. Zakharov and M. A. Shifman, *JETP Lett.* **22**, 55 (1975) [*Pisma Zh. Eksp. Teor. Fiz.* **22**, 123 (1975)]; M. A. Shifman, A. I. Vainshtein and V. I. Zakharov, *Nucl. Phys.* **B120**, 316 (1977).
31. M. A. Shifman, A. I. Vainshtein and V. I. Zakharov, *Sov. Phys. JETP* **45**, 670 (1977) [*Zh. Eksp. Teor. Fiz.* **72**, 1275 (1977)].
32. M. Kobayashi and T. Maskawa, *Prog. Theor. Phys.* **49**, 652 (1973).
33. S. Weinberg, *Phys. Rev. Lett.* **37**, 657 (1976).
34. F. J. Gilman and M. B. Wise, *Phys. Lett.* **B83**, 83 (1979).

35. F. J. Gilman and M. B. Wise, *Phys. Rev.* **D20**, 2392 (1979).
36. B. Guberina and R. D. Peccei, *Nucl. Phys.* **B163**, 289 (1980).
37. F. J. Gilman and M. B. Wise, *Phys. Lett.* **B93**, 129 (1980).
38. F. J. Gilman and M. B. Wise, *Phys. Rev.* **D27**, 1128 (1983).
39. J. Bijnens and M. B. Wise, *Phys. Lett.* **B137**, 245 (1984).
40. J. M. Flynn and L. Randall, *Phys. Lett.* **B224**, 221 (1989) [*Phys. Lett.* **B235**, 412 (1989)].
41. G. Altarelli *et al.*, *Phys. Lett.* **B99**, 141 (1981); *Nucl. Phys.* **B187**, 461 (1981).
42. J. Kambor, J. Missimer and D. Wyler, *Phys. Lett.* **B261**, 496 (1991).
43. J. Bijnens, P. Dhonte and F. Persson, hep-ph/0205341.
44. D. E. Groom *et al.*, *Eur. Phys. J.* **C15**, 1 (2000).
45. A. Alavi-Harati et al. (KTeV), *Phys. Rev. Lett.***83**, 22 (1999); V. Fanti et al. (NA48), *Phys. Lett.* **B465**, 335 (1999); H. Burkhart et al. (NA31), *Phys. Lett.* **B206**, 169 (1988); G.D. Barr et al. (NA31), *Phys. Lett.* **B317**, 233 (1993); L.K. Gibbons et al. (E731), *Phys. Rev. Lett.* **70**, 1203 (1993)
46. R. Kessler, "Recent KTeV results", hep-ex/0110020 .
47. L. Iconomidou-Fayard, NA48 Collaboration, hep-ex/0110028.
48. J. Bijnens and J. Prades, *J. High Energy Phys.* **0001**, 002 (2000) [hep-ph/9911392].
49. A. J. Buras, hep-ph/9806471, Les Houches lectures.
50. Refs. [27,28,30,34,35,36,39,40] and M. Lusignoli, *Nucl. Phys.* **B325**, 33 (1989);
51. A. J. Buras and P. H. Weisz, *Nucl. Phys.* **B333**, 66 (1990); A. J. Buras *et al.*, *Nucl. Phys.* **B370**, 69 (1992), Addendum-ibid. **B375**, 501 (1992); A. J. Buras, M. Jamin and M. E. Lautenbacher, *Nucl. Phys.* **B400**, 75 (1993) [hep-ph/9211321]; A. J. Buras *et al.*, *Nucl. Phys.* **B400**, 37 (1993) [hep-ph/9211304];
    M. Ciuchini *et al.*, *Nucl. Phys.* **B415**, 403 (1994) [hep-ph/9304257].
52. G. Buchalla, A. J. Buras and M. E. Lautenbacher, *Rev. Mod. Phys.* **68**, 1125 (1996) [hep-ph/9512380].
53. J. Bijnens and J. Prades, *J. High Energy Phys.* **9901**, 023 (1999) [hep-ph/9811472].
54. J. Bijnens and J. Prades, *J. High Energy Phys.* **0006**, 035 (2000) [hep-ph/0005189].
55. C. T. Sachrajda, hep-ph/0110304; G. Martinelli, hep-ph/0110023.
56. M. A. Shifman, A. I. Vainshtein and V. I. Zakharov, *Nucl. Phys.* **B147**, 385 (1979); *Nucl. Phys.* **B147**, 448 (1979).
57. W. A. Bardeen, A. J. Buras and J.-M. Gérard, *Phys. Lett.* **B192**, 138 (1987); *Nucl. Phys.* **B293**, 787 (1987).
58. T. Hambye *et al.*, *Phys. Rev.* **D58**, 014017 (1998) [hep-ph/9802300]; T. Hambye, G. O. Köhler and P. H. Soldan, *Eur. Phys. J.* **C10**, 271 (1999) [hep-ph/9902334]; T. Hambye *et al.*, *Nucl. Phys.* **B564**, 391 (2000) [hep-ph/9906434].
59. J. Bijnens, C. Bruno and E. de Rafael, *Nucl. Phys.* **B390**, 501 (1993) [hep-ph/9206236]; J. Bijnens, *Phys. Rept.* **265**, 369 (1996) [hep-ph/9502335] and references therein.
60. J. Bijnens and J. Prades, *Phys. Lett.* **B342**, 331 (1995) [hep-ph/9409255];

Nucl. Phys. **B444**, 523 (1995) [hep-ph/9502363].

61. J. Bijnens, E. Pallante and J. Prades, *Nucl. Phys.* **B521**, 305 (1998) [hep-ph/9801326].

62. M. Knecht, S. Peris and E. de Rafael, *Phys. Lett.* **B457**, 227 (1999) [hep-ph/9812471]; S. Peris and E. de Rafael, *Phys. Lett.* **B490**, 213 (2000) [hep-ph/0006146]. E. de Rafael, arXiv:hep-ph/0109280.

63. A. Pich and E. de Rafael, *Nucl. Phys.* **B358**, 311 (1991); V. Antonelli *et al.*, *Nucl. Phys.* **B469**, 143 (1996) [hep-ph/9511255]; V. Antonelli *et al.*, *Nucl. Phys.* **B469**, 181 (1996) [hep-ph/9511341]; S. Bertolini, J. O. Eeg and M. Fabbrichesi, *Nucl. Phys.* **B476**, 225 (1996) [hep-ph/9512356]; S. Bertolini *et al.*, *Nucl. Phys.* **B514**, 93 (1998) [hep-ph/9706260]; S. Bertolini, J. O. Eeg and M. Fabbrichesi, *Phys. Rev.* **D63**, 056009 (2001) [hep-ph/0002234]; M. Franz, H. C. Kim and K. Goeke, *Nucl. Phys.* **B562**, 213 (1999) [hep-ph/9903275].

64. G. Ecker *et al.*, *Phys. Lett.* **B477**, 88 (2000) [hep-ph/9912264].

65. E. Pallante and A. Pich, *Phys. Rev. Lett.* **84**, 2568 (2000) [hep-ph/9911233]; *Nucl. Phys.* **B592**, 294 (2001) [hep-ph/0007208].

66. T. Das, G. S. Guralnik, V. S. Mathur, F. E. Low and J. E. Young, *Phys. Rev. Lett.* **18**, 759 (1967).

67. J. F. Donoghue and E. Golowich, *Phys. Lett.* **B315**, 406 (1993) [hep-ph/9307263]; *Phys. Lett.* **B478**, 172 (2000) [hep-ph/9911309];
V. Cirigliano and E. Golowich, *Phys. Lett.* **B475**, 351 (2000) [hep-ph/9912513];
V. Cirigliano, J. F. Donoghue and E. Golowich, *J. High Energy Phys.* **0010**, 048 (2000) [hep-ph/0007196];
V. Cirigliano and E. Golowich, *Phys. Rev.* **D65**, 054014 (2002) [hep-ph/0109265].

68. M. Knecht, S. Peris and E. de Rafael, *Phys. Lett.* **B508**, 117 (2001) [hep-ph/0102017] and private communication.

69. S. Narison, *Nucl. Phys.* **B593**, 3 (2001) [hep-ph/0004247]; *Nucl. Phys. Proc. Suppl.* **96**, 364 (2001) [hep-ph/0012019].

70. V. Cirigliano, J. F. Donoghue, E. Golowich and K. Maltman, *Phys. Lett.* **B522**, 245 (2001) [hep-ph/0109113].

71. J. Bijnens, E. Gamiz and J. Prades, *J. High Energy Phys.* **0110**, 009 (2001) [hep-ph/0108240].

72. A. Donini *et al.*, *Phys. Lett.* **B470**, 233 (1999)

ELECTROMAGNETIC FORM FACTOR OF THE PION

H. LEUTWYLER

Institute for Theoretical Physics, University of Bern,
Sidlerstr. 5, CH-3012 Bern, Switzerland
E-mail: leutwyler@itp.unibe.ch

The Standard Model prediction for the magnetic moment of the muon requires a determination of the electromagnetic form factor of the pion at high precision. It is shown that the recent progress in $\pi\pi$ scattering allows us to obtain an accurate representation of this form factor on the basis of the data on $e^+e^- \to \pi^+\pi^-$. The same method also applies to the form factor of the weak vector current, where the data on the decay $\tau \to \pi^-\pi^0\,\nu_\tau$ are relevant. Unfortunately, however, the known sources of isospin breaking do not explain the difference between the two results. The discrepancy implies that the Standard Model prediction for the magnetic moment of the muon is currently subject to a large uncertainty.

1. Motivation: magnetic moment of the muon

The fabulous precision reached in the measurement of the muon magnetic moment [1] allows a thorough test of the Standard Model. The prediction that follows from the Dirac equation, $\mu = e\,\hbar/2m_\mu$, only holds to leading order in the expansion in powers of the fine structure constant α. It is customary to write the correction in the form

$$\mu = \frac{e\,\hbar}{2\,m_\mu}\,(1+a) \ . \tag{1}$$

Schwinger was able to calculate the term of first order in α, which stems from the triangle graph in fig. 1a and is universal [2],

$$a = \frac{\alpha}{2\pi} + O(\alpha^2) \ . \tag{2}$$

The contributions of $O(\alpha^2)$ can also unambiguously be calculated, except for the one from hadronic vacuum polarization, indicated by the graph in fig. 1e. It is analogous to the contributions generated by leptonic vacuum polarization in figs. 1b, 1c and 1d, but involves quarks and gluons instead of leptons. All of these graphs may be viewed as arising from vacuum

24

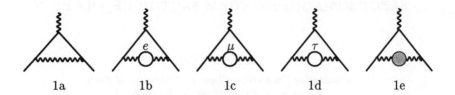

1a   1b   1c   1d   1e

polarization in the photon propagator:

$$D_{\mu\nu}(q) = \frac{g_{\mu\nu}\, Z}{q^2\{1 + \Pi(q^2)\}} + \text{gauge terms} \ .$$

The expansion of the self energy function $\Pi(t)$ in powers of α starts with

$$\Pi(t) = \alpha\, \Pi^{(0)}(t) + \alpha^2\, \Pi^{(1)}(t) + \dots$$

It is normalized by $\Pi(0) = 0$. The leading term can be pictured as

$$\Pi^{(0)}(t) = \quad + \quad + \quad + \quad$$

As shown in ref. [3], the modification of the Schwinger formula (2) that is generated by vacuum polarization can be represented in compact form:[a]

$$a = \frac{\alpha}{\pi} \int_0^1 dx\, \frac{1-x}{1+\Pi(t_x)} \ , \qquad t_x = -\frac{x^2 m_\mu^2}{1-x} \ . \tag{3}$$

Expanding this formula in powers of α, we obtain

$$a = \frac{\alpha}{\pi} \int_0^1 dx\, (1-x) - \frac{\alpha^2}{\pi} \int_0^1 dx\, (1-x)\, \Pi^{(0)}(t_x) + O(\alpha^3)$$

The first term indeed reproduces the Schwinger formula (2), which corresponds to graph 1a. The term linear in $\Pi^{(0)}(t)$ accounts for graphs 1b to 1e. The contribution involving the square of $\Pi^{(0)}(t)$ describes the one-particle reducible graphs with two bubbles, etc.

---

[a]The formula only makes sense in the framework of the perturbative expansion [4]. The contribution generated by an electron loop, for instance, grows logarithmically at large momenta and tends to $-\infty$ in the spacelike region,

$$\Pi_e^{(0)}(t) = -\frac{1}{3\pi} \ln\frac{(-t)}{M_1^2} + O\left(\frac{M_e^2}{t}\right) \ , \qquad M_1 = e^{\frac{5}{6}} M_e \ .$$

Hence $1 + \alpha\, \Pi_e^{(0)}(t)$ contains a zero in the vicinity of $t = -\exp(3\pi/\alpha)\, M_1^2 \simeq -10^{555}$ GeV^2 (graphs 1c, 1d and 1e push the zero towards slightly smaller values). At academically high energies, the photon propagator thus develops a "Landau pole", reflecting the fact that the U(1) factor of the Standard Model does not give rise to an asymptotically free gauge theory. In the present context, however, this phenomenon is not relevant – we are concerned with the low energy structure of the Standard Model.

The vacuum polarization due to a lepton loop is given by ($\ell = e, \mu, \tau$)

$$\Pi_\ell^{(0)}(t) = \frac{t}{3\pi} \int_{4M_\ell^2}^\infty ds \frac{R_\ell(s)}{s(s-t)} \;, \tag{4}$$

$$R_\ell(s) = \sqrt{1 - \frac{4M_\ell^2}{s}} \left(1 + \frac{2M_\ell^2}{s}\right) \;.$$

The hadronic contribution cannot be calculated analytically, but it can be expressed in terms of the cross section of the reaction $e^+e^- \to$ hadrons. More precisely, the leading term in the expansion of this cross section in powers of α,

$$\sigma_{e^+e^- \to h} = \alpha^2 \sigma_{e^+e^- \to h}^{(0)} + O(\alpha^3) \;,$$

is relevant. In terms of this quantity the expression reads

$$\Pi_h^{(0)}(t) = \frac{t}{3\pi} \int_{4M_\pi^2}^\infty ds \frac{R_h(s)}{s(s-t)} \;, \tag{5}$$

$$R_h(s) = \frac{3s}{4\pi} \sigma_{e^+e^- \to h}^{(0)}(s) \;.$$

At low energies, where the final state necessarily consists of two pions, the cross section is given by the square of the electromagnetic form factor of the pion,

$$R_h(s) = \frac{1}{4} \left(1 - \frac{4M_\pi^2}{s}\right)^{\frac{3}{2}} |F(s)|^2 \;, \qquad s < 9\,M_\pi^2 \;. \tag{6}$$

Numerically, the contribution from hadronic vacuum polarization to the magnetic moment of the muon amounts to $a_{\mathrm{hvp}} \simeq 700 \times 10^{-10}$. This is a small fraction of the total, $a = 11\,659\,203\,(8) \times 10^{-10}$ [1], but large compared to the experimental uncertainty: a determination of a_{hvp} to about 1% is required for the precision of the Standard Model prediction to match the experimental one. Since the contribution from hadronic vacuum polarization is dominated by the one from the two pion states, this means that the pion form factor is needed to an accuracy of about half a percent.

2. Comparison of leptonic and hadronic contributions

Graphically, the formula (6) amounts to

There are three differences between the pionic loop integral and those belonging to the lepton loops:

- the masses are different
- the spins are different
- the pion is composite – the Standard Model leptons are elementary

The compositeness manifests itself in the occurrence of the form factor $F(s)$, which generates an enhancement: at the ρ peak, $|F(s)|^2$ reaches values of order 45. The remaining difference in the expressions for the quantities $R_\ell(s)$ and $R_h(s)$ in eqs. (4) and (6) originates in the fact that the leptons carry spin $\frac{1}{2}$, while the spin of the pion vanishes. Near threshold, the angular momentum barrier suppresses the function $R_h(s)$ by three powers of momentum, while $R_\ell(s)$ is proportional to the first power. The suppression largely compensates the enhancement by the form factor – by far the most important property is the mass: in units of 10^{-10}, the contributions due to the e, μ and τ loops are 59040.6, 846.4 and 4.2, respectively, to be compared with the 700 units from hadronic vacuum polarization. The latter is comparable to the one from the muon – in accordance with the fact that the masses of pion and muon are similar.

3. Pion form factor

In the following, I disregard the electromagnetic interaction – the discussion concerns the properties of the form factor in QCD. I draw from ongoing work carried out in collaboration with Irinel Caprini, Gilberto Colangelo, Simon Eidelman, Jürg Gasser and Fred Jegerlehner.

The systematic low energy analysis of the form factor based on chiral perturbation theory [5] has been worked out to two loops [6]. This approach, however, only covers the threshold region. The range of validity of the representation can be extended to higher energies by means of dispersive methods [7], which exploit the constraints imposed by analyticity and unitarity. Our approach is very similar to the one of de Trocóniz and Yndurain [8]. For a thorough discussion of the mathematical framework, I refer to Heyn and Lang [9].

We represent the form factor as a product of three functions that account for the prominent singularities in the low energy region:

$$F(s) = G_2(s) \times G_3(s) \times G_4(s) \ . \tag{7}$$

The index has to do with the number of pions that generate the relevant discontinuity: two in the case of G_2, three for G_3 and four or more for G_4.

The first term represents the familiar Omnès factor that describes the branch cut due to $\pi^+\pi^-$ intermediate states (states with two neutral pions do not contribute, because the matrix element $\langle \pi^0 \pi^0 \, \text{out} \, | \, j^\mu \, | 0 \rangle$ vanishes, on account of Bose statistics). The corresponding branch point singularity is of the type $\operatorname{Im} G_2(s) \sim (s - s_2)^{\frac{3}{2}}$, with $s_2 = 4M_\pi^2$. The Watson final state interaction theorem implies that, in the elastic region, $4M_\pi^2 < s < 9M_\pi^2$, the phase of the form factor is given by the P-wave phase shift of the elastic scattering process $\pi^+\pi^- \to \pi^+\pi^-$. Denoting this phase shift by $\delta(s)$, the explicit expression for the Omnès factor reads:

$$G_2(s) = \exp\left\{ \frac{s}{\pi} \int_{4M_\pi^2}^{\infty} \frac{dx\, \delta(x)}{x\,(x - s)} \right\} . \tag{8}$$

The function $G_3(s)$ contains the singularities generated by 3π intermediate states: $G_3(s)$ is analytic except for a cut starting at $s_3 = 9M_\pi^2$, with a branch point singularity of the type $\operatorname{Im} G_3(s) \sim (s - s_3)^4$. If isospin symmetry were exact, the form factor would not contain such singularities: in the limit $m_u = m_d$, the term $G_3(s)$ is equal to 1. Indeed, isospin is nearly conserved, but the occurrence of a narrow resonance with the proper quantum numbers strongly enhances the effects generated by isospin breaking: the form factor contains a pole close to the real axis,

$$G_3(s) = 1 + \epsilon\, \frac{s}{s_\omega - s} + \dots \qquad s_\omega = (M_\omega - \tfrac{1}{2} i\, \Gamma_\omega)^2 . \tag{9}$$

This implies that, in the vicinity of $s = M_\omega^2$, the form factor rapidly varies, both in magnitude and in phase. The pole term cannot stand by itself because it fails to be real in the spacelike region. We replace it by a dispersion integral with the proper behaviour at threshold, but this is inessential: in the experimental range, the representation for $G_3(s)$ that we are using can barely be distinguished from the pole approximation (9).

Isospin breaking also affects the scattering amplitude. In particular, it gives rise to the inelastic reaction $2\pi \to 3\pi$, with an amplitude proportional to $m_u - m_d$. Hence unitarity implies that, in the region $9M_\pi^2 < s < 16M_\pi^2$, the elasticities of the partial waves are less than 1. Numerically, the effect is tiny, however, because it is of second order in $m_u - m_d$. To a very high degree of accuracy, the first two terms in eq. (7) thus account for all singularities below $s_4 = 16M_\pi^2$ – the function $G_4(s)$ is analytic in the plane cut from s_4 to ∞. Phase space strongly suppresses the strength of the corresponding branch point singularity: $\operatorname{Im} G_4(s) \propto (s - s_4)^{\frac{9}{2}}$. A significant discontinuity due to inelastic channels only manifests itself for $s > s_{in} = (M_\omega + M_\pi)^2$.

We analyze the background term $G_4(s)$ by means of a conformal mapping. The transformation

$$z = \frac{\sqrt{s_{in} - s_1} - \sqrt{s_{in} - s}}{\sqrt{s_{in} - s_1} + \sqrt{s_{in} - s}} \tag{10}$$

maps the s-plane cut along $s > s_{in}$ onto the unit disk in the z-plane, so that the Taylor series expansion in powers of z converges on the entire physical sheet, irrespective of the value of the arbitrary parameter s_1. We truncate this series after the first few terms, thus approximating the function $G_4(s)$ by a low order polynomial in z.

4. Roy equations

The crucial element in the above representation is the phase $\delta(s)$. The main difference between our analysis and the one in ref. [8] concerns the input used to describe the behaviour of this phase. In fact, during the last two or three years, our understanding of the $\pi\pi$ scattering amplitude has made a quantum jump. As a result of theoretical work [10–12], the low energy behaviour of the S- and P-waves is now known to an amazing accuracy – to my knowledge, $\pi\pi$ scattering is the only field in strong interaction physics where theory is ahead of experiment.

The method used to implement the requirements of analyticity, unitarity and crossing symmetry is by no means new. As shown by Roy more than 30 years ago [13], these properties of the scattering amplitude subject the partial waves to a set of coupled integral equations. These equations involve two subtraction constants, which may be identified with the two S–wave scattering lengths a_0^0, a_0^2. If these two constants are given, the Roy equations allow us to calculate the scattering amplitude in terms of the imaginary parts above the "matching point" $E_m = 0.8\,\mathrm{GeV}$. The available experimental information suffices to evaluate the relevant dispersion integrals, to within small uncertainties [10,12]. In this sense, a_0^0, a_0^2 represent the essential parameters in low energy $\pi\pi$ scattering.

As will be discussed in some detail in the next section, chiral symmetry predicts the values of the two subtraction constants and thereby turns the Roy equations into a framework that fully determines the low energy behaviour of the $\pi\pi$ scattering amplitude. In particular, the P-wave scattering length and effective range are predicted very accurately: $a_1^1 = 0.0379(5)\,M_\pi^{-2}$ and $b_1^1 = 0.00567(13)\,M_\pi^{-4}$. The manner in which the P-wave phase shift passes through $90°$ when the energy reaches the mass of the ρ is specified within the same framework, as well as the behaviour of the two S-waves. The analysis reveals, for instance, that the isoscalar S-wave contains a pole on the second sheet and the position can be calculated rather accurately: the pole occurs at $E = M_\sigma - \frac{1}{2} i\,\Gamma_\sigma$, with $M_\sigma = 470 \pm 30\,\mathrm{MeV}$, $\Gamma_\sigma = 590 \pm 40\,\mathrm{MeV}$ [11], etc.

Many papers based on alternative approaches can be found in the literature. Padé approximants, for instance, continue to enjoy popularity and the ancient idea that the σ pole represents the main feature in the isoscalar S-wave also found new adherents recently. Crude models such as these may be of interest in connection with other processes where the physics yet remains to be understood, but for the analysis of the $\pi\pi$ scattering amplitude, they cannot compete with the systematic approach based on analyticity and chiral symmetry. In view of the precision required in the determination of the pion form factor, ad hoc models are of little use, because the theoretical uncertainties associated with these are too large.

5. Prediction for the $\pi\pi$ scattering lengths

Goldstone bosons of zero momentum do not interact: if the quark masses m_u, m_d are turned off, the S-wave scattering lengths disappear, $a_0^0, a_0^2 \to 0$. Like the mass of the pion, these quantities represent effects that arise from the breaking of the chiral symmetry generated by the quark masses. In fact, as shown by Weinberg [14], a_0^0 and a_0^2 are proportional to the square of the pion mass

$$a_0^0 = \frac{7M_\pi^2}{32\pi F_\pi^2} + O(M_\pi^4) \ , \qquad a_0^2 = -\frac{M_\pi^2}{16\pi F_\pi^2} + O(M_\pi^4) \ .$$

The corrections of order M_π^4 contain chiral logarithms. In the case of a_0^0, the logarithm has an unusually large coefficient

$$a_0^0 = \frac{7M_\pi^2}{32\pi F_\pi^2} \left\{ 1 + \frac{9}{2} \frac{M_\pi^2}{(4\pi F_\pi)^2} \ln \frac{\Lambda_0^2}{M_\pi^2} + O(M_\pi^4) \right\} \ .$$

This is related to the fact that in the channel with $I = 0$, chiral symmetry predicts a strong, attractive, final state interaction. The scale Λ_0 is determined by the coupling constants of the effective Lagrangian of $O(p^4)$:

$$\frac{9}{2} \ln \frac{\Lambda_0^2}{M_\pi^2} = \frac{20}{21} \bar{\ell}_1 + \frac{40}{21} \bar{\ell}_2 - \frac{5}{14} \bar{\ell}_3 + 2 \bar{\ell}_4 + \frac{5}{2} \ .$$

The same coupling constants also determine the first order correction in the low energy theorem for a_0^2.

The couplings $\bar{\ell}_1$ and $\bar{\ell}_2$ control the momentum dependence of the scattering amplitude at first nonleading order. Using the Roy equations, these constants can be determined very accurately [11]. The terms $\bar{\ell}_3$ and $\bar{\ell}_4$, on the other hand, describe the dependence of the scattering amplitude on the quark masses – since these cannot be varied experimentally, $\bar{\ell}_3$ and $\bar{\ell}_4$ cannot be determined on the basis of $\pi\pi$ phenomenology. The constant $\bar{\ell}_3$ specifies the correction in the Gell-Mann-Oakes-Renner relation [15],

$$M_\pi^2 = M^2 \left\{ 1 - \frac{1}{2} \frac{M^2}{(4\pi F)^2} \bar{\ell}_3 + O(M^4) \right\} \ . \tag{11}$$

Here M^2 stands for the term linear in the quark masses,

$$M^2 = (m_u + m_d) |\langle 0| \bar{u}u |0\rangle| \frac{1}{F^2} \tag{12}$$

(F and $\langle 0| \bar{u}u |0\rangle$ are the values of the pion decay constant and the quark condensate in the chiral limit, respectively). The coupling constant $\bar{\ell}_4$ occurs in the analogous expansion for F_π,

$$F_\pi = F \left\{ 1 + \frac{M^2}{(4\pi F)^2} \bar{\ell}_4 + O(M^4) \right\} . \tag{13}$$

A low energy theorem relates it to the scalar radius of the pion [16],

$$\langle r^2 \rangle_s = \frac{6}{(4\pi F)^2} \left\{ \bar{\ell}_4 - \frac{13}{12} + O(M^2) \right\} . \tag{14}$$

The dispersive analysis of the scalar pion form factor in ref. [11] leads to

$$\langle r^2 \rangle_s = 0.61 \pm 0.04 \text{ fm}^2 . \tag{15}$$

The constants $\bar{\ell}_1, \ldots \bar{\ell}_4$ depend logarithmically on the quark masses:

$$\bar{\ell}_i = \ln \frac{\Lambda_i^2}{M^2} , \qquad i = 1, \ldots, 4$$

In this notation, the above value of the scalar radius amounts to

$$\Lambda_4 = 1.26 \pm 0.14 \text{ GeV} . \tag{16}$$

Unfortunately, the constant $\bar{\ell}_3 \leftrightarrow \Lambda_3$ is not known with comparable precision. The crude estimate for $\bar{\ell}_3$ given in ref. [16] corresponds to

$$0.2 \text{ GeV} < \Lambda_3 < 2 \text{ GeV} . \tag{17}$$

It turns out, however, that the contributions from $\bar{\ell}_3$ are very small, so that the uncertainty in Λ_3 does not strongly affect the predictions for the scattering lengths. This is shown in fig. 2, where the values of a_0^0, a_0^2 predicted by ChPT are indicated as a small ellipse.

6. Experimental test

Stern and collaborators [17] pointed out that "Standard" ChPT relies on a hypothesis that calls for experimental test. Such a test has now been performed and I wish to briefly describe this development.

The hypothesis in question is the assumption that the quark condensate represents the leading order parameter of the spontaneously broken chiral symmetry. More specifically, the standard analysis assumes that the term linear in the quark masses dominates the expansion of M_π^2. According to

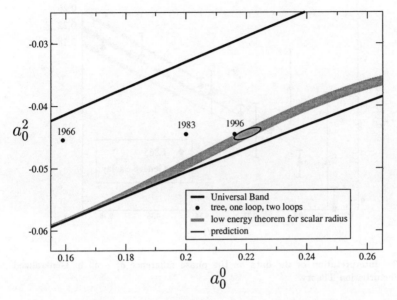

Figure 2. Prediction for the S-wave $\pi\pi$ scattering lengths

the Gell-Mann-Oakes-Renner relation (12), this term is proportional to the quark condensate, which in QCD represents the order parameter of lowest dimension. The dynamics of the ground state is not well understood. The question raised by Stern et al. is whether, for one reason or the other, the quark condensate might turn out to be small, so that the Gell-Mann-Oakes-Renner formula would fail – the "correction" might be comparable to or even larger than the algebraically leading term.

According to eq. (11), the correction is determined by the effective coupling constant $\bar{\ell}_3$. The estimate (17) implies that the correction amounts to at most 4% of the leading term, but this does not answer the question, because that estimate is based on the standard framework, where $\langle 0| \bar{u}u |0\rangle$ is assumed to represent the leading order parameter. If that estimate is discarded and $\bar{\ell}_3$ is treated as a free parameter ("Generalized" ChPT), the scattering lengths cannot be predicted individually, but the low energy theorem (14) implies that – up to corrections of next-to-next-to leading order – the combination $2a_0^0 - 5a_0^2$ is determined by the scalar radius:

$$2a_0^0 - 5a_0^2 = \frac{3M_\pi^2}{4\pi F_\pi^2} \left\{ 1 + \frac{M_\pi^2 \langle r^2 \rangle_s}{3} + \frac{41 M_\pi^2}{192\pi^2 F_\pi^2} + O(M_\pi^4) \right\} \ .$$

The resulting correlation between a_0^0 and a_0^2 is shown as a narrow strip in fig. 2 (the strip is slightly curved because the figure accounts for the

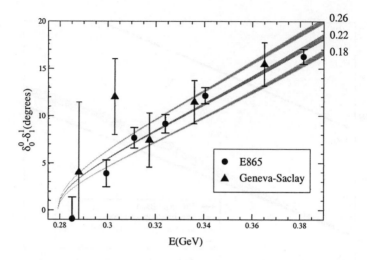

Figure 3. Interpretation of the data on the phase difference $\delta_1^1 - \delta_0^0$ in Generalized Chiral Perturbation Theory.

corrections of next-to-next-to leading order).

In view of the correlation between a_0^0 and a_0^2, the data taken by the E865-collaboration at Brookhaven [18] allow a significant test of the Gell-Mann-Oakes-Renner relation. The final state interaction theorem implies that the phase of the form factors relevant for the decay $K^+ \to \pi^+\pi^- e^+ \bar{\nu}_e$ is determined by the elastic $\pi\pi$ scattering amplitude. Conversely, the phase difference $\delta_0^0 - \delta_1^1$ can be measured in this decay. The analysis of the $4 \cdot 10^5$ events of this type collected by E865 leads to the round data points in fig. 3, taken from ref. [19] (the triangles represent the K_{e_4} data collected in the seventies). The three bands show the result obtained in Generalized ChPT for $a_0^0 = 0.18, 0.22, 0.26$, respectively. The width of the bands corresponds to the uncertainty in the prediction. A fit of the data that exploits the correlation between a_0^0 and a_0^2 yields

$$a_0^0 = 0.216 \pm 0.013 \,(\text{stat}) \pm 0.004 \,(\text{syst}) \pm 0.005 \,(\text{th}) \;\; [18] \;,$$

where the third error bar accounts for the theoretical uncertainties. The result thus beautifully confirms the prediction of ChPT, $a_0^0 = 0.220 \pm 0.005$. The agreement implies that more than 94% of the pion mass originate in the quark condensate, thus confirming that the Gell-Mann-Oakes-Renner relation is approximately valid [19]. May Generalized ChPT rest in peace.

7. Comparison of electromagnetic and weak form factors

In the theoretical limit $m_u = m_d$ and in the absence of the electromagnetic interaction, the vector current relevant for strangeness conserving semileptonic transitions is conserved. The matrix element of this current that shows up in the decay $\tau \to \pi^- \pi^0 \nu_\tau$ is then determined by the electromagnetic form factor of the pion. In reality, however, m_u differs from m_d and the radiative corrections in τ decay are different from those relevant for $e^+ e^- \to \pi^+ \pi^-$. For the anomalous moment of the muon, τ decays are of interest only to the extent that these isospin breaking effects are understood, so that the e.m. form factor can be calculated from the weak transition matrix element.

The leading isospin breaking effects are indeed well understood: those enhanced by the small energy denominator associated with ω exchange, which are described by the factor $G_3(s)$ introduced in section 3. As these do not show up in the weak transition matrix element, they must be corrected for when calculating the electromagnetic form factor from τ decays.

There is another effect that shows up in the process $e^+ e^- \to \pi^+ \pi^-$, but does not affect τ decay: vacuum polarization in the photon propagator, as illustrated by the graphs below:

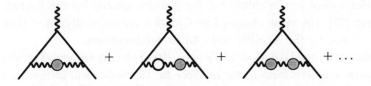

The same graphs also show up in the magnetic moment of the muon:

To avoid double counting, the data on the reaction $e^+ e^- \to \pi^+ \pi^-$ must be corrected for vacuum polarization, multiplying the cross section by the factor $|1 + \Pi(s)|^2$. In the timelike region, this factor is less than 1, so that the correction reduces the magnitude of the form factor. In the ω region, the vacuum polarization due to 3π intermediate states generates a pronounced structure in $\Pi(s)$. While below that energy, the correction is of order 1%, it reaches about 7% immediately above the ω and then decreases to about 3% towards the upper end of the range covered by the CMD2 data [20].

Unfortunately, applying the two corrections just discussed, the results for the form factor obtained with e^+e^- collisions are systematically lower than those found in τ decays [21–23]. The two phenomena mentioned above are not the only isospin breaking effects. Radiative corrections must be applied, to e^+e^- collisions [24] as well as to τ decays [25] and terms of order $m_u - m_d$ need to be estimated as well. I do not know of a mechanism, however, that could give rise to an additional isospin breaking effect of the required order of magnitude.

One way to quantify the discrepancy is to assume that the uncertainties in the overall normalization of some of the data are underestimated. Indeed, if the normalization of either the rate of the decay $\tau \to \pi^-\pi^0\nu_\tau$ or the cross section of the reaction $e^+e^- \to \pi^+\pi^-$ are treated as free parameters, the problem disappears. The renormalization, however, either lowers the ALEPH and CLEO data by about 4% or lifts the CMD2 data by this amount.

While completing this manuscript, a comparison of the τ and e^+e^- data appeared [26], where the problem is discussed in detail. One way to put the discrepancy in evidence is to compare the observed rate of the decay $\tau \to \pi^-\pi^0\nu_\tau$ with the prediction that follows from the data on the reaction $e^+e^- \to \pi^+\pi^-$ if the known sources of isospin breaking are accounted for. Using the observed lifetime of the τ, the prediction for the branching ratio of the channel $\tau \to \pi^-\pi^0\nu_\tau$ reads $B_{\mathrm{pred}} = 0.2408 \pm 0.0031$. The observed value, $B_{\mathrm{obs}} = 0.2546 \pm 0.0012$, differs from this number at the 4σ level [26].

A difference between the τ and e^+e^- results existed before, but with the new CMD2 data, where the hadronic part of vacuum polarization is now corrected for, the disagreement has become very serious. The problem also manifests itself in the values for the ρ width quoted by the Particle Data Group [27]: the value obtained by CMD2 is substantially lower than the results found by the ALEPH and CLEO collaborations.

So, unless the Standard Model fails here, either the experimental results for the electromagnetic form factor or those for the weak form factor must be incorrect. The preliminary data from KLOE appear to confirm the CMD2 results, but the uncertainties to be attached to that determination of the electromagnetic form factor yet remain to be analyzed.

Currently, the discrepancy between the e^+e^- and τ data prevents a test of the Standard Model prediction for the magnetic moment of the muon at an accuracy that would be comparable to the experimental value. The result for the contribution to the muon anomaly due to hadronic vacuum polarization depends on whether the e^+e^- data or the τ data are used to evaluate the electromagnetic form factor – according to ref. [26], the corresponding central values differ by 17.2×10^{-10}. Compared to this, the estimate $(8 \pm 4) \times 10^{-10}$ for the contribution from hadronic light-by-light scattering [28–31] is a rather precise number.[b]

8. Asymptotic behaviour

The behaviour of the form factor for large spacelike momenta can be predicted on the basis of perturbative QCD [32]:

$$F(-Q^2) = \frac{64\pi^2 F_\pi^2}{\beta_0 \, Q^2 L_Q} \left\{ 1 + B_2 \, L_Q^{-\frac{50}{81}} + B_4 \, L_Q^{-\frac{364}{405}} + O(L_Q^{-1}) \right\}^2 , \quad (18)$$

$$L_Q = \ln \frac{Q^2}{\Lambda^2} , \qquad \beta_0 = 11 - \frac{2}{3} N_f .$$

The leading asymptotic term only involves the pion decay constant. The coefficients $B_2, B_4 \ldots$ of the fractional logarithmic corrections are related to the pion distribution amplitude or null plane wave function $\psi(x, \mu)$, which is a function of the momentum fraction x of the quark and depends on the scale μ. Normalizing the wave function to the pion decay constant, the expansion in terms of Gegenbauer polynomials starts with

$$\psi(x,\mu) = 6\,x\,(1-x) \left\{ 1 + B_2 \, L_\mu^{-\frac{50}{81}} \, C_2^{\frac{3}{2}}(2x-1) + B_4 \, L_\mu^{-\frac{364}{405}} \, C_4^{\frac{3}{2}}(2x-1) + \ldots \right\}$$

$$C_2^{\frac{3}{2}}(z) = \tfrac{3}{2}\,(5z^2 - 1) , \quad C_4^{\frac{3}{2}}(z) = \tfrac{15}{8}\,(21z^4 - 14z^2 + 1) , \quad \ldots$$

The wave function cannot be calculated within perturbative QCD and the phenomenological information about the size of the coefficients $B_2, B_4, \ldots$ is meagre. It is therefore of interest to see whether the data on the form factor allow us to estimate these terms.

In the representation (7), the asymptotic behaviour of the form factor can be accounted for as follows. One first continues the asymptotic formula (18) into the timelike region and reads off the asymptotic behaviour of the phase of the form factor:

$$\phi(s) \to \pi + \frac{\pi}{\ln \frac{s}{\Lambda^2}} + \ldots$$

---

[b]The physics – in particular the sign – of the light-by-light contribution is well understood: The low energy expansion of this term is dominated by a logarithmic singularity with known residue [29,30]. In my opinion, the quoted error estimate is conservative.

If the asymptotic behaviour of the phase used for the Omnès factor agrees with this, then the Omnès formula (8) ensures that the ratio $F(s)/G_2(s)$ approaches a constant for large spacelike momenta. The value of the constant is determined by F_π and by the behaviour of the phase shift at nonasymptotic energies. This implies that the background term $G_4(s)$ tends to a known constant for large values of s, or equivalently, for $z \to -1$.

The corrections involving fractional logarithmic powers can also be accounted for with a suitable contribution to the phase. For the asymptotic expansion not to contain a term of order $s^{-1/2}$, the derivative of $G_4(s)$ with respect to z must vanish at $z = -1$. This then yields a representation of the form factor for which the asymptotic behaviour agrees with perturbative QCD, for any value of the coeffcients B_2 and B_4.

We have analyzed the experimental information with a representation of this type, including data in the spacelike region, as well as those available at large timelike momenta [33-37]. The numerical analysis yet needs to be completed and compared with the results in the literature (for a recent review and references, see for instance [38]). Our preliminary results are: If the fractional logarithmic powers are dropped ($B_2 = B_4 = 0$), we find that the asymptotic formula is reached only at academically high energies. With the value for B_2 proposed by Chernyak and Zhitnitsky, the situation improves. For the asymptotic behaviour to set in early, an even larger value of B_2 appears to be required.

This indicates that the leading asymptotic term can dominate the behaviour only for very high energies. A direct comparison of that term with the existing data, which only cover small values of s does therefore not appear to be meaningful.

9. Zeros and sum rules

Analyticity subjects the form factor to strong constraints. Concerning the asymptotic behaviour, I assume that $|\ln F(s)|$ at most grows logarithmically for $|s| \to \infty$, in any direction of the complex s-plane. This amounts to the requirement that a) for a sufficiently large value of n, the quantity $|F(s)/s^n|$ remains bounded and b) the phase of $F(s)$ at most grows logarithmically, so that the real and imaginary parts of $F(s)$ do not oscillate too rapidly at high energies. In view of asymptotic freedom, I take these properties for granted. If the form factor does not have zeros, the function

$$\psi(s) = \frac{1}{(s_2 - s)^{\frac{3}{2}}} \ln \frac{F(s)}{F(s_2)}$$

is then analytic in the cut plane and tends to zero for $|s| \to \infty$. The branch point singularity at threshold is of the type $\psi(s) \sim (s - s_2)^{-\frac{1}{2}}$. Hence $\psi(s)$ obeys the unsubtracted dispersion relation

$$\psi(s) = \frac{s}{\pi} \int_{s_2}^{\infty} dx \, \frac{\operatorname{Im} \psi(x)}{x \, (x - s)}$$

in the entire cut plane. The discontinuity across the cut is determined by the magnitude of the form factor:

$$\operatorname{Im} \psi(s) = -\frac{1}{(s - s_2)^{\frac{3}{2}}} \ln \left| \frac{F(s)}{F(s_2)} \right| \qquad s > s_2 \ .$$

Hence the above dispersion relation amounts to a representation of the form factor in terms of its magnitude in the timelike region:

$$F(s) = |F(s_2)| \exp \left\{ -\frac{(s_2 - s)^{\frac{3}{2}}}{\pi} \int_{s_2}^{\infty} \frac{dx}{(x - s_2)^{\frac{3}{2}} (x - s)} \ln \left| \frac{F(x)}{F(s_2)} \right| \right\} . \tag{19}$$

The relation implies, for instance, that the magnitude of the form factor in the timelike region also determines the charge radius.[c]

Since the value at the origin is the charge, $F(0) = 1$, the magnitude of the form factor must obey the sum rule

$$\frac{8M_\pi^3}{\pi} \int_{s_2}^{\infty} \frac{ds}{s(s - s_2)^{\frac{3}{2}}} \ln \left| \frac{F(s)}{F(s_2)} \right| = \ln |F(s_2)| \ . \tag{20}$$

A second sum rule follows from the asymptotic properties. For the quantity $\ln F(s)$ not to grow more rapidly than the logarithm of s, the function $\psi(s)$ must tend to zero more rapidly than $1/s$. Hence the magnitude of the form factor must obey the condition

$$\frac{2M_\pi}{\pi} \int_{s_2}^{\infty} \frac{ds}{(s - s_2)^{\frac{3}{2}}} \ln \left| \frac{F(s)}{F(s_2)} \right| = 0 \ . \tag{21}$$

The relations (20) and (21) are necessary and sufficient for the existence of an analytic continuation of the boundary values of $|F(s)|$ on the cut that (a) is free of zeros, (b) satisfies the condition $F(0) = 1$ and (c) behaves properly for large values of $|s|$.

The above relations only hold if the form factor does not have zeros. In the scattering amplitude, zeros necessarily occur, as a consequence of chiral symmetry – indeed, the main low energy properties of the scattering amplitude may be viewed as consequences of the Adler zeros [40]. For

---

[c]For a detailed discussion of the interrelation between the behaviour in the spacelike and timelike regions, in particular also in the presence of zeros, I refer to [39].

the form factor, however, chiral perturbation theory implies that zeros can only occur outside the range where the low energy expansion holds: For the form factor to vanish, the higher order contributions must cancel the leading term of the chiral perturbation series.

In quantum mechanics, the form factor represents the Fourier transform of the charge density. For the ground state of the hydrogen atom, for instance, the charge density of the electron cloud is proportional to the square of the wave function, which does decrease with distance, so that the corresponding form factor is positive in the spacelike region. It does not have any complex zeros, either. The wave functions of radially excited states, on the other hand, contain nodes, so that the form factor does exhibit zeros. Qualitatively, I expect the properties of the pion charge distribution to be similar to the one of the electron in the ground state of the hydrogen atom – in the null plane picture, the form factor again represents the Fourier transform of the square of the wave function [41]. In simple models such as those described in [42], the form factor is free of zeros.

The hypothesis that the form factor does not contain zeros can be tested experimentally: The sum rules (20) and (21) can be evaluated with the data on the magnitude of the form factor. The evaluation confirms that the sum rules do hold within the experimental errors, but in the case of the slowly convergent sum rule (21), these are rather large. Alternatively, we may examine the properties of the form factor obtained by fitting the data with the representation (7). By construction, the first two factors in that representation are free from zeros, but the term $G_4(s)$ may or may not have zeros. In fact, as we are representing this term by a polynomial in the conformal variable z, it necessarily contains zeros in the z-plane – their number is determined by the degree of the polynomial. The question is whether some of these occur on the physical sheet of the form factor, that is on the unit disk $|z| \leq 1$. The answer is negative: we invariably find that all of the zeros are located outside the disk. It is clear that zeros at large values of $|s|$ cannot be ruled out on the basis of experiment. In view of asymptotic freedom, however, I think that such zeros are excluded as well.

10. Conclusion

The recent progress in our understanding of $\pi\pi$ scattering provides a solid basis for the low energy analysis of the pion form factor. The main problem encountered in this framework is an experimental one: the data on the processes $e^+e^- \to \pi^+\pi^-$ and $\tau \to \pi^-\pi^0\,\nu_\tau$ are not consistent with our understanding of isospin breaking. If the data are correct, then this represents a very significant failure of the Standard Model – or at least of our understanding thereof. The discrepancy must be clarified also in order for the accuracy of the Standard Model prediction to become comparable with the fabulous precision at which the magnetic moment of the muon has been measured.

Acknowledgment

It is a great pleasure to thank Volodya Eletsky, Misha Shifman and Arkady Vainshtein for their warm hospitality. I very much profited from the collaboration with Irinel Caprini, Gilberto Colangelo, Simon Eidelman, Jürg Gasser and Fred Jegerlehner and I am indebted to Andreas Höcker, Achim Stahl, Alexander Khodjamirian and Gérard Wanders for useful comments. Part of the work reported here was carried out during the Workshop on Lattice QCD and Hadron Phenomenology, held in Seattle. I thank the Institute of Nuclear Theory, University of Washington and the Humboldt Foundation for support.

References

1. G. W. Bennett *et al.* [Muon g-2 Collaboration], hep-ex/0208001.
2. J. Schwinger, Phys. Rev. **73** (1948) 416.
3. J. Calmet, S. Narison, M. Perrottet and E. de Rafael,
   Phys. Lett. B **61** (1976) 283.
4. B. Lautrup, Phys. Lett. B **69** (1977) 109.
5. J. Gasser and H. Leutwyler, Nucl. Phys. B **250** (1985) 517.
6. G. Colangelo, M. Finkemeier and R. Urech, Phys. Rev. D **54** (1996) 4403;
   J. Bijnens, G. Colangelo and P. Talavera, JHEP **9805** (1998) 014;
   P. Post and K. Schilcher, hep-ph/0112352;
   J. Bijnens and P. Talavera, JHEP **0203** (2002) 046.
7. J. F. Donoghue, J. Gasser and H. Leutwyler, Nucl. Phys. B **343** (1990) 341;
   J. Gasser and U. Meißner, Nucl. Phys. B **357** (1991) 90;
   J. F. Donoghue and E. S. Na, Phys. Rev. D **56** (1997) 7073;
   I. Caprini, Eur. Phys. J. C **13** (2000) 471;
   A. Pich and J. Portolés, Phys. Rev. D **63** (2001) 093005;
   J. A. Oller, E. Oset and J. E. Palomar, Phys. Rev. D **63** (2001) 114009.
8. J. F. De Trocóniz and F. J. Yndurain, Phys. Rev. D **65** (2002) 093001.
9. M.F. Heyn and C.B. Lang, Z. Phys. C **7** (1981) 169.
10. B. Ananthanarayan, G. Colangelo, J. Gasser and H. Leutwyler,
    Phys. Rept. **353** (2001) 207.
11. G. Colangelo, J. Gasser and H. Leutwyler, Nucl. Phys. B **603** (2001) 125.
12. S. Descotes, N. H. Fuchs, L. Girlanda and J. Stern,
    Eur. Phys. J. C **24** (2002) 469.
13. S. M. Roy, Phys. Lett. B **36** (1971) 353.
14. S. Weinberg, Phys. Rev. Lett. **17** (1966) 616.
15. M. Gell-Mann, R. J. Oakes and B. Renner, Phys. Rev. **175** (1968) 2195.

16. J. Gasser and H. Leutwyler, Phys. Lett. B **125** (1983) 325; Annals Phys. **158** (1984) 142.
17. M. Knecht, B. Moussallam, J. Stern and N. H. Fuchs, Nucl. Phys. B **457** (1995) 513. ibid. B **471** (1996) 445.
18. S. Pislak *et al.* [BNL-E865 Collaboration], Phys. Rev. Lett. **87** (2001) 221801.
19. G. Colangelo, J. Gasser and H. Leutwyler, Phys. Rev. Lett. **86** (2001) 5008.
20. R. R. Akhmetshin *et al.* [CMD-2 Collaboration], Phys. Lett. B **527** (2002) 161.
21. R. Barate *et al.* [ALEPH Collaboration], Z. Phys. C **76** (1997) 15.
22. K. W. Edwards *et al.* [CLEO Collaboration], Phys. Rev. D **61** (2000) 072003.
23. K. Ackerstaff *et al.* [OPAL Collaboration], Eur. Phys. J. C **7** (1999) 571.
24. A. Hoefer, J. Gluza and F. Jegerlehner, Eur. Phys. J. C **24** (2002) 51.
25. V. Cirigliano, G. Ecker and H. Neufeld, Phys. Lett. B **513** (2001) 361; JHEP **0208** (2002) 002.
26. M. Davier, S. Eidelman, A. Höcker and Z. Zhang, hep-ph/0208177.
27. K. Hagiwara *et al.* [Particle Data Group], Phys. Rev. D **66** (2002) 010001.
28. M. Knecht and A. Nyffeler, Phys. Rev. D **65** (2002) 073034.
29. M. Knecht, A. Nyffeler, M. Perrottet and E. de Rafael, Phys. Rev. Lett. **88** (2002) 071802.
30. M. Ramsey-Musolf and M. B. Wise, Phys. Rev. Lett. **89** (2002) 041601.
31. E. de Rafael, hep-ph/0208251.
32. G. P. Lepage and S. J. Brodsky, Phys. Lett. B **87** (1979) 359; Phys. Rev. D **22** (1980) 2157; A. V. Efremov and A. V. Radyushkin, Phys. Lett. B **94** (1980) 245; Theor. Math. Phys. **42** (1980) 97; V. L. Chernyak and A. R. Zhitnitsky, JETP Lett. **25** (1977) 510; Sov. J. Nucl. Phys. **31** (1980) 544; G. R. Farrar and D. R. Jackson, Phys. Rev. Lett. **43** (1979) 246.
33. J. Volmer *et al.* [The Jefferson Lab F(pi) Collaboration], Phys. Rev. Lett. **86** (2001) 1713.
34. S. R. Amendolia *et al.*, Nucl. Phys. B **277** (1986) 168.
35. C. J. Bebek *et al.*, Phys. Rev. D **17** (1978) 1693.
36. D. Bollini *et al.*, Nuovo Cim. Lett. **14** (1975) 4188.
37. J. Milana, S. Nussinov and M. G. Olsson, Phys. Rev. Lett. **71** (1993) 2533.
38. J. Bijnens and A. Khodjamirian, hep-ph/0206252.
39. B. V. Geshkenbein, Yad. Fis. **9** (1969) 1932; Yad. Fis. **13** (1971) 1087; Z. Phys. C **45** (1989) 351; Phys. Rev. D **61** (2000) 033009.
40. M.R. Pennington and J. Portolés, Phys. Lett. B **344** (1995) 399.
41. H. Leutwyler, Nucl. Phys. B **76** (1974) 413.
42. W. Jaus, Phys. Rev. D **44** (1991) 2851.

MULTIPLE USES OF THE QCD INSTANTONS

E. V. SHURYAK

Department of Physics and Astronomy,
SUNY Stony Brook, NY 11794, USA
E-mail: shuryak@dau.physics.sunysb.edu

In spite of very general title, this is not a mini-review on instantons, but a sum of two unrelated new developments. The first deals with the *pion and nucleon formfactors* which were found to be calculable in a simple single-instanton approximation. The second is summary of several recent papers on production of the topological sphaleron-type gluomagnetic clusters in high energy reactions, which clarified their properties and production cross section. The paper ends with some estimates claiming that the latter process can be important for heavy ion collisions at RHIC energies.

1. Arkady and Modern Mythology

Before I proceed to physics, let me say few words about Arkady. It is not recollections about the past (some of which I contributed elsewhere) but few random thoughts which I had after several other speakers at this meeting have appealed to old and modern mythology in order to explain Arkady's role in our science. Adding a bit to those was just too tempting to resist.

Matt Strassler have made a conjecture that Arkady belongs to a long-lost tribe of *Arkadians* originating from Arcadia, presumably the best corner of the ancient Greece, with exceptionally pleasant climate and always happy inhabitants.

Indeed, sometimes one meets *Arcadians* among us: not only they themselves are happy, but so become people around them. If some need another example think e.g. about Arkady Migdal. As for the climate, the Arkady's worldline would indicate rather preference to cold climates – Siberia and Minnesota. However, my theory is that it is a *moral* climate which really matters: and as long experience (mine included) shows both places are actually quite warm in this respect.

The major question to be explained here is: *Why* are Arcadians so happy? An answer has been suggested in the talk by Frank Wilczek, who said about Arkady that " *the Force is with him*"[a]. Well, we all belong to a sect dealing with more or less mysterious forces. Let me further suggest that the *Force* (with capital F) Frank actually meant is nothing but *Deep Understanding* of what he was doing: this is the only thing which can make miracles and lead to success and happiness.

Being somewhat carried away by this mythology, let me add few more speculations. As predicted by Star Wars (made at the time of the Evil Empire was quite real) it later collapsed. That happened however not so much because some knights (scientists?) were fighting with it: in fact only few of them did. Part of the reason they were not interested in the Empire status was, I think, an apparent lack of *Force* ($\equiv$ Understanding of what they are doing) on the part of the Empire leaders. However in the back-sight it is clear that there were many things worth studying. The minds of the bosses were filled by what we would technically call a *False Vacuum*, and, when its fraction reached the critical magnitude, the whole thing imploded (without any bombs and fighter pilots but in perfect agreement with the semiclassical theory). The reason I mention it here: some amount of similar and dangerous stuff have penetrated the minds of some theorists, to the extent that quite a few (myself included) worry if its concentration may slowly approach the critical value.

2. Are all hadrons alike? Consider the Pion and Nucleon Formfactors

As many still remember, the first part of the title of this section is borrowed from Novikov, Shifman, Vainshtein and Zakharov work [1], in which it has been argued that all spin zero channels are special. In contrast to other channels, those display exceptionally strong non-perturbative effects not described by standard OPE.

My attempts [2] to relate all of those effects to each other and other phenomenological facts was the beginning of the instanton liquid model. Although these works mostly dealt first with a simple single-instanton calculations, it was clear from the onset that the main phenomenon is chiral symmetry breaking and creation of the quark condensate, and that can only be a product of the whole *ensemble* of instantons working together. Eventually, the instanton liquid was formulated with proper ensemble, and was then solved in the mean field and numerically, to *all orders in t' Hooft inter-*

---

[a]In unlikely case somebody is not familiar with Star Wars mythology, it describes a sect of knights which spend time dealing with a mysterious Force and fighting with the Evil Empire.

action, see review [3]. Over the years, I presented parts of this development in previous "advances", and would not discuss it again.

Before we go to technical details, let me remind the issue to the be discusses in the old-fasioned way. Basically, there are two kind of hadrons: some of them are *bound by confinement* and some "from within", *by short-range instanton-induced forces*. The ρ meson is an example of the former case, and the *pion* represents the latter variety. If one would like to connect their formfactors and wave functions to some Schreodinger-like equation, the effective potential would be rising potential in the former case and something like a delta function at the origin (like for the deuteron) in the latter. For convenience, taking quadratic (rather than linear) potential in the former case, one ends up with Gaussian wave function and formfactor. In the latter case a cusp at the origin would lead to a formfactor decreasing only as a power of the momentum transfer.

This leads us to the main idea: in order to tell experimentally one kind of hadrons from the other, one has to *study formfactors at sufficiently large momentum transfer* Q[b] .

Now what nucleon has to do with it? The *nucleon* as a whole is not like the pion, but 2/3 of it – the so called *scalar diquark* – is[c].

The work itself is mostly due to my graduating student Pietro Faccioli. The objective of this work was evaluation of the pion and nucleon formfactors in the instanton liquid model. We naturally wondered how good the results would be and whether they would in particular agree with new experimental results for both at relatively high $(Q \sim 1 - 3\,GeV)$ momentum transfer where the transition from non-perturbative to one-gluon exchange is supposed to happen.

What we have not anticipated at the beginning was the existence of the *window* in which the *single instanton approximation* (SIA) can be used for description of these two formfactors. If so, one can use nice analytic results rather than numerics. Second, the model dependence is maximally reduced: the results depend only on one dimensional parameter, the average instanton size, $\bar{\rho} \approx 1/3\,fm$. The calculation is independent of the properties of the instanton liquid ensemble such as the instanton-instanton interaction and the instanton density. Moreover, those results could easily be obtained 20 years ago. And, let me repeat again, the same calculation

---

[b]Unfortunately one cannot go to too large Q either, because pQCD gluon exchange lead to some universal power dependences for all hadrons. We mean here Q above the region where the pQCD asymptotics starts to work.

[c]In fact, in the 2-color QCD they are even identical.

for regular hadrons like ρ, Δ (instead of π, N) would just produce trivial zero. What it means is that those particles are expected to have *softer* formfactors, going to their perturbative high-Q asymptotics much more rapidly. Unfortunately, we cannot readily test it experimentally[d]: maybe lattice calculations will eventually do it.

A comparison of the pQCD asymptotics and the experimental data determines the momentum scale where the perturbative regime of QCD is reached. Recently, the charged pion formfactor has been measured very accurately at momentum transfers $0.6 \, GeV^2 < Q^2 < 1.6 \, GeV^2$ by the Jefferson Laboratory (JLAB) F_π collaboration [4]. Not only are the data at the highest experimentally accessible momenta very far from the asymptotic limit, but the trend is still away from it (see Fig. 1(a)).

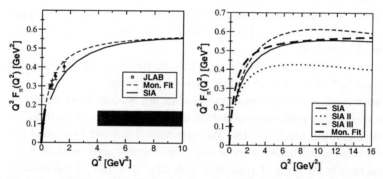

Figure 1. (a) The recent JLAB data for $Q^2 F_\pi(Q^2)$ in comparison with the asymptotic pQCD prediction (thick bar, for a typical $\alpha_s \approx 0.2 - 0.4$), the monopole fit (dashed line), and our SIA calculation (solid line). The SIA calculation is not reliable below $Q^2 \sim 1 \, GeV^2$. The solid circles denote the SLAC data. (b)The dependence of the pion formfactor $Q^2 F_\pi(Q^2)$ on the instanton size distribution. The SIA (solid) curve represents a small-size 't Hooft distribution with a Fermi distribution cutoff and lattice QCD parameters. The SIA II (dotted) curve has a different mean instanton size $\bar{\rho} = 0.47 \, fm$ (same width) and the SIA III (dashed) curve is obtained with the simplest delta distribution $n(\rho) = \bar{n} \, \delta(\rho - 1/3 \, fm)$.

In the pion channel (as well as other scalar and pseudoscalar channels) there is the so called direct instanton contribution, which can be represented by the zero-mode terms in the quark propagators. As a result the effect is enhanced (relative to e.g. the vector or axial channels) by a factor $1/(m^\star \rho)^2$, where $m^\star$ denotes the effective quark mass[e] defined and discussed in detail

---

[d]Except for the transitional formfactors like $N - \Delta$ one, which is indeed softer.

[e]Although the results we obtain are independent of the value of $m^\star$, the window of

in [5]. Numerically, the enhancement factor is about 30, and parametrically it is the inverse diluteness of the instanton ensemble. As a result, instanton-induced forces are dominant in the pion pseudoscalar correlator, starting from rather small distances. The same enhancement exists for the three-point correlator, which is related to the pion formfactor.

This feature, however, depends on the particular three-point function under investigation. For example, the enhancement is absent, when one considers the pion contribution to the axial correlator. Similarly, there is no such enhancement of the $\gamma\gamma^\star\pi^0$ neutral pion transition formfactor. The relevant instanton effects for this process are not due to (enhanced) zero modes, but are either related to non-zero mode propagators in the instanton background or to multi-instanton effects, which are suppressed by the instanton diluteness. This conclusion is nicely supported by recent CLEO measurements of this formfactor, which indeed show that the asymptotic pQCD regime is reached much earlier, at $Q^2 \sim 2\,GeV$ [6].

I appologize for not going into a long history of formfactors here, see reviews for that [7,8]. The first calculation of the pion formfactor in the Single Instanton Approximation was performed by Forkel and Nielsen [9], who complemented QCD sum rules by the instanton-induced contribution. In such an approach, however, a model description of the continuum of excitations with the quantum numbers of the pion is absolutely needed. In order to avoid such additional model dependence, Blotz and Shuryak proposed a different approach [10], in which large-sized three-point functions obtained both from simulations in the Instanton Liquid Model (ILM) were compared *directly in coordinate space*. This analysis revealed that, at large distances, the ILM results were completely consistent with the monopole fit. The radius of the pion was shown to be determined by the instanton size, and experimental value corresponding to $\rho = .35\,fm$.

The central prediction following from our SIA calculation is shown in Fig. 1(a) in comparison to the recent JLAB measurements. We find the very intriguing result that the instanton contribution to the formfactor is completely consistent with the monopole fit at intermediate momentum transfers, $2\,GeV^2 < Q^2 \lesssim 10\,GeV^2$, where the vector dominance model has no justification. It is a first microscopic study showing that all other resonances with quantum numbers of the ρ (except the ρ itself) are not seen in the pion formfactor. For large momenta transfer, $Q^2 > 20\,GeV^2$, our SIA breaks down, as it is necessary to increase the distances in order to isolate the pion ground state. At these needed distances, however, the

---

applicability of our method does depend on it.

correlation functions will become sensitive to multi-instanton effects.

We used the spatial Fourier transforms of the Euclidean three-point function and two-point function,

$$G_\mu(t, \mathbf{p} + \mathbf{q}; -t, \mathbf{p}) = \int d^3\mathbf{x}\, d^3\mathbf{y}\, e^{-i\,\mathbf{p}\cdot\mathbf{x}+i\,(\mathbf{p}+\mathbf{q})\cdot\mathbf{y}}$$

$$\times \langle 0|\, j_5(t, \mathbf{y})\, J_\mu(0, \mathbf{0})\, j_5^\dagger(-t, \mathbf{x})\, |0\rangle, \tag{1}$$

$$G(2t, \mathbf{p}) = \int d^3\mathbf{x}\, e^{i\,\mathbf{P}\cdot\mathbf{x}}\, \langle 0|\, j_5(t, \mathbf{x})\, j_5^\dagger(-t, \mathbf{0})\, |0\rangle, \tag{2}$$

where the pseudo-scalar current $j_5(x) = \bar{u}(x)\, \gamma_5\, d(x)$ excites states with the quantum numbers of the pion and $J_\mu(0)$ denotes the electro-magnetic current operator. In the large t limit (at fixed momenta), both correlation functions are dominated by the pion pole contribution and the ratio of the three-point function to the two-point function becomes proportional to the pion formfactor [11]. In the Breit frame, $\mathbf{p} = -\mathbf{q}/2$ and $Q^2 = \mathbf{q}^2$, one has simply

$$\frac{G_4(t, \mathbf{q}/2; -t, -\mathbf{q}/2)}{G(2t, \mathbf{q}/2)} \rightarrow F_\pi(Q^2). \tag{3}$$

Notice that the LHS of Eq. (3) should not depend on t, for t large enough. Below we demonstrate that, for the pion, this is achieved already for $t \sim 0.6\, fm$. The LHS of Eq. (3) is calculated in the SIA analytically, till some last integrals.

In principle, it is not obvious that such an approach is justified, as the mean Euclidean distance between two instantons in the vacuum is about $1\, fm$. Therefore, if $t \sim 1\, fm$, one would expect many instanton effects to play a non negligible role. However, two of the authors showed that the pion and nucleon three-point functions, evaluated in the SIA, agree with the results of numerical simulations in the instanton liquid model up to distances of the order of $1\, fm$ [5]. Moreover, they found that the ratio of three- to two-point function is dominated by single instanton effects to even larger distances [12]. This result enables us to reliably evaluate this ratio in the SIA up to Euclidean distances of $\sim 1.4\, fm$.

As the momentum is increased, one requires larger times in order to isolate the pion pole from the higher excitations. In sum-rule approaches, contributions from the continuum are obtained from the free and perturbative gluon exchange correlation functions. Perturbative contributions will never develop a pion pole. Indeed, we observed that the free correlators, $G_4^{(free)}$ and $G^{(free)}$, become non negligible for $|\mathbf{q}| \gtrsim 4 - 5\, GeV$. The presence of such continuum contributions destroys the time-independence of the ratio of three- to two-point functions around $t \sim 0.6\,fm$ and hence the method can no longer be used. In principle, one could extend the range of validity of our approach by increasing the time. However we recall that, when t becomes larger, multi-instanton effects are significant and the SIA breaks down. We conclude that our approach is not applicable to the study of the high momentum transfer region, $Q^2 > 20\,GeV^2$.

Since the pion formfactor may experimentally be rather accurately measured, it is instructive to ask whether such data may shed some light on the instanton size *distribution*. In Fig. 1(b), we have plotted the results of our theoretical predictions for $Q^2 F_\pi(Q^2)$ obtained for different cases of $n(\rho)$. We contrast the simplest size distribution [2], $n(\rho) \sim \delta(\rho - 1/3\,fm)$, to the results obtained from a lattice QCD parameterization. We notice that the presence of tails in the size distribution introduces only small corrections to the formfactor. Therefore, we conclude that the simplest distribution $n(\rho) \sim \delta(\rho - 1/3\,fm)$ indeed captures the relevant features for the pion formfactor at intermediate momentum transfer. We observe that our result becomes closer to the perturbative limit, if the average instanton size is larger or possibly if there is an asymmetric tail toward larger-sized instantons in the distribution.

In Fig.1(b) we also compare the results from distributions which have the same small-size limit, but cut off at different instanton sizes, $\bar{\rho} = 0.37\,fm$ and $0.47\,fm$. We observe that, throughout the entire kinematic region we have considered, our predictions are quite sensitive to the average instanton size. This implies that the asymptotic region, where the pion formfactor probes only the small-size semiclassical behavior, is not *reached within the window* where our approach is justified. So the answer is not yet 100 percent model independent: the cutoff in the instanton size (of still unknown origin) is needed.

Let us now move to the nucleon formfactors which are currently subject to a renewed experimental interest as well. At low momenta, the proton electric and magnetic form factors can be very well described by the same dipole fit, $G^p_{E(M)\,dip} = e(\mu)/(1 + Q^2/M^2_{dip})^2$, where $M_{dip} = 0.84\,rmGeV$. For larger momenta $(Q^2 \gtrsim 2\,rmGeV^2)$, however, recent measurements at JLab show that the electric form factor falls off faster than the magnetic one. On the other hand, the electric form factor of the neutron has been measured up to $Q^2 \approx 2\,rmGeV^2$ and was found to be small and positive.

Not going into general theoretical discussion (I again apologize) let me jump directly to instanton-based works. Our first paper calculated the proton electro-magnetic three-point function in coordinate space [12] both from numerical simulations in the ILM, i.e., including multi-instanton effects, and analytically in the SIA. The Green function evaluated theoretically was then compared to a phenomenological one derived from the Fourier transform of several parameterizations of the experimental data. This approach had the advantage to consider large-sized correlation functions, for which the contribution of the continuum of excitations was certainly negligible. However, such a procedure has the shortcoming that it does not allow a direct comparison to the experimental data. From a theoretical point of view, the main result was that the proton electro-magnetic three-point function is completely dominated by the contribution of a single instanton, up to surprisingly large distances of $\approx 1.8 \ rmfm$. From a phenomenological point of view, it was shown that the ILM predictions are consistent with a deviation of G_E^p from the dipole fit.

The second one, which I would briefly describe here, use SIA and are similar to the pion one discussed above. In the wall-to-wall formalism, the electric form factors can be extracted from a combination of three- to two-point functions. In particular, we choose to work in the Breit frame and consider the following spatial Fourier transform of the Euclidean three-point correlator

$$G_4(t, \mathbf{q}/2; -t, -\mathbf{q}/2) =$$

$$\int d^3\mathbf{x}\, d^3\mathbf{y}\, e^{i\,\mathbf{q}\cdot(\mathbf{x}+\mathbf{y})/2} \langle 0|\, Tr\, \eta_{sc}(t, \mathbf{y})\, J_4(0, 0)\, \bar{\eta}_{sc}(-t, \mathbf{x})\, \gamma_4\, |0\rangle. \tag{4}$$

J_4 is the fourth component of the electro-magnetic current operator and $\eta_s(x)$ is the so-called nucleon scalar current which, in the case of the proton, reads [f]

$$\eta_s(x) = \epsilon_{abc}\, [u^a(x) C\gamma_5 d^b(x)]\, u^c(x). \tag{5}$$

In this talk I focus on the *electric* form factors of the proton as it comes from maximally enhanced diagrams. The instanton contribution to the *magnetic* form factors of the nucleon can be extracted from a different combination of the three- and two-point functions [11,13], which however receives only sub-leading contributions from a single instanton. Small neutron formfactor depends on the sea contribution: see paper for discussion.

In order to access the form factors, we need to ensure that, at the largest value of the Euclidean time allowed by the SIA, the nucleon pole is reasonably isolated in both the two- and three-point correlation functions.

---

[f]We note that the nucleon scalar current contains explicitly the operator which excites a scalar ud diquark. It is this combination that couples strongly to the instanton zero modes. The corresponding operator for the neutron is obtained through the substitution $u \leftrightarrow d$.

This worked for the pion, essentially due to the large separation from its resonances. For the nucleon, this is a more delicate task, because the first resonance with the same quantum numbers, $N(1440)$, is only a few hundred MeV heavier than the ground state. This implies that larger time intervals are needed to separate their contributions. Therefore, it is a priori not guaranteed that there is a window, in which the SIA is reliable and the nucleon is isolated from its resonances.

We extract the electric form factors from the ratio

$$2\,\Lambda_{sc}^2(t)\left(\frac{\omega_{\mathbf{q}/2}(t)}{M(t)}\right)^2\frac{G_4^{p(n)}(t,\mathbf{q}/2,-t,-\mathbf{q}/2,\mathbf{t})}{(G(t,\mathbf{q}/2))^2}\to G_E^{p(n)}(Q^2),\qquad(6)$$

where $\Lambda_{sc}(t)$, $M(t)$ and $\omega_{\mathbf{q}/2}(t) \doteq \sqrt{\mathbf{q}^2/4 + M(t)^2}$ denote the values extracted from a fit of the two point function $G(t,\mathbf{q}/2)$ keeping only the nucleon contribution in the spectral decomposition. The ratio of $G_4(2t)/G(t)^2$ ensures that both correlators can be calculated reliably in the SIA. The two factors of the two-point function are needed to sustain the nucleon pole over the total Euclidean distance $2\,t$ with a necessary $t = 0.9$ fm. It corresponds to the leading order in a virial expansion in the instanton density. This procedure is at the expense of a dependence on the multi-instanton induced parameters, namely the quark effective mass m^* and the average instanton density $\bar{n}$, or equivalently a dependence on the nucleon coupling Λ_{sc} and mass. For the pion form factor, it could be achieved that such a dependence cancels in the calculation, with only the average instanton size $\bar{\rho}$ remaining.

In Fig. 2, we give the resulting proton electric form factor in the SIA, obtained for different values of the Euclidean time. We observe that our outcome is nearly independent on the chosen time interval restricted to the SIA window. This indicates a cancellation between the small contribution from the excited states to the numerator and the denominator in Eq. (6).

Summarizing, we have computed the single instanton contribution to pion and nucleon electric form factors. We have carried out the calculations analytically in a window of momentum transfer in which the single instanton approximation can be used. The range of validity of the SIA for the nucleon was found to correspond to momentum transfers $Q^2 \sim 1-4\ rmGeV^2$. At smaller Q^2, multi-instanton effects appear, and at larger Q^2, there are admixtures of excited states. Already the existence of such a window is highly non-trivial. The physical reason is that the (single) instanton induces a compact diquark structure inside the nucleon. Other baryons, such as Δ do not have such a structure and thus are expected to have much smaller form factors at intermediate Q.

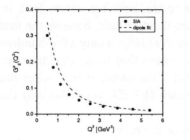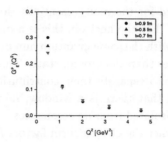

Figure 2. The proton electric form factor $G_E^p(Q^2)$ compared to monopole fit (a) and as a function of Euclidean time t (b). Although the relevant Green functions receive a contribution from the excited states of the order of 10%, for $t = 0.9$ $rmfm$, the form factor is independent of t.

3. How Quantum Mechanics of the Yang-Mills Fields may help us understand the RHIC Puzzles

As nearly all papers mentioned at this conference, the one I would like to discuss next [14] can be traced back to one of the Arkady's works. In this case it is McLerran, Vainshtein and Voloshin [15] in which these authors studied electroweak instanton-induced processes and argued that those may become strong at high energies and instanton-induced reaction may be experimentally observable. Although the latter hopes were destroyed by subsequent development [16] and left those three authors strongly disappointed, the idea itself has its own life.

Recently it was suggested that similar instanton-induced processes are important in QCD, for high energy hadronic scattering, explaining the cross section growth with energy (or the so called *"soft pomeron"* pole) [18,19]. Estimates of its parameters – the pomeron intercept and the slope – made in these works have produced reasonable results, with some additional qualitative insights (e.g. there is no odderon in this picture [19]). We will not discuss these works now.

More recent developments is rather amazing. The first, based on 3 major papers [20,21,22], have lead to much better understanding of what actually is produced and how to evaluate the cross section. Below we would like to discuss the simplest (Minkowski) part of the story.

The second part is based on two papers [14,23] devoted to the role of cluster production in heavy ion collisions at RHIC energies. It is argued that those processes may be very important since such clusters may be produced by the hundreds per collision. They should be crucial supplier of entropy and especially of light quarks, and also should generate noticeable jet quenching. It was also argued that classical description can actually be followed further in heavy ion environment due to very low pQCD cutoff in QGP inside the so called *RHIC window*.

3.1. *Quantum mechanics of the YM fields*

We all know that the *potential energy* of Yang-Mills field versus the topological Chern-Simons coordinate has the shape schematically shown in Fig.3: it is a periodic function with zeros at all *integer* points. The the minimal action path between those are *instanton*, see the line shown by the lowest dashed line).

At the moment of high energy collisions *a sudden localization* of all quantum coordinates including the topological one takes place. Although the values of the coordinates remain the same as they were prior to the collision moment, the system suddenly get real, placed *at or above the barrier* (this case is shown by the dashed line (a) in Fig.3).

(Similar phenomenon is well known perturbatively: the partons – virtual field harmonics – after collisions becomes real outgoing radiation. The on-shell vertex is described by the so called Lipatov vertex and is the basis of BFKL theory of high energy collisions in pQCD framework. Our vertex produce on-shell sphalerons.)

The same phenomenon for non-perturbative virtual fields produce remnants of the interrupted tunneling, already present in the vacuum Another possibility (shown by the dashed line (b) in Fig.3) is that a system at the collision moment is *not* under barrier, but becomes able to tunnel through it because it gets excited enough.

Whatever way the system is driven, it emerges out of the barrier via what we will call *"a turning state"*, at which momenta (gluoelectric fields) are zero. The most important ones, at the top of the potential, are relative of the sphaleron [24] solution of the electroweak theory. From there starts the real time motion outside the barrier (shown by horizontal solid lines): here the action is real and $|e^{iS}| = 1$. That means that whatever happens at this Minkowski stage has the probability 1 and cannot affect the total cross section of the process. However this part of the path is needed for understanding of the properties of the final state.

In [20] the structure of these produced objects have been studied, as well as their subsequent evolution. First of all, the shape of the potential and the corresponding turning states were obtained [20] from the *minimization of the potential energy of a static Yang-Mills fields*, consistent with two appropriate constraints:

(i) fixed value of (corrected) Chern-Simons number N_{CS} .

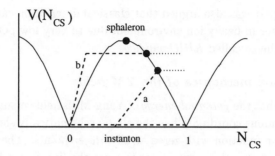

Figure 3. Schematic plot of the energy of Yang-Mills field versus the Chern-Simons number N_{cs}. It is a periodic function, with zeros at integer points. The *instanton* (shown by the lowest horizontal dashed line) is a transition between such points. However if some nonzero energy is deposited into the process during transition, the virtual paths (the dashed lines) emerges from the barrier, via the *turning points* (black circles). The later real time motion outside the barrier (shown by horizontal dotted lines) conserves energy, as the driving force is switched off. The maximal cross section corresponds to the transition around the top of the barrier, the *sphaleron*.

(ii) fixed value of the r.m.s. size R).
To find those one should search for the minimum of the following functional

$$E_{eff} = \frac{1}{2} \int B_m^2 d^3x + R(A_\mu)/\rho^2 + \kappa N_{CS}(A_\mu) \tag{7}$$

where $1/\rho^2, \kappa$ are two Lagrange multipliers. Although these two terms append YM equations and make it more complicated, an *analytical solution is found.* Skipping the details, let me only say that the energy is

$$E_{stat} = 3\pi^2(1 - \kappa^2)^2/(g^2\rho) \tag{8}$$

and (corrected) Chern-Simons number

$$\tilde{N}_{CS} = \text{sign}(\kappa)(1 - |\kappa|)^2(2 + |\kappa|)/4 \tag{9}$$

Eliminating κ, we find a potential and turning points which within accuracy of about 1 percent nicely agree with what was found from the Yung ansatz. The maximum (YM sphaleron) corresponds to $\kappa = 0$, and its energy is $3\pi^2$: the true potential however is exactly symmetric around $N_{cs} = 1/2$. We said before that it has mass about 2.5-3 GeV: we got that assuming that its size is determined by the mean radii of instantons in QCD vacuum, $\rho = 1/3$ fm 2.

Excitation of these states is studied alternatively [30] by some sample paths, described by the so called Yung ansatz [25]: although not exact, it was shown to produce very accurate description of the same states and the potential.

The main part of [30] is description (both numerical and analytic) of the explosive behavior of the turning states. after they being produced. Similar study has been made a decade ago in electroweak theory [26] for the sphaleron, where it has been found that it decays in about 51 W,Z,H quanta. The difference is: in QCD there are no Higgs scalar and its non-zero VEVs, so gluons are massless. This makes the process more *explosive* because all harmonics with different momenta move together, with the speed of light. As described in [20], we had solved it both *numerically* and *analytically* (based on work by Luescher and Schekhter [27]). I skip the technical part of it and only show few results.

At large times the solution becomes a spherical shell of weak field: it has the following energy density

$$4\pi r^2 e(r,t) = \frac{8\pi}{g^2\rho^2}(1-\kappa^2)^2 \left(\frac{\rho^2}{\rho^2+(r-t)^2}\right)^3 \qquad (10)$$

Numerical studies of the problem has been reported in [20] as well. Those are naturally more flexible than analytic and allows for more realistic initial shape of instantons and sphalerons, with exponential (rather than power-like) tails of the fields at large distances[g]. Some of the results are shown in Fig.4. The explosive behavior is very robust by itself (the left part of the figure) but the shape of the resulting energy spectrum $\omega n(\omega)$ (the right part of that figure) is dependent on the exact shape of the initial profile.

Addressing now what happens in the real world of QCD, we now have to include quarks. Here we come to the problem much debated in literature on electroweak theory in 1980's but it is not yet completely solved. In its general form, the issue is to derive an analog of the index theorem for the case when there are outgoing fields: it suppose to tell us how many fermionic levels have crossed zero (and are produced) based on some topological properties of the gauge field alone. The usual form of it, involving a change in Chern-Simons number, is obviously incorrect, as its variation is in general (and specifically for the time-dependent solutions we speak about) not an integer.

Very recent progress in this direction [22] is the derivation of explicit solution to the Dirac eqn in the background of exploding sphalerons. This solution shows how the quarks initially bound in a zero energy state of a sphaleron get accelerated into a finite-energy spectrum of emitted quarks. Naturally, this phenomenon happens simultaneously for the whole set of

---

[g]The phenomenological reasons for exponential rather than power instanton tail are discussed e.g. in [30].

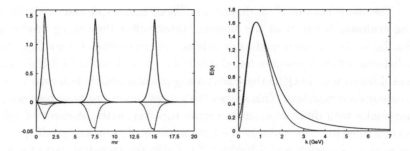

Figure 4. Results of numerical solution of YM equations. The left figure shows the energy density $r^2\epsilon(r)$ at certain times (in fm) and density of Chern-Simons number density (negative values). One can see that the profile is stabilized and no longer changes. The right figure shows the energy spectrum of final gluons, obtained by Fourier decomposition of the fields (solid line), compared to a thermal distribution with T=285 MeV (the dashed line).

light quark pairs, $\bar{u}u\bar{d}d\bar{s}s$. (We have ignored strange quark mass.) So tentatively we estimate the yield of partons per cluster to be about 3 gluons and 6 quarks-antiquarks.

3.2. How many "mini-bangs" are there in Heavy Ion Collisions ?

We have already mentioned in the Introduction that the calculation of the cross section from first principles is not yet available. The original semiclassical approach with vacuum (undeformed) instantons was pioneered by Ringwald and Espinosa [16], who noticed that multi-gluon production is more (not less as in pQCD) probable than few-gluon one. Unfortunately it is good only at low energies of partonic sub-collision, much below the sphaleron mass (10 TeV in electroweak theory and 3 GeV in QCD). The problem to find a general semi-classical answer was known as the so called *holy grail* problem. Three methods toward its solution have been proposed

(i) *Unitarization* of the multi-gluon amplitude when it becomes strong was first suggested by Zakharov and worked out by Shifman and Maggiore [16]. Basically one can treat a sphaleron as a resonance, and even the resulting expression for the cross sections in [19] looks similar to Breit-Wigner formula. This is the most worked out approach, but it still cannot guarantee the numerical values of the cross sections.

(ii) *Landau method with singular instantons* was applied by Diakonov and Petrov [17] (following some earlier works which are cited there) who were able to find the opposite limit of high energies. It follows from the comparison of the two limits, that the peak is indeed very close to the sphaleron mass, and the cross section is very close to be *first order* in instanton diluteness, not the second order as the initial probability. Unfortunately they were not able to find the solution at intermediate times which would provide the turning points of this approach.

(iii) *Classical solution on the complex time plane* [29] is another possible direction, in which a zig-zag shaped path in complex time includes classical evolution and tunneling in one common solution. Unfortunately, this interesting idea also has not been fully implemented, even for toy models with only scalar fields considered in this paper.

The method (ii) was discussed in recent paper [21] were further details of the high energy limit of the process have been found. Explicit solution for the fields lead to realization of what clusters (the turning states at t=0) are actually produced: those happen to be just rescaled YM sphalerons we discussed above. Their subsequent evolution is then (a rescaled version of) classical explosion also.

From parton-model-style phenomenology of hadronic collisions, an estimate for the cross section was developed by G. W. Carter, D. M. Ostrovsky and myself [30]. The main idea was to identify two components of the hh collisions, the color exchanges and the "color objects production", and deduce the corresponding cross sections at the partonic level. We looked at high energy NN, πN, γN, and $\gamma\gamma$ cross sections which all *increase* with energy logarithmically for $\sqrt{s} \sim 100\,GeV$

$$\sigma_{hh'}(s) = \sigma_{hh'}(s_0) + X_{hh'} \ln(s/s_0) \qquad (11)$$

We identified the two components mentioned above with these two terms, respectively, and concentrated on the last (growing) terms. We found that some universal *semi-hard* parton-parton collisions can explain all known $X_{hh'}$. Using fitted structure functions of N, π, γ and simple scaling – each gluon can be counted as 2 quarks[h] – we have expressed all of those with only one parameter, the value of the qq *cross section*. With the fitted value[i]

$$\sigma_{qq} = 1.69 \times 10^{-3} fm^2 \qquad (12)$$

we got the rising part of cross sections for 4 hadronic reactions reported in the Table in [30] well agree with data.

We may now extend our analysis to estimate the number of sphaleron-type clusters produced from excited instantons in heavy ion collisions. For

---

[h]Corresponding to SU(2) Casimir scaling, appropriate for instanton-induced reactions.
[i]Note that simple parametric estimate for this cross section, namely $\pi\rho^2 n_{inst}\rho^4$ gives the right magnitude.

central AA collisions of two Au nuclei we got the following *upper limit*[j] for the production of sphaleron-like clusters at RHIC energies

$$N_{objects} \approx 400, \tag{13}$$

In the real AA collisions thousands of outgoing secondaries include thousands of quark-anti-quark pairs, nearly all produced in the process. At low (several GeV) collision energies those are believed to be generated at hadronization stage, by breaking of the QCD strings, or color tubes. At higher energies, such as RHIC, it was long believed that the produced QGP remains during its evolution of several fm/c mostly a "hot glue", with small quark admixture. Various pQCD calculations concluded the same.

Assuming for cluster production a naive (unshadowed) upper limit one gets for central $AuAu$ collisions at RHIC about $400 * 9 = 3600$ partons may come from clusters, which is comparable to *the maximal value*, given by total entropy (observed multiplicity of hadrons). Surprisingly, the quark/gluon composition roughly matches the equilibrium value for QGP.

References

1. V. A. Novikov, M. A. Shifman, A. I. Vainshtein and V. I. Zakharov, Nucl. Phys. B **191**, 301 (1981).
2. E.V. Shuryak, Nucl. Phys. **B214** (1982) 237.
3. T. Schaefer and E.V. Shuryak, Rev. Mod. Phys. **70** (1998) 323.
4. J. Volmer *et al.* [The Jefferson Laboratory F_π Collaboration], Phys. Rev. Lett. **86** (2001) 1713.
5. P. Faccioli and E.V. Shuryak, hep-ph/0106019, Phys. Rev. **D** (2002) in press.
6. J. Gronberg *et al.*, Phys. Rev. **D57** (1998) 33.
7. V.L. Chernyak and A.R. Zhitnitsky, Phys. Rept. **112** (1984) 173.
8. G.Sterman and P. Stoler, Annu. Rev. Nucl. Part. Sci. **47** (1997) 193.
9. H. Forkel and M. Nielsen, Phys. Lett. **B345** (1997) 55.
10. A. Blotz and E.V. Shuryak, Phys. Rev. **D55** (1997) 4055.
11. T. Draper, R.M. Woloshyn, W. Wilcox, and K.F. Liu, Nucl. Phys. **B318** (1989) 319.
12. P. Faccioli and E.V. Shuryak, Phys. Rev.**D65** (2002) 076002.
13. S.J. Dong, K.F. Liu and A.G. Williams, Phys. Rev. **D58** (1998) 074504.
14. E. Shuryak, "How quantum mechanics of the Yang-Mills fields may help us understand the RHIC puzzles," arXiv:hep-ph/0205031.
15. L. D. McLerran, A. I. Vainshtein and M. B. Voloshin, Phys. Rev. D **42**, 171 (1990).

---

[j]It is the upper limit because we do not know the growing part of the hh cross sections but simply because the "objects" mentioned here can still be either gluons or colored clusters we discuss, or something else: studies of the cross section alone cannot tell the difference. This estimate is also naive: one should correct it for nuclear shadowing.

16. A.Ringwald, Nucl.Phys. **B330** (1990) 1, O.Espinosa, Nucl.Phys. **B343** (1990) 310; V.V. Khoze, A. Ringwald, Phys.Lett. **B259**:106-112, 1991 V.I.Zakharov, Nucl.Phys. **B353** (1991) 683: M.Maggiore and M.Shifman, Phys.Rev. **D46**:3550-3564,1992

17. D. Diakonov and V. Petrov, Phys. Rev. D **50**, 266 (1994) [arXiv:hep-ph/9307356].

18. D. E. Kharzeev, Y. V. Kovchegov and E. Levin, Nucl. Phys. A **690**, 621 (2001) [arXiv:hep-ph/0007182].

19. M. A. Nowak, E. V. Shuryak and I. Zahed, Phys. Rev. D **64**, 034008 (2001) [hep-ph/0012232].

20. D. M. Ostrovsky, G. W. Carter and E. V. Shuryak, "Forced Tunneling and Turning State Explosion in Pure Yang-Mills Theory," arXiv:hep-ph/0204224.

21. R. A. Janik, E. Shuryak and I. Zahed, "Prompt multi-gluon production in high energy collisions from singular Yang-Mills solutions," arXiv:hep-ph/0206005.

22. E. Shuryak and I. Zahed, "Prompt quark production by exploding sphalerons," arXiv:hep-ph/0206022.

23. E.V.Shuryak and I.Zahed, Jet Quenching by QCD Synchrotron-like radiation, in progress.

24. N.Manton, Phys.Rev.D 28 (1983) 2019; F.R.Klinkhamer and N.Manton, Phys.Rev.D30 (1984) 2212.

25. A. V. Yung, Nucl. Phys. B **297**, 47 (1988).

26. J. Zadrozny, Phys. Lett. B **284**, 88 (1992). M. Hellmund and J. Kripfganz, Nucl. Phys. B **373**, 749 (1992).

27. M.Luscher, Phys.Lett.B70 (1977) 321; B.Schechter, Phys.Rev.D16 (1977) 3015.

28. P. Levai and U. W. Heinz, Phys. Rev. C **57**, 1879 (1998) [arXiv:hep-ph/9710463].

29. V. A. Rubakov and D. T. Son, Nucl. Phys. B **424**, 55 (1994) [arXiv:hep-ph/9401257].

30. G. W. Carter, D. M. Ostrovsky and E. V. Shuryak, Phys. Rev. D **65**, 074034 (2002) [arXiv:hep-ph/0112036].

QCD SUM RULES - A WORKING TOOL FOR HADRONIC PHYSICS

ALEXANDER KHODJAMIRIAN *)

*Institut für Theoretische Teilchenphysik, Universität Karlsruhe,
D-76128 Karlsruhe, Germany*

QCD sum rules are overviewed with an emphasize on the practical applications of
this method to the physics of light and heavy hadrons.

1. Introduction

Imagine a big birthday cake for Arkady Vainshtein, each candle on that
cake corresponding to one of his outstanding contributions to the modern
particle theory. I think, a very bright and illuminating candle should then
mark QCD sum rules.

The renown papers introducing QCD sum rules [1] have been published
by Shifman, Vainshtein and Zakharov in 1979. The method, known also
under the nickname of SVZ or ITEP sum rules very soon became quite
popular in the particle theory community, especially in Russia. Not only
experienced theorists, but also many students of that time contributed to
the development of this field with important results. It was indeed a lot
of fun to start with an explicit QCD calculation in terms of quark-gluon
Feynman diagrams and end up estimating dynamical characteristics of real
hadrons. The flexibility and universality of the sum rule method allowed
one to go from one interesting problem to another, describing, in the same
framework, very different hadronic objects, from pions and nucleons to
charmonium and B mesons. Nowadays, QCD sum rules are still being
actively used, providing many important applications and representing an
important branch on the evolution tree of approximate QCD methods.

In this short overview, I start, in Sect. 2, from explaining the basic idea
of sum rules which is rooted in quantum mechanics. After that, in Sect. 3, I
outline the SVZ sum rule derivation in QCD. Some important applications

---

*) on leave from Yerevan Physics Institute, 375036 Yerevan, Armenia

and extensions of the method are listed in Sect. 4.. Furthermore, in Sect. 5 I demonstrate how QCD sum rules are used to calculate the soft contributions to the pion form factor. The light-cone version of sum rules is introduced. Many interesting applications of QCD sum rules remain outside this survey, some of them can be found in recent reviews [2,3].

2. SVZ sum rules in quantum mechanics

To grasp the basic idea of the QCD sum rule method it is sufficient to consider a dynamical system much simpler than QCD, that is quantum mechanics of a nonrelativistic particle in the potential $V(r)$. The latter has to be smooth enough at small distances and confining at large distances. The spherically-symmetrical harmonic oscillator $V(r) = m\omega^2 r^2/2$ is a good example. Evidently, having defined the potential, one is able to solve the problem *exactly* e.g., by means of the Schrödinger equation, $H\psi_n(\vec{x}) = E_n\psi_n(\vec{r})$, with the Hamiltonian $H = \vec{p}^{\,2}/2m + V(r)$, obtaining the wave functions $\psi_n(\vec{r})$ and energies E_n of all eigenstates, $n = 0, 1, \dots$.

As demonstrated in [4], it is possible to use an alternative procedure allowing one to calculate *approximately* the energy E_0 and the wave function at zero, $\psi_0(0)$ of the lowest level. The starting object is the time-evolution operator, or the Green's function of the particle $G(\vec{x}_2, \vec{x}_1; t_2 - t_1)$, taken at $\vec{x}_1 = \vec{x}_2 = 0$ and written in terms of the standard spectral representation:

$$G(\vec{x}_2 = 0, \vec{x}_1 = 0; t_2 - t_1) = \sum_{n=0}^{\infty} |\psi_n(0)|^2 e^{-iE_n(t_2 - t_1)} . \tag{1}$$

Performing an analytical continuation of the time variable to imaginary values: $t_2 - t_1 \to -i\tau$, one transforms Eq. (1) into a sum over decreasing exponents:

$$G(0, 0; -i\tau) \equiv M(\tau) = \sum_{n=0}^{\infty} |\psi_n(0)|^2 e^{-E_n \tau} . \tag{2}$$

The function $M(\tau)$ has a *dual* nature depending on the region of the variable τ. At small τ, the perturbative expansion for $M(\tau)$ is valid, and it is sufficient to retain a few first terms:

$$M^{pert}(\tau) = M^{free}(\tau) \left(1 - 4m \int_0^{\infty} r dr V(r) e^{-2mr^2/\tau} + O(V^2) + \dots \right) , \tag{3}$$

where $M^{free}(\tau) = \left(\frac{m}{2\pi\tau}\right)^{3/2}$ is the Green's function of the free particle motion. Using QCD terminology, we may call the behavior of $M(\tau)$ at small τ "asymptotically free" having in mind that it is approximated by a universal, interaction-free particle motion. Equating (2) and (3) we obtain

$$\sum_{n=0}^{\infty} |\psi_n(0)|^2 e^{-E_n \tau} \simeq M^{pert}(\tau), \qquad (4)$$

a typical *sum rule* which is valid at small τ, relating the sum over the bound-state contributions to the result of the perturbative expansion. Note that the latter includes certain "nonperturbative" or "long-distance" effects too, namely the subleading terms containing the interaction potential V.

At large τ one has a completely different picture. In the spectral representation (2) the entire sum over excited levels dies away exponentially with respect to the lowest level contribution:

$$\lim_{\tau \to \infty} M(\tau) = |\psi_0(0)|^2 e^{-E_0 \tau} . \qquad (5)$$

Thus, at large τ one encounters a typical "confinement" regime, because the lowest level parameters determining $M(\tau)$ essentially depend on the long-distance dynamics (in this case determined by $V(r)$ at large r).

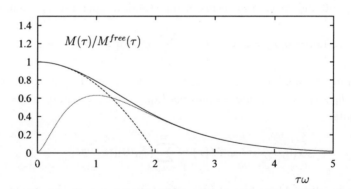

Figure 1. The analytically continued Green's function $M(\tau)$ for a particle in the oscillator $V(r) = m\omega^2 r^2/2$ normalized to $M^{free}(\tau)$ and plotted as a function of $\tau\omega$. The exact solution (solid) is compared with the perturbative calculation including the first order in V correction (dashed) and with the contribution of the lowest bound state (dotted).

An important observation made in [4] is that at intermediate values of τ both descriptions (3) and (5) are approximately valid (see Fig. 1). It is therefore possible to retain only the lowest-level contribution in the sum rule (4) allowing one to estimate both E_0 and $|\psi_0(0)|$, without actually solving the Schrödinger equation.

To further improve the quality of this determination, $M^{pert}(\tau)$ can be rewritten in a form of the integral

$$M^{pert}(\tau) = \int_0^\infty \rho^{pert}(E)e^{-E\tau}dE, \qquad (6)$$

resembling the spectral representation, so that the positive definite function $\rho^{pert}(E)$ can be called perturbative spectral density. The integral in Eq. (6) is then splitted into two parts, introducing some threshold energy $E_{th} > E_0$ and the sum over all excited states $n \geq 1$ in Eq. (4) is approximated by an integral over $\rho^{pert}(E)$ starting from this threshold:

$$\sum_{n=1}^\infty |\psi_n(0)|^2 e^{-E_n\tau} \simeq \int_{E_{th}}^\infty \rho^{pert}(E)e^{-E\tau}dE. \qquad (7)$$

The latter equation can be called a "duality" relation having in mind duality between the asymptotic-freedom regime and the spectral sum. The integral (7) is then subtracted from both sides of Eq. (4) leading to the sum rule for the lowest level:

$$|\psi_0(0)|^2 e^{-E_0\tau} = \int_0^{E_{th}} \rho^{pert}(E)e^{-E\tau}dE. \qquad (8)$$

Note that one could make use of the sum rule relations similar to Eq. (4) in an opposite way. Imagine that the interaction potential is unknown but we have a possibility to measure, for a set of low levels, their wave functions at zero separations and energies experimentally. The sum rule (4) could then be used to extract or at least constrain the potential $V(r)$.

Interestingly, quantum mechanics may also serve as a model for more complicated patterns of nonperturbative interactions, in which case sum rules have to be treated with care. An example presented in [5] is a potential containing two terms: $V(r) = V_0[(r/r_4)^4 + (r/r_{11})^{11}]$ with $r_{11} \ll r_4$, that is, a sharp short-range confining potential combined with a broader one. At $\tau \to 0$, in the perturbative expansion of $M(\tau)$ the correction due to the second term in the potential is much smaller than the correction associated with the first term. However, if one ignores the "nonperturbative effect" related to the short-distance scale r_{11}, the resulting sum rule simply reproduces the lowest level in the potential $V(r) = V_0(r/r_4)^4$. In reality, the physical picture is quite different because it is the $\sim r^{11}$ part of the potential which mainly determines the formation of bound states. The important lesson drawn from this example is: if for some reason a short-distance nonperturbative effect is missing and/or ignored, the sum rule does not work (in other words duality is violated).

Interestingly, the sum rule approach in quantum mechanics can be generalized to calculate more complicated characteristics such as the amplitudes of electric-dipole transitions between the lowest S and P levels in a given nonrelativistic potential [6]. One has to construct a three-point correlation function:

$$
\widetilde{M}(\tau_1, \tau_2) = \left\{ \int dt_3 d\vec{x}_3 \frac{\partial}{\partial|\vec{x}_2|} G(\vec{x}_2, \vec{x}_3; -i\tau_2 - t_3) \right.
$$

$$
\left. \times (\vec{e} \cdot \vec{v}_3) G(\vec{x}_3, \vec{x}_1; t_3 - (-i\tau_1)) \right\}_{\vec{x}_{1,2}=0}, \tag{9}
$$

where $\vec{v} = i(H\vec{x} - \vec{x}H)$ is the quantum-mechanical velocity operator, and $\vec{e} \cdot \vec{v}$ is the operator corresponding to the dipole radiation of a photon with polarization $\vec{e}$. The correlator (9) corresponds to the propagation of a particle in P wave (below threshold or in imaginary time) from point 2 to point 3 where a dipole photon is radiated and then further propagation to point 1 in S wave. Calculating Eq. (9) perturbatively and matching it to the double spectral sum over P and S levels, one gets a sum rule which, at intermediate values of the two variables $\tau_{1,2}$ is well approximated by the contributions of the three lowest E1 transition amplitudes ($1P \to 1S, 2S \to 1P, 2P \to 2S$).

The sum rule approach considered here is, of course not very important for quantum mechanics itself, but as we shall see in a moment, serves as a very convenient prototype for an analogous method in QCD, in the theory where no exact solution is so far available.

3. SVZ Sum rules in QCD

We now move from the safe haven of nonrelativistic quantum mechanics to QCD, a complicated theory with a rich pattern of quark-gluon and gluon-gluon interactions. At short distances, due to asymptotic freedom the theory can still be resolved. One considers a quasi-free quark propagation with calculable perturbative corrections. However, at large distances, $r \sim 1/\Lambda_{QCD}$, the QCD perturbation theory becomes inapplicable and the confinement phenomenon takes over, driven by the quark-gluon fluctuations in the QCD vacuum. As a result, quarks build coherent bound states, hadrons. In general, it is not possible to describe QCD interactions with a potential. Nevertheless, qualitatively, the pattern of quark-antiquark forces in QCD, with asymptotic freedom at small distances and formation of bound states at large distances, is very similar to the quantum-mechanical motion in the confining, oscillator-type potential considered in the previous section. It is therefore not surprising that sum rules [1] analogous to the quantum-mechanical ones exist also in QCD.

The starting object in QCD analogous to the Green's function $G(0,0,t)$ is the correlation function describing an evolution of a colorless quark-antiquark pair emitted and absorbed by external currents. A "classical example" considered in [1] is the correlation of two $j_\mu^\rho = (\bar{u}\gamma_\mu u - \bar{d}\gamma_\mu d)/2$ quark currents with the ρ^0 meson quantum numbers (isospin 1, $J^P = 1^-$):

$$\Pi_{\mu\nu}(q) = i \int d^4x e^{iqx} \langle 0|T\{j_\mu^\rho(x), j_\nu^\rho(0)\}|0\rangle . \tag{10}$$

The dispersion relation (Källen-Lehmann representation) for this correlation function contains a sum over all intermediate hadronic states, a direct analog of the spectral representation (1) :

$$\Pi_{\mu\nu}(q) = \sum_h \frac{\langle 0|j_\mu^\rho|h\rangle\langle h|j_\nu^\rho|0\rangle}{m_h^2 - q^2} + \text{subtractions} . \tag{11}$$

Note that, for brevity, I wrote the above relation in a very schematic way, including the excited ρ resonances and the continuum states with ρ quantum numbers in one discrete sum.

The Borel transformation, $\hat{B}\{1/(m_h^2 - q^2)\} \to \exp(-m_h^2/M^2)$, converts the hadronic representation (11) into a sum over decreasing exponents,

$$\hat{B}\Pi_{\mu\nu} = \sum_h \langle 0|j_\mu^\rho|h\rangle\langle h|j_\nu^\rho|0\rangle e^{-m_h^2/M^2} , \tag{12}$$

i.e., the inverse Borel variable $1/M^2$ plays essentially the same role as the auxiliary variable τ in the quantum-mechanical case. Another very important virtue of the Borel transformation is that it kills subtraction terms in the dispersion relation.

At large spacelike momentum transfers $q^2 < 0$, $Q^2 \equiv -q^2 \gg \Lambda_{QCD}^2$ (corresponding to large $M \gg \Lambda_{QCD}$ after Borel transformation) the quark-antiquark propagation described by the correlation function (10) is highly virtual, the characteristic times/distances being $x_0 \sim |\vec{x}| \sim 1/\sqrt{Q^2}$. One can then benefit from asymptotic freedom and calculate the correlation function in this region perturbatively. The corresponding diagrams up to $O(\alpha_s)$ are depicted in Fig. 2.

As first realized in [1], there are additional important effects due to the interactions with the vacuum quark and gluon fields. The latter have typically long-distance ($\sim \Lambda_{QCD}$) scales and, in first approximation, can be replaced by static fields, the *vacuum condensates*. An adequate framework to include these effects in the correlation function was developed in a form of the Wilson operator product expansion (OPE). The Borel transformed answer for the correlation function (10) reads:

$$\hat{B}\Pi_{\mu\nu}^{OPE} = \hat{B}\Pi_{\mu\nu}^{pert} + \sum_{d=3,4,\dots} \hat{B}C_{\mu\nu}^d \langle 0|O_d|0 \rangle, \tag{13}$$

where the first term on the r.h.s. corresponds to the perturbative diagrams in Fig. 2, whereas the sum contains the contributions of vacuum condensates, ordered by their dimension d. Diagrammatically, these contributions are depicted in Fig. 3. The terms with $d \leq 6$ contain the vacuum aver-

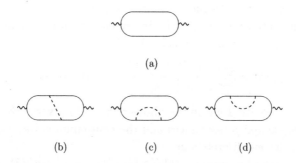

Figure 2. Diagrams determining the perturbative part of the correlation function (10): the free-quark loop (a) and the $O(\alpha_s)$ corrections (b,c,d). Solid lines denote quarks, dashed lines gluons, wavy lines external currents.

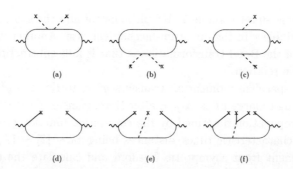

Figure 3. Diagrams corresponding to the gluon (a,b,c,), quark (d), quark-gluon (e) and four-quark (f) condensate contributions to the correlation function (10). The crosses denote the vacuum fields.

ages of the operators $O_3 = \bar{q}q$, $O_4 = G_{\mu\nu}^a G^{a\mu\nu}$, $O_5 = \bar{q}\sigma_{\mu\nu}(\lambda^a/2)G^{a\mu\nu}q$,

$O_6 = (\bar{q}\Gamma_r q)(\bar{q}\Gamma_s q)$, and $O_6^G = f_{abc}G_{\mu\nu}^a G_\sigma^{b\,\nu} G^{c\sigma\mu}$, where $q = u, d, s$ are the light-quark fields, $G_{\mu\nu}^a$ is the gluon field strength tensor, and $\Gamma_{r,s}$ denote various combinations of Lorentz and color matrices. Importantly, to compensate the growing dimension of the operator O_d, the Wilson coefficients $C_{\mu\nu}^d$ contain increasing powers of $1/Q^2$. Correspondingly, $\hat{B}C_{\mu\nu}^d$ contain powers of $1/M^2$, making it possible at large M^2 to retain in the r.h.s. of Eq. (13) only a few first condensates. Thus, at $M^2 \sim 1$ GeV^2 it is practically possible to neglect all operators with $d > 6$.

Equating at large M^2 the hadronic representation to the result of the OPE calculation we obtain the desired sum rule:

$$\sum_h \langle 0|j_\mu^\rho|h\rangle\langle h|j_\nu^\rho|0\rangle e^{-m_h^2/M^2} = \hat{B}\Pi_{\mu\nu}^{pert} + \sum_{d=3,4,..} \hat{B}C_{\mu\nu}^d\langle 0|O_d|0\rangle. \qquad (14)$$

The explicit form of this relation is [1] :

$$f_\rho^2 e^{-m_\rho^2/M^2} + \{\text{excited,continuum } \rho \text{ states}\}$$

$$= M^2\left[\frac{1}{4\pi^2}\left(1 + \frac{\alpha_s(M)}{\pi}\right) + \frac{(m_u + m_d)\langle\bar{q}q\rangle}{M^4}\right.$$

$$\left. + \frac{1}{12}\frac{\langle\frac{\alpha_s}{\pi}G_{\mu\nu}^a G^{a\mu\nu}\rangle}{M^4} - \frac{112\pi}{81}\frac{\alpha_s\langle\bar{q}q\rangle^2}{M^6}\right], \qquad (15)$$

where the decay constant of the ρ meson is defined in the standard way, $\langle\rho^0 | j_\nu^\rho | 0\rangle = (f_\rho/\sqrt{2})m_\rho\epsilon_\nu^{(\rho)*}$. In obtaining this relation the four-quark vacuum densities are factorized into a product of quark condensates. The quark-gluon and three-gluon condensates have very small Wilson coefficients and are neglected. The strong coupling α_s is taken at the scale M which is the characteristic virtuality of the loop diagrams after the Borel transformation. A more detailed derivation of this sum rule can be found, e.g. in the review [3]. The QCD vacuum condensates were recently discussed in [7].

In full analogy with quantum mechanics, there exists a SVZ region of intermediate M^2 where the ρ meson contribution alone saturates the l.h.s. of the sum rule (15). To illustrate this statement numerically, in Fig. 4 the experimentally measured f_ρ (obtained from the $\rho^0 \to e^+e^-$ width) is compared with the same hadronic parameter calculated from Eq. (15) where all contributions of excited and continuum states are neglected. One indeed observes a good agreement in the region $M^2 \sim 1$ GeV^2.

An important step to improve the sum rule (14) is to use the *quark-hadron duality* approximation. The perturbative contribution to the correlation function (the sum of Fig. 2 diagrams) is represented in the form of a

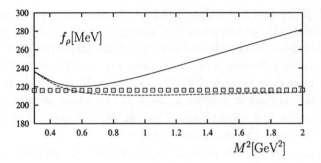

Figure 4.   The ρ meson decay constant calculated from the sum rule (15) neglecting all excited and continuum states (solid), as a function of the Borel parameter, in comparison with the experimental value (boxes). The dashed curve corresponds to an improved calculation , where the sum over excited and continuum states is estimated using quark-hadron duality with a threshold $s_0^\rho = 1.7$ GeV^2.

dispersion integral and splitted into two parts:

$$\hat{B}\Pi_{\mu\nu}^{pert} = \int\limits_0^{s_0} \rho_{\mu\nu}^{pert} e^{-s/M^2} + \int\limits_{s_0}^{\infty} \rho_{\mu\nu}^{pert} e^{-s/M^2} . \tag{16}$$

The sum over excited state and continuum contributions in Eq. (14) is approximated by the second integral over the perturbative spectral density $\rho_{\mu\nu}^{pert}$. This integral is then subtracted from both parts of Eq. (14). Correspondingly Eq. (15) is modified: the l.h.s. contains only the ρ term , and, on the r.h.s., the perturbative contribution has to be multiplied by a factor $(1 - e^{-s_0/M^2})$. The numerical result obtained from the duality improved SVZ sum rule is also shown in Fig. 4. Quark-hadron duality can independently be checked for the channels with sufficient experimental information on excited hadronic states, such as the J/ψ and ρ channels. For ψ resonances one of the first analyses of that type has been done in [8].

Importantly, not all correlation functions lead to valid QCD sum rules. The response of the QCD vacuum to the quarks and gluons "injected" by external currents crucially depends on the quantum numbers and flavour content of the current. After all, that is the main reason why hadrons are not alike [5]. In certain channels, e.g. for the correlation functions of spin zero light-quark currents, specific short-distance nonperturbative effects related to instantons are present (the so called "direct instantons"). These effects remain important even at comparatively large M^2 and are not accountable in a form of OPE. The subtle duality balance between "quasiperturbative" OPE and resonances is destroyed in such cases. Due to instantons, it is not possible, for example, to calculate the pion parameters using correlators with pseudoscalar $\bar{u}\gamma_5 d$ currents [5,9]. Models based on the instanton calculus have to be invoked (see ,e.g.,[10]).

The QCD procedure outlined in this section has indeed many similarities with the sum rule derivation in quantum mechanics. To make the analogy more transparent, in the following table I put together the main points of the two sum rule approaches: in quantum mechanics and in QCD.

Quantum mechanics	QCD
Particle in a smooth confining potential	quark-antiquark pair in QCD vacuum
Green's function $G(0,0,t)$	Correlation function $\Pi_{\mu\nu}(q^2)$
Spectral representation	Dispersion relation in q^2
Analytical continuation $to - i\tau$	Borel transformation $q^2 \to M^2$
Perturb. expansion in powers of V	OPE (Condensate expansion)
matching the lowest level to $M^{pert}(\tau)$	matching the lowest hadron to $\hat{B}\Pi(M^2)$
duality of the quasifree-motion and the spectral sum	quark-hadron duality
extracting $V(r)$ from exp. known spectral sum	extracting condensates, $m_{u,d,s,c,b}$ from exp. known spectral density
sum rule does not work if the short-distance part of the potential is ignored	sum rules do not work in the channels with direct instantons
3-point sum rules for E1-transition amplitudes	3-point sum rules for hadronic matrix elements

4. Applying and extending the method

4.1. *Baryons*

Following very successful applications of QCD sum rules in the mesonic channels [1], the next essential step was to extend the method to the baryonic sector [11,12]. Correspondingly, the correlators of specially constructed quark currents with baryon quantum numbers were considered. A well known example is the Ioffe current with the proton quantum numbers:

$$J^N(x) = \epsilon_{abc}(u^{aT}(x)\hat{C}\gamma_\mu u^b(x))\gamma_5\gamma^\mu d^c(x), \qquad (17)$$

where a, b, c are color indices and $\hat{C}$ is the charge conjugation matrix. From the QCD sum rule for the correlator $\langle 0|J_N(x)J_N^\dagger(0)|0\rangle$ an approximate formula can be obtained,

$$m_N \simeq [-(2.0)(2\pi)^2\langle 0|\bar{q}q|0\rangle(\mu = 1\text{GeV})]^{1/3}, \qquad (18)$$

relating the nucleon mass and the quark condensate density. Thus, QCD sum rules unambiguously confirm the fundamental fact that $\sim 99\%$ of the baryonic mass in the Universe is due to the vacuum condensates.

4.2. *Quark mass determination*

The sum rule relations similar to Eq. (14) are widely used to extract the fundamental QCD parameters, not only the condensates themselves but also the quark masses. One needs sufficient experimental data on hadronic parameters in a given channel (masses and decay constants of ground and excited states, experimentally fitted ansätze for continuum states) in order to saturate the hadronic part of the sum rule.

The ratios of the light (u, d, s) quark masses are predicted from the QCD chiral perturbation theory [13]:

$$\frac{m_u}{m_d} = 0.553 \pm 0.043, \quad \frac{m_s}{m_d} = 18.9 \pm 0.8, \quad \frac{2m_s}{m_u + m_d} = 24.4 \pm 1.5, \,(19)$$

(there is also a more recent estimate $m_u/m_d = 0.46 \pm 0.09$ [14]). QCD sum rules offer a unique opportunity to estimate the individual masses of u, d, s quarks. To illustrate the continuous efforts in this direction, let me mention one recent determination of the strange quark mass [15], based on the correlation function of the derivatives of the strangeness-changing vector current $j_\mu = \bar{s}\gamma_\mu q$, $q = u, d$:

$$\Pi^s(q) = i \int d^4x e^{iqx} \langle 0|T\{\partial_\mu j^\mu(x)\partial_\nu j^{\dagger\nu}(0)\}|0\rangle \,. \qquad (20)$$

The OPE answer for Π^s is proportional to $(m_s - m_q)^2 \simeq m_s^2$ turning this correlator into a very convenient object for the m_s extraction. Furthermore, the recent progress in the multiloop QCD calculations allows to reach the $O(\alpha_s^3)$ accuracy in the perturbative part of Π^s. An updated analysis of kaon S wave scattering on π, η, η' is used to reproduce the hadronic spectral density. The sum rule yields [15] for the running mass in the $\overline{\text{MS}}$ scheme: $m_s(2\text{GeV}) = 99 \pm 16$ MeV, in a good agreement with the recent lattice QCD estimates. Using the ratios (19) one obtains $m_u(2\text{GeV}) = 2.9 \pm 0.6$ MeV and $m_d(2\text{GeV}) = 5.2 \pm 0.9$ MeV. The earlier work on predicting the light-quark masses from QCD sum rules is summarized in [3,16].

The charmed quark mass determination was one of the first successful applications of the QCD sum rule approach [17,1]. The correlation function of two $\bar{c}\gamma_\mu c$ currents (in other words, the charm contribution to the photon polarization operator) was matched to its hadronic dispersion relation, where the imaginary part is simply proportional to the $e^+e^- \to charm$

cross section including ψ resonances and the open charm continuum. The lowest power moments of this sum rule at $q^2 = 0$ are well suited for m_c determination because nonpertrbative effects are extremely small. Replacing $c \to b$, $\psi \to \Upsilon$ and open charm by open beauty one obtains analogous sum rule relations for the b quark [18]. In recent years the mainstream development in the heavy quark mass determination went in another direction, employing the higher moments which are less sensitive to the experimental input above the open flavour threshold. These moments, however, demand careful treatment of Coulomb corrections [19] which is only possible in the nonrelativistic QCD (the current status of this field is reviewed in [20]). Recent precise measurements of the $e^+e^- \to hadrons$ cross section on one side and a substantial progress in the calculation of perturbative diagrams on the other side, allowed to reanalyze with a higher precision the low moments of the original SVZ sum rules for quarkonia with the following results [21] for the $\overline{\text{MS}}$ masses: $m_c(m_c) = 1.304 \pm 0.027$ GeV, $m_b = 4.209 \pm 0.05$ GeV. Another subset of charmonium sum rules (higher moments at fixed large $q^2 < 0$) was employed in [22], with a prediction for m_c in agreement with the above.

4.3. Calculation of the B meson decay constant

Having determined the condensates and quark masses from a set of experimentally proven QCD sum rules for light-quark and quarkonium systems one has an exciting possibility to predict the unknown hadronic characteristics of B meson. In the amplitudes of exclusive weak B decays the hadronic matrix elements are multiplied by poorly known CKM parameters, such as V_{ub}. QCD sum rule calculations may therefore provide a useful hadronic input for extraction of CKM parameters from data on exclusive B decays. Importantly, the theoretical accuracy of the sum rule determination can be estimated by varying the input within allowed intervals.

One of the most important parameters involved in B physics is the B meson decay constant f_B defined via the matrix element $\langle 0|\bar{u}i\gamma_5 b|B\rangle$. The calculation of f_B using QCD sum rules has a long history, a detailed review and relevant references can be found, e.g. in [3,23], I only mention the very first papers [17,24]. One usually employs the SVZ sum rule for the two-point correlator of $\bar{b}i\gamma_5 q$ currents. I will not write down this sum rule explicitly. It looks very similar to the one for f_ρ discussed in sect. 3, in a sense that the sum rule contains a (duality subtracted) perturbative part and condensate terms. The expressions for the Wilson coefficients are in this case much more complicated, especially the radiative corrections to the heavy-light loop diagrams. The recent essential update of the sum rule for

f_B is worked out in [25] taking into account the $O(\alpha_s^2)$ corrections to the heavy-light loop recently calculated in [26] and treating the b quark mass in $\overline{MS}$ scheme. The result is (for $m_b(m_b) = 4.21 \pm 0.05$ GeV): $f_B = 210 \pm 19$ MeV and $f_{B_s} = 244 \pm 21$ MeV, in a good agreement with the most recent lattice QCD determination (including dynamical sea quark effects). I think, the example of f_B determination demonstrates that QCD sum rules indeed provide a reliable analytical tool for the hadronic B physics.

4.4. Hadronic amplitudes

To complete this short survey of QCD sum rule applications, it is important to mention that this method allows to calculate various hadronic amplitudes involving more than one hadron. Let me consider, as a generic example, a calculation of the hadronic matrix element $\langle h_f(p+q)|j|h_i(p)\rangle$ of a certain quark current j with a momentum transfer q. The convenient starting object in this case is the three-point correlation function depending on two independent 4-momenta:

$$T_{fi}(p,q) = (i)^2 \int d^4x \, d^4y \, e^{-i(px+qy)} \langle 0|T\{j_f(0)j(y)j_i(x)\}|0\rangle. \qquad (21)$$

As a next step, one writes down a double dispersion relation, in the variables p^2 and $(p+q)^2$ at fixed q^2, expressed in a form similar to Eq. (11):

$$T_{fi}(p,q) = \sum_{h_f}\sum_{h_i} \frac{\langle 0|j_f|h_f\rangle\langle h_f|j|h_i\rangle\langle h_i|j_i|0\rangle}{(m_{h_f}^2 - (p+q)^2)(m_{h_i}^2 - p^2)} + \text{subtractions}, \qquad (22)$$

where the double sum includes all possible transitions between the states with h_i and h_f quantum numbers. Two independent Borel transformations in p^2 and $(p+q)^2$ applied to Eq. (22) enhance the ground-state term containing the desired matrix element and allow to get rid of subtraction terms:

$$\hat{B}_1\hat{B}_2 T_{fi} = \sum_{h_f}\sum_{h_i}\langle 0|j_f|h_f\rangle\langle h_f|j|h_i\rangle\langle h_i|j_i|0\rangle e^{-m_{h_f}^2/M_2^2 - m_{h_i}^2/M_1^2}, \qquad (23)$$

where M_1 and M_2 are the Borel variables corresponding to p^2 and $(p+q)^2$, respectively. On the other hand, the correlator (21) can be computed, in terms of perturbative and condensate contributions:

$$\hat{B}_1\hat{B}_2 T_{fi}^{OPE} = \int ds\, ds'\, \rho_{fi}^{pert}(s,s',q^2)e^{-s'/M_2^2 - s/M_1^2} + \sum_{d=3,4,..}\hat{B}_1\hat{B}_2 C_d^{fi}\langle 0|O_d|0\rangle. \qquad (24)$$

In the above, the perturbative contribution calculated from the diagram in Fig. 5a is represented in a convenient form of double spectral representation with a spectral density ρ_{fi}^{pert}. The Wilson coefficients C_d^{fi} are

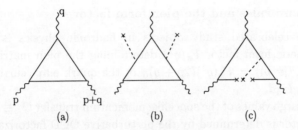

Figure 5. Contributions to the 3-point correlation function (21): (a) perturbative, ze-roth order in α_s; (b)-(c) some nonperturbative corrections.

calculated from the diagrams exemplified in Fig. 5b,c. Importantly, one does not introduce new parameters/inputs in this calculation, benefiting from the universality of quark/gluon condensates. The above expansion is valid at large spacelike external momenta: $|p^2|, |(p+q)^2| \gg \Lambda_{QCD}^2$, far from the hadronic thresholds in the corresponding channels. Accordingly, the squared momentum transfer is also kept large, $Q^2 = -q^2 \gg \Lambda_{QCD}^2$, at least for the currents containing light quarks [a]. Equating the hadronic dispersion relation to the OPE result at large M_1^2, M_2^2 and invoking quark-hadron duality one obtains the sum rule for the matrix element:

$$
f_i f_f \langle h_f | j | h_i \rangle e^{-m_{h_f}^2/M_2^2 - m_{h_i}^2/M_1^2} = \int\limits_{R(s_0, s_0')} ds\, ds' \rho_{fi}^{pert}(s, s', q^2) e^{-s'/M_2^2 - s/M_1^2}
$$

$$
+ \sum_d \hat{B}_1 \hat{B}_2 C_{fi}^d \langle 0 | O_d | 0 \rangle \,, \tag{25}
$$

where $f_i = \langle h_i | j_i | 0 \rangle$, $f_f = \langle 0 | j_f | h_f \rangle$ are the decay constants of the initial and final hadrons. The latter are calculable from two-point sum rules, or simply known from experiment. In the above, $R(s_0, s_0')$ is the quark-hadron duality domain in the s, s' plane, s_0, s_0' are the corresponding thresholds. Using 3-point correlators, the sum rules for charmonium radiative transi-tions have been derived in [28]. Another important application [29] is the pion e.m. form factor discussed in more detail in the next section.

---

[a]Hadronic matrix elements at $q^2 = 0$, e.g., the nucleon magnetic moment can also be calculated within QCD sum rule approach, using the external (background) field technique [27], with additional vacuum condensates induced by external, non-QCD fields.

5. QCD sum rules and the pion form factor

One of the celebrated study objects in hadronic physics is the pion electromagnetic form factor $F_\pi(q^2)$ determining the pion matrix element $\langle\pi(p+q)|j_\mu^{em}|\pi(p)\rangle = F_\pi(q^2)(2p+q)_\mu$ of the quark e.m. current $j_\mu^{em} = e_u\bar{u}\gamma_\mu u + e_d\bar{d}\gamma_\mu d$.

At very large values of the spacelike momentum transfer $Q^2 \equiv -q^2 \to \infty$ the form factor is determined by the perturbative QCD factorization [30]:

$$F_\pi(Q^2) = \frac{8\pi\alpha_s f_\pi^2}{9Q^2}\left|\int_0^1 du\frac{\varphi_\pi(u)}{1-u}\right|^2, \tag{26}$$

obtained by the convolution of distribution amplitudes (DA) $\varphi_\pi(u)$ of the initial and final pions (see the definition below) with the $O(\alpha_s)$ quark-gluon hard-scattering amplitude. At finite Q^2, the major problem is to estimate the "soft", $O(\alpha_s^0/Q^4)$ part of this form factor. It corresponds to an overlap of end-point configurations of the quark-parton momenta in the initial and final pions, so that the large momentum is transferred without a hard gluon exchange (the so called Feynman mechanism).

The first model-independent estimate of the soft contribution to the pion form factor was provided by QCD sum rules [29]. The three-point correlator (21) was used, with j, j_i and j_f replaced by j_μ^{em}, $j_{\nu 5}$ and $j_{\rho 5}^\dagger$, respectively, where $j_{\nu 5} = \bar{u}\gamma_\nu\gamma_5 d$ is the axial-vector current generating the pion state from the vacuum: $\langle\pi(p)|j_{\nu 5}^{(\pi)}|0\rangle = -if_\pi p_\nu$. The calculation based on OPE and condensates is valid at sufficiently large Q^2, practically at $Q^2 \sim$ 1 GeV^2. The resulting sum rule for the form factor written in the form (25) has a rather compact expression:

$$f_\pi^2 F_\pi(Q^2) = \int\limits_{R(s_0^\pi)} ds\, ds'\rho^{pert}(s,s',Q^2)e^{-\frac{s+s'}{M^2}}$$

$$+ \frac{\alpha_s}{12\pi M^2}\langle G_{\mu\nu}^a G^{a\mu\nu}\rangle + \frac{208\pi}{81M^4}\alpha_s\langle\bar{q}q\rangle^2\left(1 + \frac{2Q^2}{13M^2}\right), \tag{27}$$

where the perturbative spectral density is

$$\rho^{pert}(s,s',Q^2) = \frac{3Q^4}{4\pi^2}\frac{1}{\lambda^{7/2}}\left[3\lambda(\sigma+Q^2)(\sigma+2Q^2) - \lambda^2 - 5Q^2(\sigma+Q^2)^3\right], \tag{28}$$

with $\lambda = (\sigma+Q^2)^2 - 4ss'$ and $\sigma = s+s'$. In Eq. (27) the condensates up to $d = 6$ are included, $m_\pi = 0$, $M_1 = M_2 = M$ and $s_0 = s_0' = s_0^\pi \simeq 0.7$ GeV^2. The duality threshold is inferred from the two-point sum rule for the axial-vector channel [1]. At $Q^2 = 1 \div 3$ GeV^2, the form factor predicted from the sum rule agrees with the experimental data. E.g, compare

$F_\pi(Q^2 = 1\text{GeV}^2) \simeq 0.3$ with the most accurate CEBAF data [31] shown in Fig. 7 below. The good agreement indicates that the soft mechanism is the most important one in this region and that the $O(\alpha_s)$ hard scattering effect which should dominate at infinitely large Q^2 is still a small correction. (The latter corresponds to the perturbative gluon exchanges added to the diagram of Fig. 5a). At large Q^2 the perturbative part of the sum rule (27) has a $\sim 1/Q^4$ behavior, in full accordance with our expectation for the soft, end-point contribution to the form factor. However, the condensate contributions to $F_\pi(Q^2)$ are either Q^2-independent or grow $\sim Q^2/M^2$. A careful look at one of the relevant diagrams in Fig. 5c reveals the reason of this anomalous behavior. Using local (static field) condensate approximation, one implicitly neglects the momenta of vacuum quark/gluon fields. The external large momentum p is carried by a single quark, which, after the photon absorption, propagates with the momentum $p + q$, so that the contribution of this diagram is q^2 independent. Therefore, the truncated local condensate expansion is not an adequate approximation to reproduce the large Q^2 behavior of the pion form factor.

A possibility to calculate $F_\pi(Q^2)$ including both soft and hard scattering effects at large Q^2 [32,33], is provided by the light-cone sum rule (LCSR) approach [34] combining the elements of the theory of hard exclusive processes with the SVZ procedure.

One starts with introducing a vacuum-to-pion correlation function

$$F_{\mu\nu}(p,q) = i \int d^4x \, e^{-iqx} \langle 0|T\{j_{\mu 5}(0)j_\nu^{\text{em}}(x)\}|\pi(p)\rangle \,, \tag{29}$$

where one of the pions is put on-shell, $p^2 = m_\pi^2$, and the second one is replaced by the generating current $j_{\mu 5}$. For this correlator a dispersion relation is written, in full analogy with Eq. (11):

$$F_{\mu\nu}(p,q) = \sum_h \frac{\langle 0|j_{\mu 5}|h\rangle \langle h|j_\nu^{\text{em}}|\pi(p)\rangle}{m_h^2 - (p+q)^2} + \text{subtractions} \,. \tag{30}$$

The lowest pion-state term ($h = \pi$) in the hadronic sum,

$$F_{\mu\nu}^{(\pi)}(p,q) = \frac{if_\pi F_\pi(Q^2)(p+q)_\mu(2p+q)_\nu}{m_\pi^2 - (p+q)^2} \,, \tag{31}$$

contains the desired form factor.

At large spacelike momenta, $Q^2, |(p+q)^2| \gg \Lambda_{QCD}^2$, the correlation function (29) is dominated by small values of the space-time interval x^2, allowing one to expand the product of two currents around the light-cone $x^2 = 0$. The leading-order contribution is obtained from the diagram in Fig. 6a and consists of two parts: (1) the short-distance amplitude involving

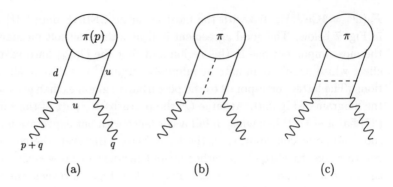

Figure 6. Light-cone expansion of the correlation function (29).

the virtual quark propagating between the points x and 0, and (2) the vacuum-to-pion matrix element of a nonlocal quark-antiquark operator, $\langle 0|\bar{u}(x)\gamma_\rho\gamma_5 d(0)|\pi\rangle$. This matrix element contains long-distance effects and is therefore not directly calculable. On the other hand, being expanded near $x^2 = 0$ it can be resolved in terms of universal distribution functions:

$$\langle 0|\bar{u}(x)\gamma_\mu\gamma_5 d(0)|\pi(p)\rangle = -ip_\mu f_\pi \int_0^1 du\, e^{-iup\cdot x}\left(\varphi_\pi(u,\mu) + x^2 g_1(u,\mu)\right)$$

$$+ f_\pi\left(x_\mu - \frac{x^2 p_\mu}{p\cdot x}\right)\int_0^1 du\, e^{-iup\cdot x} g_2(u,\mu) + \dots, (32)$$

where the terms up to $O(x^2)$ are shown explicitly. This expansion contains normalized light-cone distribution amplitudes (DA): $\varphi_\pi(u,\mu)$, $g_1(u,\mu)$, $g_2(u,\mu)$, ... , and the scale μ reflects the logarithmic dependence on x^2. Importantly, the power moments of DA, e.g., $M_n(\mu) = \int_0^1 du\, u^n \varphi_\pi(u,\mu)$, are related to the vacuum-to-pion matrix elements of local quark-antiquark operators with a definite *twist* (dimension minus Lorentz spin). For that reason, φ_π is called twist 2 DA, and, correspondingly, g_1 and g_2 are of twist 4. Thus, in the light-cone OPE one deals with a completely different pattern of long-distance effects, as compared with the local OPE considered in Sect. 3. Instead of a set of universal vacuum condensates, there is a set of DA for a given light meson, each of DA representing a series of matrix elements.

Actually, the twist 2 DA φ_π was originally introduced in the QCD analysis of hard exclusive hadronic processes [30], see e.g., Eq. (26). Some of its properties are well understood, in particular, the following expansion can be written:

$$\varphi_\pi(u,\mu) = 6u(1-u)\left(1 + \sum_{n=1} a_{2n}(\mu)C_{2n}^{3/2}(2u-1)\right), \qquad (33)$$

based on the approximate conformal symmetry of QCD with light quarks. In the above, C_{2n} are Gegenbauer polynomials and the coefficients $a_{2n}(\mu)$ determine the deviation of φ_π from its asymptotic form $6u(1-u)$. Due to the perturbative evolution, $a_{2n}(\mu)$ are logarithmically suppressed at large μ. The low-scale values of a_{2n} (and of similar coefficients for other DA) have to be considered a nonperturbative input.

The correlation function (29) calculated from the light-cone OPE represents a convolution of the pion DA and short-distance (hard scattering) amplitudes:

$$F_{\mu\nu}(p,q) = 2if_\pi p_\mu p_\nu \int\limits_0^1 du \frac{u\varphi_\pi(u,\mu)}{(1-u)Q^2 - u(p+q)^2} + \dots . \qquad (34)$$

For simplicity, only the leading order, twist 2 contribution with the relevant kinematical structure is shown. The ellipses denote the $O(\alpha_s)$ corrections (one of the diagrams is presented in Fig. 6c) and the higher twist contributions suppressed by powers of the denominator. Physically, the higher-twist corrections take into account the transverse momentum of the quark-antiquark state (e.g., the twist 4 terms in the expansion (32)) and the contributions of higher Fock states in the pion wave function (such as the quark-antiquark-gluon DA contributing via the diagram in Fig. 6b.). These two effects are related via QCD equations of motion. More details on the pion DA can be found in [35]. The factorization scale μ in Eq. (13) effectively separates the large virtualities ($> \mu^2$) in the hard scattering amplitude and the small ones ($< \mu^2$) in the pion DA.

Equating the dispersion relation (30) to the OPE result (34) at large spacelike $(p+q)^2$, one extracts the form factor $F_\pi(Q^2)$ applying the standard elements of the QCD sum rule technique, the Borel transformation in the variable $(p+q)^2$ and the quark-hadron duality. The latter reduces to a simple replacement of the lower limit in the u-integration in Eq. (34), $0 \to Q^2/(s_0^\pi + Q^2)$. The resulting sum rule [32,33] is:

$$F_\pi(Q^2) = \int\limits_{Q^2/(s_0^\pi+Q^2)}^1 du\, \varphi_\pi(u,\mu)e^{-\frac{(1-u)Q^2}{uM^2}} + F_\pi^{(tw2,\alpha_s)}(Q^2) + F_\pi^{(tw4,6)}(Q^2), \qquad (35)$$

where the leading-order twist 2 part is shown explicitly. The $1/Q^4$ behavior of Eq. (35) corresponds to the soft end-point mechanism, provided that in

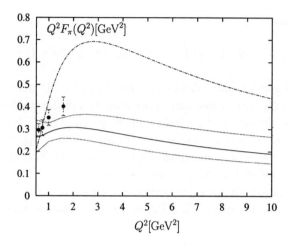

Figure 7. The pion e.m. form factor calculated from LCSR in comparison with the CEBAF data shown with points. The solid line corresponds to the asymptotic pion DA, dashed lines indicate the estimated overall theoretical uncertainty ; the dash-dotted line is calculated with the CZ model of the pion DA.

the $Q^2 \to \infty$ limit the integration region shrinks to the point $u = 1$. The $O(\alpha_s)$ part of this sum rule was calculated in [33] and is indeed small at $Q^2 \sim 1$ GeV^2. Importantly, in $F_\pi^{(\alpha_s)}(Q^2)$ one recovers the $\sim 1/Q^2$ asymptotic term corresponding to the hard perturbative mechanism, with a coefficient which, in the adopted approximation, coincides with the one in Eq. (26). The higher twist contributions to the sum rule (35) manifest the same $\sim 1/Q^4$ behavior as the leading twist. Altogether, we seem to achieve the goal. The pion form factor obtained from LCSR contains both the hard-scattering and soft (end-point) contributions, with a proper asymptotic behavior at large Q^2.

The updated LCSR prediction for $F_\pi(Q^2)$ [36] is shown in Fig. 7. One important practical use of this result is to estimate/constrain the nonasymptotic coefficients a_{2n} by fitting the sum rule (35) to the experimental data on the pion form factor. However, currently there are no sufficient data at $Q^2 > 1$ GeV^2 to constrain complicated patterns of nonasymptotic coefficients. Considering simple ones, one finds that, e.g. the asymptotic DA $\varphi_\pi(u) = 6u(1 - u)$ is not excluded, whereas the CZ -model [37] seems to be disfavored by data. Assuming that only $a_2 \neq 0$ and neglecting all other coefficients yields [36] the following range $a_2(1 \, \text{GeV}) = 0.24 \pm 0.14 \pm 0.08$, where the first error reflects the estimated theoretical uncertainty and the second one corresponds to the experimental errors.

Other LCSR applications to the physics of hard exclusive processes include the $\gamma^*\gamma\pi$ form factor[38], the kaon e.m. form factor[36], and the first attempt to calculate the nucleon form factors [39].

Furthermore, an important task of LCSR is to provide B physics with various heavy-to-light hadronic matrix elements. In particular, the sum rule for the $B \to \pi$ form factor can be obtained from a correlation function very similar to the one depicted in Fig. 6, if one replaces the virtual light quark by a b quark. The calculable short-distance part will then change considerably, but the long-distance part remains essentially the same, determined by the set of pion DA. The sum rule predictions for the $B \to \pi$ [40] and $B \to \rho$ [41] form factors are already used to extract $|V_{ub}|$ from the widths of $B \to \pi(\rho)l\nu_l$ decays. One can also employ LCSR to estimate the hadronic amplitudes for $B \to \pi\pi$ and similar decays beyond factorization [42]. A summary of the sum rule applications to the heavy flavour physics can be found in [43].

6. Conclusion

For more than twenty years, QCD(SVZ) sum rules serve as a virtual laboratory for studying the transition from short- to long-distance QCD. The practical use of this analytical method is twofold. On one hand, using sum rules for experimentally known hadronic quantities, QCD parameters such as quark masses are extracted. On the other hand, the sum rules are employed to predict unknown hadronic parameters, for example f_B, with a controllable accuracy. Finally, the example of the pion form factor calculation demonstrates that the light-cone version of QCD sum rules has a large potential in describing QCD mechanisms of exclusive hadronic transitions.

Acknowledgments

I am grateful to Misha Shifman, Arkady Vainshtein, and Misha Voloshin for hospitality and for organizing a very enjoyable and fruitful workshop. This work is supported by BMBF (Bundesministerium für Bildung und Forschung).

References

1. M. A. Shifman, A. I. Vainshtein and V. I. Zakharov, *Nucl. Phys.* B **147**, 385, 448 (1979).
2. M. Shifman, *Prog. Theor. Phys. Suppl.* **131**, 1 (1998);
   E. de Rafael, Lectures at Les Houches Summer School, Session 68, Les

Houches, France (1997), hep-ph/9802448;
A. V. Radyushkin, hep-ph/0101227.

3. P. Colangelo and A. Khodjamirian, hep-ph/0010175, published in the Boris Ioffe Festschrift 'At the Frontier of Particle Physics / Handbook of QCD', ed. by M. Shifman (World Scientific, Singapore, 2001), vol. 3, 1495-1576.

4. A. I. Vainshtein, V. I. Zakharov, V. A. Novikov and M. A. Shifman, Sov. J. Nucl. Phys. **32** (1980) 840.

5. V. A. Novikov, M. A. Shifman, A. I. Vainshtein and V. I. Zakharov, Nucl. Phys. B **191** (1981) 301.

6. A. D. Magakian, A.Khodjamirian, Sov. J. Nucl.Phys.**50** (1989) 831.

7. B. L. Ioffe, hep-ph/0207191.

8. B. V. Geshkenbein and M. S. Marinov, Yad. Fiz. **30** (1979) 1400.

9. B. V. Geshkenbein and B. L. Ioffe, Nucl. Phys. B **166** (1980) 340.

10. E. V. Shuryak, talk at this Workshop.

11. B. L. Ioffe, Nucl. Phys. B **188** (1981) 317 [Erratum-ibid. B **191** (1981) 591]; Z. Phys. C **18** (1983) 67.

12. Y. Chung, H. G. Dosch, M. Kremer and D. Schall, Phys. Lett. B **102** (1981) 175; Nucl. Phys. B **197** (1982) 55.

13. H. Leutwyler, Phys. Lett. B **378** (1996) 313.

14. G. Amoros, J. Bijnens and P. Talavera, Nucl. Phys. B **602** (2001) 87.

15. M. Jamin, J. A. Oller and A. Pich, Eur. Phys. J. C **24** (2002) 237.

16. N. Paver, Int. J. Mod. Phys. A **16S1B** (2001) 588.

17. V. A. Novikov, L. B. Okun, M. A. Shifman, A. I. Vainshtein, M. B. Voloshin and V. I. Zakharov, Phys. Rev. Lett. **38** (1977) 626 [Erratum-ibid. **38** (1977) 791]; Phys. Rept. **41** (1978) 1.

18. L. J. Reinders, H. R. Rubinstein and S. Yazaki, Nucl. Phys. B **186** (1981) 109;
S. S. Grigorian, Sov. J. Nucl. Phys. **30** (1979) 729.

19. M. B. Voloshin, Nucl. Phys. B **154** (1979) 365; Sov. J. Nucl. Phys. **36** (1982) 143.

20. A. H. Hoang, hep-ph/0204299.

21. J. H. Kuhn and M. Steinhauser, Nucl. Phys. B **619** (2001) 588.

22. B. L. Ioffe and K. N. Zyablyuk, hep-ph/0207183.

23. A. Khodjamirian and R. Ruckl, in *Heavy Flavors*, 2nd edition, eds., A.J. Buras and M. Lindner, World Scientific (1998), p. 345, hep-ph/9801443.

24. D. J. Broadhurst, Phys. Lett. B **101** (1981) 423;
T. M. Aliev and V. L. Eletsky, Sov. J. Nucl. Phys. **38** (1983) 936.

25. M. Jamin and B. O. Lange, Phys. Rev. D **65** (2002) 056005.

26. K. G. Chetyrkin and M. Steinhauser, Eur. Phys. J. C **21** (2001) 319.

27. B. L. Ioffe and A. V. Smilga, *JETP Lett.* **37**, 298 (1983); Nucl. Phys. B **232** (1984) 109;
I. I. Balitsky and A. V. Yung, Phys. Lett. B **129** (1983) 328.

28. A. Y. Khodjamirian, Phys. Lett. B **90** (1980) 460; Sov. J. Nucl. Phys. **39** (1984) 614;
V. A. Beilin and A. V. Radyushkin, Nucl. Phys. B **260** (1985) 61.

29. B. L. Ioffe and A. V. Smilga, Nucl. Phys. B **216** (1983) 373;
V. A. Nesterenko and A. V. Radyushkin, Phys. Lett. B **115** (1982) 410.

30. G. P. Lepage and S. J. Brodsky, Phys. Lett. **87B** (1979) 359; Phys. Rev. **D22** (1980) 2157;
    A. V. Efremov and A. V. Radyushkin, Phys. Lett. **B94** (1980) 245; Theor. Math. Phys. **42** (1980) 97;
    V. L. Chernyak and A. R. Zhitnitsky, JETP Lett. **25** (1977) 510; Sov. J. Nucl. Phys. **31** (1980) 544;
    G. R. Farrar and D. R. Jackson, Phys. Rev. Lett. **43** (1979) 246.
31. J. Volmer *et al.* [The Jefferson Lab F(pi) Collaboration], Phys. Rev. Lett. **86** (2001) 1713.
32. V. Braun and I. Halperin, Phys. Lett. B **328** (1994) 457.
33. V. M. Braun, A. Khodjamirian and M. Maul, Phys. Rev. D **61** (2000) 073004.
34. I. I. Balitsky, V. M. Braun and A. V. Kolesnichenko, Nucl. Phys. B **312** (1989) 509;
    V. M. Braun and I. E. Filyanov, Z. Phys. C **44** (1989) 157;
    V. L. Chernyak and I. R. Zhitnitsky, Nucl. Phys. B **345** (1990) 137.
35. V. M. Braun and I. E. Filyanov, Z. Phys. C **48** (1990) 239;
    P. Ball, JHEP**9901** (1999) 010.
36. J. Bijnens and A. Khodjamirian, hep-ph/0206252, to be published in Eur. Phys. J. C.
37. V.L. Chernyak and A.R. Zhitnitsky, Phys. Rept. **112** (1984) 173.
38. A. Khodjamirian, Eur. Phys. J. C **6** (1999) 477.
39. V. M. Braun, A. Lenz, N. Mahnke and E. Stein, Phys. Rev. D **65** (2002) 074011.
40. V. M. Belyaev, A. Khodjamirian and R. Rückl, Z. Phys. **C60** (1993) 349;
    A. Khodjamirian, R. Rückl, S. Weinzierl and O. Yakovlev, Phys. Lett. **B410** (1997) 275;
    E. Bagan, P. Ball and V. M. Braun, Phys. Lett. **B417** (1998) 154;
    P. Ball, JHEP **9809** (1998) 005;
    P. Ball and R. Zwicky, JHEP **0110** (2001) 019.
41. P. Ball and V. M. Braun, Phys. Rev. D **58** (1998) 094016.
42. A. Khodjamirian, Nucl. Phys. **B605** (2001) 558.
43. A. Khodjamirian, AIP Conf. Proc. **602** (2001) 194, hep-ph/0108205.

THE LEADING PARTICLE EFFECT IN CHARM HADROPRODUCTION FROM HEAVY QUARK RECOMBINATION

ERIC BRAATEN

Department of Physics, 174 W. 18th Ave,
The Ohio State University,
Columbus, Ohio 43210 USA
E-mail: braaten@pacific.mps.ohio-state.edu

The large asymmetry between the cross sections for charm hadrons and their antiparticles in charm hadroproduction is referred to as the leading particle effect. Heavy quark recombination is a new mechanism for heavy hadron production that generates charm asymmetries by a hard-scattering parton process. Using this mechanism, the E791 data on the $D^- - D^+$ asymmetry in $\pi^- N$ collisions can be fit reasonably well by adjusting a single multiplicative nonperturbative constant.

1. Introduction

Heavy hadron production is one of the areas of QCD phenomenology for which the agreement between theory and experiment is least satisfactory. The conventional wisdom is that perturbative methods can be used to calculate the heavy quark cross section, i.e. the cross section summed over all hadrons containing a specific heavy quark, but its decomposition into the cross sections for individual heavy hadrons is inherently nonperturbative. However large asymmetries between the cross sections for specific charm particles and their antiparticles have been observed. The most dramatic of these asymmetries has been observed in charm hadroproduction and is referred to as the *leading particle effect*. These asymmetries can be described by models for the nonperturbative hadronization of the charm quark that give a reasonable description of these asymmetries, and the conventional wisdom is that this is the best one can do. However, it would be unfortunate if these dramatic experimental results could not be related to more fundamental aspects of QCD.

2. Charm Asymmetries

The basis for the conventional wisdom on heavy hadron production is the factorization theorem of perturbative QCD for inclusive single-particle production [1,2]. The factorization formula separates the cross section into short-distance factors that can be calculated using perturbative QCD and nonperturbative long-distance factors. In the case of the production of a charm hadron D in πN collisions, the factorization formula is

$$d\sigma[\pi N \to D + X] = f_{g/\pi} \otimes f_{g/N} \otimes d\hat{\sigma}[gg \to c\bar{c}] \otimes D_{c \to D}$$
$$+ \text{ perturbative corrections}$$
$$+ \text{ power corrections.} \qquad (1)$$

At leading order in α_s, the short-distance factor is the parton cross section for $gg \to c\bar{c}$. The long-distance factor associated with the final state is the fragmentation function for $c \to D$. The perturbative corrections have been calculated to NLO in α_s [3]. At large transverse momentum, the leading power corrections to the transverse momentum distribution $d\sigma/dP_T^2$ are suppressed by powers of Λ_{QCD}/P_T. The leading power corrections to the rapidity distribution obtained by integrating over P_T are suppressed by Λ_{QCD}/m_c [4,5].

The asymmetry between the cross sections for a charm hadron D and its antiparticle $\bar{D}$ is conveniently measured by the variable

$$\alpha[D] = \frac{\sigma[D] - \sigma[\bar{D}]}{\sigma[D] + \sigma[\bar{D}]}, \qquad (2)$$

which has the range $-1 < \alpha[D] < +1$. The QCD factorization formula predicts that the asymmetry should be small. At leading order in α_s, the $D - \bar{D}$ asymmetry is predicted to be 0, because the parton process $gg \to c\bar{c}$ produces no $c - \bar{c}$ asymmetry and charge conjugation implies that the fragmentation functions satisfy $D_{c \to D} = D_{\bar{c} \to \bar{D}}$. At NLO in α_s, the parton process $qg \to qc\bar{c}$ generates a small $c - \bar{c}$ asymmetry. Provided the parton distributions of the colliding hadron h satisfy $f_{q/h} \neq f_{\bar{q}/h}$, this generates a small $D - \bar{D}$ asymmetry. However the predicted asymmetries are about an order of magnitude too small to explain the data. The large asymmetries must therefore arise from the power corrections to the factorization formula (1). Charm asymmetries are therefore important because they provide a probe of these power corrections.

The most dramatic example of a charm asymmetry is the leading particle effect that has been observed in hadroproduction experiments. In the forward direction of a colliding hadron h, the cross sections are observed to be larger for heavy hadrons D or $\bar{D}$ that have a valence quark in common with h. The valence structures of the lightest charm mesons are $D^+ = c\bar{d}$, $D^0 = c\bar{u}$, $D^- = \bar{c}d$, and $\bar{D}^0 = \bar{c}u$. The beams that have been used include $h = \pi^-$, p, and Σ^-, with valence structures $d\bar{u}$, uud, and sdd, respectively. In the forward direction of the beam, the leading particle effect implies that $\sigma[D^-] > \sigma[D^+]$ for π^-, p, or Σ^-. It also implies that the direct cross sections for D^0 and $\bar{D}^0$ satisfy $\sigma[D^0] > \sigma[\bar{D}^0]$ for π^-, $\sigma[\bar{D}^0] > \sigma[D^0]$ for p, and $\sigma[D^0] \approx \sigma[\bar{D}^0]$ for Σ^-. The most accurate data on charm asymmetries are measurements of $\alpha[D^-]$ in πN collisions by Fermilab experiment E791 [6]. The forward direction for the beam is $x_F > 0$, where x_F is the ratio of the longitudinal momentum of the D to its maximum value in the center-of-momentum (CM) frame, which has the range $-1 < x_F < +1$. The x_F distribution measured by E791 is shown in Fig. 1. The leading particle effect is the increase in $\alpha[D^-]$ as x_F approaches $+1$. Note that at $x_F \approx 0.7$, the largest values for which α has been measured, the asymmetry is $\alpha[D^-] \approx 0.7$, which corresponds to a cross section ratio $\sigma[D_-]/\sigma[D_+] \approx 6$. Thus the difference between the cross sections for D^- and D^+ is very dramatic.

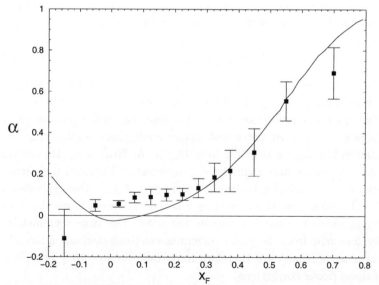

Figure 1. Data from E791 on $\alpha[D^-]$ as a function of x_F compared to the prediction from heavy quark recombination.

3. Previous Models

Predictions for charm asymmetries require some phenomenological descriptions of the hadronization process that binds the charm quark into a charm hadron. Many models have been developed to calculate the asymmetries, three of which I will describe briefly. One feature common to all these models is that the asymmetries are generated primarily by nonperturbative effects of the hadronization process. Another feature in common is that the phenomenological inputs are functions, although simplifying assumptions may be used to reduce the functions to a few parameters. These parameters generally do not scale in any simple way with the heavy quark mass. Therefore measurements of asymmetries for charm production are of little help in estimating the asymmetries for bottom production.

The *recombination model* was developed by Hwa [7] to describe the production of light hadrons and has been applied to the production of charm hadrons by Likhoded and Slabopitsky [8]. This model involves a hard perturbative process and a soft nonperturbative process:

- a hard scattering involving a parton from the hadron h creates a $c\bar{c}$ pair,
- the c recombines with a $\bar{q}$ in the remnant of h to form a D meson.

The nonperturbative inputs consist of two functions:

- the momentum distribution for the $\bar{q}$ in the remnant of h,
- a recombination function that gives the probability for the c and $\bar{q}$ to form the D meson.

The *color string model* evolved from the Lund model for hadronization in e^+e^- annihilation. Its application to heavy quark production has been developed by Norrbin and Sjostrand [9]. This model involves a hard perturbative process and a soft nonperturbative process:

- a hard scattering involving a parton from the hadron h creates a $c\bar{c}$ pair,
- a color string connecting the c to a $\bar{q}$ (or diquark) in the remnant of h decays into particles that include a D meson.

The nonperturbative inputs consist of two functions:

- the momentum distribution for the $\bar{q}$ in the remnant of h,
- the momentum distribution of the D from the decay of the color string, which is calculated using the PYTHIA event generator.

The *intrinsic charm model* was proposed by Brodsky, Hoyer, Peterson, and Sakai [10] and applied to charm asymmetries by Vogt and Brodsky [11]. The intrinsic charm mechanism proceeds through higher Fock states of a hadron h that include a $c\bar{c}$ pair, e.g. the $|d\bar{u}c\bar{c}\rangle$ Fock state for π^- or the $|uudc\bar{c}\rangle$ Fock state for p. This model involves a hard perturbative process and a soft nonperturbative process:

- a hard scattering removes a valence parton from a Fock state of the hadron h that also contains a $c\bar{c}$ pair and a valence parton $\bar{q}$,
- the c and $\bar{q}$ in the remnant of h coalesce to form the D meson.

The nonperturbative inputs consist of two functions:

- the momentum distribution for the c and $\bar{q}$ in the remnant of h,
- a coalescence function that gives the probability for the c and $\bar{q}$ to form the D meson.

4. Heavy Quark Recombination Mechanism

Heavy quark recombination (HQR) is a new mechanism for heavy hadron production that was proposed by Braaten, Jia and Mehen in 2001 [12,13,14] and is still being developed. It differs from previous models in that the asymmetry is generated by the hard-scattering process that creates the heavy quark. The mechanism involves a hard perturbative process followed by a soft nonperturbative process:

- a hard scattering involving $\bar{q}$ creates a $c\bar{c}$ pair and scatters the $\bar{q}$ into a state with small momentum in the c rest frame,
- the c and $\bar{q}$ hadronize into a D plus soft hadrons.

The nonperturbative inputs associated with the soft process are multiplicative constants that depend only on the color and spin state of the $c\bar{q}$ and scale with the heavy quark mass like $1/m_c$. Thus if the parameters are measured from asymmetries in charm production, the asymmetries in bottom production can be estimated simply by scaling the parameters by m_c/m_b. One of the diagrams that contributes to the HQR process is shown in Fig. 2.

The HQR cross section is one of the power corrections in the factorization formula (1). It is unique in that at large P_T, it gives contributions to $d\sigma/dP_T^2$ that are suppressed only by $\Lambda_{\rm QCD}m_c/P_T^2$, in contrast to other power corrections that are suppressed by at least $\Lambda_{\rm QCD}^2/P_T^2$. In the cross section integrated over P_T, it is one of the power corrections that are suppressed by $\Lambda_{\rm QCD}/m_c$. Another special feature of the HQR mechanism is that it allows the large fraction of the energy of a colliding hadron that can

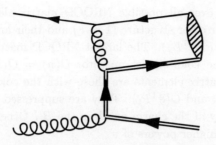

Figure 2. Diagram that contributes to heavy quark recombination into a $c\bar{q}$ meson via the parton process $\bar{q}g \to c\bar{q} + \bar{c}$.

be carried by a valence $\bar{q}$ to be used for the production process. Finally, the HQR cross section is strongly peaked in the forward $\bar{q}$ direction, which makes it plausible as an explanation for the leading particle effect.

Rather than attempting to define the HQR contribution to the cross section precisely, we will simply give an operational definition. To calculate the HQR cross section for producing the charm meson D in $\bar{q}g$ collisions,

- calculate the cross section for $\bar{q}g \to c\bar{q}(n) + \bar{c}$, where n denotes a specific color ($\underline{1}$ or $\underline{8}$) and angular momentum ($^{1}S_0$ or $^{3}S_1$) state and the c and $\bar{q}$ have momenta P and xP with $x \ll 1$,
- isolate the contribution proportional to $1/x^2$, and absorb the factor of $1/x^2$ into a nonperturbative constant $\rho[c\bar{q}(n) \to D]$.

This operational definition can be motivated from the NRQCD factorization formula for quarkonium production and also from factorization formulas for exclusive B decays.

5. Motivation from NRQCD factorization

The *NRQCD factorization formalism* allows quarkonium cross sections to be factored into perturbative parton cross sections and nonperturbative factors. The nonperturbative factors associated with the quarkonium are called NRQCD matrix elements. In the case of B_c production in gluon-gluon collisions at order α_s^4, the factorization formula is

$$d\sigma[gg \to B_c + X] = \sum_n d\hat{\sigma}[gg \to \bar{b}c(n) + b + \bar{c}] \times \langle O(n)^{B_c} \rangle, \qquad (3)$$

where the sum is over all possible NRQCD matrix elements which can be labelled by their color structure ($\underline{1}$ or $\underline{8}$) and their angular momentum quantum numbers ($^{2S+1}L_J$). The largest NRQCD matrix element for B_c is the one with the color-singlet operator $O(n) = O(\underline{1}\,^1S_0)$. The next most important matrix elements are those with the color-octet operators $O(\underline{8}\,^1S_0)$, $O(\underline{8}\,^3S_1)$, and $O(\underline{8}\,^1P_1)$. They are suppressed by v^4, where v is the relative velocity of the charm quark in the B_c. Other matrix elements are suppressed by higher powers of v.

The NRQCD factorization formula is based on the assumption that m_c is much larger than $\Lambda_{\rm QCD}$. If we decrease m_c, the NRQCD factorization formula breaks down for several reasons. First, v becomes of order 1, so there are no longer any velocity suppression factors for NRQCD matrix elements. There are also mass singularities in the form of logarithms of m_c, but they can be factored into parton distributions $f_{c/g}$ and fragmentation functions $D_{\bar{b}\to B_c}$. Some of the terms with a factor $f_{c/g}$ are suppressed only by $m_c m_b / P_T^2$ at large P_T. These are the heavy quark recombination terms. If we neglect terms suppressed by m_c^2/P_T^2, the factorization formula (3) reduces to

$$d\sigma[gg \to B_c + X] \approx d\hat{\sigma}[gg \to \bar{b} + b] \otimes D_{\bar{b}\to B_c}$$
$$+ f_{c/g} \otimes \sum_n \lim_{m_c \to 0} \left(m_c^2 \, d\hat{\sigma}[cg \to \bar{b}c(n) + \bar{c}] \right) \times \frac{1}{m_c^2} \langle O(n)^{B_c} \rangle, \qquad (4)$$

where now the sum is only over the 4 S-wave NRQCD matrix elements. The reason P-wave and higher orbital angular momentum matrix elements are suppressed at large P_T is that they involve higher dimension operators which must be compensated by higher powers of $1/P_T$ in the parton cross section $d\hat{\sigma}$. The first term on the right side of (4) is the fragmentation contribution and the last term is the HQR contribution. The correspondence with our operational definition in Section 4 is obtained by using the nonrelativistic quark model for the heavy light meson, in which case the lightcone momentum fraction of the c in the B_c is $x = m_c/(m_b + m_c)$.

6. Motivation from factorization of B decays

There has been some recent progress in deriving factorization theorems for exclusive 2-body B decays [15]. For example, in the decay $B^0 \to \pi^+\pi^-$, the decay amplitude can be factored into a short-distance part involving the weak decay $\bar{b} \to \bar{u}u\bar{d}$ and long-distance factors. The long-distance factor associated with the initial state B^0 can be expressed in terms of the distribution amplitude $\phi(x)$ for the B^0 to consist of a $\bar{b}$ and a d carrying fractions

$1 - x$ and x of its light-front momentum. This amplitude is dominated by small x of order $\Lambda_{\rm QCD}/m_b$. According to the factorization theorem, up to corrections suppressed by powers of $\Lambda_{\rm QCD}/m_b$, all nonperturbative effects associated with the B^0 wavefunction can be factored into a quantity f_+ that is proportional to the -1 moment of $\phi(x)$:

$$f_+ = \frac{F_B}{4M_B} \frac{\int_0^1 dx\,(1/x)\phi(x)}{\int_0^1 dx\,\phi(x)}. \tag{5}$$

Since the B decay constant F_B scales like $\Lambda_{\rm QCD}^{3/2}/m_b^{1/2}$ and $\langle 1/x \rangle$ scales like $m_b/\Lambda_{\rm QCD}$, it is easy to see that f_+ scales like $(\Lambda_{\rm QCD}/m_b)^{1/2}$.

There are contributions to amplitudes for exclusive B production that involve the same nonperturbative factor f_+. For example, consider the production of a B meson of momentum P via the parton process $qq \rightarrow (\bar{b}q) + b$. We take the initial and final momenta of the q to be p and xP, where x is the light-front momentum fraction of the q in the B. The wavefunction of the B is dominated by $x \sim \Lambda_{\rm QCD}/m_b$. There are diagrams such as that in Fig. 2 with $c \rightarrow \bar{b}$ and $\bar{q} \rightarrow q$ in which the momentum transferred to the q is provided by a virtual gluon of momentum $p - xP$ and invariant mass $(p - xP)^2 \approx -2xp \cdot P$, such as that in Fig. 2. Because of the factor of $1/x$ from the gluon propagator, the production amplitude is proportional to the -1 moment of x, and therefore to f_+. If the B is produced with large transverse momentum P_T, the parton cross section is suppressed by m_b^2/P_T^2 relative to that for $gg \rightarrow \bar{b} + b$. Multiplying by $f_+^2 \sim \Lambda_{\rm QCD}/m_b$, we see that this is a power correction to the B cross section that is suppressed only by $\Lambda_{\rm QCD}m_b/P_T^2$. This is a simple example of a HQR contribution to the B cross section.

The inclusive B cross section will receive contributions from the same diagrams, except that instead of the $\bar{b}q$ forming a B meson only, it can hadronize into a state consisting of the B meson plus light hadrons that are soft in the rest frame of the B. The $\bar{b}q$ need not have the color and spin quantum numbers of the B meson, because these quantum numbers can be changed by the emission of soft gluons. However, we can take the orbital angular momentum state to be S-wave, because the contributions from higher orbital angular momentum have additional factors of $\Lambda_{\rm QCD}$ in the numerator that must be compensated by additional factors of $1/P_T$ from the parton cross section and they will not contribute at order $\Lambda_{\rm QCD}m_b/P_T^2$. Thus there are only 4 possibilities for the color and angular momentum state for the $\bar{b}q$: $n = \underline{1}\,^1S_0,\,\underline{1}\,^3S_1,\,\underline{8}\,^1S_0$, and $\underline{8}\,^3S_1$. We denote the corresponding nonperturbative factor by $\rho[\bar{b}q(n) \rightarrow B]$. Thus the HQR contribution to the B cross section in gg collisions has the same form as the last term in

(4), with $c \to q$ and $\rho[\bar{b}q(n) \to B]$ proportional to $\langle O(n) \rangle / m_q^2 m_b$.

7. Application to Charm hadroproduction

To see if HQR provides a viable explanation for the charm asymmetries, we will apply it to charm hadroproduction. We will include the fragmentation contributions and the HQR contributions, both with parton cross sections calculated only to leading order in α_s. The expression we use for the cross section for the production of D^+ in the collision of hadrons h_A and h_B is

$$d\sigma[h_A h_B \to D^+ + X] \tag{6}$$
$$\approx f_{g/h_A} \otimes f_{g/h_B} \otimes d\hat{\sigma}[gg \to c + \bar{c}] \otimes D_{c \to D^+}$$
$$+ \sum_q f_{q/h_A} \otimes f_{\bar{q}/h_B} \otimes d\hat{\sigma}[q\bar{q} \to c + \bar{c}] \otimes D_{c \to D^+}$$
$$+ f_{\bar{d}/h_A} \otimes f_{g/h_B} \otimes d\hat{\sigma}[\bar{d}g \to c\bar{d}(\underline{1}\,^1S_0) + \bar{c}] \times \rho[c\bar{d}(\underline{1}\,^1S_0) \to D^+]$$
$$+ \sum_{q,n,\bar{D}} f_{q/h_A} \otimes f_{g/h_B} \otimes d\hat{\sigma}[qg \to \bar{c}q(n) + c] \times \rho[\bar{c}q(n) \to \bar{D}] \otimes D_{c \to D^+}.$$

Symmetrization under $h_A \leftrightarrow h_B$ is implied. The first two terms on the right side of (6) are the c fragmentation terms, which give no $D^- - D^+$ asymmetry. The asymmetry comes primarily from $c\bar{d}$ recombination, which is represented by the 3rd term. The asymmetry tends to be diluted by the last term, which takes into account $\bar{c}q$ recombination together with c fragmentation. The expression (6) is the cross section for direct production of D^+. We must also take into account indirect production via the feeddown from $D^{*+} \to D^+ \pi^0$.

The parton cross sections for the $c\bar{q}$ recombination process are obtained by evaluating the 5 diagrams for $\bar{q}g \to c\bar{q} + \bar{c}$ in the limit described in Section 4. The resulting expressions for the parton cross sections $d\hat{\sigma}/d\hat{t}$ are reasonably compact when expressed in terms of the variables $\hat{s}$, $T = \hat{t} - m_c^2$, $U = \hat{u} - m_c^2$, which satisfy $\hat{s} + T + U = 0$. They are given in Ref. [12]. To develop some intuition for the HQR parton cross sections, it is useful to consider the ratio with the parton cross section for the gg fusion process:

$$R = \frac{d\hat{\sigma}[\bar{q}g \to c\bar{q} + c]}{d\hat{\sigma}[gg \to c + \bar{c}]}. \tag{7}$$

This ratio has a coupling constant suppression factor of $\alpha_s(P_T)$ or $\alpha_s(m_c)$. It also has kinematic suppression factors that depend on the scattering angle θ of the $c\bar{q}$ in the CM frame. For large-angle scattering, we have

$$\theta = \pi/2: \qquad R \sim m_c^2/(P_T^2 + m_c^2). \tag{8}$$

R is therefore suppressed at large P_T by m_c^2/P_T^2. In the forward gluon direction, R is suppressed by powers of $m_c^2/\hat{s}$:

$$\theta = \pi: \qquad R \sim (m_c^2/\hat{s})^3, \qquad \text{for } c\bar{q}(\underline{1}\,{}^1S_0),$$
$$R \sim m_c^2/\hat{s}, \qquad \text{for } c\bar{q}(\underline{1}\,{}^3S_1).$$

However in the forward quark direction, there is no kinematic suppression:

$$\theta = 0: \qquad R \sim 1. \tag{9}$$

Thus the HQR cross section is strongly peaked in the forward $\bar{q}$ direction, which is a necessary condition for explaining the leading particle effect.

The HQR cross sections involve many nonperturbative factors $\rho[c\bar{q}(n) \to D]$. The light antiquark $\bar{q}$ can be $\bar{u}$, $\bar{d}$, or $\bar{s}$. The color state of the $c\bar{q}$ can be $\underline{1}$ or $\underline{8}$, and its angular momentum state can be ${}^1S_0$ or ${}^3S_1$. The heavy meson D can be one of the spin-0 states D^+, D^0, and D_s or one of the spin-1 states D^{*+}, D^{*0}, and D_s^*. However, many of these probability factors are related by approximate symmetries of QCD. $SU(3)$ flavor symmetry relates u, d, and s. Heavy quark spin symmetry relates the spin-1 state D^* to the corresponding spin-0 state D. There is also a large$-N_c$ suppression if the flavor quantum numbers of the D do not match those of the $c\bar{q}$. Taking all these into account, we can reduce the number of independent probability factors to four:

$$\rho_1 = \rho[c\bar{u}(\underline{1},{}^1S_0) \to D^0], \tag{10}$$
$$\tilde{\rho}_1 = \rho[c\bar{u}(\underline{1},{}^1S_0) \to D^{*0}],$$
$$\rho_8 = \rho[c\bar{u}(\underline{8},{}^1S_0) \to D^0],$$
$$\tilde{\rho}_8 = \rho[c\bar{u}(\underline{8},{}^1S_0) \to D^{*0}].$$

The parameter ρ_1 is the only one that receives a contribution from the exclusive final state, and therefore we expect it to be the most important. We will simplify our analysis by setting $\tilde{\rho}_1 = \rho_8 = \tilde{\rho}_8 = 0$ and considering only the single parameter ρ_1. We have already made this simplification in the 3${}^{\text{rd}}$ term on the right side of (6).

We proceed to apply the HQR mechanism to the data from E791 on the $D^+ - D^-$ asymmetry. This experiment involved a 500 GeV π^- beam on a N target. They measured the asymmetry variable $\alpha[D^+]$ defined by (2) as a function of x_F and P_T. We calculate the cross section for D^+ production using (6). The parton cross sections are calculated to leading order in α_s. We use the one-loop formula for the running coupling constant with $\Lambda_{\text{QCD}} = 200$ MeV. We take the charm quark mass to $m_c = 1.5$ GeV. There are three types of nonperturbative factors. For the parton distributions, we use the GRV 98 LO distribution for the nucleon and the GRV-P LO distribution for the π^-. For the fragmentation functions, we use the Peterson formula with shape parameter $\epsilon = 0.06$ and normalizations consistent with data on the fragmentation probabilities from LEP. In our simplified analysis, we consider only the probability factor ρ_1 defined in (10).

We determine the constant ρ_1 by fitting the E791 data on $\alpha[D^-]$ as a function of x_F and as a function of P_T. The best fit gives $\rho_1 = 0.06$. This value is consistent with the scaling $\rho_1 \sim \Lambda_{QCD}/m_c$. The resulting prediction for $\alpha[D^-]$ as a function of x_F is shown in Fig. 1. The fit is remarkably good, given that it involves only a single multiplicative parameter. The fit could be improved by allowing nonzero values of $\tilde{\rho}_1$, ρ_8, and $\tilde{\rho}_8$. We conclude that the HQR mechanism provides a viable explanation for the leading particle effect in charm hadroproduction.

8. Conclusion and Outlook

We have shown that heavy quark recombination is a viable mechanism for generating the large charm asymmetries that have been observed in fixed target experiments. This mechanism is distinguished at large P_T as a power correction to the fragmentation cross section that is suppressed only by $\Lambda_{QCD} m_c / P_T^2$. It is a very predictive mechansim: the asymmetries are generated by a hard-scattering parton process that can be calculated using perturbative QCD. It is also very economical: the nonperturbative inputs are multiplicative constants. It provides a natural explanation for the leading particle effect, because the $c\bar{q}$ recombination cross section is strongly peaked in the forward $\bar{q}$ direction.

Thus far we have considered three applications of the HQR mechanism. In the case of the $D^- - D^+$ asymmetry in $\pi^- N$ collisions, we obtain a good description of the leading particle effect by fitting a single multiplicative constant $\rho_1 \approx 0.06$ [14]. We have also applied this mechanism to charm meson production in fixed-target photoproduction experiments [13], where the asymmetries are much smaller and the experimental errors larger. Again we get a good fit to the data by adjusting a single parameter $\rho_1 \approx 0.15$, whose value is consistent within errors to the hadroproduction value. Finally we have applied the HQR mechanism to bottom meson production at the Tevatron [12]. Using the value of ρ_1 obtained by scaling the charm value by m_c/m_b, we find that the asymmetries should be tiny for rapidities $|y| < 3$, which puts them out of reach of present experiments.

There are many important applications of the HQR mechanism that remain to be done. In our analysis of charm hadroproduction, we have considered only the asymmetry for $D^- - D^+$ in $\pi^- N$ collisions for which the data is most accurate. That analysis should be extended to the other charm mesons as well as to other beams like p and Σ^-. Another interesting application is to charm meson production in ep collisions at HERA, where the ZEUS experiment has seen indications of an excess of D^* in the forward p direction [16]. Perhaps the most exciting application is to the hadroproduction of charm baryons, such as Λ_c^+, Σ_c^0, and Σ_c^{++}. Large asymmetries in these cross sections have been observed by the SELEX collaboration [17]. The most likely HQR mechanism for generating the charm baryon asymmetries is not $c\bar{q}$ recombination but cq recombination.

In conclusion, heavy quark recombination is a promising mechanism for explaining the observed asymmetries between heavy hadrons and their antiparticles. Hopefully, it represents the first step in replacing models for the hadronization of heavy quarks by a systematic treatment of the power corrections.

Acknowledgments

This work was carried out in collaboration with Yu Jia and Tom Mehen. It was supported in part by the U.S. Department of Energy under Grant DE-FG02-91-ER40690.

References

1. G. Curci, W. Furmanski and R. Petronzio, Nucl. Phys. B **175**, 27 (1980).
2. J. C. Collins and D. E. Soper, Nucl. Phys. B **194**, 445 (1982).
3. P. Nason, S. Dawson and R. K. Ellis, Nucl. Phys. B **303**, 607 (1988).
4. J. C. Collins, D. E. Soper and G. Sterman, Nucl. Phys. B **263**, 37 (1986).
5. S. J. Brodsky, J. F. Gunion and D. E. Soper, Phys. Rev. D **36**, 2710 (1987).
6. E. M. Aitala *et al.* [E791 Collaboration], Phys. Lett. B **371**, 157 (1996).
7. R. C. Hwa, Phys. Rev. D **22**, 1593 (1980).
8. A. K. Likhoded and S. R. Slabospitsky, Phys. Atom. Nucl. **62**, 693 (1999) [Yad. Fiz. **62**, 742 (1999)].
9. E. Norrbin and T. Sjostrand, Eur. Phys. J. C **17**, 137 (2000).
10. S. J. Brodsky, P. Hoyer, C. Peterson and N. Sakai, Phys. Lett. B **93**, 451 (1980).
11. R. Vogt and S. J. Brodsky, Nucl. Phys. B **478**, 311 (1996).
12. E. Braaten, Y. Jia and T. Mehen, arXiv:hep-ph/0108201.
13. E. Braaten, Y. Jia and T. Mehen, Phys. Rev. D **66**, 014003 (2002).
14. E. Braaten, Y. Jia and T. Mehen, arXiv:hep-ph/0205149.
15. M. Beneke, G. Buchalla, M. Neubert and C. T. Sachrajda, Phys. Rev. Lett. **83**, 1914 (1999); Nucl. Phys. B **591**, 313 (2000).
16. J. Breitweg *et al.* [ZEUS Collaboration], Eur. Phys. J. C **6**, 67 (1999).
17. F. G. Garcia *et al.* [SELEX Collaboration], Phys. Lett. B **528**, 49 (2002).

LIFETIMES OF HEAVY HADRONS BEYOND LEADING LOGARITHMS

ULRICH NIERSTE

Fermi National Accelerator Laboratory
*Batavia, IL60510 — 500, USA.**
E-mail: nierste@fnal.gov

The lifetime splitting between the B^+ and B_d^0 mesons has recently been calculated in the next-to-leading order of QCD. These corrections are necessary for a reliable theoretical prediction, in particular for the meaningful use of hadronic matrix elements computed with lattice QCD. Using results from quenched lattice QCD we find $\tau(B^+)/\tau(B_d^0) = 1.053 \pm 0.016 \pm 0.017$, where the uncertainties from unquenching and $1/m_b$ corrections are not included. The lifetime difference of heavy baryons Ξ_b^0 and Ξ_b^- is also discussed.

1. Introduction

In my talk I present work done in collaboration with Martin Beneke, Gerhard Buchalla, Christoph Greub and Alexander Lenz [1].

Twenty years ago the hosts of this conference showed that inclusive decay rates of hadrons containg a heavy quark can be computed from first principles of QCD. The *Heavy Quark Expansion* (HQE) technique [2] exploits the heaviness of the bottom (or charm) quark compared to the fundamental QCD scale Λ_{QCD}. In order to study the lifetime of some b-flavored hadron H containing a single heavy quark one needs to compute its total decay rate $\Gamma(H_b)$. Now the HQE is an operator product expansion (OPE) expressing $\Gamma(H_b)$ in terms of matrix elements of local $\Delta B = 0$ (B denotes the bottom number) operators, leading to an expansion of $\Gamma(H_b)$ in terms of Λ_{QCD}/m_b. In the leading order of Λ_{QCD}/m_b the decay rate of H_b equals the decay rate of a free b-quark, unaffected by the light degrees of freedom of H_b. Consequently, the lifetimes of all b-flavored hadrons are the same at this order. The dominant source of lifetime differences are weak interaction

---

*Fermilab is operated by URA under DOE contract No. DE-AC02-76CH03000.

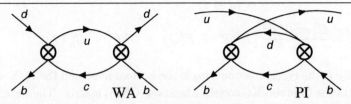

Figure 1. *Weak annihilation* (WA) and *Pauli interference* (PI) diagrams in the leading order of QCD. They contribute to $\Gamma(B_d^0)$ and $\Gamma(B^+)$, respectively. The crosses represent $|\Delta B|=1$ operators, which are generated by the exchange of W bosons. CKM-suppressed contributions are not shown.

effects between the b-quark and the light valence quark. They are depicted in Fig. 1 for the case of the B^+–B_d^0 lifetime difference. The relative size of these weak non-spectator effects to the leading free-quark decay is of order $16\pi^2(\Lambda_{QCD}/m_b)^3 = O(5$–$10\%)$. The measurement of lifetime differences among different b-flavored hadrons therefore tests the HQE formalism at the third order in the expansion parameter.

The optical theorem relates the total decay rate $\Gamma(H_b)$ to the self-energy of H_b:

$$\Gamma(H_b) = \frac{1}{2M_{H_b}}\langle H_b|T|H_b\rangle. \tag{1}$$

Here we have introduced the transition operator:

$$T = \mathrm{Im}\, i \int d^4x\, T[H(x)\, H(0)] \tag{2}$$

with the effective $|\Delta B| = 1$ Hamiltonian H describing the W-mediated decay of the b quark. The HQE amounts to an OPE applied to T which effectively integrates out the hard loop momenta (corresponding to the momenta of the final state quarks). We decompose the result as

$$T = [T_0 + T_2 + T_3]\left[1 + O(1/m_b^4)\right]$$
$$T_3 = T^u + T^d + T_{sing}. \tag{3}$$

Here T_n denotes the portion of T which is suppressed by a factor of $1/m_b^n$ with respect to T_0 describing the free quark decay. The contributions to T_3 from the weak interaction with the valence quark read

$$T^u = \frac{G_F^2 m_b^2 |V_{cb}|^2}{6\pi}\left[\ |V_{ud}|^2\left(F^u Q^d + F_S^u Q_S^d + G^u T^d + G_S^u T_S^d\right)\right.$$

$$\left. + |V_{cd}|^2\left(F^c Q^d + F_S^c Q_S^d + G^c T^d + G_S^c T_S^d\right)\right]$$

$$+ (d \to s)$$

$$T^d = \frac{G_F^2 m_b^2 |V_{cb}|^2}{6\pi} \left[F^d Q^u + F_S^d Q_S^u + G^d T^u + G_S^d T_S^u \right]. \tag{4}$$

Here G_F is the Fermi constant, m_b is the bottom mass and the V_{ij}'s are elements of the Cabibbo-Kobayashi-Maskawa (CKM) matrix. The superscript q of the coefficients F^q, F_S^q, G^q, G_S^q refers to the cq intermediate state. The leading contributions to T^u and T^d are obtained from the left and right diagram in Fig. 1, respectively. They involve the local dimension-6, $\Delta B = 0$ operators

$$Q^q = \bar{b}\gamma_\mu(1 - \gamma_5)q \, \bar{q}\gamma^\mu(1 - \gamma_5)b,$$
$$Q_S^q = \bar{b}(1 - \gamma_5)q \, \bar{q}(1 + \gamma_5)b,$$
$$T^q = \bar{b}\gamma_\mu(1 - \gamma_5)T^a q \, \bar{q}\gamma^\mu(1 - \gamma_5)T^a b,$$
$$T_S^q = \bar{b}(1 - \gamma_5)T^a q \, \bar{q}(1 + \gamma_5)T^a b, \tag{5}$$

where T^a is the generator of color SU(3). The Wilson coefficients $F^u \dots G_S^d$ contain the physics from scales above m_b and are computed in perturbation theory. The remainder T_{sing} in (3) involves additional dimension-6 operators, which are $SU(3)_F$ singlets and do not contribute to the lifetime splitting within the (B^+, B_d^0) and (Ξ_b^0, Ξ_b^-) iso-doublets. In order to predict the widths $\Gamma(B_d^0)$ and $\Gamma(B^+)$ one needs to compute the hadronic matrix elements of the operators in (5). After using the isospin relation $\langle B_d^0 | Q^{d,u} | B_d^0 \rangle = \langle B^+ | Q^{u,d} | B^+ \rangle$ the matrix elements will enter $\Gamma(B_d^0) - \Gamma(B^+)$ in isospin-breaking combinations, which are conventionally parametrized as [3,4]

$$\langle B^+|(Q^u - Q^d)|B^+\rangle = f_B^2 M_B^2 B_1, \quad \langle B^+|(Q_S^u - Q_S^d)|B^+\rangle = f_B^2 M_B^2 B_2,$$
$$\langle B^+|(T^u - T^d)|B^+\rangle = f_B^2 M_B^2 \epsilon_1, \quad \langle B^+|(T_S^u - T_S^d)|B^+\rangle = f_B^2 M_B^2 \epsilon_2. \tag{6}$$

Here f_B and M_B are decay constant and mass of the B meson, respectively. In the *vacuum saturation approximation* (VSA) one has $B_1 = 1$, $B_2 = 1 + O(\alpha_s(m_b), \Lambda_{QCD}/m_b)$ and $\epsilon_{1,2} = 0$. Corrections to the VSA results are of order $1/N_c$, where $N_c = 3$ is the number of colors.

We now find from (1) and (4):

$$\Gamma(B_d^0) - \Gamma(B^+) = \frac{G_F^2 m_b^2 |V_{cb}|^2}{12\pi} f_B^2 M_B \left(|V_{ud}|^2 \vec{F}^u + |V_{cd}|^2 \vec{F}^c - \vec{F}^d \right) \cdot \vec{B}. \tag{7}$$

Here we have introduced the shorthand notation

$$\vec{F}^q(z) = \begin{pmatrix} F^q(z) \\ F_S^q(z) \\ G^q(z) \\ G_S^q(z) \end{pmatrix}, \quad \vec{B} = \begin{pmatrix} B_1 \\ B_2 \\ \epsilon_1 \\ \epsilon_2 \end{pmatrix} \quad \text{for } q = d, u, c. \quad (8)$$

Since the hard loops involve the charm quark, the coefficient $\vec{F}^q$ depends on the ratio $z = m_c^2/m_b^2$. The minimal way to include QCD effects is the leading logarithmic approximation, which includes corrections of order $\alpha_s^n \ln^n(m_b/M_W)$, $n = 0, 1, \ldots$ in $\vec{F}^q$ in (7). The corresponding leading order (LO) calculation of the width difference in (7) involves the diagrams in Fig. 1 [2,3]. Yet LO results are too crude for a precise calculation of lifetime differences. The heavy-quark masses in (7) cannot be defined in a proper way and one faces a large dependence on unphysical renormalization scales. Furthermore, results for $B_{1,2}$ and $\epsilon_{1,2}$ from lattice gauge theory cannot be matched to the continuum theory in a meaningful way at LO. Finally, as pointed out in [3], at LO the coefficients F, F_S in (7) are anomalously small. They multiply the large matrix elements parametrized by $B_{1,2}$, while the larger coefficients G, G_S come with the small hadronic parameters $\epsilon_{1,2}$, rendering the LO prediction highly unstable. To cure these problems one must include the next-to-leading-order (NLO) QCD corrections of order $\alpha_s^{n+1} \ln^n(m_b/M_W)$.

The first calculation of a lifetime difference beyond the LO was performed for the B_s^0–B_d^0 lifetime difference [5], where $O(\alpha_s)$ corrections were calculated in the $SU(3)_F$ limit neglecting certain terms of order z. In this limit only a few penguin effects play a role. A complete NLO computation has been carried out for the lifetime difference between the two mass eigenstates of the B_s^0 meson in [6]. In particular the correct treatment of infrared effects, which appear at intermediate steps of the calculation, has been worked out in [6]. The recent computation in [1] is conceptually similar to the one in [6], except that the considered transition is $\Delta B = 0$ rather than $\Delta B = 2$ and the quark masses in the final state are different. The NLO calculation of $\Gamma(B_d^0) - \Gamma(B^+)$ involves the diagrams of Fig. 2. In [4] the NLO corrections to $\Gamma(B_d^0) - \Gamma(B^+)$ have been calculated for the limiting case $z = 0$. The corrections to this limit are of order $z \ln z$ or roughly 20%. The first NLO calculation with the complete z dependence was presented in [1] and subsequently confirmed in [7].

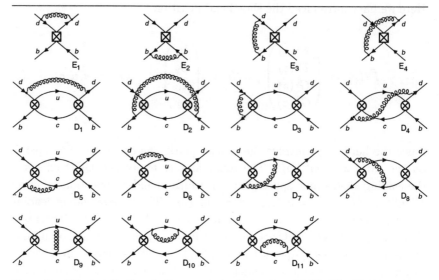

Figure 2. WA contributions in the next-to-leading order of QCD. The PI diagrams are obtained by interchanging u and d and reversing the fermion flow of the u and d lines. The first line shows the radiative corrections to $\Delta B = 0$ operators, which are necessary for the proper infrared factorization. Not displayed are the diagrams E'_3, E'_4 and D'_{3-8} which are obtained from the corresponding unprimed diagrams by left-right reflection and the reverse of the fermion flow.

2. Lifetime differences at next-to-leading order

The analytic expressions for the Wilson coefficients $F_{ij}^{u,(1)} - F_{ij}^{d,(1)} \ldots G_{S,ij}^{u,(1)} - G_{S,ij}^{d,(1)}$ are cumbersome functions of z involving dilogarithms. They depend on the renormalization scheme chosen for the $\Delta B = 0$ operators in (5) and also on the renormalization scale $\mu_0 = O(m_b)$ at which these operators are defined. These dependences properly cancel between $\vec{F}^q$ and $\vec{B}$ in physical observables like (7). When our results for $F_{ij}^{u,(1)} - F_{ij}^{d,(1)} \ldots G_{S,ij}^{u,(1)} - G_{S,ij}^{d,(1)}$ are combined with some non-perturbative computation of $B_1, \ldots \epsilon_2$, one has to make sure that the numerical values of these hadronic parameters correspond to the same renormalization scheme. Our scheme is defined by the use of dimensional regularization with $\overline{\text{MS}}$ [8] subtraction, an anticommuting γ_5 and a choice of evanescent operators preserving Fierz invariance at the loop level [9]. Choosing further $\mu_0 = m_b$ the desired lifetime ratio can be compactly written as

$$\frac{\tau(B^+)}{\tau(B_d^0)} - 1 = \tau(B^+) \left[\Gamma(B_d^0) - \Gamma(B^+) \right]$$

$$= 0.0325 \left(\frac{|V_{cb}|}{0.04}\right)^2 \left(\frac{m_b}{4.8\,\mathrm{GeV}}\right)^2 \left(\frac{f_B}{200\,\mathrm{MeV}}\right)^2 \times$$

$$\left[(1.0 \pm 0.2)\,B_1 \;+\; (0.1 \pm 0.1)\,B_2 \;-\; (18.4 \pm 0.9)\,\epsilon_1 \;+\; (4.0 \pm 0.2)\,\epsilon_2\right]. \quad (9)$$

Here $\tau(B^+) = 1.653\,\mathrm{ps}$ has been used in the overall factor.

The hadronic parameters have been computed in [10] with quenched lattice QCD using the same renormalization scheme as in the present paper. They read

$$(B_1, B_2, \epsilon_1, \epsilon_2) = (1.10 \pm 0.20,\, 0.79 \pm 0.10,\, -0.02 \pm 0.02,\, 0.03 \pm 0.01). \quad (10)$$

Inserting $|V_{cb}| = 0.040 \pm 0.0016$ from a CLEO analysis of inclusive semileptonic B decays [13], the world average $f_B = (200 \pm 30)\,\mathrm{MeV}$ from lattice calculations [14] and $m_b = 4.8 \pm 0.1\,\mathrm{GeV}$ for the one-loop bottom pole mass into (9), our NLO prediction reads

$$\frac{\tau(B^+)}{\tau(B_d^0)} = 1.053 \pm 0.016 \pm 0.017 \quad (11)$$

compared to

$$\left[\frac{\tau(B^+)}{\tau(B_d^0)}\right]_{\mathrm{LO}} = 1.041 \pm 0.040 \pm 0.013. \quad (12)$$

Here the first error is due to the errors on the coefficients and the hadronic parameters (10), and the second error is the overall normalization uncertainty due to m_b, $|V_{cb}|$ and f_B in (9). The Wilson coefficients also depend on the renormalization scale μ_1 at which the $\Delta B = 1$ operators entering the diagrams in Figs. 1 and 2 are defined. This dependence stems from the truncation of the perturbation series and diminishes order-by-order in α_s. The dependence on μ_1 is the dominant uncertainty of the LO prediction of the lifetime ratio. In Fig. 3 the μ_1-dependence of the LO and NLO predictions for $\tau(B^+)/\tau(B_d^0) - 1$ is shown. The substantial reduction of scale dependence at NLO leads to the improvement in the NLO vs. LO results in (11),(12). Note that the NLO calculation has firmly established that $\tau(B^+) > \tau(B_d^0)$, a conclusion which could not be drawn from the old LO result. The result in (11) is compatible with recent measurement from the B factories [11, 12]:

$$\frac{\tau(B^+)}{\tau(B_d^0)} = \begin{cases} 1.082 \pm 0.026 \pm 0.012 \ (\text{B\textsc{a}B\textsc{ar}}) \\ 1.091 \pm 0.023 \pm 0.014 \ (\text{BELLE}) \end{cases}$$

The calculated Wilson coefficients can also be used to predict the lifetime splitting within the iso-doublet ($\Xi_b^0 \sim bus$, $\Xi_b^- \sim bds$) with NLO precision. The corresponding LO diagrams are shown in Fig. 4. Note that

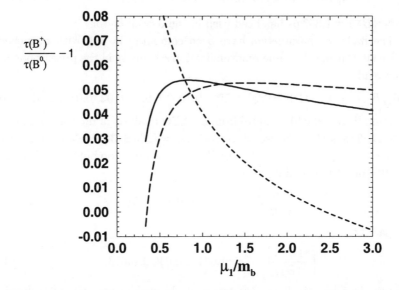

$\dfrac{\tau(B^+)}{\tau(B^0)} - 1$

Figure 3. Dependence of $\tau(B^+)/\tau(B_d^0) - 1$ on μ_1/m_b for the central values of the input parameters and $\mu_0 = m_b$. The solid (short-dashed) line shows the NLO (LO) result. The long-dashed line shows the NLO result in the approximation of [4], i.e. z is set to zero in the NLO corrections.

the role of T^u and T^d is interchanged compared to the meson case with T^u describing the Pauli interference effect. The lifetime difference between $\Lambda_b \sim bud$ and Ξ_b^0 is expected to be small, as in the case of B_s^0 and B_d^0, because it mainly stems from the small U-spin breaking effects in the matrix elements appearing at order $1/m_b^2$.

For Ξ_b's the weak decay of the valence s-quark could be relevant: the decays $\Xi_b^- \to \Lambda_b \pi^-$, $\Xi_b^- \to \Lambda_b e^- \overline{\nu}_e$ and $\Xi_b^0 \to \Lambda_b \pi^0$ are triggered by $s \to u$ transitions and could affect the total rates at the $O(1\%)$ level [15]. Once the lifetime measurements reach this accuracy, one should correct for this effect. To this end we define

$$\overline{\Gamma}(\Xi_b) \equiv \Gamma(\Xi_b) - \Gamma(\Xi_b \to \Lambda_b X) = \frac{1 - B(\Xi_b \to \Lambda_b X)}{\tau(\Xi_b)} \equiv \frac{1}{\overline{\tau}(\Xi_b)}$$

$$\text{for } \Xi_b = \Xi_b^0, \Xi_b^-, \qquad (13)$$

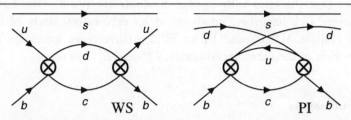

Figure 4. *Weak scattering* (WS) and PI diagrams for Ξ_b baryons in the leading order of QCD. They contribute to $\Gamma(\Xi_b^0)$ and $\Gamma(\Xi_b^-)$, respectively. CKM-suppressed contributions are not shown.

where $B(\Xi_b \to \Lambda_b X)$ is the branching ratio of the above-mentioned decay modes. Thus $\overline{\Gamma}(\Xi_b)$ is the contribution from $b \to c$ transitions to the total decay rate. In contrast to the B meson system, the matrix elements of the four operators in (5) are not independent at the considered order in Λ_{QCD}/m_b. Since the light degrees of freedom are in a spin-0 state, the matrix elements $\langle \Xi_b | 2Q_S^q + Q^q | \Xi_b \rangle$ and $\langle \Xi_b | 2T_S^q + T^q | \Xi_b \rangle$ are power-suppressed compared to those in (14) (see e.g. [2,3]). This, however, is not true in all renormalization schemes, in the $\overline{\text{MS}}$ scheme used by us $2Q_S^q + Q^q$ and $2T_S^q + T^q$ receive short-distance corrections, because hard gluons can resolve the heavy b-quark mass. A priori one can choose the renormalization of e.g. Q_S^q independently from Q^q, so that $\langle \Xi_b | 2Q_S^q + Q^q | \Xi_b \rangle = O(\Lambda_{QCD}/m_b)$ can only hold in certain renormalization schemes. This is also the case, if the operators are defined in heavy quark effective theory (HQET) rather than in full QCD. After properly taking into account these short-distance corrections, one can express the desired lifetime ratio solely in terms of two hadronic parameters defined as

$$\langle \Xi_b^0 | (Q^u - Q^d)(\mu_0) | \Xi_b^0 \rangle = f_B^2 M_B M_{\Xi_b} L_1^{\Xi_b}(\mu_0),$$
$$\langle \Xi_b^0 | (T^u - T^d)(\mu_0) | \Xi_b^0 \rangle = f_B^2 M_B M_{\Xi_b} L_2^{\Xi_b}(\mu_0). \tag{14}$$

Then one finds

$$\frac{\overline{\tau}(\Xi_b^0)}{\overline{\tau}(\Xi_b^-)} - 1 = \overline{\tau}(\Xi_b^0) \left[\Gamma(\Xi_b^-) - \Gamma(\Xi_b^0) \right]$$

$$= 0.59 \left(\frac{|V_{cb}|}{0.04} \right)^2 \left(\frac{m_b}{4.8\,\text{GeV}} \right)^2 \left(\frac{f_B}{200\,\text{MeV}} \right)^2 \frac{\overline{\tau}(\Xi_b^0)}{1.5\,\text{ps}} \times$$

$$\left[(0.04 \pm 0.01)\, L_1 - (1.00 \pm 0.04)\, L_2 \right], \tag{15}$$

with $L_i = L_i^{\Xi_b}(\mu_0 = m_b)$. For the baryon case there is no reason to expect the color-octet matrix element to be much smaller than the color-singlet ones, so that the term with L_2 will dominate the result. The hadronic parameters $L_{1,2}$ have been analyzed in an exploratory study of lattice HQET [16] for Λ_b baryons. Up to $SU(3)_F$ corrections, which are irrelevant in view of the other uncertainties, $L_i^{\Xi_b}$ and $L_i^{\Lambda_b}$ are equal.

3. Conclusions

Twenty years ago the ITEP group has developed the Heavy Quark Expansion, which allows to study inclusive decay rates of heavy hadrons in a model-free, QCD-based framework [2]. The HQE expresses these decay rates as a series in both Λ_{QCD}/m_b and $\alpha_s(m_b)$. With the advent of precision measurements of lifetimes of b-flavored hadrons at the B factories and the Tevatron correspondingly precise theory predictions are desirable. This requires the calculation of higher-order terms in the HQE. The inclusion of the α_s corrections presented in this talk is in particular mandatory for any meaningful use of hadronic matrix elements computed in lattice gauge theory. The calculated QCD corrections to the WA and PI diagrams in Figs. 1,2 allow to study the lifetime splitting within the (B^+, B_d^0) and (Ξ_b^0, Ξ_b^-) iso-doublets with NLO accuracy. It is gratifying that these corrections have been independently calulated by two groups finding agreement in their analytic expressions for the Wilson coefficients [1,7].

Current lattice calculations, which are still in a relatively early stage in this case, yield, when combined with our calculations, $\tau(B^+)/\tau(B_d^0) = 1.053 \pm 0.016 \pm 0.017$ [see (11)]. The effects of unquenching and $1/m_b$ corrections are not included in the error estimate, but the unquenching effects can well be sizable. A substantial improvement of the NLO calculation is the large reduction of perturbative uncertainty reflected in the scale dependence stemming from the $\Delta B = 1$ operators. This scale dependence had been found to be very large at leading order, preventing even an unambiguous prediction of the sign of $\tau(B^+)/\tau(B_d^0) - 1$ up to now [3].

At present the experimentally measured Λ_b lifetime falls short of $\tau(B_d^0)$ by roughly 20% [17], which has raised concerns about the applicability of the HQE to baryons. Unfortunately this interesting topic cannot yet be addressed at the NLO level for two reasons: First, $\tau(\Lambda_b)/\tau(B_d^0)$ receives contributions from the yet uncalculated $SU(3)_F$-singlet portion T_{sing} of the transition operator in (3). Second, the hadronic matrix elements entering $\tau(\Lambda_b)/\tau(B_d^0)$ involve penguin (also called 'eye') contractions of the operators in (5), which are difficult to compute. These penguin contractions are contributions to the matrix elements in which the light q and $\bar{q}$ quark fields of the operator are contracted with each other, not with the hadron's valence quarks.

Acknowledgments

I thank the organizers for the invitation to this wonderful Arkadyfest workshop. I have enjoyed stimulating discussions with many participants and look forward to future Arkadyfests, possibly on the occasions of Arkady's 70th, 80th, 90th and 100th birthdays! I am grateful to Martin Gorbahn for proofreading the manuscript.

References

1. M. Beneke, G. Buchalla, C. Greub, A. Lenz and U. Nierste, Nucl. Phys. B **639** (2002) 389 [arXiv:hep-ph/0202106].
2. M. A. Shifman and M. B. Voloshin, in: *Heavy Quarks* ed. V. A. Khoze and M. A. Shifman, Sov. Phys. Usp. **26** (1983) 387; M. A. Shifman and M. B. Voloshin, Sov. J. Nucl. Phys. **41** (1985) 120 [Yad. Fiz. **41** (1985) 187]; M. A. Shifman and M. B. Voloshin, Sov. Phys. JETP **64** (1986) 698 [Zh. Eksp. Teor. Fiz. **91** (1986) 1180]; I. I. Bigi, N. G. Uraltsev and A. I. Vainshtein, Phys. Lett. B **293** (1992) 430 [Erratum-ibid. B **297** (1992) 477]. For a recent review see: M. Voloshin, in: *B physics at the Tevatron: Run-II and Beyond*, Chapter 8, [hep-ph/0201071].
3. M. Neubert and C. T. Sachrajda, Nucl. Phys. B **483** (1997) 339. M. Beneke, G. Buchalla and I. Dunietz, Phys. Rev. D **54** (1996) 4419.
4. M. Ciuchini, E. Franco, V. Lubicz and F. Mescia, [hep-ph/0110375].
5. Y. Y. Keum and U. Nierste, Phys. Rev. D **57** (1998) 4282.
6. M. Beneke, G. Buchalla, C. Greub, A. Lenz and U. Nierste, Phys. Lett. B **459** (1999) 631.
7. E. Franco, V. Lubicz, F. Mescia and C. Tarantino, Nucl. Phys. B **633** (2002) 212 [arXiv:hep-ph/0203089].
8. W. A. Bardeen, A. J. Buras, D. W. Duke and T. Muta, Phys. Rev. D **18** (1978) 3998.
9. S. Herrlich and U. Nierste, Nucl. Phys. B **455** (1995) 39.
10. D. Becirevic, [hep-ph/0110124].
11. B. Aubert *et al.* [BABAR Collaboration], Phys. Rev. Lett. **87** (2001) 201803.
12. K. Abe [Belle Collaboration], [hep-ex/0202009].
13. D. Cassel [CLEO coll.], talk at *Lepton Photon 01*, 23-28 Jul 2001, Rome, Italy.
14. S. Ryan, [hep-lat/0111010].
15. M. B. Voloshin, Phys. Lett. B **476** (2000) 297.
16. M. Di Pierro, C. T. Sachrajda and C. Michael [UKQCD collaboration], Phys. Lett. B **468** (1999) 143.
17. D. E. Groom *et al.* [Particle Data Group Collaboration], Eur. Phys. J. C **15** (2000) 1; updated at *http://pdg.lbl.gov*.

CP-VIOLATION AND MIXING IN CHARMED MESONS

A. A. PETROV

Department of Physics and Astronomy
Wayne State University
Detroit, MI 48201, USA
E-mail: apetrov@physics.wayne.edu

The Standard Model contribution to $D^0 - \overline{D}^0$ mixing is dominated by the contributions of light s and d quarks. Neglecting the tiny effects due to b quark, both mass and lifetime differences vanish in the limit of $SU(3)_F$ symmetry. Thus, the main challenge in the Standard Model calculation of the mass and width difference in the $D^0 - \overline{D}^0$ system is to estimate the size of $SU(3)$ breaking effects. We prove that D meson mixing occurs in the Standard Model only at *second* order in $SU(3)$ violation. We consider the possibility that phase space effects may be the dominant source of $SU(3)$ breaking. We find that $y = (\Delta\Gamma)/(2\Gamma)$ of the order of one percent is natural in the Standard Model, potentially reducing the sensitivity to new physics of measurements of D meson mixing. We also discuss the possibility of observing lifetime differences and CP violation in charmed mesons both at the currently operating and proposed facilities.

1. Introduction

One of the most important motivations for studies of weak decays of charmed mesons is the possibility of observing a signal from new physics which can be separated from the one generated by the Standard Model (SM) interactions. The low energy effect of new physics particles can be naturally written in terms of a series of local operators of increasing dimension generating $\Delta C = 1$ (decays) or $\Delta C = 2$ (mixing) transitions. For $D^0 - \overline{D}^0$ mixing these operators, as well as the one loop Standard Model effects, generate contributions to the effective operators that change D^0 state into $\overline{D}^0$ state leading to the mass eigenstates

$$|D_{\frac{1}{2}}\rangle = p|D^0\rangle \pm q|\bar{D}^0\rangle, \tag{1}$$

where the complex parameters p and q are obtained from diagonalizing the $D^0 - \overline{D}^0$ mass matrix. The mass and width splittings between these eigenstates are parameterized by

$$x \equiv \frac{m_2 - m_1}{\Gamma}, \quad y \equiv \frac{\Gamma_2 - \Gamma_1}{2\Gamma}, \tag{2}$$

where $m_{1,2}$ and $\Gamma_{1,2}$ are the masses and widths of $D_{1,2}$ and the mean width and mass are $\Gamma = (\Gamma_1 + \Gamma_2)/2$ and $m = (m_1 + m_2)/2$. Since y is constructed from the decays of D into physical states, it should be dominated by the Standard Model contributions, unless new physics significantly modifies $\Delta C = 1$ interactions. On the contrary, x can receive contributions from all energy scales, so it is usually conjectured that new physics can significantly modify x leading to the inequality $x \gg y$. As we discuss later, this signal for new physics is lost if a relatively large y, of the order of a percent, is observed. It is known experimentally that $D^0 - \overline{D}^0$ mixing proceeds extremely slow, which in the Standard Model is usually attributed to the absence of superheavy quarks destroying GIM cancelations[1].

Another possible manifestation of new physics interactions in the charm system is associated with the observation of (large) CP-violation. This is due to the fact that all quarks that build up the hadronic states in weak decays of charm mesons belong to the first two generations. Since 2×2 Cabbibo quark mixing matrix is real, no CP-violation is possible in the dominant tree-level diagrams that describe the decay amplitudes. In the Standard Model CP-violating amplitudes can be introduced by including penguin or box operators induced by virtual b-quarks. However, their contributions are strongly suppressed by the small combination of CKM matrix elements $V_{cb}V_{ub}^*$. It is thus widely believed that the observation of (large) CP violation in charm decays or mixing would be an unambiguous sign for new physics. This fact makes charm decays a valuable tool in searching for new physics, since the statistics available in charm physics exepriment is usually quite large.

As in B-physics, CP-violating contributions in charm can be generally classified by three different categories: (I) CP violation in the decay amplitudes. This type of CP violation occurs when the absolute value of the decay amplitude for D to decay to a final state f (A_f) is different from the one of corresponding CP-conjugated amplitude ("direct CP-violation"); (II) CP violation in $D^0 - \overline{D}^0$ mixing matrix. This type of CP violation is manifest when $R_m^2 = |p/q|^2 = (2M_{12} - i\Gamma_{12})/(2M_{12}^* - i\Gamma_{12}^*) \neq 1$; and (III) CP violation in the interference of decays with and without mixing. This type of CP violation is possible for a subset of final states to which both D^0 and $\overline{D}^0$ can decay.

For a given final state f, CP violating contributions can be summarized in the parameter

$$\lambda_f = \frac{q}{p}\frac{\overline{A}_f}{A_f} = R_m e^{i(\phi+\delta)}\left|\frac{\overline{A}_f}{A_f}\right|, \qquad (3)$$

where A_f and $\overline{A}_f$ are the amplitudes for $D^0 \to f$ and $\overline{D}^0 \to f$ transitions respectively and δ is the strong phase difference between A_f and $\overline{A}_f$. Here ϕ represents the convention-independent weak phase difference between the ratio of decay amplitudes and the mixing matrix.

2. Present and perspective experimental constraints

Presently, experimental information about the $D^0 - \overline{D}^0$ mixing parameters x and y comes from the time-dependent analyses that can roughly be divided into two categories. First, more traditional studies look at the time dependence of $D \to f$ decays, where f is the final state that can be used to tag the flavor of the decayed meson. The most popular is the non-leptonic doubly Cabibbo suppressed decay (DCSD) $D^0 \to K^+\pi^-$. Time-dependent studies allow one to separate the DCSD from the mixing contribution $D^0 \to \overline{D}^0 \to K^+\pi^-$,

$$\Gamma[D^0(t) \to K^+\pi^-] = e^{-\Gamma t}|A_{K^-\pi^+}|^2$$
$$\times \left[R + \sqrt{R}R_m(y'\cos\phi - x'\sin\phi)\Gamma t + \frac{R_m^2}{4}(y^2 + x^2)(\Gamma t)^2 \right], \quad (4)$$

where R is the ratio of DCS and Cabibbo favored (CF) decay rates. Since x and y are small, the best constraint comes from the linear terms in t that are also *linear* in x and y. A direct extraction of x and y from Eq. (4) is not possible due to unknown relative strong phase δ of DCS and CF amplitudes[2], as $x' = x\cos\delta + y\sin\delta$, $y' = y\cos\delta - x\sin\delta$. This phase can be measured independently[3]. The corresponding formula can also be written[4] for $\overline{D}^0$ decay with $x' \to -x'$ and $R_m \to R_m^{-1}$.

Second, D^0 mixing can be measured by comparing the lifetimes extracted from the analysis of D decays into the CP-even and CP-odd final states. This study is also sensitive to a *linear* function of y via

$$\frac{\tau(D \to K^-\pi^+)}{\tau(D \to K^+K^-)} - 1 = y\cos\phi - x\sin\phi\left[\frac{R_m^2 - 1}{2}\right]. \quad (5)$$

Time-integrated studies of the semileptonic transitions are sensitive to the *quadratic* form $x^2 + y^2$ and at the moment are not competitive with the analyses discussed above.

The construction of a new tau-charm factory at Cornell (CLEO-c) will introduce new *time-independent* methods that are sensitive to a linear function of y. In particluar, one can use the fact that heavy meson pairs produced in the decays of heavy quarkonium resonances have the useful property that the two mesons are in the CP-correlated states[5]. By tagging one of the mesons as a CP eigenstate, a lifetime difference may be determined by measuring the leptonic branching ratio of the other meson. The initial $D^0\overline{D}^0$ state is prepared as

$$|D\overline{D}^0\rangle_L = \frac{1}{\sqrt{2}}\left\{|D^0(k_1)\overline{D}^0(k_2)\rangle + (-1)^L|D^0(k_2)\overline{D}^0(k_1)\rangle\right\}, \qquad (6)$$

where L is the relative angular momentum of two D mesons. There are several possible resonances at which CLEO-c will be running, for example $\psi(3770)$ where $L = 1$ and the initial state is antisymmetric, or $\psi(4114)$ where the initial $D^0\overline{D}^0$ state can be symmetric due to emission of additional pion or photon in the decay. In this scenario, the CP quantum numbers of the $D(k_2)$ can be determined. The semileptonic *width* of this meson should be independent of the CP quantum number since it is flavor specific. It follows that the semileptonic *branching ratio* of $D(k_2)$ will be inversely proportional to the total width of that meson. Since we know whether $D(k_2)$ is tagged as a (CP-eigenstate) D_+ or and D_- from the decay of $D(k_1)$ to S_σ, we can easily determine y in terms of the semileptonic branching ratios of $D_\pm$. This can be expressed simply by introducing the ratio

$$R_\sigma^L = \frac{\Gamma[\psi_L \to (H \to S_\sigma)(H \to Xl^\pm\nu)]}{\Gamma[\psi_L \to (H \to S_\sigma)(H \to X)]\,Br(H^0 \to Xl\nu)}, \qquad (7)$$

where X in $H \to X$ stands for an inclusive set of all final states. A deviation from $R_\sigma^L = 1$ implies a lifetime difference. Keeping only the leading (linear) contributions due to mixing, y can be extracted from this experimentally obtained quantity,

$$y\cos\phi = (-1)^L\sigma\frac{R_\sigma^L - 1}{R_\sigma^L}. \qquad (8)$$

The current experimental upper bounds on x and y are on the order of a few times 10^{-2}, and are expected to improve significantly in the coming years. To regard a future discovery of nonzero x or y as a signal for new physics, we would need high confidence that the Standard Model predictions lie well below the present limits. As was recently shown[6], in the Standard Model x and y are generated only at second order in $SU(3)$ breaking,

$$x, y \sim \sin^2\theta_C \times [SU(3) \text{ breaking}]^2, \qquad (9)$$

where θ_C is the Cabibbo angle. Therefore, predicting the Standard Model values of x and y depends crucially on estimating the size of $SU(3)$ breaking. Although y is expected to be determined by the Standard Model processes, its value nevertheless affects significantly the sensitivity to new physics of experimental analyses of D mixing[4].

Theoretical caluclations of x and y, as will be discussed later, are quite uncertain, and the values near the current experimental bounds cannot be ruled out. Therefore, it will be difficult to find a clear indication of physics beyond the Standard Model in $D^0 - \overline{D}^0$ mixing measurements alone. The only robust potential signal of new physics in charm system at this stage is CP violation.

CP violation in D decays and mixing can be searched for by a variety of methods. For instance, time-dependent decay widths for $D \to K\pi$ are sensitive to CP violation in mixing (see Eq.(4)). Provided that the x and y are comparable to experimental sensitivities, a combined analysis of $D \to K\pi$ and $D \to KK$ can yield interesting constraints on CP-violating parameters[4].

Most of the techniques that are sensitive to CP violation make use of the decay asymmetry,

$$A_{CP}(f) = \frac{\Gamma(D \to f) - \Gamma(\overline{D} \to \overline{f})}{\Gamma(D \to f) + \Gamma(\overline{D} \to \overline{f})} = \frac{1 - \left|\overline{A}_{\overline{f}}/A_f\right|^2}{1 + \left|\overline{A}_{\overline{f}}/A_f\right|^2}. \tag{10}$$

Most of the properties of Eq.(10), such as dependence on the strong final state phases, are similar to the ones in B-physics[7]. Current experimental bounds from various experiments, all consistent with zero within experimental uncertainties, can be found in[8].

Other interesting signals of CP-violation that are being discussed in connection with tau-charm factory measurements are the ones that are using quantum coherence of the initial state. An example of this type of signal is a decay $(D^0\overline{D}^0) \to f_1 f_2$ at $\psi(3770)$ with f_1 and f_2 being the different final CP-eigenstates with $CP|f_1\rangle = CP|f_2\rangle$. This type of signals are very easy to detect experimentally. It is easy to compute this CP-violating decay rate for the final states f_1 and f_2

$$\Gamma_{f_1 f_2} = \frac{\left(2 + x^2 - y^2\right)|\lambda_{f_1} - \lambda_{f_2}|^2 + \left(x^2 + y^2\right)|1 - \lambda_{f_1}\lambda_{f_2}|^2}{2R_m^2(1 + x^2)(1 - y^2)} \Gamma_{f_1}\Gamma_{f_2} \tag{11}$$

The result of Eq. (11) repersents a generalization of the formula given in Ref. [9]. It is clear that both terms in the numerator of Eq. (11) receive contributions from CP-violation of the type I and III, while the second term is also sensitive to CP-violation of the type II. Moreover, for a large set of the final states the first term would be additionally suppressed by $SU(3)$ symmetry. For instance, $\lambda_{\pi\pi} = \lambda_{KK}$ in the $SU(3)$ symmetry limit. It is easy to see that only the second term survives if only CP violation in the mixing matrix is retained, $\Gamma_{f_1 f_2} \propto |1 - R_m^2|^2 \propto A_m^2$. This expession is of the *second* order in CP-violating parameters. As it follows from the existing experimental constraints on rate asymmetries, CP-violating phases are quite small in charm system, regardless of whether they are produced by the Standard Model mechanisms or by some new physics contributions. In that respect, it looks unlikely that the SM signals of CP violation would be observed at CLEO-c with this observable.

While the searches for direct CP violation via the asymmetry of Eq. (10) can be done with the charged D-mesons (which are self-tagging), investigations of the other two types of CP-violation require flavor tagging of the initial state. This severely cuts the available dataset. It is therefore interesting to look for signals of CP violation that do not require identification of the initial state. One possible CP-violating signal involves the observable obtained by summing over the initial states, $\sum \Gamma_i = \Gamma_i + \overline{\Gamma}_i$ for $i = f, \overline{f}$. A CP-odd observable that can be formed out of $\sum \Gamma_i$ is an asymmetry

$$A_{CP}^U = \frac{\sum \Gamma_f - \sum \Gamma_{\overline{f}}}{\sum \Gamma_f + \sum \Gamma_{\overline{f}}}. \tag{12}$$

Note that this asymmetry does not require quantum coherence of the initial state and therefore is accessible in any D-physics experiment. The final states must be chosen such that A_{CP}^U is not trivially zero. As we shall see below, decays of D into the final states that are CP-eigenstates would result in zero asymmetry, while the final states like $K^+ K^{*-}$ or $K_S \pi^+ \pi^-$ would not. A non-zero value of A_{CP}^U in Eq. (12) can be generated by both direct and indirect CP-violating contributions. These can be separated by appropriately choosing the final states. For example, indirect CP violating amplitudes are tightly constrained in the decays dominated by the Cabibbo-favored tree level amplitudes, while singly Cabibbo suppressed amplitudes also receive contributions from direct CP violating amplitudes. Neglecting small CP-violation in the mixing matrix ($R_m \to 1$) one obtains,

$$A_{CP}^U = \frac{\Gamma_f - \overline{\Gamma}_{\overline{f}} - \Gamma_{\overline{f}} + \overline{\Gamma}_f}{\Gamma_f + \Gamma_{\overline{f}} + \overline{\Gamma}_f + \overline{\Gamma}_{\overline{f}}} + \frac{2y}{\Gamma_f + \Gamma_{\overline{f}} + \overline{\Gamma}_f + \overline{\Gamma}_{\overline{f}}}$$
$$\times \left[\cos \phi \left(Re \overline{A}_f^* A_{\overline{f}} - Re A_f^* A_{\overline{f}} \right) + \sin \phi \left(Im \overline{A}_f \overline{A}_{\overline{f}}^* + Im A_f^* A_{\overline{f}} \right) \right] \tag{13}$$

It is easy to see that, as promised, this asymmetry vanishes for the final states that are CP-eigenstates, as $\Gamma_f = \Gamma_{\overline{f}}$ and $\Gamma_f - \overline{\Gamma}_{\overline{f}} = \Gamma_{\overline{f}} - \overline{\Gamma}_f$.

3. Theoretical expectations for mixing parameters

Theoretical predictions of x and y within and beyond the Standard Model span several orders of magnitude[10]. Roughly, there are two approaches, neither of which give very reliable results because m_c is in some sense intermediate between heavy and light. The "inclusive" approach is based on the operator product expansion (OPE). In the $m_c \gg \Lambda$ limit, where Λ is a scale characteristic of the strong interactions, ΔM and $\Delta \Gamma$ can be expanded in terms of matrix elements of local operators[11]. Such calculations yield $x, y < 10^{-3}$. The use of the OPE relies on local quark-hadron duality, and on Λ/m_c being small enough to allow a truncation of the series after the first few terms. The charm mass may not be large enough for these to be good approximations, especially for nonleptonic D decays. An observation of y of order 10^{-2} could be ascribed to a breakdown of the OPE or of duality, but such a large value of y is certainly not a generic prediction of OPE analyses. The "exclusive" approach sums over intermediate hadronic states, which may be modeled or fit to experimental data[12]. Since there are cancellations between states within a given $SU(3)$ multiplet, one needs to know the contribution of each state with high precision. However, the D is not light enough that its decays are dominated by a few final states. In the absence of sufficiently precise data on many decay rates and on strong phases, one is forced to use some assumptions. While most studies find $x, y < 10^{-3}$, Refs.[12] obtain x and y at the 10^{-2} level by arguing that $SU(3)$ violation is of order unity, but the source of the large $SU(3)$ breaking is not made explicit.

In what follows we first prove that $D^0 - \overline{D}^0$ mixing arises only at *second* order in $SU(3)$ breaking effects. The proof is valid when $SU(3)$ violation enters perturbatively. This would not be so, for example, if D transitions were dominated by a single narrow resonance close to threshold[6,13]. Then we argue that reorganization of "exclusive" calculation by explicitly building $SU(3)$ cancellations into the analysis naturally leads to values of $y \sim 1\%$ if only one source of $SU(3)$ breaking (phase space) is taken into account.

The quantities M_{12} and Γ_{12} which determine x and y depend on matrix elements $\langle \overline{D}^0 | H_w H_w | D^0 \rangle$, where H_w denote the $\Delta C = -1$ part of the weak Hamiltonian. Let D be the field operator that creates a D^0 meson and annihilates a $\overline{D}^0$. Then the matrix element, whose $SU(3)$ flavor group theory properties we will study, may be written as

$$\langle 0 | D H_w H_w D | 0 \rangle . \tag{14}$$

Since the operator D is of the form $\bar{c}u$, it transforms in the fundamental representation of $SU(3)$, which we will represent with a lower index, D_i. We use a convention in which the correspondence between matrix indices and quark flavors is $(1, 2, 3) = (u, d, s)$. The only nonzero element of D_i is $D_1 = 1$. The $\Delta C = -1$ part of the weak Hamiltonian has the flavor structure $(\bar{q}_i c)(\bar{q}_j q_k)$, so its matrix representation is written with a fundamental index and two antifundamentals, H_k^{ij}. This operator is a sum of irreps contained in the product $3 \times \bar{3} \times \bar{3} = \overline{15} + 6 + \bar{3} + \bar{3}$. In the limit in which the third generation is neglected, H_k^{ij} is traceless, so only the $\overline{15}$ and 6 representations appear. That is, the $\Delta C = -1$ part of H_w may be decomposed as $\frac{1}{2}(O_{\overline{15}} + O_6)$, where

$$O_{\overline{15}} = (\bar{s}c)(\bar{u}d) + (\bar{u}c)(\bar{s}d) + s_1(\bar{d}c)(\bar{u}d) + s_1(\bar{u}c)(\bar{d}d)$$
$$- s_1(\bar{s}c)(\bar{u}s) - s_1(\bar{u}c)(\bar{s}s) - s_1^2(\bar{d}c)(\bar{u}s) - s_1^2(\bar{u}c)(\bar{d}s),$$
$$O_6 = (\bar{s}c)(\bar{u}d) - (\bar{u}c)(\bar{s}d) + s_1(\bar{d}c)(\bar{u}d) - s_1(\bar{u}c)(\bar{d}d)$$
$$- s_1(\bar{s}c)(\bar{u}s) + s_1(\bar{u}c)(\bar{s}s) - s_1^2(\bar{d}c)(\bar{u}s) + s_1^2(\bar{u}c)(\bar{d}s), \qquad (15)$$

and $s_1 = \sin\theta_C$. The matrix representations $H(\overline{15})_k^{ij}$ and $H(6)_k^{ij}$ have nonzero elements

$$
\begin{aligned}
H(\overline{15})_k^{ij}: &\qquad H_2^{13} = H_2^{31} = 1, &\qquad H_2^{12} = H_2^{21} = s_1, \\
&\qquad H_3^{13} = H_3^{31} = -s_1, &\qquad H_3^{12} = H_3^{21} = -s_1^2, \\
H(6)_k^{ij}: &\qquad H_2^{13} = -H_2^{31} = 1, &\qquad H_2^{12} = -H_2^{21} = s_1, \\
&\qquad H_3^{13} = -H_3^{31} = -s_1, &\qquad H_3^{12} = -H_3^{21} = -s_1^2.
\end{aligned}
\qquad (16)
$$

We introduce $SU(3)$ breaking through the quark mass operator M, whose matrix representation is $M_j^i = \text{diag}(m_u, m_d, m_s)$ as being in the adjoint representation to induce $SU(3)$ violating effects. We set $m_u = m_d = 0$ and let $m_s \neq 0$ be the only $SU(3)$ violating parameter. All nonzero matrix elements built out of D_i, H_k^{ij} and M_j^i must be $SU(3)$ singlets.

We now prove that $D^0 - \overline{D}^0$ mixing arises only at second order in $SU(3)$ violation, by which we mean second order in m_s. First, we note that the pair of D operators is symmetric, and so the product $D_i D_j$ transforms as a 6 under $SU(3)$. Second, the pair of H_w's is also symmetric, and the product $H_k^{ij} H_n^{lm}$ is in one of the reps which appears in the product

$$\left[(\overline{15} + 6) \times (\overline{15} + 6)\right]_S = (\overline{15} \times \overline{15})_S + (\overline{15} \times 6) + (6 \times 6)_S \qquad (17)$$
$$= (\overline{60} + \overline{24} + 15 + 15' + \bar{6}) + (42 + 24 + 15 + \bar{6} + 3) + (15' + \bar{6}).$$

A direct computation shows that only three of these representations actually appear in the decomposition of $H_w H_w$. They are the $\overline{60}$, the 42, and the $15'$ (actually twice, but with the same nonzero elements both times). So we have product operators of the form (the subscript denotes the representation of $SU(3)$)

$$DD = D_6, \qquad H_w H_w = O_{\overline{60}} + O_{42} + O_{15'}. \qquad (18)$$

Since there is no $\bar{6}$ in the decomposition of $H_w H_w$, there is no $SU(3)$ singlet which can be made with D_6, and no $SU(3)$ invariant matrix element of the form (14) can be formed. This is the well known result that $D^0 - \overline{D}^0$

mixing is *prohibited by* $SU(3)$ *symmetry*. Now consider a single insertion of the $SU(3)$ violating spurion M. The combination $D_6 M$ transforms as $6 \times 8 = 24 + \overline{15} + 6 + \overline{3}$. There is still no invariant to be made with $H_w H_w$, thus $D^0 - \overline{D}^0$ mixing is *not induced at first order in* $SU(3)$ *breaking*. With two insertions of M, it becomes possible to make an $SU(3)$ invariant. The decomposition of DMM is

$$6 \times (8 \times 8)_S = 6 \times (27 + 8 + 1) \tag{19}$$
$$= (60 + \overline{42} + 24 + \overline{15} + \overline{15}' + 6) + (24 + \overline{15} + 6 + \overline{3}) + 6.$$

There are three elements of the 6×27 part which can give invariants with $H_w H_w$. Each invariant yields a contribution to $D^0 - \overline{D}^0$ mixing proportional to $s_1^2 m_s^2$. Thus, $D^0 - \overline{D}^0$ mixing arises only at *second order* in the $SU(3)$ violating parameter m_s.

We now turn to the contributions to y from on-shell final states, which result from every common decay product of D^0 and $\overline{D}^0$. In the $SU(3)$ limit, these contributions cancel when one sums over complete $SU(3)$ multiplets in the final state. The cancellations depend on $SU(3)$ symmetry both in the decay matrix elements and in the final state phase space. While there are $SU(3)$ violating corrections to both of these, it is difficult to compute the $SU(3)$ violation in the matrix elements in a model independent manner. Yet, with some mild assumptions about the momentum dependence of the matrix elements, the $SU(3)$ violation in the phase space depends only on the final particle masses and can be computed. We estimate the contributions to y solely from $SU(3)$ violation in the phase space. We find that this source of $SU(3)$ violation can generate y of the order of a few percent.

The mixing parameter y may be written in terms of the matrix elements for common final states for D^0 and $\overline{D}^0$ decays,

$$y = \frac{1}{\Gamma} \sum_n \int [\text{P.S.}]_n \langle \overline{D}^0 | H_w | n \rangle \langle n | H_w | D^0 \rangle, \tag{20}$$

where the sum is over distinct final states n and the integral is over the phase space for state n. Let us now perform the phase space integrals and restrict the sum to final states F which transform within a single $SU(3)$ multiplet R. The result is a contribution to y of the form

$$\frac{1}{\Gamma} \langle \overline{D}^0 | H_w \left\{ \eta_{CP}(F_R) \sum_{n \in F_R} |n\rangle \rho_n \langle n| \right\} H_w | D^0 \rangle, \tag{21}$$

where ρ_n is the phase space available to the state n, $\eta_{CP} = \pm 1$ [6]. In the $SU(3)$ limit, all the ρ_n are the same for $n \in F_R$, and the quantity in braces above is an $SU(3)$ singlet. Since the ρ_n depend only on the known masses of the particles in the state n, incorporating the true values of ρ_n in the sum is a calculable source of $SU(3)$ breaking.

This method does not lead directly to a calculable contribution to y, because the matrix elements $\langle n|H_w|D^0\rangle$ and $\langle \overline{D}^0|H_w|n\rangle$ are not known. However, CP symmetry, which in the Standard Model and almost all scenarios of new physics is to an excellent approximation conserved in D decays, relates $\langle \overline{D}^0|H_w|n\rangle$ to $\langle D^0|H_w|\overline{n}\rangle$. Since $|n\rangle$ and $|\overline{n}\rangle$ are in a common $SU(3)$ multiplet, they are determined by a single effective Hamiltonian. Hence the ratio

$$y_{F,R} = \frac{\sum_{n\in F_R}\langle \overline{D}^0|\,H_w|n\rangle \rho_n \langle n|H_w\,|D^0\rangle}{\sum_{n\in F_R}\langle D^0|\,H_w|n\rangle \rho_n \langle n|H_w\,|D^0\rangle}$$
$$= \frac{\sum_{n\in F_R}\langle \overline{D}^0|\,H_w|n\rangle \rho_n \langle n|H_w\,|D^0\rangle}{\sum_{n\in F_R}\Gamma(D^0 \to n)} \tag{22}$$

is calculable, and represents the value which y would take if elements of F_R were the only channel open for D^0 decay. To get a true contribution to y, one must scale $y_{F,R}$ to the total branching ratio to all the states in F_R. This is not trivial, since a given physical final state typically decomposes into a sum over more than one multiplet F_R. The numerator of $y_{F,R}$ is of order s_1^2 while the denominator is of order 1, so with large $SU(3)$ breaking in the phase space the natural size of $y_{F,R}$ is 5%. Indeed, there are other $SU(3)$ violating effects, such as in matrix elements and final state interaction phases. Here we assume that there is no cancellation with other sources of $SU(3)$ breaking, or between the various multiplets which occur in D decay, that would reduce our result for y by an order of magnitude. This is equivalent to assuming that the D meson is not heavy enough for duality to enforce such cancellations. Performing the computations of $y_{F,R}$, we see[6] that effects at the level of a few percent are quite generic. Our results are summarized in Table 1. Then, y can be formally constructed from the individual $y_{F,R}$ by weighting them by their D^0 branching ratios,

$$y = \frac{1}{\Gamma}\sum_{F,R}y_{F,R}\left[\sum_{n\in F_R}\Gamma(D^0 \to n)\right]. \tag{23}$$

However, the data on D decays are neither abundant nor precise enough to disentangle the decays to the various $SU(3)$ multiplets, especially for the three- and four-body final states. Nor have we computed $y_{F,R}$ for all or even most of the available representations. Instead, we can only estimate individual contributions to y by assuming that the representations for which we know $y_{F,R}$ to be typical for final states with a given multiplicity, and then to scale to the total branching ratio to those final states.

The total branching ratios of D^0 to two-, three- and four-body final states can be extracted from the Review of Particle Physics[14]. Rounding to the nearest 5% to emphasize the uncertainties in these numbers, we conclude that the branching fractions for PP, $(VV)_s$-wave, $(VV)_d$-wave and $3P$ approximately amount to 5%, while the branching ratios for PV and $4P$ are of the order of 10%[6].

Final state representation		$y_{F,R}/s_1^2$	$y_{F,R}$ (%)
PP	8	-0.0038	-0.018
	27	-0.00071	-0.0034
PV	8_A	0.032	0.15
	8_S	0.031	0.15
	10	0.020	0.10
	$\overline{10}$	0.016	0.08
	27	0.04	0.19
$(VV)_s$-wave	8	-0.081	-0.39
	27	-0.061	-0.30
$(VV)_p$-wave	8	-0.10	-0.48
	27	-0.14	-0.70
$(VV)_d$-wave	8	0.51	2.5
	27	0.57	2.8
$(3P)_s$-wave	8	-0.48	-2.3
	27	-0.11	-0.54
$(3P)_p$-wave	8	-1.13	-5.5
	27	-0.07	-0.36
$(3P)_{form-factor}$	8	-0.44	-2.1
	27	-0.13	-0.64
$4P$	8	3.3	16
	27	2.2	11
	$27'$	1.9	9.2

We observe that there are terms in Eq. (23), like nonresonant $4P$, which could make contributions to y at the level of a percent or larger. There, the rest masses of the final state particles take up most of the available energy, so phase space differences are very important. One can see that y on the order of a few percent is completely natural, and that anything an order of magnitude smaller would require significant cancellations which do not appear naturally in this framework. Cancellations would be expected only if they were enforced by the OPE, or if the charm quark were heavy enough that the "inclusive" approach were applicable. The hypothesis underlying the present analysis is that this is not the case.

4. Conclusions

We proved that if $SU(3)$ violation may be treated perturbatively, then $D^0 - \overline{D}^0$ mixing in the Standard Model is generated only at second order in $SU(3)$ breaking effects. Within the exclusive approach, we identified an $SU(3)$ breaking effect, $SU(3)$ violation in final state phase space, which can be calculated with minimal model dependence. We found that phase space effects alone provide enough $SU(3)$ violation to induce $y \sim 10^{-2}$. Large effects in y appear for decays close to D threshold, where an analytic expansion in $SU(3)$ violation is no longer possible.

Indeed, some degree of cancellation is possible between different multiplets, as would be expected in the $m_c \to \infty$ limit, or between $SU(3)$ breaking in phase space and in matrix elements. It is not known how effective these cancellations are, and the most reasonable assumption in light of our analysis is that they are not significant enough to result in an order of magnitude suppression of y, as they are not enforced by any symmetry arguments. Therefore, any future discovery of a D meson width difference should not by itself be interpreted as an indication of the breakdown of the Standard Model.

At this stage the only robust potential signal of new physics in charm system is CP violation. We discussed several possible experimental observables that are sensitive to CP violation.

Acknowledgments

It is my pleasure to thank S. Bergmann, E. Golowich, Y. Grossman, A. Falk, Z. Ligeti, and Y. Nir for collaborations on the related projects. I would like to thank the organizers for the invitation to the wonderfully organized Arkadyfest workshop.

References

1. A. Datta, D. Kumbhakar, Z. Phys. C27, 515 (1985); A. A. Petrov, Phys. Rev. D56, 1685 (1997).
2. A. F. Falk, Y. Nir and A. A. Petrov, JHEP 9912, 019 (1999).
3. M. Gronau, Y. Grossman and J. L. Rosner, Phys. Lett. B 508, 37 (2001); J. P. Silva and A. Soffer, Phys. Rev. D 61, 112001 (2000). E. Golowich and S. Pakvasa, Phys. Lett. B 505, 94 (2001).

114

4. S. Bergmann, Y. Grossman, Z. Ligeti, Y. Nir, A. Petrov, Phys. Lett. B **486**, 418 (2000).
5. D. Atwood and A. A. Petrov, arXiv:hep-ph/0207165.
6. A. F. Falk, Y. Grossman, Z. Ligeti and A. A. Petrov, Phys. Rev. D **65**, 054034 (2002).
7. I. I. Bigi and A. I. Sanda, *CP violation* (Cambridge University Press, 2000).
8. D. Pedrini, J. Phys. G **27**, 1259 (2001).
9. I. I. Bigi and A. I. Sanda, Phys. Lett. B **171**, 320 (1986).
10. H. N. Nelson, in *Proc. of the 19th Intl. Symp. on Photon and Lepton Interactions at High Energy LP99* ed. J.A. Jaros and M.E. Peskin, arXiv:hep-ex/9908021.
11. H. Georgi, Phys. Lett. B297, 353 (1992); T. Ohl, G. Ricciardi and E. Simmons, Nucl. Phys. B403, 605 (1993); I. Bigi and N. Uraltsev, Nucl. Phys. B **592**, 92 (2001), for a recent review see A. A. Petrov, *Proc. of 4th Workshop on Continuous Advances in QCD*, Minneapolis, Minnesota, 12-14 May 2000, arXiv:hep-ph/0009160.
12. J. Donoghue, E. Golowich, B. Holstein and J. Trampetic, Phys. Rev. D33, 179 (1986); L. Wolfenstein, Phys. Lett. B164, 170 (1985); P. Colangelo, G. Nardulli and N. Paver, Phys. Lett. B242, 71 (1990); T.A. Kaeding, Phys. Lett. B357, 151 (1995). A. A. Anselm and Y. I. Azimov, Phys. Lett. B **85**, 72 (1979);
13. E. Golowich and A. A. Petrov, Phys. Lett. B **427**, 172 (1998).
14. D. E. Groom *et al.* [Particle Data Group Collaboration], Eur. Phys. J. C **15**, 1 (2000).

SOFT AND COLLINEAR RADIATION
AND FACTORIZATION IN PERTURBATION THEORY
AND BEYOND

E. GARDI

TH Division, CERN, CH-1211 Geneva 23, Switzerland
E-mail: Einan.Gardi@cern.ch

Power corrections to differential cross sections near a kinematic threshold are anal-
ysed by Dressed Gluon Exponentiation. Exploiting the factorization property of
soft and collinear radiation, the dominant radiative corrections in the threshold
region are resummed, yielding a renormalization-scale-invariant expression for the
Sudakov exponent. The interplay between Sudakov logs and renormalons is clarified, and the necessity to resum the latter whenever power corrections are non-
negligible is emphasized. The presence of power-suppressed ambiguities in the
exponentiation kernel suggests that power corrections exponentiate as well. This
leads to a non-perturbative factorization formula with non-trivial predictions on
the structure of power corrections, which can be contrasted with the OPE. Two
examples are discussed. The first is event-shape distributions in the two-jet region, where a wealth of precise data provides a strong motivation for the improved
perturbative technique and an ideal situation to study hadronization. The second
example is deep inelastic structure functions. In contrast to event shapes, structure
functions have an OPE. However, since the OPE breaks down at large x, it does
not provide a practical framework for the parametrization of power corrections.
Performing a detailed analysis of twist 4 it is shown precisely how the twist-2
renormalon ambiguity eventually cancels out. This analysis provides a physical
picture which substantiates the non-perturbative factorization conjecture.

1. Dressed Gluon Exponentiation

A classical application of QCD is the evaluation of semi-inclusive differential
cross sections of hard processes depending on several scales. We shall consider here cross sections that depend on a hard scale Q and an intermediate
scale W, both in the perturbative regime $Q > W \gg \Lambda$. Here Λ represents
the fundamental QCD scale. In case of a large hierarchy, $Q \gg W$, there
are large perturbative corrections containing logarithms, $\ln Q/W$. Typically, non-perturbative corrections are suppressed by powers of the *lower*
scale W. If the latter is not so large, such power corrections must be taken
into account. From a theoretical point of view, power corrections are particularly interesting because of their relation to confinement [1].

The first example is provided by event-shape distributions near the two-jet limit. Here the hard scale Q is the centre-of-mass energy and the lower one is set by the shape variable, e.g. in the case of the thrust (T), it is $W^2 = Q^2(1-T)^2$, the sum of the squared invariant masses of the two hemispheres. For $T \longrightarrow 1$ hadrons are produced in two narrow jets and large perturbative and non-perturbative corrections appear due to soft gluon radiation and hadronization. The second example is structure functions in deep inelastic scattering (DIS) at $x \longrightarrow 1$, where the hard scale is the momentum transfer $Q = \sqrt{-q^2}$ and the lower scale is the invariant mass of the hadronic system $W^2 = (p + q)^2 = Q^2(1 - x)/x$, where p and q are the momenta of the proton and γ^*, respectively. For $x \longrightarrow 1$ the recoiling quark [a] develops into a narrow jet, and large corrections appear from the jet fragmentation process.

From the outset it is clear that resummation must be applied. Two relevant types of radiative corrections can be computed to all orders: renormalons and Sudakov logs. Renormalons appear from integration over the running coupling. Using the large N_f limit, one calculates the diagrams where a *single* gluon is dressed by radiative corrections [2]. These contributions dominate at large orders and exhibit the strongest sensitivity to infrared physics. Consequently they are useful in detecting power corrections [2,3]. In the absence of an infrared cutoff, infrared renormalons make the perturbative expansion non-summable. The summation ambiguity is cancelled by non-perturbative corrections. Sudakov logs, on the other hand, emerge from *multiple* emission of soft and collinear gluons. These contributions dominate the perturbative coefficients at large x. At a fixed logarithmic accuracy, Sudakov logs can be summed to all orders [4,5], and, contrary to renormalons, they do not indicate non-perturbative corrections [6,7]. As shown schematically in fig. 1, these two classes of radiative corrections correspond to "orthogonal" sets of diagrams. Both classes are relevant; however, none of the limits considered (large orders and large x, respectively) is appropriate in the threshold region, where both Sudakov logs and power corrections are important.

The gap is closed by [7,8,9] Dressed Gluon Exponentiation (DGE). This is a resummation method that incorporates both types of diagrams by re-summing renormalons in the Sudakov exponentiation kernel. Thus, contrary to the standard approach to Sudakov resummation the result is renormalization-scale-invariant. In the Sudakov exponent, renormalons appear through the enhancement of subleading logs. At any given power of

---

[a] At large x the gluon distribution is small.

Dressed Gluon Exponentiation

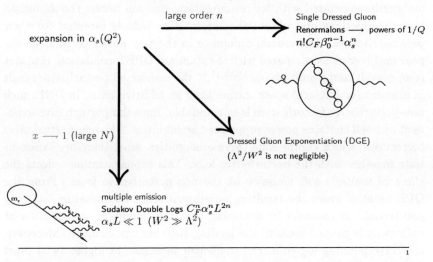

Figure 1. Different directions in summing the perturbative expansion in a two-scale problem.

the coupling α_s^n, the leading log [b], L^{n+1}, typically has a coefficient of order 1, whereas the L^k term (where $k \leq n$) appears with a larger numerical factor $\sim n!/k!$. Thus, the factorial growth in the exponent appears upon summing *all the logs*. In practice, subleading logs are not fully known and the state-of-the-art computation is restricted to a single dressed gluon (SDG) in the exponentiation kernel. In this case DGE reproduces the exact leading and next-to-leading logs (NLL) [c], but only generates a certain class of subleading logs (next-to-next-to-leading logs and beyond) that can be regarded as an approximation to these coefficients. Not much is known about the accuracy of this approximation in general. On the other hand, the factorial growth of subleading logs implies that a resummation with a fixed logarithmic accuracy [d] has a small range of validity. By construction it does not hold to power accuracy. If the latter is required, the additional resummation provided by DGE is necessary.

---

[b] $L \equiv \ln N$, where N is a moment index conjugate to $1 - x$, so N is large.

[c] NLL require using the "gluon bremsstrahlung" effective charge [10] with a 2-loop renormalization-group equation.

[d] Such resummation is derived in the $N \longrightarrow \infty$ limit with $\alpha_s(Q^2) \longrightarrow 0$, so that $\alpha_s(Q^2) \cdot \ln N$ is small.

Analysing the large-order behaviour of the Sudakov exponent and the ambiguity associated with its resummation, one can access the dominant power corrections and obtain the information that is essential for their parametrization. The crucial difference in the way power corrections appear in this case, as compared with the standard OPE formulation, is that it is an overall factor multiplying [11,12,13,14] the resummed perturbative result in moment (or Laplace) space, rather than an additive term. In DGE, such non-perturbative factorization is unavoidable, since the perturbative exponent by itself contains power suppressed ambiguities. The non-perturbative corrections, which compensate these ambiguities, must therefore exponentiate together with the perturbative logs. This exponentiation reflects the effect of multiple soft emission at the non-perturbative level. From the OPE point of view, the resulting non-perturbative factorization is highly non-trivial. It amounts to assuming that the dominant contribution at each twist is proportional to the leading-twist matrix element. Moreover, the corresponding log-enhanced coefficient functions at higher twist must coincides with that of the leading twist. The matrix element information is essentially inaccessible by perturbative methods; however the higher-twist coefficient function can be computed, allowing one to check some of these far-reaching conclusions. A first step in this direction was recently taken [15] in the context of DIS structure functions.

2. Event-shape distributions in the two-jet region

A strong motivation for the improved resummation technique as well as for a systematic study of power corrections is provided by the very precise data on event-shape distributions in e^+e^- annihilation. The goal is to have a handle on the parametrization of hadronization effects and to understand how they change depending on the observable considered. This will hopefully lead to better understanding of the hadronization process itself.

The DGE result for the single-jet mass distribution [e] is given by a Laplace integral,

$$\frac{1}{\sigma}\frac{d\sigma}{d\rho}(\rho, Q^2) = \int_c \frac{d\nu}{2\pi i} \exp\left\{\rho\nu + \ln J_\nu(Q^2)\right\}, \tag{1}$$

where c is an integration contour parallel to the imaginary axis. Here, terms that are not enhanced by logarithms of ρ were discarded. Eventually, such

---

[e]The single-jet mass distribution is used as an intermediate step in the evaluation of the thrust and the heavy-jet mass distributions. Note that this observable obtains additional non-global [16] corrections at the NLL level, which are absent in the thrust and the heavy-jet mass considered here.

terms are included by matching [5] the resummed expression to the fixed-order result, which is currently available [17] numerically to next-to-leading order (NLO). The Sudakov exponent in (1) is

$$\ln J_\nu(Q^2) = \int d\rho \, \frac{1}{\sigma} \frac{d\sigma}{d\rho}(\rho, Q^2)\Big|_{\text{SDG}} \left(e^{-\nu\rho} - 1\right)$$

$$= \frac{C_F}{2\beta_0} \int_0^\infty du \, B_\nu(u) \left(\frac{Q^2}{\bar\Lambda^2}\right)^{-u} \frac{\sin \pi u}{\pi u} \bar{A}_B(u), \qquad (2)$$

where $\bar{A}_B(u)$ depends only on the renormalization group equation for the coupling and $\bar\Lambda$ corresponds to the "gluon bremsstrahlung" effective charge [10]. The Borel function is remarkably simple:

$$B_\nu(u) = \frac{2}{u} \left(\nu^{2u} - 1\right) \Gamma(-2u) - \left(\frac{2}{u} + \frac{1}{1-u} + \frac{1}{2-u}\right) (\nu^u - 1) \, \Gamma(-u). \quad (3)$$

Using this distribution, and the assumption that the hemisphere masses are independent [f], both the thrust (we define $t \equiv 1 - T$) and the heavy-jet mass (ρ_H) distributions are readily obtained,

$$\frac{1}{\sigma} \frac{d\sigma}{dt}(t, Q^2) = \frac{d}{dt} \int_c \frac{d\nu}{2\pi i\nu} \exp\left\{\nu t + 2\ln J_\nu(Q^2)\right\} \qquad (4)$$

$$\frac{1}{\sigma} \frac{d\sigma}{d\rho_H}(\rho_H, Q^2) = \frac{d}{d\rho_H} \left[\int_c \frac{d\nu}{2\pi i\nu} \exp\left\{\nu\rho_H + \ln J_\nu(Q^2)\right\}\right]^2. \qquad (5)$$

These resummation formulae suggest a specific way in which power corrections should be included [7,9]. First of all, since renormalon ambiguities appear in the exponent, $\ln J_\nu^{\text{PT}}(Q^2) \longrightarrow \ln J_\nu^{\text{PT}}(Q^2) + \ln J_\nu^{\text{NP}}(Q^2)$, power corrections appear as a factor in Laplace space, implying factorization and exponentiation of these terms, as previously suggested [11,12,13,14]. Moreover, the particular structure of Borel ambiguities from the renormalon singularities in (3) allows one to deduce the dependence of $\ln J_\nu^{\text{NP}}(Q^2)$ on Q and ν. There are two classes of corrections: (a) *odd powers* of $\bar\Lambda\nu/Q$ from the first term in (3), which are related to large-angle soft emission; and (b) the *first two powers* of $\bar\Lambda^2\nu/Q^2$ from the second term in (3), which are associated with collinear emission. The leading corrections are (a) and they can be resummed into a shape function of a single variable [12]. The effect of the leading $\lambda_1 \bar\Lambda\nu/Q$ power correction in the exponent is to shift the entire distribution [12,13], whereas higher (odd) power corrections modify the shape of the perturbative spectrum.

---

[f]In the two-jet region, correlations between the hemispheres are suppressed perturbatively. It may play a more important role non-perturbatively [18,19]. We neglect this effect [9], still finding a good agreement with the data.

Fitting the thrust distribution in a large range of Q and t values with such a shape function provides a strong test of the approach. Good fits were obtained to the world data [7,9,20]. An example [9] is shown in fig. 2. An even more stringent test is the comparison of the extracted parameters

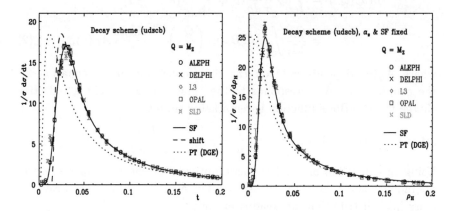

Figure 2. Left: fit to the thrust distribution data at $Q = M_Z$. The dotted line is the perturbative DGE result (principal-value regularization of the Borel sum), the dashed line shows a fit based on shifting the perturbative distribution (a single non-perturbative parameter) and the full line shows a shape-function-based fit. Right: the heavy-jet mass data compared with the predicted distribution based on the parameters fixed in the thrust fit. Here $\alpha_s^{\overline{MS}} (M_Z) = 0.1086$.

from the thrust and the heavy-jet mass distributions, both defined in the decay scheme [21] in order to minimize the effect of hadron masses. Assuming that (4) and (5) hold non-perturbatively, the power corrections to the two distributions are associated with the same exponent and are therefore simply related. The agreement in the description of the two distributions is demonstrated in fig. 1, where the full line in the right frame shows not a fit but rather a calculated distribution for heavy-jet mass, where the parameters (α_s and the shape function) are fixed by the fit to the thrust. A comparison of the leading power correction extracted from the two distributions is shown in fig. 3. In addition to DGE, the figure shows a fit based on the NLL result. Contrary to DGE, in the NLL case there is no agreement between the two. This demonstrates the necessity to resum also the factorially enhanced subleading logs, as done by DGE, when a quantitative power correction analysis is done. Note also the significant impact on the extracted value of α_s. The best fit for the thrust distribution yields $\alpha_s^{\overline{MS}} (M_Z) = 0.1086$. This value is consistent with that extracted from the average thrust upon performing renormalon resummation [22].

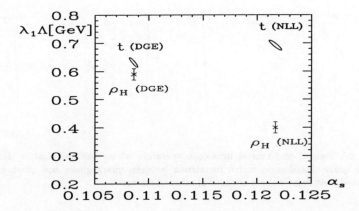

Figure 3. The leading non-perturbative correction on the scale Q/ν, extracted from the t and ρ_H distributions. Results based on DGE and on NLL resummation are shown. In each case α_s is fixed by the fit to the thrust.

3. DIS structure functions at large Bjorken x

In contrast to event-shape distributions, DIS structure functions have an OPE, by which power-suppressed contributions can be systematically identified and related to hadronic matrix elements [23]. The moments of the structure functions can be written as an expansion in inverse powers of the momentum transfer $q^2 \equiv -Q^2$,

$$\int_0^1 dx\, x^{N-1} F_2(x, Q^2) = C^{(2)}(N, \mu_F) \langle O^{(2)}(N) \rangle_{\mu_F}$$

$$+ \frac{1}{Q^2} \sum_j C_j^{(4)}(N, \mu_F) \langle O_j^{(4)}(N) \rangle_{\mu_F} + \ldots,$$

where $O_j^{(m)}$ are operators of twist m with the appropriate quantum numbers. The OPE can be expressed in terms on non-local operators [24,25,26,27] defined on the light cone $y^2 = 0$ (y is the Fourier conjugate of the γ^* momentum q), as demonstrated in fig. 4. Calculating the hadronic matrix elements of $O_j^{(m)}$ requires a full knowledge of the hadron structure. Thus, in practice these are simply parametrized and fixed by fit to experimental data. Moreover, most phenomenological analyses of structure functions are still restricted nowadays to the leading twist. On the other hand, the very existence of a non-perturbative definition puts the study of power corrections to structure functions on a firmer basis, as compared with event shapes. The OPE allows one to answer certain questions that are hard to address otherwise. In particular, within the OPE one can trace the cancellation of infrared renormalon ambiguities [2,28]. This is not only important in

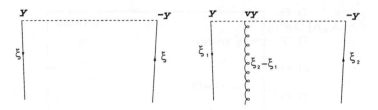

Figure 4. Twist-2 and twist-4 light-cone operators whose hadronic matrix elements are the quark distribution and a correlation between quark–gluon and quark states, respectively. At twist 4 there are several other operators.

principle, but, in fact, *essential* for any reliable measurement of higher-twist matrix elements, since the renormalon ambiguity from the resummation of $C^{(2)}(N, \mu_F)$ in (6) is of the same order as the next term in the expansion. It is well known that by studying the renormalization properties of these operators one can determine their large Q^2 scaling violation. We shall also see that by calculating the higher-twist coefficient functions one can gain some insight into which partonic configurations may be relevant to certain kinematics. Both the anomalous dimension and the information encoded into the coefficient functions of the higher twist can be useful in identifying the dominant higher-twist contributions, without making too strong assumptions concerning the matrix elements themselves.

Let us now consider F_2 in the large-x limit. In particular, consider the limit where Q^2 gets large but $W^2 = Q^2(1 - x)/x$ is held fixed at some low (yet perturbative) scale. At the level of the leading twist ($m = 2$) the analysis of the structure function simplifies, since the gluon distribution can be neglected and only valance quarks contribute. The leading-twist factorization amounts to [g] incorporating the effect of gluons softer than some scale μ_F into the hadronic matrix element $\langle O^{(2)}(N)\rangle_{\mu_F}$, whereas the effect of gluons of higher virtualities go into the coefficient functions $C^{(2)}(N, \mu_F)$. Independence of the structure function of μ_F is guaranteed by an evolution equation stating that the logarithmic dependence of the renormalized operator on μ_F is cancelled by that of the coefficient function. Moreover, radiative corrections can be factorized into a hard subprocess, a jet sub-

---

[g]The way factorization is implemented in practice (through dimensional regularization) removes infrared singularities from the coefficient functions, making them well defined order by order. However, it is not equivalent to a rigid cutoff. Infrared effects do penetrate into the coefficient functions at orders as renormalons.

process, and a soft subprocess, which are mutually incoherent [4], as shown in fig. 5. Interaction between the remnants of the target and the recoiling

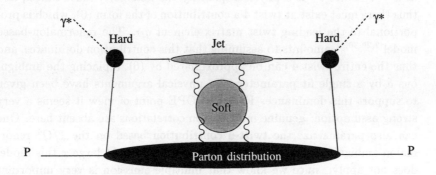

Figure 5. Factorization of DIS structure functions at large Bjorken x.

jet proceeds, at the level of the leading twist, only through the exchange of soft gluons (harder gluons contribute at higher twist), which cannot resolve the jet or the hadron structure.

Since non-perturbative corrections at large x go effectively in powers of $1/W^2$, the OPE in (6) tends to break down and needs to be summed up. At first sight this seems an incredibly complex task, in particular, since the number of parameters involved increases sharply at large N: at each twist ($m \geq 4$) the number of local matrix elements [h] grows as a power of N.

Not only the need to compensate for the renormalon ambiguity at twist 2, but also the complexity of the twist expansion, and the difficulty to use it to parametrize power corrections, have led to renormalon-based phenomenology of structure functions. Renormalons resummation in the twist-2 coefficient function, at the level of a SDG (the large-N_f limit), leads to the following ambiguity [3,29,30,31] to order $1/Q^2$,

$$\Delta \int_0^1 dx\, x^{N-2}\, F_2 =$$

---

[h] As N increases, larger distances along the light cone become relevant. This makes the formulation of the OPE in terms of non-local matrix elements attractive. Indeed, using light-cone distributions, the analysis of twist 4 at large x becomes tractable.

$$\frac{C_F}{2\beta_0} \left[4\psi(N+1) + 4\gamma + \frac{2}{1+N} + \frac{12}{2+N} - 8 - N \right] q_N \frac{\Lambda^2 \delta}{Q^2}, \qquad (6)$$

with $\delta \equiv \int_C du/(1-u)$, where the contour C is the difference between different integration contours in the Borel plane, which avoid the singularity at $u = 1$. Since F_2 is a physical quantity the ambiguity must cancel, and thus there must exist at twist 4 a contribution of the form (6), which is proportional to the leading twist matrix element q_N. The renormalon-based model [3,29,30,31] amounts to assuming that this contribution dominates, and thus the entire twist 4 can be approximated by (6), replacing the ambiguous δ by a single fit parameter. No physical arguments have been given to support this dominance. From the OPE point of view it seems a very strong assumption: genuine multiparton correlations are absent here. One can also parametrize the twist-6 contribution based on the $1/Q^4$ renormalon ambiguity in the twist-2 coefficient function. At large x this model does not apply: since we know that multiple emission is very important perturbatively, it is unlikely that power corrections will be entirely associated with a *single* dressed gluon. Instead, at large x it is more appropriate to parametrize power corrections according to DGE, taking multiple gluon emission into account. But also in this case strong assumptions are made on the higher-twist contribution. To proceed we need to clarify the meaning of the renormalon dominance assumption, but we must also identify the dominant higher-twist contribution independently of any prejudice. Both these issues were recently addressed [15] at the level of twist 4.

To clarify the meaning of the renormalon dominance assumption we must trace the cancellation of renormalon ambiguities within the OPE. The infrared renormalon ambiguity cancels against another ambiguity in the definition of higher-twist matrix elements, due to the mixing of the corresponding operators with the leading-twist operator [32,33,2]. To make use of the OPE, one must regularize both sources of ambiguity in (6):

$$C^{(2)}(N, \mu_{\rm F}) \longrightarrow C^{(2)}(N, \mu_{\rm F})\Big|_{\rm reg} + \frac{\Lambda^2 \delta_{\rm reg}}{Q^2} \qquad (7)$$

$$\langle O_j^{(4)}(N) \rangle_{\mu_{\rm F}} \longrightarrow \langle O_j^{(4)}(N) \rangle_{\mu_{\rm F}}\Big|_{\rm reg} + \langle O^{(2)}(N) \rangle_{\mu_{\rm F}} \Lambda^2 \delta_{\rm reg}^{(j)}. \qquad (8)$$

Here $\delta_{\rm reg}$ and $\delta_{\rm reg}^{(j)}$ in eqs. (7) and (8) represent the effect of changing the regularization prescription in defining the sum of the series in the twist-2 coefficient functions and the ultraviolet-divergent integrals in the renormalized twist-4 operators, respectively. A consistent regularization guarantees the cancellation of all the δ terms leading to unambiguous predictions for the structure functions to power accuracy. It is clear from (8) that the mixing of twist 4 with twist 2 must be associated with *quadratic* divergence in the renormalization of $O_j^{(4)}(N)$. Indeed, contracting the gluon in fig. 4 to one of the quark lines, there appears a loop which is quadratically divergent in the ultraviolet, while the remaining operator is the twist-2 one. An ambiguity emerges from the regularization of this loop. Summing the ambiguous ultraviolet contributions from all the relevant operators, each appearing with its own coefficient function, we recover an ambiguous expression, which is identical to (6) but has an opposite overall sign [i]. This way, when the two are summed in (6), a well-defined expression is obtained.

Note that while the renormalon ambiguity is associated with infrared scales, the presence of ambiguity in the higher twist is an ultraviolet property of the operator, which is not related to infrared physics. In conclusion, the renormalon ambiguity merely reflects the arbitrariness in separating contributions of different twists. The precise separation (or regularization of the renormalon sum) does not have any physical significance, and it should be regarded as complementary to the standard factorization used to define separately the coefficient functions and the operator matrix elements at each twist.

From this discussion it follows that the renormalon dominance assumption should be interpreted, within the OPE, as the assumption that the ultraviolet divergent contribution [32,33,2], the one which mixes under renormalization with the leading twist, dominates the higher twist. Independently of this assumption we conclude that any treatment of higher twist which fails to deal with renormalon resummation (and the corresponding regularization of higher twist) is bound to be ambiguous.

A priori, it is natural to expect that terms which mix with the leading twist will be of the same order of magnitude as other higher-twist effects. Ultraviolet dominance is the assumption that the former dominate. One can also imagine a scenario opposite to ultraviolet dominance, where the matrix elements are much larger than their ambiguous part. This point of view was adopted in the framework of QCD sum rules [1]. The success of renormalon-based phenomenology in various applications calls for reconsideration of this assumption also in the framework of the sum rules. In general, similarly to DIS higher twist, condensates should be extracted when performing renormalon resummation; they can be assigned numerical values only within a given regularization prescription for the renormalons.

Let us now return to the large-x limit and study the twist-4 contribution

---

[i]The first example where this cancellation was demonstrated is the longitudinal structure function [2,28]. Recently [15] it was demonstrated in the case of F_2.

to F_2. Examining the coefficient functions of the corresponding light-cone operators at leading order we find [15] that a significant simplification occurs at large x. This simplification is due to the fact that certain partonic configurations dominate the entire twist-4 contribution. In particular, we find that the dominating final state is that of a single energetic quark, just as at twist 2. The final state in which a quark and a gluon share the momentum is subdominant – it is suppressed by a power of $1 - x$. Thus the difference between the leading twist and twist 4 at large x is restricted to the initial states. However, also here a great simplification occurs. Representing the coefficient functions of the twist-4 operator in fig. 4 in terms of the longitudinal momentum fractions of the quarks ξ_1 and $-\xi_2$, the dominant contribution arises from the region where the gluon momentum fraction is small $\xi_2 - \xi_1 \longrightarrow 0$, since the coefficient functions are singular at this point. Consequently the Heisenberg uncertainty principle implies that the quark–gluon–quark correlator $\langle p|O_j^{(4)}(v,y)|p\rangle_{\mu_F}$ is essentially independent of the position of the gluon field on the light cone (v), and it effectively becomes a function of the light-cone separation between the quarks (py), just as the leading-twist matrix element $\langle p|O_j^{(2)}(y)|p\rangle_{\mu_F}$. In conclusion, the configurations that dominate the twist-4 contribution to F_2 at large x, in both the final and initial states, make it twist-2-like. It is therefore natural to conjecture that ultraviolet dominance indeed holds, namely that the dominant ingredient in twist 4 is the part that mixes with the leading twist.

While justifying ultraviolet dominance at large x, this picture does not apply to moderate or small values of x, and therefore it does not support the application of the renormalon model where eq. (6) is assumed to represent the N dependence of twist 4.

The essence of the simplification we identified at twist 4 at large x is that the dominant multiparton correlation measured by F_2 in such kinematics is still associated with the leading twist. This is most naturally realized through ultraviolet dominance. Gluons of momentum scale of order W which are exchanged between the jet and the remnants of the target simply cannot resolve the full multiparton correlation function. Their interaction can be associated with "power-like evolution", similarly to the way the interaction of soft gluons at the leading twist is associated with logarithmic evolution.

Assuming that ultraviolet dominance holds to all orders in the twist expansion, a non-perturbative factorization formula (valid up to perturbative and non-perturbative corrections that are suppressed by $1/N$) emerges from the leading contributions to each twist:

$$\int_0^1 dx \; x^{N-1} F_2(x, Q^2) =$$

$$= H\left(Q^2\right) J\left(Q^2/N; \mu_{\text{F}}^2\right) q_N(\mu_{\text{F}}^2) \left[1 + \kappa_1 \frac{N\Lambda^2}{Q^2} + \cdots\right] \quad (9)$$

$$= H\left(Q^2\right) J\left(Q^2/N; \mu_{\text{F}}^2\right) q_N(\mu_{\text{F}}^2) J_{\text{NP}}\left(N\bar{\Lambda}^2/Q^2\right),$$

where q_N is the twist-2 quark matrix element, H and J are the hard and jet components in the twist-2 coefficient function, and κ_i are target-dependent non-perturbative parameters. In the second line all the non-perturbative corrections on the scale Q^2/N are resummed into a shape function of a single argument, similarly to the parametrization of non-perturbative effects in event-shape distributions.

DGE can now be applied to calculate J. Using the standard $\overline{\text{MS}}$ factorization with $\mu_{\text{F}} = Q$, we obtain

$$J(Q^2/N, \mu_{\text{F}} = Q) = \exp\left\{\frac{C_F}{2\beta_0} \int_0^\infty du \, B_N(u) \left(\frac{Q^2}{\bar{\Lambda}^2}\right)^{-u} \frac{\sin \pi u}{\pi u} \bar{A}_B(u)\right\} \quad (10)$$

with [8]

$$B_N(u) = -\left(\frac{2}{u} + \frac{1}{1-u} + \frac{1}{2-u}\right) (N^u - 1) \, \Gamma(-u) - \frac{2}{u} \ln(N), \quad (11)$$

where the first term is similar to the collinear part in the event-shape case (3). Contrary to the latter, here the collinear singularity (appearing as a pole at $u = 0$) requires a subtraction. This is the rôle of the second term in (11), which can be identified as the large-N limit of the leading twist anomalous dimension. Having implemented the $\overline{\text{MS}}$ factorization, J is regular at $u = 0$ and thus has a well defined perturbative expansion. Still, it has Borel singularities at $u = 1, 2$. These will be cancelled by the non-perturbative jet function $J_{\text{NP}}(N\bar{\Lambda}^2/Q^2)$. Thus, the following ansatz suggests itself [8,15]

$$J_{\text{NP}}\left(N\bar{\Lambda}^2 Q^2\right) = \exp\left\{-\omega_1 \frac{C_F}{\beta_0} \frac{N\bar{\Lambda}^2}{Q^2} - \frac{1}{2}\omega_2 \frac{C_F}{\beta_0} \frac{N^2\bar{\Lambda}^4}{Q^4}\right\}, \quad (12)$$

where ω_i are non-perturbative parameters.

4. Conclusions

Many interesting hard processes involve kinematic thresholds. Owing to the emission of soft and collinear radiation, the corresponding differential cross-sections tend to have large perturbative and non-perturbative corrections. As a result, a naive, fixed-order perturbative treatment is insufficient. Moreover, the OPE does not apply or tends to break down. Here we shortly reviewed the case of event-shape distributions in e^+e^- annihilation, where we demonstrated the virtues of DGE both as a resummation method and as a way to study power corrections. This was followed by a deeper look into the case of DIS structure functions at large x, where the OPE was used to get additional insight into the problem. There are many other physical applications where DGE and the shape-function approach can be applied, including, for example, fragmentation functions of light and heavy quarks and Drell–Yan or heavy-boson production processes.

DGE is primarily a novel approach to resummation: the Sudakov exponent is calculated in a renormalization-scale-invariant manner by means of renormalon resummation. The criterion of a fixed-logarithmic accuracy becomes irrelevant when power corrections are being quantified: it is the subleading logs that carry the characteristic factorial growth of the coefficients, which is associated with the power corrections. Perturbatively, the additional resummation achieved by DGE with respect to the standard NLL resummation is significant. For event-shape distributions, it is $\sim 20\%$ at M_Z. In principle, power corrections cannot be quantified without renormalon resummation. Our analysis of the thrust and the heavy-jet mass shows that this has very practical implications: a consistent description of the two observables is possible only if renormalon resummation in the Sudakov exponent is performed (fig. 3). Another important consequence is the significant impact on the extracted value of α_s.

DGE and the shape-function approach can be applied in the case of DIS structure function at large x. However, here it can be contrasted with the OPE. As we have seen, a non-perturbative factorization can be consistent with the OPE, and it is supported by the OPE-based analysis. In spite of the fact that the OPE tends to break down in the large-x limit, it is very useful: a simple picture emerges from the analysis of twist 4 in terms of light-cone distributions. We have found that the dominant non-perturbative corrections at large x are associated with the formation of a narrow jet in the final state. These corrections are due to the exchange of gluons with momentum scale of the order W, which are insensitive to the details of multiparton correlations in the target. Instead of the full correlation, they measure a particular ingredient which is twist-2 like. It is therefore natural to conjecture that the dominant contributions at large x are associated with mixing with the leading twist. To $1/N$ accuracy the hadronization process of the jet involves a single target-dependent nonperturbative scale at each order in the twist expansion. These dominant corrections can be resummed into a shape function of a single argument: N/Q^2, defining a non-perturbative jet function. Thus, at large N the OPE collapses (to $1/N$ accuracy) into a factorized formula in which the leading twist is multiplied by a jet function [j]. The application of DGE is then quite natural.

Acknowledgements

It is a pleasure to thank my collaborators J. Rathsman, G.P. Korchemsky, D.A. Ross and S. Tafat for very enjoyable and fruitful collaboration. The research was supported in part by the EC program "Training and Mobility of Researchers", Network "QCD and Particle Structure", contract ERBFMRXCT980194.

References

1. M. A. Shifman, A. I. Vainshtein and V. I. Zakharov, Nucl. Phys. B **147** (1979) 385.
2. M. Beneke, *Phys. Rep.* **317** (1999) 1 [hep-ph/9807443]; M. Beneke and V. M. Braun, [hep-ph/0010208].
3. Yu. L. Dokshitzer, G. Marchesini and B. R. Webber, *Nucl. Phys.* **B469** (1996) 93 [hep-ph/9512336].
4. H. Contopanagos, E. Laenen and G. Sterman, Nucl. Phys. B **484** (1997) 303 [hep-ph/9604313].
5. S. Catani, L. Trentadue, G. Turnock and B. R. Webber, *Nucl. Phys.* **B407** (1993) 3.
6. M. Beneke and V. M. Braun, *Nucl. Phys.* **B454** (1995) 253 [hep-ph/9506452].
7. E. Gardi and J. Rathsman, *Nucl. Phys.* **B609** (2001) 123 [hep-ph/0103217].
8. E. Gardi, *Nucl. Phys.* **B622** (2002) 365 [hep-ph/0108222].
9. E. Gardi and J. Rathsman, [hep-ph/0201019], to appear in *Nucl. Phys. B*.
10. S. Catani, B. R. Webber and G. Marchesini, *Nucl. Phys.* **B349** (1991) 635.
11. G. P. Korchemsky and G. Sterman, *Nucl. Phys.* **B437** (1995) 415 [hep-ph/9411211].
12. G.P. Korchemsky and G. Sterman, Proc. 30th Rencontres de Moriond, *QCD and high energy hadronic interactions*, Les Arcs, France, 1995, ed. J. Tran Thanh Van (Editions Frontières, Gif-sur-Yvette, 1995), p. 383 [hep-ph/9505391].
13. Yu. L. Dokshitzer and B. R. Webber, *Phys. Lett.* **B404** (1997) 321 [hep-ph/9704298].

---

[j]It should be stressed that the factorization formula is not derived here from first principles. Its justification involves a strong assumption. Nevertheless, it is possible to check explicitly certain results. In particular, the formula predicts a common asymptotic behaviour of the logarithmic evolution in the large N limit for any twist. This can be verified at twist 4.

14. G. P. Korchemsky and G. Sterman, *Nucl. Phys.* **B555** (1999) 335 [hep-ph/9902341].
15. E. Gardi, G. P. Korchemsky, D. A. Ross and S. Tafat, [hep-ph/0203161], to appear in *Nucl. Phys.* **B**.
16. M. Dasgupta and G. P. Salam, *Phys. Lett.* B **512** (2001) 323 [hep-ph/0104277].
17. S. Catani and M.H. Seymour, *Phys. Lett.* **B378** (1996) 287; *Nucl. Phys.* **B485** (1997) 291; http://hepwww.rl.ac.uk/theory/seymour/nlo/
18. G. P. Korchemsky and S. Tafat, *JHEP* **0010** (2000) 010 [hep-ph/0007005].
19. A. V. Belitsky, G. P. Korchemsky and G. Sterman, *Phys. Lett.* **B515** (2001) 297 [hep-ph/0106308].
20. E. Gardi and J. Rathsman, DGESHAPE, A program for calculating the thrust and heavy-jet mass distributions using DGE, http://www3.tsl.uu.se/~rathsman/dgeshape/.
21. G. P. Salam and D. Wicke, *JHEP* **0105** (2001) 061 [hep-ph/0102343].
22. E. Gardi and G. Grunberg, *JHEP* **9911** (1999) 016 [hep-ph/9908458].
23. R. L. Jaffe, "Spin, twist and hadron structure in deep inelastic processes", [hep-ph/9602236].
24. R. L. Jaffe and M. Soldate, *Phys. Rev.* **D26** (1982) 49.
25. R. K. Ellis, W. Furmanski and R. Petronzio, *Nucl. Phys.* **B212** (1983) 29; and **B207** (1982) 1.
26. R. L. Jaffe, *Nucl. Phys.* **B229** (1983) 205.
27. I. I. Balitsky and V. M. Braun, *Nucl. Phys.* **B311** (1989) 541.
28. V. M. Braun, unpublished notes.
29. E. Stein, M. Meyer-Hermann, L. Mankiewicz and A. Schafer, *Phys. Lett.* **B376** (1996) 177 [hep-ph/9601356].
30. M. Dasgupta and B. R. Webber, *Phys. Lett.* **B382** (1996) 273 [hep-ph/9604388].
31. M. Maul, E. Stein, A. Schafer and L. Mankiewicz, *Phys. Lett.* **B401** (1997) 100 [hep-ph/9612300].
32. V. M. Braun, "Ultraviolet dominance of power corrections in QCD?", [hep-ph/9708386].
33. M. Beneke, V. M. Braun and L. Magnea, *Nucl. Phys.* **B497** (1997) 297, [hep-ph/9701309].

SECTION 2.
GENERAL ASPECTS OF QCD AND THE STANDARD MODEL

PROBING NEW PHYSICS: FROM CHARM TO SUPERSTRINGS

M. K. GAILLARD

Department of Physics, University of California
and
Theoretical Physics Group, 50A-5101, Lawrence Berkeley National Laboratory,
Berkeley, CA 94720, USA
E-mail: mkgaillard@lbl.gov

Effective theories based on experimental data provide powerful probes and tests of underlying theories in elementary particle physics. Examples within and beyond the Standard Model are discussed, including a specific model for supersymmetry breaking within the context of the weakly coupled heterotic string.

1. Introduction

This talk begins with some historical remarks about work done during the period of the inception of the Standard Model, especially areas where my own work overlapped significantly with that of Arkady Vainshtein. The remainder of the talk will be devoted to more recent work, in which Arkady's work on supersymmetry also had an influence. The common thread through the work of both periods is the interconnectedness of theory and experiment in unraveling the elementary structure of nature.

Experiment provides us with effective theories. That is, the data tell us how to write down effective Lagrangians with which we can calculate tree-level S-matrix elements that reproduce the data over some range of energy and distance scales with reasonable accuracy. When we try to take an effective theory seriously as a quantum field theory, we typically encounter difficulties (usually in the form of infinite amplitudes) that suggest a scale at which new physics must come into play. Given a hypothesis for the specifics of the new physics, detailed studies of the effective lower energy theory can test its validity. I will recall examples that contributed to the construction and study of the Standard Model and review approaches to the phenomenology of supergravity and superstring theory. A specific model for supersymmetry breaking within the context of the weakly coupled heterotic string will be discussed.

2. Effective Theories for the Standard Model

An early example of a successful effective Lagrangian is Fermi theory. A series of experiments led theorists such as Fermi, Gamov and Teller to postulate four fermion couplings to describe nuclear β-decay. Further data revealed the V-A nature of the couplings as well as a set of flavor selection rules for both strangeness-changing and strangeness-conserving (semi-)leptonic decays of hadrons. This led to the interpretation of the effective Fermi interaction as arising from the exchange of heavy, electrically charged vector bosons $W^\pm$, with the identification $G_F/\sqrt{2} = g_W^2/m_W^2$, coupled to bilinear quark currents that are Noether currents of the strong interaction. These studies provided early building blocks both of the GWS model of electroweak interactions and of the quark model and QCD.

The Fermi theory is nonrenormalizable. Attempts to treat it as a quantum theory revealed quadratic divergences which suggested that loop corrections must be effectively cut off by new physics at a scale $\Lambda_F \sim 300$ GeV. Similarly, analyses of high energy scattering amplitudes revealed a breakdown of tree unitarity at cm energies above 600 GeV. These results were the motivation for the construction of the $p\bar{p}$ colliders that ultimately produced the W, Z bosons with masses ~ 100 GeV, providing the needed cut-off. As it happens, by the time of their discovery, the underlying electroweak theory had already been developed and tested, and the measurement of the weak boson masses provided spectacular confirmation of that theory. Once the underlying theory is known, there is no impediment to treating the low energy effective Fermi theory as a full quantum theory, provided loop integration is appropriately cut off at the physical threshold for new physics, for example

$$G_F^2 \Lambda_F^2 \sim G_F^2 m_W^2 = g_W^4 m_W^{-2}. \tag{1}$$

Another example of an effective theory for the Standard Model is the low energy chiral Lagrangian for pions. The approximate $SU(2)_R \otimes SU(2)_L$ invariance of the strong interactions that was uncovered in the study of pion couplings was a crucial building block for the theory of QCD with very light u, d quarks. The chiral Lagrangian is nonrenormalizable, but its study at the quantum level, with appropriate cut-offs related to the scale of confinement, has contributed important information on quark masses and other aspects of low energy QCD matrix elements.

In the following I discuss two applications of effective theories for the Standard Model that both Arkady and I were involved in.

2.1. *The Charm Threshold*

The effective Fermi theory constructed from experimental measurements of weak (semi-)leptonic decays contained both strangeness-changing ($\bar{u}s$) and strangeness-conserving ($\bar{u}d$) quark currents. At the quantum level this leads to a semi-leptonic ($\bar{d}s$)($\bar{\mu}\mu$) Fermi coupling with effective Fermi constant

$$G_{\Delta S} \sim \sin\theta_c G_F^2 \Lambda_{\Delta S}^2, \qquad (2)$$

and consistency with experiment requires $\Lambda_{\Delta S} \sim 1$ GeV. The new threshold was provided by the GIM mechanism[1] whereby the u loop contribution was canceled up to quark mass effects by the loop contribution from the postulated charm quark c, and suggested a charm quark mass $m_c \sim$ GeV. The BIM mechanism,[2] using the same new quark to cancel gauge anomalies in the context of the GWS electroweak theory, strengthened the case for the existence of charm.

In 1973, Ben Lee and I noticed[3] that in the limit of u-c mass degeneracy, not only the amplitudes for $K \to \mu\mu$ and $K^0 \leftrightarrow \bar{K}^0$ vanish, but so do those for processes like $K \to \gamma\gamma$, that are observed to be unsuppressed. A similar obervation was made by Ernest Ma.[4] A careful analysis of these and other K-decays in the context of the GWS electroweak theory revealed that unsuppressed processes of the latter type have amplitudes $\propto \alpha G_F \ln(m_c/m_u)$, whereas suppressed processes of the former type have amplitudes $\propto \alpha G_F (m_c^2 - m_u^2)/m_W^2$. Consistency with data required $m_u^2 \ll m_c^2 \ll m_W^2$, further supporting very small u, d masses. While the calculation of $K \to \mu\mu$ is theoretically very clean, it turns out that the leading contribution, enhanced by a factor $\ln(m_W/m_c)$, cancels between W and Z exchange diagrams; the result was a rather weak limit: $m_c \le 9$ GeV. On the other hand, if one was brave enough to attempt to evaluate the matrix element for $K^0 \leftrightarrow \bar{K}^0$, a prediction on the order of a GeV for the charmed quark could be inferred and used to predict branching ratios for other rare kaon decays that had not yet been observed. Once we had done all this work, we learned from Bjorken that we had been scooped, at least in part, by our Russian colleagues Vainshtein and Kriplovich[5] who had considered the same processes and made similar inferences about the charm quark mass.

Ben and I evaluated the $K^0 \leftrightarrow \bar{K}^0$ matrix element using a simple factorization ansatz, and, from the observed value of the K_L-K_S mass difference, found $m_c \approx 1.5$ GeV, neglecting color (QCD was just beginning to emerge at that time as a candidate theory for the strong interactions). This low value worried us, since charm had not been seen in experiments (or so it was assumed; several hints[6] were indeed in the experimental literature). So we "rounded it off" to 2 GeV, until a reader[7] of the draft insisted that we really found 1.5. So we caved in, but decided to include a second evaluation using colored quarks which gave us back the seemingly safer value of 2. Of course 1.5 is the right answer even though QCD is correct. At the time we knew nothing of a third generation, and its appearance muddied the waters for a while, but since the top quark couplings to lighter quarks are very weak, its contribution to the neutral kaon mass difference is insignificant. Analyses based on a $1/N_c$ expansion suggest[8] that one should neglect the color factor, and lattice QCD calculations support[9] the naïve factorization hypothesis to a good approximation.

2.2. Penguins and the $\Delta I = \frac{1}{2}$ Rule

While the elementary couplings of the quarks and gluons of QCD become manifest at high momentum transfer $|q| \gg \Lambda_{QCD}$, at low energy's we are forced to work with an effective theory of hadrons. Here I consider weak decays. Purely leptonic processes can of course trivially be dealt with using perturbation theory. Semi-leptonic processes are also tractable because the hadronic vertex can be factored out, and the assumption that the hadronic currents are the Noether currents of flavor/chiral $SU(3)$ is highly predictive. In contrast, in nonleptonic decays, the underlying $\bar{q}q'W$ couplings are completely masked by gluon exchanges across all weak vertices. However we learned from Ken Wilson how to find the correct effective quark Lagrangian at scales much lower than m_W: the operator product expansion.[10] The dominant operator is the one of lowest dimension, in this case a four quark operator. Since gluon exchange conserves helicity, and only left-handed quarks participate in the charge-changing weak interaction, this is a V-A Fermi operator, but the I-spin structure is modified with respect to the original operator. This is because the eigenstates of the S-matrix are in fixed representations of color $SU(3)$, and these are related by Fermi statistics to the final state ud I-spin in the scattering process $us \to ud$. The scattering occurs in a $J = L = S = 0$ state, which is antisymmetric, so the color and I-spin states must have the same symmetry. These are 1) an anti-triplet of $SU(3)_c$ with $I_{ud} = 0$, which is attractive, and 2) a sextet of $SU(3)_c$ with $I_{ud} = 1$, which is repulsive. Since the initial state has $I = \frac{1}{2}$, the former is a pure $\Delta I = \frac{1}{2}$ transition and the second is a mixture of $\Delta I = \frac{1}{2}$ and $\Delta I = \frac{3}{2}$. The explicit calculation[11] gives a mild enhancement of the former and suppression of the latter such that the ratio of the effective Fermi coupling constants is $G_F^{\frac{1}{2}}/G_F^{\frac{3}{2}} \approx 5$, whereas experiment suggests that a factor of about 20 is needed, if one evaluates the amplitudes using simple factorization. Subsequently, Arkady and his collaborators pointed out[12] that we had forgotten a diagram, namely the penguin diagram, depicted in Figure 1. Although our Russian colleagues invented the diagram, its name was first introduced in a paper[13] (where we also estimated the $b\bar{b}$ photoproduction cross section using the SVZ QCD sum rules[14]) on the properties of the newly discovered B states after a dart game that ended with John Ellis required to use the word "penguin" in his next paper. (More details on the dart game can be found in Misha Shifman's contribution to a tribute[15] to Arkady.) This diagram gives an effective local four-quark operator because the q^2 in the gluon propagator cancels the q^2 in the numerator associated with the color charge radius. It contributes only to $\Delta I = \frac{1}{2}$ transitions because the penguin's head has $\Delta I = \frac{1}{2}$ ($s \to d$) and its foot has $\Delta I = 0$ since gluon exchange conserves flavor. It further has a different spin structure from the operators analyzed previously[11] because, while the penguin's head has only V-A couplings, its foot is a pure vector coupling because gluon exchange conserves parity. As a consequence, after a Fierz transformation so as to express the effective operator in terms of color singlet quark bilinears, one gets S, P interactions that can have enhanced matrix elements with respect to the usual V, A couplings.

Figure 1.   Penguin diagram.

One is still left with the thorny problem of evaluating hadronic matrix elements of the effective quark operators. This involves techniques such as chiral symmetry and lattice calculations; the latter have indicated an enhancement of penguin diagrams. The $\Delta I = \frac{1}{2}$ rule in kaon decay was discussed extensively by Bijnens.[16] To the extent that the nonrelativistic quark model is a good approximation for baryons, the $\Delta I = \frac{3}{2}$ operator discussed above cannot contribute because it has no overlap with the baryon wave function.[17] It seems likely to this observer that the undeservedly accurate ($\sim 2\%$) $\Delta I = \frac{1}{2}$ rule does not have a simple explanation, but results from the conjunction of a number of QCD effects, all of which point at least qualitatively in the right direction.

Aside from their contribution to the $\Delta I = \frac{1}{2}$ rule, the SVZ penguin diagrams have played, and continue to play, an essential role in the analysis of heavy quark decays and CP violation.

3. Is the Standard Model an Effective Theory?

Over the last twenty years experiments have continued to verify the predictions of the Standard Model to great accuracy over a wide range of energy scales. Yet the majority of theorists believe that the Standard Model itself has a limited range of validity. There are a variety of reasons for this, one being the large number of arbitrary parameters. Here I concentrate on just two, very likely connected, difficulties: the gauge hierarchy problem and the failure of the Standard Model to incorporate gravity. The gauge hierarchy problem points to a scale of new physics in the TeV region, providing the motivation for the LHC and an upgraded Tevatron. The fact that gravity couples to everything else suggests new physics at a much higher scale, the Planck scale, that (at least in the "conservative" approach taken here) will never be probed directly by experiments; probing this physics must rely on the interpretation of indirect effects in the effective low energy theory. Incorporating gravity in a fully consistent, predictive and verifiable theory, often referred to as the "Theory of Everything" (ToE) is the holy grail of elementary particle physics.

3.1. Bottom Up Approach

This approach starts from experimental data with the aim of deciphering what it implies for an underlying, more fundamental theory. One outstanding datum is the observed large gauge hierarchy, *i.e.*, the ratio of the Z mass, characteristic of the scale of electroweak symmetry breaking, to the reduced Planck scale m_P:

$$m_Z \approx 90\text{GeV} \ll m_P = \sqrt{\frac{1}{8\pi G_N}} \approx 2 \times 10^{18}\text{GeV},$$

which can be technically resolved by supersymmetry (SUSY) (among other conjectures that are by now disfavored by experiment). The conjunction of SUSY and general relativity (GR) inexorably implies supergravity (SUGRA). The absence of observed SUSY partners (sparticles) requires broken SUSY in the vacuum, and a more detailed analysis of the observed particle spectrum constrains the mechanism of SUSY-breaking in the observable sector: spontaneous SUSY-breaking is not viable, leaving soft SUSY-breaking as the only option that preserves the technical SUSY solution to the hierarchy problem. This means introducing SUSY-breaking operators of dimension three or less–such as gauge invariant masses–into the Lagrangian for the SUSY extension of the Standard Model (SM). The unattractiveness of these *ad hoc* soft terms strongly suggests that they arise from spontaneous SUSY breaking in a "hidden sector" of the underlying theory. Based on the above facts, a number of standard scenarios have emerged. These include:

• Gravity mediated SUSY-breaking, usually understood as "Minimal SUGRA" (MSUGRA), with masses of fixed spin particles set equal at the Planck scale; this scenario is typically characterized at the weak scale by

$$m_{\text{scalars}} = m_0 > m_{\text{gauginos}} = m_{\frac{1}{2}} \sim m_{\text{gravitino}} = m_{\frac{3}{2}}.$$

• Anomaly mediated SUSY-breaking,[18,19] in which $m_0 = m_{\frac{1}{2}} = 0$ classically; these models are characterized by $m_{\frac{3}{2}} >> m_0$, $m_{\frac{1}{2}}$, and typically $m_0 > m_{\frac{1}{2}}$. An exception is the Randall-Sundrum (RS) "separable potential", constructed[18] to mimic SUSY-breaking on a brane spatially separated from our own in a fifth dimension; in this scenario $m_0^2 < 0$ and m_0 arises first at two loops. More generally, the scalar masses at one loop depend on the details of Planck-scale physics.[20]

• Gauge mediated SUSY uses a hidden sector that has renormalizable gauge interactions with the SM particles. These scenarios are typically characterized by small $m_{\frac{1}{2}}$.

3.2. *Top Down Approach*

This approach starts from a ToE with the hope of deriving the Standard Model from it; the current favored candidate is superstring theory. The driving motivation is that this is at present the only known candidate for reconciling GR with quantum mechanics. Superstring theories are consistent in ten dimensions; in recent years it was discovered that all the consistent superstring theories are related to one another by dualities. These are, in my nomenclature: S-duality: $\alpha \rightarrow 1/\alpha$, and T-duality: $R \rightarrow 1/R$, where α is the fine structure constant of the gauge group(s) at the string scale, and R is a radius of compactification from dimension D to dimension $D - 1$. Figure 2 shows[21] how these dualities relate the various 10-D superstring theories to one another, and to the currently presumed ToE, M-theory. Not a lot else is known about M-theory, except that it lives in 11 dimensions and involves membranes. In Figure 2 the small circles, line, torus and cylinder represent the relevant compact manifolds in reducing D by one or two. The two $O(32)$ theories are S-dual to one another, while the $E_8 \otimes E_8$ weakly coupled heterotic string theory (WCHS) is perturbatively

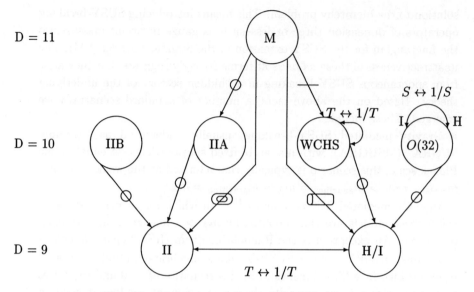

Figure 2.   M-theory according to John Schwarz.

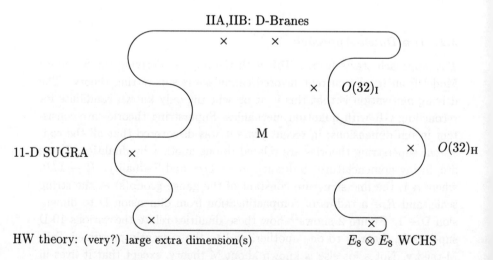

Figure 3.   M-theory according to Mike Green.

invariant[22] under T-duality. We will be specifically concentrating on this theory, and T-duality will play an important role.

Another image of M-theory, the "puddle diagram" of Figure 3, indicates[23] that all the known superstring theories, as well as $D = 11$ SUGRA, are particular limits of M-theory. Currently, there is a lot of activity in type I and II theories, or more generally in theories with branes. Similarly the Hořava-Witten (HW) scenario[24] and its inspirations have received considerable attention. If one compactifies one dimension of the 11-D limit of M-theory, one gets the HW scenario with two 10-D branes, each having an E_8 gauge group. As the radius of this 11th dimension is shrunk to zero, the WCHS scenario is recovered. This is the scenario addressed here, in a marriage of the two approaches that may serve as an illustrative example of the diversity of possible SUSY breaking scenarios.

4. The $E_8 \otimes E_8$ Heterotic String

I first recall the reasons for the original appeal of the weakly coupled $E_8 \otimes E_8$ heterotic string theory[25] compactified on a Calabi-Yau (CY) manifold[26] (or a CY-like orbifold[27]). The zero-slope (infinite string tension) limit of this superstring theory[28] is ten dimensional supergravity coupled to a supersymmetric Yang-Mills theory with an $E_8 \otimes E_8$ gauge group. To make contact with the real world, six of these ten dimensions must be unobservable in current experiments; here they are assumed to be compactified to a size of order of the reduced Planck length, 10^{-32}cm. If the topology of the extra dimensions were a six-torus, which has a flat geometry, the 8-component spinorial parameters of $N = 1$ supergravity in ten dimensions would appear as the four two-component parameters of $N = 4$ supergravity in ten dimensions. However a Calabi-Yau manifold leaves only one of these spinors invariant under parallel transport; for this manifold the group of transformations under parallel transport (holonomy group) is the $SU(3)$ subgroup of the maximal $SU(4) \cong SO(6)$ holonomy group of a six dimensional compact space. This breaks $N = 4$ supersymmetry to $N = 1$ in four dimensions. As is well known, the only phenomenologically viable supersymmetric theory at low energies is $N = 1$, because it is the only one that admits complex representations of the gauge group that are needed to describe quarks and leptons. For this solution, the classical equations of motion impose the identification of the affine connection of general coordinate transformations on the compact space (described by three complex dimensions) with the gauge connection of an $SU(3)$ subgroup of one of the E_8's: $E_8 \ni E_6 \otimes SU(3)$, resulting in $E_6 \otimes E_8$ as the gauge group in four dimensions. Since the early

1980's, E_6 has been considered the largest group that is a phenomenologically viable candidate for a Grand Unified Theory (GUT) of the Standard Model. Hence E_6 is identified as the gauge group of the "observable sector", and the additional E_8 is attributed to a "hidden sector", that interacts with the former only with gravitational strength couplings. Orbifolds, which are flat spaces except for points of infinite curvature, are more easily studied than CY manifolds, and orbifold compactifications that closely mimic the CY compactification described above, and that yield realistic spectra with just three generations of quarks and leptons, have been found.[29] In this case the surviving gauge group is $E_6 \otimes \mathcal{G}_o \otimes E_8$, $\mathcal{G}_o \in SU(3)$. The low energy effective field theory is determined by the massless spectrum, *i.e.*, the spectrum of states with masses very small compared with the scales of the string tension and of compactification. Massless particles have zero triality under an $SU(3)$ which is the diagonal of the $SU(3)$ holonomy group and the (broken) $SU(3)$ subgroup of one E_8. The ten-dimensional vector fields A_M, $M = 0, 1, \ldots 9$, appear in four dimensions as four-vectors A_μ, $\mu = M = 0, 1, \ldots 3$, and as scalars A_m, $m = M - 3 = 1, \cdots 6$. Under the decomposition $E_8 \ni E_6 \otimes SU(3)$, the E_8 adjoint contains the adjoints of E_6 and $SU(3)$, and the representation $(\mathbf{27}, \mathbf{3}) + (\overline{\mathbf{27}}, \overline{\mathbf{3}})$. Thus the massless spectrum includes gauge fields in the adjoint representation of $E_6 \otimes \mathcal{G}_o \otimes E_8$ with zero triality under each $SU(3)$, and scalar fields in $\mathbf{27} + \overline{\mathbf{27}}$ of E_6, with triality ± 1 under each $SU(3)$, together with their fermionic superpartners. The number of $\mathbf{27}$ and $\overline{\mathbf{27}}$ chiral supermultiplets that are massless depends on the detailed topology of the compact manifold. The important point for phenomenology is the decomposition under $E_6 \to SO(10) \to SU(5)$:

$$(\mathbf{27})_{E_6} = (\mathbf{16} + \mathbf{10} + \mathbf{1})_{SO(10)} = (\{\overline{\mathbf{5}} + \mathbf{10} + \mathbf{1}\} + \{\mathbf{5} + \overline{\mathbf{5}}\} + \mathbf{1})_{SU(5)}. \quad (3)$$

A $\overline{\mathbf{5}} + \mathbf{10} + \mathbf{1}$ contains one generation of quarks and leptons of the Standard Model, a right-handed neutrino and their scalar superpartners; a $\mathbf{5} + \overline{\mathbf{5}}$ contains the two Higgs doublets needed in the supersymmetric extension of the Standard Model and their fermion superpartners, as well as color-triplet supermultiplets. Thus all the states of the Standard Model and its minimal supersymmetric extension are present. On the other hand, there are no scalar particles in the adjoint representation of E_6. In conventional models for grand unification, adjoints (or one or more other representations much larger than the fundamental one) are needed to break the GUT group to the Standard Model. In string theory, this symmetry breaking can be achieved by the Hosotani, or "Wilson line", mechanism[30] in which gauge flux is trapped around "tubes" in the compact manifold, in a manner reminiscent of the Arahonov-Bohm effect. The vacuum value of the trapped flux $< \int d\ell^m A_m >$ has the same effect as an adjoint Higgs, evading the difficulties of constructing a viable Higgs sector encountered in conventional GUTS. Wilson lines reduce the gauge group in four dimensions to

$$\mathcal{G}_{obs} \otimes \mathcal{G}_{hid}, \quad \mathcal{G}_{obs} = \mathcal{G}_{SM} \otimes \mathcal{G}' \otimes \mathcal{G}_o, \quad \mathcal{G}_{SM} \otimes \mathcal{G}' \in E_6, \quad \mathcal{G}_o \in SU(3),$$
$$\mathcal{G}_{hid} \in E_8, \quad \mathcal{G}_{SM} = SU(3)_c \otimes SU(2)_L \otimes U(1)_w. \tag{4}$$

There are many other four dimensional string vacua with different features. The attractiveness of the above picture is that the requirement of $N = 1$ SUSY naturally results in a phenomenologically viable gauge group and particle spectrum. Moreover, the gauge symmetry can be broken to a product group embedding the Standard Model without the necessity of introducing large Higgs representations. In addition, the $E_8 \otimes E_8$ string theory includes a hidden sector that can provide a viable mechanism for spontaneous SUSY breaking. More specifically, if some subgroup $\mathcal{G}_a$ of $\mathcal{G}_{hid}$ is asymptotically free, with a β-function coefficient $b_a > b_{SU(3)}$, defined by the renormalization group equation (RGE)

$$\mu \frac{\partial g_a(\mu)}{\partial \mu} = -\frac{3}{2} b_a g_a^3(\mu) + O(g_a^5), \tag{5}$$

confinement and fermion condensation will occur at a scale $\Lambda_c \gg \Lambda_{QCD}$, and hidden sector gaugino condensation $< \bar{\lambda}\lambda >_{\mathcal{G}_a} \neq 0$, may induce[31] supersymmetry breaking.

To discuss supersymmetry breaking in more detail, we need the low energy spectrum resulting from the ten-dimensional gravity supermultiplet that consists of the 10-D metric g_{MN}, an antisymmetric tensor b_{MN}, the dilaton ϕ, the gravitino ψ_M and the dilatino χ. For the class of CY and orbifold compactifications described above, the zero-triality massless bosons in four dimensions are the 4-D metric $g_{\mu\nu}$, the antisymmetric tensor $b_{\mu\nu}$, the dilaton ϕ, and certain components of the tensors g_{mn} and b_{mn} that form the real and imaginary parts, respectively, of complex scalars known as moduli. (More precisely, the scalar components of the chiral multiplets of the low energy theory are obtained as functions of the scalars ϕ, g_{mn}, while the pseudoscalars b_{mn} form axionic components of these supermultiplets.) The number of moduli is related to the number of particle generations (# of **27**'s $-$ # of $\overline{\mathbf{27}}$'s). Typically, in a three generation orbifold model there are three moduli t_I; the vev's $< \text{Re} t_I >$ determine the radii of compactification of the three tori of the compact space. In some compactifications there are three other moduli u_I; the vev's $< \text{Re} u_I >$ determine the ratios of the two *a priori* independent radii of each torus. These form chiral multiplets with fermions χ_I^t, χ_I^u obtained from components of ψ_m. The 4-D dilatino

χ forms a chiral multiplet with with a complex scalar field s whose *vev* $< s >= g^{-2} - i\theta/8\pi^2$ determines the gauge coupling constant and the θ parameter of the 4-D Yang-Mills theory. The "universal" axion Ims is obtained by a duality transformation[32] from the antisymmetric tensor $b_{\mu\nu}$: $\partial_\mu \text{Im} s \leftrightarrow \epsilon_{\mu\nu\rho\sigma} \partial^\nu b^{\rho\sigma}$. Because the dilaton couples to the (observable and hidden) Yang-Mills sector, gaugino condensation induces[33] a superpotential for the dilaton superfield S (capital Greek and Roman letters denote chiral superfields, and the corresponding lower case letters denote their scalar components):

$$W(S) \propto e^{-S/b_a}. \tag{6}$$

The vacuum value $< W(S) >$ is governed by the condensation scale $\langle e^{-S/b_a} \rangle = e^{-g^{-2}/b_a} = \Lambda_c$ as determined by the RGE (5). If it is nonzero, the gravitino acquires a mass $m_{\frac{3}{2}} \propto < W >$, and local supersymmetry is broken.

5. A Runaway Dilaton?

The superpotential (6) results in a potential for the dilaton of the form $V(s) \propto e^{-2\text{Re}s/b_a}$, which has its minimum at vanishing vacuum energy and vanishing gauge coupling: $< \text{Re} s > \to \infty$, $g^2 \to 0$. This is the notorious runaway dilaton problem. The effective potential for s is determined by anomaly matching:[34] $\delta\mathcal{L}_{eff}(s,u) \longleftrightarrow \delta\mathcal{L}_{hid}(\text{gauge})$, where u, $\langle u \rangle = \langle \bar\lambda\lambda \rangle_{G_a}$, is the lightest scalar bound state of the strongly interacting, confined gauge sector. Just as in QCD, the effective low energy theory of bound states must reflect both the symmetries and the quantum anomalies of the underlying Yang-Mills theory. It turns out that the effective quantum field theory (QFT) is anomalous under T-duality. Since this is an exact symmetry of heterotic string perturbation theory, it means that the effective QFT is incomplete. This is cured by including model dependent string-loop threshold corrections[35] and a "Green-Schwarz" (GS) counter-term,[36] analogous to a similar anomaly canceling mechanism in 10-D SUGRA.[28] This introduces dilaton-moduli mixing, and the gauge coupling constant is now identified as $g^2 = 2 \langle \ell \rangle$, $\ell^{-1} = 2\text{Re}s - b\sum_I \ln(2\text{Re}t_I)$, where $b \leq b_{E_8} = 30/8\pi^2$ is the coefficient of the GS term. This term introduces a second runaway direction at strong coupling: $V \to -\infty$ for $g^2 \to \infty$. The small coupling behavior is unaffected, but the potential becomes negative for $\alpha = \ell/2\pi > .57$. This is the strong coupling regime, and nonperturbative string effects cannot be neglected; they are expected[37] to modify the Kähler potential for the dilaton, and therefore the potential

$V(\ell, u)$. It has been shown[38,39] that these contributions can indeed stabilize the dilaton.

In order to carry out the above program, one first needs to know the quantum corrections to the unconfined (i.e., at scales greater than Λ_c) Yang-Mills Lagrangian for the hidden sector, in order to do the correct anomaly matching. This is where we again encounter Arkady's work. Just as we can correctly treat the effective Fermi Lagrangian as a quantum theory only if we incorporate the physical cut-off m_W, we must include a similar physical cut-off when encountering divergences in the effective SUGRA theory from superstrings. One way to do this is using Pauli-Villars regulators with couplings that respect SUSY.[40] The required cut-offs[40,41] for regulating the coefficient of the Yang-Mills terms are (neglecting string nonperturbative corrections to the Kähler potential)

$$\Lambda_A = \Lambda(\Lambda/g)^{2(1-3q_A)}, \quad \Lambda_g = \Lambda g^{-\frac{2}{3}}, \qquad (7)$$

for matter and gauge loops, respectively, where $\Lambda = R^{-1}$ is the inverse radius of compactification, and q_A is the average modular weight for the chiral supermultiplet Φ^A. For "untwisted" matter $q_A = \frac{1}{3}$, giving the intuitive value $\Lambda_U = \Lambda = 1/R$, while the gauge-loop cut-off contains the "two-loop" factor $g^{-\frac{2}{3}}$ which assures that the anomaly is a chiral superfield. Matching[41] the field theory result with string loop calculations[35] determines the GS term for $\Phi^A = 0$, and one obtaines for the Wilson coefficient of the Yang-Mills operator

$$2\ell^{-1} - \frac{1}{16\pi^2}(C_G^a - C_M^a)\ln\ell - \frac{1}{8\pi^2}\sum_A C_A^a \ln(1 - p_A\ell) + O(|\phi^A|^2) + \Delta_a(t)(8)$$

where $\Delta_a(t)$ is a string threshold correction that is present in some orbifold models, and p_A is the coupling of $|\Phi^A|^2$ to the GS term. Neglecting Δ_a, the RGE invariant (8) can be identified with the general RGE invariant in SUSY Yang-Mills theories found by Shifman and Vainshtein[42]

$$g_a^{-2}(\mu) - \frac{1}{16\pi^2}(3C_G^a - C_M^a)\ln\mu^2 + C_G^a 8\pi^2 \ln g_a^2(\mu) + \frac{1}{8\pi^2}\sum_A C_A^a \ln Z_A^a(\mu)(9)$$

where Z_A^a are the renormalization factors for the matter fields, provided we identify

$$g_a(\mu_s) = 2\ell = (\mu_s/M_P)^2, \quad Z_A^a(\mu_s) = (1 - p_A\ell)^{-1}. \qquad (10)$$

The same boundary condition on the couplings was found in[43] where different regularization procedures were used. (The inclusion of string nonperturbative effects can modify these results; their effects were found to be negligible in the model[38] discussed below).

6. A Condensation Model for SUSY Breaking

In this section I discuss features of an explicit model[38] based on affine level one orbifolds with three untwisted moduli T_I and a gauge group of the form (4). Retaining just one or two terms of the suggested parameterizations[37] of the nonperturbative string corrections: $a_n \ell^{-n/2} e^{-c_n/\sqrt{\ell}}$ or $a_n \ell^{-n} e^{-c_n/\ell}$, the potential can be made positive definite everywhere and the parameters can be chosen to fit two data points: the coupling constant $g^2 \approx 1/2$ and the cosmological constant $\Lambda_{cos} \simeq 0$. This is fine tuning, but it can be done with reasonable (order 1) values for the parameters c_n, a_n. If there are several condensates with different β-functions, the potential is dominated by the condensate with the largest β-function coefficient b_+, and the result is essentially the same as in the single condensate case, except that a small mass is generated for the axion Ims. In this model, mass hierarchies arise from the presence of β-function coefficients; these have interesting implications for both cosmology and the spectrum of sparticles.

6.1. Modular Cosmology

The masses of the dilaton $d = \mathrm{Re}s$ and the complex t-moduli are related to the gravitino mass by[38]

$$m_d \sim \frac{1}{b_+^2} m_{\tilde{G}}, \qquad m_{t_I} \approx \frac{2\pi}{3} \frac{(b - b_+)}{(1 + b < \ell >)} m_{\tilde{G}}. \tag{11}$$

Taking $b = b_{E_8} \approx .38 \approx 10b$, gives a hierarchy of order $m_{\frac{3}{2}} \sim 10^{-15} m_{Pl} \sim 10^3$ GeV and $m_{t_I} \approx 20 m_{\frac{3}{2}} \approx 20$ TeV, $m_d \sim 10^3 m_{\frac{3}{2}} \sim 10^6$ GeV, which is sufficient to evade the late moduli decay problem[44] in nucleosynthesis.

If there is just one hidden sector condensate, the axion $a = \mathrm{Im}s$ is massless up to QCD-induced effects: $m_a \sim (\Lambda_{QCD}/\Lambda_c)^{\frac{3}{2}} m_{\frac{3}{2}} \sim 10^{-9}$ eV, and it is the natural candidate for the Peccei-Quinn axion. Because of string nonperturbative corrections to its gauge kinetic term, the decay constant f_a of the canonically normalized axion is reduced with respect to the standard result by a factor $b_+ \ell^2 \sqrt{6} \approx 1/50$ if $b_+ \approx .1 b_{E_8}$, which may be sufficiently small to satisfy the (looser) constraints on f_a when moduli are present.[45]

6.2. Sparticle Spectrum

In contrast to an enhancement of the dilaton and moduli masses, there is a suppression of gaugino masses: $m_{\frac{1}{2}} \approx b_+ m_{\frac{3}{2}}$, as evaluated at the scale Λ_c in the tree approximation. As a consequence quantum corrections can be important; for example there is an anomaly-like scenario in some regions of the (b_+, b_+^α) parameter space, where b_+^α is the hidden matter contribution to b_+. If the gauge group for the dominant condensate (largest b_a) is not E_8, the moduli t_I are stabilized at self-dual points through their couplings to twisted sector matter and/or moduli-dependent string threshold corrections, and their auxiliary fields vanish in the vacuum. Thus SUSY-breaking is dilaton mediated, avoiding a potentially dangerous source of flavor changing neutral currents (FCNC). These results hold up to the unknown couplings p_A in (8): at the scale Λ_c $m_{0A} = m_{\frac{3}{2}}$ if $p_A = 0$, while $m_{0A} = \frac{1}{2}m_{t_I} \approx 10m_{\frac{3}{2}}$ if the scalars couple with the same strength as the T-moduli: $p_A = b$. In addition, if $p_A = b$ for some gauge-charged chiral fields, there are enhanced loop corrections to gaugino masses.[46] Four sample scenarios were studied:[47] A) $p_A = 0$, B) $p_A = b$ for the superpartners of the first two generations of SM particles and $p_A = 0$ for the third, C) $p_A = b$, and D) $p_A = 0$ for the Higgs particles and $p_A = b$ otherwise. Imposing constraints from experiments and the correct electroweak symmetry-breaking vacuum rules out scenarios B and C. Scenario A is viable for $1.65 < \tan\beta < 4.5$, and scenario D is viable for all values of $\tan\beta$, which is the ratio of Higgs vev's in the supersymmetric extension of the SM. The viable range of (b_+, b_+^α) parameter space is shown[48] in Figure 4 for $g^2 = \frac{1}{2}$. The dashed lines represent the possible dominant condensing hidden gauge groups $\mathcal{G}_+ \in E_8$ with chiral matter in the coset space $E_8/\mathcal{G}_{hid}$.

6.3. *Flat Directions in the Early Universe*

Many successful cosmological scenarios–such as an epoch of inflation–require flat directions in the potential. A promising scenario for baryogenesis suggested[49] by Affleck and Dine (AD) requires in particular flat directions during inflation in sparticle field space: $< \tilde{q} >, < \tilde{\ell} > \neq 0$, where $\tilde{f}$ denotes the superpartner of the fermion f. While flat directions are common in SUSY theories, they are generally lifted[50] in the early universe by SUGRA couplings to the potential that drives inflation. This problem is evaded[51] in models with a "no-scale" structure, such as the classical potential for the untwisted sector of orbifold compactifications. Although the GS term breaks the no-scale property of the theory, quasi-flat directions can still be found. An explicit model[52] for inflation based on the effective theory described above allows dilaton stabilization within its domain of attraction with one or more moduli stabilized at the vacuum value $t_I = e^{i\pi/6}$. One of the moduli may be the inflaton. The moduli masses (11) are sufficiently large to evade the late moduli decay problem in nucleosynthesis, but unlike

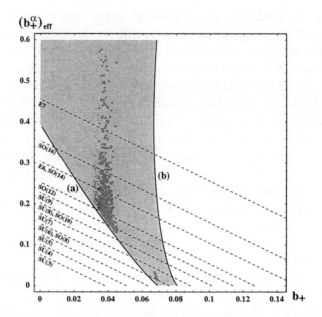

Figure 4. Viable hidden sector gauge groups for scenario A of the condensation model. The swath bounded by lines (a) and (b) is the region defined by $.1 < m_{\frac{3}{2}}/\text{TeV}, \lambda_c < 10$, where λ_c is a condensate superpotential coupling constant. The fine points correspond to $.1 \leq \Omega_d h^2 \leq .3$, and the course points to $.3 < \Omega_d h^2 \leq 1$.

the dilaton, they are insufficient to avoid a large relic LSP density without violation[53] of R-parity (a quantum number that distinguishes SM particles from their superpartners). If R-parity is conserved, this problem can be evaded if the moduli are stabilized at or near their vacuum values–or for a modulus that is itself the inflaton. It is possible that the requirement that the remaining moduli be in the domain of attraction is sufficient to avoid the problem altogether. For example, if $\text{Im} t_I = 0$, the domain of attraction near $t_I = 1$ is rather limited: $0.6 < \text{Re} t_I < 1.6$, and the entropy produced by dilaton decay with an initial value in this range might be less than commonly assumed. The dilaton decay to its true ground state may provide[54] partial baryon number dilution, which is generally needed for a viable AD scenario.

6.4. *Relic Density of the Lightest SUSY Particle (LSP)*

Two pertinent questions for SUSY cosmology are:
• Does the LSP overclose the Universe?
• Can the LSP be dark matter?

The window for LSP dark matter in the much-studied MSUGRA scenario,[55] has become ever more tiny as the Higgs mass limit has increased; in fact there is not much parameter space in which the LSP does not overclose the universe. The ratios of electroweak sparticle masses at the Plank scale determine the composition of the LSP (which must be neutral) in terms of the Bino (superpartner of the SM $U(1)$ gauge boson), the Wino (superpartner of the SM $SU(2)$ gauge boson), and the higgsino (superpartner of the Higgs boson). The MSUGRA assumption of equal gaugino masses at the Planck scale leads to a Bino LSP with rather weak couplings, resulting in little annihilation and hence the tendency to overclose the universe, except in a narrow range of parameter space where the LSP is nearly degenerate with the next to lightest sparticle (in this case a stau $\tilde{\tau}$), allowing significant coannihilation. Relaxing this assumption[48] it was found that a predominantly Bino LSP with a small admixture of Wino can provide the observed density fraction Ω_d of dark matter. In the above model,[38] this occurs in the region indicated by fine points in Figure 4. In this model the deviation from the MSUGRA scenario is due to the importance of loop corrections to small tree-level gaugino masses; in addition to a small Wino component in the LSP, its near degeneracy in mass with the lightest charged gaugino enhances coannihilation. For larger b_+ the LSP becomes pure Bino as in MSUGRA, and for smaller values it becomes Wino-dominated as in anomaly-mediated models which are cosmologically safe, but do not provide LSP dark matter, because Wino annihilation is too fast.

6.5. Realistic Orbifold Models?

Orbifold compactifications with Wilson lines needed to break E_6 to the SM gauge group generally have $b_+ \leq b \leq b_{E_8}$. An example is a model[29] with hidden gauge group $O(10)$ and $b_+ = b = b_{O(10)}$. It is clear from (11) that this would lead to disastrous modular cosmology, since the t-moduli are massless. Moreover, in typical orbifold compactifications, the gauge group $\mathcal{G}_{obs} \otimes \mathcal{G}_{hid}$ obtained at the string scale has no asymptotically free subgroup that could condense to trigger SUSY-breaking. However in many compactifications with realistic particle spectra,[29] the effective field theory has an anomalous $U(1)$ gauge subgroup, which is not anomalous at the string theory level. The anomaly is canceled[56] by a GS counterterm, similar to the GS term introduced above to cancel the modular anomaly. This results in a D-term that forces some otherwise flat direction in scalar field space to be lifted, inducing scalar vev's that further break the gauge symmetry and give masses of order Λ_D to some chiral multiplets, so that the β-function of some of the surviving gauge subgroups may be negative below the scale Λ_D, typically an order of magnitude below the string scale. The presence of such a D-term was explicitly invoked in the above-mentioned inflationary model.[52] Its incorporation into the effective condensation potential is under study.[57]

There is a large vacuum degeneracy associated with the D-term induced breaking of the anomalous $U(1)$, resulting in many massless "D-moduli" that have the potential for a yet more disastrous modular cosmology.[58] Preliminary results indicate that the D-moduli couplings to matter condensates lift the degeneracy to give cosmologically safe D-moduli masses. Although the D-term modifies the potential for the dilaton, one still obtains moduli stabilized at self-dual points giving FCNC-free dilaton dominated SUSY-breaking, an enhanced dilaton mass m_d and a suppressed axion coupling f_d. An enhancement of the ratio $m_{t_I}/m_{\frac{3}{2}}$ can result from couplings to condensates of $U(1)$-charged D-moduli, that also carry T-modular weights. However, the requirement of a viable scalar/gaugino mass ratio may impose severe restrictions on the details of the effective theory.

7. Lessons

History has taught us that high energy physics can be successfully studied using lower energy effective theories. Hopefully this will be the case for string theory. In particular I have argued that

• Quantitative studies with predictions for observable phenomena are possible within the context of the WCHS.

• Experiments can place restrictions on the underlying theory, such as the hidden sector spectrum and the couplings and modular weights of D-moduli when an anomalous $U(1)$ is present. Data can also inform us about Plank scale physics through matter couplings to the GS term and one-loop corrections to the soft SUSY-breaking scalar potential.

Finally, the SUSY-breaking scenario presented here illustrates the need for sparticle searches to avoid restrictive assumptions based on "standard scenarios" that may be misleading in the absence of concrete models.

Acknowledgments

I am grateful to my many collaborators. This work was supported in part by the Director, Office of Energy Research, Office of High Energy and Nuclear Physics, Division of High Energy Physics of the U.S. Department of Energy under Contract DE-AC03-76SF00098 and in part by the National Science Foundation under grants PHY-95-14797 and INT-9910077.

References

1. S. Glashow, J. Iliopoulos and L. Maiani, *Phys. Rev.* **D2**, 1285 (1970).
2. C. Bouchiat, J. Iliopoulos and Ph. Meyer, *Phys. Lett.* **B38**, 519 (1972).
3. M.K. Gaillard and B.W. Lee, *Phys. Rev.* **D10**, 897 (1974).
4. E. Ma, *Phys. Rev.* **D9**, 3103 (1974).
5. A.I. Vainshtein and I.B. Kriplovich, *ZhETF Pis. Red.* **18**, 141 (1973) and *Phys. Lett.* **18**, 83 (1973).
6. M.K. Gaillard, B.W. Lee and J.L. Rosner, *Rev. Mod. Phys.*, **47** 277 (1975).
7. J.-M. Gaillard, private communication.
8. A.J. Buras and J.M. Gerard, *Nucl. Phys.* **B264**, 371 (1986).
9. G.W. Kilcup, U., S.R. Sharpe, R. Gupta, A. Patel, *Phys. Rev. Lett.* **64**, 25 (1990).
10. K. Wilson, *Phys. Rev.* **179**, 1499 (1969).
11. M.K. Gaillard and B.W. Lee, *Phys. Rev. Lett.* **33**, 108 (1974); G. Altarelli and L. Maiani, *Phys. Lett.* **B52**, 351 (1974).
12. M.A. Shifman, A.I. Vainshtein and V.I. Zakharov, *Pis'ma ZhETF* **22**, (1975) 123 [*JETP Lett.* **22**, 55 (1975)] and *ZhETF* **72**, 1275 (1977) [*JETP* **45**, 670 (1977)].
13. J. Ellis, M.K. Gaillard, D.V. Nanopoulos and S. Rudaz, *Nucl. Phys.* **B131** 285 (1977).
14. M.A. Shifman, A.I. Vainshtein and V.I. Zakharov, *Phys. Lett.* **B65**, 255 (1976).
15. Fun Reading for the Participants of CAQCD2002/Arkadyfest, Theoretical Physics Institute, University of Minnesota (2002).
16. J. Bijnens, these proceedings.
17. J.C. Pati and C.H. Woo, *Phys. Rev.* **D3**, 2920 (1971).
18. L. Randall and R. Sundrum, *Nucl. Phys.* **B 557**, 557 (1999).
19. G. Giudice, M. Luty, H. Murayama and R. Rattazzi, *JHEP* **9812**, 027 (1998).
20. M.K. Gaillard and B. Nelson, *Nucl. Phys.* **B 588**, 197 (2000); P. Binétruy, M.K. Gaillard and B. Nelson, *Nucl. Phys.* **B 604**, 32 (2001).
21. J.H. Schwarz, Nucl. Phys. Proc. Suppl. B **55**, 1 (1997).
22. A. Giveon, N. Malkin and E. Rabinovici, *Phys. Lett.* B **220**, 551 (1989); E. Alvarez and M. Osorio, *Phys. Rev.* D **40**, 1150 (1989).
23. M. Green, a seminar; my apologies if the puddle diagram was invented by someone else.
24. P. Hořava and E. Witten, *Nucl. Phys.* B **460**, 506 (1996) and B **475**, 94 (1996).
25. D. Gross, J. Harvey, E. Martinec and R. Rohm, *Phys. Rev. Lett.* **54**, 502 (1985).
26. P. Candelas, G. Horowitz, A. Strominger and E. Witten, *Nucl. Phys.* B **258**, 46 (1985).
27. L.J. Dixon, V.S. Kaplunovsky and J. Louis, *Nucl. Phys.* B **329**, 27 (1990); S. Ferrara, D. Lüst, and S. Theisen, *Phys. Lett.* B **233**, 147 (1989).
28. M. Green and J. Schwarz, *Phys. Lett.* B **149**, 117 (1984).
29. L.E. Ibàñez, H.-P. Nilles and F. Quevedo, *Phys. Lett.* B **187**, 25 (1987); A. Font, L. Ibàñez, D. Lust and F. Quevedo, *Phys. Lett.* B **245**, 401 (1990); A. Font, L. E. Ibàñez, F. Quevedo and A. Sierra, *Nucl. Phys.* B **331**, 421 (1990).

152

30. Y. Hosotani, *Phys. Lett.* B **129**, 75 (1985).
31. H.P. Nilles, *Phys. Lett.* B **115**, 193 (1982).
32. E. Witten, *Phys. Lett.* B **155**, 151 (1985).
33. M. Dine, R. Rohm, N. Seiberg and E. Witten, *Phys. Lett.* B **156**, 55 (1985).
34. G. Veneziano and S. Yankielowicz, *Phys. Lett.* **113**, 231 (1982); T.R. Taylor, G. Veneziano and S. Yankielowicz, *Nucl. Phys.* B **218**, 493 (1983).
35. L.J. Dixon, V.S. Kaplunovsky and J. Louis, *Nucl. Phys.* B **355**, 649 (1991); I. Antoniadis, K.S. Narain and T.R. Taylor, *Phys. Lett.* B **267**, 37 (1991).
36. G.L. Cardoso and B.A. Ovrut, *Nucl. Phys.* B **369**, 315 (1993); J.-P. Derendinger, S. Ferrara, C. Kounnas and F. Zwirner, *Nucl. Phys.* B **372**, 145 (1992).
37. S.H. Shenker, in *Random Surfaces and Quantum Gravity*, Proceedings of the NATO Advanced Study Institute, Cargèse, France, 1990, edited by O. Alvarez, E. Marinari, and P. Windey, NATO ASI Series B: Physics Vol.262 (Plenum, New York, 1990); T. Banks and M. Dine, *Phys. Rev.* D **50**, 7454 (1994); E. Silverstein, *Phys. Lett.* B **396**, 91 (1997).
38. P. Binétruy, M. K. Gaillard and Y.-Y. Wu, *Nucl. Phys.* B**481**, 109 (1996) and B **493**, 27 (1997); *Phys. Lett.* B**412**, 228 (1997).
39. J.A. Casas, *Phys. Lett.* B**384**, 103 (1996).
40. M. K. Gaillard, *Phys. Lett.* B **342** (1995), 125; *Phys. Rev.* D **58** (1998), 105027; D 61 (2000), 084028.
41. M.K. Gaillard and T.R. Taylor, *Nucl. Phys.* B**381**, 577 (1992).
42. M.A. Shifman and A.I. Vainshtein, *Nucl. Phys.* **277**, 456 (1986).
43. V.S. Kaplunovsky and J. Louis, *Nucl. Phys.* B**444**, 191 (1995).
44. G.D. Coughlan, W. Fischler, E.W. Kolb, S. Raby and G.G. Ross, *Phys. Lett.* B**131**, (1983) 59.
45. T. Banks and M. Dine, *Nucl. Phys.* B**505**, 445 (1997).
46. M.K. Gaillard, B. Nelson and Y.Y. Wu, *Phys. Lett.* B**459**, 549 (1999).
47. M.K. Gaillard and B. Nelson, *Nucl. Phys.* B**571**, 3 (2000).
48. B. Nelson and A. Birkedal-Hansen, *Phys. Rev.* D**64**, 015008 (2001).
49. I. Affleck and M. Dine, *Nucl. Phys.* B**249**, 361 (1985).
50. M.Dine, L. Randall, and S. Thomas, *Nucl Phys.* B**458**, 291 (1996).
51. M.K. Gaillard, H. Murayama and K. A. Olive, *Phys. Lett.* B **355**, 71 (1995).
52. M.K. Gaillard, H. Murayama and D.H. Lyth, *Phys. Rev.* D **58**, 123505 (1998); M.K. Gaillard and Mike J. Cai, *Phys. Rev.* D **62**, 047901 (2000).
53. M. Kawasaki, T. Moroi, and T. Yanagida, *Phys. Lett.* B **370**, 52 (1996).
54. M.K. Gaillard, B.A. Campbell, M.K. Gaillard, H. Murayama and K.A. Olive, *Nucl. Phys.* B **538**, 351 (1999).
55. J. Ellis, T. Falk, K.A. Olive and M. Schmitt, *Phys. Lett.* B **388**, 97 (1996) and *Phys. Lett.* B **413**, 355 (1997); J. Ellis, T. Falk, G. Ganis and K.A. Olive *Phys. Rev.* D **62**, 075010 (2000).
56. M. Dine, N. Seiberg and E. Witten, *Nucl. Phys.* B **289**, 589 (1987); J. Attick, L. Dixon and A. Sen, *Nucl. Phys.* B **292**, 109 (1987); M. Dine, I. Ichinose and N. Seiberg, *Nucl. Phys.* B **293**, 253 (1988).
57. M.K. Gaillard and Joel Giedt, hep-th/0204100 and work in progress.
58. M.K. Gaillard and Joel Giedt, *Phys. Lett.* B **479**, 308 (2000).

EXPERIMENTAL CHALLENGES FOR QCD - THE PAST AND THE FUTURE

HARRY J. LIPKIN

Department of Particle Physics, Weizmann Institute of Science,
Rehovot, Israel
E-mail: ftlipkin@ckever.weizmann.ac.il

School of Physics and Astronomy, Raymond and Beverly Sackler Faculty of
Exact Sciences,
Tel Aviv University, Tel Aviv, Israel

High Energy Physics Division, Argonne National Laboratory,
Argonne, IL 60439-4815, USA
E-mail: lipkin@hep.anl.gov

The past leaves the surprising experimental successes of the simple constituent quark model to be expained by QCD. The future opens the way to new insight into QCD from heavy flavor experiments

1. Introduction

QCD is supposed to explain everything about Hadron Physics - But How?

QED is supposed to explain everything about Superconductivity .

Will explaining Hadron Spectrocopy from QCD be as difficult as explaining Superconductivity from QED?

The Past leaves us with many experimental regularities which await explanation by QCD. The future offers many new experimental opportunities to learn about QCD from heavy flavvor physics. .

1.1. *The Past - pre-QCD questions - Challenges for QCD*

1.1.1. *What is a Hadron?*

Present attempts to describe hadrons recall the story of the blind men and the elephant[1] . Each investigation finds one particular property of hadrons and many contradictory conclusions arise that are all correct,

(1) A pion is a Goldstone Boson and a proton is a Skyrmion,

153

(2) A pion is two-thirds of a proton. The simple quark model prediction $\sigma_{tot}(\pi^-p) \approx (2/3) \cdot \sigma_{tot}(pp)$ [2,3] still fits experimental data better than 7% up to 310 Gev/c[4];

(3) The a_1 is a $q\bar{q}$ pair in a $^3P_1$ state similar to other ^3P states: scalar and tensor (a_2)

(4) The a_1 is the chiral partner of the ρ coupled similarly to the W.

(5) The η and η' are orthogonal linear combinations of the same strange and nonstrange ground state wave functions

(6) The η and η' contain other components like glueballs or radial excitations

(7) Mesons and Baryons are made of the same quarks. Describing both as simple composites of asymptotically free quasiparticles with a unique effective mass value predicts hadron masses, magnetic moments and hyperfine splittings[5,6,7].

(8) Lattice QCD can give all the answers,

(9) Lattice calculations disagree on whether the H dibaryon is bound and offer no hope of settling this question until much bigger lattices are available[1].

1.1.2. *What is a good hadronic symmetry? Many contradictions*

(1) Light (uds) SU(3) symmetry and Heavy Quark symmetry (cbt) are good;

(2) Light (uds) SU(3) symmetry is bad. All nontrivial hadron states violate SU(3). All light V, A and T mesons have good isospin symmetry with flavor mixing in (u.d) space and no $s\bar{s}$ component; e.g. ρ, ω.

(3) The s-quark is a heavy quark. Flavor mixing in mass eigenstates predicted by SU(3) is not there. Most nontrivial strange hadron states satisfy (scb) heavy quark symmetry with no flavor mixing.; e.g. ϕ, ψ, Υ.

(4) Light (uds) SU(3) symmetry is the basis of Cabibbo theory of weak interactions and gives excellent description of hyperon decays.

(5) Violation of the Gottfried sum rule shows the proton sea is not isoscalar.

(6) Isospin symmetry requires a proton with an isovector sea to have a component with a valence neutron and a charged sea

(7) SU(3) requires proton with isovector sea to have a component with a valence hyperon and a strange sea to satisfy Cabibbo theory for vector current.

(8) Experiment shows SU(3) symmetry manifestly broken in proton sea
(9) No consistent explanation of Gottfried violation, strange sea suppression breaking SU(3) and Cabibbo theory requiring good SU(3).
(10) Why are the ω and ρ degenerate while the η and π are not? Is there a symmetry beyond SU(3) that forbids octet-singlet splitting for vectors but not for pseudoscalars??

1.1.3. Why do the constituent quark model and the OZI rule work so well?

Surprising agreement with experiment from simple Sakharov-Zeldovich model (1966) having quarks with effective masses and hyperfine interaction. Nambu's (1966) Colored quarks with gauge gluons gave mass spectrum with only qqq and $\bar{q}$ bound states.

The topological quark-line OZI rule does not follow from any symmetry and predicts experiments successfully without any solid theoretical justification.

1.2. The Future - Heavy Flavor Decays Give New Insight

(1) Weak Decays need hadron models and QCD to interpret decays, but have too many diagrams, too many free parameters
(2) No rigorous QCD results for FSI and strong phases
(3) Too many decay modes, too much data. Need phenomenologists to choose data for analysis
(4) Experimental results from B and Charm factories that defy conventional wisdom can provide clues to new physics and inadequacies in hadron models.

2. The Past - The constituent quark model and other pre-QCD Challenges for QCD

2.1. The Sakharov-Zeldovich 1966 Quark model (SZ66)

2.1.1. The Model

Andrei Sakharov, a pioneer in quark-hadron physics asked in 1966 "Why are the Λ and Σ masses different? They are made of the same quarks". Sakharov and Zeldovich[5]. assumed a quark model for hadrons with a flavor dependent linear mass term and hyperfine interaction,

$$M = \sum_i m_i + \sum_{i>j} \frac{\vec{\sigma}_i \cdot \vec{\sigma}_j}{m_i \cdot m_j} \cdot v_{ij}^{hyp} \tag{1}$$

where m_i is the effective mass of quark i, $\vec{\sigma}_i$ is a quark spin operator and v_{ij}^{hyp} is a hyperfine interaction with different strengths but the same flavor dependence for qq and $\bar{q}q$ interactions.

This model can be considered analogous to the BCS description of superconductivity. The constituent quarks are quasiparticles of unknown structure with a background of a condensate. They have effective masses not simply related to the bare current quark masses, and somehow including all effects of confinement and other flavor independent potentials. The only contribution to hadron masses not already included is a flavor-dependent two-body hyperfine interaction inversely proportional to the product of these same effective quark masses. Hadron magnetic moments are described simply by adding the contributions of the moments of these constituent quarks with Dirac magnetic moments having a scale determined by the same effective masses. The model describes low-lying excitations of a complex system with remarkable success.

2.1.2. Striking Results and Predictive Power

Sakarov and Zeldovich already in 1966 obtained two relations between meson and baryon masses in remarkable agreement with experiment. Both the mass difference $m_s - m_u$ between strange and nonstrange quarks and their mass ratio m_s/m_u have the same values when calculated from baryon masses and meson masses[5]

$$\langle m_s - m_u \rangle_{Bar} = M_\Lambda - M_N = 177\,\text{MeV} \tag{2}$$

$$\langle m_s - m_u \rangle_{mes} = \frac{3(M_{K^*} - M_\rho) + M_K - M_\pi}{4} = 180\,\text{MeV} \tag{3}$$

$$\left(\frac{m_s}{m_u}\right)_{Bar} = \frac{M_\Delta - M_N}{M_{\Sigma^*} - M_\Sigma} = 1.53 = \left(\frac{m_s}{m_u}\right)_{Mes} = \frac{M_\rho - M_\pi}{M_{K^*} - M_K} = 1.61 \tag{4}$$

Further extension of this approach led to two more relations for $m_s - m_u$ when calculated from baryon masses and meson masses[6,7]. and to three magnetic moment predictions with no free parameters[8,9]

$$\langle m_s - m_u \rangle_{mes} = \frac{3M_\rho + M_\pi}{8} \cdot \left(\frac{M_\rho - M_\pi}{M_{K^*} - M_K} - 1\right) = 178\,\text{MeV}. \tag{5}$$

$$\langle m_s - m_u \rangle_{Bar} = \frac{M_N + M_\Delta}{6} \cdot \left(\frac{M_\Delta - M_N}{M_{\Sigma^*} - M_\Sigma} - 1\right) = 190\,\text{MeV}. \tag{6}$$

$$\mu_\Lambda = -0.61\,\text{n.m.} = \mu_\Lambda = -\frac{\mu_p}{3} \cdot \frac{m_u}{m_s} = -\frac{\mu_p}{3}\frac{M_{\Sigma^*} - M_\Sigma}{M_\Delta - M_N} = -0.61\,\text{n.m.} \tag{7}$$

$$-1.46 = \frac{\mu_p}{\mu_n} = -\frac{3}{2} \qquad (8)$$

$$\mu_p + \mu_n = 0.88\,\text{n.m.} = \frac{M_p}{3m_u} = \frac{2M_p}{M_N + M_\Delta} = 0.865\,\text{n.m.} \qquad (9)$$

Also in 1966 Levin and Frankfurt[2] noted a remarkable systematics in hadron-nucleon total cross sections indicating that mesons and baryons were made of the same basic building blocks. The analysis supporting their ratio of 3/2 between baryon-nucleon and nucleon-nucleon cross sections has been refined[3] and consistently confirmed by new experiments[4]. QCD calculations have not yet explained such remarkably successful simple constituent quark model results. A search for new experimental input to guide us is therefore of interest.

2.2. The A.....Z or OZI rule and QCD

2.2.1. OZI for light quarks

No rigorous QCD derivation has yet been found for this flavor-topology rule arising also in duality diagrams of Regge phenomenology where leading t-channel exchanges are dual to s-channel resonances and in more modern planar quark diagrams in large N_c QCD. It has been repeatedly confirmed in a large variety of experimental results and theoretical analyses for strong interaction three-point and four-point functions, beginning with the first controversial prediction relating final states in completely different isospin and flavor-SU(3) multiplets unrelated by any known symmetry.

$$\sigma(K^- p \to \Lambda \rho^o) = \sigma(K^- p \to \Lambda \omega) \qquad (10)$$

Is this connected with the $\omega - \rho$ degeneracy?

2.2.2. OZI for heavy flavors - Why is J/ψ narrow?

One diagram $J/\psi \to 3G \to$ light hadrons fits the narrowness but $J/\psi \to D\bar{D} \to$ light hadrons is larger and neglected along with ad hoc forbidden "hairpin diagrams". Hand waving explanations with cancellations give predictive power which can be tested with future experimental data and still challenge QCD for explanation.

2.3. Problems with the η and η'

The pseudoscalars are conventionally decribed by adding an additional mass contribution to the SU(3) singlet state, thus breaking U(3) while conserving

SU(3) and leaving SU(3) breaking as entirely due to quark mass differences.. The dynamical origin of this additional singlet contribution is still unclear and controversial, with some models attributing it the annihilation of an $q\bar{q}$ pair into gluons or instantons and no reason to limit the mixing to only ground state $q\bar{q}$ wave functions. Admixtures of radial excitations and glueballs have been considered.

2.4. How to go beyond SZ66 with QCD

Many approaches are being investigated to use QCD in the description of hadron spectroscopy[10]. The complexity of QCD calculations necessitates the introduction of ad hoc approximations and free parameters to obtain results, thus losing the simplicity of the constituent quark model, with its ability to make many independent predictions with very few parameters, There is also a tendency to lose some of the good results of the constituent quark model; namely

- The universal treatment of mesons and baryons made of the same quarks
- The spin dependence of hadron masses as a hyperfine interaction
- The appearance of the same effective quark masses in hadron masses, spin splittings and magnetic moments
- The systematic regularities relating meson-nucleon and baryon-nucleon cross sections

While none of these results can be considered to have a firm theoretical foundation based on QCD, it is difficult simply to dismiss the striking agreement with experiment and the successful predictive power as purely purely accidental.

3. The Future - Heavy Flavor Physics Gives New Insight

Weak Decays need hadron models and QCD to interpret decays, but have too many diagrams and too many free parameters. Use of flavor topology can simplify analyses on one hand and challenge QCD to explain them if they work.

3.1. Experimental systematics challenging conventional wisdom

3.1.1. Universality of vector dominance couplings

The large branching ratios observed[11] for the appearance of the $a_1(1260)^\pm$ in all quasi-two-body decays $D \to a_1(1260)^\pm X$ and $B \to a_1(1260)^\pm X$ are comparable to those observed for $\pi^\pm X$ and $\rho^\pm X$. No decays to the other p-wave mesons are within an order of magnitude of these values; e.g the difference between the a_1 and the a_2. All 24 B decays of the form $B \to \bar{D}W^+ \to \bar{D}M^+$, where M can denote $a_1, \rho, \pi, \ell^+\nu_\ell, D_s, D_s^*$, are dominant with branching ratios above 0.3%. Other B-decay modes have upper limits in the 10^{-4} ball park, including the absence with significant upper limits of neutral decays $B^o \to \bar{D}^o M^o$ which are coupled by strong final state interactions to $B^o \to D^- M^+$.

These experimental systematics suggested a "vector-dominance" model[12] where the initial hadron state i decays to a final state f by emitting a $W^\pm$ which then hadronizes into an a_1^+, ρ^+ or π^+, along with a universality relation,

$$[if\pi] \equiv \frac{BR[i \to f\pi^+]}{BR[i \to f\rho^+]} \approx \left|\frac{W^+ \to \pi^+}{W^+ \to \rho^+}\right|^2$$

$$[ifa] \equiv \frac{BR[i \to fa_1(1260)^+]}{BR[i \to f\rho^+]} \approx \left|\frac{W^+ \to a_1^+}{W^+ \to \rho^+}\right|^2$$

for all states i and f with corrections for phase space.

$$[D^+\bar{K}^o\pi] \approx [D^oK^-\pi] \approx [B^oD^-\pi] \approx [B^oD^{*-}\pi] \approx [B^+\bar{D}^o\pi] \approx [B^+\bar{D}^{*o}\pi]$$

$$.44 \pm .17 \approx .35 \pm .09 \approx .38 \pm .08 \approx .41 \pm .20 \approx .40 \pm .06 \approx .30 \pm .07$$

$$[D^+\bar{K}^o a] \approx [D^oK^- a] \approx [B^oD^- a] \approx [B^oD^{*-} a] \approx [B^+\bar{D}^o a] \approx [B^+\bar{D}^{*o} a]$$

$$1.2 \pm .5 \approx .68 \pm .12 \approx .8 \pm .4 \approx 1.9 \pm 1.0 \approx .37 \pm .30 \approx 1.2 \pm .4$$

The a_1 data have large errors, b ut the experimental ratios $[ifa]$ are all consistent with 0.7, and more than order of magnitude higher than other upper limits

$$[D^oK^- a_2^+] < 0.019 \pm 0.002; \quad [D^+\bar{K}^o a_2^+] < 0.045 \pm 0.017$$

That such widely different decays should agree so well is impressive and suggests further investigation. e.g. reducing the experimental errors and looking for more decay modes like $D_s^+ \to \phi a_1$, $D_s^+ \to \omega a_1$, $D^+ \to K^{*0} a_1$ and $D^0 \to K^{*-} a_1$.

3.1.2. Vector-Dominance Decays of the B_c

The B_c meson is identified against a large combinatorial background by decay modes including a J/ψ. Vector dominance decay modes including the J/ψ are expected to have relatively large branching ratios. These include: $J/\psi\rho^+$, $J/\psi a_1^+$, $J/\psi\pi^+$, $J/\psi D_s^*$, $J/\psi D_{s1A}$, and $J/\psi D_s$. The corresponding modes with a ψ' instead of a J/ψ are expected to have comparable branching ratios.

3.1.3. Puzzles in Singly-Suppressed Charm Decays

Two Cabibbo suppressed D^+ decay modes have anomalously high branching ratios which are not simply explained by any model[13].

$$BR[D^+ \to K^*(892)^+\bar{K}^o] = 3.2 \pm 1.5\% \tag{11}$$

$$BR[D^+ \to K^*(892)^+\bar{K}^*(892)^o] = 2.6 \pm 1.1\% \tag{12}$$

These are the same order as corresponding Cabibbo allowed branching ratios

$$BR[D^+ \to \rho^+\bar{K}^o] = 6.6 \pm 2.5\% \tag{13}$$

$$BR[D^+ \to \rho^+\bar{K}^*(892)^o] = 2.1 \pm 1.3\% \tag{14}$$

The dominant tree diagrams for these corresponding allowed and suppressed decays differ only in the weak vertices $c \to W^+ + s \to \rho^+ + s$ and $c \to W^+ + s \to K^*(892)^+ + s$ and have the same hadronization of the strange quark s and spectator $\bar{d}$. These diagrams should show the expected Cabibbo suppression which is not observed.

All standard model diagrams that can contribute to these anomalously enhanced decays (11-12) are related by symmetries to a very similar diagrams for one of the following decay modes which show the expected Cabibbo suppression

$$BR[D^+ \to K^+\bar{K}^*(892)^o] = 0.42 \pm 0.05\% \tag{15}$$

$$BR[D^o \to K^*(892)^+K^-] = 0.35 \pm 0.08\% \tag{16}$$

$$BR[D^o \to K^*(892)^-K^+] = 0.18 \pm 0.01\% \tag{17}$$

$$BR[D^o \to K^*(892)^o\bar{K}^o] < 0.08\% \tag{18}$$

$$BR[D^o \to \bar{K}^*(892)^oK^o] < 0.16\% \tag{19}$$

$$BR[D^o \to K^*(892)^o\bar{K}^*(892)^o] = 0.14 \pm 0.05\% \tag{20}$$

There is no simple diagram that enhances the suppressed diagrams (11-12) without also enhancing others that show no experimental enhancement.

It is therefore of interest to check the branching ratios for the transitions (11-12) and reduce the errors. Using the present data we find:

$$BR[D^+ \to K^*(892)^+\bar{K}^o]+BR[D^+ \to K^*(892)^+\bar{K}^*(892)^o] = 5.8\pm1.9\%(21)$$

This is still large even at two standard deviations. If the large branching ratios are confirmed with smaller errors, there may be good reason to look for a new physics explanation.

3.1.4. *Anomalously high η' in charmless strange B decays*

The large experimental branching ratio[11] $BR(B^+ \to K^+\eta') = 6.5 \pm 1.7 \times 10^{-5}$ as compared with $BR(B^+ \to K^+\eta) < 1.4 \times 10^{-5}$ and $BR(B^+ \to K^o\pi^+) = 2.3 \pm 1.1 \times 10^{-5}$ still has no completely satisfactory explanation and has aroused considerable controversy[14] Also the large inclusive $B^+ \to K^+\eta'X$ .branching ratio is equally puzzling.

A parity selection rule provides a clear experimental method to distinguish between two proposed explanations with different flavor topologies

1. The OZI-forbidden hairpin diagram[15] predicts a universal parity-independent enhancement for all final states arising from the flavor singlet component of the η' [16,17]..

2. Parity-dependent interference between diagrams producing the η' via its strange and nonstrange components[15] predicts a large η'/η ratio for even parity final states like $K\eta$ and $K\eta'$ the reverse for odd parity states like $K^*(892)\,\eta$ and $K^*\,\eta'$[14]. This selection rule agrees with experiment, although so far the $K^*\,\eta$ has been seen and the $K^*\,\eta'$ has not.

3.2. *Predictions from simple easily-tested assumptions*

3.2.1. *The Flavor-Topology OZI rule and QCD*

Two predictions which challenge QCD if they agree with experimenmt.

$$BR(B^\pm \to K^\pm\omega) = BR(B^\pm \to K^\pm\rho^o) \tag{22}$$

Because the ρ^o and ω mesons both come only from $\bar{u}u$ this prediction requires only exclusion of hairpin diagram topology and holds even in presence of strong final state rescattering via all other quark-gluon diagrams.

$$\tilde{\Gamma}(B^\pm \to K^\pm\phi) = \tilde{\Gamma}(B^\pm \to K^o\rho^\pm) \tag{23}$$

This prediction also assumes the SU(3) flavor symmetry relation between strange and nonstrange pair production

162

3.2.2. *The "inactive spectator" approach*

Many interesting predictions that can be checked experimentally and challenge QCD if they work follow from a simple flavor-topology rule[18]. The spectator quark line must flow continuously from initial to final state, emitting and absorbing gluons freely, but not undergoing annihilation or pair creation. One example arises in B decays to final states containing charmonium. The first prediction forbids all decays without the spectator flavor in the final state :

$$A[B_d \to J/\psi M(\bar{q}s)] = 0 = A[B_s \to J/\psi M(\bar{q}d)] \tag{24}$$

$$A(B_s \to J/\psi \rho^o) = A(B_s \to J/\psi \omega) = A(B_d \to J/\psi \phi) = 0 \tag{25}$$

where $M(\bar{q}s)$ and $M(\bar{q}d)$ denote any $\bar{q}q$ meson with these constituents. If this selection rule holds all other decays described by two amplitudes

$$B(\bar{b}q) \to J/\psi \bar{d}q \to J/\psi M(\bar{d}q) \tag{26}$$

$$B(\bar{b}q) \to J/\psi \bar{s}q \to J/\psi M(\bar{s}q) \tag{27}$$

Decay is product of $\bar{b} \to J/\psi \bar{d}$ or $\to J/\psi \bar{s}$ decay and hadronization function h

$$
\begin{aligned}
A[B_d \to J/\psi M^0(\bar{s}d)] &= A(\bar{b} \to J/\psi \bar{s}) \cdot h[\bar{s}d \to M^0(\bar{s}d)] \\
A[B_s \to J/\psi M^0(\bar{d}s)] &= A(\bar{b} \to J/\psi \bar{d}) \cdot h[\bar{d}s \to M^0(\bar{d}s)] \\
A[B_d \to J/\psi M^0(\bar{d}d)] &= A(\bar{b} \to J/\psi \bar{d}) \cdot h[\bar{d}d \to M^0(\bar{d}d)] \\
A[B_s \to J/\psi M^0(\bar{s}s)] &= A(\bar{b} \to J/\psi \bar{s}) \cdot h[\bar{s}s \to M^0(\bar{s}s)] \\
A[B^+ \to J/\psi M^+(\bar{s}u)] &= A(\bar{b} \to J/\psi \bar{s}) \cdot h[\bar{s}d \to M^+(\bar{s}u)] \\
A[B^+ \to J/\psi M^+(\bar{d}u)] &= A(\bar{b} \to J/\psi \bar{d}) \cdot h[\bar{d}d \to M^+(\bar{d}u)]
\end{aligned} \tag{28}
$$

Decays into charge-conjugate strange final states differ only by weak interaction vertex and kinematic and form factor differences induced by $B_d - B_s$ mass difference. For any partial wave L in a vector-vector final state, we can write

$$A(B_s \to J/\psi \bar{K}^{*0})_L = F^L_{CKM} \cdot A(B_d \to J/\psi K^{*0})_L \tag{29}$$

If the weak transition is $\bar{b} \to \bar{c} + W^+ \to \bar{c} + c + \bar{q}$ as in the dominant tree diagram.

$$F^L_{CKM} = \frac{A_L(\bar{b} \to J/\psi \bar{d})}{A_L(\bar{b} \to J/\psi \bar{s})} \approx V_{cd}/V_{cs} \tag{30}$$

If $F^L_{CKM} \neq V_{cd}/V_{cs}$ other contributions are indicated.

Additional SU(3) assumption gives

$$A_L(B_d \to J/\psi \rho^o) = A_L(B_d \to J/\psi \omega) = \frac{A_L(B_s \to J/\psi \bar{K}^{*0})}{\sqrt{2}} \quad (31)$$

$$A_L(B_s \to J/\psi \phi) = A_L(B_d \to J/\psi K^{*0}) \quad (32)$$

$$A_L(B_s \to J/\psi \rho^o) = A_L(B_s \to J/\psi \omega) = A_L(B_d \to J/\psi \phi) = 0 \quad (33)$$

3.2.3. How to test $\eta - \eta'$ mixing

If the $\eta - \eta'$ system satisfies the standard mixing,

$$|\eta\rangle = |\eta_n\rangle \cos\phi - |\eta_s\rangle \sin\phi; \quad |\eta'\rangle = |\eta_n\rangle \sin\phi + |\eta_s\rangle \cos\phi \quad (34)$$

Then

$$r_d = \frac{p_{\eta'}^3 \Gamma(\bar{B}^0 \to J/\psi\eta)}{p_\eta^3 \Gamma(\bar{B}^0 \to J/\psi\eta')} = \cot^2\phi; \quad r_s = \frac{p_{\eta'}^3 \Gamma(\bar{B}_s \to J/\psi\eta)}{p_\eta^3 \Gamma(\bar{B}_s \to J/\psi\eta')} = \tan^2\phi \quad (35)$$

$$R\eta = \frac{p_{B\eta}^3 \Gamma(\bar{B}^0 \to J/\psi\eta)}{p_{Bs\eta}^3 \Gamma(\bar{B}_s \to J/\psi\eta)} \cdot \frac{p_{BsK}^3 \Gamma(\bar{B}_s \to J/\psi K^0)}{p_{BK}^3 \Gamma(\bar{B}^0 \to J/\psi\bar{K}^0)} = \cot^2\phi \quad (36)$$

$$R'_\eta = \frac{p_{B\eta'}^3 \Gamma(\bar{B}^0 \to J/\psi\eta')}{p_{Bs\eta'}^3 \Gamma(\bar{B}_s \to J/\psi\eta')} \cdot \frac{p_{BsK}^3 \Gamma(\bar{B}_s \to J/\psi K^0)}{p_{BK}^3 \Gamma(\bar{B}^0 \to J/\psi\bar{K}^0)} = \tan^2\phi \quad (37)$$

$$r = \sqrt{r_d r_s} = 1; \quad R_B = \sqrt{R_\eta R'_\eta} = 1 \quad (38)$$

Any large deviation of r or R_B from 1 would indicate evidence of non standard $\eta - \eta'$ mixing[18].

3.2.4. SU(3) Relations between Cabibbo-Favored and Doubly-Cabibbo Suppressed D^o decays

The SU(3) transformation $d \leftrightarrow s$, also called a Weyl reflection or a U-spin reflection relates Cabibbo-favored $\leftrightarrow$ doubly-cabibbo suppressed charm decays

$$d \leftrightarrow s; \quad K^+ \leftrightarrow \pi^+; \quad K^- \leftrightarrow \pi^-; \quad D^+ \leftrightarrow D_s;$$
$$D^o \leftrightarrow D^o; \quad K^+\pi^- \leftrightarrow K^-\pi^+ \quad (39)$$

If strong interaction final state interactions conserve SU(3) the only SU(3) breaking occurs in the CKM matrix elements.

A simple test of this SU(3) symmetry is

$$tan^4\theta_c = \frac{BR(D^o \to K^+\pi^-)}{BR(D^o \to K^-\pi^+)} = \frac{BR[D^o \to K^*(892)^+\rho^-]}{BR[D^o \to K^*(892)^-\rho^+]} \quad (40)$$

These relations involve only branching ratios and are easily tested . They involve no phases and only branching ratios of decay modes all expected to be comparable to the observed DCSD $D^o \to K^+\pi^-$. A similar relation .

$$tan^4\theta_c = \frac{BR[D^o \to K^+a_1(1260)^-]}{BR[D^o \to K^-a_1(1260)^+]} \tag{41}$$

may be subject to a different type of SU(3) breaking. . A weak vector dominance form factor can enhance

$$D^o(c\bar{u}) \to (s\bar{u} \to K^-)_S \cdot (u\bar{d} \to a_1^+)_W \to K^-a_1^+ \tag{42}$$

where the subscripts S and W denote strong and weak form factors.

No such enhancement should occur in

$$D^o(c\bar{u}) \to (d\bar{u} \to a_1^-)_S \cdot (u\bar{s} \to K^+)_W \to a_1^-K^+ \tag{43}$$

If the SU(3) breaking is really due to the difference between products of weak axial and strong kaon form factors and vice versa, the SU(3) relation invlving the a_1 can be expected to be strongly broken and replaced by the inequality

$$\frac{BR[D^o \to K^-a_1(1260)^+]}{BR(D^o \to K^-\pi^+)} \gg \frac{BR[D^o \to K^+a_1(1260)^-]}{BR(D^o \to K^+\pi^-)} \tag{44}$$

3.2.5. *A problem with strong phases*

.

The $d \leftrightarrow s$ interchange SU(3) transformation also predicts[19] $D^o \to K^+\pi^-$ and $D^o \to K^-\pi^+$ have the same strong phases. This has been shown to be in disagreement with experiment[20] showing SU(3) violation.

But the $K^+\pi^-$ and $K^-\pi^+$ final states are charge conjugates of one another and strong interactions conserve charge conjugation. SU(3) can be broken in strong interactions without breaking charge conjugation only in the quark - hadron form factors arising in hadronization transitions like

$$D^o(c\bar{u}) \to (s\bar{u} \to K^-)_S \cdot (u\bar{d} \to a_1^+)_W$$
$$\to K^-a_1^+ \to K^-\pi^+ \tag{45}$$

$$D^o(c\bar{u}) \to (d\bar{u} \to a_1^-)_S \cdot (u\bar{s} \to K^{*+})_W$$
$$\to a_1^-K^+ \to \pi^-K^+ \tag{46}$$

with the SU(3) breaking given by the inequality (44).

The a_1 and π wave functions are very different and not related by SU(3). The $K^\mp a_1^\pm \to K^\mp\pi^\pm$ transition can proceed via ρ exchange

3.2.6. *SU(3) relations between D^+ and D_s decays*

Both of the following ratios of branching ratios

$$\frac{BR(D_s \to K^+ K^+ \pi^-)}{BR(D_s \to K^+ K^- \pi^+)} \approx \frac{BR(D^+ \to K^+ \pi^+ \pi^-)}{BR(D^+ \to K^- \pi^+ \pi^+)} \approx O(tan^4 \theta_c) \qquad (47)$$

are ratios of a doubly Cabibbo forbidden decay to an allowed decay and should be of order $tan^4 \theta_c$. The SU(3) transformation $d \leftrightarrow s$ takes the two ratios (47) into the reciprocals of one another. SU(3) requires the product of these two ratios to be EXACTLY $tan^8 \theta_c$[21].

$$\frac{BR(D_s \to K^+ K^+ \pi^-)}{BR(D_s \to K^+ K^- \pi^+)} \cdot \frac{BR(D^+ \to K^+ \pi^+ \pi^-)}{BR(D^+ \to K^- \pi^+ \pi^+)} = tan^8 \theta_c \qquad (48)$$

Most obvious SU(3)-symmetry-breaking factors cancel out in this product; e.g. phase space. Present data[11] show

$$\frac{BR(D^+ \to K^+ \pi^- \pi^+)}{BR(D^+ \to K^- \pi^+ \pi^+)} \approx 0.65\% \approx 3 \times tan^4 \theta_c \qquad (49)$$

Then SU(3) predicts

$$\frac{BR(D_s \to K^+ K^+ \pi^-)}{BR(D_s \to K^+ K^- \pi^+)} \approx \frac{tan^4 \theta_c}{3} \approx 0.07\%. \qquad (50)$$

If this SU(3) prediction is confirmed experimentally some new dynamical explanation will be needed for the order of magnitude difference between effects of the final-state interactions in D^+ and D_s decays.

If the final state interactions behave similarly in D_s and D^+ decays, the large violation of SU(3) will need some explanation.

New physics enhancing the doubly suppressed decays might produce a CP violation observable as a charge asymmetry in the products of above the two ratios; i.e between the values for D^+ and D_s decays and for D^- and $\bar{D}_s$ decays.

An obvious caveat is the almost trivial SU(3) breaking arising from resonances in the final states. But sufficient data and Dalitz plots should enable including these effects. In any case the SU(3) relation and its possible violations raise interesting questions which deserve further theoretical and experimental investigation. Any really large SU(3)-breaking final state interactions that we don't understand must cast serious doubts on many SU(3) predictions.

Acknowledgments

This work was supported in part by the U.S. Department of Energy, Division of High Energy Physics,

Contract W-31-109-ENG-38.and in part by a grant from the United States-Israel Binational Science Foundation (BSF), Jerusalem, Israel and by the Basic Research Foundation administered by the Israel Academy of Sciences and Humanities

References

1. Harry J. Lipkin, in Proceedings of the 6th Conference on the Intersections of Particle and Nuclear Physics, Big Sky, Montana, May (1997) Edited by T. W. Donnelly, AIP Conference Proceedings no. 412, (1997) p.504
2. E. M. Levin and L. L. Frankfurt, Zh. Eksperim. i. Theor. Fiz.-Pis'ma Redakt (1965) 105; JETP Letters (1965) 65
3. H.J. Lipkin and F. Scheck, Phys. Rev. Lett. 16 (1966) 71
4. Harry J. Lipkin, Physics Letters B335 (1994) 500
5. Ya. B. Zeldovich and A.D. Sakharov, Yad. Fiz 4 (1966)395; Sov. J. Nucl. Phys. 4 (1967) 283
6. I. Cohen and H. J. Lipkin, Phys. Lett. 93B, (1980) 56
7. Harry J. Lipkin, Phys. Lett. B233 (1989) 446; Nuc. Phys. A507 (1990) 205c
8. A. De Rujula, H. Georgi and S.L. Glashow, Phys. Rev. D12 (1975) 147
9. Harry J. Lipkin, Nucl. Phys. A478, (1988) 307c
10. Harry J. Lipkin, In "Hadron 99" Proceedings of the Eighth International Conference on Hadron Spectroscopy, Beijing, China, August 24-38 (1999) Edited by W. G. Li, Y. Z. Huang and B. S. Zou, Nuclear Physics A675 (2000) 443c
11. Particle Data Group, Eur. Phys. J. C 15 (2000) 1
12. Harry J. Lipkin, hep-ph/0011228, Physics Letters B 515 (2001) 81
13. F. E. Close and H. J. Lipkin hep-ph/0208217
14. Harry J. Lipkin, In Proceedings of the 2nd International Conference on B Physics and CP Violation, Honolulu, Hawaii, 24-27 March 1997 Edited by T. E. Browder et al, World Scientific, (1998) p.436 hep-ph/9708253
15. H. J. Lipkin, Phys. Lett. B433, 117 (1998)
16. D. Atwood and A. Soni, Phys. Lett. B405, 150 (1997) [hep-ph/9704357].
17. I. Halperin and A. Zhitnitsky, Phys. Rev. Lett. 80, 438 (1998) [hep-ph/9705251].
18. Alakabaha Datta, Harry J.Lipkin and Patrick J. O'Donnell, hep-ph-0111336, Physics Letters B 529 (2002) 93
19. L. Wolfenstein, Phys. Rev. Lett. 75 (1995) 2460
20. S. Bergmann, Y.Grossman, Z. Ligeti, Y.Nir and A.A.Petrov, hep - ph/0005181
21. Harry J. Lipkin and Zhi-zhong Xing, Phys. Lett. B450 (1999) 405

POSITION SPACE INTERPRETATION FOR GENERALIZED
PARTON DISTRIBUTIONS

M. BURKARDT

Dept. of Physics, New Mexico State University, Las Cruces, NM 88003, USA
E-mail: burkardt@nmsu.edu*

For an unpolarized target, the generalized parton distribution $H_q(x, 0, t)$ is related
to the distribution of partons in impact parameter space. The transverse distortion
of this distribution for a transversely polarized target is described by $E_q(x, 0, t)$.

1. Introduction

Generalized parton distributions (GPDs) [1] have attracted significant interest
since it has been recognized that they can not only be probed in deeply virtual
Compton scattering experiments but can also be related to the orbital angular
momentum carried by quarks in the nucleon [2]. However, remarkably little
is still known about the physical interpretation of GPDs, and one may ask
the question: *suppose, about 10-15 years from now, after a combined effort
from experiment, simulation and theory, we know how GPDs look like for the
nucleon. What will we have learned about the structure of the nucleon?* Of
course, we will have learned something about the orbital angular momentum
carried by the quarks [2], but is that all there is? In these notes, I will discuss
another interesting piece of information that can be extracted from GPDs,
namely *how partons are distributed in the transverse plane.*

In nonrelativistic quantum mechanics, the physics of form factors is illu-
cidated by transforming to the center of mass frame and by interpreting the
Fourier transform of form factors as charge distributions in that frame.

GPDs [1] are the form factors of the same operators [light cone correlators
$\hat{O}_q(x, \mathbf{0}_\perp)$] whose forward matrix elements also yield the usual (forward) parton
distribution functions (PDFs). For example, the unpolarized PDF $q(x)$ can be

---
*this work was supported by the doe (de-fg03-95er40965)

expressed in the form[a]

$$q(x) = \langle p, \lambda | \hat{O}_q(x, \mathbf{0}_\perp) | p, \lambda \rangle, \qquad (1)$$

while the GPDs $H_q(x, \xi, t)$ and $E_q(x, \xi, t)$ are defined as

$$\langle p', \lambda' | \hat{O}_q(x, \mathbf{0}_\perp) | p, \lambda \rangle = \frac{1}{2\bar{p}^+} \bar{u}' \left(\gamma^+ H_q(x, \xi, t) + i \frac{\sigma^{+\nu}\Delta_\nu}{2M} E_q(x, \xi, t) \right) u, \qquad (2)$$

where $\Delta = p' - p$, $2\bar{p} = p + p'$, $t = \Delta^2$, $2\bar{p}^+\xi = \Delta^+$, and

$$\hat{O}_q(x, \mathbf{b}_\perp) = \int \frac{dx^-}{4\pi} \bar{q}\left(-\frac{x^-}{2}, \mathbf{b}_\perp \right) \gamma^+ q\left(\frac{x^-}{2}, \mathbf{b}_\perp \right) e^{ixp^+x^-}. \qquad (3)$$

In the case of form factors, non-forward matrix elements provide information about how the charge (i.e. the forward matrix element) is distributed in position space. By analogy with form factors, one would therefore expect that the additional information (compared to PDFs) contained in GPDs helps to understand how the usual PDFs are distributed in position space [3]. Of course, since the operator $\hat{O}_q(x, \mathbf{0}_\perp)$ already 'filters out' quarks with a definite momentum fraction x, Heisenberg's uncertainty principle does not allow a simultaneous determination of the partons' longitudinal position, but determining the distributions of partons in impact parameter space is conceiveable. Making these intuitive expectation more precise (e.g. what is the 'reference point', 'are there relativistic corrections', 'how does polarization enter', 'is there a strict probability interpretation') will be the main purpose of these notes.

2. Impact parameter dependent PDFs

In nonrelativistic quantum mechanics, the Fourier transform of the form factor yields the charge distribution in the center of mass frame. In general, the concept of a center of mass has no analog in relativistic theories, and therefore the position space interpretation of form factors is frame dependent.

The infinite momentum frame (IMF) plays a distinguished role in the physical interpretation of regular PDFs as momentum distributions in the IMF. It is therefore natural to attempt to interpret GPDs in the IMF. This task is facilitated by the fact that there a is Galilean subgroup of transverse boosts in the IMF, whose generators are defined as

$$B_x \equiv M^{+x} = (K_x + J_y)/\sqrt{2} \qquad\qquad B_y \equiv M^{+y} = (K_y - J_x)/\sqrt{2}, \qquad (4)$$

---

[a]We will suppress the scale (i.e. Q^2) dependence of these matrix elements for notational convenience. In the end, the $\perp$ 'resolution' will be limited by $1/Q$.

where $M_{ij} = \varepsilon_{ijk} J_k$, $M_{i0} = K_i$, and $M^{\mu\nu}$ is the familiar generator of Lorentz transformations. The commutation relations between $\mathbf{B}_\perp$ and other Poincaré generators

$$[J_3, B_k] = i\varepsilon_{kl} B_l \qquad\qquad [P_k, B_l] = -i\delta_{kl} P^+$$
$$[P^-, B_k] = -iP_k \qquad\qquad [P^+, B_k] = 0 \qquad (5)$$

where $k, l \in \{x, y\}$, $\varepsilon_{xy} = -\varepsilon_{yx} = 1$, and $\varepsilon_{xx} = \varepsilon_{yy} = 0$, are formally identical to the commutation relations among boosts/translations for a nonrelativistic system in the plane provided we make the identification [4]

$$\begin{aligned}
\mathbf{P}_\perp &\longrightarrow \text{momentum in the plane} & P^+ &\longrightarrow \text{mass} \\
L_z &\longrightarrow \text{rotations around } z\text{-axis} & P^- &\longrightarrow \text{Hamiltonian} \quad (6) \\
\mathbf{B}_\perp &\longrightarrow \text{generator of boosts in the } \perp \text{ plane.}
\end{aligned}$$

Because of this isomorphism it is possible to transfer a number of results and concepts from nonrelativistic quantum mechanics to the infinite momentum frame. For example, for an eigenstate of P^+, one can define a (transverse) *center of momentum* (CM)

$$\mathbf{R}_\perp \equiv -\frac{\mathbf{B}_\perp}{p^+} = \frac{1}{p^+} \int dx^- \int d^2\mathbf{x}_\perp T^{++} \mathbf{x}_\perp, \qquad (7)$$

where $T^{\mu\nu}$ is the energy momentum tensor. Like its nonrelativistic counterpart, it satisfies $[J_3, R_k] = i\varepsilon_{kl} R_l$ and $[P_k, R_l] = -i\delta_{kl}$. These simple commutation relations enable us to form simultaneous eigenstates of $\mathbf{R}_\perp$ (with eigenvalue $\mathbf{0}_\perp$), P^+ and J_3

$$\left| p^+, \mathbf{R}_\perp = \mathbf{0}_\perp, \lambda \right\rangle \equiv N \int d^2\mathbf{p}_\perp \left| p^+, \mathbf{p}_\perp, \lambda \right\rangle, \qquad (8)$$

where N is some normalization constant, and λ corresponds to the helicity when viewed from a frame with infinite momentum. For details on how these IMF helicity states are defined, as well as for their relation to usual rest frame states, see Ref. [5].

In the following we will use the eigenstates of the $\perp$ center of momentum operator[b] (8) to define the concept of a parton distributions in impact parameter space [c]

$$q(x, \mathbf{b}_\perp) \equiv \left\langle p^+, \mathbf{R}_\perp = \mathbf{0}_\perp, \lambda \right| \hat{O}_q(x, \mathbf{b}_\perp) \left| p^+, \mathbf{R}_\perp = \mathbf{0}_\perp, \lambda \right\rangle. \qquad (9)$$

---

[b]Note that the Galilei invariance in the IMF is crucial for being able to construct a useful CM concept.
[c]In Ref. [6], wave packets were used in order to avoid states that are normalized to δ functions. The final results are unchanged. This was also verified in Ref. [8].

The impact parameter dependent PDFs defined above (9) are the Fourier transform of H_q [6,7,8] (without relativistic corrections!) [d]

$$q(x, \mathbf{b}_\perp) = \frac{|N|^2}{(2\pi)^2} \int d^2\mathbf{p}_\perp \int d^2\mathbf{p}'_\perp \left\langle p^+, \mathbf{0}_\perp, \lambda \left| \hat{O}_q(x, \mathbf{b}_\perp) \right| p^+, \mathbf{0}_\perp, \lambda \right\rangle \quad (10)$$

$$= \frac{|N|^2}{(2\pi)^2} \int d^2\mathbf{p}_\perp \int d^2\mathbf{p}'_\perp H_q(x, -(\mathbf{p}'_\perp - \mathbf{p}_\perp)^2) e^{i\mathbf{b}_\perp \cdot (\mathbf{p}_\perp - \mathbf{p}'_\perp)}$$

$$= \int \frac{d^2\mathbf{\Delta}_\perp}{(2\pi)^2} H_q(x, -\mathbf{\Delta}_\perp^2) e^{-i\mathbf{b}_\perp \cdot \mathbf{\Delta}_\perp}$$

and its normalization is $\int d^2\mathbf{b}_\perp q(x, \mathbf{b}_\perp) = q(x)$. Furthermore, $q(x, \mathbf{b}_\perp)$ has a probabilistic interpretation. Denoting $\tilde{b}_s(k^+, \mathbf{b}_\perp)$ $[\tilde{d}_s(k^+, \mathbf{b}_\perp)]$ the canonical destruction operator for a quark [antiquark] with longitudinal momentum k^+ and $\perp$ position $\mathbf{b}_\perp$, one finds [7]

$$q(x, \mathbf{b}_\perp) = \begin{cases} \sum_s \left| \tilde{b}_s(xp^+, \mathbf{b}_\perp) |p^+, \mathbf{0}_\perp, \lambda\rangle \right|^2 \geq 0 & \text{for } x > 0 \\ -\sum_s \left| \tilde{d}_s^\dagger(xp^+, \mathbf{b}_\perp) |p^+, \mathbf{0}_\perp, \lambda\rangle \right|^2 \leq 0 & \text{for } x < 0 \end{cases} \quad (11)$$

For large x, one expects $q(x, \mathbf{b}_\perp)$ to be not only small in magnitude (since $q(x)$ is small for large x) but also very narrow (localized valence core!). In particular, the $\perp$ width should vanish as $x \to 1$, since $q(x, \mathbf{b}_\perp)$ is defined with the $\perp$ CM as a reference point. A parton representation for $\mathbf{R}_\perp$ (7) is given by $\mathbf{R}_\perp = \sum_{i \in q, g} x_i \mathbf{r}_{i,\perp}$, where x_i ($\mathbf{r}_{i,\perp}$) is the momentum fraction ($\perp$ position) of the ith parton, and for $x = 1$ the position of active quark coincides with the $\perp$ CM.

In order to gain some intuition for the kind of results that one might expect for impact parameter dependent PDFs, we consider a simple model

$$H_q(x, 0, -\mathbf{\Delta}_\perp^2) = q(x) e^{-a\mathbf{\Delta}_\perp^2 (1-x) \ln \frac{1}{x}}. \quad (12)$$

The precise functional form in this ansatz should not be taken too seriously, and the model should only be considered a simple parameterization which is consistent with both Regge behavior at small x and Drell-Yan-West duality at large x. A straightforward Fourier transform yields (Fig. 1)

$$q(x, \mathbf{b}_\perp) = q(x) \frac{1}{4\pi(1-x) \ln \frac{1}{x}} e^{-\frac{\mathbf{b}_\perp^2}{4a(1-x) \ln \frac{1}{x}}}. \quad (13)$$

---

[d] A similar interpretation exists for $\tilde{H}_q(x, 0, t)$ in terms of impact parameter dependent polarized quark distributions $\Delta q(x, \mathbf{b}_\perp) = \int \frac{d^2\mathbf{\Delta}_\perp}{(2\pi)^2} \tilde{H}_q(x, 0, -\mathbf{\Delta}_\perp^2) e^{-i\mathbf{\Delta}_\perp \cdot \mathbf{b}_\perp}$.

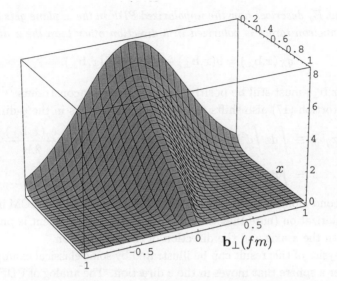

Figure 1. Impact parameter dependent parton distribution $q(x, \mathbf{b}_\perp)$ for the model (13).

3. Position Space Interpretation for $E(x, 0, -\mathbf{\Delta}_\perp^2)$

While both $H(x, 0, t)$ and $\tilde{H}(x, 0, t)$ are diagonal in helicity, $E(x, 0, t)$ contributes only to helicity flip matrix elements. In fact for $p^+ = p^{+\prime}$ [9] [10]

$$\int \frac{dx^-}{4\pi} e^{ip^+x^- x} \langle P{+}\Delta, \uparrow | \bar{q}(0) \, \gamma^+ q(x^-) | P, \uparrow \rangle = H(x, 0, -\mathbf{\Delta}_\perp^2) \qquad (14)$$

$$\int \frac{dx^-}{4\pi} e^{ip^+x^- x} \langle P{+}\Delta, \uparrow | \bar{q}(0) \, \gamma^+ q(x^-) | P, \downarrow \rangle = -\frac{\Delta_x - i\Delta_y}{2M} E(x, 0, -\mathbf{\Delta}_\perp^2).$$

Since E is off diagonal in helicity, it will therefore only have a nonzero expectation value in states that are *not* eigenstates of helicity, i.e. if we search for a probabilistic interpretation for $E(x, 0, t)$ we need to look for it in states that are superpositions of helicity eigenstates. For this purpose, we consider the state $|X\rangle \equiv |p^+, \mathbf{R}_\perp = 0, X\rangle \equiv (|p^+, \mathbf{R}_\perp = 0, \uparrow\rangle + |p^+, \mathbf{R}_\perp = 0, \downarrow\rangle)/\sqrt{2}$. In this state, we find for the (unpolarized) impact parameter dependent PDF

$$q_X(x, \mathbf{b}_\perp) \equiv \langle X | O_q(x, \mathbf{b}_\perp) | X \rangle \qquad (15)$$

$$= \int \frac{d^2\mathbf{\Delta}_\perp}{(2\pi)^2} \left[H_q(x, 0, -\mathbf{\Delta}_\perp^2) + \frac{i\Delta_y}{2M} E_q(x, 0, -\mathbf{\Delta}_\perp^2) \right] e^{-i\mathbf{b}_\perp \cdot \mathbf{\Delta}_\perp}$$

Upon introducing the Fourier transform of E_q

$$e_q(x, \mathbf{b}_\perp) \equiv \int \frac{d^2\mathbf{\Delta}_\perp}{(2\pi)^2} E_q(x, 0, -\mathbf{\Delta}_\perp^2) e^{-i\mathbf{b}_\perp \cdot \mathbf{\Delta}_\perp} \qquad (16)$$

172

we find that E_q describes how the unpolarized PDF in the $\perp$ plane gets distorted when the nucleon target is polarized in a direction other than the z direction

$$q_X(x,\mathbf{b}_\perp) = q(x,\mathbf{b}_\perp) - \frac{1}{2M}\frac{\partial}{\partial b_y}e_q(x,\mathbf{b}_\perp). \tag{17}$$

Since $q_X(x,\mathbf{b}_\perp)$ must still be positive further positivity constraints [7,11] follow. Above distortion (17) also shifts the $\perp$CM of the partons in the y-direction

$$\langle x_q b_y^q \rangle \equiv \int dx \int d^2\mathbf{b}_\perp x b_y^q q_X(x,\mathbf{b}_\perp) = \int dx \int d^2\mathbf{b}_\perp x \frac{e_q(x,\mathbf{b}_\perp)}{2M}$$

$$= \int dx\, x \frac{E_q(x,0,0)}{2M}, \tag{18}$$

i.e. the second moment of E_q describes the displacement of the $\perp$CM in a state with $\perp$ polarization (note that the direction of the displacement is perpendicular to both the z axis and the direction of the polarization.

The physics of this result can be illustrated by some classical example (Fig. 2) Consider a sphere that moves in the z direction. The analog of PDFs in this example are the distributions of momenta in the z direction from particles on the sphere as seen by an observer at rest. The distribution of these momenta in the transverse plane is the same regardless whether the sphere is nonrotating or spinning around the z axis (in either direction) because spinning around the z axis does not modify the momenta in the z direction.[e] However, if the sphere spins around the x-axis while still moving along the z-axis, the longitudinal momentum distribution of particles on the sphere as seen by an observer at rest changes dependent on the transverse position of the observer since on one side of the z axis the rotational motion adds to the translatory motion and on the other side it subtracts.

Of course, the nucleon is not just a rotating sphere but this simple example illustrate the physics of why the parton distribution in the transverse plane gets distorted and is no longer invariant under rotations around the z-axis when the nucleon is not longitudinally polarized. And in the nucleon it is the Fourier transform of $E(x,0,-\mathbf{\Delta}_\perp^2)$, which describes this distortion.

Of course, the net (summed over all quark flavors plus glue) displacement of the $\perp$CM vanishes $\sum_{i\in q,g}\langle x_i b_y^i \rangle = 0$ [12]. The resulting $\perp$ dipole moment is given by

$$d_q^y \equiv \int dx \int d^2\mathbf{b}_\perp b_y q_X(x,\mathbf{b}_\perp) = \int dx \int d^2\mathbf{b}_\perp \frac{e_q(x,\mathbf{b}_\perp)}{2M} = \int dx \frac{E_q(x,0,0)}{2M}$$

$$= \frac{\kappa_q(0)}{2M}. \tag{19}$$

---

[e]One may also take this example as a simple illustration as to why the (unpolarized) impact parameter dependent PDFs are the same for nucleons with $\lambda = \uparrow$ and $\lambda = \downarrow$.

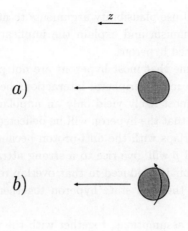

Figure 2. Comparison of a non-rotating sphere that moves in the z direction with a sphere that spins at the same time around the z axis and a sphere that spins around the x axis When the sphere spins around the x axis, the rotation changes the distribution of momenta in the z direction (adds/subtracts to velocity for $y > 0$ and $y < 0$ respectively). For the nucleon the resulting modification of the (unpolarized) momentum distribution is described by $E(x, 0, t)$.

κ_q is the contribution from flavor q to the anomalous Dirac moment $F_2(0)$. In order to get some feeling for the order of magnitude, we consider a very simple model where only $q = u, d$ contribute to $F_2(0)$, one finds for example $\kappa_d \approx -2$ and therefore a mean displacement of d quarks of by about $0.2 fm$. For u quarks the effect is about half as large and in the opposite direction. Again one can use the semiclassical example (Fig. 2) to uderstand why the distortion may be in different directions for different flavors since for different flavor the orbital angular momentum may be parallel or anti-parallel to the nucleon spin.

3.1. Implications for Transverse Hyperon Polarization

The transverse distortion of impact parameter dependent PDFs for states with transverse polarization should have many interesting implications for the production of hadrons with transverse polarization as one can see from the the reactions $P + P \longrightarrow Y + X$ or $P + \bar{P} \longrightarrow Y + X$ with $Y \in \{\Lambda, \Sigma, \Xi\}$ at high energies. We will use plausibility arguments to motivate a rather simple general reaction mechanism and explain the implications for the transverse polarization of produced hyperons.

First we will assume that most hyperons are not produced in central collisions, but instead in more peripheral interactions because at high energy a central collision will most likely yield only an unpolarized background. Secondly, we will assume that the hyperon will be deflected in the direction where the initial proton overlaps with the anti-proton because most reaction mechanisms between p and $\bar{p}$ will give rise to a strong attraction. Finally, we will assume that the $s\bar{s}$ pair is produced in that overlap region and that the produced s quarks form the final state hyperon together with the 'rest' of the initial proton.

These very simple assumptions, together with the transverse distortion of the strange quark cloud in a transversely polarized hyperon will favor hyperons with a specific transverse polarization: Consider first the case where the hyperon has been deflected to the left (Fig. 3), where we look in the direction of the outgoing hyperon. Based on our model assumptions above, the s quark has been produced on the left side of the hyperon, which we assume has a positive strange anomalous magnetic moment κ_s^Y (the case $\kappa_s^Y < 0$ yields the opposite effect). If the final state hyperon is polarized 'up' (w.r.t. the reaction plane) then the s quark distribution in the hyperon is distorted to the right, i.e. away from the reaction zone, while it is distorted to the left (towards the reaction zone) if the polarization is down (Fig. 4). Clearly the second possibility yields a better overlap between the intermediate state and the final state and we would therefore expect a polarization 'down' in this case. The polarization is reversed when the hyperon is deflected to the right (because then the reaction zone is on its right side) and it is also reversed when the sign of κ_s^Y is reversed. These simple considerations lead to the prediction that the polarization direction in this reaction is determined by

$$\vec{P}_Y \sim -\kappa_s^Y \, \vec{p}_P \times \vec{p}_Y. \tag{20}$$

Before we can compare this remarkably simple prediction to experimental data, we need to determine κ_s^Y for various hyperons. Using $SU(3)$ flavor symmetry, one finds

$$\kappa_s^\Lambda = \kappa^p + \kappa_s^p = 1.79 + \kappa_s^p$$
$$\kappa_s^\Sigma = \kappa^p + 2\kappa^n + \kappa_s^p = -2.03 + \kappa_s^p$$
$$\kappa_s^\Xi = 2\kappa^p + \kappa^n + \kappa_s^p = 1.67 + \kappa_s^p. \tag{21}$$

Although the exact value for the strange magnetic moment of the nucleon is not known, it is unlikely to be on the same order as κ^p or κ^n, i.e. Eq. (21) tells

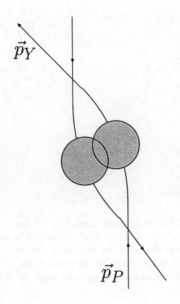

Figure 3. $P + P(\bar{P}) \longrightarrow Y + \bar{Y}$ where the incoming P (from bottom) is deflected to the left during the reaction. The $s\bar{s}$ pair is assumed to be produced roughly in the overlap region, i.e. on the left 'side' of the Y.

us that the strange anomalous magnetic moment of the Λ and the Ξ hyperon is positive, while it is negative for the Σ, yielding for the polarizations

$$\vec{P}_\Lambda \sim -\vec{p}_P \times \vec{p}_Y$$
$$\vec{P}_\Sigma \sim +\vec{p}_P \times \vec{p}_Y$$
$$\vec{P}_\Xi \sim -\vec{p}_P \times \vec{p}_Y \qquad (22)$$

which agrees with the experimentally observed pattern of polarizations (P w.r.t. $\vec{p}_P \times \vec{P}_Y$)

$$0 < P_{\Sigma^0} \approx P_{\Sigma^-} \approx P_{\Sigma^+} \approx -P_\Lambda \approx -P_{\Xi^0} \approx -P_{\Xi^-} \qquad (23)$$

in hyperon production reactions.[f]

Although we illustrated the effect here in the example of $P + P(\bar{P}) \longrightarrow Y + X$, it should be clear that the effect should also apply to many other hyperon (as well as other hadron) production reactions, but a complete discussion of

---

[f]For a recent discussion of polarization in such reactions, see Ref. [13] and references therein.

176

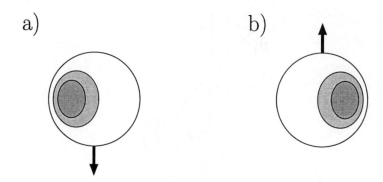

Figure 4. Schematic view of the transverse distortion of the s quark distribution (in grayscale) in the transverse plane for a transversely polarized hyperon with $\kappa_s^Y > 0$. The view is (from the rest frame) into the direction of motion (i.e. momentum into plane) for a hyperon that moves with a large momentum. In the case of spin down (a), the s-quarks get distorted towards the left, while the distortion is to the right for the case of spin up (b).

this subject would go beyond the scope of this section which should serve only as a simple illustration of the striking consequences that the transverse distortion of the parton distribution in transversely polarized may have.

References

1. D. Müller et al., Fortschr. Phys. **42**, 101 (1994); X. Ji, J. Phys. G24 (1998) 1181; A.V. Radyushkin, Phys. Rev. D56 (1997) 5524; K. Goeke et al., Prog. Part. Nucl. Phys. 47 (2001) 401.
2. X. Ji, Phys. Rev. Lett. **78** (1997) 610.
3. J.P. Ralston and B. Pire, hep-ph/0110075; J.P. Ralston, B. Pire and R.V. Buniy, hep-ph/0206074.
4. J. Kogut and D.E. Soper, Phys. Rev. D **1** (1970) 2901.
5. D.E. Soper, Phys. Rev. D 5 (1972) 1956.
   Phys. Lett. B **460**, 204 (1999); Eur. Phys. J. **C8**, 409 (1999); M. Vanderhaeghen, Eur. Phys. J. **A8**, 455 (2000); R. Jakob, hep-ph/0010158; P.Kroll, hep-ph/0011280;
6. M. Burkardt, Phys. Rev. D **62** (2000) 071503.
7. M. Burkardt, proceedings of the workshop on *Lepton Scattering, Hadrons and QCD*, Eds. W.Melnitchouk et al., Adelaide, March 2001; hep-ph/0105324; M. Burkardt, *submitted to Int. Journ. Mod. Phys. A*; hep-ph/0207047.
8. M. Diehl, hep-ph/0205208.
9. M. Diehl, Eur. Phys. J. C19, 485 (2001); P.V. Pobylitsa, hep-ph/0201030.
10. M. Diehl *et al.*, Nucl. Phys. B **596**, 33 (2001).
11. P.V. Pobylitsa, hep-ph/0204337.
12. S.J. Brodsky et al., Nucl. Phys. B **593**, 311 (2001).
13. S.M. Troshin and N.E. Tyurin, hep-ph/0201267.

RENORMALONS AS DILATATION MODES IN THE FUNCTIONAL SPACE

A. BABANSKY AND I. BALITSKY

Physics Department, Old Dominion University, Norfolk VA 23529
and Theory Group, Jefferson Lab, Newport News VA 23606
E-mail: babansky@jlab.org, balitsky@jlab.org

There are two sources of the factorial large-order behavior of a typical perturbative series. First, the number of the different Feynman diagrams may be large; second, there are abnormally large diagrams known as renormalons. It is well known that the large combinatorial number of diagrams is described by instanton-type solutions of the classical equations. We demonstrate that from the functional-integral viewpoint the renormalons do not correspond to a particular configuration but manifest themselves as dilatation modes in the functional space.

It is well known that the perturbative series in a typical quantum field theory is at best asymptotic: the coefficients in front of a typical perturbative expansion grow like $n!$ where n is the order of the perturbation series. There are two sources of the $n!$ behavior which correspond to two different situations. In the first case all Feynman diagrams are ~ 1 but their number is large ($\sim n!$) [1] (for a review, see e.g. [2]). In the second case, we have just one diagram but it is abnormally big $- \sim n!$ (the famous 't Hooft renormalon [3]). The first type of factorial behavior is not specific to a field theory; for example, it can be studied in the anharmonic-oscillator quantum mechanics. On the contrary, renormalon singularities can occur only in field theories with running coupling constant.

It is convenient to visualize the large-order behavior of perturbative series using the 't Hooft picture of singularities[3]. Consider Adler's function related to the polarization operator in (Euclidean) QCD

$$D(q^2) = 4\pi^2 q^2 \frac{d}{dq^2} \frac{1}{3q^2} \Pi(q^2), \tag{1}$$

$$\Pi(q^2) = \int dx e^{iqx} \int D\bar{\psi} D\psi DA j_\mu(x) j_\mu(0) e^{-S_{\text{QCD}}}$$

Suppose we write down $D(q^2)$ as a Borel integral

$$D(\alpha_s(q)) = \int_0^\infty dt D(t) e^{-\frac{4\pi}{\alpha_s(q)}t}. \tag{2}$$

The divergent behavior of the original series $D(\alpha_s(q))$ is encoded in the singularities of its Borel transform shown in Fig.1. The ultraviolet (UV)

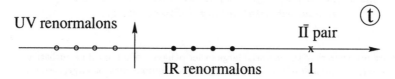

Figure 1.   't Hooft picture of singularities in the Borel plane for QCD.

renormalons located at $t = -\frac{1}{b}, \frac{2}{b}, \frac{3}{b}...$ ($b = 11 - \frac{2}{3}n_f$) come from the regions of hard momenta in Feynman diagrams; the infrared (IR) renormalons placed at $t = \frac{2}{b}, \frac{3}{b}...$ come from the region of soft momenta (for a review of the renormalons, see ref.[4] and references therein). The instanton singularities are located at $t = 1, 2, 3...$ and they correspond to the large number of graphs. (Actually, the first topologically trivial classical configuration, which contributes to the divergence of perturbation theory, is not an instanton itself but a weakly coupled instanton-antiinstanton pair [5]). The main result of this paper is that unlike the instanton-type singularities, the renormalons do not correspond to a particular configuration but manifest themselves as dilatation modes in the functional space.

The interpretation of renormalons as dilatation modes is based upon the similarity of the functional integral in the vicinity of a valley in the functional space[6] to the Borel representation (2). At first we will consider the quantum mechanical example without renormalons and then demonstrate that in a field theory the same integral along the valley leads to renormalon singularities. With QCD in view, we consider the double-well anharmonic oscillator described by the functional integral

$$\int D\phi e^{-\frac{S(\phi)}{g^2}}, \quad S(\phi) = \frac{1}{2} \int dt \left(\dot{\phi}^2 + \phi^2(1-\phi)^2\right) \tag{3}$$

The large-order behavior in this model is determined by the instanton-antiinstanton $(I\bar{I})$ configuration. The $I\bar{I}$ valley for the double-well system may be chosen as

$$f_\alpha(t - \tau) = \frac{1}{2} \tanh \frac{t - \tau + \alpha}{2} - \frac{1}{2} \tanh \frac{t - \tau - \alpha}{2} \tag{4}$$

It satisfies the valley equation[67]

$$\frac{12}{\xi^2} w_\alpha(t) f'_\alpha(t) = L_\alpha(t) \tag{5}$$

where $\xi \equiv e^\alpha$, $f'_\alpha \equiv \frac{\partial}{\partial \alpha} f_\alpha$, and $L_\alpha(t) = \frac{\delta S}{\delta \phi}\Big|_{\phi = f_\alpha(t)}$. Here $w_\alpha(t) = \frac{\xi}{4} \sinh \alpha (\cosh \alpha \cosh t + 1)^{-1}$ is the measure in the functional space so $(f, g) \equiv \int dt w_\alpha(t) f(t) g(t)$. The valley (4) connects two classical solutions: the perturbative vacuum at $\alpha = 0$ and the infinitely separated $I\bar{I}$ pair at $\alpha \to \infty$. The $I\bar{I}$ separation α and the position of the $I\bar{I}$ pair τ are the two collective coordinates of the valley. In order to integrate over the small fluctuations near (4) we insert two δ-functions $\delta(\phi(t) - f_\alpha(t - \tau), \dot{f}_\alpha(t - \tau))$ and $\delta(\phi(t) - f_\alpha(t - \tau), f'_\alpha(t - \tau))$ restricting the integrations along the two collective coordinates, make a shift $\phi(t) \to \phi(t) + f_\alpha(t - \tau)$, expand in quantum deviations $\phi(t)$ and perform the (gaussian) integration in the first nontrivial order in perturbation theory. After the shift, the the linear term in the exponent $\int dt \phi(t) L_\alpha(t)$ is disabled due to the valley equation (5) so the functional integral (3) for the vacuum energy reduces to

$$T \int d\alpha \int D\phi (f_\alpha, L_\alpha)(\dot{f}_\alpha, \dot{f}_\alpha)$$
$$\delta(\phi, f'_\alpha)\delta(\phi, \dot{f}_\alpha) e^{-\frac{1}{g^2}\left[S_\alpha + \frac{1}{2}\int dt \phi(t) \Box_\alpha \phi(t)\right]} + O(g^2). \tag{6}$$

Here T (the total "volume" in one space-time dimension) is the result of trivial integration over τ, $\Box_\alpha = -\partial^2 + 1 - 6 f_\alpha(1 - f_\alpha)$ is the operator of second derivative of the action and

$$S_\alpha \equiv S(\xi) = \frac{6\xi^4 - 14\xi^2}{(\xi^2 - 1)^2} - \frac{17}{3} + \frac{12\xi^2 + 4}{(\xi^2 - 1)^3} \ln \xi \tag{7}$$

is the action of $I\bar{I}$ valley. Performing Gaussian integrations one gets

$$T \frac{1}{g^2} \int_0^\infty d\alpha \, e^{-\frac{1}{g^2} S_\alpha} F(\alpha), \tag{8}$$

$$F(\alpha) = \frac{(\det \Box_\alpha)^{-1/2} (f'_\alpha, f'_\alpha)(\dot{f}_\alpha, \dot{f}_\alpha)}{\left(f'_\alpha, \Box_\alpha^{-1} w_\alpha f'_\alpha\right)^{1/2} \left(\dot{f}_\alpha, \Box_\alpha^{-1} w_\alpha \dot{f}_\alpha\right)^{1/2}}. \tag{9}$$

At $\alpha \to \infty$ we have the widely separated I and $\bar{I}$. In this case, the determinant of the $I\bar{I}$ configuration factorizes into a product of two one-instanton determinants (with zero modes excluded) so $F(\alpha) \to$ const at $\alpha \to \infty$. The divergent part of the integral (8) corresponds to the second iteration of the one-instanton contribution to the vacuum energy and therefore it must be subtracted from the $I\bar{I}$ contribution to E_{vac}:

$$g^2 E_{vac}(g^2) = \int d\alpha \left(e^{-\frac{1}{g^2}S_\alpha} F(\alpha) - e^{-\frac{1}{g^2}2S_I} F(\infty) \right) \tag{10}$$

where S_I is the one-instanton action. Since S_α is a monotonous function of α one can invert Eq. (7) and obtain

$$g^2 E_{vac}(g^2) = \int_0^{2S_I} dS e^{-\frac{1}{g^2}S} F(S) \left(\frac{1}{S - 2S_I} \right)_+ \tag{11}$$

which has the desired form of the Borel integral (1) for the vacuum energy. Thus the leading singularity for $E_{vac}(S)$ is $F(2S_I)/(S - 2S_I)$. Note that our semiclassical calculation does not give the whole answer for the Borel transform of vacuum energy $(F(S) \neq E_{vac}(S)$ in general) since we threw away the higher quantum fluctuations around the $I\bar{I}$ pair in Eq. (6) yet it determines the leading singularity in the Borel plane.

The situation with the instanton-induced asymptotics of perturbative series in a field theory such as QCD is pretty much similar with one notable exception: in QCD there is an additional dimensional parameter ρ – the overall size of the $I\bar{I}$ configuration. The classical $I\bar{I}$ action does not depend on this parameter but the quantum determinant does, leading to the replacement

$$e^{-\frac{1}{g^2}S} \to e^{-\frac{1}{g^2(\rho)}S} \tag{12}$$

so the Borel integrand have the following generic form

$$F(S) \sim \int d\rho e^{-\frac{1}{g^2(\rho)}S} \bar{F}(\rho). \tag{13}$$

The divergence of this integral at either large or small ρ determines the position of the renormalon-type singularities of $F(S)$. We'll demonstrate (by purely dimensional analysis) that $\bar{F}(\rho) \sim \rho^{-5}$ at $\rho \to \infty$ and $\bar{F}(\rho) \sim \rho$ at $\rho \to \infty$ leading to the IR renormalon at $S = \frac{32\pi^2}{b}$ and the UV renormalon at $S = -\frac{16\pi^2}{b}$, respectively.

The $I\bar{I}$ valley in QCD[7] can be chosen as a conformal transformation of the spherical configuration

$$A_\mu^s(x) = -i \left(\sigma_\mu \bar{x} - x_\mu \right) x^{-2} f_\alpha(t) \tag{14}$$

with $t = \ln x^2/d^2$ where d is an arbitrary scale. (We use the notations $x \equiv x_\mu \sigma_\mu$, $\bar{x} \equiv x_\mu \bar{\sigma}_\mu$ where $\sigma_\mu = (1, -i\vec{\sigma})$, $\bar{\sigma}_\mu = (1, i\vec{\sigma})$). To obtain the $I\bar{I}$ configuration with arbitrary sizes ρ_1, ρ_2 and separation R one performs shift $x \to x - a$, inversion $x \to \frac{d^2}{x^2}x$ and second shift $x \to x - x_0$. The resulting valley $A^v_\mu(x - x_0)$ is the sum of the I with size ρ_1 and $\bar{I}$ with size ρ_2 in the singular gauge in the maximum attractive orientation plus the additional term which is small at large $I\bar{I}$ separations (the explicit expression can be found in [8]). The action of the $I\bar{I}$ valley is equal to the action of the spherical configuration (14) which is proportional to (7):

$$S^v(z) = 48\pi^2 S(\xi), \quad \xi = z + \sqrt{z^2 - 1} \tag{15}$$

where the "conformal parameter" z is given by

$$z = (\rho_1^2 + \rho_2^2 + R^2)/(2\rho_1\rho_2). \tag{16}$$

Let us find now the polarization operator (1) in the valley background. The collective coordinates are the sizes of instantons ρ_i, separation R, overall position x_0 and the orientation in the color space (the valley of a general color orientation has the form $\bar{O}_{ab}A^v_b$ where $\bar{O}$ is an arbitrary $SU(3)$ matrix). The structure of Gaussian result for the polarization operator is

$$\int_0^\infty \frac{d\rho_1 d\rho_2}{\rho_1^5 \rho_2^5} d^4R d^4x_0 d\bar{O}\Pi^v(q)\frac{1}{g^{17}}e^{-\frac{S^v(z)}{g^2}}\Delta(\rho_i, R) \tag{17}$$

where 17 is a number of the collective coordinates. Here $\Pi^v(q)$ is a Fourier transform of $\Pi^v(x) = (\sum e_q^2)\mathrm{Tr}\gamma_\mu G(x,0)\gamma_\mu G(0,x)$ where $G(x,y)$ is the Green function in the valley background. The factor $\Delta(\rho_i, R)$ in Eq. (17) is the quantum determinant - the result of Gaussian integrations near the $I\bar{I}$ valley (cf. eq. (9)). For our purposes, it is convenient to introduce the conformal parameter z and the average size $\rho = \sqrt{\rho_1\rho_2}$ as the collective coordinates in place of ρ_1 and ρ_2. We have

$$\int dz \frac{d\rho}{\rho^9} d^4R d^4x_0 \Pi^v(q)\frac{1}{g^{17}(\rho)}F(z, R^2/\rho^2)e^{-\frac{S^v(z)}{g^2(\rho)}} \tag{18}$$

where $F(z, R^2/\rho^2)$ includes $\theta(z - 1 - \frac{R^2}{2\rho^2})$ (see Eq. (16)). We have included in F the trivial integral over color orientation which gives the volume of $SU(3)$ group.

The main effect of the quantum determinant Δ is the replacement of the bare coupling constant g^2 in Eq. (17) by the effective coupling constant $g^2(\rho)$ in Eq. (18) so the remaining function F is the (dimensionless) function of the ratio R^2/ρ^2 and the conformal parameter. This is almost evident from the renormalizability of the theory since the only dimensional parameters are ρ and R. Formally, one can prove that rescaling of the configuration by factor λ (so that $\rho \to \lambda\rho$ and $R \to \lambda R$) leads to multiplication of the determinant by factor $\lambda^{bS^v(z)/8\pi^2}$ due to conformal anomaly (see e.g. the book [9]).

Consider the singularities of the integral (18). The function $F(z, R^2/\rho^2)$ is non-singular since the singularity in Φ ($\equiv$ singularity in Δ) would mean a non-existing zero mode in quantum determinant. Moreover, the integration over R is finite due to $\theta(z - 1 - \frac{R^2}{2\rho^2})$ which means that the only source of singularity at finite z is the divergence of the ρ integral at either large or small ρ. (At $z \to \infty$ in a way similar to the derivation of Eq. (11) we obtain the first instanton-type singularity located at $t = 1^{10}$).

Let us demonstrate that the singularity of the integral (18) at large $\rho \gg \frac{1}{q}$ corresponds to the IR renormalon. The polarization operator $\Pi(x)$ in the background of the large-scale vacuum fluctuation reduces to[11]

$$\langle \Pi(x) \rangle_A \to -\frac{\sum e_q^2}{64\pi^4} \left(\frac{G^2(0)}{x^2} + c\alpha_s G^3(0) \ln x^2 + ... \right) \tag{19}$$

where $G^2 \equiv 2\text{Tr}G_{\xi\eta}G_{\xi\eta}$, $G^3 \equiv 2\text{Tr}G_{\xi\eta}G_{\eta\sigma}G_{\sigma\xi}$, and c is an (unknown) constant. (The coefficient in front of G^3 vanishes at the tree level[12]). Consider the leading term in this expansion; since the field strength for the $I\bar{I}$ valley configuration depends only on $x - x_0$

$$\int d^4x_0 \text{Tr}G_{\xi\eta}^v(0)G_{\xi\eta}^v(0) = 4S^v(z) \tag{20}$$

the intergal (18) reduces to

$$\frac{1}{q^2} \int_1^\infty dz \int_{1/q}^\infty \frac{d\rho}{\rho^9} d^4R g^{-17}(\rho) e^{-\frac{S^v(z)}{g^2(\rho)}} F(z, \frac{R^2}{\rho^2}) \tag{21}$$

where we have included the factor $\frac{1}{36} \sum e_q^2 S^v(z)$ in F. The (finite) integration over R can be performed resulting in an additional dimensional factor ρ^4: $\int d^4R F(z, R^2/\rho^2) = \rho^4 \Phi(z)$ where the function Φ is dimensionless so it can depend only on z. We get

$$\frac{1}{q^2} \int_1^\infty dz \int_{1/q}^\infty \frac{d\rho}{\rho^5} g^{-17}(\rho) e^{-\frac{1}{g^2(\rho)}S^v(z)} \Phi(z) \tag{22}$$

Inverting Eq. (7) we can write the corresponding contribution to Adler's function as an integral over the valley action ($t \equiv \frac{S}{16\pi^2}$):

$$D(q^2) \simeq \frac{1}{3q^4} \int_0^1 dt \int_{1/q}^\infty \frac{d\rho}{\rho^5} g^{-17}(\rho) e^{-\frac{4\pi}{\alpha_s(\rho)}t} \Phi(t)$$

$$= \frac{1}{q^4} \int_0^1 dt \int_{1/q}^\infty \frac{d\rho}{\rho^5} e^{-\frac{4\pi}{\alpha_s(\rho)}t} \Psi(t) \tag{23}$$

where $\Psi(t) = \frac{1}{3\sqrt{\pi}(4\pi)^{17}} \int_0^t dt'(t-t')^{-1/2}\Phi^{(9)}(t')$ after nine integrations by parts and a Laplace transformation (we neglect terms $\sim e^{-\frac{4\pi}{\alpha_s}}$ corresponding to the $I\bar{I}$ singularity). At the one-loop level we obtain

$$D(t) \simeq \Psi(t) \int_{1/q}^{\infty} \frac{d\rho}{q^4\rho^5}(q^2\rho^2)^{bt} = \frac{1}{2-bt} \tag{24}$$

which is the first IR renormalon. Strictly speaking, we should integrate over ρ only up to $\rho < \Lambda_{QCD}^{-1}$ since at $\rho \sim \Lambda_{QCD}^{-1}$ our valley is melted within large-scale fluctiuations populating the QCD vacuum (cf. ref. [13]). At the one-loop level this leads to the replacement $(2-bt)^{-1} \to (2-bt)^{-1}(1-(\Lambda^2/q^2)^{2-bt})$ so we get

$$D(q^2) = D^{\mathrm{pert}}(q^2) + D^{\mathrm{non-pert}}(q^2) \simeq \int_0^1 dt\frac{f(t)}{2-bt}\left[e^{-\frac{4\pi}{\alpha_s(q)}t} - \frac{\Lambda^4}{q^4}\right]. \tag{25}$$

Both D^{pert} and $D^{\mathrm{non-pert}}$ have the singularity $1/(2-bt)$ which cancels in their sum. If we adopt the principal value prescription for integral over t, the non-perturbative part is a real mumber (divided by q^4) which contributes to the phenomenological power correction $\frac{\pi}{3q^4}\alpha_s\langle G_{\mu\nu}^2\rangle$ (cf. ref.[14]).

At the two-loop level we get

$$D^{\mathrm{pert}}(t) \simeq \Psi(t) \int_{1/q}^{\infty} \frac{d\rho}{q^4\rho^5}(q^2\rho^2)^{bt}\left(\alpha_s(\rho)/\alpha_s(q)\right)^{\frac{2b't}{b}} \tag{26}$$

where $b' = 51 - \frac{19}{3}n_f$. Using integration by parts it is easy do demonstrate that an extra $\alpha_s(\rho)$ does not change the singularity (cf. Eq. (23) while an extra $\alpha_s(q)$ shifts it by one power of $2-bt$. Thus, at the two-loop level the first IR renormalon is a branching point singularity $(t-\frac{2}{b})^{-1-4\frac{b'}{b^2}}$[14]. The second term in the expansion (19) gives the second renormalon singularity located at $t = \frac{3}{b}$ and higher terms of the expansion of the polarization operator (19) will give the subsequent renormalons at $t = \frac{4}{b}, \frac{5}{b}$, etc.

Next we demonstrate that the divergence of the integral (18) at small ρ leads to the UV renormalon. In order to find the polarization operator in the valley background in this case we recall that a very small valley is an inversion of the very large spherical configuration (14). For a very large configuration we can use the formula (19) (in the coordinate space) and obtain

$$\Pi^v(x) \to \frac{\sum e_q^2}{192\pi^4}\left[\frac{d^8G_s^2(a)}{y^4x_0^4x^2}\left\{1 - \frac{4(x_0y)^2}{x_0^2y^2}\right\}\right.$$

$$\left. +c\alpha_s\frac{d^{12}G_s^3(a)}{y^6x_0^6}\left\{\left(3\ln\frac{x^2}{y^2x_0^2}-1\right)\frac{2(x_0y)^2}{x_0^2y^2}+1\right\}\right] \tag{27}$$

where $y \equiv x - x_0$. Here $G^s_{\mu\nu}(a)$ is the field strength of the spherical configuration (14) calculated at at $x = a$ ($\Leftrightarrow$ field strength at the origin of the configuration with center a). The integration over x and x_0 yields

$$\int dx dx_0 e^{iqx} \Pi^v_{\mu\nu}(x) = c' \alpha_s(q) d^{12} G^3_s(a) q^4 \ln^2 q^2 \qquad (28)$$

Note that the first $\sim G^2_s$ term in Eq. (27) vanishes. Rewriting d and a in terms of z, ρ, and R we obtain

$$d^{12} G^3_s(a) = \rho^6 \bar{G}(z, R^2/\rho^2) \qquad (29)$$

where the dimensionless function $\bar{G}$ is non-singular (the explicit expressions can be found in [8]). Performing the integration over R we obtain the analog of the Eq. (23) for the UV renormalon

$$D(q^2) \simeq q^2 \alpha_s(q) \int_0^1 dt \int_0^{1/q} d\rho \rho \ln^2 q^2 \rho^2 e^{-\frac{4\pi}{\alpha_s(\rho)} t} \tilde{\Psi}(t) \qquad (30)$$

which corresponds to

$$D(t) \simeq \tilde{\Psi}(t) q^2 \int_0^{1/q} d\rho^2 (\ln q^2 \rho^2)(q^2 \rho^2)^{bt} \left(\frac{\alpha_s(\rho)}{\alpha_s(q)}\right)^{\frac{2b't}{b}} \qquad (31)$$

(extra $\alpha_s(q)$ in Eq. (30) is compensated by one power of $\ln q^2 \rho^2$). The integral over ρ diverges at $t = -\frac{1}{b}$ and gives the double pole at the one-loop level just as in the perturbative analysis. Subsequent terms in the expansion (19) correspond to UV renormalons located at $t = -\frac{2}{b}, -\frac{3}{b}$.... It should be mentioned that the Eq. (31) does not reproduce the strength of the first UV renormalon at the two-loop level[15]. The reason is that in Eq. (19) we have neglected the anomalous dimensions of the operators $\sim (\alpha_s(q)/\alpha_s(\rho))^{\frac{7}{b}}$. Such factors can change the strength of the singularity. For the IR renormalon this does not matter since the operator G^2 is renorm-invariant ($\gamma = 0$) and for the subsequent renormalons we can easily correct our results by corresponding γ's. For the UV renormalons, we do not know how to use the conformal invariance with the anomalous dimensions included.

It is very important to note that we actually never use the explicit form of the valley configuration. What we have really used are the three facts: (I) the fact that the rescaling of the vacuum fluctuation with an action S by a factor λ multiplies the determinant by $\lambda^{bS/8\pi^2}$ leading to the formula (12), (II) the expansion of the polarization operator (19) in slow varying fields, and (III) the conformal invariance of QCD at the tree level (for the UV renormalon we wrote down the small-size valley as an inversion of a large-scale spherical configuration). All of these properties hold true for an arbitrary vacuum fluctuation so we could take an arbitrary valley and arrive at the same results Eq. (26) and Eq. (31). It means that our result about the renormalon singularity coming from dilatation mode in the functional space is general.

Just as the conventional approach, our method yields the position and strength of the IR renormalon singularity but not the coefficient in front of it (this coefficient determines the numerical value of asymptotics of perturbative series for $R_{e^+e^-\to\text{hadrons}}$). To go beyond that and find the coefficient would require the integration over all possible valleys.

Acknowledgments

This work was supported by the US Department of Energy under contracts DE-AC05-84ER40150 and DE-FG02-97ER41028.

References

1. L.N. Lipatov, *Sov. Phys. JETP* **44** (1976) 1055; **45** (1977) 216.
2. Large order behavior of perturbation theory ed. J.C. Le Guillou and J. Zinn-Justin, North-Holland (1990).
3. G 't Hooft, in *The why's of Subnuclear Physics* (Erice, 1977), ed A. Zichichi (Plenum, New York, 1977)
4. M.Beneke, *Phys. Reports* **317** (1999) 1.
5. E.B. Bogomolny and V.A. Fateyev, *Phys. Lett.* **71B** (1977) 93.
6. I. Balitsky and A.V. Yung, *Phys. Lett.* **168B** (1986) 113; *Nucl. Phys.B* **274** (1986) 475
7. A.V. Yung, *Nucl. Phys.B* **297** (1988) 47.
8. I. Balitsky and A. Schafer, *Nucl. Phys.B* **404** (1993) 639.
9. A.S. Schwarz, Quantum field theory and topology, Springer Verlag, 1993
10. I. Balitsky, *Phys. Lett.* **273B** (1991) 282.
11. M.A. Shifman, A.I. Vainshtein, and V.I. Zakharov, *Nucl. Phys.B* **147** (1979) 1.
12. M.S. Dubovikov and A.V. Smilga, *Nucl. Phys.B* **185** (1981)109.
13. M.A. Shifman, A.I. Vainshtein, and V.I. Zakharov, *Nucl. Phys.B* **165** (1980) 45.
14. A.H. Mueller, *Nucl. Phys.B* **250** (1985) 327.
15. M. Beneke, V.M.Braun, and N. Kivel, *Phys. Lett.* **457B** (1999)147.

DYNAMICS OF QCD IN A STRONG MAGNETIC FIELD

V.A. MIRANSKY*

*Department of Applied Mathematics,
University of Western Ontario ,
London, Ontario N6A 5B7, Canada*

QCD in a strong magnetic field yields an example of a rich, sophisticated and controllable dynamics.

1. Prologue

On the first day of this wonderful meeting, I decided to change the topic of my talk. Already during the first Session, I felt it necessary to present something special, something connected with Arkady. So, I decided to talk about the dynamics of QCD in a strong magnetic field. This talk, it seemed to me, was appropriate indeed. The point is that the last time when I interacted with Arkady in person was in April 1994, in the Institute for Theoretical Physics in Santa Barbara. At that time, we, Valery Gusynin, Igor Shovkovy, and myself, had just finished our first work [1] in the series of papers concerning the role of a magnetic field in dynamical symmetry breaking in 2+1 and 3+1 dimensional field theories (actually, at that time, we discussed only the 2+1 dimensional case). So, it seemed to be appropriate to "continue" that discussion at the Arkadyfest in Minneapolis 8 years later.

I was lucky with this decision. At the time of its presentation, this work had not been yet completed. During my talk, Arkady raised a question concerning the dynamics of the magnetic catalysis in QCD with a large number of colors, $N_c \to \infty$. The answer to this question, Igor Shovkovy and I got the next day, helped to finish the work [2].

2. Introduction

Since the dynamics of QCD is extremely rich and complicated, it is important to study this theory under external conditions which provide a controllable dynamics. On the one hand, this allows one to understand better the vacuum structure and Green's functions of QCD, and, on the other hand, there can exist interesting applications of such models in themselves. The well known examples are hot QCD and QCD with a large baryon density.

Studies of QCD in external electromagnetic fields had started long ago [3,4]. A particularly interesting case is an external magnetic field. Using the Nambu-Jona-Lasinio (NJL) model as a low energy effective theory for QCD, it was shown that a magnetic field enhances the spontaneous chiral symmetry breakdown. The understanding of this phenomenon had remained obscure until a universal role of a magnetic field as a catalyst of chiral symmetry breaking was established in Refs. [1,5]. The general result states that a constant magnetic field leads to the generation of a fermion dynamical mass (i.e., a gap in the one-particle energy spectrum) even at the weakest attractive interaction between fermions. For this reason, this phenomenon was called the magnetic catalysis. The essence of the effect is the dimensional reduction $D \to D - 2$ in the dynamics of fermion pairing in a magnetic field. In the particular case of weak coupling, this dynamics is dominated by the lowest Landau level (LLL) which is essentially $D - 2$ dimensional [1,5].

The phenomenon of the magnetic catalysis was studied in gauge theories, in particular, in QED [6,7] and in QCD [8,9]. In the recent work [9], it has been suggested that the dynamics underlying the magnetic catalysis in QCD is weakly coupled at sufficiently large magnetic fields. Here we will consider this dynamical problem rigorously, from first principles. In fact, we show that, at sufficiently strong magnetic fields, $|eB| \gg \Lambda_{QCD}^2$, there exists a consistent truncation of the Schwinger-Dyson (gap) equation which leads to a reliable asymptotic expression for the quark mass m_q. Its explicit form reads:

$$m_q^2 \simeq 2C_1 |e_q B| (c_q \alpha_s)^{2/3} \exp\left[-\frac{4N_c \pi}{\alpha_s (N_c^2 - 1) \ln(C_2/c_q \alpha_s)} \right], \qquad (1)$$

where e_q is the electric charge of the q-th quark and N_c is the number of colors. The numerical factors C_1 and C_2 equal 1 in the leading approximation that we use. Their value, however, can change beyond this approximation and we can only say that they are of order 1. The constant c_q is defined as follows:

$$c_q = \frac{1}{6\pi} (2N_u + N_d) \left| \frac{e}{e_q} \right|, \qquad (2)$$

where N_u and N_d are the numbers of up and down quark flavors, respectively. The total number of quark flavors is $N_f = N_u + N_d$. The strong coupling α_s in the last equation is related to the scale $\sqrt{|eB|}$, i.e.,

$$\frac{1}{\alpha_s} \simeq b \ln \frac{|eB|}{\Lambda_{QCD}^2}, \quad \text{where} \quad b = \frac{11N_c - 2N_f}{12\pi}. \tag{3}$$

We should note that in the leading approximation the energy scale $\sqrt{|eB|}$ in Eq. (3) is fixed only up to a factor of order 1.

The central dynamical issue underlying this dynamics is the effect of screening of the gluon interactions in a magnetic field in the region of momenta relevant for the chiral symmetry breaking dynamics, $m_q^2 \ll |k^2| \ll |eB|$. In this region, gluons acquire a mass M_g of order $\sqrt{N_f \alpha_s |e_q B|}$. This allows to separate the dynamics of the magnetic catalysis from that of confinement.

Since the background magnetic field breaks explicitly the global chiral symmetry that interchanges the up and down quark flavors, the chiral symmetry in this problem is $SU(N_u)_L \times SU(N_u)_R \times SU(N_d)_L \times SU(N_d)_R \times U^{(-)}(1)_A$. The $U^{(-)}(1)_A$ is connected with the current which is an anomaly free linear combination of the $U^{(d)}(1)_A$ and $U^{(u)}(1)_A$ currents. The generation of quark masses breaks this symmetry spontaneously down to $SU(N_u)_V \times SU(N_d)_V$ and, as a result, $N_u^2 + N_d^2 - 1$ gapless Nambu-Goldstone (NG) bosons occur. In Sec. 4, we derive the effective action for the NG bosons and calculate their decay constants and velocities.

The present analysis is heavily based on the analysis of the magnetic catalysis in QED done by Gusynin, Miransky, and Shovkovy [7]. A crucial difference is of course the property of asymptotic freedom and confinement in QCD. In connection with that, our second major result is the derivation of the low energy effective action for gluons in QCD in a strong magnetic field. The characteristic feature of this action is its anisotropic dynamics. In particular, the strength of static (Coulomb like) forces along the direction parallel to the magnetic field is much larger than that in the transverse directions. Also, the confinement scale in this theory is much less than that in QCD without a magnetic field. This features imply a rich and unusual spectrum of light glueballs in this theory.

A special and interesting case is QCD with a large number of colors, in particular, with $N_c \to \infty$ (the 't Hooft limit). The question about it was raised by Arkady during my talk. This theory is considered in Sec. 6.

3. Magnetic catalysis in QCD

We begin by considering the Schwinger-Dyson (gap) equation for the quark propagator. It has the following form:

$$G^{-1}(x,y) = S^{-1}(x,y) + 4\pi\alpha_s\gamma^\mu \int G(x,z)\Gamma^\nu(z,y,z')D_{\nu\mu}(z',x)d^4z d^4z' \quad (4)$$

where $S(x,y)$ and $G(x,y)$ are the bare and full fermion propagators in an external magnetic field, $D_{\nu\mu}(x,y)$ is the full gluon propagator and $\Gamma^\nu(x,y,z)$ is the full amputated vertex function. Since the coupling α_s related to the scale $|eB|$ is small, one might think that the rainbow (ladder) approximation is reliable in this problem. However, this is not the case. Because of the (1+1)-dimensional form of the fermion propagator in the LLL approximation, there are relevant higher order contributions [6,7]. Fortunately one can solve this problem. First of all, an important feature of the quark-antiquark pairing dynamics in QCD in a strong magnetic field is that this dynamics is essentially abelian. This feature is provided by the form of the polarization operator of gluons in this theory. The point is that the dynamics of the quark-antiquark pairing is mainly induced in the region of momenta k much less than $\sqrt{|eB|}$. This implies that the magnetic field yields a dynamical ultraviolet cutoff in this problem. On the other hand, while the contribution of (electrically neutral) gluons and ghosts in the polarization operator is proportional to k^2, the fermion contribution is proportional to $|e_q B|$ [7]. As a result, the fermion contribution dominates in the relevant region with $k^2 \ll |eB|$.

This observation implies that there are three, dynamically very different, scale regions in this problem. The first one is the region with the energy scale above the magnetic scale $\sqrt{|eB|}$. In that region, the dynamics is essentially the same as in QCD without a magnetic field. In particular, the running coupling decreases logarithmically with increasing the energy scale there. The second region is that with the energy scale below the magnetic scale but much larger than the dynamical mass m_q. In this region, the dynamics is abelian like and, therefore, the dynamics of the magnetic catalysis is similar to that in QED in a magnetic field. At last, the third region is the region with the energy scale less than the gap. In this region, quarks decouple and a confinement dynamics for gluons is realized.

Let us first consider the intermediate region relevant for the magnetic catalysis. As was indicated above, the important ingredient of this dynamics is a large contribution of fermions to the polarization operator. It is large because of an (essentially) 1+1 dimensional form of the fermion propagator in a strong magnetic field. Its explicit form can be obtained by modifying appropriately the expression for the polarization operator in QED in a magnetic field [7]:

$$P^{AB,\mu\nu} \simeq \frac{\alpha_s}{6\pi}\delta^{AB}\left(k_{\|}^{\mu}k_{\|}^{\nu} - k_{\|}^2 g_{\|}^{\mu\nu}\right)\sum_{q=1}^{N_f}\frac{|e_q B|}{m_q^2}, \quad |k_{\|}^2| \ll m_q^2, \quad (5)$$

$$P^{AB,\mu\nu} \simeq -\frac{\alpha_s}{\pi}\delta^{AB}\left(k_{\|}^{\mu}k_{\|}^{\nu} - k_{\|}^2 g_{\|}^{\mu\nu}\right)\sum_{q=1}^{N_f}\frac{|e_q B|}{k_{\|}^2}, \quad m_q^2 \ll |k_{\|}^2| \ll |eB|, \quad (6)$$

where $g_{\|}^{\mu\nu} \equiv \mathrm{diag}(1,0,0,-1)$ is the projector onto the longitudinal subspace, and $k_{\|}^{\mu} \equiv g_{\|}^{\mu\nu}k_{\nu}$ (the magnetic field is in the x^3 direction). Similarly, we introduce the orthogonal projector $g_{\perp}^{\mu\nu} \equiv g^{\mu\nu} - g_{\|}^{\mu\nu} = \mathrm{diag}(0,-1,-1,0)$ and $k_{\perp}^{\mu} \equiv g_{\perp}^{\mu\nu}k_{\nu}$ that we shall use below. Notice that quarks in a strong magnetic field do not couple to the transverse subspace spanned by $g_{\perp}^{\mu\nu}$ and $k_{\perp}^{\mu}$. This is because in a strong magnetic field only the quark from the LLL matter and they couple only to the longitudinal components of the gluon field. The latter property follows from the fact that spins of the LLL quarks are polarized along the magnetic field (see the second paper in [5]).

The expressions (5) and (6) coincide with those for the polarization operator in the $1+1$ dimensional *massive* QED (massive Schwinger model) [10] if the parameter $\alpha_s|e_q B|/2$ here is replaced by the dimensional coupling α_1 of QED_{1+1}. As in the Schwinger model, Eq. (6) implies that there is a massive resonance in the $k_{\|}^{\mu}k_{\|}^{\nu} - k_{\|}^2 g_{\|}^{\mu\nu}$ component of the gluon propagator. Its mass is

$$M_g^2 = \sum_{q=1}^{N_f}\frac{\alpha_s}{\pi}|e_q B| = (2N_u + N_d)\frac{\alpha_s}{3\pi}|eB|. \quad (7)$$

This is reminiscent of the pseudo-Higgs effect in the $(1+1)$-dimensional massive QED. It is not the genuine Higgs effect because there is no complete screening of the color charge in the infrared region with $|k_{\|}^2| \ll m_q^2$. This can be seen clearly from Eq. (5). Nevertheless, the pseudo-Higgs effect is manifested in creating a massive resonance and this resonance provides the dominant forces leading to chiral symmetry breaking.

Now, after the abelian like structure of the dynamics in this problem is established, we can use the results of the analysis in QED in a magnetic field [7] by introducing appropriate modifications. The main points of the analysis are: (i) the so called improved rainbow approximation is reliable in this problem provided a special non-local gauge is used in the analysis, and (ii) for a small coupling α_s (α in QED), the relevant region of momenta in this problem is $m_q^2 \ll |k^2| \ll |eB|$. We recall that in the improved rainbow approximation the vertex $\Gamma^\nu(x, y, z)$ is taken to be bare and the gluon propagator is taken in the one-loop approximation. Moreover, as we argued above, in this intermediate region of momenta, only the contribution of quarks to the gluon polarization tensor (6) matters. [It is appropriate to call this approximation the "strong-field-loop improved rainbow approximation". It is an analog of the hard-dense-loop improved rainbow approximation in QCD with a nonzero baryon density]. As to the modifications, they are purely kinematic: the overall coupling constant in the gap equation α and the dimensionless combination $M_\gamma^2/|eB|$ in QED have to be replaced by $\alpha_s(N_c^2 - 1)/2N_c$ and $M_g^2/|e_qB|$, respectively. This leads us to the expression (1) for the dynamical gap.

After expressing the magnetic field in terms of the running coupling, the result for the dynamical mass takes the following convenient form:

$$m_q^2 \simeq 2C_1 \left|\frac{e_q}{e}\right| \Lambda_{QCD}^2 (c_q \alpha_s)^{2/3} \exp\left[\frac{1}{b\alpha_s} - \frac{4N_c\pi}{\alpha_s(N_c^2 - 1)\ln(C_2/c_q\alpha_s)}\right]. \quad (8)$$

As is easy to check, the dynamical mass of the u-quark is considerably larger than that of the d-quark. It is also noticeable that the values of the u-quark dynamical mass becomes comparable to the vacuum value $m_{dyn}^{(0)} \simeq 300$ MeV only when the coupling constant gets as small as 0.05.

Now, by trading the coupling constant for the magnetic field scale $|eB|$, we get the dependence of the dynamical mass on the value of the external field. The numerical results are presented in Fig. 1 [we used $C_1 = C_2 = 1$ in Eq. (8)]. As one can see in Fig. 1, the value of the quark gap in a wide window of strong magnetic fields, $\Lambda_{QCD}^2 \ll |eB| \lesssim (10 \text{ TeV})^2$, remains smaller than the dynamical mass of quarks $m_{dyn}^{(0)} \simeq 300$ MeV in QCD without a magnetic field. In other words, the chiral condensate is partially *suppressed* for those values of a magnetic field. The explanation of this, rather unexpected, result is actually simple. The magnetic field leads to the mass M_g (7) for gluons. In a strong enough magnetic field, this mass becomes larger than the characteristic gap Λ in QCD without a magnetic field (Λ, playing the role of a gluon mass, can be estimated as a few times larger than Λ_{QCD}). This, along with the property of the asymptotic freedom (i.e., the fact that α_s decreases with increasing the magnetic field), leads to the suppression of the chiral condensate.

This point also explains why our result for the gap is so different from that in the NJL model in a magnetic field [3]. Recall that, in the NJL model, the gap logarithmically (i.e., much faster than in the present case) grows with a magnetic field. This is the related to the assumption that both the dimensional coupling constant $G = g/\Lambda^2$ (with Λ playing a role similar

192

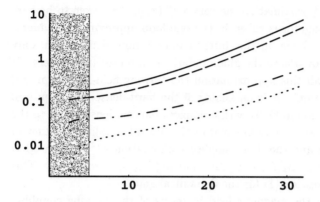

Figure 1. The dynamical masses of quarks as functions of $\ln(|eB|/\Lambda_{QCD}^2)$ for $N_c = 3$ and two different values of $N_f = N_u + N_d$: (i) masses of u-quark (solid line) and d-quark (dash-dotted line) for $N_u = 1$ and $N_d = 2$; (ii) masses of u-quark (dashed line) and d-quark (dotted line) for $N_u = 2$ and $N_d = 2$. The result may not be reliable in the weak magnetic field region (shaded) where some of the approximations break. The values of masses are given in units of $\Lambda_{QCD} = 250$ MeV.

to that of the gluon mass in QCD), as well as the scale Λ do not dependent on the value of the magnetic field. Therefore, in that model, in a strong enough magnetic field, the value of the chiral condensate is overestimated.

The picture which emerges from this discussion is the following. For values of a magnetic field $|eB| \lesssim \Lambda^2$ the dynamics in QCD should be qualitatively similar to that in the NJL model. For strong values of the field, however, it is essentially different, as was described above. This in turn suggests that there should exist an intermediate region of fields where the dynamical masses of quarks decreases with increasing the background magnetic field.

4. Effective action of NG bosons

The presence of the background magnetic field breaks explicitly the global chiral symmetry that interchanges the up and down quark flavors. This is related to the fact that the electric charges of the two sets of quarks are different. However, the magnetic field does not break the global chiral symmetry of the action completely. In particular, in the model with the N_u up quark flavors and the N_d down quark flavors, the action is invariant under the chiral symmetry $SU(N_u)_L \times SU(N_u)_R \times SU(N_d)_L \times SU(N_d)_R \times U^{(-)}(1)_A$. The $U^{(-)}(1)_A$ is connected with the current which is an anomaly free linear combination of the $U^{(d)}(1)_A$ and $U^{(u)}(1)_A$ currents.

[The $U^{(-)}(1)_A$ symmetry is of course absent if either N_d or N_u is equal to zero].

The global chiral symmetry of the action is broken spontaneously down to the diagonal subgroup $SU(N_u)_V \times SU(N_d)_V$ when dynamical masses of quarks are generated. In agreement with the Goldstone theorem, this leads to the appearance of $N_u^2 + N_d^2 - 1$ number of the NG gapless excitations in the low-energy spectrum of QCD in a strong magnetic field. Notice that there is also a pseudo-NG boson connected with the conventional (anomalous) $U(1)_A$ symmetry which can be rather light in a sufficiently strong magnetic field.

Now, in the chiral limit, the general structure of the low energy action for the NG bosons could be easily established from the symmetry arguments alone. First of all, such an action should be invariant with respect to the space-time symmetry $SO(1,1) \times SO(2)$ which is left unbroken by the background magnetic field [here the $SO(1,1)$ and the $SO(2)$ are connected with Lorentz boosts in the $x_0 - x_3$ hyperplane and rotations in the $x_1 - x_2$ plane, respectively]. Besides that, the low-energy action should respect the original chiral symmetry $SU(N_u)_L \times SU(N_u)_R \times SU(N_d)_L \times SU(N_d)_R \times U^{(-)}(1)_A$. These requirements lead to the following general form of the action:

$$
L_{NG} \simeq \frac{f_u^2}{4} \mathrm{tr} \left(g_\parallel^{\mu\nu} \partial_\mu \Sigma_u \partial_\nu \Sigma_u^\dagger + v_u^2 g_\perp^{\mu\nu} \partial_\mu \Sigma_u \partial_\nu \Sigma_u^\dagger \right)
$$

$$
+ \frac{f_d^2}{4} \mathrm{tr} \left(g_\parallel^{\mu\nu} \partial_\mu \Sigma_d \partial_\nu \Sigma_d^\dagger + v_d^2 g_\perp^{\mu\nu} \partial_\mu \Sigma_d \partial_\nu \Sigma_d^\dagger \right)
$$

$$
+ \frac{\tilde{f}^2}{4} \left(g_\parallel^{\mu\nu} \partial_\mu \tilde{\Sigma} \partial_\nu \tilde{\Sigma}^\dagger + \tilde{v}^2 g_\perp^{\mu\nu} \partial_\mu \tilde{\Sigma} \partial_\nu \tilde{\Sigma}^\dagger \right). \tag{9}
$$

The unitary matrix fields $\Sigma_u \equiv \exp\left(i \sum_{A=1}^{N_u^2-1} \lambda^A \pi_u^A / f_u \right)$, $\Sigma_d \equiv \exp\left(i \sum_{A=1}^{N_d^2-1} \lambda^A \pi_d^A / f_d \right)$, and $\tilde{\Sigma} \equiv \exp\left(i\sqrt{2}\tilde{\pi}/\tilde{f} \right)$ describe the NG bosons in the up, down, and $U^{(-)}(1)_A$ sectors of the original theory. The decay constants $f_u, f_d, \tilde{f}$ and transverse velocities $v_u, v_d, \tilde{v}$ can be calculated by using the standard field theory formalism. The results for the $N_u^2 + N_d^2 - 2$ NG bosons in the up and down sectors, assigned to the adjoint representation of the $SU(N_u)_V \times SU(N_d)_V$ symmetry, are [2]

$$
f_u^2 = \frac{N_c}{6\pi^2} |eB|, \tag{10}
$$

$$
f_d^2 = \frac{N_c}{12\pi^2} |eB|, \quad v_q = 0. \tag{11}
$$

The remarkable fact is that the decay constants are nonzero even in the limit when the dynamical masses of quarks approach zero. The reason of that is the $1+1$ dimensional character of this dynamics. A similar situation takes place in color superconductivity: in that case the $1+1$ dimensional character of the dynamics is provided by the Fermi surface.

Notice that the transverse velocities of the NG bosons are equal to zero. This is also a consequence of the $1+1$ dimensional structure of the quark propagator in the LLL approximation. The point is that quarks can move in the transverse directions only by hopping to higher Landau levels. Taking into account higher Landau levels would lead to nonzero velocities suppressed by powers of $|m_q|^2/|eB|$. In fact, the explicit form of the velocities was derived in the weakly coupled NJL model in an external magnetic field [see Eq. (65) in the second paper of Ref. [5]]. It is

$$v_{u,d}^2 \sim \frac{|m_{u,d}|^2}{|eB|} \ln \frac{|eB|}{|m_{u,d}|^2} \ll 1. \tag{12}$$

A similar expression should take place also for the transverse velocities of the NG bosons in QCD.

The decay constant $\tilde{f}$ of the singlet NG boson connected with the spontaneous breakdown of the $U^{(-)}(1)_A$ is [2]

$$\tilde{f}^2 = \frac{(N_d f_u + N_u f_d)^2}{N_f^2} = \frac{(\sqrt{2}N_d + N_u)^2 N_c}{12\pi^2 N_f^2}|eB|. \tag{13}$$

Its transverse velocity is of course zero in the LLL approximation.

5. Anisotropic confinement of gluons

Let us now turn to the infrared region with $|k| \lesssim m_d$, where all quarks decouple (notice that we take here the smaller mass of d quarks). In that region, a pure gluodynamics realizes. However, its dynamics is quite unusual. The point is that although gluons are electrically neutral, their dynamics is strongly influenced by an external magnetic field, as one can see from expression (5) for their polarization operator. In a more formal language, while quarks decouple and do not contribute into the equations of the renormalization group in that infrared region, their dynamics strongly influence the boundary (matching) conditions for those equations at $k \sim m_d$. A conventional way to describe this dynamics is the method of the low energy effective action. This low effective action was derived in Ref. [2]. Here we will discuss its main properties.

The low energy effective action is relevant for momenta $|k| \lesssim m_d$. Notice the following important feature of the action: its "bare" coupling constant g, related to the scale m_d, coincides with the value of the vacuum QCD coupling related to the scale $\sqrt{|eB|}$ (and *not* to the scale m_d). This is because g is determined from the matching condition at $|k| \sim m_d$, the lower border of the intermediate region $m_d \lesssim |k| \lesssim \sqrt{|eB|}$, where, because of the pseudo-Higgs effect, the running of the coupling is essentially frozen. Therefore the "bare" coupling g indeed coincides with the value of the vacuum QCD coupling related to the scale $\sqrt{|eB|}$: $g = g_s$. Since this value is much less that that of the vacuum QCD coupling related to the scale m_d, this implies that the confinement scale λ_{QCD} of the action should be much less than Λ_{QCD} in QCD without a magnetic field.

Actually, this consideration somewhat simplifies the real situation. Since the LLL quarks couple to the longitudinal components of the polarization operator, only the effective coupling connected with longitudinal gluons is frozen. For transverse gluons, there should be a logarithmic running of their effective coupling. It is clear, however, that this running should be quite different from that in the vacuum QCD. The point is that the time like gluons are now massive and their contribution in the running in the intermediate region is severely reduced. On the other hand, because of their negative norm, just the time like gluons are the major players in producing the antiscreening running in QCD (at least in covariant gauges). Since now they effectively decouple, the running of the effective coupling for the transverse gluons should slow down. It is even not inconceivable that the antiscreening running can be transformed into a screening one. In any case, one should expect that the value of the transverse coupling related to the matching scale m_d will be also essentially reduced in comparison with that in the vacuum QCD. Since the consideration in this section is rather qualitative, we adopt the simplest scenario with the value of the transverse coupling at the matching scale m_d also coinciding with g_s.

The interaction potential between two static quarks in this theory at "short" distances $r \sim m_d^{-1}$ reads:

$$V(x, y, z) \simeq \frac{g_s^2}{4\pi\sqrt{z^2 + \epsilon(x^2 + y^2)}}, \tag{14}$$

where the dielectric constant $\epsilon = 1 + \frac{\alpha_s}{6\pi} \sum_{q=1}^{N_f} |e_q B|/m_q^2$ is very large. Because of the dielectric constant, this Coulomb like interaction is anisotropic in space: it is suppressed by a factor of $\sqrt{\epsilon}$ in the transverse directions compared to the interaction in the direction of the magnetic field.

The potential (14) corresponds to the classical, tree, approximation which is good only in the region of distances much smaller than the confinement radius $r_{QCD} \sim \lambda_{QCD}^{-1}$. Deviations from this interaction are determined by loop corrections. The analysis of the loop expansion leads to the following estimate of the new confinement scale λ_{QCD} in QCD in a strong magnetic field:

$$\lambda_{QCD} \simeq m_d \left(\frac{\Lambda_{QCD}}{\sqrt{|eB|}} \right)^{b/b_0}, \qquad (15)$$

where $b = (11N_c - 2N_f)/12\pi$ and $b_0 = 11N_c/12\pi$. Therefore, in a strong magnetic field, λ_{QCD} is much less than Λ_{QCD}.

The hierarchy $\lambda_{QCD} \ll \Lambda_{QCD}$ is intimately connected with a somewhat puzzling point that the pairing dynamics decouples from the confinement dynamics despite it produces quark masses of order Λ_{QCD} or less [for a magnetic field all the way up to the order of $(10 \text{ TeV})^2$]. The point is that these masses are heavy in units of the new confinement scale λ_{QCD} and the pairing dynamics is indeed weakly coupled.

6. Arkady question: QCD with a large number of colors

I did not discuss the case of QCD with a large number of colors in my talk. The question about it was raised by Arkady during the talk (he actually asked not myself, the speaker, but my coauthor Igor Shovkovy who was sitting next to him). Igor and myself got the answer the next day.

Just a look at expression (7) for the gluon mass is enough to recognize that the dynamics in this limit is very different from that considered in the previous sections. Indeed, as is well known, the strong coupling constant α_s is proportional to $1/N_c$ in this limit. More precisely, it rescales as $\alpha_s = \frac{\bar{\alpha}_s}{N_c}$, where the new coupling constant $\bar{\alpha}_s$ remains finite as $N_c \to \infty$ ('t Hooft limit). Then, expression (7) implies that the gluon mass goes to zero in this limit. This in turn implies that the appropriate approximation in this limit is not the improved rainbow approximation but the rainbow approximation itself, when *both* the vertex and the gluon propagator in the SD equation (4) are taken to be bare. This leads to the following expression for the dynamical mass of quarks [2]:

$$m_q^2 = C|e_q B| \exp\left[-\pi \left(\frac{\pi N_c}{(N_c^2 - 1)\alpha_s} \right)^{1/2} \right], \qquad (16)$$

where the constant C is of order one. The confinement scale λ_{QCD} is close to Λ_{QCD} in this case [2].

It is natural to ask how large N_c should be before the expression (16) becomes reliable. One can show [2] that the threshold value N_c^{thr} grows rapidly with the magnetic field [$N_c^{thr} \gtrsim 100$ for $|eB| \gtrsim (1\text{GeV})^2$]. Expression (16) for the quark mass is reliable for the values of N_c of the order of N_c^{thr} or larger. Decreasing N_c below N_c^{thr}, one comes to expression (1).

7. Conclusion

QCD in a strong magnetic field yields an example of a rich, sophisticated and (that is very important) controllable dynamics.

Acknowledgments

I am grateful to the organizers of this Symposium for their warm hospitality. I acknowledge support from the Natural Sciences and Engineering Research Council of Canada.

References

* On leave of absence from Bogolyubov Institute for Theoretical Physics, 252143, Kiev, Ukraine.
1. V. P. Gusynin, V. A. Miransky, and I. A. Shovkovy, Phys. Rev. Lett. **73**, 3499 (1994); Phys. Rev. D **52**, 4718 (1995).
2. V. A. Miransky and I. A. Shovkovy, arXiv:hep-ph/0205348 (to appear in Phys. Rev. D).
3. S. Kawati, G. Konisi, and H. Miyata, Phys. Rev. D **28**, 1537 (1983); S. P. Klevansky and R. H. Lemmer, Phys. Rev. D **39**, 3478 (1989); H. Suganuma and T. Tatsumi, Annals Phys. **208**, 470 (1991).
4. S. Schramm, B. Muller, and A. J. Schramm, Mod. Phys. Lett. A **7**, 973 (1992).
5. V. P. Gusynin, V. A. Miransky, and I. A. Shovkovy, Phys. Lett. B **349**, 477 (1995). Nucl. Phys. B **462**, 249 (1996).
6. V. P. Gusynin, V. A. Miransky, and I. A. Shovkovy, Phys. Rev. D **52**, 4747 (1995); C. N. Leung, Y. J. Ng, and A. W. Ackley, Phys. Rev. D **54**, 4181 (1996); D. S. Lee, C. N. Leung, and Y. J. Ng, Phys. Rev. D **55**, 6504 (1997); *ibid.* D **57**, 5224 (1998). D. K. Hong, Y. Kim, and S. J. Sin, Phys. Rev. D **54**, 7879 (1996). V. P. Gusynin and A. V. Smilga, Phys. Lett. B **450**, 267 (1999). J. Alexandre, K. Farakos and G. Koutsoumbas, Phys. Rev. D **64**, 067702 (2001).
7. V. P. Gusynin, V. A. Miransky, and I. A. Shovkovy, Phys. Rev. Lett. **83**, 1291 (1999); Nucl. Phys. B **563**, 361 (1999).
8. I. A. Shushpanov and A. V. Smilga, Phys. Lett. B **402**, 351 (1997); N. O. Agasian and I. A. Shushpanov, Phys. Lett. B **472**, 143 (2000). V. C. Zhukovsky, V. V. Khudyakov, K. G. Klimenko, and D. Ebert, JETP Lett. **74**, 523 (2001) [Pisma Zh. Eksp. Teor. Fiz. **74**, 595 (2001)]; D. Ebert, V. V. Khudyakov, V. C. Zhukovsky, and K. G. Klimenko, Phys. Rev. D **65**, 054024 (2002).
9. D. Kabat, K. Lee, and E. Weinberg, Phys. Rev. D **66**, 014004 (2002) .
10. J. Schwinger, Phys. Rev. **125**, 397 (1962); S. R. Coleman, R. Jackiw, and L. Susskind, Annals Phys. **93**, 267 (1975); S. R. Coleman, Annals Phys. **101**, 239 (1976).

LIGHT-FRONT QUANTIZATION OF GLUODYNAMICS REVISITED

V. T. KIM

St. Petersburg Nuclear Physics Institute,
Gatchina 188300, Russia
E-mail: kim@pnpi.spb.ru

G. B. PIVOVAROV

Institute for Nuclear Research,
7a 60th October Anniversary Prospect,
Moscow, 117312 Russia
E-mail: gbpivo@ms2.inr.ac.ru

J. P. VARY

Department of Physics and Astronomy,
Iowa State University, Ames, Iowa 50011, USA
E-mail: jvary@iastate.edu

We review recent advances in the light-front quantization of gluodynamics. A definition is given to the DLCQ Hamiltonian for $SU(2)$ gludynamics. In particular, it is explained that a global contribution to the Wilson loop along the compactified light-like direction features this Hamiltonian as a degree of freedom.

1. Introduction

In 1966 Weinberg demonstrated [1] that going over to the infinite momentum frame removes many of the diagrams of the old-fashioned perturbation theory. The outcome of his consideration is a shortcut to Feynman integrals in a representation, where momentum loop integrals are represented by the integrals over the transverse momenta (i.e., space-like components of the momenta transverse to a singled out space-like direction), and over Feynman parameters. In this, Feynman parameters have the meaning of the fractions of the total longitudinal (i.e., light-like) momentum that flow along the lines of the diagrams. These integrals run from zero to unit. The longitudinal momentum conserves in the vertices. This is the reason the vacuum diagrams of the old-fashioned perturbation theory disappear in the infinite momentum frame: there is no way to conserve the positive longitudinal momentum in a vertex if the external momentum vanishes.

In [2], it was realized that the same result could be obtained as an outcome of a specific quantization scheme of the corresponding field theory, where a light-like direction is used as dynamical time, and commutation relations for the fields are set on a light-front, at a fixed light-like time variable. In this way, a nonperturbative meaning can be given to the transformation of Feynman integrals to the parametric form.

Next step was made in [3], where the connection to the Dirac paper [4] on the forms of relativistic dynamics was made, compactification of the light-like direction transverse to the light-like time was suggested, and the problem of the zero modes was pointed out. In [5] it was advocated that the formulation with the compactified light like-direction can be used to define nonperturbatively field theories. Here, it was pointed out that the Hamiltonian of the theory is block-diagonal in this formulation, each block represents a Hamiltonian of quantum mechanics with the number of particles bounded above. The formulation with a compactified light-like direction was called the discretized light-cone quantization (DLCQ) in [6].

Let us explain the block-diagonal structure of the DLCQ Hamiltonian using perturbation theory. The compactification of the light-like direction $x^- \equiv (x^0 - x^3)/\sqrt{2}$ replaces the loop integrals over the longitudinal momenta with finite sums (the sums are finite, because the longitudinal momentum is nonnegative and cannot exceed the external longitudinal momentum). The external longitudinal momentum is quantized: $P_- = (2\pi K)/L$, where L is the size of the system in the longitudinal direction. The positive integer K above is called the harmonic resolution. Discard the terms in the sums whose lines carry zero longitudinal momentum. Consider any cross section of a diagram of the old-fashioned perturbation theory that cuts between the initial and final state. The number of the lines that are cut across is bounded above by K. As each line corresponds to a propagating particle, the number of the particles is bounded above by the harmonic resolution K. Evidently, the number of interactions and the number of Feynman diagrams are unbounded for a finite K. When we diagonalize a block of the DLCQ Hamiltonian, we sum up an infinite series of Feynman diagrams. (We point out here that each of the diagrams we may be able to sum up in this way is distorted by the compactification of the longitudinal direction, the distortion is small at large L and K.)

The doubtful point above is the discard of the lines with the zero longitudinal momentum. Naively, it is justified, because when we approximate an integral with a sum, any finite number of terms can be dropped, and the integral can still be restored in the continuum limit. Evidently, this reasoning depends on the properties of the integrand. These terms we discarded above constitute the contribution of the zero modes. Keeping them would destroy the above reasoning, discarding them is dangerous. This is the problem of the zero modes.

A consistent solution to this problem may be approached with a functional integral formulation. The zero modes (the Fourier modes of the fields with the zero longitudinal momentum) should be integrated out [7].

In any case, the DLCQ program starts with a canonical quantization of the classical field theory with the x^--direction compactified, and x^+ taken as time. Even this step is nontrivial for the non-Abelian field theories. This is the case because the time derivative (i.e., the derivative over x^+) of the fields is involved in the Lagrangian in a way obstructing the Hamiltonian interpretation.

To see this, compare with a scalar field theory. In this case, the time derivative enters the Lagrangian in the combination $\int dx^- \partial_- \phi \partial_+ \phi$. Make a Fourier transform of the fields over x^-, rescale the Fourier modes, and cast the above term in the form $\sum_{k_- > 0} i a^\dagger(k_-) \dot{a}(k_-)$, where $a(k_-)$ are the rescaled Fourier modes, and the overdot denotes the derivative over x^+. In this way, we find $a^\dagger$, a as canonical variables.

For gluodynamics, we have the components A_-, A_+, $A_\perp$ of the gluon field. There is no time derivatives of A_+ in the Lagrangian. It is a Lagrange multiplier; variation over it yields Gauss law. The Lagrangian is quadratic in time derivatives of A_-, i.e., in the Hamiltonian formulation, one of the canonical momentum variables is proportional to the time derivative of A_-. The problem we pointed out above is related to the time derivatives of the transverse components of the gauge field. They enter the Lagrangian in the combination $(D_- A_\perp) \dot{A}_\perp$, where D_- is the covariant derivative. The complication is that the covariant derivative above depends on the component A_-. Therefore, the canonical variables cannot be obtained with a linear transformation of the transverse components of the gluon field, in contrast to the case of the scalar field.

The problem disappears in the light-cone gauge $A_- = 0$ [8]. However, for DLCQ this gauge is inaccessible, because the direction x^- is comactified in this case. Indeed, for x^- compactified, the Wilson loop $W \equiv \mathrm{TrP} \exp(ig \int dx^- A_-)/N$ is gauge invariant, and A_- cannot vanish if W deviates from unit (N above is the rank of the gauge algebra). Still, we can take the gauge where A_- is diagonal and independent of x^-. In this gauge, the canonical variables can be obtained with Fourier transform over x^-, as it was possible above for the case of the scalar field. The only difference is that the rescaling of the Fourier modes to the canonical variables depends now on A_-.

This way of canonical formulation of DLCQ gluodynamics was considered in [9]. In this paper, it was concluded that the off-diagonal transverse gluons cannot exist in the DLCQ formulation, because their presence contradicts the Gauss law. This conclusion is unexpected, because there are definitely solutions to the classical field equations with nonzero off-diagonal transverse gluon fields. The issue was reconsidered in [10]. In this paper, it was found that the off-diagonal gluons do not contradict the Gauss law if we allow a contribution to A_- dependent only on x^+ (i.e., independent not only of x^-, but also of $x^\perp$). This component constitutes a degree of freedom not restricted by the Gauss law. The Abelian charge of the off-diagonal gluons depends on the value this degree of freedom takes. This dependence makes it possible to remove the contradiction with the Gauss law, because the Abelian charges of the off-diagonal gluons are involved in the Gauss law.

In this way, we obtain a quantum mechanics where transverse gluons move over the transverse plane interacting with one another and with a global degree of freedom that does not depend on the location on the transverse plane. The global degree of freedom (which is related to a global contribution to the Wilson loop) has its own dynamics: the Hamiltonian contains a square of its velocity, and a potential depending on the global variable.

In the next section, we present the DLCQ Hamiltonian of $SU(2)$ gluodynamics. In the concluding section, we discuss the prospects of practical use of this Hamiltonian.

2. DLCQ Hamiltonian of $SU(2)$ Gluodynamics

The action of $SU(2)$ gluodynamics is

$$S_{glue} = \int dx^+ dy \left[\frac{1}{2} \langle E_+ | E_+ \rangle + \langle E_I | F_{+I} \rangle - \frac{1}{4} \langle F_{IJ} | F_{IJ} \rangle \right]. \tag{1}$$

Here we use the scalar product

$$\langle A | B \rangle = \int_{-\frac{L}{2}}^{\frac{L}{2}} dx^- 2\mathrm{Tr}(A^\dagger B). \tag{2}$$

The normalizing factor 2 makes the Pauli matrixes orthonormal at unit length of the x^- circle. The integration over y runs above over the transverse plane. We use above separate notations for some of the components of the gauge field strength tensor: $E_+ \equiv F_{-+}$, $E_I \equiv F_{-I}$. The upper case

Latin indixes enumerate the coordinates of the transverse plane. The field strength tensor is $F_{\mu\nu} \equiv (\partial_\mu A_\mu^a - \partial_\nu A_\mu^a - ig f^{abc} A_\mu^b A_\nu^c)\sigma^a/2$, where σ^a are the Pauli matrixes.

With the above scalar product, the covariant derivative D_- is anti-Hermitean, where $D_- B \equiv \partial_- B - ig[A_-, B]$. We use extensively expansions of the fields over the eigenfunctions χ_k of D_-, $iD_-\chi_k = \omega_k \chi_k$. We call it non-Abelian Fourier transform. The formulation of a non-Abelian gauge theory with a compactified dimension in terms of the Fourier coefficients of the non-Abelian Fourier transform is considered in [11]. Our notations here are mostly in agreement with [11]. We skip here the discussion of the non-Abelian Fourier transform, in particular, of the structure of the spectrum of D_-. We only give an explicit formula for the eigenvalues: $\omega_k = 2\pi n_k/L + g\sigma_k\alpha/\sqrt{L}$, where n_k is an integer, and σ_k is a sign: $\sigma_k = -1, 0, 1$. Further discussion of the non-Abelian Fourier transform can be found in [11].

We use the gauge in which χ_k are independent of x^+ and $x^\perp$, and $A_- = \alpha\chi_0$ (χ_0 is the eigenfunction with zero eigenvalue, $iD_-\chi_0 = 0$). The α depends on x^+ and $x^\perp$. The eigenfunctions χ_k form a complete orthonormal set: $\langle A|B \rangle = \sum_k (A^k)^\dagger B^k$, where the superscripts k denote the projections onto the eigengunctions: $A^k \equiv \langle \chi_k|A \rangle$, and the sum over $k = (n_k, \sigma_k)$ comprises the sums over the integer n_k and the sign σ_k. In fact, our consideration can be performed without any gauge fixing. The mild complications related to a treatment without gauge fixing are considered in [11]. Hereafter, we use a simplified consideration in the above gauge.

The first step in the derivation of the DLCQ Hamiltoninan for $SU(2)$ gluodynamics is to rewrite the above action of gluodynamics in terms of the Fourier modes $\Psi_{+,I}^k \equiv \langle \chi_k|A_{+,I} \rangle$, $k \neq 0$, $G_{+,I}^k = \langle \chi_0|A_{+,I} \rangle$, and $\alpha = \langle \chi_0|A_- \rangle$. The field strengths are expressed in these terms as follows:

$$E_+^0 = -\dot{\alpha}, \quad E_+^k = -i\omega_k \Psi_+^k, \tag{3}$$

$$E_I^0 = -\partial_I \alpha, \quad E_I^k = -i\omega_k \Psi_I^k, \tag{4}$$

and

$$F_{+I}^0 = \mathcal{F}_{+I} - i\frac{g}{\sqrt{L}} \sum_k f(\sigma_k - \sigma_{-k}) \Psi_+^k \Psi_I^{-k}, \tag{5}$$

$$F_{+I}^{k \neq 0} = \mathcal{D}_+ \Psi_I^k - \mathcal{D}_I \Psi_+^k - $$
$$-i\frac{g}{\sqrt{L}} \sum_{k_1} f(\sigma_{k_1} - \sigma_{k-k_1}) \Psi_+^{k_1} \Psi_I^{k-k_1}. \tag{6}$$

There are similar explicit expressions for F_{IJ}. The above sums over k do not include the term $k = 0$. Hereafter, we single out the zero mode terms

$k = 0$. The zero mode terms form above the Abelian field strengths, and the Abelian covariant derivatives:

$$\mathcal{F}_{ij} \equiv \partial_i G_j - \partial_j G_i, \tag{7}$$

$$\mathcal{D}_i \Psi_j^k \equiv \partial_i \Psi_j^k - i \frac{g \sigma_k}{\sqrt{L}} G_k \Psi_j^k \tag{8}$$

The lower case Latin indixes stand here for $+, I$. Note that the zero modes G_i play the role of an Abelian gauge field, and Ψ_i^k are the (charged) vector fields in this Abelian gauge theory.

The objects above that differ in notations from their analogs in [11] are

$$\omega_k \equiv \frac{2\pi n_k}{L} + \frac{\sigma_k g}{\sqrt{L}} \alpha, \tag{9}$$

and

$$f(\sigma) \equiv \epsilon(\sigma)(-1)^{|\sigma|}, \tag{10}$$

where the ϵ function returns the sign of its argument, and vanishes with the argument.

We single out the terms in S_{glue} containing the time derivatives, and G_+:

$$S_{glue} = \int dx^+ dy \left[\frac{\dot{\alpha}^2}{2} - \dot{\alpha} \partial_I G_I + \sum_k i \omega_k \Psi_I^{-k} \dot{\Psi}_I^k + G_+ G \right] - \int dx^+ H', \tag{11}$$

where

$$G \equiv -\Delta_\perp \alpha + \frac{g}{\sqrt{L}} \sum_k \omega_k \sigma_k \Psi_I^{-k} \Psi_I^k, \tag{12}$$

and H' is defined by comparison with (1).

Go over to the first order formalism replacing S_{glue} with an equivalent action:

$$S_{glue} = \int dx^+ dy \left[\pi \dot{\alpha} - \frac{(\pi + \partial_I G_I)^2}{2} + \sum_k i \omega_k \Psi_I^{-k} \dot{\Psi}_I^k + G_+ G \right] - \int dx^+ H'. \tag{13}$$

To see that this is indeed equivalent, vary over π, and substitute the result, $\pi = \dot{\alpha} - \partial_I G_I$, back in the action. From the above expression for π, its gauge transformation is $\pi' = \pi - \Delta_\perp \phi$, where ϕ is the parameter of the Abelian gauge transformations. Therefore, we can chose a gauge where $\pi = p/\sqrt{V_\perp}$ with $p \equiv \int dx^\perp \pi / \sqrt{V_\perp}$ ($V_\perp$ is the volume of the transverse plane). In this gauge

$$S_{glue} = \int dx^+ p\dot{q} + \int dx^+ dy \left[\sum_k i \omega_k \Psi_I^{-k} \dot{\Psi}_I^k + G_+ G \right] - \int dx^+ H, \tag{14}$$

where

$$q \equiv \int \frac{dx^\perp}{\sqrt{V_\perp}}\, \alpha, \tag{15}$$

and

$$H = \frac{p^2}{2} + \int dy \frac{(\partial_I G_I)^2}{2} + H'. \tag{16}$$

The same result can be achieved without gauge fixing by a corresponding transformation of the Ψ_I-variables.

Let us point out that the label k above runs over the following set:

$$k \in \mathcal{K}, \; \mathcal{K} = \{(n,\sigma)|(n=0) \to (\sigma \neq 0)\}. \tag{17}$$

Here n is any integer, and σ is either $+1$, -1, or 0. In words, k is a pair of an integer and a sign (the sign can vanish), and if the integer in the pair vanishes, the sign is not vanishing. Recall that k labels the nonvanishing eigenvalues of iD_-. The above structure of $\mathcal{K}$ is a feature of $SU(2)$. It is however a general property valid for any unitary gauge group that there is a reflection operation defined on $\mathcal{K}$. Namely, for any k there is a $-k$. The corresponding eigenvalue is reflected, $\omega_{-k} = -\omega_k$, and the Fourier coefficient is complex conjugated, $\Psi^{-k} = (\Psi^k)^\dagger$. These properties of the spectrum of iD_- are parallel to the corresponding properties of the spectrum of, say, $i\partial_-$.

We will use the set of indexes $|\mathcal{K}|$ obtained from $\mathcal{K}$ by identifying the elements related by the above reflection operation:

$$|\mathcal{K}| = \{(n,\sigma)|(n \geq 0)\&((n=0) \to (\sigma = +1))\}. \tag{18}$$

The index $|k|$ belongs to $|\mathcal{K}|$ and is obtained from the corresponding $k \in \mathcal{K}$ by a choice of sign ($|k| = k$, or $|k| = -k$).

Let us go over to new variables, $A_I^{|k|}$, replacing Ψ_I^k. This change of variables is governed by the relation

$$\Psi_I^k = \frac{\theta(\omega_k)}{\sqrt{2|\omega_k|}} A_I^{|k|} + \frac{\theta(-\omega_k)}{\sqrt{2|\omega_k|}} (A_I^{|k|})^\dagger, \tag{19}$$

where the dagger denotes the complex conjugation.

A straightforward check shows that $A_I^{|k|}$, $(A_I^{|k|})^\dagger$, and p, q form the set of canonical variables of the light-front gluodynamics.

The variables $A_I^{|k|}$, $(A_I^{|k|})^\dagger$ are not independent. They satisfy a first class constraint (Gauss Law):

$$\int dx^\perp \sum_{|k|} \epsilon(\omega_k) \sigma_k (A_I^{|k|})^\dagger A_I^{|k|} = 0. \tag{20}$$

This constraint generates global (independent of the transverse coordinate) changes of the phase of the creation-annihilation operators. Upon quantization, we implement it as a condition on the states.

The Hamiltonian of the formulation is given by (16). It is quadratic in the nondynamical variables Ψ_+^k and G_I. They are nondynamical because there are no time derivatives of these fields in the action. The crucial point is that the corresponding quadratic form is nondegenerate, i.e., the classical equations allow us to express these fields in terms of the dynamical variables. One should be careful in integrating Ψ_+^k and G_I out. In the functional integral formulation, integrating them out is reduced to replacing them with the classical solutions in terms of the dynamical variables, and multiplying the integrand of the functional integral with the corresponding determinant. The determinant depends on the dynamical variables, and cannot be neglected. We consider this problem in [7].

Another point is in order. The Hamiltonian (16) depends on α (the zero mode of A_-). We express it in terms of the dynamical variables with (15), and with the Gauss law $G = 0$ (G is defined in (12)).

This completes the description of the DLCQ Hamiltonian of $SU(2)$ gluodynamics.

3. Conclusion

The relation of the DLCQ Hamiltonian to conventional formulation is far from obvious. One of the open questions is the structure of the ultraviolet renormalization in DLCQ. As we argue in [7], to consider the ultraviolet renormalization in DLCQ formulation, the light-front time should be discretized, and diagonalization of the transfer matrix should be considered.

An important lead is a comparison with standard perturbation theory. Most promising is the comparison of the ultraviolet renormalization in DLCQ with the standard perturbative renormalization. DLCQ boils down to an infrared regularization. We hope that it does not influence the ultraviolet properties, and the standard ultraviolet renormalization can be used to renormalize the DLCQ formulation.

In conclusion, in this paper, we gave a definition to the DLCQ Hamiltonian for $SU(2)$ gluodynamics.

Acknowledgments

GP is grateful to the organizers of the Symposium and Workshop "Continuous Advances in QCD 2002/Arkadyfest" for the possibility to present these results. This work is supported in part by the Russian Foundation for Basic Research, grant No. 00-02-17432.

References

1. S. Weinberg, *Phys. Rev.* **150**, 1313 (1966).
2. J. B. Kogut and D. E. Soper, *Phys. Rev.* **D1**, 2901 (1970).
3. T. Maskawa and K. Yamawaki, *Prog. Theor. Phys.* **56**, 270 (1976).
4. P. A. M. Dirac, *Rev. Mod. Phys.* **21**, 392 (1949).
5. H. C. Pauli and S. J. Brodsky, *Phys. Rev.* **D32**, 1993 (1985).
6. T. Eller, H. C. Pauli and S. J. Brodsky, *Phys. Rev.* **D35**, 1493 (1987).
7. V. T. Kim, G. B. Pivovarov and J. P. Vary, *Work in Progress*.
8. E. Tomboulis, *Phys. Rev.* **D8**, 2736 (1973).
9. V. A. Franke, Yu. V. Novozhilov and E. V. Prokhvatilov, *Lett. Math. Phys.* **5**, 437 (1981).
10. V. T. Kim, V. A. Matveev, G. B. Pivovarov and J. P. Vary, *Phys. Rev.* **D63**, 054009 (2001) [arXiv:hep-th/9910157].
11. G. B. Pivovarov and J. P. Vary, *Phys. Rev.* **D65**, 125004 (2002) [arXiv:hep-th/0111286].

ON MIXED PHASES IN GAUGE THEORIES

V.L. CHERNYAK

Budker Institute for Nuclear Physics
630090 Novosibirsk, Russia
E-mail: chernyak@inp.nsk.su

In many gauge theories at different values of parameters entering Lagrangian, the vacuum is dominated by coherent condensates of different mutually non-local fields (for instance, by condensates of electric or magnetic charges, or by various dyons). It is argued that the transition between these "dual to each other" phases proceeds through an intermediate "mixed phase", having qualitatively different features. The examples considered include: ordinary YM, $N = 1$ SYM, $N = 1$ SQCD, and broken $N = 2$ SYM and SQCD.

To Arkady,
my friend and teacher

1. $SU(N_c)$ - YM at $\theta \neq 0$ and dyons

The physics of this theory, and in particular the vacuum energy density $\overline{E}_{vac}(\theta, N_c)$, is supposed to be periodic in $\theta \to \theta + 2k\pi$. On the other hand, the standard large N_c-counting rules imply ($b_o = 11/3$):

$$\overline{E}_{vac} = -N_c^2 \frac{b_o}{4} \Lambda^4 F(\theta/N_c) , \qquad (1)$$

with $F(z \to 0) = 1 - c_1 z^2 + c_2 z^4 + \dots$ [a]   It was first pointed out by E. Witten [1] (see also [2] for a similar behaviour in the "stringy-YM" theory) that the N_c-dependence in Eq.(1) and periodicity in $\theta \to \theta + 2\pi$ imply together that the function $F(z)$ should be nonanalytic in its argument. So, for instance, instead of Eq.(1), the more explicit form of dependence of $\overline{E}_{vac}$ on θ will rather look as:

$$\overline{E}_{vac} = -N_c^2 \frac{b_o}{4} \Lambda^4 \left\{ min \sum_k f\left(\frac{\theta + 2k\pi}{N_c}\right) \right\} , \qquad (2)$$

---

[a]The numerical coefficient c_1 is positive, but it is a dynamical quantity and can not be determined from general considerations alone.

with $f(z)$ being the "normal" analytic function.

The qualitative behaviour of the curve $\overline{E}_{vac}(\theta, N_c)$ looks as follows. First, it is symmetric in $\theta \to -\theta$ and periodic in $\theta \to \theta + 2k\pi$. Further, it has minimum at $\theta = 0$ and begins to increase with increasing $\theta > 0$, as it follows from general considerations of the Euclidean functional integral determining this theory. It reaches its maximal value at $\theta = \pi$. The curve itself is continuous at this point, but there is a cusp so that $\overline{E}_{vac}$ begins to decrease in a symmetric way in the interval $\pi < \theta < 2\pi$, reaching the same minimal value at $\theta = 2\pi$.

As for the qualitative behaviour of the topological charge density, $\overline{P}_{vac}(\theta, N_c)$, it follows from the relation: $\overline{P}_{vac}(\theta, N_c) \sim d\overline{E}_{vac}/d\theta$, and looks as follows. First, it is antisymmetric in $\theta \to -\theta$ and periodic in $\theta \to \theta + 2k\pi$. So, it is zero at $\theta = 0$ and increases with θ reaching its maximum value at $\theta = \pi$. There is a discontinuity at this point, so that the curve jumps to the same but negative value as θ overshoots π, and increasing in a symmetric way reaches zero at $\theta = 2\pi$.

The above described nonanalytic (cusped) behaviour of $\overline{E}_{vac}(\theta, N_c)$ along the real θ-axis agrees, in particular, with the asymptotic behaviour of $\overline{E}_{vac}(\theta, N_c)$ at large imaginary values of θ, $i\theta/N_c = \tilde{\theta}/N_c \gg 1$, obtained in [3]:

$$\overline{E}_{vac}(\theta, N_c) \sim -N_c^2 \Lambda^4 \exp\left\{\frac{4}{b_o}\frac{\tilde{\theta}}{N_c}\right\}. \tag{3}$$

It is seen from Eq.(3) that $\overline{E}_{vac}(\theta, N_c)$ is not naturally periodic at $\tilde{\theta} \to \tilde{\theta} \pm 2i\pi$. Rather, it implies that periodic $\overline{E}_{vac}(\theta, N_c)$ is analytic in the strip $-\pi < Re\,\theta < \pi$ in the complex θ-plane, and is glued then periodically strip by strip.

The natural physical interpretation explaining the origin of the above described cusped behaviour of $\overline{E}_{vac}(\theta)$ along the real θ-axis has been proposed in [3], and looks as follows.

Let us suppose the "standard" picture of the confinement mechanism to be valid, i.e. those of the dual superconductor. By this we imply here the dynamical mechanism with composite (naturally adjoint) Higgs field which determines the formation of $U(1)^{N_c-1}$ from the original $SU(N_c)$, $SU(N_c) \to U(1)^{N_c-1}$, and besides the $U_i(1)$-magnetically charged excitations (monopoles) condense. We will be interested to trace the qualitative behaviour of this vacuum state in its dependence of θ. For this, it will be sufficient to consider the "first" U(1)-charge only with its monopoles, the dual photon and corresponding $g^\pm$ gluons as if it were the SU(2) theory, because θ is $SU(N_c)$-singlet and all other U(1) charges will behave the same way under variation of θ.

As has been shown by E. Witten [4], the pure monopole M=(magnetic charge $=1$, electric charge$=0$) at $\theta = 0$ turns into the dyon with charges $d_1^\theta = (1, \theta/2\pi)$ at $\theta \neq 0$. So, the coherent condensate of monopoles in the vacuum at $\theta = 0$ turns into the condensate of d_1^θ-dyons as θ starts to deviate from zero, and the vacuum energy density begins to increase for this reason.

It is a specific property of our system that there are two types of condensates made of the dyons and antidyons with the charges: $\{(1, 1/2); (-1, -1/2)\}$ and $\{(1, -1/2); (-1, 1/2)\}$, and having the same energy density. This can be seen, for instance, as follows. Let us start from the pure monopole condensate at $\theta = 0$ and let us move anticlockwise along the path: $\theta = 0 \to \theta = \pi$. The vacuum state will consist of $(1, \frac{1}{2})$ - dyons and $(-1, -\frac{1}{2})$ - antidyons. Let us move now clockwise along the path: $\theta = 0 \to \theta = -\pi$. The vacuum state will consist now of $(1, -\frac{1}{2})$ - dyons and $(-1, \frac{1}{2})$ - antidyons. Because the vacuum energy density is even under $\theta \to -\theta$, these two vacuum states are degenerate. [b]

Besides, these two states belong to the same world as they are reachable one from another through a barrier, because there are electrically charged gluons, $g^\pm = (0, \pm 1)$, which can recharge these $(1, \pm 1/2)$ - dyons into each other. In contrast, the two vacuum states, $|\theta\rangle$ and $|-\theta\rangle$ at $\theta \neq 0, \pi$ are unreachable one from another and belong to different worlds, as there is no particles in the spectrum capable to recharge the $(1, \pm\theta/2\pi)$ - dyons into each other.

Thus, the vacuum state becomes twice degenerate at $\theta = \pi$, so that the "level crossing" (in the form of rechargement: $\{d_1 = (1, 1/2), \bar{d}_1 = (-1, -1/2)\} \to \{d_2 = (1, -1/2), \bar{d}_2 = (-1, 1/2)\}$) can take place if this will lower the energy density at $\theta > \pi$. And indeed it lowers, and this leads to a casp in $\bar{E}_{vac}(\theta)$. At $\theta > \pi$ the vacuum is filled now with the coherent condensate of new dyons with the charges: $d_2^\theta = (1, -1 + \theta/2\pi)$, $\bar{d}_2^\theta = (-1, 1 - \theta/2\pi)$. As θ increases further, the electric charge of these d_2^θ - dyons decreases, and the vacuum energy density decreases with it. Finally,

---

[b]The existence of two vacuum states at $\theta = \pi$ does not follow from the symmetry considerations alone, like $\bar{E}_{vac}(\theta) = \bar{E}_{vac}(-\theta)$ and $\bar{E}_{vac}(\theta) = \bar{E}_{vac}(\theta + 2\pi k)$. It is sufficient to give a counterexample. So, let us consider the SU(2) Yang-Mills together with the Higgs doublet with large vacuum condensate. In this case the θ-dependence of the vacuum energy density is due to a rare quasiclassical gas of instantons, and is $\sim \cos(\theta)$. All the above symmetry properties are fulfilled, but there is only one vacuum state at $\theta = \pi$. (See also the end of this section.)

at $\theta = 2\pi$ the d_2^θ-dyons (which were the (1,-1)-dyons at $\theta = 0$) become pure monopoles, and the vacuum state becomes exactly as it was at $\theta = 0$, i.e. the same condensate of pure monopoles and antimonopoles.

We emphasize that, as it follows from the above picture, it is wrong to imagine the vacuum state at $\theta = 2\pi$ as, for instance, a condensate of dyons with the charges (1,-1), degenerate in energy with the pure monopole condensate at $\theta = 0$. [c]

Physically, the above rechargement process will appear as a typical first order phase transition. After θ overshoots π, in a space with the coherent condensate of $d_1 = (1, 1/2)$ - dyons and $\bar{d}_1 = (-1, -1/2)$ - antidyons, the bubbles will appear with the coherent condensate of $d_2 = (1, -1/2)$ - dyons and $\bar{d}_2 = (-1, 1/2)$- antidyons deep inside each bubble, and with a transition region surface (domain wall) through which the averaged densities of two type dyons interpolate smoothly. These bubbles expand then over all the space through the rechargement process $d_1 + \bar{d}_1 \rightarrow d_2 + \bar{d}_2$ occuring on a surface of each bubble. This rechargement can be thought as going through a copious "production" of charged gluon pairs $g^+ g^-$, so that the underlying processes are: $[d_1 = (1, 1/2)] + [g^- = (0, -1)] \rightarrow [d_2 = (1, -1/2)]$ and $[\bar{d}_1 = (-1, -1/2)] + [g^+ = (0, 1)] \rightarrow [\bar{d}_2 = (-1, 1/2)]$.

Some analogy with the simplest Schwinger model may be useful at this point, in connection with the above described rechargement process. Let

---

[c]In this respect, the widely used terminology naming the two singularity points $u = \pm\Lambda^2$ on the $N = 2$ $SU(N_c = 2)$ SYM moduli space as those where monopoles and respectively dyons become massless, is not quite adequate (and may be dangerous for this reason, leading to wrong conclusions). It is based on *quantum numbers* $\bar{n} = (n_m, n_e)$ of corresponding fields, and these quantum numbers are always the same independently of the point of the moduli space we are staying in, and are not direct physical observables. In contrast, the standard physical terminology is based on *charges* $\bar{g} = (g_m, g_e)$ which are the direct physical observables because, by definition, the Coulomb interaction of two particles is proportional to product of their charges, not quantum numbers. In distinction from *quantum numbers*, the values of *charges* depend on the point of the moduli space, due to Witten's effect.

To illustrate, let us start from the vacuum $u = \Lambda^2$ where, by definition, the massless particles are pure monopoles and let us move, for instance, along a circle to the point $u = -\Lambda^2$. On the way, the former massless monopole increases its mass because it becomes the $d_1^\theta = (1, \theta(u)/2\pi)$ - dyon (here $\theta(u)/2\pi = Re\,\tau(u)$). At the same time, the former massive $d_2^o = (1, -1)$ - dyon diminishes its mass as it becomes the $d_2^\theta = (1, -1 + \theta(u)/2\pi)$ - dyon. When we reach the point $u = -\Lambda^2$, i.e. $\theta(u) = 2\pi$, the former dyon becomes massless just because it becomes the pure monopole here. So, an observer living in the world with $u = -\Lambda^2$ will also see the massless monopoles (not dyons, and this is distinguishable by their Coulomb interactions between themselves and with other dyons), exactly as those living in the world with $u = \Lambda^2$.

us consider first the pure QED_2 without finite mass charged particles, and let us put two infinitely heavy "quarks" with the charges $\pm\theta/2\pi$ (in units of some e_o) at the edges of our space. It is well known [5] that this is equivalent to introducing the θ-angle into the QED_2 Lagrangian. As a result, there is the empty vacuum at $\theta = 0$, and the long range Coulomb "string" at $\theta \neq 0$. The vacuum energy density behaves as: $\overline{E}_{vac}(\theta) = C_o e_o^2 \theta^2$, $C_o = const$, at any $0 \leq \theta < \infty$.

Let us add now some finite mass, $m \gg e_o$, and of unit charge e_o field ϕ to the Lagrangian. When there are no external charges, this massive charged field can be integrated out, resulting in a small charge renormalization. But when the above quarks are introduced, the behaviour of $\overline{E}_{vac}(\theta)$ becomes nontrivial. The charge of the external quark tends to $1/2$ as θ approaches π. As θ overshoots π it becomes preferable to produce a pair of ϕ- particles, $\phi^+ \phi^-$, from the vacuum. They separate so that to recharge the external quarks: $(\pm 1/2) \rightarrow (\mp 1/2)$ (without changing the volume energy), and the external charges become equal $(\theta/2\pi - 1)$ and $(-\theta/2\pi + 1)$ at $\theta > \pi$. As a result of this rechargement, there appears a cusp in $\overline{E}_{vac}(\theta)$ and it begins to decrease at $\theta > \pi$, so that the former "empty" vacuum is reached at $\theta = 2\pi$. Therefore, the behaviour of $\overline{E}_{vac}(\theta)$ will be: $\overline{E}_{vac}(\theta) = C_o e_o^2 \cdot \{\min_k (\theta + 2\pi k)^2\}$, so that $\overline{E}_{vac}(\theta) = C_o e_o^2 \theta^2$ at $0 \leq \theta \leq \pi$, and $\overline{E}_{vac}(\theta) = C_o e_o^2 (2\pi - \theta)^2)$ at $\pi \leq \theta \leq 2\pi$.

Let us return however to our dyons. The above described picture predicts also a definite qualitative behaviour of the topological charge density, $\overline{P}(\theta)$. At $0 < \theta < \pi$, i.e. in the condensate of the $d_1^\theta = (1, \theta/2\pi)$ - dyons and $\bar{d}_1^\theta = (-1, -\theta/2\pi)$ - antidyons, the product of signs of the magnetic and electric charges is positive for both d_1^θ - dyons and $\bar{d}_1^\theta$ - antidyons. Thus, these charges give rise to the correlated field strengths: $\vec{E} \parallel \theta\vec{H}$, $\vec{E} \cdot \vec{H} > 0$, and both species contribute a positive amount to the mean value of the topological charge density, so that $\overline{P}_1(\theta) > 0$ and grows monotonically with θ in this interval following increasing electric charge $\sim \theta/2\pi$ of the dyon.

On the other side, at $\pi < \theta < 2\pi$, i.e. in the condensate of the $d_2^\theta = (1, -1 + \theta/2\pi)$ - dyons and $\bar{d}_2^\theta = (-1, 1 - \theta/2\pi)$ - antidyons, the product of signs of the magnetic and electric charges is negative for both d_2^θ - dyons and $\bar{d}_2^\theta$ - antidyons. Thus, both species contribute a negative amount to $\overline{P}_2(\theta)$, such that: $\overline{P}_2(\theta) = -\overline{P}_1(2\pi - \theta)$, and $\overline{P}(\theta)$ jumps reversing its sign at $\theta = \pi$ due to rechargement.

On the whole, it is seen that the cusped behaviour of $\overline{E}_{vac}(\theta)$ and discontinuous behaviour of $\overline{P}(\theta)$ appear naturally in this picture of the confinement mechanism in $SU(N_c)$ - YM theory, and are exactly the same that are expected from simplest general considerations and were described in the beginning of this section.

Clearly, at $0 \leq \theta < \pi$ the condensate made of only the $d_1^{\theta} = (1, \theta/2\pi)$ - dyons (recalling also for a possible charged gluon pair production) can screen the same type $d_k^{\theta} = [\, const\, (1, \theta/2\pi) + (0, k)\,]$ - test dyon only ($k = 0, \pm 1, \pm 2, \ldots$; and the same for the d_2^{θ} - dyons at $\pi < \theta \leq 2\pi$). So, the heavy quark-antiquark pair will be confined at $\theta \neq \pi$.

New nontrivial phenomena arise at $\theta = \pi$. Because there are two degenerate states, i.e. the condensates of $(1, \pm\frac{1}{2})$ - dyons (and antidyons), a "mixed state" configuration becomes possible with, for instance, each condensate filling a half of space only, and with the domain wall interpolating between them. The simplest reasonings about the energy scales involved in this domain wall are as follows. The masses of relevant gluons $g^{\pm} = (0, \pm 1)$ and both dyons $(1, \pm 1/2)$ coexisting together in the bulk of the domain wall are naturally $\simeq \Lambda_{YM}$, and so of the same size will be increase in energy density. Besides, there are $(N_c - 1)$ independent $U_i(1)$ charges. On the whole, therefore, the domain wall tension will be $T \sim N_c \Lambda_{YM}^2$, while its typical width will be $\sim 1/\Lambda_{YM}$.

Physically, the above domain wall represents "a smeared rechargement", i.e. smeared over space interpolation of electrically charged degrees of freedom between their corresponding vacuum values, resulting in a smooth variation of the averaged densities of both type dyons $(1, \pm 1/2)$ through the domain wall. Surprisingly, there is no confinement inside the bulk of such domain wall.

The reason is as follows. Let us take the domain wall interpolating along the z-axis, so that at $z \to -\infty$ there is the main coherent density of $d_1 = (1, 1/2)$ - dyons, and at $z \to \infty$, - that of $d_2 = (1, -1/2)$ - dyons. As we move from the far left to the right, the density of d_1-dyons decreases and there is also a smaller but increasing incoherent density of d_2-dyons. This small amount of d_2-dyons is "harmless", in the sense that its presence does not result in the screening of the corresponding charge. The reason is clear: the large coherent density of d_1-dyons keeps the d_2-dyons on the confinement, so that they can not move freely and appear only in the form of rare and tightly connected neutral pairs $\bar{d}_2 d_2$, with different pairs fluctuating independently of each other. As we are going further to the right, the density of these neutral pairs grows and their typical size increases (although they are still on the confinement), because the main density of d_1-dyons decreases. Finally, at some distance from the centre of the wall the percolation takes place, i.e. the d_2-dyons form a continuous coherent network and become released, so that the individual d_2-dyon can travel freely to arbitrary large distances (in the transverse xy-plane) but only within its network. And in this percolated region the coherent network of d_1-dyons still survives, so that *these two coherent networks coexist in space and form the new "mixed phase" with qualitatively different properties.*

This is a general feature, and each time when there will coexist coherent condensates of two mutually non-local fields, they will try to keep each other on the confinement, and will resemble the above described case.

The above mixed phase shares some features in common with the mixed state of the type-II superconductor in the external magnetic field. The crucial difference is that the magnetic flux is sourceless inside the superconductor, while in the above described mixed phase there are real dual to each other charges, each type living within its network.

As we move further to the right, the density of d_2-dyons continue to increase while those of d_1 continue to decrease. Finally, at the symmetric distance to the right of the wall centre the "inverse percolation" takes place, i.e. the coherent network of d_1-dyons decays into separate independently fluctuating neutral droplets whose average density (and size) continue to decrease with increasing z. Clearly, the picture on the right side repeats in a symmetric way those on the left one, with the d_1 and d_2 dyons interchanging their roles. [d]

Let us consider now the heavy test quark put inside the bulk of the domain wall, i.e. inside the mixed phase region. Clearly, this region has the properties of the "double Higgs phase". Indeed, because the (two-dimensional) charges of two dyons, $(1, 1/2)$ and $(1, -1/2)$, are linearly independent, polarizing itself appropriately this system of charges will screen any external charge put inside, and the quark one in particular.

Finally, if the test quark is put far from the bulk of the wall, the string will originate from this point making its way toward the wall, and will be screened inside the mixed phase (i.e. the double Higgs) region. However, as is clear from the above explanations, the electric string can not be stretched between two such domain walls.

---

[d]Evidently, if we replace the above domain wall with fixed $\theta = \pi$ by the domain wall of the light axion field $a(z)$, $m_a \ll \Lambda_{YM}$, interpolating between $a = 0$ at $z \to -\infty$ and $a = 2\pi$ at $z \to \infty$ with $a(z = 0) = \pi$, all the properties will remain the same in the bulk, i.e. at $|z| \lesssim (several)\,\Lambda_{YM}^{-1}$. The main difference will be that the condensates of $d_1 = (1, 1/2)$ and $d_2 = (1, -1/2)$ - dyons will turn into condensates of pure monopoles at $|z| \gg 1/m_a$.

Let us point out finally that the assumption about the confinement property of the $SU(N_c)$ YM theory is not a pure guess, as the above discussed nonanalytic (i.e. cusped) behaviour of the vacuum energy density, $\overline{E}_{vac}(\theta, N_c)$, is a clear evidence for a phase transition at some finite temperature. Indeed, at high temperatures the θ - dependence of the free energy density in the gluon plasma is under control and is : $\sim T^4(\Lambda/T)^{N_c b_o} \cos(\theta)$, due to a rare gas of instantons. It is important for us here that it is perfectly analytic in θ, and that the form of its θ dependence is T-independent, i.e. it remains to be $\sim \cos(\theta)$ when the temperature decreases. On the opposite side at $T = 0$, i.e. in the confinement phase, the θ - dependence is nonanalytic and, clearly, this nonanalyticity survives at small temperatures as there are no massless particles in the spectrum. So, there should be a phase transition (confinement - deconfinement) at some critical temperature, $T_c \simeq \Lambda$, where the θ - dependence changes qualitatively.

2. $N = 1$ $SU(N_c)$ SYM

In this theory the residual non-anomalous discrete axial symmetry is broken spontaneously, so that there are N_c vacuum states differing by the phase of the gluino condensate [6], [7]:

$$\langle 0| \lambda\lambda |0\rangle_k \sim N_c\Lambda^3 \exp\left\{i\frac{2\pi k}{N_c}\right\} . \tag{4}$$

Besides, it is widely believed that this theory is confining, similarly to the usual YM- theory. In what follows, we will suppose that the confinement mechanism here is the same as in the previous section, i.e. those of the dual superconductor. Our purpose in this section will be to describe qualitatively the physical properties of domain walls interpolating between the above vacua and, in particular, their ability to screen the quark charge [8].

For this, let us consider the effective theory obtained by integrating out all degrees of freedom except for the composite chiral field $\Omega = (W_\alpha^2/32\pi^2 N_c)$, $\Omega = (\lambda\lambda, \ldots, \theta^2(GG + iG\tilde{G}))/32\pi^2 N_c = (\rho\exp\{i\phi\}, \ldots)$.

Because the field $N_c\phi$ in SYM is the exact analog of θ in the ordinary YM, the physical interpretation and qualitative behaviour of $\overline{E}_{vac}(\theta)$ in the YM- theory described in the previous section can be transfered now to SYM, with only some evident changes:

a) $\overline{E}_{vac}(\theta) \to U(N_c\phi)$, and it is not the vacuum energy density now but rather the potential of the field ϕ;

b) if we start with the condensate of pure monopoles at $\phi = 0$, the rechargement $d_1^\phi = (1, N_c\phi/2\pi) \to d_2^\phi = (1, -1 + N_c\phi/2\pi)$ and the cusp in $U(N_c\phi)$ will occur now at $\phi = \pi/N_c$, so that at $\phi = 2\pi/N_c$ we will arrive at the next vacuum with the same pure monopole condensate but with shifted phase of the gluino condensate.

Let us consider now the domain wall interpolating along z-axis between two nearest vacua with $k = 0$ and $k = 1$, so that $\phi(z) \to 0$ at $z \to -\infty$, and $\phi(z) \to 2\pi/N_c$ at $z \to \infty$. There is a crucial difference between this case and those described just above where the field ϕ was considered as being space-time independent, i.e. $\phi(z) = const$. The matter is that the system can not behave now in a way described above (which allowed it to have a lowest energy $U(N_c\phi)$ at each given value of $\phi(z) = \phi = const$): i.e. to be the pure coherent condensate of d_1^ϕ - dyons at $0 \le \phi < \pi/N_c$, the pure coherent condensate of d_2^ϕ - dyons at $\pi/N_c < \phi \le 2\pi/N_c$, and to recharge suddenly at $\phi = \pi/N_c$. The reason is that the fields corresponding to electrically charged degrees of freedom also become functions of z at $q = \int dz \, [d\phi_{dw}(z)/dz] \ne 0$. So, they can not change abruptly now at $z = 0$ where $\phi_{dw}(z)$ goes through π/N_c, because their kinetic energy will become infinitely large in this case. Thus, the transition will be smeared necessarily.

The qualitative properties of the domain wall under consideration here will be similar to those described in the previous section. The main difference is that θ was fixed at π in sect.1, while $N_c\phi_{dw}(z)$ varies here between its limiting values, and the electric charges of dyons follow it.

So, at far left there will be a large coherent condensate of $d_1^\phi = (1, N_c\phi/2\pi)$-dyons (pure monopoles at $z \to -\infty$), and a small incoherent density of $d_2^\phi = (1, -1 + N_c\phi/2\pi)$-dyons. [e] The d_2^ϕ-dyons can not move freely in this region as they are on the confinement and appear as rare and tightly connected neutral pairs $\overline{d}_2^\phi d_2^\phi$ only. Therefore, their presence does not result in the screening of the corresponding charge. As we move to the right, the density of d_1^ϕ-dyons decreases while those of d_2^ϕ - increases. These last move more and more freely, but are still on the confinement. Finally, their density reaches a critical value at $z = -z_o$, so that a percolation takes place and the d_2^ϕ-dyons form a continuous coherent network within which the individual d_2^ϕ-dyons can move freely to any distance (in the transverse xy-plane). At the same time, there still sirvives a coherent condensate of d_1^ϕ-dyons, which still can freely move individually within their own network.

At the symmetrical point $z = z_o$ to the right of the domain wall centre at $z = 0$, the "inverse percolation" takes place, so that the connected network of d_1^ϕ-dyons decays into separate independently fluctuating neutral droplets,

---

[e]Other possible dyons play no role in the transition we consider, and we will ignore them.

whose density (and size) decreases with further increasing z. At large z we arrive at the vacuum state with a large coherent condensate of monopoles (former $(1, -1)$-dyons at large negative z).

Now, let us consider what happens when a heavy quark is put inside the bulk of the domain wall. The crucial point is that there is a mixture of all four dyon and antidyon species (of all $N_c - 1$ types): $d_1^\phi = (1, N_c\phi/2\pi)$, $\bar{d}_1^\phi = (-1, -N_c\phi/2\pi)$, $d_2^\phi = (1, -1 + N_c\phi/2\pi)$ and $\bar{d}_2^\phi = (-1, 1 - N_c\phi/2\pi)$ in this percolated region, with each dyon moving freely inside its coherent network. So, this region has the properties of "the double Higgs phase", as here both the d_1^ϕ and d_2^ϕ-dyons are capable to screen corresponding charges. And because the charges of d_1^ϕ and d_2^ϕ-dyons are linearly independent, polarizing itself appropriately this mixture of dyons will screen any test charge put inside, the heavy quark one in particular.

If the test quark is put at far left (right) of the wall, the string will originate from this point making its way towards a wall, and will disappear inside the bulk of the wall, i.e. in the mixed phase region where the string flux will be screened. The above described explanation of the physical phenomena resulting in quark string ending in the wall differs from both, those described by E. Witten in [9] and those proposed by I. Kogan, A. Kovner and M. Shifman in [10] (see also the footnote "c").

3. $N = 1$ $SU(2)$, $N_F = 1$ SQCD

As previously, we will imply here that there is confinement of electric charges in the $N = 1$ pure SYM- theory (see previous sections). Then, there will be three phases in this $N = 1$ $SQCD$ - theory, depending on the value of m - the mass parameter of the quark [8].

At small $m \ll \Lambda$, there will be the usual electric Higgs phase, with the large quark condensate $\langle \overline{Q}Q \rangle \sim (\Lambda^5/m)^{1/2}$, and light quark composite fields $(\overline{Q}Q)$ with masses $\sim m$. The effective Lagrangian for these fields is those of Affleck-Dine-Seiberg [11]. The heavy magnetically charged excitations (monopoles) will be confined, and so the monopoles will appear as rare and tightly bound neutral pairs only, with different pairs fluctuating independently of each other. [f]

---

[f]That there are monopoles in this theory at $m \ll \Lambda$ can be seen as follows. First, let us consider the effective Lagrangian obtained by integrating out degrees of freedom with energy scales $\mu \geq \mu_o$, $\mu_o = (\Lambda^5/m)^{1/4}$. These include the instanton contributions, as the typical instanton size is just $\bar{\rho} \sim \langle \overline{Q}Q \rangle^{-1/2} \sim 1/\mu_o$. The instanton will add the Affleck-Dine-Seiberg term $\Lambda^5/(\overline{Q}Q)$ to the original superpotential. Now, the so obtained

With increasing m the quark condensate and monopole mass decrease, while the density of monopole pairs (and their typical size) increases. At some value $m = c_1 \Lambda$ the percolation of the monopole droplets takes place, so that in the interval $c_1 \Lambda < m < c_2 \Lambda$ there will be the mixed phase (or equivalently, "the double Higgs phase") with two infinite size connected coherent networks of monopoles and quarks, with their densities (averaged over the scale of distances $R \gg 1/\Lambda$) being constant over the space, and following only the value of m.

There will be screening rather than confinement (although the difference between these two becomes to a large extent elusory here) of any test charge in this interval of m.

Finally, at $m = c_2 \Lambda$ the "inverse percolation" of the coherent electric quark condensate takes place, so that it decays into independently fluctuating neutral droplets whose density and typical size decrease with increasing m. At $m \gg \Lambda$ we arrive at the $N = 1$ SYM - theory with $\Lambda_{YM} = m^{1/6} \Lambda^{5/6}$ and with heavy $(m \gg \Lambda_{YM})$ quarks which are confined.

4.   Broken $N = 2$ $SU(2)$ SYM

Let us recall the famous solution of this theory by N. Seiberg and E. Witten, with the low energy Lagrangian (at small $\mu \ll \Lambda$) [12]:

$$L = \int d^4\theta \, \{ M^\dagger e^V M + \overline{M}^\dagger e^{-V} \overline{M} + Im\,(A_D A^\dagger / 4\pi) \} - \frac{i}{16\pi} \int d^2\theta \, \tau \, W_\alpha^2 +$$

$$+ \int d^2\theta \, \{ \sqrt{2} \, \overline{M} M A_D + \mu U \} + h.c. \tag{5}$$

Here M is the monopole field. Because it was not integrated yet, the terms entering the Lagrangian in Eq.(5) (τ, etc.)   do not contain the monopole loop contributions and have no singularity at $\langle U \rangle \to \Lambda^2$.

---

effective Lagrangian is the appropriate one to look, in particular, for a possible string solution because the characteristic distances involved in the string formation are larger than $\bar{\rho}$.

At the classical level, there will the solution for the Abrikosov-Nielsen-Olesen like string with the magnetic flux. But because the quarks are in the fundamental representation, the gauge group is $SU(2)$ which is simply connected and there are no truly uncontractable strings in this theory.   This implies that the above classical string will break up on account of quantum tunneling effects. Physically, this break up will be realized through the production of a pair of magnetically charged "particle" and "antiparticle", with their subsequent separation along the string axis to screen the external infinitely heavy monopoles at the string ends. So, these magnetically charged particles should be present in the excitation spectrum of this theory (even if they are not well formed).

Below, it will be convenient for us to consider A, τ and U in Eq.(5) as functions of the field A_D which has zero vacuum expectation value. The vacuum state we are dealing with is at $\langle U \rangle = \Lambda^2$, with $\tau \sim O(1)$ and $Im\,(A_D A^\dagger/4\pi) \sim A_D A_D^\dagger$ in Eq.(5) at small μ.

How the effective Lagrangian for these fields can look if μ is not small in comparison with Λ ? Because at $\mu \gg \Lambda$ the degrees of freedom which have been integrated out were heavy (in particular, the charged Higgs fields, with their masses $\sim \mu$), the $N = 2$ supersymmetry will be broken explicitly and μ- dependence will penetrate the effective Lagrangian. At the same time, it is not difficult to see that due to: a) holomorphicity; b) R-charge conservation (with the R-charge of μ equal two); c) the known limit at $\mu \ll \Lambda$, the additional μ- dependence can not appear in the F - terms, and so will appear in the D - terms only.

So, in terms with no more than two space-time derivatives, the only difference with Eq.(5) will be the appearance of Z - factors of the monopole and A_D - fields, Z_M and Z_H. Besides, for our qualitative considerations below, it will be sufficient to replace in these Z_i - factors the possible dependence on the monopole and U - fields by their vacuum values, $(M^\dagger M/U) \to \mu/\Lambda$, and to consider them as c-numbers: $Z_i = Z_i(\mu/\Lambda)$, with $Z_i(0) = 1$. Let us denote by L_μ the so obtained Lagrangian.

Our purpose now is to consider L_μ at large values of μ, $\mu \gg \Lambda$. Recalling that the original theory was $N = 2$ SYM broken by the mass term of the Higgs fields, we are ensured that at $\mu \gg \Lambda$ the Higgs fields become heavy, with their masses $m_H \to \mu$, and decouple. So, we end up with $N = 1$ SYM with the scale parameter: $\Lambda_{YM} = \mu^{1/3}\Lambda^{2/3}$, and this is the only scale of this theory.

On the other hand, one obtains from L_μ at $\mu \gg \Lambda$ that [g] Z_H stays intact, $Z_H(\mu/\Lambda \gg 1) \sim 1$, in order to have $m_H \sim \mu$, while the values of the dual photon and monopole masses look as

$$m_{\tilde{\gamma}}^2 \sim Z_M \left(\frac{\mu}{\Lambda}\right) \langle 0|\overline{M}M|0\rangle \sim Z_M \left(\frac{\mu}{\Lambda}\right) \mu\Lambda, \quad m_M^2 \sim Z_M^{-2} \left(\frac{\mu}{\Lambda}\right) \Lambda^2. \quad (6)$$

Let us combine now Eq.(6) with the additional assumption: *there is no*

<hr>

[g]In the above described set up, there is no need for the function $U(A_D)$ in the $\mu\,U(A_D)$-term in L_μ to be exactly the Seiberg-Witten function. For our purposes and for simplicity, it will be sufficient to keep only three first terms (i.e. constant, linear and quadratic in A_D) in the expansion of $U(A_D)$ in powers of A_D, to ensure that the adjoint Higgs becomes heavy, $m_H \sim \mu$, and decouples.

massless particles in the spectrum of $N=1$ SYM. Then, this requires [h] :

$$Z_M\left(\frac{\mu}{\Lambda}\right) \sim \left(\frac{\Lambda}{\mu}\right)^{1/3} \quad at \quad \mu \gg \Lambda. \tag{7}$$

It follows now from Eqs.(6),(7):

$$m_{\tilde{\gamma}} \sim m_M \sim \Lambda_{YM} \quad at \quad \mu \gg \Lambda, \tag{8}$$

i.e. *both the dual photon and monopole survive in the spectrum of the* $N = 1$ SYM. This is nontrivial in the sense that one of them or both could become heavy and decouple at $\mu \gg \Lambda$.

As for the value of the monopole condensate, it depends clearly on the normalization of the monopole field. In the presence of the monopole Z_M-factor, the normalization $\langle 0|\overline{M}M|0\rangle \sim \mu\Lambda$ is not the natural one. The appropriate normalization is: $\langle 0|\overline{N}N|0\rangle = \langle 0|(\sqrt{Z_M}\ \overline{M})(\sqrt{Z_M}\ M)|0\rangle$, and it has the right scale: $\langle 0|\overline{N}N|0\rangle \sim \Lambda_{YM}^2$.

On the whole, the above described results imply that there is no phase transition in the broken $N = 2$ SYM theory when going from small $\mu \ll \Lambda$ to large $\mu \gg \Lambda$, and they give a strong support to the widely accepted expectation that the $N = 1$ SYM theory is confining, with the confining mechanism those of the dual superconductor. In other words, when going from $\mu \ll \Lambda$ to $\mu \gg \Lambda$ in the broken $N = 2$ *SYM* - theory, the "external" adoint Higgs of this theory decouples and its role is substituted smoothly by the "internal" composite adjoint Higgs of the $N = 1$ *SYM* - theory.

5. Broken $N = 2$ $SU(2)$, $N_F = 1$ $SQCD$

The solution of the unbroken $N = 2$ theory has been given by N. Seiberg and E. Witten [13]. The original superpotential of the broken $N = 2 \rightarrow N = 1$ theory has the form:

$$W = m\,\overline{Q}Q + h\sqrt{2}\,\overline{Q}\,\frac{\tau^a}{2}\phi^a\,Q + \mu U, \tag{9}$$

where the quark fields Q and $\overline{Q}$ are in the **2** and $\overline{\mathbf{2}}$ representations of the colour group $SU(2)$, and ϕ is the adjoint Higgs field. The unbroken $N = 2$ SUSY corresponds to $\mu = 0$ and $h = 1$.

The properties of this $N = 2$ - broken theory have been considered previously in [14 - 16]. The most detailed description has been given recently by A. Gorsky, A. Vainshtein and A. Yung in [17], and we use widely the

---

[h]If $(\mu/\Lambda)^{1/3} \cdot Z_M(\mu/\Lambda) \to 0$, the dual photon will be massless (on the scale Λ_{YM}, i.e. $m_{\tilde{\gamma}}/\Lambda_{YM} \to 0$), while if $(\mu/\Lambda)^{1/3} \cdot Z_M(\mu/\Lambda) \to \infty$ the monopole will be massless.

results of this paper below. For our purposes, we will deal with the special case of light quarks weakly coupled to the Higgs field:

$$m \ll \Lambda, \quad \sigma = h \left(\frac{\Lambda}{m}\right)^{3/2} \ll 1, \tag{10}$$

where Λ is the scale parameter of the original fundamental theory ($\Lambda = 1$ in what follows). Under the conditions of Eq.(10), one vacuum state decouples and there remain two physically equivalent vacuum states. So, it will be sufficient to deal with one of them where the condensates of original fields take the values [17]:

$$\langle \overline{Q} Q \rangle \sim \mu m^{-1/2}; \quad \langle \phi^2 \rangle \sim m^{1/2}; \quad \langle \lambda\lambda \rangle \sim \mu m^{1/2}, \tag{11}$$

while the condensate of the monopole field is

$$\langle \overline{M} M \rangle \sim \mu m^{1/4}. \tag{12}$$

Under the conditions of Eq.(10), the only freedom remained is the relative value of μ and m, and the phase and physical content of this theory depend essentially on this. Indeed:

a) at sufficiently small μ; the quarks $Q\,(\overline{Q})$ are "heavy" and decouple, we are in the pure $N = 2$ SYM - theory with $\Lambda_{eff}^{(1)} = m^{1/4}\Lambda^{3/4}$, broken by the small μU - term. The vacuum is the Seiberg-Witten vacuum, i.e. the dominant condensate is the Higgs one, $\langle U \rangle$, leading to $SU(2) \rightarrow U(1)$, with $W^{\pm}$ masses $\sim m^{1/4}$. The lightest particles are the dual photon $\tilde{\gamma}$, its $N = 2$ partner A_D and monopole M, with small masses $\sim \mu^{1/2} m^{1/8}$. The monopole field condenses in the low energy $U(1)$ - theory, resulting in the confinement of electric charges. We will call this phase the magnetic one.

b) at sufficiently large μ; the Higgs field ϕ is heavy and decouples, we are in $N = 1$, $N_F = 1$ SQCD with $\Lambda_{eff}^{(2)} = \mu^{2/5}\Lambda^{3/5}$ and with light quark composite $\overline{Q}Q$- fields, with their masses $\sim m \ll \Lambda_{eff}^{(2)}$. Here, the large quark condensate $\langle \overline{Q}Q \rangle$ dominates, $SU(2)$ is broken completely and there is confinement of magnetically charged excitations (monopoles), see the footnote "f". The low energy effective Lagrangian is those of Affleck-Dine-Seiberg [11]. We will call this phase the electric one.

So, unlike the examples considered in previous sections, at the conditions given by Eq.(10) we have a good control here over the phases of our theory in both limiting cases of small and large values of μ, and these phases are dual to each other and dominated by condensates of mutually non-local fields. Our purpose now is to trace in more detail the transition between the magnetic and electric phases at some value $\mu \simeq \mu_o$, when going from small $\mu \ll \mu_o$ to large $\mu \gg \mu_o$ values of μ. We expect that this transition proceeds through the formation of the mixed phase in some region $c_1 \mu_o \lesssim \mu \lesssim c_2 \mu_o$ (with $c_1 < c_2$, but parametrically both $c_1 \sim c_2 \sim O(1)$). [i]

In the magnetic phase region $0 \leq \mu \lesssim \mu_o$ we will proceed in the same way as in the previous section, by retaining only the lightest fields of the dual photon $\tilde{\gamma}$, A_D and monopole M. All quark fields, in particular, are integrated out. Although, see Eq.(10), $h \to 0$ and quarks do not interact directly with the Higgs fields, they interact with massive charged gluons and gluinos and will give corrections in powers of $\sim (W_\alpha^2/m^3) \sim (\langle\lambda\lambda\rangle/m^3) \sim (\mu/m^{5/2})$. Further, being integrated out, the massive gluons and gluinos will transmit these corrections to the monopole Z_M-factor: $Z_M = Z_M(\mu/m^{5/2})$. So, the quarks really decouple only at $\mu \ll m^{5/2}$ where $Z_M(\mu/m^{5/2} \to 0) \to 1$, while at $\mu > m^{5/2}$ the quarks influence the physics and $Z_M \neq 1$.

Similarly, in the electric phase region $\mu_o \lesssim \mu$, after integrating out the Higgs and gauge fields, the quark Z_Q-factor will obtain corrections in powers of $(\langle\lambda\lambda\rangle/\mu^3) \sim (m^{1/2}/\mu^2)$, so that $Z_Q = Z_Q(m^{1/4}/\mu)$ and the heavy Higgs field really decouples only at $\mu \gg m^{1/4}$: $Z_Q(m^{1/4}/\mu \to 0) \to 1$. At $\mu_o \lesssim \mu < m^{1/4}$ the adjoint Higgs field still influences the physics, and $Z_Q \neq 1$.

It is not difficult to see that with the only choice:

$$Z_M^o \equiv Z_M\left(\frac{\mu_o}{m^{5/2}}\right) = \frac{m^{1/4}}{\mu_o}, \quad Z_Q^o \equiv Z_Q\left(\frac{m^{1/4}}{\mu_o}\right) = \frac{m}{\mu_o}, \quad (13)$$

all particle masses and all (properly normalized) condensates are matched in the transition region $\mu \simeq \mu_o$, as it should be:

$$M_H \sim M_M \sim M_Q \sim \mu_o; \quad M_{W^\pm} \sim M_\gamma \sim M_{\tilde{\gamma}} \sim m^{1/4} > \mu_o, \quad (14)$$

$$\langle\phi^2\rangle \sim \langle(\sqrt{Z_Q^o}\,\overline{Q})(\sqrt{Z_Q^o}\,Q)\rangle \sim \langle(\sqrt{Z_M^o}\,\overline{M})(\sqrt{Z_M^o}\,M)\rangle \sim m^{1/2}. \quad (15)$$

The nontrivial fact is that the number of matching conditions in Eqs. (14),(15) is larger than the choice of only two numbers in Eq.(13).

---

[i]Here and in other supersymmetric theories, the condensates of chiral superfields are always simple smooth functions of chiral parameters *in the whole parameter space*, see for instance Eqs.(11), (12). This does not contradict to a possibility for a system to be in qualitatively different (dual to each other) phases at different values of parameters, because these condensates are not order parameters for these phases. Rather, the masses of the direct and dual photon look more like the order parameters in the electric and magnetic phases respectively.

Unfortunately, the value of μ_o remained undetermined, we know only that $m^{5/2} < \mu_o < m^{1/4}$. It can be obtained supposing a kind of duality at the transition point $\mu = \mu_o$ (see Eq.(13)):

$$Z_M^o \sim 1/Z_Q^o \quad \to \quad \mu_o \sim m^{5/8}, \tag{16}$$

which looks reasonable. In this case Z_M behaves like $Z_M = [1 + (\mu/m^{5/2})]^{1/5}$ in the magnetic phase region $0 < \mu \lesssim m^{5/8}$, and Z_Q behaves like $Z_Q = [1 + (m^{1/4}/\mu)]^{-1}$ in the electric phase region $m^{5/8} \lesssim \mu$.

As was pointed out above, the example considered in this section has an advantage that we have a good control over the properties of the magnetic and electric phases at both sides, $\mu \ll \mu_o$ and $\mu \gg \mu_o$, of the transition region at $\mu \sim \mu_o$. As for the properties of the mixed phase in the transition region, they are similar to those described in previous sections. In short:

1) At $\mu \ll c_1 \mu_o$ the monopole condensate dominates, the quarks are confined and there are rare incoherent fluctuations of small size quark-antiquark pairs.

2) The density and the typical size of these pairs increases with increasing μ.

3) These quark bags (or strings) percolate at $\mu = c_1 \mu_o$. At $c_1 \mu_o \lesssim \mu \lesssim c_2 \mu_o$ the system is in the mixed phase, where two infinite size connected networks of electric and magnetic strings (or bags) coexist. In this region (parametrically): quark condensate $\sim$ monopole condensate, quark mass $\sim$ monopole mass, photon mass $\sim$ dual photon mass, etc.

4) At $\mu = c_2 \mu_o$ the quark condensate takes over, and the connected coherent magnetic condensate decays into separate independently fluctuating droplets (depercolation), whose density and typical size decrease with increasing μ.

5) At $\mu \gg c_2 \mu_o$ the magnetically charged excitations are heavy and confined into rare small size neutral pairs.

6.   Summary

As has been argued on a number of examples above, the mixed phases exist with their properties qualitatively different from those of pure phases. And the appearance of mixed phases is not an exception but rather a typical phenomenon in various gauge theories, both supersymmetric and ordinary.

Conclusion

Dear Arkady, be healthy and happy!

Acknowledgements

I am grateful to the Organizing Committee of "QCD-2002/Arkadyfest" for a kind hospitality and support.

References

1. E. Witten, *Nucl. Phys.* **B149**, 285 (1979); *Ann. Phys. (N.Y.)* **128**, 363 (1980)
2. E. Witten, *Phys. Rev. Lett.* **81**, 2862 (1998); [hep-th/**9807109**]
3. V. Chernyak, Preprint BINP 98-61 (1998) (revised) [hep-th/**9808092**]
4. E. Witten, *Phys. Lett.* **B86**, 283 (1979)
5. S. Coleman, Ann. Phys. (N.Y.) **101** (1976) 239
6. G. Veneziano and S. Yankielowicz, *Phys. Lett.* **B113**, 321 (1982)
7. M.A. Shifman and A.I. Vainshtein, *Nucl. Phys.* **B296**, 445 (1988)
8. V. Chernyak, *Phys. Lett.* **B450**, 65 (1999) [hep-th/**9808093**]
9. E. Witten, *Nucl. Phys.* **B507**, 658 (1997); [hep-th/**9706109**]
10. I.I. Kogan, A. Kovner and M. Shifman, *Phys. Rev.* **D57**, 5195 (1998); [hep-th/**9712046**]
11. I. Affleck, M. Dine and N. Seiberg, *Phys. Lett.* **B137**, 187 (1984)
12. N. Seiberg and E. Witten, *Nucl. Phys.* **B426**, 19 (1994) [hep-th/**9407087**]
13. N. Seiberg and E. Witten, *Nucl. Phys.* **B431**, 484 (1994) [hep-th/**9408099**]
14. K. Intrilligator and N. Seiberg, *Nucl. Phys.* **B431**, 551 (1994) [hep-th/**9408155**]
15. S. Elitzur, A. Forge, A. Giveon and E. Rabinovici, *Phys. Lett.* **B353**, 79 (1995) [hep-th/**9504080**]; *Nucl. Phys.* **B459**, 160 (1996) [hep-th/**9509130**]
16. S. Elitzur, A. Forge, A. Giveon, K. Intrilligator and E. Rabinovici, *Phys. Lett.* **B379**, 121 (1996) [hep-th/**9603051**]
17. A. Gorsky, A. Vainshtein and A. Yung, *Nucl. Phys.* **B584**, 197 (2000) [hep-th/**0004087**]

RESHUFFLING THE OPE:
DELOCALIZED OPERATOR EXPANSION

R. HOFMANN

Max-Planck-Institut für Physik,
Werner-Heisenberg-Institut
Föhringer Ring 6
80805 München,
Germany
E-mail: ralfh@mppmu.mpg.de

A generalization of the operator product expansion for Euclidean correlators of gauge invariant QCD currents is presented. Each contribution to the modified expansion, which is based on a delocalized multipole expansion of a perturbatively determined coefficient function, sums up an infinite series of local operators. On a more formal level the delocalized operator expansion corresponds to an *optimal* choice of basis sets in the dual spaces which are associated with the interplay of perturbative and nonperturbative N-point correlations in a distorted vacuum. A consequence of the delocalized expansion is the running of condensates with the external momentum. Phenomenological evidence is gathered that the gluon condensate, often being the leading nonperturbative parameter in the OPE, is indeed a function of resolution. Within a model calculation of the nonperturbative corrections to the ground state energy of a heavy quarkonium system it is shown exemplarily that the convergence properties are better than those of the OPE. Potential applications of the delocalized operator expansion in view of estimates of the violation of local quark-hadron duality are discussed.

1. Introduction

1.1. *Some historical remarks*

Once again: Happy Birthday, Arkady! This is a very lively conference that I am happy to attend.

In this talk I would like to report about work done jointly with André Hoang. The subject is to formulate a framework for a generalization of Wilson's operator product expansion (OPE), to confront this framework with experiment, and to demonstrate its usefulness in some model calculations.

As a warm-up I remind you of some of the OPE applications. The OPE is a powerful tool for the treatment of conformally invariant field theories where its convergence is explicitly shown. Unfortunately, this feature does not survive in realistic, four-dimensional field theories. Here the expansion is believed to be asymptotic at best.

The construction of effective theories like HQET or of the effective Hamiltonians of electroweak theory are crucially relying on the OPE which determines effective vertices by a perturbative matching of amplitudes with those of the fundamental theory. Heavy fields are effectively integrated out in this procedure (see [1] and references therein).

The OPE turned out to be important for a semi-perturbative derivation of the exact β function in $N = 1$ super Yang-Mills theory [2].

And, finally, there is a major application in QCD sum rules [3] where the "theoretical side" of a dispersion relation for the correlator of gauge invariant currents is represented in terms of an OPE. The occurence of power corrections to the low-order in α_s perturbative result can be argued from the summation of bubble graphs inducing so-called infrared renormalons, and formal contact to the OPE can be made [4]. A practical version of the OPE was proposed by Shifman, Vainshtein, and Zakharov in [3] where nonperturbative factors in the power corrections, the QCD condensates, are viewed as universal parameters of the QCD vacuum to be determined by experiment. Due to the universality of the condensates this method is rather predictive in practice.

1.2. *Problems of the OPE in QCD*

Despite its many successes in the framework of the QCD sum rule method there are some unanswered questions concerning the convergence properties of the OPE. Answering these questions very likely resolves the problems connected with local quark-hadron duality (see [5] in references therein) which can be formulated as follows.

Given analyticity of the correlator away from the positive real axis and invoking the optical theorem the discontinuity of the correlator at time-like momenta should yield the spectral function of the hadronic channel which corresponds to the QCD current in the correlator. This procedure, however, fails badly if a "practical" OPE with a small number of power corrections is used. Whereas the experimentally measured spectral function shows resonance wiggles at moderate time-like momenta a continuation of the OPE generates a smooth behavior down to small momenta. This may not harm the sum rule method since the relevant quantity there is a spectral *integral*, that is, an average over resonances which agrees quite well with the average over the associated discontinuity of the OPE. However, in applications, where the local properties of OPE discontinuities are needed [6], the violation of local quark-hadron duality can induce large errors.

Another concern, which is connected with the violation of local quark-hadron duality, is the asymptotic nature of the expansion. Asymptotic behavior can be argued from estimates of operator averages using the instanton gas approximation [3] or by appealing to the renormalon idea. From the OPE itself, where naively an expansion in powers of Λ_{QCD}/Q is assumed, there is no way of estimating the critical dimension at which the expansion ceases to approximate. In the framework of a delocalized expansion it was argued in [7] that the knowledge about nonperturbative, gauge invariant correlation functions would allow for such an estimate if the decay of effective correlation length with mass dimension is sufficiently fast.

The central theme of this talk is a nonlocal generalization of the OPE which has the potential to cure the above short-comings.

2. Delocalized operator expansion

The OPE of the correlator of a current j reads

$$i \int d^4x\, e^{iqx} \langle Tj(x)j(0)\rangle = Q^2 \sum_{d=0} \sum_{l=1}^{l_d} c_{dl}(Q^2)\langle O_{dl}(0)\rangle\,, \quad (Q^2 = -q^2 > 0)\,,$$

(1)

where d, l run over the dimension and the field content of the local operator O_{dl}, respectively. The Wilson coefficient c_{dl} is perturbatively calculable. The expansion (1) contains reducible chains of operators, such as

$$G^2, GD^2G, GD^4G, GD^6G\dots\,,$$
$$\bar{q}q, \bar{q}D_\mu\gamma_\mu q, \bar{q}D^2q, \bar{q}D^2D_\mu\gamma_\mu q\,,$$

(2)

D_μ denotes the gauge covariant derivative. Here "reducible" refers to the fact that each chain can be obtained from a Taylor expansion of an associated gauge invariant 2-point correlator which is not decomposible into gauge invariant factors. Examples, which correspond to the chains in (2), are the gluonic field strength correlator [8]

$$g_{\mu\nu\kappa\lambda}(x) \equiv tr\left\langle g^2 G_{\mu\nu}(x)\, Pe^{ig\int_0^x dz_\mu A_\mu(z)}\, G_{\kappa\lambda}(0) Pe^{ig\int_x^0 dz_\mu A_\mu(z)}\right\rangle$$

(3)

and the bilocal quark "condensate"

$$q(x) \equiv \left\langle \bar{q}(x)\, Pe^{ig\int_0^x dz_\mu A_\mu(z)}\, q(0)\right\rangle\,,$$

(4)

where P denotes path ordering. For lattice measurements the path connecting the points 0 and x is taken to be a straight line. This should capture the largest scale Λ in the nonperturbative decay of the correlator. The general case of N-point correlations will be addressed below. The effect of a delocalized expansion is a partial summation of such a chain in each order of the momentum expansion of the perturbative coefficient. This leads to an improvement of the convergence properties.

Let me make this explicit for the case of 2-point correlations in a 1-dimensional Euclidean world [9]. The contribution to the OPE, which arises from this, can formally be written as

$$
\int_{-\infty}^{\infty} dx\, f(x)\, g(x) = \int dx\, dy\, f(y)\, \delta(x - y)\, g(x)
$$

$$
= \int dx\, dy\, f(y) \left[\sum_{n=0}^{\infty} \frac{(-1)^n}{n!}\, y^n \delta^{(n)}(x) \right] g(x)
$$

$$
= \sum_{n=0}^{\infty} \left[\int dy\, f(y)\, y^n \right] \left[\int dx\, \frac{(-1)^n}{n!} \delta^{(n)}(x)\, g(x) \right]
$$

$$
\equiv \sum_{n=0}^{\infty} f_n(\infty)\, g_n(\infty)\,.
\tag{5}
$$

Here $f(x)$ is associated with perturbative correlations whereas $g(x)$ describes nonlocal, nonperturbative effects. The integral expressions in the first and second square bracket correspond to local Wilson coefficients and operators of the like of (2), respectively. From (5) it is seen that the part of the OPE being generated by 2-point correlations is induced by a bilinear form $(\cdot, \cdot) \equiv \int dx \cdot \times \cdot$. The expansion (5) corresponds to a specific choice of basis $\{\tilde{e}_n(\infty), e_n(\infty)\}$ in the associated dual space of square integrable functions where

$$
\tilde{e}_n(\infty) = y^n \qquad \text{and} \qquad e_n(\infty) = \frac{(-1)^n}{n!}\, \delta^{(n)}(x)\,.
\tag{6}
$$

and we have orthonormality

$$
(e_n, \tilde{e}_m) = \delta_{nm}\,.
\tag{7}
$$

From Eq. (5) it follows that

$$
f(x) = \sum_{n=0}^{\infty} \left[\int dy\, f(y)\, y^n \right] \frac{(-1)^n}{n!}\, \delta^{(n)}(x)\,.
\tag{8}
$$

This corresponds to an expansion of the perturbative factor $f(x)$ into degenerate, that is, zero-width multipoles. It is suggestive that an expansion into multipoles of width $\sim \Omega^{-1}$, which is comparable with $\Delta_f \sim Q^{-1}$, converges better (see Fig. 1). Like the OPE this expansion is controlled by

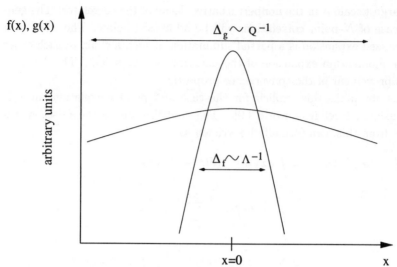

f(x), g(x)

arbitrary units

$\Delta_g \sim Q^{-1}$

$\Delta_f \sim \Lambda^{-1}$

x=0   x

Figure 1.   Schematic drawing of the short-distance function $f(x)$ and the nonperturbative 2-point-correlator $g(x)$ illustrating the scale hierarchy $\Delta_f/\Delta_g \sim \Lambda/Q \ll 1$.

powers of $\frac{\Delta_f}{\Delta_g} \sim \frac{\Lambda}{Q}$, and it breaks down if $\Delta_f \sim Delta_g$. The expansion into finite-width multipoles corresponds to a change of basis in dual space. Rather than expanding into the basis $\{\tilde{e}_n(\infty), e_n(\infty)\}$ one may expand into $\{\tilde{e}_n(\Omega), e_n(\Omega)\}$ where

$$\tilde{e}_n^{\Omega}(x) = \frac{H_n(\Omega x)}{(2\,\Omega)^n}, \quad e_n^{\Omega}(x) \equiv \frac{\Omega^{n+1}}{\sqrt{\pi\, n!}}\, H_n(\Omega x)\, e^{-\Omega^2 x^2} \tag{9}$$

We then have

$$\int dx\, f(x)\, g(x) = \sum_{n=0}^{\infty} f_n(\Omega)\, g_n(\Omega). \tag{10}$$

In (9) H_n denotes the nth Hermite polynomial. An expansion into any other basis that allows for a sensible multipole expansion of $f(x)$ is possible, and the restriction to a Gaussian based expansion is just for definiteness. Note that (9) reduces to (6) in the limit $\Omega \to \infty$. According to (9) the transformations, which link a basis with resolution parameter Ω to a basis with Ω', form a 1-parameter group. The moments $f_n(\Omega)$ in (10) have finite expansions into the Wilson coefficients $f_n(\infty)$

$$f_0(\Omega) = f_0(\infty), \quad f_2(\Omega) = f_2(\infty) - \frac{1}{2\Omega^2} f_0(\infty), \cdots. \tag{11}$$

Chosing a particular basis and a value for Ω is analogous to a choice of regularization and normalization point in pure perturbation theory. In

practice (11) enables to consider perturbative renormalization and delocalized operator expansion (DOE) separately. The expansion of $g(\Omega)$ into
local condensates is *infinite*, for example

$$g_0(\Omega) = g_0(\infty) + \frac{1}{2\Omega^2} g_2(\infty) + \cdots . \qquad (12)$$

To evaluate sums like in (12) one would have to know infinitely many
anomalous dimensions and condensates. On the other hand, one may be
more pragmatic and start with a (lattice inspired) model for $g(x)$ which is
believed to contain the information about all $g_n(\infty)$. To *each* order $\frac{\Lambda}{\Omega}$ one
may then derive evolution equations for $f_n(\Omega)$ and $g_n(\Omega)$ which sum these
contributions to *all* orders [9].

Setting $\Omega = Q$, the parametric organization of the expansion in $\frac{\Lambda}{Q}$ up
to an overall factor is

$$f_0(Q) \times g_0(Q) \sim \left(\frac{\Lambda}{Q}\right)^0 \times \{\left(\frac{\Lambda}{Q}\right)^0 + \left(\frac{\Lambda}{Q}\right)^2 + \cdots\}$$

$$f_2(Q) \times g_2(Q) \sim \left(\frac{\Lambda}{Q}\right)^2 \times \{\left(\frac{\Lambda}{Q}\right)^0 + \left(\frac{\Lambda}{Q}\right)^2 + \cdots\}$$

$$\vdots \qquad \qquad (13)$$

The generalization to K-dimensional Euclidean space is straightforward [
9]. It also allows for a treatment of N-point correlations in 4-dimensional
space. In particular, the $n = 0$ term of the contribution to the DOE from
2-gluon correlations yields the local Wilson coefficient times a "runnning
gluon condensate" $(Q = \Omega)$.

3. Applications

3.1. *Gluonic field strength correlator*

To have a well-motivated model for $g(x)$ for the case of 2-gluon correlations
we appeal to a parametrization of the gluonic field strength correlator (3)
as it was proposed in [8]

$$g_{\mu\nu\kappa\lambda}(x) \equiv (\delta_{\mu\kappa}\delta_{\nu\lambda} - \delta_{\mu\lambda}\delta_{\nu\kappa})[D(x^2) + D_1(x^2)] +$$

$$(x_\mu x_\kappa \delta_{\nu\lambda} - x_\mu x_\lambda \delta_{\nu\kappa} + x_\nu x_\lambda \delta_{\mu\kappa} - x_\nu x_\kappa \delta_{\mu\lambda})\frac{\partial D_1(x^2)}{\partial x^2} . \quad (14)$$

In a lattice measurement [10] the scalar functions D and D_1 were
parametrized as

$$D(x^2) = A_0 \exp[-|x|/\lambda_A] + \frac{a_0}{|x|^4} \exp[-|x|/\lambda_a] ,$$

$$D_1(x^2) = A_1 \exp[-|x|/\lambda_A] + \frac{a_1}{|x|^4} \exp[-|x|/\lambda_a], \qquad (15)$$

where the purely exponentially decaying terms are attributed to purely nonperturbative dynamics. The measurements at various quark masses for dynamical light quarks indicate that for $|x| \leq \lambda_A$ the nonperturbative contribution to D dominates the nonperturbative contribution to D_1 implying that the tensor structure of $g_{\mu\nu\kappa\lambda}$ is identical to that of the corresponding local operator obtained by letting $x \to 0$. This feature leads to great simplifications.

3.2. Extraction of the running gluon condensate

Using the DOE instead of the OPE, the running gluon condensate can be extracted from experiment in channels where 2-gluon correlations are dominant. We have done this for the $V + A$ correlator and the charmonium system. For a more detailed presentation see [9].

3.2.1. Sum rules for the V+A correlator

The spectral function for light quark production in the $V + A$ channel has been remeasured recently from hadronic τ decays for $q^2 \leq m_\tau^2$ by Aleph [11] and Opal [12]. In the OPE the associated current correlator is dominated by the gluon condensate, and the dimension 6 power corrections that are not due to a double covariant derivative in the local gluon condensate are suppressed. The corresponding currents are $j_\mu^{L/R} = \bar{u}\gamma_\mu(1 \pm \gamma_5)d$ and the relevant correlator in the chiral limit reads

$$i \int d^4x \, e^{iqx} \left\langle T j_\mu^L(x) j_\nu^R(0) \right\rangle = (q_\mu q_\nu - q^2 g_{\mu\nu}) \, \Pi^{V+A}(Q^2), \qquad Q^2 = -q^2.$$

$$(16)$$

Since the correlator $\Pi(Q^2)^{V+A}$ is cutoff-dependent itself, we investigate the Adler function

$$D(Q^2) \equiv -Q^2 \frac{\partial \, \Pi^{V+A}(Q^2)}{\partial \, Q^2} = \frac{Q^2}{\pi} \int_0^\infty ds \frac{\mathrm{Im} \, \Pi^{V+A}(s)}{(s + Q^2)^2}. \qquad (17)$$

For the corresponding experimental V+A spectral function we have used the Aleph measurement in the resonance region up to 2.2 GeV^2. For the continuum region above 2.2 GeV^2 we used 3-loop perturbation theory for $\alpha_s(M_Z) = 0.118$, and we have set the renormalization scale μ to Q. We remark that the pion pole of the axial vector contribution has to be taken into account in order to yield a consistent description in terms of the OPE

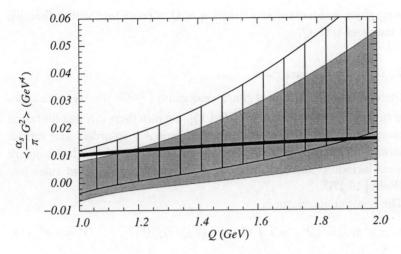

Figure 2. The running gluon condensate as a function of Q when extracted from the Adler function. The grey area represents the allowed region using perturbation theory at $O(\alpha_s^3)$ and the striped region using perturbation theory at $O(\alpha_s^2)$. The thick solid line denotes the running gluon condensate ($\Omega = Q$) from the lattice-inspired ansatz.

for asymptotically large Q [13]. We have checked that the known perturbative contributions to the Adler function show good convergence properties. For Q between 1.0 and 2.0 GeV and setting $\mu = Q$, the 3-loop (two-loop) corrections amount to 5% (7%) and 0.5% (1.4%), respectively. On the other hand, the $O(\alpha_s)$ correction to the Wilson coefficient of the gluon condensate are between -16% and -6% for μ between 0.8 and 2.0 GeV. We have compared the local dimension 6 contributions contained in the running gluon condensate with the sum of all dimension 6 terms in the OPE as they were determined in [13]. We again found that the corresponding dimension 6 contributions have equal sign and roughly the same size.

Our result for the running gluon condensate as a function of Q is shown in Fig. 2. for $1\,\text{GeV} \le Q \le 2$ GeV. The uncertainties are due to the experimental errors in the spectral function and a variation of the renormalization scale μ in the range $Q \pm 0.25$ GeV. The strong coupling has been fixed at $\alpha_s(M_Z) = 0.118$. We have restricted our analysis to the range $1\,\text{GeV} < Q < 2$ GeV because for $Q \lesssim 1$ GeV perturbation theory becomes unreliable and for $Q \gtrsim 2$ GeV the experimentally unknown part of the spectral function at $s \ge 2.2\,\text{GeV}^2$ is being probed. The thick black line in Fig. 2 shows the running gluon condensate obtained from the lattice-inspired model of the last section for $\Omega = Q$. We find that our result for the running gluon condensate as a function of Q is consistent with a function that increases with Q.

3.2.2. Charmonium sum rules

The determination of the local gluon condensate $\left\langle \frac{\alpha_s}{\pi} G^2 \right\rangle(\infty)$ from charmonium sum rules was pioneered in Refs. [15]. By now there is a vast literature on computations of Wilson coefficients to various loop-orders and various dimensions of the power corrections in the correlator of two heavy quark currents including updated analyses of the corresponding sum rules (see e.g. Refs. [16,17]).

The relevant correlator is

$$(q_\mu q_\nu - q^2 g_{\mu\nu}) \Pi^c(Q^2) = i \int d^4x \, e^{iqx} \left\langle T j_\mu^c(x) j_\nu^c(0) \right\rangle, \qquad Q^2 = -q^2, (18)$$

where $j_\mu \equiv \bar{c}\gamma_\mu c$, and the n-th moment is defined as

$$M_n = \frac{1}{n!} \left(-\frac{d}{dQ^2} \right)^n \Pi^c(Q^2) \Bigg|_{Q^2=0}. \tag{19}$$

Assuming analyticity of Π^c in Q^2 away from the negative, real axis and employing the optical theorem, the nth moment can be expressed as a dispersion integral over the charm pair cross section in e^+e^- annihilation,

$$M_n = \frac{1}{12\pi^2 Q_c^2} \int \frac{ds}{s^{n+1}} \frac{\sigma_{e^+e^- \to c\bar{c}+X}(s)}{\sigma_{e^+e^- \to \mu^+\mu^-}(s)}, \tag{20}$$

where s is the square of the c.m. energy and $Q_c = 2/3$. We consider the ratio [15]

$$r_n \equiv \frac{M_n}{M_{n-1}} \tag{21}$$

and extract the running gluon condensate as a function of n from the equality of the theoretical ratio using Eq. (19) and the ratio based on Eq. (20) determined from experimental data.

For the experimental moments we use the compilation presented in Ref. [18], where the spectral function is split into contributions from the charmonium resonances, the charm threshold region, and the continuum. For the latter the authors of Ref. [18] used perturbation theory since no experimental data are available for the continuum region. We have assigned a 10% error for the spectral function in the continuum region. For the purely perturbative contribution of the theoretical moments we used the compilation of analytic $O(\alpha_s^2)$ results from Ref. [18] and adopted the $\overline{\text{MS}}$ mass definition $\overline{m}_c(\overline{m}_c)$ (for any renormalization scale μ). We have used

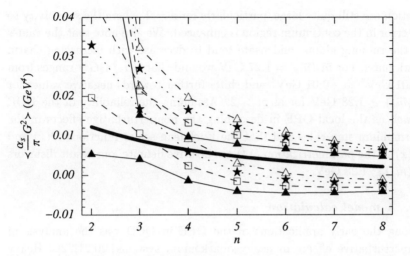

Figure 3. The running gluon condensate as a function of n when extracted from the ratio of moments r_n for $\overline{m}_c(\overline{m}_c) = 1.23$ (white triangles), 1.24 (black stars), 1.25 (white squares) and 1.26 GeV (black triangles). The area between the upper and lower symbols represent the uncertainties. The thick solid line indicates the running gluon condensate as it is obtained from the lattice-inspired model for $g(x)$.

the one-loop expression for the Wilson coefficient of the gluon condensate, and we have checked that for $n \leq 8$ the perturbative $O(\alpha_s^2)$ corrections do not exceed 50% of the $O(\alpha_s)$ corrections for μ between 1 and 4 GeV.

Our result for the running gluon condensate as a function of n is shown in Fig. 3. For each value of $\overline{m}_c(\overline{m}_c)$ and n the area between the upper and lower symbols represents the uncertainty due to the experimental errors of the moments and variations of $\alpha_s(M_Z)$ between 0.116 and 0.120 and of μ between 1 and 3 GeV. We see that the running gluon condensate appears to be a decreasing function of n. Since the exact width of the short-distance function f for the moments is unknown we estimate it using physical arguments. For large n, i.e. in the nonrelativistic regime, the width is of the order of the quark c.m. kinetic energy mv^2, which scales like m/n because the average quark velocity in the n-th moment scales like $1/\sqrt{n}$. [19] For small n the relevant short-distance scale is just the quark mass. Therefore, we take $\Omega = 2m_c/n$ as the most appropriate choice of the resolution scale. In Fig. 3 the running gluon condensate is displayed for $\lambda_A^{-1} = 0.7$ GeV and, exemplarily, $A_0 = 0.04$ (i.e. $\langle(\alpha_s/\pi)G^2\rangle_{\text{lat}}(\infty) = 0.024$ GeV^4) and for $\Omega = (2.5 \text{ GeV})/n$ as the thick solid line. We find that the n-dependence of the lattice inspired running gluon condensate and our fit result from the charmonium sum rules are consistent. However, the uncertainties of our extraction are still quite large, particularly for $n = 2$, where the sensitivity to the error in the continuum region is enhanced. We also note that the values for the running gluon condensate tend to decrease with increasing charm quark mass. For $\overline{m}_c(\overline{m}_c) = 1.27$ GeV we find that $\langle (\alpha_s/\pi)G^2 \rangle$ ranges from -0.01 GeV^4 to $+0.01$ GeV^4 and shifts further towards negative values for $\overline{m}_c(\overline{m}_c) \gtrsim 1.28$ GeV for all $n > 2$. Assuming the reliability of the DOE as well as the local OPE in describing the nonperturbative effects in the charmonium sum rules and that the Euclidean scalar functions $D(x)$ and $D_1(x)$ in the parametrization (14) are positive definite, our result disfavors $\overline{m}_c(\overline{m}_c) \gtrsim 1.28$ GeV.

3.3. A model calculation

Among the early applications of the OPE in QCD was the analysis of nonperturbative effects in heavy quarkonium systems [20,21,22]. Heavy quarkonium systems are nonrelativistic quark-antiquark bound states for which there is the following hierarchy of the relevant physical scales m (heavy quark mass), mv (relative momentum), mv^2 (kinetic energy) and $\Lambda_{\rm QCD}$:

$$m \gg mv \gg mv^2 \gg \Lambda_{\rm QCD}. \tag{22}$$

Thus the spatial size of the quarkonium system $\sim (mv)^{-1}$ is much smaller than the typical dynamical time scale $\sim (mv^2)^{-1}$ effectively rendering the problem 1-dimensional. In this section we demonstrate the DOE for the nonperturbative corrections to the $n^{2s+1}L_j = 1^3S_1$ ground state. We adopt the local version of the multipole expansion (OPE) for the expansion in the ratios of the scales m, mv and mv^2. The resolution dependent expansion (DOE) is applied with respect to the ratio of the scales mv^2 and $\Lambda_{\rm QCD}$. The former expansion amounts to the usual treatment of the dominant perturbative dynamics by means of a nonrelativistic two-body Schrödinger equation. The interaction with the nonperturbative vacuum is accounted for by two insertions of the local $\mathbf{xE}$ dipole operator, $\mathbf{E}$ being the chromoelectric field [20]. The chain of VEV's of the two gluon operator with increasing numbers of covariant derivatives times powers of quark-antiquark octet propagators [20], i.e. the expansion in Λ/mv^2, is treated using the DOE.

At leading order in the local multipole expansion with respect to the scales m, mv, and mv^2 the expression for the nonperturbative corrections to the ground state energy reads

$$E^{np} = \int_{-\infty}^{\infty} dt \, f(t) \, g(t), \tag{23}$$

where

$$f(t) = \frac{1}{36} \int \frac{dq_0}{2\pi} e^{iq_0(it)} \int d^3\mathbf{x} \int d^3\mathbf{y}\, \phi(x)\, (\mathbf{xy})\, G_O\left(\mathbf{x},\mathbf{y}, -\frac{k^2}{m} - q_0\right) \phi(y), \tag{24}$$

with

$$G_O\left(\mathbf{x},\mathbf{y}, -\frac{k^2}{m}\right) = \sum_{l=0}^{\infty} (2l+1)\, P_l\left(\frac{\mathbf{xy}}{xy}\right) G_l\left(x,y, -\frac{k^2}{m}\right),$$

$$G_l\left(x,y, -\frac{k^2}{m}\right) = \frac{mk}{2\pi} (2kx)^l (2ky)^l\, e^{-k(x+y)}$$

$$\sum_{s=0}^{\infty} \frac{L_s^{2l+1}(2kx)\, L_s^{2l+1}(2ky)\, s!}{(s+l+1 - \frac{m\alpha_s}{12k})(s+2l+1)!},$$

$$\phi(x) = \frac{k^{3/2}}{\sqrt{\pi}} e^{-kx}, \qquad k = \frac{2}{3} m\alpha_s. \tag{25}$$

The term G_O is the quark-antiquark octet Green-function [22], and ϕ denotes the ground state wave function. The functions P_n and L_n^k are the Legendre and Laguerre polynomials, respectively. We note that t is the Euclidean time. This is the origin of the term it in the exponent appearing in the definition of the function f. In Fig. 4 the function $f(t)$ (solid line) is displayed for $m = 5\,\text{GeV}$ and $\alpha_s = 0.39$. Since the spatial extension of the quarkonium system is neglected and the average time between interactions with the vacuum is of the order of the inverse kinetic energy, the characteristic width of f is of order $(mv^2)^{-1} \sim (m\alpha_s^2)^{-1}$. As a comparison we have also displayed the function $[\int dt' f(t')\tilde{e}_0^\Omega(t')]e_0^\Omega(t)$ for $\Omega = k^2/m$ (dashed line), which is the leading term in the delocalized multipole expansion of f. The values of the first few local multipole moments $f_n(\infty)$, which correspond to the local Wilson coefficients, read

$$f_0(\infty) = 1.6518\, \tfrac{m}{36\,k^4}, \qquad f_2(\infty) = 1.3130\, \tfrac{m^3}{36\,k^8},$$

$$f_4(\infty) = 7.7570\, \tfrac{m^5}{36\,k^{12}}, \qquad f_6(\infty) = 30.492\, \tfrac{m^7}{36\,k^{16}}, \tag{26}$$

$$f_8(\infty) = 4474.1\, \tfrac{m^9}{36\,k^{20}}, \qquad f_{10}(\infty) = 262709.3\, \tfrac{m^{11}}{36\,k^{24}}, \dots$$

The term $f_0(\infty)$ agrees with Ref. [21,22] and $f_2(\infty)$ with Ref. [14]. The results for $f_{n>2}(\infty)$ are new.

Let us compare the local expansion of E^{np} with the resolution-dependent expansion using the basis functions of Eq. (9). For the nonperturbative gluonic field strength correlator we use a lattice inspired model

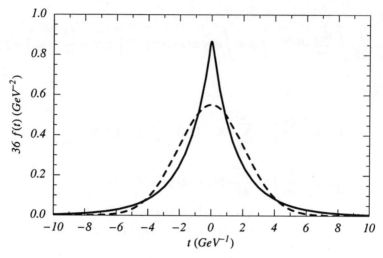

Figure 4. The perturbative short-distance function $f(t)$ for $m = 5$ GeV and $\alpha_s = 0.39$ (solid line) and the leading term in the delocalized multipole expansion of $f(t)$ for $\Omega = k^2/m$ (dashed line).

of the form

$$g(t) = 12\,A_0 \exp\left(-\sqrt{t^2 + \lambda_A^2}/\lambda_A + 1\right),$$

$$A_0 = 0.04 \text{ GeV}^4, \qquad \lambda_A^{-1} = 0.7 \text{ GeV}, \tag{27}$$

This model has an exponential large-time behavior and a smooth behavior for small t. The local dimension 4 gluon condensate in this model is

$$\left\langle \frac{\alpha_s}{\pi} G_{\mu\nu}^a G_{\mu\nu}^a \right\rangle = \frac{6\,A_0}{\pi^2} = 0.024 \text{ GeV}^4. \tag{28}$$

In Tab. 1 we have displayed the exact result and the first four terms of the resolution-dependent expansion of E^{np} for the quark masses $m = 5, 25, 45, 90, 175$ GeV and for $\Omega = \infty$ and $\Omega = k^2/m$. For each value of the quark mass the strong coupling has been fixed by the relation $\alpha_s = \alpha_s(k)$. Note that the series are all asymptotic, i.e. there is no convergence for any resolution. The local expansion ($\Omega = \infty$) is quite badly behaved for smaller quark masses because for $k^2/m \lesssim \lambda_A^{-1}$ any local expansion is meaningless. In particular, for $m = 5$ GeV the subleading dimension 6 term is already larger than the parametrically leading dimension 4 term. This is consistent with the size of the dimension 6 term based on a phenomenological estimate of the local dimension 6 condensate. For quark masses, where $k^2/m \gtrsim \lambda_A^{-1}$, the local expansion is reasonably good. However, for finite

Table 1. Nonperturbative corrections to the heavy quarkonium ground state level at leading order in the multipole expansion with respect to the scales m, mv and mv^2 for various quark masses m based on the model in Eq. (27). Displayed are the exact result and the first few orders in the DOE for $\Omega = \infty$ and $\Omega = k^2/m$. The numbers are rounded off to units of 0.1, 0.01 or 0.001 MeV.

					$\Omega = \infty$		$\Omega = k^2/m$	
m (GeV)	α_s	k^2/m (MeV)	E^{np} (MeV)	n	$f_n g_n$ (MeV)	$\sum_{i=0}^{n} f_i g_i$ (MeV)	$f_n g_n$ (MeV)	$\sum_{i=0}^{n} f_i g_i$ (MeV)
5	0.39	0.338	24.8	0	38.6	38.6	24.2	24.2
				2	−65.7	−27.2	−3.9	20.3
				4	832.7	805.5	12.1	32.4
				6	−35048.0	−34242.4	−43.1	−10.8
25	0.23	0.588	12.6	0	16.0	16.0	12.8	12.8
				2	−9.0	7.0	−1.2	11.5
				4	37.8	44.8	2.6	14.1
				6	−526.6	−481.8	−6.7	7.4
45	0.19	0.722	4.9	0	5.9	5.9	5.0	5.0
				2	−2.2	3.7	−0.4	4.6
				4	6.1	9.8	0.7	5.3
				6	−56.4	−46.6	−1.4	3.8
90	0.17	1.156	1.05	0	1.15	1.15	1.07	1.07
				2	−0.17	0.98	−0.04	1.02
				4	0.18	1.16	0.04	1.07
				6	−0.65	0.51	−0.07	1.00
175	0.15	1.750	0.245	0	0.258	0.258	0.249	0.249
				2	−0.016	0.242	−0.005	0.244
				4	0.008	0.250	0.003	0.247
				6	−0.012	0.237	−0.003	0.244

resolution $\Omega = k^2/m$, the size of higher order corrections is considerably smaller than in the local expansion for all quark masses, and the series appears to be much better behaved. The size of the order n term is suppressed by approximately a factor 2^{-n} as compared to the order n term in the local expansion. We find explicitly that terms in the series with larger n decrease more quickly for finite resolution scale as compared to the local expansion. One also observes that even in the case $k^2/m < \lambda_A^{-1}$, where the leading term of the local expansion overestimates the exact result, the leading term in the delocalized expansion for $\Omega = k^2/m$ agrees with the exact result within a few percent. It is intuitively clear that this feature is a general property of the delocalized expansion.

We would like to emphasize the above calculation is not intended to provide a phenomenological determination of nonperturbative corrections to the heavy quarkonium ground state energy level, but rather to demonstrate the DOE within a specific model. For a realistic treatment of the nonperturbative contributions in the heavy quarkonium spectrum a model-independent analysis should be carried out. In addition, higher orders in the local multipole expansion with respect to the ratios of scales m, mv and mv^2 should be taken into account, which have been neglected here. These corrections might be substantial, in particular for smaller quark masses. However, having in mind an application to the bottomonium spectrum, we believe that our results for the expansion in Λ/mv^2 indicate that going beyond the leading term in the OPE for the bottomonium ground state is probably meaningless and that calculations based on the DOE with a suitable choice of resolution are more reliable.

4. Summary and Outlook

In this talk I have presented a generalization of the expansion of the correlator of gauge invariant currents into local operator averages. It is built on appropriate choices of basis sets spanning the dual spaces which correspond to the interplay between perturbative and nonperturbative N-point correlations in the Euclidean propagation of a gauge invariant current. It was demonstrated that this framework, when applied to the $V + A$ and charmonium systems, yields an experimentally confirmed running gluon condensate. In a model calculation for the nonperturbative corrections to the ground state of a heavy quarkonium superior convergence properties of the DOE were obtained when compared with the OPE.

The framework has a wealth of potential applications. For example, it can be used to improve the scaling relation for decay constants of heavy-light mesons as it is obtained in HQET (see [9] for an attempt). The issue of the violation of local quark-hadron duality by the OPE can be re-addressed [23], and a scheme can be developed to estimate duality violation effects. This is of particular importance for the determination of standard model parameters from mixing and decay of B mesons [6].

Acknowledgments

I would like to thank the organizers for providing the atmosphere for a very stimulating conference. Helpful comments on the manuscript by A. Hoang are gratefully acknowledged.

References

1. A. J. Buras [arXiv:hep-ph/ph/9806471].

2. M. A. Shifman, A. I. Vainshtein, Nucl. Phys. **B277**, 456 (1986).

3. M. A. Shifman, A. I. Vainshtein and V. I. Zakharov, Nucl. Phys. B **147**, 385 (1979). Nucl. Phys. B **147**, 448 (1979).

4. A. H. Mueller, Nucl. Phys. B **250**, 327 (1985)
   M. Beneke, Phys. Rept. **317**, 1 (1999) [arXiv:hep-ph/9807443].

5. I. I. Y. Bigi and N.Uraltsev, Int. J. Mod. Phys.**A16**, 5201 (2001) [arXiv:hep-ph/0106346].

6. M. Beneke et al., Phys. Lett. B **459**, 631 (1999) [arXiv:hep-ph/9808385], [arXiv:hep-ph/0202106].

7. R. Hofmann, Phys. Lett. B **520**, 257 (2001)[arXiv:hep-ph/0109007].

8. H. G. Dosch and Y. A. Simonov, Phys. Lett. B **205**, 339 (1988).

9. A. H. Hoang and R. Hofmann [arXiv:hep-ph/0206201].

10. M. D'Elia, A. Di Giacomo and E. Meggiolaro, Phys. Lett. B **408**, 315 (1997) [arXiv:hep-lat/9705032].

11. R. Barate et al. [ALEPH Collaboration], Eur. Phys. J. C **4**, 409 (1998).

12. K. Ackerstaff et al. [OPAL Collaboration], Eur. Phys. J. C **7**, 571 (1999) [arXiv:hep-ex/9808019].

13. E. Braaten, S. Narison and A. Pich, Nucl. Phys. B **373**, 581 (1992).

14. A. Pineda, Nucl. Phys. B **494**, 213 (1997) [arXiv:hep-ph/9611388].

15. V. A. Novikov, L. B. Okun, M. A. Shifman, A. I. Vainshtein, M. B. Voloshin and V. I. Zakharov, Phys. Rev. Lett. **38**, 626 (1977) [Erratum-ibid. **38**, 791 (1977)]; Phys. Lett. B **67**, 409 (1977).

16. S. N. Nikolaev and A. V. Radyushkin, Nucl. Phys. B **213**, 285 (1983).

17. S. N. Nikolaev and A. V. Radyushkin, Phys. Lett. B **124**, 243 (1983).

18. J. H. Kuhn and M. Steinhauser, Nucl. Phys. B **619**, 588 (2001) [arXiv:hep-ph/0109084].

19. A. H. Hoang, Phys. Rev. D **59**, 014039 (1999) [arXiv:hep-ph/9803454].

20. M. B. Voloshin, Nucl. Phys. B **154**, 365 (1979).

21. H. Leutwyler, Phys. Lett. B **98**, 447 (1981).

22. M. B. Voloshin, Sov. J. Nucl. Phys. **35**, 592 (1982) [Yad. Fiz. **35**, 1016 (1982)]; Sov. J. Nucl. Phys. **36**, 143 (1982) [Yad. Fiz. **36**, 247 (1982)].

23. R. Hofmann, Nucl. Phys. **B623**, 301 (2002) [arXiv:hep-ph/0109008].

SECTION 3.
GAUGE DYNAMICS AT HIGH
TEMPERATURE AND DENSITY

WHAT QCD TELLS US ABOUT NATURE

FRANK WILCZEK

Center for Theoretical Physics
Laboratory for Nuclear Science and Department of Physics
Massachusetts Institute of Technology
Cambridge, MA 02139
E-mail: wilczek@mit.edu

I discuss why QCD is our most perfect physical theory and visit a few of its current frontiers. Finally I draw some appropriate conclusions.

Arkady Vainshtein and the ITEP→Minnesota group have made QCD the primary focus of their research from the earliest days of the subject right up to the present. They have contributed mightily to our understanding of the theory itself, of its phenomenological applications, and of its lessons for quantum field theory in general. It therefore seems appropriate on this occasion to help celebrate Arkady's birthday with an appreciation of QCD.

1. QCD is our most perfect physical theory

Here's why:

1.1. *It embodies deep and beautiful principles*

These are, first of all, the general principles of quantum mechanics, special relativity, and locality, which lead one to relativistic quantum field theory[1]. In addition, we require invariance under the nonabelian gauge symmetry $SU(3)$, the specific matter content of quarks – six spin-1/2 Dirac fermions that are color triplets – and renormalizability. These requirements determine the theory completely, up to a very small number of continuous parameters as discussed below.

Deeper consideration reduces the axioms further. Theoretical physicists have learned the hard way that consistent, nontrivial relativistic quantum field theories are difficult to construct, due to the infinite number of degrees of freedom (per unit volume) needed to construct local fields, which tends to bring in ultraviolet divergences. To construct a relativistic quantum theory, one typically introduces at intermediate stages a cutoff, which spoils the locality or relativistic invariance of the theory. Then one attempts to remove the cutoff while adjusting the defining parameters, to achieve a finite, cutoff-independent limiting theory. Renormalizable theories are those for which this can be done, order by order in a perturbation expansion around free field theory. This formulation, while convenient for mathematical analysis, obviously begs the question whether the perturbation theory converges (and in practice it never does).

A more straightforward procedure, conceptually, is to regulate the theory as a whole by discretizing it, approximating space-time by a lattice[2]. This spoils the continuous space-time symmetries of the theory. Then one attempts to remove dependence on the discretization by refining it, while if necessary adjusting the defining parameters, to achieve a finite limiting theory that does not depend on the discretization, and does respect the space-time symmetries. The redefinition of parameters is necessary, because in refining the discretization one is introducing new degrees of freedom. The earlier, coarser theory results from integrating out these degrees of freedom, and if it is to represent the same physics it must incorporate their effects, for example in vacuum polarization.

In this procedure, the big question is whether the limit exists. It will do so only if the effects of integrating out the additional short-wavelength modes, that are introduced with each refinement of the lattice, can be captured accurately by a re-definition of parameters already appearing in the theory. This, in turn, will occur only in a straightforward way if these modes are weakly coupled. (Another simple possibility is that short-distance modes of different types cancel in vacuum polarization. This is what occurs in supersymmetric theories. Other types of ultraviolet fixed points are in principle possible, but difficult to imagine or investigate.) But this is true, if and only if the theory is asymptotically free.

One can investigate this question, i.e., whether the couplings decrease to zero with distance, or in other words whether the theory is asymptotically free, within weak coupling perturbation theory[3]. One finds that only nonabelian gauge theories with simple matter content, and no nonrenormalizable couplings, satisfy this criterion[4]. Supersymmetric versions of these theories allow more elaborate, but still highly constrained, matter content.

Summarizing the argument, only those relativistic field theories which are asymptotically free can be argued in a straightforward way to exist. And the only asymptotically free theories in four space-time dimensions involve nonabelian gauge symmetry, with highly restricted matter content. So the axioms of gauge symmetry and renormalizability are, in a sense, gratuitous.

They are implicit in the mere *existence* of nontrivial interacting quantum field theories.

Thus QCD is a member of a small aristocracy: the closed, consistent embodiments of relativity, quantum mechanics, and locality. Within this class, it is among the least affected members.

1.2. *It provides algorithms to answer any physically meaningful question within its scope*

As I just discussed, QCD can be constructed by an explicit, precisely defined discretization and limiting procedure. This provides, in principle, a method to compute any observable, in terms that could be communicated to a Turing machine.

In fact marvelous things can be accomplished, in favorable cases, using this direct method. For a stirring example, see Figure 4 below.

The computational burden of the direct approach is, however, heavy at best. When one cannot use importance sampling, as in addressing such basic questions as calculating scattering amplitudes or finding the ground state energy at finite baryon number density, it becomes totally impractical.

For these reasons various improved perturbation theories continue to play an enormous role in our understanding and use of QCD. The most important and well-developed of these, directly based on asymptotic freedom, applies to hard processes and processes involving heavy quarks[5]. It is what is usually called "perturbative QCD", and leads to extremely impressive results as exemplified by Figures 1–3, below. The scope of these methods is continually expanding, to include additional "semi-hard" processes, as will be discussed in many talks at the Conference. When combined with some additional ideas, they allow us to address major questions regarding the behavior of the theory at high temperatures and large densities, as I'll touch on below.

Chiral perturbation theory[6], which is based on quite different aspects of QCD, is extremely useful in discussing low-energy processes, though it is difficult to improve systematically. The ITEP or QCD sum rules have proved so useful and widely applicable that the seminal papers are among the most frequently cited in the history of high-energy physics, and have rained glory and prizes upon their authors[7,8]. A proper discussion of this, or of many other approaches each of which offers some significant insight (traditional nuclear physics, bag model, Reggeism, sum rules, large N, Skyrme model, etc.) would not be appropriate here.

I would however like to mention a perturbation theory which I think is considerably underrated, that is strong coupling perturbation theory[9]. It leads to a simple, appealing, and correct understanding of confinement[10], and even its existing, crude implementations provide a remarkably good caricature of the low-lying hadron spectrum. It may be time to revisit this approach, using modern algorithms and computer resources.

1.3. *Its scope is wide*

There are significant applications of QCD to nuclear physics, accelerator physics, cosmology, extreme astrophysics, unification, and natural philosophy. I'll say just a few words about each, in turn.

Nuclear physics Understanding atomic nuclei was of course the original goal of strong interaction physics. In principle QCD provides answers to all its questions. But in practice QCD has not superseded traditional nuclear physics within its customary domain. The relationship between QCD and traditional nuclear physics is in some respects similar to the relationship between QED and chemistry. The older disciplines retain their integrity and independence, because they tackle questions that are exceedingly refined from the point of view of the microscopic theories, involving delicate cancellations and competitions that manufacture small net energy scales out of much larger gross ones. QCD offers many insights and suggestions, however, of which we will hear much at this Conference. There is also an emerging field of extreme nuclear physics, including the study of nuclei with hard probes and heavy-ion collisions, where the influence of QCD is decisive.

Accelerator physics Most of what goes on at high-energy accelerators is described by QCD. This application has been so successful, that experimenters no longer speak of "tests of QCD", but of "QCD backgrounds"! Two- and even three-loop calculations of such "backgrounds" are in urgent demand. What can one add to that sincere testimonial?

Cosmology Because of asymptotic freedom, hadronic matter becomes not impenetrably complex, but rather profoundly simpler, under the extreme conditions predicted for the early moments of the Big Bang. This stunning simplification has opened up a large and fruitful area of investigation.

Extreme astrophysics The physics of neutron star interiors, neutron star collisions, and collapse of very massive stars involves extreme nuclear physics. It should be, and I believe that in the foreseeable future it will be, firmly based on microscopic QCD.

Unification and Natural philosophy See below.

1.4. *It contains a wealth of phenomena*

Let me enumerate some major ones: radiative corrections, running couplings, confinement, spontaneous (chiral) symmetry breaking, anomalies, instantons. Much could be said about each of these, but I will just add a few words about the first. The Lamb shift in QED is rightly celebrated as a triumph of quantum field theory, because it shows quantitatively, and beyond reasonable doubt, that loop effects of virtual particles are described by the precise, intricate rules of that discipline. But in QCD, we probably have by now 50 or so cases where two- and even three-loop effects are needed to do justice to experimental results – and the rules are considerably more intricate!

1.5. *It has few parameters ...*

A straightforward accounting of the parameters in QCD would suggest 8: the masses of six quarks, the value of the strong coupling, and the value of the P and T violating θ parameter. The fact that there are only a small finite number of parameters is quite profound. It is a consequence of the constraints of gauge invariance and renormalizability (or alternatively, as we saw, existence, by way of asymptotic freedom).

In reality there are not 8 parameters, but only 6. g is eliminated by dimensional transmutation[11]. This means, roughly stated, that because the coupling runs as a function of distance, one cannot specify a unique numerical value for it. It will take any value, at some distance or other. One can put (say) $g(l) \equiv 1$, thereby determining a length scale l. What appeared to be a choice of dimensionless coupling, is revealed instead to be a choice of unit of length, or equivalently (with $\hbar = c = 1$) of mass. Only the dimensionless *ratio* of this mass to quark masses can enter into predictions for dimensionless quantities. So what appeared to be a one-parameter family of theories, with different couplings, turns out to be a single theory measured using differently calibrated meter-sticks.

The θ term is eliminated, presumably, by the Peccei-Quinn mechanism[12]. To assure us of this, it would be very nice to observe the quanta associated with this mechanism, namely axions[13]. In any case, we know for sure that the θ term is very small. For purposes of strong interaction physics, within QCD itself, we can safely set it to zero, invoking P or T symmetry, and be done with it.

1.6. ... *or none*

This economy of parameters would already be quite impressive, given the wealth of phenomena described. However if we left it at that we would be doing a gross injustice to QCD, and missing one of its most striking features.

To make my point, let me call your attention to a simplified version of QCD, that I call "QCD Lite". QCD Lite is simply QCD truncated to contain just two flavors of quarks, both of which are strictly massless, and with the θ parameter set to zero. These choices are natural in the technical sense, since they can be replaced by symmetry postulates. Indeed, assuming masslessness of the quarks is tantamount to assuming exact $SU(2) \times SU(2)$ chiral symmetry, and $\theta = 0$ is tantamount to assuming the discrete symmetries P or T. (Actually, once we have set the quark masses to zero, we can dial away θ by a field redefinition.)

Now there are two especially remarkable things about QCD Lite[14]. The first is that it is a theory which *contains no continuous free parameters at all*. Its only inputs are the numbers 3 (colors) and 2 (flavors). The second is that it *provides an excellent semi-quantitative theory of hadronic matter.*

Indeed, in reality the strange and heavier quarks have very little influence on the structure or masses of protons, neutrons, atomic nuclei, pions, rho mesons, ... Leaving them out would require us to abridge, but not to radically revise, the Rosenfeld Tables. This is proved by the remarkable quantitative success, at the 5-10% level, of lattice gauge theory in the 'quenched' approximation[15]. For in this approximation, the influence of the heavier quarks on the lighter ones is systematically ignored.

The only major effect of putting the u and d masses to zero is to make the pions, which are already quite light by hadronic standards, strictly massless. The perturbation to the proton mass, for instance, can be related using chiral perturbation theory to the value of the so-called σ term, a directly measurable quantity[16]. When this is done, one finds that the u and d quark masses are responsible for only about 5% of the proton mass.

Thus QCD Lite provides a truly remarkable realization of John Wheeler's program, "Getting Its From Bits". For here we encounter an extremely rich and complex class of physical phenomena – including, in principle, nuclear and particle spectra – that can be calculated, accurately and without ambiguity, using as sole inputs the numbers 3 and 2.

1.7. *It is true*

I will not waste a lot of words on this, showing instead a few pictures, each worth many thousands of words.

Figure 1[17] displays graphically that many independent types of experiments at different energy scales have yielded determinations of the strong coupling constant, all consistent with the predicted running. The overall accuracy and consistency of this phenomenology is reflected in the precision with which this coupling is determined, to wit 5% (at the Z mass). A remarkable feature of the theory is that a wide range of possible values for the coupling at relatively small energies focuses down to quite a narrow range at the highest accessible energies. Thus any "reasonable" choice of the scale at which the coupling becomes numerically large leads, within a few per cent, to a unique value of the coupling at the Z mass. Our successful QCD predictions for high energy experiments have no wiggle-room!

While Figure 1 is impressive, it does not do complete justice to the situation. For several of the experimental 'points' each represents a summary of hundreds of independent measurements, any one of which might have invalidated the theory, and which display many interesting features. Figures 2 and 3 partially ameliorate the omission. Figure 2[18] shows some of the experimental data on deep inelastic scattering – all subsumed within the 'DIS' point in Figure 1 – unfolded to show the complete Q^2 and x dependence. Our predictions[5] for the pattern of evolution of structure functions with Q^2 –decrease at large x, increase at small x – are now confirmed in great detail, and with considerable precision. Particularly spectacular is the rapid growth at small x. This was predicted[19], in the form now observed, very early on. However, even at the time we realized this rise could not continue forever. The proliferating partons begin to form a dense system, and eventually one must cease to regard them as independent. There is a very interesting many-body problem developing here, which seems ripe for experimental and theoretical investigation, and may finally allow us to make contact between microscopic QCD and the remarkably successful Regge-pole phenomenology.

Figure 3[20] displays a comparison of the experimental distribution of jet energies, in 3-jet events, with the QCD prediction. Shown is the energy fraction of the second hardest jet, compared to its kinematic maximum. For a detailed explanation, see[20]. The rise at $x_2 \to 1$ reflects the singularity of soft gluon bremsstrahlung, and matches the prediction of QCD (solid line). For comparison, the predictions for hypothetical scalar gluons are shown by the dotted line. This is as close to a direct measurement of the core interaction of QCD, the basic quark-gluon vertex, as you could hope to see. The other piece of this Figure displays a related but more sophisticated comparison, using the Ellis-Karliner angle.

Finally Figure 4[21] shows the comparison of the QCD predictions to the spectrum of low-lying hadrons. Unlike what was shown in the previous two Figures, and most of the points in Figure 1, this tests the whole structure of the theory, not only its perturbative aspect. The quality of the fit is remarkable. Note that only one adjustable parameter (the strange quark mass) and one overall choice of normalization go into the calculation. Otherwise it's pure "Its From Bits". Improvements due to enhanced computing power and to the use of domain wall quarks[22], that more nearly respect chiral symmetry, are on the horizon.

Since there seems to be much confusion (and obfuscation) on the point, let me emphasize an aspect of Figure 4 that ought to be blindingly clear: what you don't see in it. You don't see massless degrees of freedom with long-range gauge interactions, nor parity doublets. That is, confinement and chiral symmetry breaking are simply true facts about the solution of QCD, that emerge by direct calculation. The numerical work has taken us way beyond abstract discussion of these features.

1.8. *It lacks flaws*

Finally, to justify the adverb in "*most* perfect" I must briefly recall for you some prominent flaws in our other best theories of physics, which QCD does not share.

Quantum electrodynamics is of course extremely useful – incomparably more useful than QCD – and successful in practice. But there is a worm in its bud. It is not asymptotically free. Treated outside of perturbation theory, or extrapolated to extremely high energy, QED becomes internally inconsistent. Modern electroweak theory shares many of the virtues of QED, but it harbors the same worm, and in addition contains many loose ends and continuous free parameters. General relativity is the deepest and most beautiful theory of all, but it breaks down in several known circumstances, producing singularities that have no meaningful interpretation within the theory. Nor does it mesh seamlessly with the considerably better tested and established framework we use for understanding the remainder of physics. Specifically, general relativity is notoriously difficult to quantize. Finally, it begs the question of why the cosmological term is zero, or at least fantastically small when measured in its natural units. Superstring theory promises to solve some and conceivably all of these difficulties, and to provide a fully integrated theory of Nature, but I think it is fair to say that in its present form superstring theory is not defined by clear principles, nor does it provide definite algorithms to answer questions within its claimed scope, so there remains a big gap between promise and delivery.

2. Breaking New Ground in QCD, 1: High Temperature

The behavior of QCD at high temperature, and low baryon number density, is relevant to cosmology – indeed, it describes the bulk of matter filling the Universe, during the first few seconds of the Big Bang – and to the description of both numerical and physical experiments. There are ambitious experimental programs planned for RHIC, and eventually LHC, to probe this physics. It is also, I think, intrinsically fascinating to ask – what happens to empty space, if you keep adding heat?

The equilibrium thermodynamics of QCD at finite temperature (and zero chemical potential) is amenable to direct simulation, using the techniques of lattice gauge theory. Figure 5[23] does not quite represent the current, rapidly evolving, state of the art, but it does already demonstrate some major qualitative points.

The chiral symmetry breaking condensate, clearly present (as previously advertised) at zero temperature, weakens and seems to be gone by $T \sim 150$ Mev. Likewise at these temperatures there is a sizable increase in the value of the Polyakov loop, indicating that the force between distant color sources has considerably weakened. Furthermore the energy density increases rapidly, approaching the value one would calculate for an ideal gas of quarks and gluons. The pressure likewise increases rapidly, but lags somewhat behind the ideal gas value.

All these phenomena indicate that at these temperatures and above a description using quarks and gluons as the degrees of freedom is much simpler and more appropriate than a description involving ordinary hadrons. Indeed, the quarks and gluons appear to be quasi-free. That is what one expects, from asymptotic freedom, for the high-energy modes that dominate the thermodynamics.

It is an interesting challenge to reproduce the pressure analytically[24]. Since the only scale in the problem is the large temperature, if one can organize the calculation so as to avoid infrared divergences, asymptotic freedom will legitimize a weak coupling treatment. Even more interesting would be to do this by a method that also works at finite density, since the equation of state is of great interest for astrophysics and is *not* accessible numerically.

There is no doubt, in any case, that QCD predicts the existence of a quark-gluon plasma phase, wherein its basic degrees of freedom, normally hidden, come to occupy center stage.

While transition to a quark-gluon plasma at asymptotically high temperatures is not unexpected, the abruptness and especially the *precocity* of the change is startling. Below 150 Mev the only important hadronic degrees of freedom are the pions. Why does this rather dilute pion gas suddenly go berserk?

The change is enormous, quantitatively. The pions represent precisely 3 degrees of freedom. The free quarks and gluons, with all their colors, spins, and antiparticles, represent 52 degrees of freedom.

There are many ideas for detecting signals of quark-gluon plasma formation in heavy ion collisions, which you will be hearing much of. I would like discuss briefly a related but more focused question, on which there has been dramatic progress recently[25]. This is the question, whether there is a true phase transition in QCD accessible to experiment.

One might think that the answer is obviously "yes", since there are striking differences between ordinary hadronic matter and the quark-gluon plasma. This is not decisive, however. Let me remind you that the dissociation of ordinary atomic gases into plasmas is not accompanied by a phase transition, even though these states of matter are very different (so different, that at Princeton they are studied on separate campuses). Similarly, confinement of quarks is believed to go over continuously, at high temperature, into screening – certainly, no one has demonstrated the existence of an order parameter to distinguish between them.

What about chiral symmetry restoration? For massless quarks, there is a definite difference between the low-temperature phase of broken chiral symmetry and the quasi-free phase with chiral symmetry restored, so there must be a phase transition. A rather subtle analysis using the renormalization group indicated that for two massless quarks one might have a second-order phase transition, while for three massless quarks it must be first-order[26]. This is the pattern observed in numerical simulations.

The real world has two very light quarks and one (the strange quark) whose mass is neither clearly small nor clearly large compared to basic QCD scales. Here, then, a sharp question emerges. If the strange quark is effectively heavy, and the other quarks are taken strictly massless, we should have a second-order transition. If the strange quark is effectively light, we should have a first-order transition. Which is the case, for the physical value of the strange quark mass? Although this has been a controversial question, there seems to be an emerging consensus among lattice gauge theorists that it is second-order.

Unfortunately, this means that with small but finite u and d quark masses we will not have a sharp phase transition at all, but only a crossover. And not a particularly sharp one, at that. For while we are ordinarily encouraged to treat these masses as small perturbations, they are responsible for the pion masses, which are far from negligible at the temperatures under discussion. So the relevant correlation length never gets very large.

Nevertheless it was interesting to point out[27] that in the $m_s - T$ plane one could naturally connect the first- and second- order behaviors, as in Figure 6. The line of first-order transitions ends at a tricritical point. This is a true critical point, with diverging correlation lengths and large fluctuations. The first-order line, since it is a locus of discontinuities, and therefore the existence of a tricritical point where it terminates, are features which survive the small perturbation due to nonzero light quark masses. All these statements can be tested against numerical simulation.

Stephanov, Rajagopal and Shuryak[25] have brought the subject to a new level of interest, taking off from the simple but brilliant observation that one expects similar behavior in the $\mu - T$ plane. The big advantage of this is, that while m_s is not a control parameter one can vary experimentally, the chemical potential μ is. They have proposed quite specific, characteristic signatures for passage near this transition in the thermal history of a fireball, such as might be obtained in heavy ion collisions[28]. The signatures involve enhanced fluctuations and excess, nonthermal production of low-energy pions.

I believe it ought to be possible to refine the prediction, by locating the tricritical point theoretically. For while it is notoriously difficult to deal with large chemical potentials at small temperature numerically, there are good reasons to be optimistic about high temperatures and relatively small chemical potentials, which is our concern here[29].

If all these strands can be brought together, it will be a wonderful interweaving of theory, experiment, and numerics.

3. Breaking New Ground in QCD, 2: High Density

The behavior of QCD at high baryon number density, and low temperature, is of direct interest for describing neutron star interiors, neutron star collisions, and events near the core of collapsing stars. Unfortunately, it has proved quite difficult to calculate this behavior directly numerically using lattice gauge theory techniques. This is because in the presence of a chemical potential the functional integral for the partition function is no longer positive definite (or even real) configuration by configuration, so importance sampling fails, and the calculation converges only very slowly.

On the other hand, there has been remarkable progress on this problem over the last year or two using analytical techniques. This has shed considerable new light on many aspects of QCD. We have new, fully calculable mechanisms for confinement and chiral symmetry breaking[30]. Amazingly, we find that two famous, historically influential "mistakes" from the prehistory of QCD – the Han-Nambu[26] assignment of integer charge to quarks, and the Sakurai[32] model of vector mesons as Yang-Mills fields[33] – emerge from the microscopic theory at high density. And we find that in the slightly idealized version of QCD with three degenerate light quarks, there need be no phase transition separating the calculable high density phase from (the appropriate version of) nuclear matter[34]!

For a detailed review of these developments, see[35].

Why might we expect QCD to become analytically tractable at high density? At the crudest heuristic level, it is a case of asymptotic freedom meets the fermi surface.

Let us suppose, optimistically, that a weak coupling treatment is going to be appropriate, and see where it leads.

If the coupling is weak and the density large, our first approximation to the ground state is large fermi balls for all the quarks. Due to the Pauli exclusion principle, the modes deep within the ball will be energetically costly to excite, and the important low-energy degrees of freedom will be the modes close to the fermi surface. But these modes will have large momentum. Thus their interactions, generically, will either hardly deflect them, or will involve large momentum transfer. In the first case we don't care, while the second involves a weak coupling, due to asymptotic freedom.

On reflection, one perceives two big holes in this argument. First, it doesn't touch the gluons. They remain massless, with singular interactions and strong couplings in the infrared that do not appear to be under control. Second, as we learn in the theory of superconductivity, the fermi surface is generically unstable, even at weak coupling. This is because pairs of particles (or holes) of equal and opposite momenta are low-energy excitations which can all scatter into one another. Thus one is doing highly degenerate perturbation theory, and in that circumstance even a small coupling can have large qualitative effects.

Fortunately, our brethren in condensed matter physics have taught us how to deal with the second problem, and its proper treatment also cures the first. There is an attractive interaction between quarks on the opposite sides of the fermi surface, and they pair up and condense. In favorable cases – and in particular, for three degenerate or nearly degenerate flavors – this color superconductivity produces a gap for all the fermion excitations, and also gives mass to all the gluons.

Thus a proper weak-coupling treatment automatically avoids all potential infrared divergences, and our optimistic invocation of asymptotic freedom provides, at asymptotically high density, its own justification.

4. Breaking New Ground in QCD, 3: Unification

The different components of the standard model have a similar mathematical structure, all being gauge theories. Their common structure encourages the speculation that they are different facets of a more encompassing gauge symmetry, in which the different strong and weak color charges, as well as electromagnetic charge, would all appear on the same footing. The multiplet structure of the quarks and leptons in the standard model fits beautifully into small representations of unification groups such as $SU(5)$ or $SO(10)$. There is the apparent difficulty, however, that the coupling strengths of the different standard model interactions are widely different, whereas the symmetry required for unification requires that they share a common value.

The running of couplings suggests an escape from this impasse[36]. Since the strong, weak, and electromagnetic couplings run at different rates, their inequality at currently accessible scales need not reflect the ultimate state of affairs. We can imagine that spontaneous symmetry breaking – a soft effect – has hidden the full symmetry of the unified interaction. What is really required is that the fundamental, bare couplings be equal, or in more prosaic terms, that the running couplings of the different interactions should become equal beyond some large scale.

Using simple generalizations of the formulas derived and tested in QCD, which are none other than the ones experimentally validated in Figure 1, we can calculate the running of all the couplings, to see whether this requirement is met. In doing so one must make some hypothesis about the spectrum of virtual particles. If there are additional massive particles (or, better, fields) that have not yet been observed, they will contribute significantly to the running of couplings once the scale exceeds their mass.

Let us first consider the default assumption, that there are no new fields beyond those that occur in the standard model. The results of this calculation are displayed in Figure 7[37]. Considering the enormity of the extrapolation this works remarkably well, but the accurate experimental data indicates unequivocally that something is wrong.

There is one particularly attractive way to extend the standard model, by including supersymmetry. Supersymmetry cannot be exact, but if it is only mildly broken (so that the superpartners have masses $\lesssim 1$ TeV) it can help explain why radiative corrections to the Higgs mass parameter, and thus to the scale of weak symmetry breaking, are not enormously large.

In the absence of supersymmetry power counting would indicate a hard, quadratic dependence of this parameter on the cutoff. Supersymmetry removes the most divergent contribution, by canceling boson against fermion loops. If the masses of the superpartners are not too heavy, the residual finite contributions due to supersymmetry breaking will not be too large. The minimal supersymmetric extension of the standard model, then, makes definite predictions for the spectrum of virtual particles starting at 1 TeV or so. Since the running of couplings is logarithmic, it is not extremely sensitive to the unknown details of the supersymmetric mass spectrum, and we can assess the impact of supersymmetry on the unification hypothesis quantitatively. The results, as shown in Figure 8[38], are quite encouraging.

A notable result of the unification of couplings calculation, especially in its supersymmetric form, is that the unification occurs at an energy scale which is enormously large by the standards of traditional particle physics, perhaps approaching 10^{16-17} GeV. From a phenomenological viewpoint, this is fortunate. The most compelling unification schemes merge quarks, antiquarks, leptons, and antileptons into common multiplets, and have gauge bosons mediating transitions among all these particle types. Baryon number violating processes almost inevitably result, whose rate is inversely proportional to the fourth power of the gauge boson masses, and thus to the fourth power of the unification scale. Only for such large values of the scale is one safe from experimental limits on nucleon instability.

From a theoretical point of view the large scale is fascinating because it brings us, from the internal logic of particle physics, to the threshold of quantum gravity.

I find it quite remarkable that the logarithmic running of couplings, discovered theoretically and now amply verified within QCD, permits a meaningful quantitative discussion of these extremely ambitious and otherwise thinly rooted ideas, and even allows us to discriminate between different possibilities (especially, SUSY vs. non-SUSY).

5. Lessons: The Nature of Nature

Since QCD is our most perfect example of a fundamental theory of Nature, it is appropriate to use it as a basis for drawing broad conclusions about how Nature works, or, in other words, for "natural philosophy".

Let me do this by listing some adjectives we might use to describe the theory; the implication being, that these adjectives therefore describe Nature herself:

Alien As has been the case for all fundamental physical theories since Galileo, QCD is formulated in abstract mathematical terms. In particular, there are no hints of moral concepts or purposes. Nor, in thousands of rigorous experiments, have we encountered any signs of active intervention in the unfolding of the equations according to permanent laws.

Simple In its appropriate, natural language, QCD can be written in one line, using only symbols that cleanly embody its conceptual basis.

Beautiful That she achieves so much with such economy of means, marks Nature as a skillful artist. She plays with symmetries, creating and destroying them in varied, fascinating ways.

Weird Quantum mechanics is notoriously weird, and QCD incorporates it in its marrow. Less remarkable, but to me no less weird, is the need to define QCD through a limiting procedure. This would seem to be a rather difficult and inefficient way to run a Universe.

Comprehensible QCD wonderfully illustrates Einstein's remark, "The most incomprehensible thing about Nature is that it is comprehensible." One hundred years ago people did not know there were such things as an atomic nuclei and a strong interaction; just over fifty years ago the pion and kaon were discovered; just over twenty-five years ago the strong interaction problem still seemed hopelessly intractable. That we, collectively, have got from there to here so quickly, against overwhelming odds, is an extraordinary achievement. It is a tribute to our culture, and to the glory of the human mind. And it is there that where we must locate, for now, the most incomprehensible thing about Nature.

Acknowledgments

The gist of this work originally appeared as hep-ph/9907340, and a version was published in Nucl. Phys. A663 (2000), pp. 3–20. This work is supported in part by the US Dept. of Energy (DOE) under cooperative research agreement #DF-FC02-94ER40818.

References

1. F. Wilczek, Rev. Mod. Phys. 71, S85-S95 (1999).
2. K. Wilson, Phys. Rev. D10, 2445 (1975).
3. S. Coleman and D. Gross, Phys. Rev. Lett. 31, 851 (1973).
4. D. Gross and F. Wilczek, Phys. Rev. Lett. 30, 1343 (1973); H. Politzer, Phys. Rev. Lett. 30, 1346 (1973).
5. D. Gross and F. Wilczek, Phys. Rev. D8, 3633 (1973), Phys. Rev. D9, 980 (1974); H. Georgi and H. Politzer, Phys. Rev. D9, 416 (1974).
6. Reviewed in H. Leutwyler, hep-ph/9609465.
7. M.A. Shifman, A.I. Vainshtein, and V.I. Zakharov, Nucl. Phys. B147, 385, 448 (1979); V.A. Novikov, M.A. Shifman, A.I. Vainshtein, and V.I. Zakharov, Nucl. Phys. B191, 301 (1981); V.A. Novikov, L.B. Okun, M.A. Shifman, A.I.

258

Vainshtein, and V.I. Zakharov, Phys. Rept. 41, 1 (1978).

8. P Colangelo, A. Khodjamirian, "QCD Sum Rules, A Modern Perspective" in *At the Frontiers of Particle Physics: Handbook of QCD*, vol. 3, M. Shifman, ed. (World Scientific, Singapore, 2001).

9. T. Banks, S. Raby, L. Susskind, J. Kogut, D. Jones, P. Scharbach, and D. Sinclair, Phys. Rev. 15, 1111 (1977).

10. M. Creutz, Phys. Rev. D21, 2308 (1980).

11. S. Coleman and E. Weinberg, Phys. Rev. D7, 1888 (1973).

12. R. Peccei and H. Quinn, Phys. Rev. D16, 1791 (1977).

13. S. Weinberg, Phys. Rev. Lett. 40, 223 (1978); F. Wilczek, Phys. Rev. Lett. 40, 279 (1978).

14. F. Wilczek, Nature, 397, 303–306 (1999).

15. S. Aoki et al., Phys. Rev. Lett. 84 (2000) 238–241.

16. J. Gasser, H. Leutwyler and M. Sainio, Phys. Lett. B253, 252, 260 (1991).

17. M. Schmelling, preprint MPI-H-V39, hep-ex/9701002. Talk given at the 28th International Conference on High-energy Physics (ICHEP96), Warsaw, Poland (1996).

18. R.D. Ball, hep-ph/9609309. Summary Talk of WG1, in proceedings of DIS96, Rome, April 1996.

19. A. de Rujula, S. Glashow, H. Politzer, S. Treiman, F. Wilczek and A. Zee, Phys. Rev. D10, 2216 (1974).

20. S. Bethke and J.E. Pilcher, Ann. Rev. Nucl. Part. Sci., 42, 251–289, (1992)

21. R. Burkhalter, Nucl. Phys. Proc. Suppl. 73 (1999) 3–15.

22. D. Kaplan, Phys. Lett. B288, 342 (1992), Nucl. Phys. B30 (Proc. Suppl.) 597 (1993).

23. S. Gottlieb, et al., Phys. Rev. D47, 3619 (1993).

24. J. Andersen, E. Braaten, and M. Strickland, Phys. Rev. D61 (2000) 014017; J.-P. Blaizot, E. Iancu and A. Rebhan, Phys. Rev. Lett. 83 (1999) 2906–2909.

25. M. Stephanov, K. Rajagopal and E. Shuryak, Phys. Rev. Lett. 81, 4816 (1998). hep-ph/9806219.

26. R. Pisarski and F. Wilczek, Phys. Rev. D29, (1984).

27. F. Wilczek, Int. J. Mod. Phys A7 3911 (1992).

28. M. Stephanov, K. Rajagopal and E. Shuryak, Phys. Rev. D60 (1999) 114028.

29. M. Alford, A. Kapustin and F. Wilczek, Phys. Rev. D59, 054502 (1999).

30. M. Alford, K. Rajagopal and F. Wilczek, Nucl. Phys. B537, 443–458 (1999).

31. M. Han and Y. Nambu, Phys. Rev. 139B, 1006 (1965).

32. J. Sakurai, Annals of Physics 11, 1 (1960).

33. C. N. Yang and R. Mills, Phys. Rev. 96, 191 (1954).

34. T. Schaefer and F. Wilczek, Phys. Rev. Lett. 82, 3956 (1999).

35. F. Wilczek and K. Rajagopal, "The Condensed Matter Physics of QCD" in *At the Frontiers of Particle Physics: Handbook of QCD*, vol. 3, M. Shifman, ed. (World Scientific, Singapore, 2001).

36. For a short and a long review of the subject, see respectively S. Dimopoulos, S. Raby, and F. Wilczek,Physics Today 44, 25 (1991); K. Dienes, Physics Reports 287, 447–525 (1997)

37. P. Langacker and M. Luo, Phys. Rev. D44, 817–822 (1991).

38. Figure 8 courtesy of K. Dienes, CERN Theory Group.

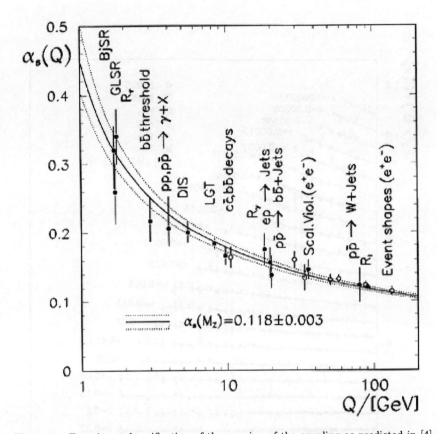

Figure 1. Experimental verification of the running of the coupling as predicted in [4]. The determinations, running from left to right, are from: corrections to Bjorken sum rule, corrections to Gross-Llewellyn-Smith sum rule, hadronic width of τ lepton, $b\,\bar{b}$ threshold production, prompt photon production in pp and $\bar{p}$ collisions, scaling violation in deeply inelastic scattering, lattice gauge theory calculations for heavy quark spectra, heavy quarkonium decays, shape variables chacterizing jets at different energies (white dots), total e^+e^- annihilation cross section, jet production in semileptonic and hadronic processes, energy dependence of photons in Z decay, W production, and electroweak radiative corrections. From [17].

260

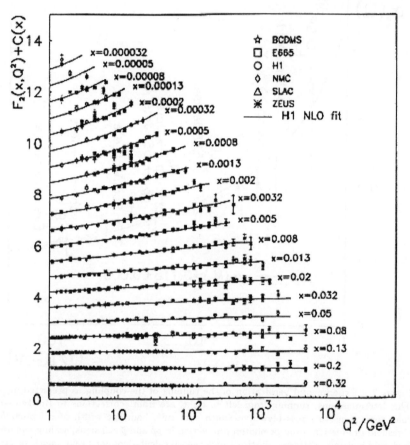

Figure 2. Evolution of structure function F_2, as measured and compared with QCD predictions (solid lines). From [18].

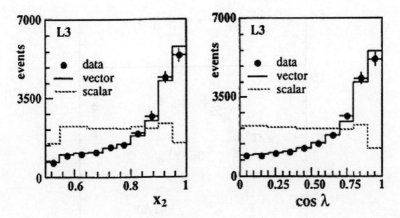

Figure 3. Energy and angular characterization of three-jet events, testing the basic quark-gluon vertex very directly. From [20].

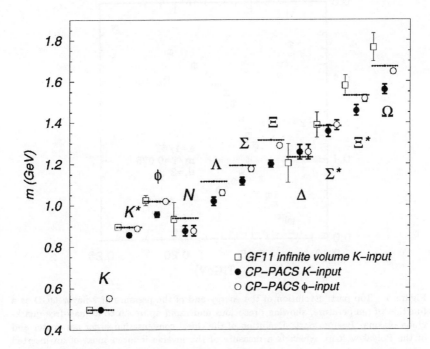

Figure 4. Comparison of the hadronic spectrum with first-principles calculations from QCD, using techniques of lattice gauge theory. From [21].

262

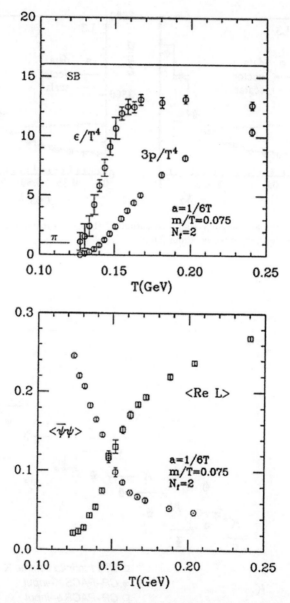

Figure 5. Top part: Evolution of the energy and of the pressure of 2-flavor QCD as a function of temperature, showing precocious and rapid approach to a quasi-free quark-gluon plasma. Bottom part: Evolution of the chiral condensation order parameter and of the Polyakov loop, which is a measure of the inverse induced mass of an inserted color source, and vanishes in a confined phase. One sees clear signals of chiral symmetry restoration and deconfinement[23].

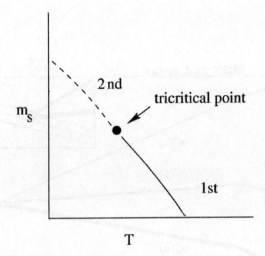

Figure 6. Connecting the second- and first- order chiral symmetry restoration transitions for two, respectively three, light quark flavors. The end of the first-order line is a tricritical point. A similar diagram may be valid for fixed strange quark mass and varying chemical potential.

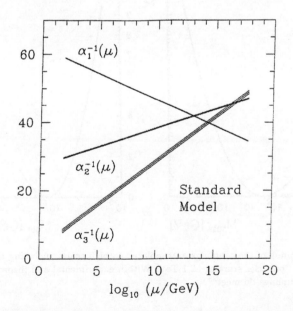

Figure 7. Running of the couplings extrapolated toward very high scales, using just the fields of the standard model. The couplings do not quite meet. Experimental uncertainties in the extrapolation are indicated by the width of the lines[37].

264

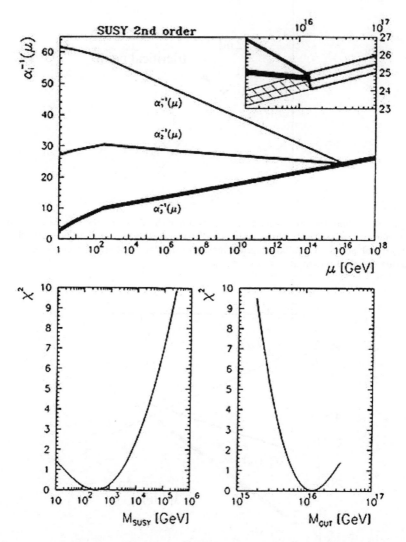

Figure 8. Running of the couplings extrapolated to high scales, including the effects of supersymmetric particles starting at 1 TeV. Within experimental and theoretical uncertainties, the couplings do meet[38].

DIRAC SPECTRA AND REAL QCD AT NONZERO
CHEMICAL POTENTIAL

D. TOUBLAN[1] AND J.J.M. VERBAARSCHOT[2]

[1] *Department of Physics, UIUC,*
1110 West Green Street, Champaign-Urbana, Il 61801, USA
E-mail:toublan@uiuc.edu

[2] *Department of Physics and Astronomy, SUNY at Stony Brook,*
Stony Brook, NY 11794, USA
E-mail: verbaarschot@tonic.physics.sunysb.edu

We show that QCD Dirac spectra well below Λ_{QCD}, both at zero and at nonzero chemical potential, can be obtained from a chiral Lagrangian. At nonzero chemical potential Goldstone bosons with nonzero baryon number condense beyond a critical value. Such superfluid phase transition is likely to occur in any system with a chemical potential with the quantum numbers of the Goldstone bosons. We discuss the phase diagram for one such system, QCD with two colors, and show the existence of a tricritical point in an effective potential approach.

1. Introduction

For strongly interacting quantum field theories such as QCD a complete nonperturbative analysis from first principles is only possible by means of large scale Monte Carlo simulations. Therefore, partial analytical results in some parameter domain of the theory are extremely valuable, not only to provide additional insight in the numerical calculations, but also as an independent check of their reliability. This has been our main motivation for analyzing such domains. The principle idea we have been pursuing is based on chiral perturbation theory [1,2]: because of confinement and the spontaneous breaking of chiral symmetry, the low-energy chiral limit of QCD is a theory of weakly interacting Goldstone bosons which are described by a chiral Lagrangian that is completely determined by the symmetries of QCD. This idea can be applied to the QCD Dirac spectrum which can be extracted from the valence quark mass dependence of the chiral condensate. The valence quark mass is not a physical parameter of the QCD partition function and can be chosen in a domain where the valence quark mass dependence of the QCD partition function can be described to an arbitrary

265

accuracy by a corresponding chiral Lagrangian. If the Compton wavelength of Goldstone bosons containing only valence quark masses is much larger than the size of the box the low-energy effective theory simplifies even much further [3]. Then only the zero momentum component of the Goldstone fields has to be taken into account so that the valence quark mass dependence of the QCD partition function is given by a unitary matrix integral. This idea was first applied to the QCD partition function [4] with quark masses of order $m \sim 1/V\langle\bar{\psi}\psi\rangle$ (with $\langle\bar{\psi}\psi\rangle$ the chiral condensate and V the volume of space-time). However, we emphasize that for physical values of the quark masses and volumes, a part of the Dirac spectrum, as probed by the valence quark mass, is always in this mesoscopic domain of QCD. More precisely, using the Gell-Mann-Oakes-Renner relation (with F the pion decay constant), in the domain

$$\frac{m_v\langle\bar{\psi}\psi\rangle}{F^2} \ll 1/\sqrt{V} \quad \text{and} \quad V^{1/4}\Lambda_{\text{QCD}} \gg 1, \tag{1}$$

the kinetic term in the chiral Lagrangian can be ignored and the valence quark mass dependence of the QCD partition function reduces to a unitary matrix integral [3,5]. This integral is equivalent to a chiral Random Matrix Theory in the limit of large matrices [6,7]. The second condition ensures that excitations of the order Λ_{QCD} decouple from the low-energy sector of the partition function.

At nonzero baryon chemical potential the Dirac spectrum is scattered in the complex plane. However, at a sufficiently small nonzero baryon chemical potential and finite physical quark masses, the Dirac spectrum in the phase of broken chiral symmetry is still described by a partition function of Goldstone bosons containing valence quarks [8]. In order to eliminate the fermion determinant containing the valence quarks, one has to calculate the valence quark mass dependence of the chiral condensate in the limit of a vanishing number of valence quarks. The existence of this limit requires the introduction of conjugate antiquarks [9,10], resulting in the appearance of Goldstone bosons with nonzero baryon number containing only valence quarks. They condense if the chemical potential exceeds their mass. In terms of the Dirac spectrum this phase transition is visible as a sharp boundary of the locus of the eigenvalues.

Such phase transition to a Bose condensed phase is likely to occur in any theory with a chemical potential with the quantum numbers of Goldstone bosons. For example, for QCD with two fundamental colors [11,12] or for adjoint QCD with two or more colors [12], the lightest baryon is a Goldstone boson. A transition to a Bose condensed phase occurs for a chemical potential larger than the mass of this boson. Other examples are pion condensation, which may occur for a nonzero isospin chemical potential [13], and kaon condensation which may occur for a nonzero strangeness chemical potential [14]. If the mass of the Goldstone bosons and the chemical potential are both well below $\Lambda_{\rm QCD}$, such phase transition can be described in terms of a chiral Lagrangian. We have analyzed such Lagrangian for QCD with two colors at nonzero temperature and chemical potential [12,15,16]. In an effective potential approach we have found a tricritical point [16] in agreement with recent lattice QCD simulations [17].

We start this lecture by discussing QCD Dirac spectra at zero chemical potential and explaining its description in terms of a chiral Lagrangian. In section 3 we analyze QCD Dirac spectra at nonzero chemical potential. The phase diagram of QCD with two colors at nonzero temperature and chemical potential is discussed in section 4 and concluding remarks are made in section 5.

2. Dirac Spectrum at Zero Chemical Potential

The Euclidean QCD Dirac operator is given by

$$iD = \gamma_\mu(\partial_\mu + iA_\mu), \tag{2}$$

where the γ_μ are the Euclidean gamma matrices and the A_μ are $SU(N_c)$ valued gauge fields. The Dirac spectrum for a fixed gauge field configuration is obtained by solving the eigenvalue equation

$$iD\phi_k = i\lambda_k\phi_k. \tag{3}$$

In a regularization scheme with a finite number of N eigenvalues, the average spectral density is defined by

$$\rho(\lambda) = \langle \sum_{k=1}^{N} \delta(\lambda - \lambda_k) \rangle_{\rm QCD}, \tag{4}$$

where the average $\langle \cdots \rangle_{\rm QCD}$ is over gauge field configurations weighted by the Euclidean QCD action. As a result of the averaging we expect that $\rho(\lambda)$ will be a smooth function of λ. Because of the involutive automorphism $\gamma_5 iD\gamma_5 = -iD$ the Dirac operator can always be represented in block-form as

$$iD = \begin{pmatrix} 0 & iW \\ iW^\dagger & 0 \end{pmatrix}. \tag{5}$$

If W is a square matrix the nonzero eigenvalues of iD occur in pairs $\pm\lambda_k$. For nonzero topological charge the total number of zero eigenvalues is given by the difference of the the number of right-handed modes and left-handed modes. In that case, the matrix W is a rectangular matrix with the absolute value of the difference between the number of rows and columns equal to the topological charge. For very large values of λ the Dirac spectrum converges to the free Dirac spectrum so that the spectral density given by $\rho(\lambda) \sim V\lambda^3$. The smallest nonzero eigenvalue, $\lambda_{\min}$, is of the order of the average level spacing and is thus given by

$$\lambda_{\min} = \Delta\lambda = \frac{1}{\rho(0)}. \tag{6}$$

2.1. *Spontaneous Chiral Symmetry Breaking and Eigenvalue Correlations*

The chiral condensate is given by

$$
\begin{aligned}
\langle \bar{\psi}\psi \rangle &= \lim_{\Lambda\to\infty} \lim_{m\to 0} \lim_{V\to\infty} \frac{1}{V} \left\langle \mathrm{Tr}\frac{1}{iD+m} \right\rangle_{\mathrm{QCD}} \\
&= \lim_{\Lambda\to\infty} \lim_{m\to 0} \lim_{V\to\infty} \frac{1}{V} \int_0^\Lambda \frac{2m\rho(\lambda)}{\lambda^2 + m^2}.
\end{aligned} \tag{7}
$$

The limit $m \to 0$ is taken before $\Lambda \to \infty$ to eliminate divergent contributions from the ultraviolet part of the Dirac spectrum (the ultraviolet cutoff, Λ, may also appear in the spectral density). Because of spontaneous breaking of chiral symmetry, the limit $V \to \infty$ cannot be interchanged with the limit $m \to 0$ in (7). If the chiral condensate is nonzero the limits $m \to 0^+$ and $m \to 0^-$ have opposite signs. This can only happen if $\rho(0) \sim V$. If we expand the spectral density as

$$\rho(\lambda) = \rho(0^+) + a_1|\lambda| + a_2\lambda^2 + \cdots, \tag{8}$$

we obtain Banks-Casher formula [18]

$$\langle \bar{\psi}\psi \rangle = \lim_{V\to\infty} \frac{\pi\rho(0^+)}{V}. \tag{9}$$

In this article we avoid taking limits by mainly focusing on finite values of m, V and Λ.

Let us now consider the QCD partition function $Z(m_f)$,

$$Z(m_f) = \langle \prod_f \prod_k (i\lambda_k + m_f) \rangle_{\mathrm{YM}}, \tag{10}$$

where $\langle \cdots \rangle_{\mathrm{YM}}$ denotes averaging with respect to the Yang-Mills action. Because in the thermodynamic limit the derivative of the partition function with respect to m_f has a discontinuity across the imaginary axis, we expect that its zeros are also located on the imaginary axis as well and, for finite volume, are spaced as $1/V$. This average can also be written as an average over the joint eigenvalue distribution

$$\rho(\lambda_1, \lambda_2, \cdots) \equiv \langle \delta(\lambda_1 - \lambda_1^A)\delta(\lambda_2 - \lambda_2^A)\cdots\rangle_{\text{QCD}}, \tag{11}$$

where λ_k^A are the eigenvalues of the Dirac operator for a given gauge field configuration A. This results in

$$Z(m_f) = \int \rho(\lambda_1, \lambda_2, \cdots) \prod_f \prod_k (i\lambda_k + m). \tag{12}$$

If the eigenvalues are uncorrelated the joint eigenvalue distribution factorizes into one-particle distributions

$$\rho(\lambda_1, \lambda_2, \cdots) = \rho_1(\lambda_1)\rho_1(\lambda_2)\cdots \tag{13}$$

and the partition function is the product of N identical factors. For example, for $N_f = 1$, in the sector of zero topological charge, we obtain

$$Z(m) = (\langle \lambda^2 \rangle_1 + m^2)^N, \tag{14}$$

where $\langle \cdots \rangle_1$ is the average with respect to the one particle distribution (which in this case is the average spectral density of the QCD Dirac operator). Therefore $Z(m)$ is a smooth function as m crosses the imaginary axis along the real axis and chiral symmetry is not broken. We conclude that the absence of eigenvalue correlations implies that chiral symmetry is not spontaneously broken, or conversely, if chiral symmetry is broken spontaneously the eigenvalues of the Dirac operator are necessarily correlated. The question we wish to answer is what are these correlations.

2.2. Low Energy Limit of QCD

Because of confinement the chiral limit of QCD at low energy is a theory of weakly interacting Goldstone bosons. For small values of the quark masses m_f and chemical potentials μ_f the QCD partition function coincides with a partition function of Goldstone bosons:

$$Z_{\text{QCD}}(m_f, \mu_f, \theta) \sim Z_{\text{Gold}}(m_f, \mu_f, \theta), \tag{15}$$

where θ is the vacuum θ-angle. Up to phenomenological coupling constants, the mass dependence of Z_{Gold} is completely determined by the symmetries and transformation properties of the QCD partition function. In particular, both partition functions have the same low mass expansion. Equating the coefficients of powers of the quark masses leads to sum-rules for the inverse Dirac eigenvalues [4]. To derive them we consider the Fourier components of the θ dependence which are just the partition function in a given sector of topological charge,

$$Z_\nu(\cdots) = \frac{1}{2\pi} \int_0^{2\pi} d\theta e^{i\nu\theta} Z(\cdots,\theta). \tag{16}$$

As an example, let us consider the case $N_f = 1$, $\mu_f = 0$ and $\nu = 0$. In this case there are no Goldstone bosons and the the mass dependence of the partition function for $\theta = 0$ is given by

$$Z \sim e^{V\Sigma(m+m^*)/2}. \tag{17}$$

The θ dependence id obtained from the substitution $m \to me^{i\theta}$. For the sector of zero topological charge we thus find

$$\langle(\lambda^2 + m^2)\rangle_{\nu=0} \sim \frac{1}{2\pi} \int_0^{2\pi} d\theta e^{mV\Sigma \cos\theta}$$

$$= 1 + \frac{1}{4}m^2 V^2 \Sigma^2 + \cdots. \tag{18}$$

This result in the sum rule [4]

$$\left\langle \sum_{\lambda_k > 0} \frac{1}{\lambda_k^2} \right\rangle_{\nu=0} = \frac{V^2 \Sigma^2}{4}. \tag{19}$$

In fact, an infinite number of sum rules can be derived for the partition function of QCD and QCD-like theories with spontaneous chiral symmetry breaking [4,19,20,21,22]. Nevertheless, these sum rules are not sufficient to determine the Dirac spectrum.

2.3. Resolvent

In order to derive the QCD Dirac spectrum we introduce the resolvent

$$G(z) = \frac{1}{V} \left\langle \text{Tr} \frac{1}{z + iD} \right\rangle_{\text{QCD}}. \tag{20}$$

Here, z is a complex 'valence quark mass' which does not occur inside the fermion determinant that is included in the average. The spectral density is obtained from the discontinuity of the resolvent across the imaginary axis,

$$\rho(\lambda) = \frac{1}{2\pi}(G(i\lambda + \epsilon) - G(i\lambda - \epsilon))$$

$$= \frac{1}{2\pi}(G(i\lambda + \epsilon) + G(-i\lambda + \epsilon)). \tag{21}$$

The resolvent can be obtained [23,24,25,26,27,28,29,30] from the generating function $Z_{\text{spect}}(z, z', m_f)$,

$$G(z) = \frac{1}{V}\partial_z \, Z_{\text{spect}}(z, z', m_f)|_{z'=z}, \qquad (22)$$

with

$$Z_{\text{spect}}(z, z', m_f) = \left\langle \frac{\det(iD + z)}{\det(iD + z')} \prod_f \det(iD + m_f) \right\rangle_{\text{YM}}. \qquad (23)$$

The variable z is a parameter that probes the Dirac spectrum and can be chosen arbitrary small. For z, z', $m_f \ll \Lambda_{\text{QCD}}$ this partition function can be approximated arbitrarily well by a chiral Lagrangian which is completely determined by the symmetries of the QCD partition function. In addition to fermionic quarks, this partition function also contains bosonic ghost quarks. The corresponding chiral Lagrangian therefore includes both bosonic and fermionic Goldstone bosons with masses given by $2\text{Re}(z)\Sigma/F^2$, $\text{Re}(z + z')\Sigma/F^2$, $\text{Re}(z + m_f)\Sigma/F^2$, etc., as given by the usual Gell-Mann-Oakes-Renner relation.

The inverse fermion determinant can be written as a convergent bosonic integral provided that $\text{Re}(z) > 0$:

$$\frac{1}{\det(iD + z)} = \int d\phi d\phi^* e^{-\phi^*(iD+z)\phi}. \qquad (24)$$

The convergence requirements restrict the possible symmetry transformations of the partition function. For example the axial $U(1)$ transformation which in the fermionic case is given by

$$\psi_R \to e^{i\theta}\psi_R, \qquad \psi_L \to e^{-i\theta}\psi_L$$
$$\bar{\psi}_R \to e^{-i\theta}\psi_R, \qquad \bar{\psi}_L \to e^{+i\theta}\psi_L, \qquad (25)$$

would violate the complex conjugation structure of the bosonic integral with $\bar{\phi}_R = \phi_L^*$ and $\bar{\phi}_L = \phi_R^*$. Instead, the allowed $U_A(1)$ transformation is

$$\phi_R \to e^s\phi_R, \qquad \phi_L \to e^{-s}\phi_L,$$
$$\bar{\phi}_R \to e^{-s}\phi_R, \qquad \bar{\phi}_L \to e^s\phi_L, \qquad (26)$$

with s a real parameter. The Goldstone manifold is therefore *not* given by the super-unitary group $SU(N_f+1|1)$ but rather by its complexified version that reflects the convergence requirements of the bosonic axial transformations [29,30]. We will denote this manifold by $\hat{SU}(N_f+1|1)$ and an explicit parameterization for the simplest case, $N_f = 0$, will be given below. Vector flavor symmetry transformations are consistent with the complex conjugation properties of the bosonic integral. This symmetry group is thus given by $SU(N_f + 1|1)$.

In the chiral limit the mass dependence of generating function (23) can be obtained from a chiral Lagrangian determined by its symmetries and transformation properties. It is given by

$$L = \mathrm{Str}\partial_\mu U \partial_\mu U^{-1} - \frac{1}{2}\langle \bar{\psi}\psi \rangle \mathrm{Str}(M(U + U^{-1})), \tag{27}$$

and the corresponding partition function reads

$$Z_\nu = \int_{\hat{U}(N_f + 1|1)} dU(x)\mathrm{Sdet}^\nu U_0 e^{-\int d^4 x L}. \tag{28}$$

Because ν is the global topological charge only the zero momentum component of U, denoted by U_0, appears in the argument of the superdeterminant. In the chiral limit, QCD is flavor symmetric so that the kinetic term of the chiral Lagrangian should be flavor symmetric as well. Therefore, the pion decay constant of the extended flavor symmetry is the same as in QCD. The mass matrix is given by $M = \mathrm{diag}(m, \cdots, m, z, z')$.

If $z \ll m_c \equiv F^2/\langle \bar{\psi}\psi \rangle \sqrt{V}$ the fluctuations of the zero momentum modes are much larger that the fluctuations of the nonzero momentum modes, which then can be ignored in the calculation of the resolvent. More physically, this condition means that the Compton wavelength of Goldstone bosons containing ghostquarks with mass z or z' is much larger than the size of the box. In condensed matter physics, the energy scale m_c is known as the Thouless energy and has been related to the inverse diffusion time of an electron through a disordered sample [31].

In the Dirac spectrum we therefore can distinguish three different energy scales, the smallest eigenvalue $\lambda_{\min}$, the Thouless energy m_c and the QCD scale Λ_{QCD}. On mass scales well below Λ_{QCD} the mass dependence of the QCD partition function is given by the chiral Lagrangian. For mass scales well below the Thouless energy only the zero momentum modes have to be taken into account. However, for masses *not* much larger than $\lambda_{\min}$, a perturbative calculation breaks down and the group integrals have to be performed exactly. An interesting possibility is if $\lambda_{\min}$ and m_c coincide which may lead to critical statistics [32].

In the zero momentum limit, it is straightforward to calculate the integrals over the superunitary group. The simplest case is the quenched case ($N_f = 0$) where U can be parameterized as

$$U = \begin{pmatrix} e^{i\theta} & \alpha \\ \beta & e^s \end{pmatrix}, \tag{29}$$

with α and β are Grassmann variables, $\theta \in [0, 2\pi]$ and $s \in \langle -\infty, \infty \rangle$. In terms of the rescaled variable $u = zV\langle \bar{\psi}\psi \rangle$, one obtains the resolvent

$$\frac{G(u)}{\langle \bar{\psi}\psi \rangle} = u(K_a(u)I_a(u) + K_{a-1}(u)I_{a+1}(u)) + \frac{\nu}{u}, \qquad (30)$$

where $a = N_f + |\nu|$. From the definitions of the modified Bessel functions it is clear that the compact/noncompact parameterization of the superunitary group is essential. The microscopic spectral density is obtained from the discontinuity of the resolvent and is given by

$$\rho_s(\zeta) = \frac{\rho(\zeta/V\langle \bar{\psi}\psi \rangle)}{V\langle \bar{\psi}\psi \rangle} = \frac{\zeta}{2}(J_a^2(\zeta) - J_{a+1}(\zeta)J_{a-1}(\zeta)) + \nu\delta(\zeta), \qquad (31)$$

where $\zeta = \lambda V\langle \bar{\psi}\psi \rangle$.

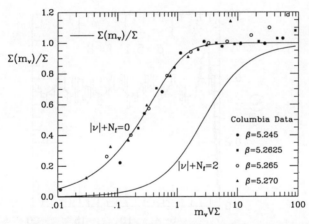

Figure 1. The valence quark mass dependence of the chiral condensate $\Sigma(m_v)$ plotted as $\Sigma(z/m_v)/\Sigma$ versus $m_v V\Sigma$. The dots and squares represent lattice results by the Columbia group [33] for values of β as indicated in the label of the figure. (Figure adapted from ref. [3]).

2.4. Lattice Results

The properties of the Dirac spectrum have been analyzed in many lattice QCD simulations [33,3,34,35,36,37,38,39,40,41,42,43,44,45,46,47,48] [49,50,51,52,53,54] and have been found to be in complete agreement with the conclusions of the previous section. We only show three representative examples.

In Fig. 1 we show the valence mass dependence of the chiral condensate as calculated by the Columbia group [33]. In this figure the valence quark mass is denoted by m_v and $\Sigma = \langle \bar{\psi} \psi \rangle$. Our variable u in (30) is thus given by $m_v V \Sigma$ and $\Sigma(m_v)$ should be identified with $G(u)$. The reason that the lattice data agree with the quenched approximation is that the sea-quark masses in the lattice calculation are much larger than the valence masses. The topological charge is zero because the instanton zero modes are completely mixed with the nonzero modes due to the lattice discretization.

Because the valence quark mass dependence agrees with (30) the corresponding lattice QCD microscopic spectral density should agree with (31). This was shown by two independent calculations [42,43]. In fig. 2 we show results for an 8^4 lattice with quenched staggered fermions [42].

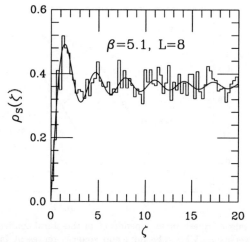

Figure 2. The microscopic spectral density for quenched QCD with three colors. The solid curve represents the analytical result (31) for $a = 0$. (Figure adapted from ref. [42])

In Fig. 3 we show the disconnected chiral susceptibility defined by

$$\chi^{\mathrm{disc}}(m_v) = \frac{1}{N} \left\langle \sum_{k,l=1}^{N} \frac{1}{(i\lambda_k + m_v)(i\lambda_l + m_v)} \right\rangle - \frac{1}{N} \left\langle \sum_{k=1}^{N} \frac{1}{i\lambda_k + m_v} \right\rangle^2 . (32)$$

This quantity can be obtained from the two-point spectral correlation function but can also be directly computed in chPT [55,56,29,54]. The dashed curve represents the result obtained from taking into account only the zero momentum modes whereas the solid curve is obtained from a perturbative one-loop calculation. Also in this figure $\Sigma = \langle \bar{\psi} \psi \rangle$. This figure clearly demonstrates the existence of a domain where a perturbative calculation can be applied to the zero momentum sector of the theory.

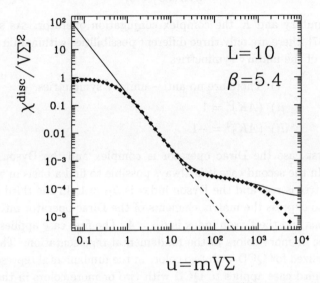

$$u = m V \Sigma$$

Figure 3. The disconnected susceptibility for quenched SU(3) with staggered fermions (solid points). The solid curve represents the prediction from chPT, and the dashed one is the exact result for the zero momentum approximation to the chiral susceptibility. (Note the dashed line is hidden by the data points for $u < 10$.) (Figure taken from ref. [54].)

2.5. *Chiral Random Matrix Theory*

Correlations of Dirac eigenvalues on the scale of the average level spacing are completely determined by the zero mode part of the partition function which only includes the mass term and the topological term of the chiral Lagrangian. This raises the question of what is the most symmetric theory that can be reduced to this partition function. The answer is chiral Random Matrix Theory in the limit of large matrices. This theory is invariant under and additional $U_R(n) \times U_L(n + \nu)$ group (with $n \times (n + \nu)$ the size of the nonzero blocks of the Dirac matrix). Because of this much larger symmetry group, all correlation function of the eigenvalues can be obtained analytically, often in a much simpler way than by means of the supersymmetric generating functions for the resolvent.

Before defining chiral Random Matrix Theories, we have to introduce the Dyson index of the Dirac operator. It is defined as the number of independent degrees of freedom per matrix element and is is determined by the anti-unitary symmetries of the Dirac operator. They are of the form

$$[AK, iD] = 0, \tag{33}$$

with A unitary and K the complex conjugation operator. As shown by Dyson [57], there are only three different possibilities within an irreducible subspace of the unitary symmetries

i) There are no anti − unitary symmetries,

ii) $(AK)^2 = 1$,

iii) $(AK)^2 = -1$. $\tag{34}$

In the first case the Dirac operator is complex and the Dyson index is $\beta_D = 2$. In the second case it is always possible to find a basis in which the Dirac matrix is real and the Dyson index is $\beta_D = 1$. In the third case it is possible to express the matrix elements of the Dirac operator into selfdual quaternions and the Dyson index is $\beta_D = 4$. The first case applies to QCD with three or more colors in the fundamental representation. The second case is realized for QCD with two colors in the fundamental representation, and the third case applies to QCD with two or more colors in the adjoint representation.

Chiral Random Matrix Theory is a Random Matrix Theory with the global symmetries of the QCD partition function. It is defined by the partition function [6,7]

$$Z_\nu(m_1, \cdots, m_f) = \int dW \prod_{f=1}^{N_f} \det{}^\nu(iD + m_f) e^{-\text{tr} V(WW^\dagger)}, \tag{35}$$

where the Random Matrix Theory Dirac operator is defined by

$$iD = \begin{pmatrix} 0 & iW \\ iW^\dagger & 0 \end{pmatrix}, \tag{36}$$

and W is an $n \times (n + \nu)$ matrix so that iD has exactly ν zero eigenvalues. In general the probability potential is a finite order polynomial. However, one can show [58,59,60,61,62,63] that correlations on the scale of the average level spacing do not depend on the details of this polynomial and the same results can be obtained much simpler from the Gaussian case. Depending on the Dyson index of the Dirac operator we have three different possibilities, the matrix elements of W are real, complex or self-dual quaternion for $\beta_D = 1, 2, 4$, respectively. The corresponding Gaussian ensembles are known as the chiral Gaussian Orthogonal Ensemble (chGOE), the chiral Gaussian Unitary Ensemble (chGUE) and the chiral Gaussian Symplectic

Ensemble (chGSE), in this order. Together with the Wigner-Dyson Ensembles and four ensembles that can be applied to superconducting systems, these ensembles can be classified according to the Cartan classification of large symmetric spaces [64].

The results (30) and (31) quoted in the previous section were obtained first by means of standard Random Matrix Theory methods [55,3].

3. Dirac Spectra at Nonzero Chemical Potential

Quenched lattice QCD Dirac spectra at $\mu \neq 0$ were first obtained numerically in the pioneering paper by Barbour et al. [65] and have since then been studied in several other works [66,67,68,69]. Since the Dirac operator has no hermiticity properties at $\mu \neq 0$ its spectrum is scattered in the complex plane. However, it was found [65] that for not too large values of the chemical potential the spectrum is distributed homogeneously inside an oval shape with a width proportional to μ^2. In this section we will explain these results in terms of a chiral Lagrangian for phase quenched QCD at nonzero chemical potential.

3.1. *Spectra of Nonhermitian Operators*

The spectral density of a nonhermitian operator is defined by

$$\rho(\lambda) = \left\langle \sum_k \delta(\mathrm{Re}(\lambda - \lambda_k))\delta(\mathrm{Im}(\lambda - \lambda_k)) \right\rangle_{\mathrm{QCD}}$$

$$= \frac{1}{\pi}\partial_{z^*}G(z), \tag{37}$$

where the resolvent $G(z)$ is defined by

$$G(z) = \frac{1}{V}\left\langle \sum_k \frac{1}{z - \lambda_k} \right\rangle_{\mathrm{QCD}}. \tag{38}$$

Often it is useful to interpret the real and imaginary parts of the resolvent as the electric field in the plane at point z from charges located at λ_k.

Since the fermion determinant is invariant for multiplication of the Dirac operator by an unimodular matrix, one could analyze the spectrum of various Dirac operators. The Dirac operator that is of interest is the one with eigenvalues that are related to an observable. For example, the Dirac operator in a chiral representation has the structure

$$iD = \begin{pmatrix} 0 & iW + \mu \\ iW^\dagger + \mu & 0 \end{pmatrix}. \tag{39}$$

In terms of its eigenvalues, the chiral condensate is given by

$$\langle \bar{\psi}\psi \rangle = \langle \frac{1}{V} \sum_k \frac{1}{m + i\lambda_k} \rangle. \tag{40}$$

If we are interested in the baryon number, on the other hand, we consider the Dirac operator

$$iD_\mu = \begin{pmatrix} iW & m \\ m & iW^\dagger \end{pmatrix}, \tag{41}$$

which satisfies the relation $\det(iD_\mu + \mu) = \det(iD + m)$. In terms of its eigenvalues μ_k the baryon density is given by

$$n_B = \left\langle \frac{1}{V} \sum_k \frac{1}{\mu + i\mu_k} \right\rangle_{\text{QCD}}. \tag{42}$$

Finally, let us consider QCD at nonzero isospin chemical potential. In this case the fermion determinant is given by

$$\det \begin{pmatrix} m & iW + \mu_I \\ iW^\dagger + \mu_I & m \end{pmatrix} \det \begin{pmatrix} m & iW - \mu_I \\ iW^\dagger - \mu_I & m \end{pmatrix}, \tag{43}$$

which can be rewritten as the determinant of the antihermitian matrix

$$\begin{pmatrix} 0 & 0 & -m & iW - \mu_I \\ 0 & 0 & iW^\dagger - \mu_I & -m \\ m & iW + \mu_I & 0 & 0 \\ iW^\dagger + \mu_I & m & 0 & 0 \end{pmatrix}. \tag{44}$$

In terms of its eigenvalues $i\pi_k$, the pion condensate is given by

$$\langle \pi \rangle = \left\langle \frac{1}{V} \sum_k \frac{1}{j_\pi + i\pi_k} \right\rangle_{\text{QCD}}, \tag{45}$$

where j_π is the source term for the pion condensate.

3.2. Low Energy Limit of Phase Quenched QCD

The generating function for the quenched Dirac spectrum is given by the replica limit ($N_f \to 0$) of phase quenched QCD partition function [10] defined by

$$\begin{aligned} Z &= \langle [\det(iD + z + \mu\gamma_0)\det{}^*(iD + z + \mu\gamma_0)]^{N_f} \rangle_{\text{QCD}} \\ &= \langle [\det(iD + z + \mu\gamma_0)\det(iD + z^* - \mu\gamma_0)]^{N_f} \rangle_{\text{QCD}}. \end{aligned} \tag{46}$$

Since this is a partition function of quarks and conjugate anti-quarks we can have Goldstone bosons with nonzero baryon number. For a chemical potential equal to half the pion mass we thus expect a phase transition to a Bose condensed phase. For a quark mass much less than Λ_{QCD} this phase transition can be described completely in terms a chiral Lagrangian. In nonhermitian Random Matrix Theory, the technique to determine the spectral density by analyzing a corresponding Hermitian ensemble is known as Hermitization [70,71],

The chiral Lagrangian is again determined by the symmetries and the transformation properties of the QCD partition function. These can be made more explicit if we rewrite the fermion determinant as

$$\det \begin{pmatrix} M_1 & d + B_R \\ -d^\dagger + B_L & M_2 \end{pmatrix}, \tag{47}$$

where $M_1 = M_2 = \mathrm{diag}(z, \cdots, z, z^*, \cdots, z^*)$ and $B_L = B_R = \mathrm{diag}(\mu, \cdots, \mu, -\mu, \cdots, -\mu)$. For $z = \mu = 0$ our theory is invariant under $SU_L(2N_f) \times SU_R(2N_f)$. For $z \neq 0$ and $\mu \neq 0$ this invariance can be restored if the the mass and chemical potential matrices are transformed as [2,8,13]

$$M_1 \to V_R M_1 V_L^{-1}, \qquad B_R \to V_R B_R V_R^{-1}, \tag{48}$$
$$M_2 \to V_L M_1 V_R^{-1}, \qquad B_L \to V_L B_L V_L^{-1}. \tag{49}$$

However, since $B_{R(L)}$ are a vector fields we can achieve local covariance by transforming them according to

$$B_L \to V_L(\partial_0 + B_L)V_L^{-1},$$
$$B_R \to V_R(\partial_0 + B_R)V_R^{-1}. \tag{50}$$

In the effective Lagrangian local covariance is obtained by replacing the derivatives in the kinetic term by a covariant derivative given by [2]

$$\partial_\nu \Sigma \to \nabla_\nu \Sigma \equiv \partial_\nu \Sigma - B_L \Sigma + \Sigma B_R. \tag{51}$$

This results in the chiral Lagrangian

$$L = \frac{F^2}{4} \mathrm{Tr} \nabla_\nu \Sigma \nabla_\nu \Sigma^\dagger - \frac{G}{2} \mathrm{Tr}(M_1 \Sigma^\dagger + M_2 \Sigma). \tag{52}$$

In our mean field analysis to be discussed below we only need the static part of this Lagrangian which is given by [8]

$$L^{stat} = \frac{F^2}{4} \mu^2 \mathrm{Tr} B_R \Sigma B_L \Sigma^\dagger - \frac{G}{2} \mathrm{Tr}(M_1 \Sigma^\dagger + M_2 \Sigma). \tag{53}$$

3.3. *Mean Field Analysis*

In this subsection we describe the mean field analysis [8] of the static Lagrangian (53). In phase quenched QCD, baryonic Goldstone modes contain a quark with mass z and a conjugate antiquark with mass z^*. According to the GOR relation their mass is given by

$$M^2 = \frac{(z + z^*)G}{2F^2}.$$ (54)

If the chemical potential is less than $M/2$ only the vacuum state contributes to the QCD partition function. This results in

$$Z = e^{V(z+z^*)G}.$$ (55)

We then find the following result for the resolvent and the spectral density

$$G(z) = G, \qquad \rho(z) = 0, \qquad \text{for} \qquad \mu < M/2.$$ (56)

For $\mu > M/2$ the baryonic Goldstone modes condense resulting a nontrivial vacuum field which can be obtained from a mean field analysis. The mass term and the chemical potential term in the static Lagrangian are respectively minimized by[*]

$$\Sigma_c = \begin{pmatrix} 1 & 0 \\ 0 & 1 \end{pmatrix}, \quad \text{and} \quad \Sigma_d = \begin{pmatrix} 0 & 1 \\ 1 & 0 \end{pmatrix}.$$ (57)

A natural ansatz for the minimum of the static Lagrangian (53) is thus given by

$$\Sigma = \Sigma_c \cos \alpha + \Sigma_d \sin \alpha.$$ (58)

An effective potential for α is obtained by substituting this ansatz into the static Lagrangian. It is given by

$$L(\alpha) = \mu^2 F^2 N_f (\cos^2 \alpha - \sin^2 \alpha) - G N_f (z + z^*) \cos \alpha.$$ (59)

This potential is minimized for $\bar{\alpha}$ given by

$$\begin{aligned} \mu^2 < \tfrac{G(z+z^*)}{4F^2} : & \quad \sin \bar{\alpha} = 0, & L(\bar{\alpha}) = -G N_f (z + z^*), \\ \mu^2 > \tfrac{G(z+z^*)}{4F^2} : & \quad \cos \bar{\alpha} = \tfrac{G(z+z^*)}{4\mu^2 F^2}, & L(\bar{\alpha}) = -\tfrac{G^2 N_f (z+z^*)^2}{8F^2 \mu^2}. \end{aligned}$$ (60)

From the free energy at the minimum we easily derive the resolvent and the spectral density (see Fig. 4 in units with $2\mu^2 F^2/G = 1$)

$$\begin{aligned} \mu^2 < \tfrac{G(z+z^*)}{4F^2} : & \quad G(z) = G, & \rho(\lambda) = 0, \\ \mu^2 > \tfrac{G(z+z^*)}{4F^2} : & \quad G(z) = \tfrac{G^2(z+z^*)}{F^2}, & \rho(\lambda) = \tfrac{G^2}{4F^2 \mu^2}. \end{aligned}$$ (61)

---

[*]The minimum Σ_d is not unique which leads to massless Goldstone bosons in the condensed phase.

We conclude that the Dirac eigenvalues are distributed homogeneously inside a strip with width $\sim \mu^2$ in agreement with the numerical simulations [65]. For a discussion of correlations of eigenvalues of a nonhermitian operator we refer to the specialized literature [74,75,70,77,76,78,79].

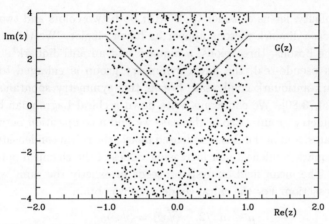

Figure 4.   The distribution of eigenvalues of the Dirac operator in the complex z-plane. The resolvent given by eq. (61) is represented by the dotted curve.

In [13] this analysis was applied to the problem of QCD at finite isospin density with a partition function that coincides with the phase quenched QCD partition function (46) [72]. In that reference [13] it was also shown that the ansatz (58) is a true minimum of the static Lagrangian. However, although we believe that it is an absolute minimum, this has not yet been shown.

4. Real QCD and Nonzero Chemical Potential

The analysis of the previous section can be repeated for any theory with a chemical potential with the quantum numbers of the Goldstone bosons. Both for QCD with two colors in the fundamental representation and for QCD with two or more colors in the adjoint representation, a baryon has quark number two and is a boson. For broken chiral symmetry some of these baryonic states are Goldstone bosons so that Bose-Einstein condensation is likely to occur if the baryon chemical potential surpasses the mass of the Goldstone bosons. For QCD with three or more colors in the fundamental representation we expect a similar low energy behavior if we introduce a chemical potential for isospin [72,13] or strangeness [14] leading to pion condensation or kaon condensation, respectively. Below we only discuss QCD with two colors.

4.1. QCD with $N_c = 2$

For simplicity, let us consider QCD with both two colors and two flavors. In that case diquark mesons appear as flavor singlet. We thus have five Goldstone bosons, three pions, a diquark and an anti-diquark[†]. Because $SU(2)$ is pseudo-real, the flavor symmetry group is enlarged to $SU(4)$. The quark-antiquark condensate breaks this symmetry spontaneously to $Sp(4)$ [20,73,80]. We can again write down a chiral Lagrangian based on this symmetry group. Also in this case we find a competition between two condensates, and in the Bose condensed phase, the chiral condensate rotates into a diquark condensate for increasing values of the chemical potential as in (58). The mean field analysis proceeds in exactly the same way as in previous section. For the chiral condensate we obtain

$$\mu < m_\pi/2, \quad \langle \bar{\psi}\psi \rangle = \langle \bar{\psi}\psi \rangle_0,$$

$$\mu > m_\pi/2, \quad \langle \bar{\psi}\psi \rangle = \langle \bar{\psi}\psi \rangle_0 \frac{m_\pi^2}{4\mu^2}. \tag{62}$$

In Fig. 5 we show that our predictions agree with lattice simulations by Hands et al. [81]. The simulations were done for a $4^3 \times 8$ lattice for $SU(2)$ in the adjoint representation and staggered fermions which is in the same symmetry class as QCD with two colors in the fundamental representation. A similar type of agreement was found by several other groups [82,83,84,85,86]

Results for QCD with two colors in the fundamental representation obtained in [84] are shown in Fig. 6. Again we find good agreement with the mean field results (62). Furthermore, if we plot the same data versus $m_\pi^2/4\mu^2$ the curve reminds us of the resolvent for QCD in phased quenched QCD after transforming the z dependence of the resolvent at fixed μ into a μ dependence at fixed z (see Fig. 4). Since the condensate can be interpreted as the electric field at the quark mass due to charges at the position of the eigenvalues we have no eigenvalues for $\mu > m_\pi/2$ and for a narrow strip along the $m_\pi^2/4\mu^2$ axis. In the remaining region the Dirac eigenvalues are distributed homogeneously. The absence of eigenvalues close to the $m_\pi^2/4\mu^2$

[†]For QCD in the adjoint representation, the diquarks appear as flavor triplet. For two flavors this results in three pions, three diquarks and three anti-diquarks in agreement with spontaneous symmetry breaking according to $SU(4) \to O(4)$

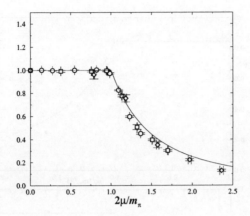

$2\mu/m_\pi$

Figure 5. The chiral condensate versus $\mu/(m_\pi/2)$ for QCD with two colors in the adjoint representation (Figure taken from ref. [81]).

axis is a signature[87] of $\beta_D = 4$. Indeed, such behavior has been identified both numerically [87] and analytically [77].

4.2. Beyond Mean Field

One of the recurring questions in the study of phase transitions is the stability of the mean field analysis. In the following, we carry out a next-to-leading order study of the second order phase transition found at the mean-field level. Additional details can be found in [15,16]. We will concentrate on the free energy of the Bose condensed phase close to the mean-field critical chemical potential $\mu_c = M/2$, with the leading order pion mass given by the GOR relation: $M^2 = Gm_q/F^2$.

The chiral Lagrangian (52) contains the two operators that have the lowest dimension in momentum space and that are invariant under local flavor transformations. There are, of course, many operators of higher dimension that fulfill these symmetry constraints. One has to introduce a systematic power counting to account for their relative importance [2]. Our power-counting scheme is the same as the one used in chiral perturbation theory, extended to include the chemical potential: $p \sim \mu \sim M \sim \sqrt{m_q}$, where p is a Goldstone momentum. The leading order chiral Lagrangian contains all the operators of order p^2 that fulfill the symmetry constraints. The next-to-leading order chiral Lagrangian contains all suitable operators of order p^4. In general, for any N_f, there are ten such operators that contribute to the free energy. They are made out of traces of $\nabla_\nu \Sigma \nabla_\nu \Sigma^\dagger$

284

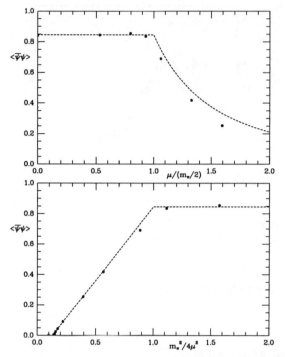

Figure 6. The chiral condensate versus $\mu/(m_\pi/2)$ (upper) and versus $m_\pi^2/4\mu^2$ (lower). The dashed curves in the lower figure are drawn to guide the eye and in the lower figure they represent the mean field result (62) (Data points taken from ref. [84]).

and of $M_1\Sigma^\dagger + M_2\Sigma$. The next-to-leading order chiral Lagrangian can be written as

$$L^{(4)} = \sum_{i=0}^{9} L_i O_i. \tag{63}$$

At next-to-leading order, that is p^4, one has to take into account the one-loop diagrams from the leading-order chiral Lagrangian (52), as well as the tree diagrams from the next-to-leading order Lagrangian (63). In this perturbative scheme, three Feynman diagrams contribute to the free energy at next-to-leading order (see fig. 7). The one-loop diagram is divergent in four dimensions. The theory can be renormalized by introducing renormalized coupling constants

$$L_i \to L_i^r(\Lambda) + \gamma_i \left[\frac{1}{4} - \Gamma(-d/2)\right] \frac{\Lambda^{d-4}}{(4\pi)^{d/2}}, \tag{64}$$

where Λ is the renormalization scale and γ_i are numbers that can depend

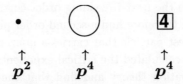

Figure 7. Feynman diagrams that enter into the free energy at next-to-leading order. The dot denotes the contribution from L (52), and the boxed 4 the contribution from $L^{(4)}$ (63). The order in the momentum expansion is also given under each diagram.

N_f [2,15]. The renormalization can be carried out order by order in the perturbation theory. It does not depend on the chemical potential [15].

The main technical difficulty at next-to-leading order comes from the computation of the one-loop diagram in the Bose condensed phase: Some modes are mixed [12,15]. Because of this mixing, one-loop integrals may be quite complicated. However, we notice that the angle α that appears in (58) can be used as an order parameter of the Bose condensed phase. Since we want to study the free energy of that phase near the mean-field critical chemical potential $\mu_c = M/2$, it is sufficient to compute the one-loop integrals for small α, and μ close to $M/2$. The free energy is then given by

$$\frac{\Omega}{M^2F^2} \sim \text{cst} - \left(a_2 + (2+a_3)(\frac{\mu}{M} - \frac{1}{2})\right)\alpha^2 + \left(\frac{1}{8} - a_4\right)\alpha^4 + \dots \quad (65)$$

The coefficients a_i come from the next-to-leading order corrections. They are numbers that can be expressed in terms of the renormalized coupling constants (64). Their general form is given by

$$a_i = \left(\sum_{k=0}^{9} b_{ik}L_k^r(\Lambda) - \frac{1}{32\pi^2}\sum_{k=0}^{9} b_{ik}\gamma_k \ln\frac{M^2}{\Lambda^2}\right)\frac{M^2}{F^2}. \quad (66)$$

They do not depend on the renormalization scale Λ and can be evaluated from the L_K^r which can in principle be obtained from lattice simulations. They are expected to be small (of the order of 0.05 in 3-color QCD [2]).

The free energy (65) can be analyzed in the same way as a Landau-Ginzburg model. The coefficient of α^4 is positive. Therefore, there is a second order phase transition when the coefficient of α^2 vanishes. We thus find that the critical chemical potential at next-to-leading order is given by

$$\mu_c = \frac{1}{2}M(1 - a_2) = \frac{1}{2}m_\pi^{\text{NLO}}, \quad (67)$$

where m_π^{NLO} is the mass of the Goldstones at next-to-leading order and at zero μ. It is remarkable that the next-to-leading order shift in μ_c corresponds exactly to the next-to-leading order correction of m_π. At next-to-leading order, we therefore find a second order phase transition at half the mass of the lightest particle that carries a nonzero baryon charge.

We have also calculated the critical exponents at next-to-leading order in chiral perturbation theory and find that they are still given by their mean-field values [15]. Since $d = 4$ is the critical dimension beyond which mean field exponents become valid, this is not entirely surprising. From the form of the propagators, we conjecture that the critical exponents are given by mean-field theory at any (finite) order in perturbation theory.

In summary, we find that the next-to-leading order corrections are only marginal. The main picture obtained from the mean-field analysis is still valid at next-to-leading order: A second order phase transition at $\mu_c = m_\pi/2$ with mean-field critical exponents separates the normal phase from a Bose condensed phase.

4.3. Nonzero Temperature

At the one-loop level in chiral perturbation theory, the influence of the temperature on the second-order phase transition can also be studied [16,88]. In order to study the phase transition at nonzero T and μ, we compute the free energy of the Bose condensed phase close the critical chemical potential $\mu_c = m_\pi/2$ at $T = 0$. The temperature dependence of the free energy is solely contained in the 1-loop diagram in fig. 7. Since we are only interested in the behavior of the free energy of the Bose condensed phase near the phase transition, it is sufficient to compute it for small α, small T, and μ close to $m_\pi/2$. This procedure again leads to a free energy that can be analyzed as a usual Landau-Ginzburg model. The minimum of the free energy is given by

$$\frac{\partial \Omega}{\partial \alpha} = 0 \Rightarrow \begin{cases} \alpha = 0 \\ -c_2 + c_4 \alpha^2 + c_6 \alpha^4 = 0, \end{cases} \tag{68}$$

where c_i are coefficients that can be computed exactly. For instance, we get that $c_2 = (32\sqrt{2\pi^3}F_\pi^2 - \zeta(\frac{3}{2})\sqrt{m_\pi T^3})^2$ for $N_f = 2$ [16]. We find that at

$$\mu_{\text{tri}} = \frac{m_\pi}{2} + \frac{m_\pi^3}{6\sqrt{3}\zeta^2(3/2)F_\pi^2}\left(1 - \frac{\zeta(1/2)\zeta(3/2)}{4\pi}\right)^{3/2},$$

$$T_{\text{tri}} = 2m_\pi \frac{4\pi - \zeta(1/2)\zeta(3/2)}{3\zeta^2(3/2)}, \tag{69}$$

both c_2 and c_4 vanish. Therefore, we find that the second-order phase transition line given by $c_2 = 0$, that is

$$\mu_{\text{sec}}(T) = \frac{m_\pi}{2} + \frac{1}{32F_\pi^2}\sqrt{\frac{m_\pi^3 T^3}{2\pi^3}}\zeta(3/2), \tag{70}$$

ends at μ_{tri}. For a larger chemical potential, the phase transition is of first order. The second-order phase transition line (70) is in complete agreement with the semi-classical analysis of a dilute Bose gas in the canonical ensemble. This phase diagram has been confirmed by lattice simulations [82,85].

5. Conclusions

Below Λ_{QCD} the QCD Dirac spectrum both at zero and at a sufficiently small nonzero chemical potential is described completely by a suitable chiral Lagrangian. Below the Thouless energy, i.e. the scale for which the Compton wavelength of the Goldstone bosons is equal to the size of the box, the Dirac spectrum can be obtained from the zero momentum part of this theory. This matrix integral can also be derived from a chiral Random Matrix Theory with the global symmetries of the QCD partition function. Therefore, below the Thouless energy, the correlations of QCD Dirac eigenvalues are given by chiral Random Matrix Theory. At nonzero chemical potential the Dirac eigenvalues are located inside a strip in the complex plane. Going inside this strip the chiral condensate rotates into a superfluid Bose-Einstein condensate. A very similar phase transition is found for any system with a chemical potential with the quantum numbers of the Goldstone bosons. We have discussed in detail the phase diagram for QCD with two colors. Because of the Pauli-Gürsey symmetry diquarks appear as Goldstone bosons in this theory. We have analyzed the phase diagram of this theory to one-loop order and have found a tricritical point in the chemical potential-temperature plane. All our results are in agreement with recent lattice QCD simulations.

Acknowledgments

Arkady Vainshtein is thanked for being a long lasting inspirational force of our field and the TPI is thanked for its hospitality. D.T. is supported in part by "Holderbank"-Stiftung and by NSF under grant NSF-PHY-0102409. This work was partially supported by the US DOE grant DE-FG-88ER40388.

288

References

1. S. Weinberg, Phys. Rev. Lett. 18 (1967) 188; Phys. Rev. 166 (1968) 1568; Physica A 96 (1979) 327.
2. J. Gasser and H. Leutwyler, Ann. Phys. 158 (1984) 142; J. Gasser and H. Leutwyler, Nucl. Phys. B 250 (1985) 465; H. Leutwyler, Ann. Phys. 235 (1994) 165.
3. J.J.M. Verbaarschot, Phys. Lett. B 368 (1996) 137.
4. H. Leutwyler and A.V. Smilga, Phys. Rev. D 46 (1992) 5607.
5. J.C. Osborn and J.J.M. Verbaarschot, Phys. Rev. Lett. 81 (1998) 268; Nucl. Phys. B 525 (1998) 738.
6. E.V. Shuryak and J.J.M. Verbaarschot, Nucl. Phys. A 560 (1993) 306.
7. J.J.M. Verbaarschot, Phys. Rev. Lett. 72 (1994) 2531.
8. D. Toublan and J. J. Verbaarschot, Int. J. Mod. Phys. B 15 (2001) 1404.
9. V.L. Girko, *Theory of random determinants* Kluwer Academic Publishers, Dordrecht, 1990.
10. M. A. Stephanov, Phys. Rev. Lett. 76 (1996) 4472.
11. J.B. Kogut, M.A. Stephanov, and D. Toublan, Phys. Lett. B 464 (1999) 183.
12. J.B. Kogut, M.A. Stephanov, D. Toublan, J.J.M Verbaarschot, and A. Zhitnitsky, Nucl. Phys. B 582 (2000) 477.
13. D. T. Son and M. A. Stephanov, Phys. Rev. Lett. 86 (2001) 592.
14. J. B. Kogut and D. Toublan, Phys. Rev. D 64 (2001) 034007.
15. K. Splittorff, D. Toublan and J. Verbaarschot, Nucl. Phys. B 620 (2002) 290.
16. K. Splittorff, D. Toublan and J. Verbaarschot, hep-ph/0204076.
17. J. B. Kogut, D. Toublan and D. K. Sinclair, Phys. Lett. B 514 (2001) 77. J. B. Kogut, D. Toublan and D. K. Sinclair, hep-lat/0205019.
18. T. Banks and A. Casher, Nucl. Phys. B 169 (1980) 103.
19. J.J.M. Verbaarschot, Phys. Lett. B329 (1994) 351.
20. A. Smilga and J. J. Verbaarschot, Phys. Rev. D51 (1995) 829.
21. P. H. Damgaard, Phys. Lett. B425 (1998) 151.
22. K. Zyablyuk, hep-ph/9911300.
23. E. Brézin, Lec. Notes Phys. 216 (1984) 115.
24. K. Efetov, Adv. Phys. 32 (1983) 53.
25. A. Morel, J. Physique 48 (1987) 1111.
26. J.J.M. Verbaarschot, H.A. Weidenmüller, and M.R. Zirnbauer, Phys. Rep. 129 (1985) 367.
27. C. Bernard and M. Golterman, Phys. Rev. D49 (1994) 486; C. Bernard and M. Golterman, hep-lat/9311070; M.F.L. Golterman, Acta Phys. Polon. B25 (1994).
28. M. F. Golterman and K. C. Leung, Phys. Rev. D 57 (1998) 5703.
29. J. Osborn, D. Toublan and J. Verbaarschot, Nucl. Phys. **B540** (1999) 317.
30. P.H. Damgaard, J.C. Osborn, D. Toublan, and J.J.M. Verbaarschot, Nucl. Phys. B 547 (1999) 305.
31. B.L. Altshuler, I.Kh. Zharekeshev, S.A. Kotochigova and B.I. Shklovskii, Zh. Eksp. Teor. Fiz. 94 (1988) 343.
32. A. Garcia-Garcia and J. Verbaarschot, Nucl. Phys. B 586 (2000) 668; A. Garcia-Garcia and J. Verbaarschot, cond-mat/0204151.
33. S. Chandrasekharan and N. Christ, Nucl. Phys. B (Proc. Suppl.) 47 (1996)

527.

34. M.Á. Halász and J.J.M. Verbaarschot, Phys. Rev. Lett. 74 (1995) 3920; M.Á. Halász, T. Kalkreuter, and J.J.M. Verbaarschot, Nucl. Phys. B (Proc. Suppl.) 53 (1997) 266.

35. R. Pullirsch, K. Rabitsch, T. Wettig, and H. Markum, Phys. Lett. B 427 (1998) 119.

36. B.A. Berg, H. Markum, and R. Pullirsch, Phys. Rev. D 59 (1999) 097504.

37. R.G. Edwards, U.M. Heller, J. Kiskis, and R. Narayanan, Phys. Rev. Lett. 82 (1999) 4188.

38. R.G. Edwards, U.M. Heller, J. Kiskis, and R. Narayanan, Phys. Rev. Lett. 82 (1999) 4188.

39. P.H. Damgaard, R.G. Edwards, U.M. Heller, and R. Narayanan, Phys. Rev. D 61 (2000) 094503; Nucl. Phys. B (Proc. Suppl.) 47 (1996) 527.

40. M.E. Berbenni-Bitsch, S. Meyer, A. Schäfer, J.J.M. Verbaarschot, and T. Wettig, Phys. Rev. Lett. 80 (1998) 1146.

41. J.-Z. Ma, T. Guhr, and T. Wettig, Eur. Phys. J. A 2 (1998) 87.

42. P.H. Damgaard, U.M. Heller, and A. Krasnitz, Phys. Lett. B 445 (1999) 366.

43. M. Göckeler, H. Hehl, P.E.L. Rakow, A. Schäfer, and T. Wettig, Phys. Rev. D 59 (1999) 094503.

44. R. Edwards, U. Heller, and R. Narayanan, Phys. Rev. D 60 (1999) 077502.

45. M.E. Berbenni-Bitsch, S. Meyer, and T. Wettig, Phys. Rev. D 58 (1998) 071502.

46. F. Farchioni, I. Hip, C.B. Lang, and M. Wohlgenannt, Nucl. Phys. B 549 (1999) 364; Nucl. Phys. B (Proc. Suppl.) 73 (1999) 939.

47. M.E. Berbenni-Bitsch, A.D. Jackson, S. Meyer, A. Schäfer, J.J.M. Verbaarschot, and T. Wettig, Nucl. Phys. B (Proc. Suppl.) 63 (1998) 820.

48. P.H. Damgaard, U.M. Heller, R. Niclasen, and K. Rummukainen, Phys. Rev. D 61 (2000) 014501.

49. F. Farchioni, I. Hip, and C.B. Lang, Phys. Lett. B 471 (1999) 58; Nucl. Phys. B (Proc. Suppl.) 83-84 (2000) 482.

50. M. Schnabel and T. Wettig, hep-lat/9912057.

51. F. Farchioni, P. de Forcrand, I. Hip, C.B. Lang, and K. Splittorff, Phys. Rev. D 62 (2000) 014503.

52. P.H. Damgaard, U.M. Heller, R. Niclasen, and K. Rummukainen, hep-lat/0003021.

53. M.E. Berbenni-Bitsch, M. Göckeler, T. Guhr, A.D. Jackson, J.-Z. Ma, S. Meyer, A. Schäfer, H.A. Weidenmüller, T. Wettig, and T. Wilke, Phys. Lett. B 438 (1998) 14; M.E. Berbenni-Bitsch, M. Göckeler, S. Meyer, A. Schäfer, and T. Wettig, Nucl. Phys. B (Proc. Suppl.) 73 (1999) 605.

54. M.E. Berbenni-Bitsch, M. Göckeler, H. Hehl, S. Meyer, P.E.L. Rakow, A. Schäfer, and T. Wettig, Phys. Lett. B 466 (1999) 293; Nucl. Phys. B (Proc. Suppl.) 83-84 (2000) 974.

55. J.J.M. Verbaarschot and I. Zahed, Phys. Rev. Lett. 70 (1993) 3852.

56. D. Toublan and J. J. Verbaarschot, Nucl. Phys. B 603 (2001) 343.

57. F.J. Dyson, J. Math. Phys. 3 (1962) 140, 157, 166, 1199.

58. E. Brézin, S. Hikami, and A. Zee, Nucl. Phys. B 464 (1996) 411.

59. A. Jackson, M. Sener, J. Verbaarschot, Nucl. Phys. B 479 (1996) 707.

60. G. Akemann, P.H. Damgaard, U. Magnea, and S. Nishigaki, Nucl. Phys. B 487 (1997) 721.
61. A. Jackson, M. Sener, J. Verbaarschot, Nucl. Phys. B 506 (1997) 612.
62. T. Guhr and T. Wettig, Nucl. Phys. B 506 (1997) 589.
63. E. Kanzieper and V. Freilikher, Phys. Rev. Lett. 78 (1997) 3806; Phys. Rev. E 55 (1997) 3712; cond-mat/9809365.
64. M.R. Zirnbauer, J. Math. Phys. 37 (1996) 4986; F.J. Dyson, Comm. Math. Phys. 19 (1970) 235.
65. I. Barbour, N. Behihil, E. Dagotto, F. Karsch, A. Moreo, M. Stone and H. Wyld, Nucl. Phys. B275 (1986) 296; M.-P. Lombardo, J. Kogut and D. Sinclair, Phys. Rev. D54 (1996) 2303.
66. C. Baillie, K.C. Bowler, P.E. Gibbs, I.M. Barbour, and M. Rafique, Phys. Lett. B 197 (1987) 195.
67. H. Markum, R. Pullirsch, and T. Wettig, Phys. Rev. Lett. 83 (1999) 484.
68. I.M. Barbour, S.E. Morrison, E.G. Klepfish, J.B. Kogut, and M.-P. Lombardo, Nucl. Phys. (Proc. Suppl.) A 60 (1998) 220.
69. S. Hands, I. Montvay, S. Morrison, M. Oevers, L. Scorzato and J. Skullerud, Eur. Phys. J. C 17 (2000) 285.
70. K.B. Efetov, Phys. Rev. Lett. 79 (1997) 491; Phys. Rev. B 56 (1997) 9630.
71. J. Feinberg and A. Zee, Nucl. Phys. B 504 (1997) 579; Nucl. Phys. B 501 (1997) 643.
72. M. Alford, A. Kapustin and F. Wilczek, Phys. Rev. D 59 (1999) 054502.
73. D. Toublan and J. J. Verbaarschot, Nucl. Phys. B 560 (1999) 259.
74. J. Ginibre, J. Math. Phys. 6 (1965) 440.
75. Y. Fyodorov, B. Khoruzhenko, and H.-J. Sommers, Phys. Lett. A 226 (1997) 46; Phys. Rev. Lett. 79 (1997) 557; Ann. Ins. H. Poincaré, 68 (1998) 449.
76. P.J. Forrester, Phys. Rep. 301 (1998) 235.
77. A. V. Kolesnikov and K. B. Efetov, Waves Random Media 9 (1999) 71.
78. G. Akemann, Phys. Rev. Lett. 89 (2002) 072002; G. Akemann, hep-th/0204246; G. Akemann, hep-th/0206086.
79. A. Garcia-Garcia, S. Nishigaki and J. Verbaarschot, cond-mat/0202151.
80. S . Hands, J. B. Kogut, M. P. Lombardo and S. E. Morrison, Nucl. Phys. B 558 (1999) 327.
81. S. Hands, I. Montvay, S. Morrison, M. Oevers, L. Scorzato and J. Skullerud, Eur. Phys. J. C 17 (2000) 285.
82. J. B. Kogut, D. Toublan and D. K. Sinclair, Phys. Lett. B 514 (2001) 77.
83. J. B. Kogut and D. K. Sinclair, hep-lat/0202028.
84. J. B. Kogut, D. K. Sinclair, S. J. Hands and S. E. Morrison, Phys. Rev. D 64 (2001) 094505.
85. J. B. Kogut, D. Toublan and D. K. Sinclair, Phys. Lett. B 514 77 (2001).
86. R. Aloisio, A. Galante, V. Azcoiti, G. Di Carlo and A. F. Grillo, hep-lat/0007018.
87. M. Halasz, J. Osborn and J. Verbaarschot, Phys. Rev. D 56 (1997) 7059.
88. J. Gasser and H. Leutwyler, Phys. Lett. B 184 (1987) 83; Phys. Lett. B 188 (1987) 477; P. Gerber and H. Leutwyler, Nucl. Phys. B 321 (1989) 387; A. Schenk, Nucl. Phys. B 363 (1991) 97; D. Toublan, Phys. Rev. D 56 (1997) 5629.

QUARK COLOR SUPERCONDUCTIVITY AND THE COOLING OF COMPACT STARS

I. A. SHOVKOVY AND P. J. ELLIS

School of Physics and Astronomy,
University of Minnesota,
Minneapolis, MN 55455, USA

The thermal conductivity of the color-flavor locked phase of dense quark matter is calculated. The dominant contribution to the conductivity comes from photons and Nambu-Goldstone bosons associated with the breaking of baryon number, both of which are trapped in the quark core. Because of their very large mean free path the conductivity is also very large. The cooling of the quark core arises mostly from the heat flux across the surface of direct contact with the nuclear matter. As the thermal conductivity of the neighboring layer is also high, the whole interior of the star should be nearly isothermal. Our results imply that the cooling time of compact stars with color-flavor locked quark cores is similar to that of ordinary neutron stars.

1. Introduction

At sufficiently high baryon density the nucleons in nuclear matter should melt into quarks so that the system becomes a quark liquid. It should be weakly interacting due to asymptotic freedom [1], however, it cannot be described as a simple Fermi liquid. This is due to the nonvanishing attractive interaction in the color antitriplet quark-quark channel, provided by one-gluon exchange, which renders a highly degenerate Fermi surface unstable with respect to Cooper pairing. As a result the true ground state of dense quark matter is, in fact, a color superconductor [2].

Recent phenomenological [3] and microscopic studies [4,5,6,7] have confirmed that quark matter at a sufficiently high density undergoes a phase transition into a color superconducting state. Phenomenological studies are expected to be appropriate to intermediate baryon densities, while microscopic approaches are strictly applicable at asymptotic densities where perturbation theory can be used. It is remarkable that both approaches concur that the superconducting order parameter (which determines the gap Δ in the quark spectrum) lies between 10 and 100 MeV for baryon densities existing in the cores of compact stars.

At realistic baryon densities only the three lightest quarks can participate in the pairing dynamics. The masses of the quarks are much smaller than the baryon chemical potential, thus, to a good approximation, all three flavors participate equally in the color condensation. The ground state is then the so-called color flavor locked (CFL) phase [8]. The original gauge symmetry $SU(3)_c$ and the global chiral symmetry $SU(3)_L \times SU(3)_R$ break down to a global diagonal "locked" $SU(3)_{c+L+R}$ subgroup. Because of the Higgs mechanism the gluons become massive and decouple from the infrared dynamics. The quarks also decouple because large gaps develop in their energy spectra. The breaking of the chiral symmetry leads to the appearance of an octet of pseudo-Nambu-Goldstone (NG) bosons (π^0, $\pi^\pm$, $K^\pm$, K^0, $\bar{K}^0$, η). In addition an extra NG boson ϕ and a pseudo-NG boson η' appear in the low energy spectrum as a result of the breaking of global baryon number symmetry and approximate $U(1)_A$ symmetry, respectively.

The low energy action for the NG bosons in the limit of asymptotically large densities was derived in Refs. [9,10]. By making use of an auxiliary "gauge" symmetry, it was suggested in Ref. [11] that the low energy action of Refs. [9,10] should be modified by adding a time-like covariant derivative to the action of the composite field. Under a favorable choice of parameters, the modified action predicted kaon condensation in the CFL phase. Some unusual properties of such a condensate were discussed in Ref. [12].

While the general structure of the low energy action in the CFL phase can be established by symmetry arguments alone [9], the values of the parameters in such an action can be rigorously derived only at asymptotically large baryon densities [10,11]. Thus, in the most interesting case of intermediate densities existing in the cores of compact stars, the details of the action are not well known. For the purposes of the present paper, however, it suffices to know that there are 9 massive pseudo-NG bosons and one massless NG boson ϕ in the low energy spectrum. If kaons condense [11] an additional NG boson should appear. These NG bosons should be relevant for the kinetic properties of dense quark matter.

It has been found [13] that neutrino and photon emission rates for the CFL phase are very small so that they would be inefficient in cooling the core of a neutron star. The purpose of the present investigation is to determine quantitatively the thermal conductivity of the CFL phase of dense quark matter in order to see whether it can significantly impact the cooling rate. We shall argue that the temperature of the CFL core, as well as the neighboring neutron layer which is in contact with the core, falls quickly due to the very high thermal conductivities on both sides of the interface. In fact, to a good approximation, the interior of the star is isothermal. A noticeable gradient of the temperature appears only in a relatively thin surface layer of the star where a finite flux of energy is carried outwards by photon diffusion. A more complete account of this work is to be found in Ref. [14].

2. Thermal conductivity

A detailed understanding of the cooling mechanism of a compact star with a quark core is not complete without a study of thermal conductivity effects in the color superconducting quark core. The conductivity, as well as the other kinetic properties of quark matter in the CFL phase, is dominated by the low energy degrees of freedom. It is clear then that at all temperatures of interest to us, $T \ll \Delta$, it is crucial to consider the contributions of the NG bosons. In addition, there may be an equally important contribution due to photons; this is discussed in Sec. 4. Note that, at such small temperatures, the gluon and the quark quasiparticles become completely irrelevant. For example, a typical quark contribution to a transport coefficient would be exponentially suppressed by the factor $\exp(-\Delta/T)$.

Let us start from the general definition of the thermal conductivity as a characteristic of a system which is forced out of equilibrium by a temperature gradient. In response to such a gradient transport of heat is induced. Formally this is described by the following relation:

$$u_i = -\kappa \partial_i T, \tag{1}$$

where u_i is the heat current, and κ is the heat conductivity. As is clear from this relation, the heat flow would persist until a state of uniform temperature is reached. The higher the conductivity, the shorter the time for this relaxation.

In the linear response approximation, the thermal conductivity is given in terms of the heat current correlator by a Kubo-type formula. We derive the expression for the heat (energy) current carried by a single (pseudo-) NG boson field φ. The corresponding Lagrangian density reads

$$L = \frac{1}{2} \left(\partial_0 \varphi \partial_0 \varphi - v^2 \partial_i \varphi \partial_i \varphi - m^2 \varphi^2 \right) + \ldots, \tag{2}$$

where the ellipsis stand for the self-interaction terms as well as interactions with other fields. Notice that we introduced explicitly the velocity parameter v. In microscopic studies of color superconducting phases, which are valid at very large densities, this velocity is equal to $1/\sqrt{3}$ for all (pseudo-) NG bosons. It is smaller than 1 because Lorentz symmetry is broken due to the finite value of the quark chemical potential. By making use of the above Lagrangian density, we derive the following expression for the heat current:

$$u_i = \frac{\partial L}{\partial(\partial^i \varphi)} \partial_0 \varphi = v^2 \partial_i \varphi \partial_0 \varphi. \tag{3}$$

This definition leads to an expression for the heat conductivity in terms of the corresponding correlator [15]:

$$\kappa_{ij} = -\frac{i}{2T} \lim_{\Omega \to 0} \frac{1}{\Omega} \left[\Pi_{ij}^R(\Omega + i\epsilon) - \Pi_{ij}^A(\Omega - i\epsilon) \right], \tag{4}$$

where, in the Matsubara formalism,

$$\Pi_{ij}(i\Omega_m) = v^4 T \sum_n \int \frac{d^3 k}{(2\pi)^3} k_i k_j i\Omega_n (i\Omega_n + i\Omega_{n-m})$$

$$\times S(i\Omega_n, \vec{k}) S(i\Omega_{n-m}, \vec{k}). \tag{5}$$

Here $\Omega_n \equiv 2\pi n T$ is the bosonic Matsubara frequency, and $S(i\Omega_n, \vec{k})$ is the propagator of the (pseudo-) NG boson. In general, the propagator should have the following form:

$$S(\omega, \vec{k}) = \frac{1}{(\omega + i\Gamma/2)^2 - v^2 \vec{k}^2 - m^2}, \tag{6}$$

where the width parameter $\Gamma(\omega, \vec{k})$ is related to the inverse lifetime (as well as the mean free path) of the boson. In our calculation, it is very convenient to utilize the spectral representation of the propagator,

$$S(i\Omega_n, \vec{k}) = \frac{1}{\pi} \int_{-\infty}^{\infty} \frac{d\omega A(\omega, \vec{k})}{i\Omega_n - \omega}. \tag{7}$$

Then the conductivity is expressed through the spectral function $A(\omega, \vec{k})$ as follows:

$$\kappa_{ij} = \frac{v^4}{2\pi T^2} \int_{-\infty}^{\infty} \frac{\omega^2 d\omega}{\sinh^2 \frac{\omega}{2T}} \int \frac{d^3 k}{(2\pi)^3} k_i k_j A^2(\omega, \vec{k}). \tag{8}$$

By making use of the explicit form of the propagator in Eq. (6), we see that the spectral function of the (pseudo-) NG boson is

$$A(\omega, \vec{k}) = \frac{\omega \Gamma}{(\omega^2 - e_k^2 - \Gamma^2/4)^2 + \omega^2 \Gamma^2}, \tag{9}$$

where $e_k \equiv \sqrt{v^2 k^2 + m^2}$. Because of the rotational symmetry of the system, the conductivity is characterized by a single scalar quantity κ which is introduced as follows: $\kappa_{ij} = \kappa \delta_{ij}$. The explicit expression for this scalar function reads

$$\kappa = \frac{1}{48\sqrt{2}\pi^2 v \Gamma T^2} \int_0^{\infty} \frac{\omega d\omega}{\sinh^2 \frac{\omega}{2T}} \left(\sqrt{X^2 + \omega^2 \Gamma^2} + X \right)^{3/2}, \tag{10}$$

where we introduced the notation $X \equiv \omega^2 - m^2 - \Gamma^2/4$. For our purposes it will be sufficient to consider the conductivity in the limit of a small width, $\Gamma \to 0$. This is because the (pseudo-) NG bosons in the CFL quark matter are weakly interacting. Thus, we derive the following approximate relation:

$$\kappa = \frac{1}{24\pi^2 v T^2 \Gamma} \int_m^\infty \frac{d\omega \omega}{\sinh^2 \frac{\omega}{2T}} \left(\omega^2 - m^2 \right)^{3/2}. \tag{11}$$

At small temperature, $T \ll m$, this result is further approximated by

$$\kappa \simeq \frac{m^{5/2}\sqrt{T}}{2\sqrt{2}\pi^{3/2} v \Gamma} e^{-m/T}. \tag{12}$$

This demonstrates clearly that the contributions of heavy pseudo-NG bosons to the thermal conductivity are suppressed. The largest contribution comes from the massless NG boson ϕ for which the thermal conductivity is

$$\kappa_\phi = \frac{4T^3}{3\pi^2 v \Gamma_\phi} \int_0^\infty \frac{x^4 dx}{\sinh^2 x} = \frac{2\pi^2 T^3}{45 v \Gamma_\phi}. \tag{13}$$

In order to calculate κ_ϕ the width Γ_ϕ is required, or equivalently the mean free path ℓ_ϕ since $\ell_\phi \equiv \bar{v}/\Gamma_\phi$, where $\bar{v}$ is the average thermal velocity of the NG bosons. This will be discussed in the next section.

3. Mean free path of the NG boson

As we have remarked, the contribution of massive pseudo-NG bosons to the thermal conductivity is suppressed. In the CFL phase of quark matter, however, there is one truly massless NG boson ϕ which should therefore give the dominant contribution to the heat conductivity. The interactions of ϕ with the CFL matter leads to a finite value for its mean free path. Since this boson is a composite particle there is always a non-zero probability at finite temperature for its decay into a pair of quark quasiparticles. It is natural to expect that such a process is strongly suppressed at small temperatures, $T \ll \Delta$. This is confirmed by a direct microscopic calculation in the region of asymptotic densities which yields a decay width [16]:

$$\Gamma_{\phi \to qq}(k) \simeq \frac{5\sqrt{2}\pi v k}{4(21 - 8\ln 2)} \exp\left(-\sqrt{\frac{3}{2}} \frac{\Delta}{T} \right). \tag{14}$$

If this were the only contribution, then the order of magnitude of the mean free path of the NG boson would be

$$\ell_{\phi \to qq} \sim \frac{v}{T} \exp\left(\sqrt{\frac{3}{2}} \frac{\Delta}{T} \right). \tag{15}$$

This grows exponentially with decreasing temperature. For example, if $\Delta \simeq 50$ MeV and $T \lesssim 1.5$ MeV, the mean free path is 30 km or more. This scale is a few times larger than the typical size of a compact star.

The decay channel of the NG bosons into quarks is not the only contribution to the mean free path. They can also scatter on one another. The corresponding amplitude is of order k^4/μ^4 [17] which gives a cross section of $\sigma_{\phi\phi} \simeq T^6/\mu^8$, yielding the following contribution to the width:

$$\Gamma_{\phi\phi} = v\sigma_{\phi\phi}n_\phi \sim \frac{T^9}{\mu^8} , \qquad (16)$$

where n_ϕ is the equilibrium number density of the NG bosons [14]. At small temperatures the scattering contribution in Eq. (16) dominates the width. This leads to a mean free path

$$\ell_{\phi\phi} \sim \frac{\mu^8}{T^9} \approx 8 \times 10^5 \frac{\mu_{500}^8}{T_{\mathrm{MeV}}^9} \text{ km.} \qquad (17)$$

Here we defined the following dimensionless quantities: $\mu_{500} \equiv \mu/(500 \text{ MeV})$ and $T_{\mathrm{MeV}} \equiv T/(1 \text{ MeV})$. Both $\ell_{\phi\phi}$ and $\ell_{\phi\to qq}$ depend very strongly on temperature, however the salient point is that they are both larger than the size of a compact star for temperatures T_{MeV} of order 1.

We define $\tilde{T}$ to be the temperature at which the massive NG bosons decouple from the system. This is determined by the mass of the lightest pseudo-NG boson for which it is not presently possible to give a reliable value. Different model calculations [10,11,18] produce different values which can range as low as 10 MeV. Thus, conservatively, we choose $\tilde{T} \simeq 1$ MeV. Then the mean free path of the NG boson is comparable to or even larger than the size of a star for essentially all temperatures $T \lesssim \tilde{T}$. It is also important to note that the mean free path is very sensitive to temperature changes. In particular, at temperatures just a few times higher than $\tilde{T}$ the value of ℓ may already become much smaller than the star size. This suggests that, during the first few seconds after the supernova explosion when the temperatures remain considerably higher than $\tilde{T}$, a noticeable temperature gradient may exist in the quark core. This should relax very quickly because of the combined effect of cooling (which is very efficient at $T \gg \tilde{T}$) and diffusion. After that almost the whole interior of the star would become isothermal.

Before concluding this section, we point out that the geometrical size of the quark phase limits the mean free path of the NG boson since the scattering with the boundary should also be taken into account. It is clear from simple geometry that $\ell \sim R_0$, where R_0 is the radius of the quark core.

4. Photon contributions

Now, let us discuss the role of photons in the CFL quark core. It was argued in Ref. [13] that the mean free path of photons is larger than the typical size of a compact star at all temperatures $T \lesssim \tilde{T}$. One might conclude therefore that all photons would leave the stellar core shortly after the core becomes transparent. If this were so the photons would be able to contribute neither to the thermodynamic nor to the kinetic properties of the quark core. However the neighboring neutron matter has very good metallic properties due to the presence of a considerable number of electrons. As is known from plasma physics, low frequency electromagnetic waves cannot propagate inside a plasma. Moreover, an incoming electromagnetic wave is reflected from the surface of such a plasma [19]. In particular, if Ω_p is the value of the plasma frequency of the nuclear matter, then all photons with frequencies $\omega < \Omega_p$ are reflected from the boundary. This effect is similar to the well known reflection of radio waves from the Earth's ionosphere.

The plasma frequency is known to be proportional to the square root of the density of charge carriers and inversely proportional to the square root of their mass. It is clear therefore that the electrons, rather than the more massive protons, will lead to the largest value of the plasma frequency in nuclear matter. Our estimate for the value of this frequency is

$$\Omega_p = \sqrt{\frac{4\pi e^2 Y_e \rho}{m_e m_p}} \simeq 4.7 \times 10^2 \sqrt{\frac{\rho Y_e}{\rho_0}} \text{ MeV}, \tag{18}$$

where the electron density $n_e = Y_e \rho / m_p$ is given in terms of the nuclear matter density ρ and the proton mass m_p. Also m_e denotes the electron mass, $Y_e \simeq 0.1$ is the number of electrons per baryon, and $\rho_0 \approx 2.8 \times 10^{14}$ g cm^{-3} is equilibrium nuclear matter density.

Since Ω_p is more than 100 MeV, essentially all thermally populated electromagnetic waves at $T \lesssim \tilde{T}$ will be reflected back into the core region. In a way the boundary of the core looks like a good quality mirror with some leakage which will allow a thermal photon distribution to build up and stay. Thus photons will be *trapped* in such a core surrounded by a nuclear layer. Notice that the transparency of the core is reached only after the temperature drops substantially below $\tilde{T} \simeq 2m_e$, i.e., when the density of thermally excited electron-positron pairs becomes very small.

Now, since photons are massless they also give a sizable contribution to the thermal conductivity of the CFL phase. The corresponding contribution κ_γ will be similar to the contribution of massless NG bosons in Eq. (13). Since the photons move at approximately the speed of light [20] ($v \simeq 1$ at the densities of interest) and they have two polarization states, we obtain

$$\kappa_\gamma = \frac{4\pi^2 T^3}{45\Gamma_\gamma} . \tag{19}$$

Since the thermal conductivity is additive the total conductivity of dense quark matter in the CFL phase is given by the sum of the two contributions:

$$\kappa_{CFL} = \kappa_\phi + \kappa_\gamma \simeq \frac{2\pi^2}{9} T^3 R_0, \tag{20}$$

where for both a photon and a NG boson the mean free path $\ell \sim R_0$. This yields the value

$$\kappa_{CFL} \simeq 1.2 \times 10^{32} T_{\mathrm{MeV}}^3 R_{0,\mathrm{km}} \ \mathrm{erg} \ \mathrm{cm}^{-1} \mathrm{sec}^{-1} \mathrm{K}^{-1}, \tag{21}$$

where $R_{0,\mathrm{km}}$ is the quark core radius measured in kilometers. The value of κ_{CFL} is many orders of magnitude larger than the thermal conductivity of regular nuclear matter in a neutron star [21].

5. Stellar cooling

In discussing the cooling mechanism for a compact star we have to make some general assumptions about the structure of the star. We accept without proof that a quark core exists at the center of the star. This core stays in direct contact with the neighboring nuclear matter. From this nuclear layer outwards the structure of the star is essentially the same as an ordinary neutron star. The radius of the core is denoted by R_0, while the radius of the whole star is denoted by R.

A detailed analysis of the interface between the quark core and the nuclear matter was made in Ref. [22]. A similar analysis might also be very useful for understanding the mechanism of heat transfer from one phase to the other. Here we assume that direct contact between the phases is possible, and that the temperature is slowly varying across the interface.

Now, let us consider the physics that governs stellar cooling. We start from the moment when the star is formed in a supernova explosion. Immediately after the explosion many high-energy neutrinos are trapped inside the star. After about 10 to 15 seconds most of them escape from the star by diffusion. The presence of the CFL quark core could slightly modify the rate of such diffusion [23,24,25]. By the end of the deleptonization process, the temperature of the star will have risen to a few tens of MeV. Then, the star cools down relatively quickly to about $\tilde{T}$ by the efficient process of neutrino emission. It is unlikely that the quark core would greatly affect the time scale for this initial cooling stage. An ordinary neutron star would continue to cool by neutrino emission for quite a long time even after that [26]. Here we discuss how the presence of the CFL quark core affects the cooling process of the star after the temperature drops below $\tilde{T}$.

Our result for the mean free path of the NG boson demonstrates that the heat conductivity of dense quark matter in the CFL phase is very high. For example, a temperature gradient of 1 MeV across a core of 1 km in size is washed away by heat conduction in a very short time interval of order $R_{0,\mathrm{km}}^2/v\ell(T) \simeq 6 \times 10^{-4}$ sec. In deriving this estimate, we took into account the fact that the specific heat and the heat conductivity in the CFL phase are dominated by photons and massless NG bosons and that $\ell \sim R_0$. In addition, we used the classical relation $\kappa = \bar{v} c_v \ell/3$, where c_v is the specific heat; this can be shown to hold in the present context [14]. Since heat conduction removes a temperature gradient in such a short time interval, it is clear that, to a good approximation, the quark core is isothermal at all temperatures $T \lesssim \tilde{T}$.

The heat conductivity of the neighboring nuclear matter is also known to be very high because of the large contribution from degenerate electrons which have a very long mean free path. It is clear, then, that both the quark and the nuclear layers should be nearly isothermal with equal values of the temperature. When one of the layers cools down by any mechanism, the temperature of the other will adjust almost immediately due to the very efficient heat transfer on both sides of the interface.

Now consider the order of magnitude of the cooling time for a star with a CFL quark core. One of the most important components of the calculation of the cooling time is the thermal energy of the star which is the amount of energy that is lost in cooling. There are contributions to the total thermal energy from both the quark and the nuclear parts of the star. The dominant amount of thermal energy in the CFL quark matter is stored in photons and massless NG bosons. Numerically, its value is [14]

$$E_{CFL}(T) \simeq 2.1 \times 10^{42} R_{0,\mathrm{km}}^3 T_{\mathrm{MeV}}^4 \text{ erg.} \qquad (22)$$

The thermal energy of the outer nuclear layer is provided mostly by degenerate neutrons. The corresponding numerical estimate is [27]:

$$E_{NM}(T) \simeq 8.1 \times 10^{49} \frac{M - M_0}{M_\odot} \left(\frac{\rho_0}{\rho}\right)^{2/3} T_{\mathrm{MeV}}^2 \text{ erg,} \qquad (23)$$

where M is the mass of the star, M_0 is the mass of the quark core and $M_\odot$ is the mass of the Sun. It is crucial to note that the thermal energy of the quark core is negligible in comparison to that of the nuclear layer.

The second important component that determines stellar cooling is the luminosity which describes the rate of energy loss due to neutrino and photon emission. Typically, the neutrino luminosity dominates the cooling of young stars when the temperatures are still higher than about 10 keV and after that the photon diffusion mechanism starts to dominate. Photon and neutrino emission from the CFL quark phase is strongly suppressed at low temperatures [13]. The neighboring nuclear layer, on the other hand, emits neutrinos quite efficiently. As a result, it cools relatively fast in the same way as an ordinary neutron star. The nuclear layer should be able to emit not only its own thermal energy, but also that of the quark core which constantly arrives by the very efficient heat transfer process. The analysis of the cooling mechanism is greatly simplified by the fact that the thermal energy of the quark core is negligible compared to the energy stored in the nuclear matter. By taking this into account, we conclude that the cooling time of a star with a quark core is essentially the same as for an ordinary neutron star provided that the nuclear layer is not extremely thin.

6. Conclusions

Our analysis shows that the thermal conductivity of CFL color superconducting dense quark matter is very high for typical values of the temperature found in a newborn compact star. This is a direct consequence of the existence of the photon and the massless NG boson ϕ whose mean free paths are very large. Note that the photons are trapped in the core because of reflection by the electron plasma in the neighboring nuclear matter. The NG bosons are also confined to the core since they can only exist in the CFL phase.

It is appropriate to mention that the (pseudo-) NG bosons and photons should also dominate other kinetic properties of dense quark matter in the CFL phase. For example, the shear viscosity should be mostly due to photons and the same massless NG bosons associated with the breaking of baryon number. The electrical conductivity, on the other hand, would be mostly due to the lightest *charged* pseudo-NG boson, i.e., the K^+. Thus, in the limit of small temperatures, $T \to 0$, the electrical conductivity will be suppressed by a factor $\exp(-m_{K^+}/T)$.

Since the neutrino emissivity of the CFL core is strongly suppressed, the heat is transferred to the outer nuclear layer only through direct surface contact. While both the core and the outer layer contribute to the heat capacity of the star, it is only the outer layer which is capable of emitting this heat energy efficiently in the form of neutrinos. The combination of these two factors tends to extend the cooling time of a star. However, because of the very small thermal energy of the quark core, the time scale for cooling could be noticeably longer than that for an ordinary neutron star only if the outer nuclear layer was very thin. (Note that, while little is known about the properties of thin boundary layers outside CFL matter, it is possible that photon emission from this layer might quickly drain the relatively small amount of CFL thermal energy.) Thus it appears that the cooling of stars with not too large CFL quark cores will differ little from the cooling of typical neutron stars. A similar conclusion has been reached for stars with regular, non-CFL quark interiors [28].

In passing it is interesting to speculate about the possibility that a bare CFL quark star made entirely of dense quark matter could exist. If it were possible, it would look like a transparent dielectric [29]. Our present study suggests such a star would also have very unusual thermal properties. Indeed, if the star has a finite temperature $T \lesssim \tilde{T}$ after it was created, almost all of its thermal energy would be stored in the NG bosons. Notice that all the photons would leave the star very soon after transparency set in because the star is assumed to have no nuclear matter layer. The local interaction as well as the self-interaction of the NG bosons is very weak so that we argued in Sec. 3 that their mean free path would be limited only by the geometrical size of the star. This suggests that, since photon and neutrino emission inside the CFL phase is strongly suppressed [13], the only potential source of energy loss in the bare CFL star would be the interaction of the NG bosons at the stellar boundary and photon emission there. It is likely, therefore, that such stars might be very dim and might even be good candidates for some of the baryonic dark matter in the Universe.

Acknowledgments

This work was supported by the U.S. Department of Energy Grant No. DE-FG02-87ER40328.

References

1. J.C. Collins and M.J. Perry, *Phys. Rev. Lett.* **34**, 1353 (1975).
2. B. C. Barrois, *Nucl. Phys.* **B129**, 390 (1977); S. C. Frautschi, in *Hadronic Matter at Extreme Energy Density*, edited by N. Cabibbo and L. Sertorio (Plenum, New York, 1980); D. Bailin and A. Love, *Phys. Rep.* **107**, 325 (1984).
3. M. G. Alford, K. Rajagopal and F. Wilczek, *Phys. Lett.* **B422**, 247 (1998); R. Rapp, T. Schäfer, E. V. Shuryak and M. Velkovsky, *Phys. Rev. Lett.* **81**, 53 (1998).
4. D. T. Son, *Phys. Rev.* **D59**, 094019 (1999); R. D. Pisarski and D. H. Rischke, *Phys. Rev. Lett.* **83**, 37 (1999).
5. T. Schafer and F. Wilczek, *Phys. Rev.* **D60**, 114033 (1999); D. K. Hong, V. A. Miransky, I. A. Shovkovy and L. C. R. Wijewardhana, *ibid.* **D61**, 056001

302

(2000); **D62**, 059903(E) (2000); R. D. Pisarski and D. H. Rischke, *ibid.* **D61**, 051501 (2000).

6. S. D. Hsu and M. Schwetz, *Nucl. Phys.* **B572**, 211 (2000); W. E. Brown, J. T. Liu and H. C. Ren, *Phys. Rev.* **D61**, 114012 (2000).

7. I. A. Shovkovy and L. C. R. Wijewardhana, *Phys. Lett.* **B470**, 189 (1999); T. Schäfer, *Nucl. Phys.* **B575**, 269 (2000).

8. M. Alford, K. Rajagopal and F. Wilczek, *Nucl. Phys.* **B537**, 443 (1999); note that the 2SC phase involving just two quark flavors is thought to be absent in compact stars, see M. Alford and K. Rajagopal, *JHEP* **0206**, 031 (2002).

9. R. Casalbuoni and R. Gatto, *Phys. Lett.* **B464**, 111 (1999).

10. D. T. Son and M. A. Stephanov, *Phys. Rev.* **D61**, 074012 (2000); **D62**, 059902(E) (2000).

11. P. F. Bedaque and T. Schäfer, *Nucl. Phys.* **A697**, 802 (2002); D. B. Kaplan and S. Reddy, *Phys. Rev.* **D65**, 054042 (2002).

12. V. A. Miransky and I. A. Shovkovy, *Phys. Rev. Lett.* **88**, 111601 (2002); T. Schafer, D. T. Son, M. A. Stephanov, D. Toublan and J. J. Verbaarschot, *Phys. Lett.* **B522**, 67 (2001).

13. P. Jaikumar, M. Prakash and T. Schäfer, astro-ph/0203088.

14. I. A. Shovkovy and P. J. Ellis, *Phys. Rev.* **C66**, 015802 (2002).

15. E. J. Ferrer, V. P. Gusynin and V. de la Incera, cond-matt/0203217.

16. V. P. Gusynin and I. A. Shovkovy, *Nucl. Phys.* **A700**, 577 (2002).

17. D. T. Son, hep-ph/0204199.

18. T. Schäfer, hep-ph/0201189.

19. P. A. Sturrock, *Plasma Physics* (Cambridge University Press, Cambridge, 1994).

20. D. F. Litim and C. Manuel, *Phys. Rev.* **D64**, 094013 (2001).

21. J. M. Lattimer, K. A. Van Riper, M. Prakash and M. Prakash, *Astrophys. J.* **425**, 802 (1994).

22. M. G. Alford, K. Rajagopal, S. Reddy and F. Wilczek, *Phys. Rev.* **D64**, 074017 (2001).

23. G. W. Carter and S. Reddy, *Phys. Rev.* **D62**, 103002 (2000).

24. A. W. Steiner, M. Prakash and J. M. Lattimer, *Phys. Lett.* **B509**, 10 (2001).

25. S. Reddy, M. Sadzikowski and M. Tachibana, nucl-th/0203011.

26. M. Prakash, J. M. Lattimer, J. A. Pons, A. W. Steiner and S. Reddy, *Lect. Notes Phys.* **578**, 364 (2001).

27. S. L. Shapiro and S. A. Teukolsky, *Black Holes, White Dwarfs, and Neutron Stars: The Physics of Compact Objects* (Wiley, New York, 1983).

28. D. Page, M. Prakash, J. M. Lattimer and A. W. Steiner, *Phys. Rev. Lett.* **85**, 2048 (2000).

29. K. Rajagopal and F. Wilczek, *Phys. Rev. Lett.* **86**, 3492 (2001).

COLOR SUPERCONDUCTIVITY FROM MAGNETIC
INTERACTION INDUCED BY FLOW EQUATIONS

E. GUBANKOVA

Center for Theoretical Physics,
Massachusetts Institute of Technology,
Cambridge, MA 02139

Using flow equations, we derive an effective quark-quark interaction and obtain the coupled set of gap equations for the condensates of the CFL phase of massless $N_f = 3$ dense QCD. We find two different sources of the infrared cutoff in magnetic interaction. When the penetration depth of the magnetic field inside the superconductor is less than the coherence length of quark-quark bound state, our results for the gap agree with those of Son [4]. In the other case, we obtain parametric enhancement of the gap on the coupling constant.

1. Introduction

Quark matter at sufficiently high density is a color superconductor. It has been known for some time [1], however the value of the gap was too small to be of any practical interest. Recent revision of the subject brought much larger gap [2] than previously thought. The consistent numerical value of the gap has been obtained using phenomenological four-fermion interactions [3] and from QCD one-gluon exchange [4], though with different parametric dependence on the coupling. Many microscopic calculations of the gap in $N_f = 2$ [5,6] and in $N_f = 3$ [7] have been done after that.

Microscopic studies rely on the fact that due to asymptotic freedom QCD becomes a weakly interacting theory at high densities and perturbation theory can be used. However, high density quark matter cannot be described as a simple Fermi gas. As known from the BSC theory of superconductivity, the Fermi surface is unstable to an arbitrarily weak attractive interaction. In QCD attractive interaction is mediated by one-gluon exchange between quarks in color antitriplet state.

In the language of renormalization group, BCS instability with respect to Cooper pairing is associated with breaking down the conventional perturbation theory in the vicinity of the Fermi surface. This signals formation of a gap which regulates the infrared divergent behavior. It is in full analogy with electron superconductivity in metals. The difference between the two is that the attractive interaction via gluon exchange is already present in QCD rather than being induced by the solid state lattice as in metal. Therefore, properties of a gluon propagating in high density quark matter are essential in calculating the color superconducting gap. It has been suggested by Son (the first reference in [4]), that the gap is dominated by magnetic long range exchanges, and that the infrared behavior is regulated by dynamic screening in quark plasma. We show that there are two characteristic scales in superconducting medium: the penetration depth of magnetic field inside the superconductor, δ, and the coherence length of the bound quarks (or size of the qq Cooper pair), ξ. The prevailing energy $1/\delta$ or $1/\xi$ regulates the long range behavior of magnetic field in color superconductor.

The purpose of the present work is to obtain an effective microscopic low energy theory of quasiparticles and holes in the vicinity of the Fermi surface. We shall argue that depending on the state of superconductor, there are two different sources of the infrared cutoff in magnetic interaction. At small temperature, the superconductor is of a Pipperd type, $\delta \ll \xi$, and magnetic gluon is damped by the quark medium according to Landau damping mechanism. As we increase the temperature, the quark plasma 'dilutes' and the penetration depth of magnetic field grows. It turns out that usually the superconducting condensate melts slower than the penetration depth grows [12], corresponding to London type of superconductor, $\xi \ll \delta$. In this case magnetic gluon scatters over diquark Cooper pairs and damped by the superconducting medium. Magnetic interaction is regulated by the inverse coherence length, or roughly by the superconducting gap. In these two limiting cases we obtain for the gap the same parametric dependence on coupling constant, though different numerical factors.

We apply flow equations to the Coulomb gauge QCD Hamiltonian with $N_f = 3$ at nonzero quark density. Flow equation method [10], which is a synthesis of the perturbative renormalization group and the many-body technique, has been successfully applied to the problem of electron superconductivity in metals [11]. The main idea of the method is to eliminate interactions which mix states with different particle content (or number of particles) and obtain an effective theory which conserves particle number in each (Fock) sector. This procedure is performed in sequence for matrix elements with large energy differences down to more degenerate states. In high density QCD, all nonabelian contributions are suppressed, therefore, the quark-gluon coupling is the only term which mediates hoping between states of high- (away from Fermi surface) and low- (near Fermi surface) energies. Eliminating quark-gluon coupling by flow equations we virtually move towards the Fermi surface, in analogy to perturbative renormalization group scaling. Decoupling different energy scales corresponds to integrating out fast moving modes - quarks in the Dirac sea, and obtaining an effective theory for slow modes - quarks close to Fermi surface. As mentioned above, in this scaling down we encounter two characteristic scales: penetration depth of magnetic field and the coherence length of qq bound state. Depending which scale is bigger, we obtain different mechanisms of the infrared cutoff in magnetic interaction and corresponding gaps.

2. Effective microscopic Hamiltonian for color superconductivity

We start with the Coulomb gauge QCD Hamiltonian, $\nabla \cdot A = 0$, with $N_f = 3$ at nonzero quark densities given by

$$H = H_0 + H_I = H_0 + H_{inst} + H_{dyn} , \tag{1}$$

where H_0 is the free Hamiltonian, H_{inst} is the instantaneous interaction describing static properties, and H_{dyn} is the dynamical interaction involving the gluon propagation. The free part is the kinetic energy of quarks and gluons

$$H_0 = \int dx \bar{\psi}(x) \left(-i\gamma \cdot \nabla - \mu\gamma_0 + m \right) \psi(x)$$

$$+ \, \mathrm{Tr} \int dx \left(\mathbf{\Pi}^2(x) + \mathbf{B}_A^2(x) \right) , \tag{2}$$

where the non-abelian magnetic field is $\mathbf{B} = B_i = \nabla_j A_k - \nabla_k A_j + g[A_j, A_k]$, and its abelian part is represented by $\mathbf{B}_A$. The degrees of freedom are the transverse perturbative gluon field $\mathbf{A} = A^a T^a$ ($A \equiv A^{PT}$), its conjugate momentum $\mathbf{\Pi}$, and the quark field in the Coulomb gauge. The instantaneous interaction is given by

$$H_{inst} = -\frac{1}{2} \int dx dy \bar{\psi}(x)\gamma_0 T^a \psi(x) V_{inst}(|x - y|)\bar{\psi}(y)\gamma_0 T^a \psi(y) , \tag{3}$$

where the leading order kernel is a Coulomb potential defined, together with its Fourier transform, by

$$V_{inst}(r) = -\frac{\alpha_s}{r} , \quad V_{inst}(q) = -\frac{g^2}{q^2} , \tag{4}$$

with $\alpha_s = g^2/4\pi$. The dynamical interaction includes the minimal quark-gluon coupling, V_{qg}, and the non-abelian three- and four-gluon interactions,

V_{gg}, i.e. $H_{dyn} = V_{qg} + V_{gg}$, where

$$V_{qg} = -g \int d\boldsymbol{x} \bar{\psi}(\boldsymbol{x}) \boldsymbol{\gamma} \cdot \boldsymbol{A}(\boldsymbol{x}) \psi(\boldsymbol{x})$$

$$V_{gg} = \text{Tr} \int d\boldsymbol{x} \left(\boldsymbol{B}^2(\boldsymbol{x}) - \boldsymbol{B}_A^2(\boldsymbol{x}) \right) . \tag{5}$$

As mentioned in the introduction, at high quark densities diagrams including the non-abelian interactions are suppressed, henceforth V_{gg} is not considered.

Following the work of Alford, Rajagopal and Wilczek, [3], we assume, due to quark interactions, the diquark condensation in scalar $< \psi^T C \gamma_5 \psi >$ and pseudoscalar $< \psi^T C \psi >$ channels, or equivalently $< bb >\neq 0$ and $< b^\dagger b^\dagger >\neq 0$. In other words, perturbative vacuum $|0\rangle$ is not a ground state for the system, instead it is the BCS vacuum $|\Omega\rangle$ containing condensates of diquarks. The Fock space is constructed from this vacuum using quasiparicle operators $b^\dagger$ and $d^\dagger$

$$\psi(\boldsymbol{x}) = \sum_s \int \frac{d\boldsymbol{k}}{(2\pi)^3} [u(\boldsymbol{k},s)b(\boldsymbol{k},s) + v(-\boldsymbol{k},s)d^\dagger(-\boldsymbol{k},s)]e^{i\boldsymbol{k}\boldsymbol{x}}$$

$$\boldsymbol{A}(\boldsymbol{x}) = \sum_a \int \frac{d\boldsymbol{k}}{(2\pi)^3} \frac{1}{\sqrt{2\omega(\boldsymbol{k})}} [a(\boldsymbol{k},a) + a^\dagger(-\boldsymbol{k},a)]e^{i\boldsymbol{k}\boldsymbol{x}}$$

$$\boldsymbol{\Pi}(\boldsymbol{x}) = -i \sum_a \int \frac{d\boldsymbol{k}}{(2\pi)^3} \sqrt{\frac{\omega(\boldsymbol{k})}{2}} [a(\boldsymbol{k},a) - a^\dagger(-\boldsymbol{k},a)]e^{i\boldsymbol{k}\boldsymbol{x}} , \tag{6}$$

where $b|\Omega\rangle = d|\Omega\rangle = 0$, and the gluon part has trivial vacuum $a|0\rangle = 0$ with $\omega(\boldsymbol{k}) = |\boldsymbol{k}|$. All descrete numbers (helicity, color, and flavor for the quarks and color for the gluons) are collectively denoted as s and a, respectively. The gluon operators $a = a^{IA}(\boldsymbol{k}) = \sum_{\lambda=1,2} \varepsilon^I(\boldsymbol{k},\lambda) a^A(\boldsymbol{k},\lambda)$ are transverse, i.e. $\boldsymbol{k} \cdot a^A(\boldsymbol{k}) = 0$ and the polarization sum is $D_{IJ}(\boldsymbol{k}) = \sum_{\lambda=1,2} \varepsilon_I(\boldsymbol{k},\lambda)\varepsilon_J(\boldsymbol{k},\lambda) = \delta_{IJ} - \hat{k}_I \hat{k}_J$. In the massless basis, $m = 0$, $u(\boldsymbol{k}) = v(\boldsymbol{k}) = (L(\boldsymbol{k}), R(\boldsymbol{k}))$ with the left Weyl spinor $L(\boldsymbol{k}) = (-\sin(\Theta(\boldsymbol{k})/2)\exp(-i\phi(\boldsymbol{k})), \cos(\Theta(\boldsymbol{k})/2))$.

Then, in the mean field approximation, the Hamiltonian Eq. (1) can be written in terms of creation/annihilation operators as

$$H = \int \frac{d\boldsymbol{k}}{(2\pi)^3} (k-\mu) \, b^\dagger(\boldsymbol{k})_\alpha^i b(\boldsymbol{k})_\alpha^i + \int \frac{d\boldsymbol{k}}{(2\pi)^3} (\mu-k) \, b^\dagger(\boldsymbol{k})_\alpha^i b(\boldsymbol{k})_\alpha^i$$

$$+ \int \frac{d\boldsymbol{k}}{(2\pi)^3} (k+\mu) \, d^\dagger(\boldsymbol{k})_\alpha^i d(\boldsymbol{k})_\alpha^i + \int \frac{d\boldsymbol{k}}{(2\pi)^3} k \, a^\dagger(\boldsymbol{k})a(\boldsymbol{k})$$

$$+ \int \frac{d\boldsymbol{p}}{(2\pi)^3} \Delta_{ij}^{\alpha\gamma}(\boldsymbol{p}) T_{\alpha\beta}^A T_{\gamma\delta}^A e^{-i\phi(\boldsymbol{P})} \left(b(\boldsymbol{p})_\beta^i b(-\boldsymbol{p})_\delta^j + d^\dagger(\boldsymbol{p})_\beta^i d^\dagger(-\boldsymbol{p})_\delta^j \right) + c.c.$$

$$+ \int \frac{dk}{(2\pi)^3} \frac{dp}{(2\pi)^3} \frac{3}{4} W(k,p) e^{i\phi(k)} e^{-i\phi(p)}$$

$$\times\ b^\dagger(k)^i_\alpha T^A_{\alpha\beta} b(p)^i_\beta b^\dagger(-k)^j_\gamma T^A_{\gamma\delta} b(-p)^j_\delta + \cdots$$

$$+ \int \frac{dk}{(2\pi)^3} \frac{dp}{(2\pi)^3}\ g(k,p,k-p)$$

$$\times\ b^\dagger(k)^i_\alpha T^A_{\alpha\beta} b(p)^i_\beta \frac{a^{IA}(k-p)}{2\,|k-p|} (u^\dagger(k)\alpha^I u(p)) + \cdots, \tag{7}$$

we explicitly display color (α, β), flavor (i, j) and polarization I indices. The first three terms describe kinetic energies of particles, holes and antiparticles, respectively. The diquark condensate Δ and the effective quark-quark interaction W are unknown parameters, and dots show that there are other possible terms which we ignored.

Color-superconducting condensate $\Delta^{ij}_{\alpha\gamma}$ is a $N_c \times N_c$ matrix in fundamental color space $(\alpha, \gamma = 1, ..., N_c)$, and a $N_f \times N_f$ matrix in flavor space $(i, j = 1, ..., N_f)$. It can be parametrized for $N_f = 3$ as [3]

$$\Delta^{ij}_{\alpha\gamma}(p) = 3\left(\frac{1}{3}\left[\Delta_8(p) + \frac{1}{8}\Delta_1(p)\right]\delta^i_\alpha \delta^j_\gamma + \frac{1}{8}\Delta_1(p)\delta^i_\gamma \delta^j_\alpha \right), \tag{8}$$

where Δ_1 and Δ_8 are the gaps in the singlet and octet channels, respectively. We will also use the parametrization in terms of symmetric, $\Delta_{(6,6)}$, and antisymmetric, $\Delta_{(\bar{3},\bar{3})}$, in color and flavor gap funtions, $\Delta^{ij}_{\alpha\gamma} = \Delta_{(6,6)}(\delta^i_\alpha \delta^j_\gamma + \delta^i_\gamma \delta^j_\alpha) + \Delta_{(\bar{3},\bar{3})}(\delta^i_\alpha \delta^j_\gamma - \delta^i_\gamma \delta^j_\alpha)$. The connection with the eigenvalue gaps is given by $\Delta_{(\bar{3},\bar{3})} = 1/2(\Delta_8 - 1/4\Delta_1)$, and $\Delta_{(6,6)} = 1/2(\Delta_8 + 1/2\Delta_1)$.

Replacing color α and flavor i indices with a single color-flavor index ρ, $\Delta^{ij}_{\alpha\gamma}$ is diagonal in the CFL basis [9]

$$b(k)^i_\alpha = \sum_\rho \frac{\lambda^\rho_{i\alpha}}{\sqrt{2}}\ b(k)^\rho, \tag{9}$$

where λ^ρ are the Gell-Mann matrices for $\rho = 1, ..., 8$ and $\lambda^9_{i\alpha} = \sqrt{2/3}\delta^i_\alpha$ for $\rho = 9$, and strictly speaking we should have used different letter notation for the CFL operators b^ρ. Then, $\Delta^{ij}_{\alpha\gamma}(p)b(p)^i_\beta b(-p)^j_\delta T^A_{\alpha\beta} T^A_{\gamma\delta} = 1/2\sum_\rho \Delta_\rho(p)b(p)^\rho b(-p)^\rho$, and the Hamiltonian Eq. (7) is given in the CFL basis by

$$H = \sum_{k,\rho}|k-\mu|\ b^\dagger_\rho(k)b_\rho(k) + \sum_{k,\rho}(k+\mu)\ d^\dagger_\rho(k)d_\rho(k) + \sum_k\ k\ a^\dagger(k)a(k)$$

$$+ \frac{1}{2}\sum_{p,\rho}\Delta_\rho(p)e^{-i\phi(p)}\left(b_\rho(p)b_\rho(-p) + d^\dagger_\rho(p)d^\dagger_\rho(-p)\right) + c.c.$$

$$+ \frac{3}{4} \sum_{k,p,\rho,\rho'} W^{\rho\rho'}(k,p) \frac{1}{2} \lambda_{i\alpha}^{\rho} \lambda_{j\gamma}^{\rho} \frac{1}{2} \lambda_{i\beta}^{\rho'} \lambda_{j\delta}^{\rho'} T_{\alpha\beta}^{A} T_{\gamma\delta}^{A} e^{i\phi(k)} e^{-i\phi(p)}$$

$$\times \, b_{\rho}^{\dagger}(k) b_{\rho'}(p) b_{\rho}^{\dagger}(-k) b_{\rho'}(-p) + \ldots$$

$$+ \sum_{k,p,\rho,\rho'} g^{\rho\rho'}(k,p,k-p) \frac{1}{2} \lambda_{i\alpha}^{\rho} T_{\alpha\beta}^{A} \lambda_{i\beta}^{\rho'}$$

$$\times \, b_{\rho}^{\dagger}(k) b_{\rho'}(p) \frac{a^{IA}(k-p)}{2\,|k-p|} (u^{\dagger}(k) \alpha^{I} u(p)) + \ldots . \tag{10}$$

where we used notation $\sum_{k} = \int dk/(2\pi)^3$. As in [3], we change basis to creation/annihilation operators y and z for quasiparticles (quasiholes) and quasiantiparticles, respectively,

$$y_{\rho}(k) = \cos(\Theta_{\rho}^{y}(k)) b^{\rho}(k) + \sin(\Theta_{\rho}^{y}(k)) \exp(i\xi_{\rho}^{y}(k)) b_{\rho}^{\dagger}(-k)$$
$$z_{\rho}(k) = \cos(\Theta_{\rho}^{z}(k)) b^{\rho}(k) + \sin(\Theta_{\rho}^{z}(k)) \exp(i\xi_{\rho}^{z}(k)) b_{\rho}^{\dagger}(-k), \tag{11}$$

with $y_{\rho}(k)|0\rangle = z_{\rho}(k)|0\rangle = 0$. In order to absorb the condensate term into a new free Hamiltonian, we choose [3] $\cos(2\Theta_{\rho}^{y}(k)) = |k-\mu|/E_{\rho}^{(-)}(k)$, $\cos(2\Theta_{\rho}^{z}(k)) = |k+\mu|/E_{\rho}^{(+)}(k)$, and $\sin(2\Theta_{\rho}^{y}(k)) = \Delta_{\rho}(k)/E_{\rho}^{(-)}(k)$, $\sin(2\Theta_{\rho}^{z}(k)) = \Delta_{\rho}(k)/E_{\rho}^{(+)}(k)$; $\xi_{\rho}^{y}(k) = \phi(k) + \pi$, $\xi_{\rho}^{z}(k) = -\phi(k)$. As the result of this transformation, a new free Hamiltonian is given by

$$\tilde{H}_0 = \sum_{k,\rho} E_{\rho}^{(-)}(k) \, b_{\rho}^{\dagger}(k) b_{\rho}(k) + \sum_{k,\rho} E_{\rho}^{(+)}(k) \, d_{\rho}^{\dagger}(k) d_{\rho}(k)$$

$$+ \sum_{k} \omega(k) \, a^{\dagger}(k) a(k), \tag{12}$$

where $E_{\rho}^{(-)} = \sqrt{(k-\mu)^2 + \Delta_{\rho}(k)^2}$, $E_{\rho}^{(+)} = \sqrt{(k+\mu)^2 + \Delta_{\rho}(k)^2}$, and $\omega(k) = k$. The effective Hamiltonian Eq. (10) is given by

$$H = \tilde{H}_0 + V_{qq} + V_{qg}, \tag{13}$$

with transformations of Eq. (11) made in the diquark interaction V_{qq} and the quark-gluon coupling V_{qg}, which are the final two terms in Eq. (10). In the next section, our aim is to determine the unknown parameters of the effective Hamiltonian Eq. (13) using flow equations.

3. Flow equations

Flow equations are written for the unknown functions $W^{\rho\rho'}, g^{\rho\rho'}, \Delta_{\rho}$ of the effective Hamiltonian Eq. (13). Calculations are performed using y, z variables and the vacuum $|0\rangle$. Similar calculations have been done in [8].

The first order flow equations $dV_{qg}/dl = [\eta, \tilde{H}_0]$, $\eta = [\tilde{H}_0, V_{qg}]$ eliminate the quark-gluon coupling V_{qg}. The generator of the transformation is given by

$$\eta = \sum_{k,p,\rho,\rho'} \eta^{\rho\rho'}(k,p,k-p)\frac{1}{2}\lambda_{i\alpha}^\rho T_{\alpha\beta}^A \lambda_{i\beta}^{\rho'}$$

$$\times b_\rho^\dagger(k)b_{\rho'}(p)\frac{a^{IA}(k-p)}{2\,|k-p|}(u^\dagger(k)\alpha^I u(p))\,, \tag{14}$$

then for the generator $\eta^{\rho\rho'}$ and coupling $g^{\rho\rho'}$ functions flow equations are written as

$$\frac{dg^{\rho\rho'}}{dl} = -(E_\rho(k) - E_{\rho'}(p) - \omega(k-p))^2\ g^{\rho\rho'}$$

$$\eta^{\rho\rho'} = (E_\rho(k) - E_{\rho'}(p) - \omega(k-p))\ g^{\rho\rho'}\,. \tag{15}$$

The solution of Eq. (15) reads

$$g^{\rho\rho'}(l) = g(0)\exp\left(-(E_\rho(k) - E_{\rho'}(p) - \omega(k-p))^2\ l\right)\,, \tag{16}$$

where $g(0) = g$ is the bare coupling constant. As $l \to \infty$ the coupling is eliminated as long as the states in the exponent are not degenerate.

Using the generator Eq. (14) and the quark-gluon coupling Eq. (10), we write the second order flow equations for the effective quark-quark interaction, $dV_{qq}/dl = [\eta, V_{qg}]_{two-body}$, and for the diquark self-energy, $d\Sigma_q/dl = [\eta, V_{qg}]_{one-body}$. Flow equation for the diquark interaction is given by

$$\frac{dV_{qq}}{dl} = -\sum_{k,p,\rho\,rho'} b_\rho^\dagger(k)b_\rho^\dagger(-k)b_{\rho'}(p)b_{\rho'}(-p)\frac{1}{2}\lambda_{i\alpha}^\rho\lambda_{j\gamma}^\rho\frac{1}{2}\lambda_{i\beta}^{\rho'}\lambda_{j\delta}^{\rho'}T_{\alpha\beta}^A T_{\gamma\delta}^A$$

$$(u^\dagger(k)\alpha^I u(p))(u^\dagger(-k)\alpha^J u(-p))\frac{D_{IJ}(k-p)}{2\omega(k-p)} \tag{17}$$

$$\left(\eta^{\rho\rho'}(k,p,k-p)g^{\rho'\rho}(p,k,k-p) + \eta^{\rho'\rho}(p,k,k-p)g^{\rho\rho'}(k,p,k-p)\right)\,.$$

Integrating the flow equation Eq. (17) over $l = [0,\infty)$, with the final value $V_{qq} = V_{qq}(l = \infty)$, and adding the instantaneous interaction Eq. (3), we get

$$V_{qq} = -\sum_{k,p,\rho,\rho'} b_\rho^\dagger(k)b_\rho^\dagger(-k)b_{\rho'}(p)b_{\rho'}(-p)\frac{1}{2}\lambda_{i\alpha}^\rho\lambda_{j\gamma}^\rho\frac{1}{2}\lambda_{i\beta}^{\rho'}\lambda_{j\delta}^{\rho'}T_{\alpha\beta}^A T_{\gamma\delta}^A$$

$$\left(V^{\rho\rho'}(k,p)(u^\dagger(k)\alpha^I u(p))(u^\dagger(-k)\alpha^J u(-p))D_{IJ}(k-p)\right.$$

$$\left.+V(k,p)(u^\dagger(k)u(p))(u^\dagger(-k)u(-p))\right)\,, \tag{18}$$

with

$$V^{\rho\rho'}(k,p) = -\frac{g^2}{2}\frac{1}{(E^\rho(k) - E^{\rho'}(p))^2 + \omega_M(k-p)^2}$$

$$V(k,p) = \frac{g^2}{2}\frac{1}{\omega_E(k-p)^2}, \tag{19}$$

where we have used solutions for $\eta^{\rho\rho'}$ and $g^{\rho\rho'}$, Eqs. (15,16). Energies for magnetic and electric gluons, ω_M and ω_E respectively, are specified further. Keeping only left-left components, matrix elements are given by

$$(u^\dagger(k)\alpha^I u(p))(u^\dagger(-k)\alpha^J u(-p))D_{IJ}(k-p) =$$

$$(-1)e^{-i\phi(p)}e^{i\phi(k)}\left(-\frac{3 - \hat{k}\cdot\hat{p}}{2} + \frac{1 - \hat{k}\cdot\hat{p}}{2}\frac{(k+p)^2}{(k-p)^2}\right)$$

$$(u^\dagger(k)u(p))(u^\dagger(-k)u(-p)) = (-1)e^{-i\phi(p)}e^{i\phi(k)}\left(\frac{1 + \hat{k}\cdot\hat{p}}{2}\right), \tag{20}$$

where we used Fierz transform $(\bar{u}(k)O^A u(p))(\bar{u}(-k)O^B u(-p)) = \epsilon\sum_{CD}f^{AB}_{CD}(\bar{u}_C(p)O^C u(-p))(\bar{u}(-k)O^D u_C(k))$, where $\epsilon = -1$ for $O^A, O^B = \gamma^\mu$, and $C\bar{u}^T = u_C$, $u^T C = \bar{u}_C$. For the above matrix elements the coefficients f in Fierz transform $O^A \times O^B \to O^C \times O^D$ are given by $\gamma_0 \times \gamma_0 \to 1/4(1\times 1 - \gamma_5 \times \gamma_5) + 1/4\sigma_{\mu\nu} \times \sigma^{\mu\nu}$, and $\gamma_i \times \gamma^j \to \delta_i^j 1/4(1\times 1 - \gamma_5 \times \gamma_5) + \delta_i^j 1/4\sigma_{\mu\nu} \times \sigma^{\mu\nu} + 1/4(\sigma_{i\mu} \times \sigma^{j\mu} + \sigma_{\mu i} \times \sigma^{\mu j} - \sigma_{i\mu} \times \sigma^{\mu j} - \sigma_{\mu i} \times \sigma^{j\mu})$, where $\mu,\nu = (0, i = 1,2,3)$, $\sigma_{\mu\nu} = 1/2[\gamma_\mu, \gamma_\nu]$. The diquark condensates are 1×1-pseudoscalar $(\bar{u}_C u)$, $\gamma_5 \times \gamma_5$-scalar $(\bar{u}_C\gamma u)$, and $\sigma_{\mu\nu} \times \sigma^{\mu\nu}$-vector $(\bar{u}_C\sigma_{\mu\nu}u)$. Performing algebra with λ-matrices in Eq. (18), we get

$$V_{qq} = -\sum_{k,p}(-1)e^{-i\phi(p)}e^{i\phi(k)}$$

$$\left(b_8^\dagger(k)b_8^\dagger(-k)b_1(p)b_1(-p)F^{81}(k,p) + b_1^\dagger(k)b_1^\dagger(-k)b_8(p)b_8(-p)F^{18}(k,p)\right.$$

$$\left. - b_8^\dagger(k)b_8^\dagger(-k)b_8(p)b_8(-p)2F^{88}(k,p)\right), \tag{21}$$

where we introduced

$$F^{\rho\rho'} = V^{\rho\rho'}_{dyn}(k,p)\left(\frac{3 - \hat{k}\cdot\hat{p}}{2}\right) + V_{inst}(k,p)\left(\frac{1 + \hat{k}\cdot\hat{p}}{2}\right), \tag{22}$$

and

$$V^{\rho\rho'}_{dyn}(k,p) = \frac{2g^2}{3}\frac{1}{(E^\rho(k) - E^{\rho'}(p))^2 + \omega_M(k-p)^2}$$

$$V_{inst}(k,p) = \frac{2g^2}{3}\frac{1}{\omega_E(k-p)^2}. \tag{23}$$

Dynamically generated by flow equations interaction describes magnetic gluon exchange, and the instantaneous interaction -electric gluon exchange. Comparing Eq. (21) with Eq. (10) for V_{qq}, the effective quark-quark interaction is given by

$$W^{\rho\rho'}(\boldsymbol{k},\boldsymbol{p}) = -F^{\rho\rho'}(\boldsymbol{k},\boldsymbol{p}) = -V_{dyn}^{\rho\rho'}(\boldsymbol{k},\boldsymbol{p}) - V_{inst}^{\rho\rho'}(\boldsymbol{k},\boldsymbol{p}), \qquad (24)$$

in the collinear limit $\hat{k}\cdot\hat{p} = 1$. This interaction is attractive in the singlet and octet channels. Note that $W^{11} = 0$, and condensation in singlet channel is driven by W^{81}.

Dynamical magnetic interaction, Eq. (23), has the form $-1/(q^2+\delta E^2)$ instead of $-1/(q^2-\delta E^2)$, that is produced by the equal time perturbation theory, where δE is the enegry difference of in- and out-going quarks. The latter has the pole,therefore usually δE is neglected near the Fermi surface. Our interaction is regular, and δE will play an important role.

In order to incorporate effects of the dense quark medium, we include polarization operators for electric and magnetic gluons, modifying the gluon single energy as done in the work of Pisarski and Rischke [5]

$$\frac{1}{\omega_M(\boldsymbol{k}-\boldsymbol{p})^2} = \frac{1}{2}\left(\frac{(\boldsymbol{k}-\boldsymbol{p})^4}{(\boldsymbol{k}-\boldsymbol{p})^6 + M^4(E(\boldsymbol{k})+E(\boldsymbol{p}))^2)}\right.$$
$$\left.+ \frac{(\boldsymbol{k}-\boldsymbol{p})^4}{(\boldsymbol{k}-\boldsymbol{p})^6 + M^4(E(\boldsymbol{k})-E(\boldsymbol{p}))^2)}\right)$$
$$\frac{1}{\omega_E(\boldsymbol{k}-\boldsymbol{p})^2} = \frac{1}{(\boldsymbol{k}-\boldsymbol{p})^2 + 3m_g^2}, \qquad (25)$$

where $M^2 = (3\pi/4)m_g^2$, $m_g^2 = N_f g^2\mu^2/(6\pi)^2$. Eq. (25) is used in the complete diquark interaction Eq. (23).

Quark self-energy has $b^\dagger b$ and $bb + b^\dagger b^\dagger$ components associated with normal and anomalous propagation, respectively. For the anomalous propagation self-energy the flow equation is given by

$$\frac{d\Sigma_{qq}}{dl} = -\sum_{\boldsymbol{k},\boldsymbol{p},\rho,\rho'}\frac{1}{2}\lambda_{i\alpha}^\rho\lambda_{j\gamma}^\rho\frac{1}{2}\lambda_{i\beta}^{\rho'}\lambda_{j\delta}^{\rho'}T_{\alpha\beta}^A T_{\gamma\delta}^A$$

$$(< b_\rho^\dagger(\boldsymbol{k})b_\rho^\dagger(-\boldsymbol{k}) > b_{\rho'}(\boldsymbol{p})b_{\rho'}(-\boldsymbol{p}) + < b_{\rho'}(\boldsymbol{p})b_{\rho'}(-\boldsymbol{p}) > b_\rho^\dagger(\boldsymbol{k})b_\rho^\dagger(-\boldsymbol{k}))$$

$$(u^\dagger(\boldsymbol{k})\alpha^I u(\boldsymbol{p}))(u^\dagger(-\boldsymbol{k})\alpha^J u(-\boldsymbol{p}))\frac{D_{IJ}(\boldsymbol{k}-\boldsymbol{p})}{2\omega(\boldsymbol{k}-\boldsymbol{p})}$$

$$\left(\eta^{\rho\rho'}(\boldsymbol{k},\boldsymbol{p},\boldsymbol{k}-\boldsymbol{p})g^{\rho'\rho}(\boldsymbol{p},\boldsymbol{k},\boldsymbol{k}-\boldsymbol{p}) + \eta^{\rho'\rho}(\boldsymbol{p},\boldsymbol{k},\boldsymbol{k}-\boldsymbol{p})g^{\rho\rho'}(\boldsymbol{k},\boldsymbol{p},\boldsymbol{k}-\boldsymbol{p})\right)$$

$$+ (Terms \sim < dd >, < d^\dagger d^\dagger >). \qquad (26)$$

We neglect the contribution of antiparticles, since it is regular near the Fermi surface, $< dd > \sim \sin(2\Theta^z) \sim \Delta/2\mu$, and does not change much the condensate of particles. The anomalous propagator for particles is given by

$$< b_\rho^\dagger(k)b_{\rho'}^\dagger(-k) > = -\frac{1}{2}\sin(2\Theta_\rho(k))\exp(-i\xi_\rho(k)), \qquad (27)$$

where the average is calculated in $|0\rangle$ vacuum.

Integrating over $l = [0, 1/\lambda^2]$, where λ is the UV cut-off, and adding the self-energy instantaneous term comming from normal-ordering Eq. (3) in the $|0\rangle$ vacuum, we get

$$\Sigma_{qq}(\lambda) = \sum_{k,p,\rho,\rho'} \frac{1}{2}\lambda_{i\alpha}^\rho\lambda_{j\gamma}^\rho\frac{1}{2}\lambda_{i\beta}^{\rho'}\lambda_{j\delta}^{\rho'}T_{\alpha\beta}^A T_{\gamma\delta}^A < b_\rho^\dagger(k)b_\rho^\dagger(-k) > b_{\rho'}(p)b_{\rho'}(-p)$$

$$\left(V^{\rho\rho'}(k,p)(u^\dagger(k)\alpha^I u(p))(u^\dagger(-k)\alpha^J u(-p))D_{IJ}(k-p) \right.$$

$$\left. + V(k,p)(u^\dagger(k)u(p))(u^\dagger(-k)u(-p)) \right) R^{\rho\rho'}(k,p;\lambda) + c.c., \qquad (28)$$

where $V^{\rho,\rho'}$ and V are given in Eq. (19), and the the UV regulator is given by

$$R^{\rho\rho'}(k,p;\lambda) = \exp\left(-[(E^\rho(k) - E^{\rho'}(p))^2 + \omega_M(k-p)^2]/\lambda^2\right), \qquad (29)$$

Change in sign appear due to adopting conventional scaling up as in the perturbation theory, instead of scaling down in flow equations. Performing calculations with λ-matrices, we have

$$\Sigma_{qq} = \sum_{k,p}(-1)e^{-i\phi(p)}e^{i\phi(k)}\left(b_1(p)b_1(-p) < b_8^\dagger(k)b_8^\dagger(-k) > F^{81}(k,p) \right.$$

$$+ b_8(p)b_8(-p)[< b_1^\dagger(k)b_1^\dagger(-k) > F^{18}(k,p) - < b_8^\dagger(k)b_8^\dagger(-k) > 2F^{88}(k,p)] \left. \right)$$

$$\times R^{\rho\rho'}(k,p;\lambda) + c.c.. \qquad (30)$$

Comparing Eq. (30) with Eq. (10) for the gap functions, we obtain the system of gap equations

$$\Delta_1(p) = 8G^{81}(p)$$
$$\Delta_8(p) = G^{18}(p) - 2G^{88}(p), \qquad (31)$$

where

$$G^{\rho\rho'}(p) = -\frac{1}{4}\int \frac{dk}{(2\pi)^3}\frac{1}{2}\sin(2\Theta_\rho(k))F^{\rho\rho'}(k,p)R^{\rho\rho'}(k,p;\lambda), \qquad (32)$$

with $F^{\rho\rho'}$ is given by Eq. (22), and $R^{\rho\rho'}$ by Eq. (29).

4. Solving the system of gap equations

We solve approximately the system of gap equations Eq. (31) in two limiting cases of the Pipperd and London type superconductor. Magnetic interaction Eq. (23) is regulated by the term generated by flow equations, with dispersion $q \sim E$, and by the magnetic gluon polarization operator, Eq. (25), with dispersion $q \sim E^{1/3}$. When the first term dominates, superconductor is of London type, the second term-Pipperd type.

We take instead of the smooth regulator R, Eq. (29), a sharp cut-off, $|k - \mu| \le \delta$. Near the Fermi surface the integration measure is given by $\int dk = 2\pi\mu^2 \int_{-\delta}^{\delta} d(k-\mu) \int_{-1}^{1} d\cos\theta$, and since the integral is even in $(k-\mu)$ one has $\int_{\delta}^{\delta} \to 2\int_0^{\delta}$. In the denominator of F, Eq. (22), we approximate $(k - p)^2 = 2\mu^2(1 - \cos\theta)$.

Performing the θ-integration, we get for the Pipperd superconductor:

$$G^{\rho\rho'}(p) = -\frac{1}{4}\bar{g}^2 \int_0^{\delta} \frac{d(k-\mu)}{E_\rho(k)} \Delta_\rho(k) \frac{1}{2} \ln\left(\frac{b^2\mu^2}{|E(k)^2 - E(p)^2|}\right) , \quad (33)$$

where

$$\bar{g} = \frac{g}{3\sqrt{2\pi}} \; ; \; b_M = 32\pi \left(\frac{2}{N_f g^2}\right) \; , \; b_{M+E} = 256\pi^4 \left(\frac{2}{N_f g^2}\right)^{5/2} , \quad (34)$$

and superscript M and $M + E$ means that only magnetic or magnetic and electric components are taken. For the London superconductor:

$$G^{\rho\rho'}(p) = -\frac{1}{4}\bar{g}^2 \int_0^{\delta} \frac{d(k-\mu)}{E_\rho(k)} \Delta_\rho(k) \frac{1}{2} \ln\left(\frac{b^2\mu^2}{(E_\rho(k) - E_{\rho'}(p))^2}\right) , \quad (35)$$

where

$$\bar{g} = \frac{g}{\sqrt{6\pi}} \; ; \; b_M = 2 \; , \; b_{M+E} = 4\pi \left(\frac{2}{N_f g^2}\right)^{1/2} . \quad (36)$$

Substituting $G^{\rho\rho'}$ Eqs. (33,35) into Eq. (31), we have the system of Eliashberg type of equations. Different effective coupling $\bar{g}$ is obtained because of different dispersion law in the two cases. The coupling $\bar{g}$ defines the exponent of the solution for the gap Δ and b the preexponential factor.

If we are interested only in the correct exponential dependence, we can neglect ρ dependence by the quark energies in the anomalous quark propagator (sin factor in G) and in the gluon propagator (under the ln). Indeed, when estimating the momentum-independent gap, double logarithm arises from the region $\Delta \ll k \ll \mu$ where the integral reduces to

$$\Delta \sim \bar{g}^2 \int_{\Delta}^{\delta} \frac{d(k-\mu)}{(k-\mu)} \ln\left(\frac{b\mu}{|k-\mu|}\right) \Delta , \quad (37)$$

with the solution $\Delta \sim b\mu \exp(-\sqrt{2}/\bar{g})$; i.e. double logarithm responsible for the exponential solution does not depend on the quark energies. To calculate the preexponential factor correctly, one should make the rescaling in the integral as suggested by Schäfer [7].

In order to convert the integral gap equation into differential one, we split the logarithm in $G^{\rho\rho'}$ as suggested by Son [4]

$$\ln\left(\frac{b\mu}{E(p)}\right)\theta(p-k) + \ln\left(\frac{b\mu}{E(k)}\right)\theta(k-p), \tag{38}$$

that gives the same expression in two cases, Eq. (33) and Eq. (35). Coupled gap equations Eq. (31) decouple for the antitriplet and sixtet in color and flavor gaps, introduced after Eq. (8),

$$\Delta_{(\bar{3},\bar{3})}(p) = \bar{g}^2 \ln\left(\frac{b\mu}{E(p)}\right) \int_0^{(p-\mu)} \frac{d(k-\mu)}{E(k)} \Delta_{(\bar{3},\bar{3})}(k)$$

$$+ \bar{g}^2 \int_{(p-\mu)}^{\delta} \frac{d(k-\mu)}{E(k)} \ln\left(\frac{b\mu}{E(k)}\right) \Delta_{(\bar{3},\bar{3})}(k)$$

$$\Delta_{(6,6)}(p) = -\frac{\bar{g}^2}{2} \ln\left(\frac{b\mu}{E(p)}\right) \int_0^{(p-\mu)} \frac{d(k-\mu)}{E(k)} \Delta_{(6,6)}(k)$$

$$- \frac{\bar{g}^2}{2} \int_{(p-\mu)}^{\delta} \frac{d(k-\mu)}{E(k)} \ln\left(\frac{b\mu}{E(k)}\right) \Delta_{(6,6)}(k). \tag{39}$$

Introducing the variable, as by Pisarski and Rischke [5],

$$x = \ln\left(\frac{2b\mu}{p - \mu + E(p)}\right), \tag{40}$$

the integratiom measure is given by $d(k-\mu)/E(k) = dx/x$, and the integral equations, Eq. (39), reduce to differential equations

$$\frac{d^2}{dx^2}\Delta_{(\bar{3},\bar{3})} + \bar{g}^2 \Delta_{(\bar{3},\bar{3})} = 0$$

$$\frac{d^2}{dx^2}\Delta_{(6,6)} - \frac{\bar{g}^2}{2}\Delta_{(6,6)} = 0, \tag{41}$$

with initial conditions at the Fermi surface $d\Delta/dx(x = x_0) = 0$ and $\Delta(x = x_0) = \Delta_0$; also away from the Fermi surface $\Delta(x = 0) = 0$. The negative sign in the second equation means the repulsion in the sixtet channel. As function of momentum there is a trivial solution in the sixtet channel. In the antitriplet channel, the solution reads

$$\Delta_{(\bar{3},\bar{3})} = \Delta_0 \sin(\bar{g}x), \quad \Delta_0 = 2b\mu \exp\left(-\frac{\pi}{2\bar{g}}\right), \tag{42}$$

where $\bar{g}$ and b are given in Eqs. (34,36).

5. Conclusions

Using flow equations, we derived an effective quark-quark interaction and obtained the coupled set of gap equations for the condensates of the CFL phase of massless $N_f = 3$ dense QCD. Diquark interaction, generated dynamically by flow equations, is a long-range magnetic gluon exchange regulated by two different sources in the infrared region. One term describes the retardation effects of a magnetic gluon caused by Landau damping in dense quark-gluon plasma (normal phase). At small temperature, far below the melting of superconducting condensate $T \ll T_0$, Landau damping is the dominant mechanism in the infrared, that corresponds to the Pipperd type of superconductor. The other term describes retardation due to propagating of a magnetic gluon in a superconducting matter, i.e. through the multiple scattering of a gluon at diquark Cooper pairs. This mechanism is dominant at temperatures close to a melting point of superconducting condensate, $T - T_0 \ll T_0$, and corresponds to the London type of superconductor.

We obtain approximate analytical solutions of the gap equations in these two limiting cases. The dominant contribution to the condensate comes in the color antitriplet, flavor antitriplet channel. The color sextet, flavor sextet contribution is small but non-zero. In the color and flavor antisymmetric channel, we obtain parametric enhacement of the London type condensate

$$\Delta_0 \sim \exp(-\sqrt{\frac{3}{2}}\frac{\pi^2}{g})\,, \tag{43}$$

versus the Pipperd type condensate

$$\Delta_0 \sim \exp(-\frac{3}{\sqrt{2}}\frac{\pi^2}{g})\,. \tag{44}$$

The same conclusion holds for $N_f = 2$. Numerically, the $(\bar{3}, \bar{3})$ codensates are almost the same for the London and Pipperd type 2SC superconductors, $N_f = 2$. However, the condensate of the London type superconductor is sufficiently larger than of the Pipperd superconductor in the CFL phase, $N_f = 3$. This has important implications, which regime is energetically favorable and will actually be realized in neutron stars.

Independence of the gap on number of flavors for magnetic field and slow dependence on N_f for the sum of magnetic and electric fields in London type superconductors works in favor that the CFL phase will be observed in neutron stars, and that there is no window for the 2SC phase [13].

Our calculations have been done at zero temperature. Rigorous calculations of the condensates which include calculations of the penetration

316

depth of magnetic field and coherence length at non-zero temperature are necessary. Qualitatively it will not affect our results, since our conclusions rely on similar trends known in metal superconductors, which are derived from Landau-Ginzburg theory and are proven experimentally [12].

We included all possible condensates, $< qCq >$, $< qC\gamma_5 q >$ and $< qC\sigma_{\mu\nu}q >$, which do not mix left and right components, that accounted for the angle dependence in the gap equations. It is interesting to add the instanton-induced 't Hooft interaction and to consider simultaneously the chiral condensation in $\bar{q}q$ channel. Generalization to a non-zero strange quark mass, and adding kaon condensates is important for neutron star phenomenology.

References

1. B. C. Barrois, *Nucl. Phys.* **B129**, 390 (1977); S. C. Frautschi, in *Hadronic Matter at Extreme Energy Density*, edited by N. Cabibbo and L. Sertorio (Plenum, New York, 1980); D. Bailin and A. Love, *Phys. Rep.* **107**, 325 (1984).
2. M. G. Alford, K. Rajagopal and F. Wilczek, *Phys. Lett.* **B422**, 247 (1998); R. Rapp, T. Schäfer, E. V. Shuryak and M. Velkovsky, *Phys. Rev. Lett.* **81**, 53 (1998).
3. M. Alford, K. Rajagopal and F. Wilczek, *Nucl. Phys.* **B537**, 443 (1999);
4. D. T. Son, *Phys. Rev.* **D59**, 094019 (1999); R. D. Pisarski and D. H. Rischke, *Phys. Rev. Lett.* **83**, 37 (1999).
5. T. Schafer and F. Wilczek, *Phys. Rev.* **D60**, 114033 (1999); D. K. Hong, V. A. Miransky, I. A. Shovkovy and L. C. R. Wijewardhana, *ibid.* **D61**, 056001 (2000); **D62**, 059903(E) (2000); R. D. Pisarski and D. H. Rischke, *ibid.* **D61**, 051501 (2000); *ibid.* **D61** 074017 (2000).
6. S. D. Hsu and M. Schwetz, *Nucl. Phys.* **B572**, 211 (2000); W. E. Brown, J. T. Liu and H. C. Ren, *Phys. Rev.* **D61**, 114012 (2000).
7. K. Rajagopal and E. Shuster, *Phys. Rev.* **D62**, 085007 (2000); I. A. Shovkovy and L. C. R. Wijewardhana, *Phys. Lett.* **B470**, 189 (1999); T. Schäfer, *Nucl. Phys.* **B575**, 269 (2000).
8. E. Gubankova, hep-ph/0112213.
9. D. T. Son and M. A. Stephanov, *Phys. Rev.* **D61**, 074012 (2000); **D62**, 059902(E) (2000).
10. F. Wegner, *Ann. Phys. (Leipzig)* **3**, 77 (1994); *Physics Reports* **348**, 77 (2001).
11. P. Lenz and F. Wegner, cond-mat/9604087; for a recent work see I. Grote, E. Koerding, F. Wegner, cond-mat/0106604.
12. A. A. Abrikosov, L. P. Gorkov, and I. E. Dzyaloshinski, "Methods of quantum field theory in statistical physics," Dover publications, Inc., New York.
13. M. Alford and K. Rajagopal, *JHEP* **0206**, 031 (2002).

DOMAIN WALLS AND STRINGS IN DENSE QUARK MATTER

ARIEL R. ZHITNITSKY

Department of Physics and Astronomy, University of British Columbia,
Vancouver, BC V6T 1Z1, Canada

I discuss several types of domain walls and global strings which occur in colour superconducting quark matter due to the spontaneous violation of relevant $U(1)$ and discrete symmetries. These include the baryon $U(1)_B$, approximate axial $U(1)_A$ symmetries as well as an approximate $U(1)_Y$ symmetry arising from kaon condensation in colour-flavour locking phase. In this talk I concentrate on discussions of K strings due to their interesting internal structures. Specifically, I demonstrate that under some conditions the global $U(1)_Y$ symmetry may not be restored inside the string, in contrast with the standard expectations. Instead, K^+ condensation occurs inside the core of the string if a relevant parameter $\cos\theta_{K^0} \equiv m_{K^0}^2/\mu_{\text{eff}}^2$ is larger than some critical value $\theta_{K^0} \geq \theta_{\text{crit}}$. If this phenomenon happens, the $U(1)_Y$ strings become superconducting and may considerably influence the magnetic properties of dense quark matter, in particular in neutron stars.

1. Introduction

Domain walls and strings are common examples of topological defects which are present in various field theories. Domain walls are configurations of fields related to the nontrivial mapping $\pi_0(M)$, while topologically stable strings are due to the mapping $\pi_1(M)$ where M is manifold of degenerate vacuum states. While it is generally believed that there are no domain walls or strings in the Standard Model due to the triviality of the corresponding mappings, such objects may exist in extensions to the Standard Model, or in phases where the symmetries of the Standard Model are broken.

Topological defects that occur in extensions to the Standard Model may play an important role in cosmology, however, the focus of this talk will be strings and domain walls that exist in high density matter where the symmetries of the Standard Model are broken. As was recently discussed [1,2,3,4,5], in the regime of high baryon densities when the chemical potential μ is much larger than the QCD scale $\mu \gg \Lambda_{\text{QCD}}$, QCD supports domain walls and strings of different types.

There existence and long life time of the 2π-domain walls (such as $U(1)_A$

domain wall[1] or K domain wall[2]) is based on the following facts: 1) an explicit violation of the relevant symmetry such as $U(1)_A$ or $U(1)_Y$ is small; 2) the $U(1)_A$ or $U(1)_Y$ symmetry is spontaneously broken due to the diquark condensate or K condensate correspondingly; 3) the decay constants of the pseudoscalar Goldstone bosons are large and their masses are small at large μ.

There existence of topological string configurations is based on simple topological arguments which suggest that if some $U(1)$ symmetry is spontaneously broken, the global string would appear as the result of this symmetry breaking pattern. The important point is that the asymptotic freedom of QCD allows us to assert the existence of these domain walls and strings in the high baryon density regime, and the properties of these objects can be determined by controllable weak-coupling calculations. Indeed, in the last few years it has been demonstrated that the ground state of QCD at high density is unstable due to the formation of diquark condensate [6,7] (see [8] for a review). In this new ground state various symmetries, which were present at zero baryon density, are spontaneously broken. Specifically, I discuss here the three flavor $N_f = 3$ color flavor locking (CFL) phase. In this case both the exact $U(1)_B$ and the approximate axial $U(1)_A$ symmetries of QCD are spontaneously broken, leading to the formation of $U(1)_A$ and $U(1)_B$ global strings described in [3] and $U(1)_A$ domain wall described in[1]. If, in addition to the CFL phase, kaon condensation also occurs as argued in [9,10,11,12] then the hypercharge symmetry $U(1)_Y$ is also spontaneously broken. If this is indeed the case, one more type of domain wall[2] and string[3] related to the spontaneously broken $U(1)_Y$ global symmetry will also appear.

I can not review in this talk all topological defects I mentioned above due to the limited time; therefore, I refer to original literature for the detail discussions on this subject. Instead, in this talk I limit myself by describing only one specific topological object from the extended list of possible topological excitations described above. Namely, I will discuss $U(1)_Y$ strings which have very unusual properties. I follow[5] in this presentation.

It was demonstrated in[4] that the internal core structure of the $U(1)_Y$ string could be very different from the $U(1)_A$ and $U(1)_B$ global strings discussed earlier [3]. Namely, it was argued that the relevant symmetry may not be restored in the core: in most known cases, in particular in magnetic vortices in a conventional superconductor, the $U(1)$ symmetry is restored inside the core. If this is the case, the $U(1)_Y$ string becomes superconducting with the core having a K^+ condensate. The fact that such unusual behavior, in principle, may occur in the theory of superconducting cosmic strings and in quantum field theory in general, has been known for quite a while [13,14,15]. Even more than that, such a behavior has been observed experimentally when the laboratory experiments on ^3He provided us with strong evidence for defect core transition in the interior of vortices which appear in the superfluid $^3He - B$ phase (see [16] for a review). Still, such a behavior of the vortex core is considered as an exception rather than a common phenomenon in physics.

The goal of this talk is twofold: 1) understand the phenomenon of core transition in the interior of vortices qualitatively, using some analytical methods; 2) make quantitative estimates for phenomenologically relevant parameters in the CFL phase when the transition does occur.

The fact that there should be some kind of phase transition in the string core as a function of the external parameters can be understood from the following simple arguments. If the quark mass difference $m_d - m_u$ is relatively large, than there would be only a $U(1)$ (rather than $SU(2)_I$) symmetry to be broken. The standard topological arguments suggest that in this case, the K^0 string would be a topologically stable configuration with the restoration of the corresponding $U(1)$ symmetry inside the core (which is a typical situation). If $m_d - m_u$ is exactly zero, such that the isotopical $SU(2)_I$ symmetry is exact, then symmetry arguments suggest that both K^0 and K^+ fields condense, and no global stable strings are possible. From these two limiting cases, it is clear that there should be some intermediate region that somehow interpolates (as a function of $m_d - m_u$) between these two cases. Indeed, as we discuss below in detail, the way how this interpolation works is the following. For relatively large $m_d - m_u$ nothing unusual happens: the K^0 string has a typical behavior with K^0 condensation outside the core, and with restoration of the symmetry inside the core. At some finite magnitude of $m_d - m_u$, an instability arises through the condensation of K^+-field inside of the core of the string. As the magnitude of $m_d - m_u$ decreases, the size of the core becomes larger and larger with nonzero values of both K^0 and K^+ condensates inside the core. Finally, at $m_d = m_u$ the core of the string (with nonzero condensates K^0 and K^+) fills the entire space, in which case the meaning of the string is completely lost, and we are left with the situation when $SU(2)_I$ symmetry is exact: no stable strings are possible.

Given this argument, we would expect that there must be some transition region where the $SU(2)_I$ symmetry is broken to some degree below which K^+-condensation will occur inside the core. The point at which this occurs will be estimated in this letter. We will show that K^+-condensation only occurs above a certain point $\theta_{K^0} > \theta_{crit}$, where parametrically θ_{crit} is given by $\sin(\theta_{\text{crit}}/2) \simeq$ constant $(\Delta/m_s)\sqrt{(m_d - m_u)/m_s}$, with $\Delta \sim 100$ MeV being the superconducting gap and m_s is the strange quark mass.

This talk is organized as follows. I give a brief overview of the properties of the mixed CFL-K^0 phase of high density QCD in section 2. Section 3 will discuss the issue of the stability of global K^0 strings as a function of the parameter $(m_d - m_u)$. In this section I will calculate θ_{crit} where K^+ condensation occurs inside of the global strings. I end with concluding remarks.

2. The CFL+K^0 phase of high density QCD

It is well known that the ground state of $N_f = 3$, $N_c = 3$ QCD exhibits the Cooper pairing phenomenon as in conventional superconductivity [6,7]. The corresponding condensates in the CFL phase take the approximate form:

$$\langle q_{L\alpha}^{ia} q_{L\beta}^{jb} \rangle^* \sim \epsilon_{\alpha\beta\gamma}\epsilon^{ij}\epsilon^{abc}X_c^\gamma,$$
$$\langle q_{R\alpha}^{ia} q_{R\beta}^{jb} \rangle^* \sim \epsilon_{\alpha\beta\gamma}\epsilon^{ij}\epsilon^{abc}Y_c^\gamma, \tag{1}$$

where L and R represent left and right handed quarks, α, β, and γ are the flavor indices, i and j are spinor indices, a, b, and c are color indices, and X_c^γ and Y_c^γ are complex color-flavor matrices describing the Goldstone bosons. In order to describe the light degrees of freedom in a gauge invariant way, one introduces the color singlet field Σ

$$\Sigma_\gamma^\beta = XY^\dagger = \sum_c X_c^\beta Y_\gamma^{c*}, \tag{2}$$

such that the leading terms of the effective Lagrangian in terms of Σ take the form [10]

$$L_{\text{eff}} = \frac{f_\pi^2}{4}\text{Tr}\left[\nabla_0\Sigma\nabla_0\Sigma^\dagger - v_\pi^2\partial_i\Sigma\partial_i\Sigma^\dagger\right] + 2A\left[\det(M)\text{Tr}(M^{-1}\Sigma + h.c.\right],\tag{3}$$

$$\nabla_0\Sigma = \partial_0\Sigma + i\left(\frac{MM^\dagger}{2p_f}\right)\Sigma - i\Sigma\left(\frac{M^\dagger M}{2p_f}\right),$$

where the matrix $\Sigma = \exp(i\pi^a\lambda^a/f_\pi)$ describes the octet of Goldstone bosons with the $SU(3)$ generators λ^a normalized as $\text{Tr}(\lambda^a\lambda^b) = 2\delta^{ab}$. The quark mass matrix in Eq.(3) is given by $M = \text{diag}(m_u, m_d, m_s)$. Finally, we neglect the electromagnetic interactions in the expression (3) but keep the isospin violation $\sim (m_d - m_u)$, which is an appropriate approximation for the physically relevant case when the baryon density is not very high [12]. The constants f_π, v_π and A have been calculated in the leading perturbative approximation and are given by [17,18]:

$$f_\pi^2 = \frac{21 - 8\ln 2}{18}\frac{\mu^2}{2\pi^2}, \qquad v_\pi^2 = \frac{1}{3}, \qquad A = \frac{3\Delta^2}{4\pi^2}. \tag{4}$$

Recently, it was realized in [9,10,11,12] that the ground state of the theory may be different from the pure CFL phase for a physical value of the strange quark ($m_s \gg m_u, m_d$): condensation of the K^0 and K^+ mesons would lower the free energy of the system by reducing the strange quark content. Specifically, it has been argued that kaon condensation would occur in the CFL phase if $m_s \geq 60$ MeV. This means that $\Sigma = 1$ is no longer the global minimum of the free energy; instead, some rotated value of Σ describes the ground state in this case. In what follows we consider the realistic case when the isospin symmetry is not exact, $m_d > m_u$, such that K^0 condensation occurs. The appropriate expression for Σ describing the K^0 condensed ground state in this case can be parameterized as:

$$\Sigma = \begin{pmatrix} 1 & 0 & 0 \\ 0 & \cos\theta_{K^0} & \sin\theta_{K^0} e^{-i\phi} \\ 0 & -\sin\theta_{K^0} e^{i\phi} & \cos\theta_{K^0} \end{pmatrix}, \tag{5}$$

where ϕ describes the corresponding Goldstone mode and θ_{K^0} describes the strength of the kaon condensation with [10]:

$$\cos\theta_{K^0} = \frac{m_0^2}{\mu_{\text{eff}}^2}, \quad \mu_{\text{eff}} \equiv \frac{m_s^2}{2p_F}, \quad m_0^2 \equiv a m_u(m_d + m_s), \quad a = \frac{4A}{f_\pi^2} = \frac{3\Delta^2}{\pi^2 f_\pi^2}. \tag{6}$$

In order for this to be satisfied, we must have $m_0 < \mu_{\text{eff}}$. If kaon condensation occurs and $\theta_{K^0} \neq 0$, an additional $U(1)$ symmetry is broken. This brings us to the next section where will discuss the consequences of this symmetry breaking.

3. Global $K-$ strings

We follow our logic outlined in the Introduction and first consider global K^0 strings when they are topologically stable. This corresponds to the approximation when the splitting between K^0 and all other degrees of freedom is relatively large such that we can neglect in our effective description all fields except K^0. After that we analyze situation when the K^0 and K^+ masses are degenerate such that the K strings become unstable. Finally we introduce a small explicit isospin violation into our description $\sim (m_d - m_u)$ in order to analyze the stability/instability issue for this physically relevant case. The important global characteristic of the K^0 string, the string tension α, with logarithmic accuracy is determined by the pion decay constant $\alpha \sim f_\pi^2$ as discussed in [3], and it is not sensitive to the internal structure of the core. The subject of this talk is the analysis of the core structure of the K^0 strings.

3.1. *Topologically stable $U(1)$ strings in $CFL + K^0$ phase*

We start by considering the following effective field theory which describes a single complex K^0 field. This corresponds to the case of spontaneously broken $U(1)$ symmetry

$$L_{\text{eff}}(K^0) = |(\partial_0 + i\mu_{\text{eff}})K^0|^2 - v_\pi^2|\partial_i K^0|^2 - m_0^2|K^0|^2 - \lambda|K^0|^4. \quad (7)$$

We fix all parameters of this effective theory by comparing the amplitude of the K^0 field to the result obtained from the theory described in the previous section. We neglect all other degrees of freedom at this point (see the discussion at the end of this section). If $\mu_{\text{eff}} > m_0$ the kaon field acquires a nonzero vacuum expectation value $\langle K^0 \rangle = \eta/\sqrt{2}$ where

$$\eta^2 = \frac{\mu_{\text{eff}}^2 - m_0^2}{\lambda} = \frac{\mu_{\text{eff}}^2}{\lambda}(1 - \cos\theta_{K^0}). \quad (8)$$

For $\mu_{\text{eff}} > m_{K^0}$ it is more convenient to represent the effective Lagrangian in the familiar form of a Mexican hat type potential:

$$L_{\text{eff}} = |\partial_0 K^0|^2 - v_\pi^2|\partial_i K^0|^2 - \lambda\left(|K^0|^2 - \frac{\eta^2}{2}\right)^2. \quad (9)$$

This is a text-book Lagrangian with spontaneously broken global $U(1)$ symmetry which admits topologically stable global string solutions for which we will make the following ansatz, $K^0_{\text{string}} = \frac{\eta}{\sqrt{2}}f(r)e^{in\phi}$, where n is the winding number of the string (we will take $n = 1$ in what follows), ϕ is the azimuthal angle in cylindrical coordinates, and $f(r)$ is a yet to be determined profile function which obeys the following ordinary differential equation:

$$\frac{1}{r}\frac{d}{dr}\left(r\frac{df(r)}{dr}\right) - \frac{f(r)}{r^2} = \frac{\mu_{\text{eff}}^2(1 - \cos\theta_{K^0})}{v_\pi^2}(f(r)^2 - 1)f(r), \quad (10)$$

with the boundary conditions $f(0) = 0$ and $f(\infty) = 1$. A numerical solution of this equation can be easily found and confirms that the global string is stable and the $U(1)$ symmetry is restored inside of the core, $f(r = 0) = 0$, as expected.

3.2. *Unstable $SU(2)_I \times U(1)_Y \rightarrow U(1)$ strings in CFL/K^0 phase*

Our next task is an analysis of the K string when isospin is an exact symmetry (i.e. $m_u = m_d$), such that the symmetry pattern breaking is $SU(2)_I \times U(1)_Y \rightarrow U(1)$. In this case, based on topological arguments, we know that the K string is unstable. We want to analyze the stability issue in detail to understand this phenomenon on the quantitative level.

To simplify things we start with the effective Lagrangian which only includes a single complex kaon doublet $\Phi = (K^+, K^0)$. As discussed in [10], this is an appropriate approach to discuss kaon condensation in the CFL phase if all other degrees of freedom are much heavier. As before (see Eqs. (7,9)) we can represent the effective Lagrangian in the form of a Mexican hat type potential,

$$L_{\text{eff}}(\Phi) = |\partial_o \Phi|^2 - v_\pi^2 |\partial_i \Phi|^2 - \lambda \left(|\Phi|^2 - \frac{\eta^2}{2} \right)^2 . \tag{11}$$

All parameters here are defined in the same way as in (6,8,9) and the Φ field acquires a non-zero vacuum expectation value which we assume takes the form $\langle \Phi \rangle = (0, \frac{\eta}{\sqrt{2}})$. The time independent equations of motion for K^0 and K^+ are given by:

$$v_\pi^2 \nabla^2 K^+ = 2\lambda(|K^+|^2 + |K^0|^2 - \frac{\eta^2}{2})K^+, \tag{12}$$

$$v_\pi^2 \nabla^2 K^0 = 2\lambda(|K^+|^2 + |K^0|^2 - \frac{\eta^2}{2})K^0. \tag{13}$$

For the K^0 string solution we will make the following ansatz:

$$K^0_{\text{string}} = \frac{\eta}{\sqrt{2}} f(r) e^{in\phi}, \quad K^+ = 0, \tag{14}$$

such that $f(r)$ is the solution of Eq. (10) with the boundary conditions $f(0) = 0$ and $f(\infty) = 1$. Without calculations, from topological arguments we know that although the solution (14) satisfies the equation of motion (12, 13), it is an unstable solution. The source of instability can be seen as follows. We follow the standard procedure and expand the energy in the K^0 string background to quadratic order in K^+ and K^0 modes:

$$E(K^0 = K^0_{\text{string}} + \delta K^0, \quad K^+) \approx E_{\text{string}} + \delta E. \tag{15}$$

We know that the K^0 string itself is a stable configuration, therefore, the δK^0 modes cannot have negative eigenvalues which would correspond to the instability. Therefore, we concentrate only on the "dangerous" modes related to K^+ fluctuations, in which case δE is given by:

$$\delta E = \int d^2 r \left(v_\pi^2 |\nabla K^+|^2 + \mu_{\text{eff}}^2 (1 - \cos\theta_{K^0})(f(r)^2 - 1)|K^+|^2 \right), \tag{16}$$

where $f(r)$ is the solution of Eq. (10) with the boundary conditions $f(0) = 0$ and $f(\infty) = 1$. If δE is a positive quantity, then the K^0 string is absolutely stable and K^+ modes do not destroy the string configuration. If δE is negative, this means that this is a direction in configurational space where the K^0 string decays. The standard procedure is to expand K^+ field in Fourier modes:

$$K^+ = \sqrt{2} f_\pi \sum_m g_m(r) e^{im\phi}. \tag{17}$$

Now we have the K^+ field in terms of the dimensionless Fourier components $g_m(r)$. Setting $m = 0$ in the above expansion in order to analyze the lowest energy δE_0 contribution in (16), we arrive at:

$$\delta E_0 = 2 f_\pi^2 \int d^2 r \left(v_\pi^2 (\frac{\partial g_0(r)}{\partial r})^2 + \mu_{\text{eff}}^2 (1 - \cos\theta_{K^0}) \cdot (f^2(r) - 1) g_0^2(r) \right). \tag{18}$$

In order to have dimensionless coordinates and fields, we will perform the following change of variables, $\tilde{r} = \gamma r$, where $\gamma \equiv \mu_{\text{eff}}/v_\pi$. This change of variables sets the string width in $f(\tilde{r})$ to be $\tilde{r}_{\text{core}} \sim 1$. Equation (18) now reads:

$$\delta E_0 = 2 f_\pi^2 v_\pi^2 \int d^2\tilde{r}\, g_0(\tilde{r}) \hat{O} g_0(\tilde{r}), \quad \hat{O} = -\frac{1}{\tilde{r}} \frac{d}{d\tilde{r}} (\tilde{r} \frac{d}{d\tilde{r}}) + \lambda'(f^2(\tilde{r}) - 1), \tag{19}$$

where $\lambda' = (1 - \cos\theta_{K^0})$. The problem is reduced to the analysis of the two-dimensional Schrodinger equation for a particle in an attractive potential $V(r) = -(1 - \cos\theta_{K^0})(1 - f^2(\tilde{r}))$ with $f(\tilde{r})$ being the solution of Eq. (10) with the boundary conditions $f(0) = 0$ and $f(\infty) = 1$. Such a potential is negative everywhere and approaches zero at infinity. As is known from standard quantum mechanics [19], for an arbitrarily weak potential well there is always a negative energy bound state in one and two spatial dimensions; in three dimensions a negative energy bound state may not exist. For the two dimensional case (the relevant problem in our case) the lowest energy level of the bound state is always negative. As a consequence, the string (14) is an unstable solution of the classical equation of motion, the result we expected from the beginning from topological arguments. The instability manifests itself in the form of a negative energy bound state solution of the corresponding two-dimensional Schrodinger equation (19) irrespective of the magnitudes of the parameters.

3.3. K^+-condensation in the core of K^0 strings in $CFL + K^0$ phase

The issue of the stability or instability of K^0 strings reviewed in the previous section is highly sensitive to the degree of symmetry present in the Lagrangian describing the K^0/K^+ system. If the $SU(2)_I$ symmetry is strongly broken, the K^0 strings will be absolutely (topologically) stable as discussed in subsection A. If the $SU(2)_I$ symmetry remains unbroken, the K^0 strings will always be unstable as discussed in subsection B. Now we introduce an explicit symmetry breaking parameter δm^2 into (11) fixed by the original Lagrangian (3) such that our simplified version of the system (only $\Phi = (K^+, K^0)$ fields are taken into account) has the form

$$L_{\text{eff}}(\Phi) = |\partial_0 \Phi|^2 - v_\pi^2 |\partial_i \Phi|^2 - \lambda \left(|\Phi|^2 - \frac{\eta^2}{2} \right)^2 - \delta m^2 \Phi^\dagger \tau_3 \Phi, \qquad (20)$$

where $\delta m^2 \equiv \frac{a}{2} m_s (m_d - m_u)$ We anticipate that, as the symmetry breaking parameter δm^2 in Eq. (20) becomes sufficiently large, a stable K^0 string with restored $U(1)$ symmetry in the core, $f(r = 0) = 0$, must be reproduced, as discussed in subsection A. When the symmetry breaking parameter δm^2 in becomes sufficiently small, one should eventually reach a point where K^+ instability occurs, and it is energetically favorable for a K^+ condensate to be formed inside the core of the string.

In this subsection we calculate the critical value of θ_{K^0} when the K^+ instability occurs, and a K^+ condensate does form in the string core. In addition to this, we will obtain an estimate of the absolute value of the K^+-condensate at the center of the core of the string ($r = 0$). We treat K^+ field as a small perturbation in the K^0 background field. Such an approach is not appropriate when K^+-condensation is already well-developed in which case both fields K^0 and K^+ must be treated on the same footing. However, this approach is quite appropriate when one studies the transition from the phase where K^+ background field is zero to the region where it becomes nonzero.

To begin, we will expand the energy in the constant K^0 string background to quadratic order in K^+, i.e.: $E \approx E_{\text{string}} + \delta E$, where δE is given by

$$\delta E = \delta E_0 + 2 f_\pi^2 v_\pi^2 \int d^2 \bar{r} \, g_0(\bar{r})[\epsilon] g_0(\bar{r}), \quad \epsilon \equiv \frac{a}{2} \frac{m_s(m_d - m_u)}{\mu_{\text{eff}}^2}, \qquad (21)$$

with δE_0 defined by eq. (19). The only difference between this expression and Eq. (19) describing the instability of the string in case of exact isospin symmetry, is the presence of the term $\sim \epsilon$ in Eq. (21). The problem of determining when K^+-condensation occurs is now reduced to solving the Schrodinger type equation $\hat{O} g_0 = \hat{E} g_0$. From the previous discussions we know that $\hat{E}$ for the ground state is always negative. However, to insure the instability with respect to K^+-condensation one should require a relatively large negative value i.e. $\hat{E} + \epsilon < 0$. It can not happen for arbitrary weak coupling constant $\sim (1 - \cos \theta_{K^0})$ when θ_{K^0} is small. However, it does happen for relatively large θ_{K^0}. To calculate the minimal critical value

θ_{crit} when K^+-condensation develops, one should calculate the eigenvalue $\hat{E}$ as a function of parameter θ_{K^0} and solve the equation $\hat{E}(\theta_{\text{crit}}) + \epsilon = 0$. For very small coupling constant $\lambda' = (1 - \cos\theta_{K^0}) \to 0$ the bound state energy is negative and exponentially small, $\hat{E} \sim -e^{-\frac{1}{\lambda'}}$. However, for realistic parameters of μ, Δ, m_s, m_u, m_d the parameter ϵ is not very small and we expect that in the region relevant for us the bound state energy $\hat{E}$ is the same order of magnitude as the potential energy $\sim \lambda'$. In this case we estimate θ_{crit} from the following conditions $-\hat{E}(\theta_{\text{crit}}) \sim \lambda' \sim (1 - \cos\theta_{\text{crit}}) \sim \epsilon$ with the result which can be parametrically represented as

$$\sin\frac{\theta_{\text{crit}}}{2} \sim \text{const}\frac{\Delta}{\pi f_\pi}\sqrt{\frac{m_s(m_d - m_u)}{\mu_{\text{eff}}^2}} \sim \text{const}\frac{\Delta}{m_s}\sqrt{\frac{(m_d - m_u)}{m_s}}, \qquad (22)$$

where we have neglected all numerical factors in order to explicitly demonstrate the dependence of θ_{crit} on the external parameters. The limit of exact isospin symmetry, which corresponds to $m_d \to m_u$ when the string becomes unstable, can be easily understood from the expression (22). Indeed, in the case that the critical parameter $\theta_{\text{crit}} \to 0$ becomes an arbitrarily small number the K^+ instability would develop for arbitrarily small $\theta_{K^0} > 0$. The region occupied by the K^+ condensate at this point is determined by the behavior of lowest energy mode g_0 at large distances, $g_0(\tilde{r} \to \infty) \sim \exp(-\hat{E}\tilde{r})$ such that a typical $\tilde{r} \sim (m_d - m_u)^{-1} \to \infty$ as expected.

In order to make a quantitative, rather than qualitative estimation of the critical value θ_{crit} when $\hat{E}(\theta_{\text{crit}}) + \epsilon = 0$, we discretize the operator $\hat{O}$ and solve the problem numerically, with the boundary conditions $g_0(\infty) = 0$ and $g_0(0) = \text{constant}$. Varying the condensation angle θ_{K^0}, we see that a negative eigenvalue $\hat{E} + \epsilon < 0$ appears when $\theta_{K^0} > \theta_{\text{crit}} \approx 53°$. For our parameters we use $m_u = 5$ MeV, $m_d = 8$ MeV, $m_s = 150$ MeV, $\mu = 500$ MeV, and $\Delta = 100$ MeV which gives $\epsilon \equiv am_s(m_d - m_u)/(2\mu_{\text{eff}}^2) \simeq 0.1$.

All these discussions were quite general because they were based on the symmetry and topological properties of the system rather than on a specific form of the interaction. However, the numerical estimates presented above was derived by using a concrete form of the effective Lagrangian (20) describing the lightest K^+, K^0 degrees of freedom. The question arises how sensitive our numerical results are when the form of the potential changes. To formulate the question in a more specific way, let us remind the reader that, in general, the effective Lagrangian describing the Goldstone modes can be represented in many different forms as long as symmetry properties are satisfied. The results for the amplitudes describing the interaction of the Goldstone particles do not depend on a specific representation used. A well-known example of this fact is the possibility of describing the π meson properties by using a linear σ model as well as a non-linear σ model (and many other models which satisfy the relevant symmetry breaking pattern). The results remain the same if one discusses the local properties of the theory (such as $\pi\pi$ scattering length) when the π meson is considered as a small quantum fluctuation rather than a large background field . It may not be the case when π represents a large background field in which case some numerical difference between different representations of the effective Lagrangian may occur. Roughly speaking, the source of the difference is an inequality $\pi(x) \neq \sin\pi(x)$ for large global background fields such as a string solution which is the subject of this talk.

Having this in mind, we repeated similar numerical estimates discussed above for the original effective Lagrangian (3) where Goldstone fields represented in the exponential form rather than in form determined by the effective Lagrangian (20). As before, in these estimates we considered exclusively K^+ modes which are the energetically lowest modes and which can potentially destabilize the system presented by the background field of the K^0 string. This approximation is justified as long as the K^+ field is the lowest massive excitation in the system when a K^0 condensate develops. Also, the typical size of the core must be larger than the inverse gap Δ^{-1}, i.e. $r_{\text{core}} \simeq v_\pi/\mu_{\text{eff}} \gg \Delta^{-1}$ in order to maintain color superconductivity inside the core. We assume this is the case. Our numerical results suggest, that the critical parameters are not very sensitive to the specifics of the Lagrangian such that θ_{crit} is close to our previous numerical estimates. Therefore, most likely, the real world (with our parameters $\theta_{K^0} \simeq 70^o$) case corresponds to $\theta_{K^0} > \theta_{\text{crit}}$ and thus, a K^+ condensate does develop inside the core of the K^0 strings.

Conclusion

We have demonstrated that, within the CFL+K^0 phase of QCD, K^+ condensation does occur within the core of global K^0 strings if some conditions are met: $\theta_{K^0} > \theta_{\text{crit}}$. We presented two estimates for θ_{crit}: an analytical one which gives a qualitative explanation of the phenomenon, as well as numerical one for the physically relevant parameters realized in nature. Our results suggest that if a CFL+K^0 phase is realized in nature, it is likely that K^0 strings form together with K^+ condensation inside the core, in which case the strings become superconducting strings (see [13] for a more complete description of superconducting strings and their properties). As is known, the K string can not exist by itself, rather it is always attached to the K domain wall[2]. In such a case one should expect that a more

328

complicated object like a vorton[20] exists. A vorton which spins is a stationary ring of vortex whose string tension(as well as the tension due to the attached domain wall) gets balanced by angular momentum caused by trapped worldsheet charges. It remains to be seen whether such K vortons can have any impact on observable effects (they might, for example, affect glitches: sudden increases in the rotation frequency ω of neutron stars by as much as $\Delta\omega/\omega \sim 10^{-6}$, or the magnetic field structure and evolution). However, such speculations is subject for a different talk.

References

1. D. T. Son, M. A. Stephanov, and A. R. Zhitnitsky, Phys.Rev.Lett. **86**, 3955 (2001), [hep-ph/0012041]; Phys. Lett. **B510**, 167 (2001), [hep-ph/0103099].
2. D. T. Son, [hep-ph/0108260].
3. M. M. Forbes and A. R. Zhitnitsky, Phys. Rev. D **65**, 085009 (2002), [hep-ph/0109173].
4. D. B. Kaplan and S. Reddy, Phys. Rev. Lett. **88**, 132302 (2002) [arXiv:hep-ph/0109256].
5. K. B. W. Buckley and A. Zhitnitsky [hep-ph/0204064]
6. M. Alford, K. Rajagopal, and F. Wilczek, Phys. Lett. **B422**, 247 (1998); Nucl. Phys. **B537**, 443 (1999),
7. R. Rapp, T. Schäfer, E. V. Shuryak, and M. Velkovsky, Phys. Rev. Lett. **81**, 53 (1998); Annals Phys. **280**, 35 (2000).
8. K. Rajagopal and F. Wilczek, [hep-ph/0011333].
9. T. Schäfer, Phys.Rev.Lett. **85** 5531 (2000), [nucl-th/0007021].
10. P. F. Bedaque and T. Schäfer, Nucl. Phys. **A697**, 802 (2002), [hep-ph/0105150].
11. D. B. Kaplan and S. Reddy, Phys. Rev. **D65**, 054042 (2002), [hep-ph/0107265].
12. P. F. Bedaque, Phys.Lett. **B524**, 137 (2002), [nucl-th/0110049].
13. E. Witten, Nucl. Phys. **B249**, 557 (1985).
14. C. T. Hill, H. M. Hodges, and M. S. Turner, Phys. Rev. **D37**, 263 (1988).
15. M. Axenides and L. Perivolaropoulos, Phys.Rev. **D56**, 1973 (1997). [hep-ph/9702221]. M. Axenides, L. Perivolaropoulos, and M. Trodden, Phys.Rev. **D58**, 083505 (1998). [hep-ph/9801232]. B. Carter, R. H. Brandenberger, and A. C. Davis, [hep-ph/0201155].
16. M. M. Salomaa and G. E. Volovic, Rev. Mod. Phys. **59**, 533 (1987).
17. D. T. Son and M. A. Stephanov, Phys. Rev. **D61**, 074012 (2000), [hep-ph/9910491]. Erratum: Phys. Rev. **D62**, 059902 (2000), [hep-ph/0004095].
18. S. R. Beane, P. F. Bedaque, and M. J. Savage, Phys. Lett. **B483**, 131 (2000), [hep-ph/0002209].
19. L. D. Landau and E. M. Lifshitz, *Quantum mechanics, non-relativistic theory*, Pergamon Press, Oxford, 1977.
20. R.L.Davis, Phys. Rev. **D38**, 3722 (1988)

FINITE-TEMPERATURE BEHAVIOR OF THE (2+1)D GEORGI-GLASHOW MODEL WITH AND WITHOUT QUARKS

DMITRI ANTONOV

INFN-Sezione di Pisa, Università degli studi di Pisa, Dipartimento di Fisica,
Via Buonarroti, 2 - Ed. B - 56127 Pisa, Italy
E-mail: antonov@df.unipi.it

(2+1)-dimensional Georgi-Glashow model and its $SU(N)$-generalization are explored at nonzero temperatures and in the regime when the Higgs boson is not
infinitely heavy. The finiteness of the Higgs-boson mass leads to various novel effects. Those include the appearance of two separate phase transitions and of the
upper bound on the parameter of the weak-coupling approximation, necessary to
maintain the stochasticity of the Higgs vacuum. The modification of the finite-
temperature behavior of the model emerging due to the introduction of massless
quarks is also discussed.

1. Introduction. The model

Since the second half of the seventies [1], (2+1)-dimensional Georgi-
Glashow model is known as an example of the theory allowing for an
analytic description of confinement. However, confinement in the Georgi-
Glashow model is typically discussed in the limit of infinitely large Higgs-
boson mass, when the model is reduced to compact QED. In the present
talk, we shall review various novel effects stemming from the finiteness of
that mass. The main emphasis in this discussion will be payed to the modification of the finite-temperature properties of the Georgi-Glashow model,
as well as of its $SU(N)$-generalization.

The Euclidean action of the (2+1)D Georgi-Glashow model reads

$$S = \int d^3x \left[\frac{1}{4g^2} \left(F^a_{\mu\nu} \right)^2 + \frac{1}{2} \left(D_\mu \Phi^a \right)^2 + \frac{\lambda}{4} \left((\Phi^a)^2 - \eta^2 \right)^2 \right]. \quad (1)$$

Here, the Higgs field Φ^a transforms by the adjoint representation, $D_\mu \Phi^a \equiv \partial_\mu \Phi^a + \varepsilon^{abc} A^b_\mu \Phi^c$. Next, λ is the Higgs coupling constant of dimensionality
[mass], η is the Higgs v.e.v. of dimensionality [mass]$^{1/2}$, and g is the elec-
tric coupling constant of the same dimensionality. At the one-loop level,

the sector of the theory (1) containing dual photons and Higgs bosons is represented by the following partition function [2]:

$$\mathcal{Z} = 1 + \sum_{N=1}^{\infty} \frac{\zeta^N}{N!} \left(\prod_{i=1}^{N} \int d^3 z_i \sum_{q_i = \pm 1} \right) \times$$

$$\times \exp \left[-\frac{g_m^2}{8\pi} \sum_{\substack{a,b=1 \\ a \neq b}}^{N} \left(\frac{q_a q_b}{|\mathbf{z}_a - \mathbf{z}_b|} - \frac{e^{-m_H |\mathbf{z}_a - \mathbf{z}_b|}}{|\mathbf{z}_a - \mathbf{z}_b|} \right) \right] \equiv \int \mathcal{D}\chi \mathcal{D}\psi e^{-S}, \quad (2)$$

where

$$S = \int d^3 x \left[\frac{1}{2}(\partial_\mu \chi)^2 + \frac{1}{2}(\partial_\mu \psi)^2 + \frac{m_H^2}{2}\psi^2 - 2\zeta e^{g_m \psi} \cos(g_m \chi) \right] \equiv$$

$$\equiv \int d^3 x \mathcal{L}[\chi, \psi | g_m, \zeta]. \quad (3)$$

The partition function (2) describes the grand canonical ensemble of monopoles with the account for their Higgs-mediated interaction. In Eqs. (2) and (3), χ is the dual-photon field, and the field ψ accounts for the Higgs field, whose mass reads $m_H = \eta\sqrt{2\lambda}$. Note that from Eq. (2) it is straightforward to deduce that when m_H formally tends to infinity, one arrives at the conventional sine-Gordon theory of the dual-photon field [1] describing the compact-QED limit of the model. Next, in the above equations, g_m stands for the magnetic coupling constant related to the electric one as $g_m g = 4\pi$, and the monopole fugacity ζ has the form $\zeta = \frac{m_W^{7/2}}{g}\delta\left(\frac{\lambda}{g^2}\right)e^{-4\pi m_W \epsilon/g^2}$. In this formula, $m_W = g\eta$ is the W-boson mass, and $\epsilon = \epsilon(\lambda/g^2)$ is a certain monotonic, slowly varying function, $\epsilon \geq 1$, $\epsilon(0) = 1$ [3], $\epsilon(\infty) \simeq 1.787$ [4]. As far as the function δ is concerned, it is determined by the loop corrections. In what follows, we shall work in the standard weak-coupling regime $g^2 \ll m_W$, which parallels the requirement that η should be large enough to ensure the spontaneous symmetry breaking from $SU(2)$ to $U(1)$. The W-boson mass will thus play the role of the UV cutoff in the further analysis.

2. The model at finite temperature beyond the compact-QED limit

In the discussion of finite-temperature properties of the Georgi-Glashow model in the present and next Sections, we shall mainly follow Ref. [5].

At finite temperature $T \equiv 1/\beta$, one should supply the fields χ and ψ with the periodic boundary conditions in the temporal direction, with the period equal to β. Because of that, the lines of magnetic field emitted by a monopole cannot cross the boundary of the one-period region and consequently, at the distances larger than β, should go almost parallel to this boundary, approaching it. Therefore, monopoles separated by such distances interact via the 2D Coulomb potential, rather than the 3D one. Since the average distance between monopoles in the plasma is of the order $\zeta^{-1/3}$, we see that at $T \gtrsim \zeta^{1/3}$, the monopole ensemble becomes two-dimensional. Owing to the fact that ζ is exponentially small in the weak-coupling regime under discussion, the idea of dimensional reduction is perfectly applicable at the temperatures of the order of the critical temperature of the Berezinsky-Kosterlitz-Thouless (BKT) phase transition [6] (see e.g. Ref. [7] for a review) in the monopole plasma, which is equal to $g^2/2\pi$ [8] [a]. Up to exponentially small corrections, this temperature is unaffected by the finiteness of the Higgs-boson mass. This can be seen from the expression for the mean squared separation in the monopole-antimonopole molecule,

$$\langle L^2 \rangle = \frac{\int\limits_{|\mathbf{x}|>m_W^{-1}} d^2\mathbf{x}|\mathbf{x}|^{2-\frac{8\pi T}{g^2}} \exp\left[\frac{4\pi T}{g^2}K_0\left(m_H|\mathbf{x}|\right)\right]}{\int\limits_{|\mathbf{x}|>m_W^{-1}} d^2\mathbf{x}|\mathbf{x}|^{-\frac{8\pi T}{g^2}} \exp\left[\frac{4\pi T}{g^2}K_0\left(m_H|\mathbf{x}|\right)\right]},$$

where K_0 denotes the modified Bessel function. Disregarding the exponential factors in the numerator and denominator of this equation, we obtain $\langle L^2 \rangle \simeq \frac{4\pi T - g^2}{2m_W^2(2\pi T - g^2)}$, that yields the above-mentioned value of the BKT critical temperature $g^2/2\pi$. Besides that, it is straightforward to see that in the weak-coupling regime under study, the value of $\sqrt{\langle L^2 \rangle}$ is exponentially smaller than the characteristic distance in the monopole plasma, $\zeta^{-1/3}$, i.e., molecules are very small-sized with respect to that distance.

The factor β at the action of the dimensionally-reduced theory, $S_{\mathrm{d.-r.}} = \beta \int d^2x \mathcal{L}[\chi, \psi|g_m, \zeta]$, can be removed [and this action can be cast to the original form of eq. (3) with the substitution $d^3x \to d^2x$] by the obvious rescaling: $S_{\mathrm{d.-r.}} = \int d^2x \mathcal{L}\left[\chi^{\mathrm{new}}, \psi^{\mathrm{new}}|\sqrt{K}, \beta\zeta\right]$. Here, $K \equiv g_m^2 T$, $\chi^{\mathrm{new}} = \sqrt{\beta}\chi$, $\psi^{\mathrm{new}} = \sqrt{\beta}\psi$, and in what follows we shall denote for brevity χ^{new}

---

[a]Note that due to the T-dependence of the strength of the monopole-antimonopole interaction, which is a consequence of the dimensional reduction, the BKT phase transition in the 3D Georgi-Glashow model is inverse with respect to the standard one of the 2D XY model. Namely, monopoles exist in the plasma phase at the temperatures below the BKT critical one and in the molecular phase otherwise.

and ψ^{new} simply as χ and ψ, respectively. Averaging then over the field ψ with the use of the cumulant expansion we arrive at the following action:

$$S_{\text{d.-r.}} \simeq \int d^2x \left[\frac{1}{2}(\nabla\chi)^2 - 2\xi \cos\left(g_m\sqrt{T}\chi\right) \right] -$$

$$-2\xi^2 \int d^2x d^2y \cos\left(\sqrt{K}\chi(\mathbf{x})\right) \mathcal{K}^{(2)}(\mathbf{x}-\mathbf{y}) \cos\left(\sqrt{K}\chi(\mathbf{y})\right). \quad (4)$$

In this expression, we have disregarded all the cumulants higher than the quadratic one, and the limits of applicability of this so-called bilocal approximation will be discussed below. Further, in Eq. (4), $\mathcal{K}^{(2)}(\mathbf{x}) \equiv e^{K D_{m_H}^{(2)}(\mathbf{x})} - 1$, where $D_{m_H}^{(2)}(\mathbf{x}) \equiv K_0(m_H|\mathbf{x}|)/2\pi$ is the 2D Yukawa propagator, and $\xi \equiv \beta\zeta e^{\frac{K}{2}D_{m_H}^{(2)}(0)}$ denotes the monopole fugacity modified by the interaction of monopoles via the Higgs field. Clearly, in the compact-QED limit (when m_H formally tends to infinity) $D_{m_H}^{(2)}(0)$, being equal to $\int \frac{d^2p}{(2\pi)^2} \frac{1}{p^2+m_H^2}$, vanishes already before doing the integration, and $\xi \to \beta\zeta$, as it should be. In the general case, when the mass of the Higgs field is moderate and does not exceed m_W, we obtain

$$\xi \propto \exp\left[-\frac{4\pi}{g^2}\left(m_W\epsilon + T\ln\left(\frac{e^\gamma}{2}c\right) \right) \right].$$

Here, we have introduced the notation $c \equiv m_H/m_W$, $c < 1$, and $\gamma \simeq 0.577$ is the Euler constant, so that $\frac{e^\gamma}{2} \simeq 0.89 < 1$. We see that the modified fugacity remains exponentially small, provided that

$$T < -\frac{m_W\epsilon}{\ln\left(\frac{e^\gamma}{2}c\right)}. \quad (5)$$

This constraint should be updated by another one, which would provide the convergence of the cumulant expansion applied in course of the average over ψ. Were the cumulant expansion divergent, this fact would signal that the Higgs vacuum loses its normal stochastic properties and becomes a coherent one. In order to get the new constraint, notice that the parameter of the cumulant expansion reads $\xi I^{(2)}$, where $I^{(2)} \equiv \int d^2x \mathcal{K}^{(2)}(\mathbf{x})$. Evaluation of the integral $I^{(2)}$ yields [5]:

$$I^{(2)} \simeq \frac{2\pi}{m_H^2} \left[\frac{1}{2}\left(c^2 - 1 + \left(\frac{2}{e^\gamma}\right)^{\frac{8\pi T}{g^2}} \frac{1 - c^{2-\frac{8\pi T}{g^2}}}{1 - \frac{4\pi T}{g^2}} \right) + e^{\frac{a}{e}} - 1 + \frac{a}{e} \right]. \quad (6)$$

(Note that at $T \to g^2/4\pi$, $\frac{1-c^{2-\frac{8\pi T}{g^2}}}{1-\frac{4\pi T}{g^2}} \to -2\ln c$, i.e., $I^{(2)}$ remains finite.)
In the derivation of this expression, the parameter $a \equiv 4\pi\sqrt{2\pi T}/g^2$ was

assumed to be of the order of unity. That is because the temperatures we are working at are of the order of the BKT critical one, $g^2/2\pi$. Due to the exponential term in Eq. (6), the violation of the cumulant expansion may occur at high enough temperatures [that parallels the above-obtained constraint (5)]. The most essential, exponential, part of the parameter of the cumulant expansion thus reads

$$\xi I^{(2)} \propto \exp\left[-\frac{4\pi}{g^2}\left(m_W\epsilon + T\ln\left(\frac{e^\gamma}{2}c\right) - T\frac{\sqrt{2\pi}}{e}\right)\right].$$

Therefore, the cumulant expansion converges at the temperatures obeying the inequality

$$T < \frac{m_W\epsilon}{\frac{\sqrt{2\pi}}{e} - \ln\left(\frac{e^\gamma}{2}c\right)},$$

which updates the inequality (5). On the other hand, since we are working in the plasma phase, i.e., $T \leq g^2/2\pi$, it is enough to impose the following upper bound on the parameter of the weak-coupling approximation, g^2/m_W:

$$\frac{g^2}{m_W} < \frac{2\pi\epsilon}{\frac{\sqrt{2\pi}}{e} - \ln\left(\frac{e^\gamma}{2}c\right)}.$$

Note that although this inequality is satisfied automatically at $\frac{e^\gamma}{2}c \sim 1$, since it then takes the form $\frac{g^2}{m_W} < \sqrt{2\pi}e\epsilon$, this is not so for the Bogomolny-Prasad-Sommerfield (BPS) limit, $c \ll 1$. Indeed, in such a case, we have $\frac{g^2}{m_W}\ln\left(\frac{2}{ce^\gamma}\right) < 2\pi\epsilon$, that owing to the logarithm is however quite feasible.

3. Critical temperatures of the deconfining phase transition

The deconfining phase transition in the model under study occurs when the density of monopoles becomes equal to the one of W-bosons. Up to inessential subleading corrections, it takes place when the exponent of the monopole fugacity is equal to that of the fugacity of W-bosons [9]. [Another way to understand why the phase transition occurs when the two fugacities are equal to each other is to notice that once this happens, the thickness of the string confining two W's (which is proportional to $\xi^{-1/2}$) becomes equal to the average distance between the W's (proportional to (fugacity of W's)$^{-1/2}$). This qualitative result was also confirmed by the RG analysis performed in Ref. [9].] The density of W's approximately equals (see Ref. [5] for details) $\frac{3m_W T}{\pi}e^{-m_W \beta}$, where we have taken into account that the temperatures of our interest are much smaller than m_W in the weak-coupling regime, since they should not exceed $g^2/2\pi$. Then, in the compact-QED limit, $\xi \to \beta\zeta$ and $T_c = \frac{g^2}{4\pi\epsilon(\infty)}$ [9]. In the general case under discussion, $c < 1$, we obtain the two following distinct values of critical temperatures:

$$T_{1,2} = g^2\epsilon\frac{1 \pm \sqrt{1 - \frac{b}{\pi\epsilon^2}}}{2b}. \tag{7}$$

Here, $b \equiv -\frac{g^2}{m_W}\ln\left(\frac{e^\gamma}{2}c\right)$, $b > 0$, and the indices 1,2 refer to the smaller and the larger temperatures, respectively. The degenerate situation $T_1 = T_2 = g^2/2\pi\epsilon$ then corresponds to $b = \pi\epsilon^2$, and, since $\epsilon \geq 1$, $T_{1,2} \leq g^2/2\pi$ in this case, as it should be. In particular, in the BPS limit, $\epsilon = 1$, and the deconfining phase transition takes place together with the monopole BKT one. Obviously, at any other $b < \pi\epsilon^2$, $T_1 \neq T_2$, i.e., there exist two separate phase transitions. (Note that the existence of the upper bound for b is quite natural, since in the weak-coupling regime and aside from the BPS limit, b is definitely bounded from above.) The existence of two phase transitions means that at $T = T_1$, molecules of W-bosons start dissociating, while at $T = T_2$, this process is completed. In another words, accounting for the interaction of monopoles via the Higgs field opens a possibility for the existence of a new (metastable) phase at $T \in (T_1, T_2)$. This is the phase, where both the gas of W-molecules and W-plasma are present.

An elementary analysis shows that for $\pi(2\epsilon - 1) < b \leq \pi\epsilon^2$, $T_2 < g^2/2\pi$ [and $T_2 = g^2/2\pi$ at $b = \pi(2\epsilon - 1)$]. At the values of b lying in this interval, the phase transition corresponding to the critical temperature T_2 thus may occur. In the BPS limit, T_2 can only be equal to $g^2/2\pi$, that corresponds to the above-discussed case when both critical temperatures coincide with the one of the monopole BKT phase transition. In the same way, for any $b \leq \pi\epsilon^2$, $T_1 \leq g^2/2\pi$, and, in particular, $T_1 = g^2/2\pi$ only in the BPS limit, when $\epsilon = 1$. Therefore, the phase transition corresponding to the temperature T_1 always takes place. Also an elementary analysis shows that for any $\epsilon > 0$ (and, in particular, for the realistic values $\epsilon \geq 1$) and $b < \pi\epsilon^2$, $T_1 > g^2/4\pi\epsilon$ (and consequently $T_2 > g^2/4\pi\epsilon$ as well). Since $\epsilon < \epsilon(\infty)$, we conclude that both phase transitions always occur at the temperatures which are larger than that of the phase transition in the compact-QED limit.

Obviously, the RG analysis, performed in Ref. [9] for the compact-QED limit remains valid, but with the replacement $\beta\zeta \to \xi$. In particular, the deconfining phase transition corresponds again to the IR unstable fixed point, where the exponents of the W-fugacity $\mu \propto e^{-mw\beta}$ and of the monopole fugacity ξ become equal to each other [that yields the above-obtained critical temperatures (7)]. One can further see that the initial condition $\mu_{in} < \xi_{in}$ takes place, provided that the initial temperature, T_{in}, is either smaller than T_1 or lies between T_2 and $g^2/2\pi$. For these ranges of T_{in}, the temperature starts decreasing according to the RG equation $dt/d\lambda = \pi^2 \bar{a}^4 \left(\mu^2 - t^2\xi^2\right)$. In this equation, $t = 4\pi T/g^2$, λ is the evolution parameter, $\bar{a}$ is some parameter of the dimensionality [length], and for the comparison of μ and ξ the preexponent t^2 is again immaterial. Then, in the case $T_{in} < T_1$, the situation is identical to the one discussed in Ref. [9], namely μ becomes irrelevant and decreases to zero. Indeed, from the evolution equation for μ, the following equation for $d\mu/dt$, by virtue of which one can determine the sign of this quantity, stems:

$$\frac{d\mu}{dt} = \frac{\mu \left(2 - \frac{1}{t}\right)}{\pi^2 \bar{a}^4 \left(\mu^2 - t^2\xi^2\right)}.$$

One can see from this equation that if the evolution starts at $T_{in} \in (g^2/8\pi, T_1)$, μ temporaly increases until the temperature is not equal to $g^2/8\pi$, but then it nevertheless starts vanishing together with the temperature. However, by virtue of the same evolution equations we see that at $T_{in} \in (T_2, g^2/2\pi)$, the situation is now different. Indeed, in that case, μ is not decreasing, but rather increasing with the decrease of the temperature (since $d\mu/dt < 0$ at $T > T_2$), until it reaches some value $\mu_* \sim e^{-mw/T_2}$. Once we are in the region $T < T_2$, the temperature starts increasing again, that together with the change of the sign of $d\mu/dt$ leads to the increase of μ, and so on. Thus, we see that μ_* is the stable local maximum of μ for such initial conditions.

4. Including massless quarks

Let us consider the extension of the model (1) by the fundamental dynamical quarks, which are supposed to be massless: $\Delta S = -i \int d^3 x \bar{\psi} \vec{\gamma} \vec{D} \psi$. In this formula, $D_\mu \psi = \left(\partial_\mu - ig\frac{\tau^a}{2} A_\mu^a\right)\psi$, $\bar{\psi} = \psi^\dagger \beta$, where the Euclidean Dirac matrices are defined as $\vec{\gamma} = -i\beta\vec{\alpha}$ with $\beta = \begin{pmatrix} 1 & 0 \\ 0 & -1 \end{pmatrix}$ and $\vec{\alpha} = \begin{pmatrix} 0 & \vec{\tau} \\ \vec{\tau} & 0 \end{pmatrix}$. Our discussion in the present Section will further follow Ref. [10]. In that paper, it has been shown that at the temperatures higher than the BKT one, quark zero modes in the monopole field lead to the additional attraction between a monopole and an antimonopole in the molecule. In particular, when the number of these modes (equal to the number of massless flavors) is sufficiently large, the molecule shrinks so that its size becomes of the order of the inverse W-boson mass. Another factor which determines the size of the molecule is the characteristic range of localization of zero modes. Namely, it can be shown that the stronger zero modes are localized in the vicinity of the monopole center, the smaller molecular size is. Let us consider the case when the Yukawa coupling of quarks with the Higgs field vanishes, and originally massless quarks do not acquire any mass. This means that zero modes are maximally delocalized. We shall see that in the case of one flavor, such a weakness of the quark-mediated interaction of monopoles opens a possibility for molecules to undergo the phase transition into the plasma phase at the temperature comparable with the BKT one.

It is a well known fact that in 3D, 't Hooft-Polyakov monopole is actually an instanton [1]. Owing to this fact, we can use the results of Ref. [11] on the quark contribution to the effective action of the instanton-antiinstanton molecule in QCD. Referring the reader for details to Ref. [10], we shall present here the final expression for the effective action, which reads $\Gamma = 2N_f \ln|a|$. Here, $a = \left\langle \psi_0^{\bar{M}} \left| g\vec{\gamma}\frac{\tau^a}{2}\vec{A}^{a\,M} \right| \psi_0^M \right\rangle$ is the matrix element of the monopole field $\vec{A}^{a\,M}$ taken between the zero modes $\left|\psi_0^M\right\rangle$, $\left|\psi_0^{\bar{M}}\right\rangle$ of the operator $-i\vec{\gamma}\vec{D}$ defined at the field of a monopole and an antimonopole, respectively. The dependence of $|a|$ on the molecular size R can be straightforwardly found and reads $|a| \propto \int d^3r / \left(r^2 \left|\vec{r} - \vec{R}\right| \right) = -4\pi\ln(\mu R)$, where μ stands for the IR cutoff.

At finite temperature, in the dimensionally-reduced theory, the usual Coulomb interaction of monopoles [b] $R^{-1} = \sum\limits_{n=-\infty}^{+\infty} \left(\mathcal{R}^2 + (\beta n)^2\right)^{-1/2}$, i.e., it goes over into $-2T\ln(\mu\mathcal{R})$, where $\mathcal{R}$ denotes the absolute value of the 2D vector $\vec{\mathcal{R}}$. As far as the novel logarithmic interaction, $\ln(\mu R) = \sum\limits_{n=-\infty}^{+\infty} \ln\left[\mu\left(\mathcal{R}^2 + (\beta n)^2\right)^{1/2}\right]$, is concerned, it transforms into $\pi T\mathcal{R} + \ln\left[1 - \exp(-2\pi T\mathcal{R})\right] - \ln 2$. Accounting for both interactions, we eventually arrive at the following expression for the mean squared separation $\left\langle L^2 \right\rangle$ in the molecule as a function of T, g, and N_f:

---

[b]Without the loss of generality, we consider the molecule with the temporal component of $\vec{R}$ equal to zero.

$$\langle L^2 \rangle = \frac{\int\limits_{m_W^{-1}}^{\infty} d\mathcal{R}\mathcal{R}^{3-\frac{8\pi T}{g^2}} \left[\pi T\mathcal{R} + \ln\left[1 - \exp(-2\pi T\mathcal{R})\right] - \ln 2\right]^{-2N_f}}{\int\limits_{m_W^{-1}}^{\infty} d\mathcal{R}\mathcal{R}^{1-\frac{8\pi T}{g^2}} \left[\pi T\mathcal{R} + \ln\left[1 - \exp(-2\pi T\mathcal{R})\right] - \ln 2\right]^{-2N_f}}.$$

At large $\mathcal{R}$, $\ln 2 \ll \pi T\mathcal{R}$ and $\left|\ln\left[1 - \exp(-2\pi T\mathcal{R})\right]\right| \simeq \exp(-2\pi T\mathcal{R}) \ll \pi T\mathcal{R}$. Hence, we see that $\langle L^2 \rangle$ is finite at $T > (2 - N_f)g^2/4\pi$, that reproduces the standard result $g^2/2\pi$ at $N_f = 0$. For $N_f = 1$, the plasma phase is still present at $T < g^2/4\pi$, whereas for $N_f \geq 2$ the monopole ensemble may exist only in the molecular phase at any temperature larger than $\zeta^{1/3}$. Clearly, at $N_f \gg \max\left\{1, 4\pi T/g^2\right\}$, $\sqrt{\langle L^2 \rangle} \to m_W^{-1}$, which means that such a large number of zero modes shrinks the molecule to the minimal admissible size.

Let us finally comment on what happens to the real deconfining phase transitions at $N_f = 1$, if the Higgs boson is not infinitely heavy. Comparing the above-obtained critical temperatures (7) with the BKT critical temperature $g^2/4\pi$, we see that T_2 is always larger than $g^2/4\pi$, so that the second phase transition always takes place at $T = g^2/4\pi$. As far as the first phase transition is concerned, one can see that $T_1 < g^2/4\pi$ at $b < 4\pi(\epsilon - 1)$ for any $\epsilon \in [1, \epsilon(\infty)]$, and $T_1 = g^2/4\pi$ at $b = 4\pi(\epsilon - 1)$.

5. Properties of the $SU(N)$-case

In this Section, we shall discuss some results of Ref. [12] which concern the peculiarities of the $SU(N)$-case. The $SU(N)$-generalization of the action (3), stemming from the $SU(N)$ Georgi-Glashow model, has the form

$$S = \int d^3x \left[\frac{1}{2}(\nabla\vec{\chi})^2 + \frac{1}{2}(\nabla\psi)^2 + \frac{m_H^2}{2}\psi^2 - 2\zeta e^{g_m\psi}\sum_i \cos\left(g_m\vec{q}_i\vec{\chi}\right)\right]. \quad (8)$$

Here, $\sum\limits_i \equiv \sum\limits_{i=1}^{N(N-1)/2}$, and $\vec{q}_i$'s are the positive root vectors of the group $SU(N)$. As well as the field $\vec{\chi}$, these vectors are $(N-1)$-dimensional. Note that the $SU(3)$-version of the action (8), which incorporates the effects of the Higgs field, has been discussed in Ref. [13]. The compact-QED limit of the $SU(N)$-case has been studied in Refs. [14] and [15]. The string representation of the compact-QED limit has been explored for the $SU(3)$-case in Ref. [16]. Here, similarly to all the above-mentioned papers, we assume that W-bosons corresponding to different root vectors have the same masses.

338

Straightforward integration over ψ then yields the following equation:

$$S \simeq \int d^3x \left[\frac{1}{2}(\nabla\vec{\chi})^2 - 2\bar{\zeta}\sum_i \cos\left(g_m\vec{q}_i\vec{\chi}\right) \right] -$$

$$-2\bar{\zeta}^2 \int d^3x d^3y \sum_{i,j} \cos\left(g_m\vec{q}_i\vec{\chi}(\mathbf{x})\right) \mathcal{K}^{(3)}(\mathbf{x} - \mathbf{y}) \cos\left(g_m\vec{q}_j\vec{\chi}(\mathbf{y})\right). \qquad (9)$$

In this equation, $\mathcal{K}^{(3)}(\mathbf{x}) \equiv e^{g_m^2 D_{m_H}^{(3)}(\mathbf{x})} - 1$, where $D_{m_H}^{(3)}(\mathbf{x}) \equiv e^{-m_H|\mathbf{x}|}/(4\pi|\mathbf{x}|)$ stands for the Higgs-field propagator, and $\bar{\zeta} \equiv \zeta e^{\frac{g_m^2}{2} D_{m_H}^{(3)}(0)}$ denotes the modified fugacity. The square of the Debye mass of the field $\vec{\chi}$ can be derived from Eq. (9) by virtue of the formula $\sum_i q_i^\alpha q_i^\beta = \frac{N}{2}\delta^{\alpha\beta}$ and reads $m_D^2 = g_m^2\bar{\zeta}N\left[1 + \bar{\zeta}I^{(3)}N(N-1)\right]$, where $I^{(3)} \equiv \int d^3x\mathcal{K}^{(3)}(\mathbf{x})$. Note that this formula is valid also for the standard case $N = 2$ and reproduces the $SU(3)$-result of the compact-QED limit [16], [13] $m_D^2 = 3g_m^2\zeta$.

The new parameter of the cumulant expansion, $\bar{\zeta}I^{(3)}N(N-1)$, will be exponentially small provided that at $x \sim 1/2$,

$$\epsilon(x) > \frac{1}{2}\left[3e^{-\sqrt{2x}} + \frac{g^2}{2\pi m_W}\ln(N(N-1))\right].$$

Setting in this inequality $x = 1/2$ and recalling that [c] $\epsilon(1/2) < \epsilon(\infty) \simeq 1.787$, we obtain the following upper bound on N, which is necessary (although not sufficient) to provide the convergence of the cumulant expansion: $N(N-1) < e^{15.522 m_W/g^2}$. Clearly, in the weak-coupling regime under study, this bound is exponentially large, that allows N to be large enough too.

In the finite-temperature theory, owing to the fact that the root vectors have the unit length, the critical temperature of the monopole BKT phase transition remains the same as in the $SU(2)$-case, $g^2/2\pi$. In the case $m_H \sim m_W$, one can perform the RG analysis by integrating out high-frequency modes of the fields $\vec{\chi}$ and ψ (see the second paper of Ref. [12] for details). In particular, one can derive the leading correction to the BKT RG flow (determining the dependence of $\beta\zeta$ on K) in powers of $1/(m_H a)^2$. Here, a is the correlation radius, which diverges with an essential singularity at $T \to g^2/2\pi - 0$ as [d] $a(\tau) \sim \exp(\text{const}/\sqrt{\tau})$, $\tau = (g^2/2\pi - T)/(g^2/2\pi)$,

---

[c]Similarly to Ref. [15], we assume here that the function ϵ is one and the same for any N.

[d]In the molecular phase, the correlation radius becomes infinite due to the short-rangeness of the molecular fields.

while m_H evolves very weakly in the critical region (namely, according to the equation $\frac{|dm_H^2|}{m_H^2} \ll \frac{da}{a}$) and therefore can be treated as a constant. In particular, from the modified RG flow it follows that the corrections to the BKT fixed point, $K_{\text{cr.}} = 8\pi$, $(\beta\zeta)_{\text{cr.}} = 0$, vanish. This confirms once more the result of Section 2 that the finiteness of the Higgs-boson mass does not change the critical temperature of the BKT phase transition. As far as the zeroth-order [in the parameter $1/(m_H a)^2$] part of the free-energy density is concerned, it scales as a^{-2} and therefore remains continuous in the critical region. Moreover, the first-order correction can be shown to be continuous as well.

However, an essential difference of the $SU(N)$-case, $N > 2$, from the $SU(2)$-one does exist. Namely, while in the $SU(2)$-case the RG invariance is exact (modulo the negligibly small high-order terms of the cumulant expansion applied to the average over ψ), in the $SU(N)$-case it is only approximate, even in the compact-QED limit of the model. In fact, in course of integration over high-frequency modes, new terms appear in the action, and the RG invariance holds only modulo the approximation $\sum_{i,j} \mathcal{O}_{ij} \cos\left[(\vec{q}_i - \vec{q}_j)\vec{\chi}\right] \simeq \sum_i \mathcal{O}_{ii}$. Within this approximation, the RG flow of the $SU(N)$-model is indeed identical to the one of the $SU(2)$-case, since all the N-dependence can then be removed upon the appropriate rescaling of fields and couplings.

6. Conclusions

In this talk, we have discussed various properties of the (2+1)D Georgi-Glashow model at finite temperature, in the weak-coupling limit. First, we have explored the consequences of accounting for the Higgs field to the deconfining phase transition in this model. To this end, this field was not supposed to be infinitely heavy, as it takes place in the compact-QED limit of the model. Owing to that, the Higgs field starts propagating and, in particular, produces the additional interaction of monopoles in the plasma. Although this effect does not change the critical temperature of the monopole BKT phase transition, it modifies the monopole fugacity and also leads to the appearance of the novel terms in the action of the dual-photon field. The cumulant expansion applied in the course of the average over the Higgs field is checked to be convergent, provided that the weak-coupling approximation is implied in a certain sense. Namely, the parameter of the weak-coupling approximation should be bounded from above by a certain function of masses of the monopole, W-boson, and the Higgs field.

It has been demonstrated that although in the compact-QED limit there exists only one critical temperature of the phase transition, in general there exist two distinct critical temperatures. We have discussed the dependence of these temperatures on the parameters of the Georgi-Glashow model. In particular, both critical temperatures turn out to be larger than the one of the phase transition in the compact-QED limit. Besides that, it has been demonstrated that the smaller of the two critical temperatures always does not exceed the critical temperature of the monopole BKT phase transition. As far as the larger critical temperature is concerned, the range of parameters of the Georgi-Glashow model has been found, where it also does not exceed the BKT one. The situation when there exist two phase transitions implies that at the smaller of the two critical temperatures, W-molecules start dissociating, while at the larger one all of them are dissociated completely. This means that in the region of temperatures between the critical ones, the gas of W-molecules coexists with the W-plasma.

From the RG equations, it follows that the presence of the second (larger) critical temperature leads to the appearance of a novel stable value of the W-fugacity. This value is reached if one starts the evolution in the region where the temperature is larger than the above-mentioned critical one, and the density of W's is smaller than the one of monopoles. The resulting stable value is nonvanishing (i.e., W's at that point are still of some importance), that is the opposite to the standard situation, which takes place if the evolution starts at the temperatures smaller than the first critical one.

Next, we have found the change of critical temperature of the monopole BKT phase transition in the presence of massless dynamical quarks which interact with the Higgs boson only via the gauge field. It has been shown that for $N_f = 1$, this temperature becomes twice smaller than the one in the absence of quarks, whereas for $N_f \geq 2$ it becomes exponentially small, namely of the order $\zeta^{1/3}$. The latter effect means that this number of quark zero modes, which strengthen the attraction of a monopole and an antimonopole in the molecule, becomes enough for the support of the molecular phase at any temperature exceeding that exponentially small one. Therefore, for $N_f \geq 2$, no fundamental matter (including dynamical quarks themselves) can be confined at such temperatures by means of the monopole mechanism. As far as the real deconfining phase transition at $N_f = 1$ and at the finite Higgs-boson mass is concerned, we have seen that the larger of the two critical temperatures always exceeds the BKT one. Therefore, the second phase transition always takes place together with the monopole BKT phase transition. At the same time, the smaller of the two critical temperatures does not exceed the BKT one, only provided the following inequality holds: $\frac{g^2}{m_W} \ln \left(\frac{e^\gamma}{2} \frac{m_W}{m_H} \right) \leq 4\pi(\epsilon - 1)$.

We have further investigated the general case of $SU(N)$ (2+1)D Georgi-Glashow model in the weak-coupling limit, at $N \geq 2$. There has been found the upper bound for N, which is necessary (although not sufficient) to provide the convergence of the cumulant expansion applied in course of the average over the Higgs field. This bound is a certain exponent of the ratio of the W-boson mass to the squared electric coupling constant. It is therefore an exponentially large quantity in the weak-coupling regime, that yields an enough broad range for the variation of N. The Debye mass of the dual photon at arbitrary N has also been found.

Finally, we have discussed the influence of the Higgs field to the RG flow in the $SU(N)$-version of the Georgi-Glashow model at finite temperature. In this analysis, the Higgs-field mass was supposed to be large, namely of the order of the W-boson one, but not infinite, as it takes place in the compact-QED limit of the model. The leading correction to the coupling $g_m^2 T$ then turns out to be inversly proportional to the second power of the correlation radius and therefore it vanishes at the BKT critical point. This point is one and the same for any $N \geq 2$, since the root vectors of the group $SU(N)$ (along which monopole charges are distributed) have the unit length. As far as the Higgs mass itself is concerned, the RG equation for it shows that this quantity evolves so weakly in the vicinity of the critical point that it can be treated as a constant with a high accuracy. Next, according to the respective RG equation, the free-energy density remains continuous in the critical region, even with the account for the leading Higgs-inspired correction. It has finally been mentioned that contrary to the $SU(2)$-case, in the $SU(N)$-model at $N > 2$, the RG invariance holds only approximately, even in the compact-QED limit. Within this approximation, the RG flow in the $SU(N)$-model is identical to the one of the $SU(2)$-case. In particular, this fact confirms the above statement that the BKT critical point is universal for any N.

Acknowledgments

The author is grateful to Dr. N.O. Agasian, in collaboration with whom the papers [10] and [13] have been written, and to Profs. A. Di Giacomo and I.I. Kogan for useful discussions. This work has been supported by INFN and partially by the INTAS grant Open Call 2000, Project No. 110. And last but not least, the author is grateful to the organizers of the Symposium and Workshop "Continuous Advances in QCD 2002/ARKADYFEST (honoring the 60th birthday of Prof. Arkady Vainshtein)" (Theoretical Physics

342

Institute of the University of Minnesota, Minneapolis, 17-23 May 2002) for an opportunity to present these results in a very pleasant and stimulating atmosphere.

References

1. A.M. Polyakov, *Nucl. Phys.* **B120**, 429 (1977).
2. K. Dietz and Th. Filk, Nucl. Phys. **B164**, 536 (1980).
3. M.K. Prasad and C.M. Sommerfield, Phys. Rev. Lett. **35**, 760 (1975); E.B. Bogomolny, Sov. J. Nucl. Phys. **24**, 449 (1976).
4. T.W. Kirkman and C.K. Zachos, Phys. Rev. **D24**, 999 (1981).
5. D. Antonov, Phys. Lett. **B535**, 236 (2002).
6. V.L. Berezinsky, Sov. Phys. JETP **32**, 493 (1971); J.M. Kosterlitz and D.J. Thouless, J. Phys. **C6**, 1181 (1973); J.M. Kosterlitz, J. Phys. **C7**, 1046 (1974).
7. J. Zinn-Justin, *Quantum Field Theory and Critical Phenomena* (2nd edn., Oxford Univ. Press, New York, 1993).
8. N.O. Agasian and K. Zarembo, Phys. Rev. **D57**, 2475 (1998).
9. G. Dunne, I.I. Kogan, A. Kovner, and B. Tekin, JHEP **01**, 032 (2001); I.I. Kogan, A. Kovner, and B. Tekin, JHEP **03**, 021 (2001); Phys. Rev. **D63**, 116007 (2001); JHEP **05**, 062 (2001); I.I. Kogan, A. Kovner, and M. Schvellinger, JHEP **07**, 019 (2001); I.I. Kogan and A. Kovner, hep-th/0205026.
10. N.O. Agasian and D. Antonov, Phys. Lett. **B530**, 153 (2002).
11. C. Lee and W.A. Bardeen, Nucl. Phys. **B153**, 210 (1979).
12. D. Antonov, Mod. Phys. Lett. **A17**, 279 (2002); ibid. **A17**, 851 (2002).
13. N.O. Agasian and D. Antonov, JHEP **06**, 058 (2001).
14. S.R. Wadia and S.R. Das, Phys. Lett. **B106**, 386 (1981); Erratum-ibid. **B108**, 435 (1982).
15. N.J. Snyderman, Nucl. Phys. **B218**, 381 (1983).
16. D. Antonov, Europhys. Lett. **52**, 54 (2000).

SECTION 4.

TOPOLOGICAL FIELD
CONFIGURATIONS, DYNAMICS IN
SUPERSYMMETRIC MODELS, AND
THEORETICAL ISSUES

(DIS)ASSEMBLING COMPOSITE SUPERSYMMETRIC SOLITONS

ADAM RITZ

*Department of Applied Mathematics and Theoretical Physics,
Centre for Mathematical Sciences, University of Cambridge,
Wilberforce Rd., Cambridge CB3 0WA, UK
Email: a.ritz@damtp.cam.ac.uk*

Many supersymmetric theories which possess BPS states exhibit co-dimension one curves of marginal stability (CMS) in parameter or moduli space across which the BPS spectrum changes discontinuously. It follows that to obtain a global picture of the spectrum we require control over aspects of the near-CMS dynamics. Fortunately, the questions of interest concerning the spectrum simplify in this regime to tractable problems in nonrelativistic quantum mechanics. In this talk, we review the features of this system focusing on the interaction potential between BPS states and the delocalization mechanism via which states are removed from the spectrum. We discuss two examples in detail; kink states in $\mathcal{N}=2$ theories in $D=2$, and dyonic states in $\mathcal{N}=2$ theories in $D=4$.

1. Introduction

Questions about the spectrum of massive states are central, but often difficult to answer, within many physically interesting field theories. The presence of extended supersymmetry assists in this regard by allowing us to focus on representations preserving some fraction of the supersymmetry of the vacuum. The corresponding states inherit special properties, such as kinematically determined mass spectra and stability, from the underlying supersymmetry algebra. This presents us with an opportunity to examine the structure of this part of the massive spectrum quantitatively in interesting classes of theories.

The generic features of these Bogomol'nyi-Prasad-Sommerfield (BPS) states $|\mathbf{n}\rangle$ are well known. In a sector of the spectrum specified by the conserved charges $\mathbf{n} = \{n_i\}$, the mass (or tension) saturates the Bogomol'nyi bound [1], and is given in terms of the central charge $\mathcal{Z}|\mathbf{n}\rangle = \mathcal{Z}_\mathbf{n}|\mathbf{n}\rangle$. It will be sufficient here to restrict to cases in which $\mathcal{Z}_\mathbf{n}$ is complex, so that

$$M_\mathbf{n} = |\mathcal{Z}_\mathbf{n}|, \tag{1}$$

from which it follows that the state preserves some supersymmetry [2,3]. Consequently, BPS states lie in short supermultiplets and are generically stable with respect to decay into two constituents, $|\mathbf{n}_1\rangle$ and $|\mathbf{n}_2\rangle$, with $\mathbf{n} = \mathbf{n}_1 + \mathbf{n}_2$. More precisely, since the central charges are linear in $\{\mathbf{n}_i\}$,

$$M_\mathbf{n} = |\mathcal{Z}_{\mathbf{n}_1} + \mathcal{Z}_{\mathbf{n}_2}| \leq M_{\mathbf{n}_1} + M_{\mathbf{n}_2}. \tag{2}$$

However, the central charges generically depend on the moduli (and/or parameters) in the theory, and so there can be special submanifolds where a BPS state is only marginally stable, i.e. $M_\mathbf{n} = M_{\mathbf{n}_1} + M_{\mathbf{n}_2}$. This submanifold is characterized by the alignment of the central charges of the constituents, which amounts to the vanishing of the relative phase ω,

$$\omega(\text{CMS}) = 0, \quad \text{where} \quad e^{i\omega} \equiv \frac{\bar{\mathcal{Z}}_1 \mathcal{Z}_2}{|\mathcal{Z}_1 \mathcal{Z}_2|}. \tag{3}$$

These *curves of marginal stability* (CMS) in the moduli (and/or parameter) space are then of considerable interest as they are the only regimes where a discontinuous change in the spectrum of BPS states is possible.

Historically, the discontinuities in the BPS spectrum associated with the transition across CMS curves, were studied first in the context of BPS kinks in $\mathcal{N}=2$ theories in 1+1D [4,5]. However, as is apparent from (2), this is a generic feature within theories of extended supersymmetry where the central charges have more than a single real component. Consequently, related spectral transitions have since been observed and studied within the particle spectrum of $\mathcal{N}=2$ and $\mathcal{N}=4$ super Yang-Mills (SYM) [6–12], in the spectrum of domain walls in $\mathcal{N}=1$ theories [13,14], and within string theory, particularly with regard to IIB string junctions [15,16] and the D-brane spectrum in compactifications on manifolds containing nontrivial cycles [17–21]. In many cases there are links between these examples as much of the structure is prescribed by the level of supersymmetry.

At first sight, the presence of a CMS in moduli space may seem a hindrance to obtaining a clear picture of the BPS spectrum, negating some of the nice features alluded to above. While this is true to a certain extent, their presence is in fact crucial to many of the most interesting features of gauge and string theories, particularly the dualities which provide insight into strong coupling dynamics. Moreover, the transitions in the BPS spectrum can often be understood using quite powerful symmetry arguments. In particular, the existence and position of CMS curves is straightforwardly deduced from the BPS mass formula and the moduli dependence of the central charges $\mathcal{Z}_\mathbf{n}$. The question then boils down to whether a discontinuity in the spectrum does occur, and if so on which side of the CMS the BPS states actually exist, or indeed whether they exist at all. In some cases, such as $\mathcal{N}=2$ SYM [9,10,12], this issue can be settled without a detailed dynamical calculation.

More generally, one would like a simple dynamical means to understand the transitions in the spectrum on crossing CMS curves. The linear dependence of $\mathcal{Z}_n$ on the conserved charges implies that a natural approach is to study the interactions between "primary" states, namely those which are the lightest carrying a given charge. One can then hope to build up the full spectrum in terms of bound states of these primary constituents. In this talk, we review some recent work in collaboration with M. Shifman, A. Vainshtein*, and M. Voloshin [24,25] which addresses this problem. In actual fact the near-CMS regime turns out to exhibit important simplifications, and one can abstract the dynamical question of interest to a question of bound state formation in (supersymmetric) quantum mechanics (SQM). The problem is nonrelativistic sufficiently close to the CMS where the binding energy of a particular state becoming marginally stable necessarily tends to zero, as follows from (2). The transition across the CMS is associated with a *delocalization* [24] of the composite BPS configuration into two (or more) primary BPS constituents.

This dynamical behavior has the bonus of providing us with a tractable weakly interacting regime regardless of the nature of the underlying dynamics. Thus we can use the near-CMS regime to map out the full spectrum of BPS states at weak- and also at strong-coupling. It follows that the central quantity to obtain is the static interaction potential between primary states, which will in general depend on the moduli and parameters of the theory. In Sect. 2 we will discuss in some detail the generic constraints on this potential which are imposed by the residual supersymmetry of the reduced dynamics. We then turn to some applications. In Sect. 3 we consider marginal stability for composite kink states in $\mathcal{N}=2$ theories in 1+1D focusing on the importance of quantum corrections in the near-CMS regime [24] (see also [26]). In Sect. 4, we turn to dyonic states in $\mathcal{N}=4$ and $\mathcal{N}=2$ SYM in 3+1D where to answer questions of stability it is sufficient to determine the leading long range potential determined by massless electrostatic, magnetostatic, and scalar exchanges [25] (see also [27–36,19,20]). We utilize this potential to consider the transitions in the BPS spectrum as one breaks $\mathcal{N}=4$ to $\mathcal{N}=2$ supersymmetry. We finish with some concluding

---

*On the occasion of this conference, celebrating Arkady Vainshtein's 60th birthday, it is appropriate to recall that his contribution in this field extends right back to the original paper of Bogomol'nyi [1], which was stimulated by the earlier work of Bogomol'nyi and Vainshtein [22] on the stability of vortices, and their knowledge of the application of first-order equations in this system within the condensed matter literature [23].

remarks in Sect. 5.

2. BPS bound state constraints

Before diving into the details, it proves useful to consider the generic constraints that supersymmetry imposes on the form of the interaction potential near the CMS. We will focus exclusively now on finite mass BPS particle-like soliton states. So, if we define the two natural dimensionful scales in the bound state problem, the binding energy and the reduced mass,

$$E_{\text{bind}} = M_{\mathbf{n}_1} + M_{\mathbf{n}_2} - M_{\mathbf{n}}, \qquad M_r = \frac{M_{\mathbf{n}_1} M_{\mathbf{n}_2}}{M_{\mathbf{n}_1} + M_{\mathbf{n}_2}}, \qquad (4)$$

then, near the CMS we find that they are related by the phase ω,

$$E_{\text{bind}} = \frac{1}{2} M_r \omega^2 + \mathcal{O}(\omega^4). \qquad (5)$$

It follows that if we consider the dynamics of two BPS particles $|\mathbf{n}_1\rangle$ and $|\mathbf{n}_2\rangle$ sufficiently near the CMS for the composite state $|\mathbf{n}\rangle = |\mathbf{n}_1 + \mathbf{n}_2\rangle$, then the binding energy E_{bind} (and, consequently, the kinetic energy) can be made parametrically small. i.e. by moving near the CMS we have,

$$\epsilon \equiv \frac{E_{\text{bind}}}{M_r} \sim \frac{1}{2} \omega^2 \ll 1, \qquad (6)$$

This amounts to a restriction that we keep closer to the CMS than to points in moduli space where the primary states are massless. However, this is not a significant constraint as such submanifolds are of higher co-dimension than the CMS curves. Therefore, we can always consider a regime near the CMS where the dynamics describing the composite state is manifestly nonrelativistic, i.e. where $\epsilon \ll 1$. In this regime it is legitimate to ignore the full microscopic theory and study the effective (supersymmetric) quantum mechanics associated with the collective coordinates of the BPS states. It is important that this argument relies only on having a new small parameter $\epsilon \ll 1$ and is therefore quite general; it applies whether or not the underlying theory is strongly or weakly coupled.

The generic leading order form of the static potential between two sources (described by complex fields) separated by a distance r has the form,

$$V_{\text{LO}}(r) = E_{\text{bind}} + \eta_{12}^i K_d(m_i, r) + \cdots \qquad (7)$$

where the exchanges are limited to those states of lowest mass m_i which couple to the constituents, and $K_d(m_i, r)$ is the corresponding position space propagator for these states in the transverse space of dimension d. We conclude that the leading structure of (7) is uniquely fixed up to the (real) residue η_{12}^i, which we can write a little more explicitly in a complex basis as $\eta_{12} = \text{Re}(\eta_1 \bar{\eta}_2)$, in terms of the (complex) trilinear couplings with the constituents η_1 and η_2. It is here that supersymmetry provides additional constraints, and indeed prescribes the general form of the potential. Schematically, the relevant component of the superalgebra which acts trivially on the putative bound state has the form (suppressing Lorentz indices)

$$\{Q, Q\} = M - |\mathcal{Z}| = H_{\text{SQM}}, \tag{8}$$

where $M - |\mathcal{Z}| \to 0$ on approach to the CMS. It will be sufficient here to assume that the lowest energy configurations are determined by a spherically symmetric potential so that the dynamical system is effectively one-dimensional. Dropping kinetic terms on boths sides of (8), we see that the structure of the potential

$$\{Q, Q\}|_{p=0} = V_{\text{SQM}}(r) = V_{\text{Bose}}(r) + V_{\text{Fermi}}(r), \tag{9}$$

is correspondingly constrained[†] In particular, the bosonic terms, normalized so that $V_{\text{SQM}}(r = \infty) = E_{\text{bind}}$, take the form,

$$V_{\text{Bose}}(r) \sim \frac{M_r}{2} g_{rr} (\omega - K(r))^2, \tag{10}$$

where we have allowed for a nontrivial metric g_{rr} such that $g_{rr} \to 1$ as $r \to \infty$, while $K(r)$ is a radial function, with $K(r \to \infty) \to 0$. Our notation for $K(r)$ reflects the fact that by expansion of (10) this function bears a relation to the position space propagator in (7). Specifically, if we consider a situation where the radial dependence of the metric g_{rr} can be ignored, and the fermionic corrections are of higher order as $r \to \infty$, then

$$V_{\text{LO}}(r) \sim \frac{1}{2} M_r \omega^2 - \omega M_r K(r) + \cdots \tag{11}$$

which, on comparison with (7), relates the residue η_{12} to the phase ω in the near-CMS regime. This result, however, need not be generic. In particular, the fermionic quantum corrections in $V_{\text{Fermi}}(r)$ arising from cross-terms in the anti-commutator (8) amount (depending on dimensionality) to "spin"-dependent interactions whose importance near the CMS, depending on the model, may not be negligible [24] as we discuss below.

[†]As follows from (9), the primary quantity of interest is really the supercharge Q, from which the potential follows directly from the structure of the superalgebra. However, for clarity we will focus our attention here on the potential itself, but one should bear in mind that full use of supersymmetry is made by studying Q itself [24].

The form of the BPS mass formula indicates that as we move near a CMS curve, bound states are formed via an attractive interaction that can be made arbitrarily small. To determine the resulting constraints on the potential, we need to be a little more specific and it is natural to introduce the codimension d ($d = D - 1$ for the finite energy solitons that we focus on here) of the BPS states as a means of classification. Following the arguments above, we are concerned with quantum mechanical potentials that support very weakly bound states as we approach the CMS. The strength of the radial potential $V(r)$ is given by the characteristic dimensionless parameter v,

$$v \equiv M_r \int dr\, r(-V(r)), \tag{12}$$

where we normalize $V(r \to \infty) \to 0$ which denotes the continuum threshold.

The bound state problem [37] then bifurcates at $d = 2$, where for $d \leq 2$ a bound state will exist for *any* finite $v > 0$, while for $d > 2$ bound states exist only for $v \geq v_{\mathrm{crit}}(d) > 0$. For $d < 2$, the dimensionless binding energy,

$$\epsilon(v) \propto v^{\alpha_d}, \qquad \text{for} \quad \text{d} < 2 \tag{13}$$

scales to zero with α_d a positive dimension-dependent exponent. For the critical co-dimension $d = 2$, the binding energy is exponentially suppressed as $v \to 0$,

$$\epsilon(v) \propto \exp\left(-\frac{\mathrm{const}}{v}\right), \qquad \text{for} \quad \text{d} = 2. \tag{14}$$

For $d > 2$, the existence of the lower threshold $v_{\mathrm{crit}}(d)$ is significant as it allows us to conclude that, in general, long range forces will be involved in forming near CMS bound states. Only in this case can the coefficient of the binding component of the potential tend to zero on the CMS, as follows from (11), while keeping $v \geq v_{\mathrm{crit}}$, since the integral in (12) will diverge provided the coefficient is nonzero.

To illustrate the constraints this imposes, focus on soliton states in 3+1D, namely $d = 3$. We conclude that bound state formation near the CMS depends only on long range forces, and is therefore insensitive to detailed issues concerning short range interactions between the cores of the BPS states. Consequently, we can simplify the effective description by treating the primary states as point-like particles interacting only via long range forces. This is clearly an abstraction but is nonetheless sufficient for answering questions about the presence of bound states. In this case, it is not difficult to convince oneself that not only is it consistent for the coefficient of the leading order potential to vanish on the CMS, it *must* vanish. This follows by noting that Coulombic systems associated with an attractive $1/r$-type potential,

$$V(r) = -\frac{\alpha}{4\pi r}, \tag{15}$$

possess, in the limit where $\alpha \to 0$, towers of closely spaced bound states, only the lowest of which can be BPS saturated. In contrast, we know from the BPS mass formula that on the CMS the lowest level in the tower must reach the continuum. This can only happen if the effective coupling α, which is a function of the moduli, vanishes on the CMS.

A corollary of these constraints is that the composite BPS configuration, viewed as a bound state of primary constituents, must delocalize on approach to the CMS. A previously known example of such delocalization, at the classical level of static equilibrium configurations, is provided by the semiclassical system of two dyons with charges under different U(1) subgroups of the Cartan subalgebra in $\mathcal{N}=2$ SYM with gauge group SU(3) [31–34,38–40], as discussed from this standpoint in [24]. In this system, besides the Coulombic potential (15) with a coupling α which vanishes on the CMS, there is also a subleading repulsive $1/r^2$ potential, related by supersymmetry to the fermionic "loop corrections" in (9), which remains finite on the CMS. This leads to a diverging equilibrium separation on approach to the CMS, which becomes infinite on the CMS itself. We will illustrate this feature in several examples in the following two sections.

3. $\mathcal{N}=2$ theories in two dimensions

To make these arguments more concrete, we turn first to $\mathcal{N}=2$ theories in 1+1D exhibiting BPS kink solitons. For a generic supersymmetric Landau-Ginzburg-type Lagrangian of the form

$$\mathcal{L} = \int d^2\theta d^2\bar{\theta}\mathcal{K}(X_i, \overline{X}_i) + 2\mathrm{Re}\int d^2\theta\mathcal{W}(X_i), \tag{16}$$

which we assume to possess several isolated massive vacua, a Bogomol'nyi bound exists on the energy of static configurations. This energy corresponds to the kink mass and has the form,

$$M \geq |\mathcal{Z}|, \qquad \mathcal{Z} = 2\Delta\mathcal{W}, \tag{17}$$

where $\Delta\mathcal{W}$ is the topological charge, the difference between the values of the superpotential in the two vacua between which the soliton interpolates. Such solitons are therefore topologically stable.

At the semi-classical level, BPS configurations satisfy the first-order Bogomol'nyi equations [1],

$$g_{\bar{j}k}\partial_z X^k = e^{i\gamma}\partial_{\bar{j}}\overline{\mathcal{W}}, \tag{18}$$

where z is the spatial coordinate, and saturate the bound in (17). Note that γ is the phase of the central charge, i.e. $e^{i\gamma} = \Delta\mathcal{W}/|\Delta\mathcal{W}|$.

If we write the corresponding $\mathcal{N}=2$ superalgebra in the rest frame

$$\{Q_\alpha^i, Q_\beta^j\} = 2\delta^{ij}\delta_{\alpha\beta}M + 2\sigma_{3\alpha\beta}\sigma_3^{ij}|\mathcal{Z}| \tag{19}$$

it follows that for $M = |\mathcal{Z}|$, Q_1^2 and Q_2^1 act trivially and thus $1/2$ the supersymmetry of the vacuum is preserved.

There are two crucial features in this regard. Firstly, it follows from the BPS equations that

$$\partial_z\mathcal{W} = e^{i\gamma}g^{i\bar{j}}\partial_i\mathcal{W}\partial_{\bar{j}}\overline{\mathcal{W}} \equiv e^{i\gamma}\|\partial\mathcal{W}\|^2, \tag{20}$$

and thus the kink profile describes a straight line in the $\mathcal{W}$-plane connecting the two vacua [4]. Hence, for a solution interpolating between two specific vacua, we can introduce a fiducial coordinate $t \in [0, 1]$ so that

$$\mathcal{W}(X_a(t)) = (1 - t)\mathcal{W}_0 + t\mathcal{W}_1 \tag{21}$$

interpolates the superpotential along the wall trajectory, and this equation provides an implicit solution for the BPS kink profile as a function of the fiducial coordinate t. Secondly, there is definition of soliton number [5] which is preserved under deformations of D-terms. Since we consider massive vacua, one can expand the superpotential to quadratic order about each of the vacua, and the set of all linearized solutions to (18) forms a cycle Δ_a in field-space diffeomorphic to a sphere. A weighted soliton number S_{ab} for two specific vacua is then given by the intersection number of the associated cycles [5]

$$S_{ab} = \Delta_a \cdot \Delta_b \tag{22}$$

This counts solitons weighted by $(-1)^F$, and is consequently an index [41], independent of D-term deformations. The relevance of this result here is that it determines the kink spectrum via the topology of paths connecting vacua in the punctured W-plane (with the vacua excised). A kink is formally described by any path homotopic to a straight line on this surface connecting the two vacua. As we vary moduli the vacua may move around and, as follows from (17), CMS curves arise when three vacua become aligned. The spectrum may then change, and the change in intersection numbers is determined by the Picard-Lefshetz monodromy [5].

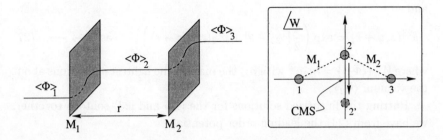

Figure 1. A composite kink configuration, with the two constituents separated by a distance r (some dimensions have been added for clarity). The CMS for the composite arises when the three vacua are aligned in the W-plane as shown on the right.

It is this transition as vacua pass through co-linear (CMS) configurations in the W-plane that we would like to describe in more physical terms. Rather than focusing on a particular model, as in [24], we will consider a theory of a single chiral superfield Φ described by a general Landau-Ginzburg superpotential $\mathcal{W}(\Phi)$. We will assume that there are at least three massive vacua, $\{\langle\Phi_1\rangle, \langle\Phi_2\rangle, \langle\Phi_3\rangle\}$ with moduli that allow us to move the vacua through CMS curves for the corresponding BPS kinks.

Consider the configuration shown in Fig. 1, where the $|13\rangle$ kink arises as a weakly bound state of the primary kinks $|12\rangle$ and $|23\rangle$. To determine the leading order interaction between the primary states, a convenient approach is to calculate the correction to the mass of kink $|23\rangle$, M_{23}, in the background of kink $|12\rangle$ (for an alternative approach see [26]). We have

$$M_{23}(r) = 2|W(\langle\Phi_3\rangle) - W(\langle\Phi_2\rangle + \delta^{12} + \delta^{23})|, \tag{23}$$

where the perturbation to Φ about vacuum $\langle\Phi_2\rangle$, $\delta = \Phi - \langle\Phi_2\rangle$, is small away from both primary kinks. Thus, with kink $|12\rangle$ situated at point x and kink $|23\rangle$ at point $x + r$, with x and r large, we have

$$M_{23}(r) = |\mathcal{Z}_{23}| - 2\mathrm{Re}\left(e^{-i\gamma_{23}}\delta^{12}(x)\delta^{23}(x+r)W''(\langle\Phi_2\rangle)\right) + \cdots, \tag{24}$$

where $\delta^{12}(x)$ and $\delta^{23}(x+r)$ denote the linearized solutions for the $|12\rangle$ and $|23\rangle$ kinks expanded about the vacuum $\langle\Phi_2\rangle$. The linearized Bogomol'nyi equations have the form (absorbing the metric, assumed to be smooth, into a rescaling of the transverse coordinate z),

$$\partial_z\delta = e^{i\gamma}\bar{\delta}\bar{W}''(\Phi_2), \tag{25}$$

and the relevant linearised solutions about vacuum $\langle\Phi_2\rangle$ are given by

$$\delta^{12}(z,x) = \exp\left(\frac{i}{2}[\gamma_{12} - \theta - \pi] - m_2(z - x)\right), \quad \text{as } z \to \infty \tag{26}$$

and

$$\delta^{23}(z, x + r) = \exp\left(\frac{i}{2}[\gamma_{23} - \theta] + m_2(z - x - r)\right), \quad \text{as } z \to -\infty \quad (27)$$

where $W''(\langle\Phi_2\rangle) = m_2 e^{i\theta}$ with m_2 the mass of the lightest excitations about the vacuum $\langle\Phi_2\rangle$.

Putting the linearized solutions for the $|12\rangle$ and $|23\rangle$ solitons together, we have from (24) the leading order potential

$$V(r) = E_{\text{bind}} + 2m_2 \sin\frac{\omega}{2}\exp(-m_2 r) + \cdots, \quad (28)$$

where $\omega = \gamma_{23} - \gamma_{12}$ is the relative phase between the two central charges. Thus, near the CMS, we can simplify this to the form

$$V_{\text{LO}}(r) = \frac{1}{2}M_r\omega^2 + m_2\omega\exp(-m_2 r) + \cdots, \quad (29)$$

using the fact that the binding energy scales as ω^2 near the CMS.

From this leading order potential we are led to conclude [24,26] that as $\omega \to 0$ on the CMS (with $\omega < 0$ so that the potential is attractive) the first two terms in the potential vanish while the subleading terms will in general remain finite. Thus the equilibrium separation will diverge,

$$r_{\text{class}} \sim -\ln|\omega| \to \infty, \quad (30)$$

with the bound state delocalizing. On the other side of the CMS, where $\omega > 0$, the potential is repulsive and this bound state does not exist.

3.1. Quantum corrections

While the picture outlined above is apparently consistent, one might wonder why the leading order term in the potential was found to vanish, given that for this system the analysis of Section 2 would impose no such constraint. In actual fact quantum corrections, associated with the fermionic content of the model, are very important and become increasingly so near the CMS.

To illustrate this in the example above, let us return to the superalgebra. Near the CMS, there is a natural decoupling of the form

$$\{Q_{\text{REL}}\} \otimes \{Q_{\text{CM}}\}, \quad (31)$$

between the generators which describe the relative dynamics of the primary states, and consequently act trivially on the BPS bound state, and those generators which describe the center of mass and always act nontrivially. For $\{Q_{\text{REL}}\}$ we have

$$(Q_1^2)^2 = (Q_{\dot{1}}^2)^2 = M - |Z| \ll M, \quad (32)$$

while for $\{Q_{CM}\}$,

$$(Q_1^1)^2 = (Q_2^2)^2 = M + |Z| \approx 2|Z|, \tag{33}$$

and we can decouple this sector of the superalgebra as it simply acts to restore the full multiplet structure.

The presence or otherwise of bound states is determined by the existence of zero energy ground states in the $\mathcal{N} = 2$ SQM associated with (32). The generic structure of $\mathcal{N} = 2$ SQM was first described by Witten [42], and we can realize the algebra as follows in terms of a quantum mechanical superpotential $\mathcal{W}_{QM}(r)$,

$$Q_1^2 = \frac{1}{\sqrt{2M_r}} \left[p\sigma_1 + \mathcal{W}'_{QM}(r)\sigma_2 \right],$$

$$Q_2^1 = \frac{1}{\sqrt{2M_r}} \left[p\sigma_2 - \mathcal{W}'_{QM}(r)\sigma_1 \right], \tag{34}$$

so that the Hamiltonian is given by

$$\mathcal{H}_{SQM} = M - |Z| = \frac{1}{2M_r} \left[p^2 + (\mathcal{W}'_{QM}(r))^2 + \sigma_3 \mathcal{W}''_{QM}(r) \right], \tag{35}$$

where the second term is the classical potential, and the final term is a quantum correction of $\mathcal{O}(\hbar)$.

Identifying the quantum mechanical superpotential $\mathcal{W}_{QM}(r)$ in general requires a detailed analysis of the model at hand, as in [24]. However, we can easily extract the general structure near the CMS by comparison with the leading order classical potential in (29). We find

$$\mathcal{W}_{QM}(r) = \omega(M_r r) - \exp(-m_2 r), \tag{36}$$

and thus near the CMS,

$$V_{SQM}(r) = \frac{1}{2} M_r \omega^2 + m_2 \left[\omega \pm \frac{m_2}{2M_r} \right] \exp(-m_2 r) + \cdots \tag{37}$$

The additional quantum correction to the leading term thus becomes important near the CMS where it remains finite.

Despite this correction, the classical equilibrium position survives as the maximum of the ground state wavefunction, which is easily determined from (34),

$$|\Psi_0\rangle = \exp(\mathcal{W}_{QM}(r))|+\rangle \xrightarrow{r \to \infty} \exp(\omega M_r r)|+\rangle + \cdots \tag{38}$$

where $|\pm\rangle$ are the eigenvectors of σ_3. In the limit $r \to \infty$ we have extracted only the leading exponential behaviour. It is apparent that this wavefunction is normalizable on only one side of the CMS. For $\omega < 0$ it is the unique ground state and when tensored with the center-of-mass component of the algebra describes a two-component BPS multiplet of composite states – the $|13\rangle$ state. However, this state delocalizes when $\omega = 0$, and for $\omega > 0$ the ground state is given by the doubly-degenerate 1st excited state $|\Psi_1\rangle$, which leads to a four-component non-BPS multiplet for the bound state.

While the qualitative picture of "classical" delocalization of the composite bound state apparently survives quantization more–or–less unscathed in this (generic) scenario, it is important to realize that this is not always so. One can also consider scenarios (see e.g. [24]) in which the entire superpotential (36) is of $\mathcal{O}(\omega)$ and thus vanishes on the CMS leading to an attractive domain for the composite on either side, and an additional modulus – the relative separation – on the CMS itself. In this case the classical equilibrium separation may vanish for any finite ω. It is then crucial to take quantum corrections into account to observe that the wavefunction, although still peaked at $r = 0$, flattens out as $\omega \to 0$ reflecting once again the delocalization of the bound state. This may have, as an observable effect, an internal spin-rearrangement of the multiplet on crossing the CMS [24].

4. Dyon states in $\mathcal{N}=2$ theories in 3+1D

Having considered an example of low co-dimension, $d = 1$, we now turn to an example with $d = 3$, namely dyonic states in $\mathcal{N} = 2$ gauge theories in 3+1D. As described in Section 2, the presence of long-range forces in this system will allow us to treat the dyons as point-like charges and moreover the leading $1/r$ potential does not receive quantum corrections from the fermionic terms. Thus, for the purpose of analysing the structure of the BPS spectrum, there are important simplifications in this case and we can focus on calculating the dominant $1/r$ potential determined by massless exchange [25].

Following the work of Seiberg and Witten [6], the massless sector of the theory on the Coulomb branch is known exactly and this is sufficient to determine the leading interactions between BPS states even in the strong coupling regime. Here, for simplicity, we will focus on the case of gauge group SU(2) (see [25] for the extension to higher rank gauge groups) and consider the changes in the spectrum as one breaks $\mathcal{N}=4$ to $\mathcal{N}=2$ SYM.

We begin by recalling that for $\mathcal{N} = 2$ SYM with gauge group SU(2), broken to U(1) on the Coulomb branch parametrized by the complex adjoint Casimir u, the low energy degrees of freedom comprise an $\mathcal{N} = 2$ abelian vector multiplet. Different duality frames are natural for this multiplet in different regions of the moduli space. At weak coupling, for large u, this multplet is naturally realized in terms of the photon and a complex scalar A, while in the strong coupling region with light monopoles, it can be realized in terms the dual magnetic photon and the corresponding scalar superpartner A_D. The corresponding vevs $(a(u), a_D(u))$ form a section of an SL(2,$\mathbb{Z}$) bundle over the moduli space, whose explicit form [6] we will not need here.

The expectation value of the central charge $\mathcal{Z}_\mathbf{n}$ entering the $\mathcal{N} = 2$ supersymmetry algebra is a linear combination of conserved U(1) charges,

$$\mathcal{Z}_\mathbf{n} = \sqrt{2}\left[n_E a(u) + n_M a_D(u)\right] + s_h m. \tag{39}$$

n_E and n_M refer respectively to integer electric and magnetic quantum numbers, while s_h denotes the integral flavor quantum numbers associated with the U(1) symmetries of any additional hypermultiplets of mass m.

4.1. *The Long Range Potential*

The leading $1/r$ potential is determined by massless exchange, and follows from treating the BPS dyonic states as point particles carrying electric, magnetic, and scalar charges. The corresponding interactions are fixed by $\mathcal{N}=2$ supersymmetry. Specifically, we describe the massive BPS particles via the hypermultiplet pair $(X, \tilde{X})$ of $\mathcal{N} = 1$ chiral superfields. Strictly speaking, this assumes a maximal spin of $j = 1/2$. Nonetheless, the results for the Coulombic potential we will obtain are valid for arbitrary j as the interactions are spin-independent.

The part of the hypermultiplet low energy effective Lagrangian describing the interactions with the massless fields, has the form

$$\mathcal{L}_{\text{hyp}} = \sum_X \int d^2\theta d^2\bar\theta \left[\bar{X} e^{n_E V} e^{n_M V_D} X + \tilde{X} e^{-n_E V} e^{-n_M V_D} \bar{\tilde{X}}\right]$$
$$+ 2 \sum_X \text{Re}\left[\int d^2\theta\, \tilde{X} \mathcal{Z}_X(A) X\right], \tag{40}$$

where n_E, n_M are the electric and magnetic charges of the field X, the superfield V_D is dual to the vector superfield V which describes the photon, and $\mathcal{Z}_X(A)$ is given by Eq. (39) in which $\mathcal{Z}$ is viewed as a function of A.

The normalization of the corresponding propagators for the photon and the scalar A are given (by $\mathcal{N}=2$ SUSY) in terms of the scalar metric,

$$g_{a\bar{a}} = \frac{4\pi}{\text{Im}\,\tau(u)}, \tag{41}$$

determined by the low energy (moduli dependent) U(1) gauge coupling $\tau(u)$. The normalization of the propagators for the dual photon and its

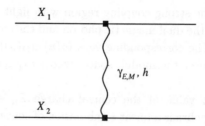

Figure 2. The tree level massless exchanges leading to the long range potential between the BPS states X_1 and X_2. $\gamma_{E,M}$ refers to the electrostatic and magnetostatic terms, while h denotes the scalar field quanta.

superpartner A_D follow from a duality rotation, which amounts to a change of basis in (41) [25].

We can now determine the long range interaction between two static BPS particles due to the massless exchanges shown schematically in Fig. 2. The $1/r$ potential is obtained as the sum of three terms,

$$V(r) = V_E(r) + V_M(r) + V_S(r), \tag{42}$$

referring to electrostatic V_E, magnetostatic V_M, and scalar V_S exchange, in correspondence with the contributions to the tree level diagram in Fig. 2 where γ and $h = A - a$ are the intermediate massless particles.

The final result takes the form [25]

$$V(r) = \frac{1}{8\pi r} \mathrm{Re} \left\{ (Q_1 \cdot \bar{Q}_2) (1 - e^{i\omega}) \right\} \tag{43}$$

where we have defined the complex charges, $Q = \partial \mathcal{Z} / \partial a = \sqrt{2} (n_E + n_M \tau)$, bearing in mind that $\mathrm{Re}(Q)$ is the physical "fractional electric charge", while $Q_1 \cdot \bar{Q}_2 \equiv g_{a\bar{a}} Q_1 \bar{Q}_2$ is the inner product on the moduli space, and finally ω is once again the relative phase between the two central charges. The first term corresponds to the sum of electrostatic and magnetostatic exchanges, corrected to account for the Witten effect [43] which in this context implies a mixing between the fields V and V_D. The second term in (43) describes the scalar exchange, generalising the classical discussion of Montonen and Olive [44]. The Yukawa interaction in this case is fixed by supersymmetry and follows from Taylor-expanding the central charge in (40).

We see that the scalar exchange differs from the gauge exchange only by the presence of the phase factor $e^{i\omega}$, so the potential vanishes on the CMS where $\omega = 0$ as anticipated in Section 2. Furthermore, we observe that (43) is manifestly reparametrization invariant with respect to the choice of the exchanged scalar field, due to its dependence only on the inner product $Q_1 \cdot \bar{Q}_2$ which is necessarily basis-independent. In other words, our choice of the exchanged scalar as a was purely for convenience and the result would be the same if we had chosen any other duality-related scalar, such as a_D, which is an important consistency check.

The validity of the nonrelativistic potential relies on being near the CMS, where the binding energy is small, and it is natural to expand (43) in this regime. Retaining terms of $\mathcal{O}(\omega^2)$ we find,

$$V(r) = -\frac{\omega}{r}\langle q_1, q_2\rangle + \frac{\omega^2}{16\pi r}\mathrm{Re}\left(Q_1 \cdot \bar{Q}_2\right) + \mathcal{O}(\omega^3), \tag{44}$$

where $\langle q_1, q_2\rangle = n_E^1 n_M^2 - n_E^2 n_M^1$ is the symplectic (or Dirac–Schwinger–Zwanziger) product of the charges. Note that the leading linear term is independent of the Kähler metric, and it follows immediately that, for mutually nonlocal primary states, the bound state is stable for

$$\omega\langle q_1, q_2\rangle > 0. \tag{45}$$

This stability criterion has also appeared in other contexts with the same level of supersymmetry [17,19].

In the subsections that follow we will consider some illustrative examples which exhibit the basic phenomena of interest.

4.2. BPS spectrum in $\mathcal{N}=4$ SYM

Let us consider $\mathcal{N}=2$ SU(2) gauge theory with one adjoint hypermultiplet. The expression for the central charge is given in (39) where the adjoint hypermultiplet (of mass m_h) is described by two superfields Φ_s for which $s = \pm 1$, while all pure $\mathcal{N}=2$ fields have s-charge zero. This UV finite theory represents $\mathcal{N}=4$ SU(2) SYM at $m_h = 0$. When m_h is finite the theory flows in the infrared to $\mathcal{N}=2$ SU(2) SYM, where m_h plays the role of a UV cut off while the infrared scale Λ is $\Lambda^4 = 4\,m_h^4 \exp\left(2\pi i\,\tau_0\right)$.

Defining $\mathcal{N} = 4$ SU(2) SYM as the limit $m_h \to 0$ implies that vevs of the hypermultiplet fields Φ_s must vanish. On the Coulomb branch of the $\mathcal{N} = 4$ theory it is always possible to pass to such an orientation by global SU(4) rotations in the space of scalar fields. There is no running of the coupling and the vevs are given by their classical values $a = \sqrt{2u}$, $a_D = \tau_0 a$. The central charge then takes the classical form,

$$\mathcal{Z}_{\{n_E, n_M\}} = \sqrt{2}\,a\left(n_E + n_M\,\tau_0\right). \tag{46}$$

The general expression for the long range potential in this case results in

$$V(r) = -\frac{1 - \cos\omega}{r\,\mathrm{Im}\,\tau_0}\,|Q_1|\,|Q_2| \tag{47}$$

for the interaction between two BPS particles with generic electric and magnetic charges $\{n_E, n_M\}$. This expression is formally valid for any nonzero a, where the primary states are massive, and we observe that the potential (47) vanishes if the central charges are aligned, i.e. $\omega = 0$, while it is attractive otherwise.

In order to study the spectrum, consider first the interaction of the particles $\pm\{1, 0\}$ and $\pm\{0, 1\}$. These are the lightest states with nonvanishing electric and magnetic charges, which guarantees their stability, and thus we will take these as "primary" constituents. The potential (47) leads to attraction in the channels $\pm\{1, 1\}$, $\pm\{1, -1\}$ but vanishes in the $\pm\{2, 0\}$, $\pm\{0, 2\}$ channels. Therefore the potential leads to bound states with quantum numbers $\pm\{1, 1\}$ and $\pm\{1, -1\}$. Although we are unable to use the nonrelativistic approximation to calculate the binding energy – it is of the same order as the reduced mass – we know that the ground states in the $\pm\{1, 1\}$, $\pm\{1, -1\}$ channels are indeed BPS saturated from the corresponding dyon solutions at weak coupling. These solutions are known to compose BPS multiplets with maximal spin $j = 1$, the short $\mathcal{N}=4$ multiplet with 16 states.

Given that interactions between the primary states are either attractive, i.e. between $\pm\{1, 0\}$ and $\pm\{0, 1\}$, or vanish, we can deduce the full spectrum of stable bound states in the following way. For a generic configuration with charges $\{n_E, n_M\}$, it follows that we can arrange the constituents into k non-interacting subgroups, which can then dissociate without any cost in energy, only if the charges have a common divisor k, i.e. they have the form $\{n_E, n_M\} = k\{n'_E, n'_M\}$ for some integer k. Under the assumption that the bound states we find are indeed BPS, which we can verify in the semiclassical region for a restricted subset of charges and spins, we then deduce that the stable BPS states form the set $\{n_E, n_M\}$ where n_E and n_M are co-prime. In this way we obtain a spectrum of BPS states which is in agreement with known semiclassical results [45], and indeed with the predictions of SL(2,$\mathbb{Z}$) duality [44–46].

4.3. Bound states and delocalization at strong coupling in $\mathcal{N}=2$ SYM

In $\mathcal{N}=4$ SYM, since the gauge coupling is a marginal parameter, we can choose to stay in the weak coupling region. However, the approach described here is not limited to the weak coupling regime as we can make use of the Seiberg-Witten solution for $\mathcal{N}=2$ SU(2) SYM to determine the long range interactions even when the microscopic gauge theory is strongly coupled.

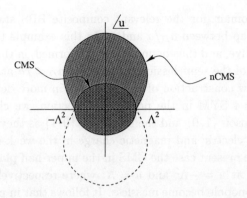

Figure 3. The BPS curve of marginal stability (CMS) is shown in the u-plane for SU(2) SYM, along with the non-BPS CMS (nCMS) where the potential vanishes between certain states which both possess magnetic charge. As an example, the bold contour bounds the exterior stability domain for the non-BPS state with charges $\{1,2\}$.

We will restrict our attention to the upper-half plane in moduli space, $\text{Im}(u) > 0$, thus avoiding the cuts involved in specifying the sections $(a(u), a_D(u))$, which can be chosen to lie along the real axis. We now consider again the interaction between two BPS particles with generic electric and magnetic charges $\{n_E, n_M\}$. The long range interaction vanishes when $\omega = 0$, which is easily seen to hold throughout the moduli space when $\{n_E^1, n_M^1\}$ and $\{n_E^2, n_M^2\}$ are parallel. This means that when the total electric (n_E) and magnetic (n_M) charges are *not* co-prime the situation is the same as it was in $\mathcal{N}=4$; the state dissociates into noninteracting BPS particles. However, if $\{n_E^1, n_M^1\}$ and $\{n_E^2, n_M^2\}$ are not parallel the phase ω becomes moduli dependent. Indeed, for $|\omega| \ll 1$ (which requires $\text{Re}(\bar{Z}_1 Z_2) > 0$), we have from Eq. (3) that

$$\omega = 2\text{Im}(\bar{a}\, a_D) \frac{\langle q_1, q_2 \rangle}{M_1 M_2} + \mathcal{O}(\omega^2)\,, \tag{48}$$

from which it follows that the CMS curve is defined by [7]

$$\text{Im}\,(\bar{a}(u)\, a_D(u)) = 0\,, \tag{49}$$

which is sketched in Fig. 3. Substituting the expression (48) for ω into Eq. (44) we find the near-CMS potential

$$V_{\text{CMS}}(r) = -\frac{2\text{Im}(\bar{a}\, a_D)}{r} \frac{\langle q_1, q_2 \rangle^2}{M_1 M_2} + \mathcal{O}(\omega^2)\,. \tag{50}$$

The sign of the potential is determined by the sign of $\text{Im}(\bar{a}\, a_D)$ independently of the charges, provided $\langle q_1, q_2 \rangle \neq 0$. Thus the problem of deducing the stability domain for the relevant composite BPS states amounts to knowing the map between a_D/a and u. In this example the sign is easily seen to be positive, and thus bound states are formed, in the exterior region by making use of the semiclassical limit where $a_D = \tau a$ at large $|u|$.

Let us follow construction of the spectrum in more detail. In our discussion of $\mathcal{N} = 4$ SYM in the previous subsection, we chose as primary states the W boson $\{1,0\}$ and the monopole $\{0,1\}$ as they are the lightest states carrying electric and magnetic charge in the weak coupling region. However, in the present case the CMS in the upper-half plane connects two singular points at $u = -\Lambda^2$ and $u = \Lambda^2$ where respectively a $\{1,1\}$ dyon and a $\{0,1\}$ monopole become massless. It follows that in passing to strong coupling the W boson is replaced by the $\{1,1\}$ dyon as the lightest state carrying electric charge, which should then be chosen as a primary state.

The W boson must then be understood as a bound state [6,7,9] and indeed we can consider the two particle system of $\{0,-1\}$ and $\{1,1\}$ which has the appropriate quantum numbers to form W^+. The curve (49) is the CMS for this bound state, so we arrive at the scenario described above; the W boson exists as a bound state only in the exterior region of the CMS, its wavefunction swells on approach to the CMS, the system becomes delocalized, and the bound state does not exist in the interior. In other words, at large moduli, $|u| \gg |\Lambda|^2$, we have a point-like W boson with a small admixture of the dyon pair. As we move close to the CMS this pair becomes dominant, and the picture of the W boson as a bound state becomes appropriate.

We can build up the spectrum of BPS states following the procedure outlined for $\mathcal{N}=4$ SYM, with the new ingredients that we take $\{0,1\}$ and $\{1,1\}$ as primary constituents (in the upper half-plane), and remember that the potential is moduli-dependent. For a given composite configuration, we ask whether it is possible to arrange the primary constituents into subgroups in such a way that the energy is minimized when, for example, two of these subgroups are at infinite separation. If so, then no bound state is formed, while if the answer is negative, then we are guaranteed that a bound state is formed with those charges at that point in the moduli space. Consider first the possible composite configurations with magnetic charge ± 1. Following the above procedure, making use of the potential (50), we obtain all the dyons $\pm\{k,1\}$ with $k = 2,3,\ldots$ and $k = -1,-2,\ldots$ as bound states in the exterior region to the CMS, while we find no bound states in the interior region. The fact that these states are BPS is easily verified semiclassically. Together with the $W^\pm$ bosons these BPS particles form the well-known stable BPS spectrum [6,9] in the exterior region of the

CMS.

4.4. Non-BPS bound states in $\mathcal{N}=2$ SYM

We now proceed further and consider composite configurations with magnetic charge 2. We will focus on the simplest example corresponding to a system with charges $\{1,2\}$. We again take the states $\{0,1\}$ and $\{1,1\}$ as primary in the upper half plane. In this case, $\mathrm{Re}\left(\bar{\mathcal{Z}}^1 \mathcal{Z}^2\right) < 0$ on the curve (49), implying $\omega = \pi$ and we cannot use the potential (50), as there is no CMS for this state according to our definition above. However, we can still study the full Coulombic potential,

$$V_{\{1,2\}}(r) = \frac{1}{r\,\mathrm{Im}\,\tau}\,\mathrm{Re}\left\{\tau\,(1+\bar{\tau})\left(1 - \frac{a_D^*\,(a+a_D)}{|a_D|\,|a+a_D|}\right)\right\}. \tag{51}$$

In the weak coupling region, where $|u| \gg |\Lambda|^2$, we have $a_D = \tau a$ and this potential matches the $\mathcal{N}=4$ potential given in Eq. (47). It is attractive, and one finds a bound state descending from the $\mathcal{N}=4$ BPS state with magnetic charge two.

The appearance of states with magnetic charge 2 in $\mathcal{N}=2$ SYM appears puzzling at first sight. It follows from the structure of the Seiberg-Witten solution, for example, that no such BPS states are possible [9,10]. The natural resolution is that this is a *non-BPS* state. At large u, in the classical approximation, this state can be viewed as lying in a 1/2-BPS multiplet of $\mathcal{N}=4$ supersymmetry, or equivalently in a degenerate set of $\mathcal{N}=2$ BPS multiplets. In this approximation the soliton mass is given by the central charge. However, part of the $\mathcal{N}=4$ SUSY preserved by this state corresponds to generators carrying a nonzero value of s, i.e. involving fields from the $\mathcal{N}=2$ adjoint hypermultiplet. When quantum corrections, associated with the nonzero mass for the adjoint hypermultiplet, are taken into account one finds that the degeneracy allows the state to leave the BPS bound as a long representation of $\mathcal{N}=2$ supersymmetry [47].

Quantum corrections therefore lift the mass $M_{\{1,2\}}$ above the BPS bound $|\mathcal{Z}_{\{1,2\}}|$. Nonetheless, the presence of an attractive potential means that this mass is still less than the sum of the constituent masses,

$$\left|\mathcal{Z}_{\{1,2\}}\right| < M_{\{1,2\}}^{\mathrm{non-BPS}} < \left|\mathcal{Z}_{\{0,1\}}\right| + \left|\mathcal{Z}_{\{1,1\}}\right|. \tag{52}$$

The potential (51) is of course also valid at strong coupling and we can enquire as to the fate of these non-BPS states in this region. In fact we find that in the upper-half plane the potential vanishes on a new curve, somewhat outside the CMS, defined by,

$$\text{Re}\left\{\tau\left(1+\bar{\tau}\right)\left(1-\frac{a_D^*\left(a+a_D\right)}{|a_D||a+a_D|}\right)\right\}=0, \qquad (53)$$

which in [25] we called a *non-BPS curve of marginal stability* (nCMS). This curve is plotted in Fig. 3, and it is important to emphasize that, unlike the CMS, this curve has a purely dynamical definition as the submanifold on which the leading $1/r$ potential vanishes. Interestingly, although we cannot calculate the mass of the composite state on this curve there are compelling arguments that the nCMS curve corresponds to the point where the mass of the composite reaches the threshold where the second inequality in (52) is saturated, i.e. when

$$M_{\{1,2\}}^{\text{non}-\text{BPS}}(\text{nCMS})=\left|\mathcal{Z}_{\{0,1\}}\right|+\left|\mathcal{Z}_{\{1,1\}}\right|. \qquad (54)$$

If this were not the case, it is clear that the state would either disappear from the spectrum while stable (i.e. $M_{\{1,2\}}^{\text{non}-\text{BPS}}<\left|\mathcal{Z}_{\{0,1\}}\right|+\left|\mathcal{Z}_{\{1,1\}}\right|$), or would remain in the spectrum when genuinely unstable to decay (i.e. $M_{\{1,2\}}^{\text{non}-\text{BPS}}>\left|\mathcal{Z}_{\{0,1\}}\right|+\left|\mathcal{Z}_{\{1,1\}}\right|$). Thus it is natural to conclude that the nCMS corresponds to the threshold condition (54), while further progress leads to a repulsive potential and the non-BPS state disappears from the spectrum.

It is important to emphasize that the position of the nCMS curve is not corrected by higher derivative terms in the effective action, since it is defined in terms of the $1/r$ potential between BPS states, which is determined purely by the two-derivative sector of the low energy effective theory. One useful consequence of this is that the threshold condition (54) also receives no corrections. Consequently, although we do not know the mass of the non-BPS bound state for generic moduli, we *do* know its value precisely on the nCMS curve as it is given by the sum of the masses of its two BPS constituents. Moreover, near this curve the constraint (6) can always be satisfied and thus the nonrelativistic approximation is valid.

To complete this picture, we note that the $\{1,2\}$ state also exists in the lower-half plane, for large $|u|$, and in this case the light constituents at strong coupling are $\pm\{0,1\}$ and $\pm\{-1,1\}$. The state $\{1,2\}$ is then a four-particle configuration comprised as follows:

$$\{1,2\}\Longleftrightarrow\{1,-1\}+3\{0,1\}. \qquad (55)$$

Noting that the monopoles do not interact, while the potential for $\{1,-1\}$ and $\{0,1\}$ is attractive outside, and repulsive inside, the CMS in the lower-half plane, we deduce that a bound state exists in the full region exterior to the CMS in the lower-half plane. Given that we have argued that this state must be non-BPS, it is clear that the scenario we have found here corresponds to the case where both inequalities in (52) are saturated at the same point, implying that the CMS and the nCMS coincide. Thus, precisely on the CMS we should only find BPS multiplets. It follows that the transition to the CMS must lead to some degeneracy as the state $\{1,2\}$ lies in a long multiplet arbitrarily close to this point. Fortunately, this is indeed consistent with the delocalized four-particle configuration that follows from (55).

One might wonder about the global structure of the marginal stability curve for the $\{1,2\}$ state, which at first sight looks inconsistent with the residual $\mathbf{Z}_2$ symmetry $u \rightarrow -u$ of the theory. However, the precise symmetry includes a CP conjugation due to the fact that one must shift the basepoint used to define the charges, and thus the reflection of the curve applies instead to the state $\{-1,2\}$. Moreover, the appearance of non-BPS states of this type, descending from higher charge states in $\mathcal{N}=4$ SYM, has been conjectured before in the literature from the string web construction [48]. Furthermore, the existence criteria for these non-BPS states, and also their stability domains, as deduced here [25] agrees well with earlier arguments of Bergman [48] using the string junction formalism.

These examples provide sufficient information on the interaction channels for us to deduce the presence, and stability domains, for higher charge composite particles. In particular, we learn that all $\{n_E, n_M\}$ states with co-prime n_E and n_M – descending from $\mathcal{N}=4$ – represent non-BPS states stable in the semiclassical region. These states are removed from the spectrum in the strong coupling regime as we cross the appropriate nCMS or CMS curve.

5. Concluding Remarks

The relative simplicity of the interaction between BPS states near marginal stability curves makes this an ideal regime in which to address questions concerning the structure of the bound state spectrum. In this talk, we dwelt on two examples, namely BPS kinks in 1+1D and dyonic states in 3+1D. However, it is clear that there are many other systems which should be amenable to this approach. As emphasized in the preceding section, one can also make use of the interaction potential to deduce certain features of the non-BPS spectrum as well – something over which supersymmetry would naively appear to have little control.

As an example, we can consider the interaction of strings (or vortices) in a supersymmetric variant of the abelian Higgs model. Ignoring the world-volume dynamics, on general grounds we can write down the leading interaction in the form (7),

$$V_{\text{LO}}(r) \sim \eta_{12} K_0(m_\gamma, r) \sim \eta_{12} \frac{\exp(-m_\gamma r)}{\sqrt{m_\gamma r}} + \cdots \tag{56}$$

where $K_0(m_\gamma, r)$ is a modified Bessel function (the massive propagator in 2+1D), and the scale is set by the photon mass m_γ. To fix the leading order potential, one then requires knowledge of the dependence of the trilinear couplings $\eta^i(n)$ on the topological charge n of the vortex, and any additional moduli. This is an interesting, and apparently nontrivial, question in itself [49] (see [50] for a recent analysis).

Another issue under closer scrutiny [47] concerns terms of higher order in $1/r$ in the full interaction potential between dyonic BPS states (see also [36,20]). Although unimportant, as argued in Sect. 2, for generic questions of bound state formation near the CMS, these bosonic and fermionic corrections are required in order to understand in detail the restructuring of the supermultiplets that takes place on breaking $\mathcal{N}=4$ to $\mathcal{N}=2$ as these terms are, in contrast to the $1/r$ potential, spin-dependent.

Acknowledgments

It is a pleasure to thank my coworkers, Misha Shifman, Misha Voloshin, and especially Arkady Vainshtein for a highly stimulating and fruitful collaboration, and also for the invitation to present this work at such a lively and enjoyable meeting. I'd also like to acknowledge helpful discussions with P. Argyres, J. Gauntlett, K. Lee, R. Portugues, P. Yi, and A. Yung.

References

1. E. B. Bogomolny, Sov. J. Nucl. Phys. **24**, 449 (1976) [Yad. Fiz. **24**, 861 (1976)].
2. E. Witten and D. Olive, Phys. Lett. **B78**, 97 (1978).
3. H. Osborn, Phys. Lett. B **83**, 321 (1979).
4. P. Fendley, S. D. Mathur, C. Vafa and N. P. Warner, Phys. Lett. B **243**, 257 (1990); E. R. Abraham and P. K. Townsend, Nucl. Phys. B **351**, 313 (1991).
5. S. Cecotti and C. Vafa, Commun. Math. Phys. **158**, 569 (1993) [hep-th/9211097].
6. N. Seiberg and E. Witten, Nucl. Phys. **B426**, 19 (1994) [hep-th/9407087]; Nucl. Phys. **B431**, 484 (1994) [hep-th/9408099].
7. U. Lindstrom, F. Gonzalez-Rey, M. Rocek and R. von Unge, Phys. Lett. **B388**, 581 (1996) [hep-th/9607089]; A. Fayyazuddin, Mod. Phys. Lett. **A10**, 2703 (1995) [hep-th/9504120].
8. M. Henningson, Nucl. Phys. **B461**, 101 (1996) [hep-th/9510138].

9. F. Ferrari and A. Bilal, Nucl. Phys. **B469**, 387 (1996) [hep-th/9602082].

10. A. Bilal and F. Ferrari, Nucl. Phys. **B480**, 589 (1996) [hep-th/9605101]; Nucl. Phys. **B516**, 175 (1998) [hep-th/9706145].

11. A. Brandhuber and S. Stieberger, Nucl. Phys. **B488**, 199 (1997) [hep-th/9610053].

12. A. Bilal and F. Ferrari, Nucl. Phys. **B516**, 175 (1998) [hep-th/9706145].

13. A. Smilga and A. Veselov, Phys. Rev. Lett. **79**, 4529 (1997) [hep-th/9706217]; Nucl. Phys. **B515**, 163 (1998) [hep-th/9710123]; Phys. Lett. **B428**, 303 (1998) [hep-th/9801142].

14. A. V. Smilga, Phys. Rev. **D58**, 065005 (1998) [hep-th/9711032].

15. O. Bergman, Nucl. Phys. **B525**, 104 (1998) [hep-th/9712211].

16. O. Bergman and A. Fayyazuddin, Nucl. Phys. **B531**, 108 (1998) [hep-th/9802033];
Nucl. Phys. **B535**, 139 (1998) [hep-th/9806011];
A. Mikhailov, N. Nekrasov, and S. Sethi, Nucl. Phys. **B531**, 345 (1998) [hep-th/9803142].

17. D. Joyce, *On counting special Lagrangian 3-spheres*, hep-th/9907013;
S. Kachru and J. McGreevy, Phys. Rev. **D61**, 026001 (2000) [hep-th/9908135].

18. M. R. Douglas, B. Fiol and C. Romelsberger, arXiv:hep-th/0002037;
M. R. Douglas, B. Fiol and C. Romelsberger, *The spectrum of BPS branes on a noncompact Calabi-Yau*, hep-th/0003263;
P. S. Aspinwall and M. R. Douglas, JHEP **0205**, 031 (2002) [arXiv:hep-th/0110071].

19. F. Denef, JHEP **0008**, 050 (2000) [arXiv:hep-th/0005049];
F. Denef, B. Greene and M. Raugas, JHEP **0105**, 012 (2001) [arXiv:hep-th/0101135];
F. Denef, arXiv:hep-th/0107152.

20. F. Denef, arXiv:hep-th/0206072.

21. B. Fiol and M. Marino, JHEP **0007**, 031 (2000) [hep-th/0006189];
B. Fiol, *The BPS spectrum of $N = 2$ $SU(N)$ SYM and parton branes*, hep-th/0012079.

22. E. B. Bogomolny and A. I. Vainshtein, Sov. J. Nucl. Phys. **23**, 588 (1976).

23. see e.g. P.G. de Gennes, *Superconductivity of Metals and Alloys*, (Benjamin, New York, 1966).

24. A. Ritz, M. Shifman, A. Vainshtein and M. Voloshin, Phys. Rev. D **63**, 065018 (2001) [hep-th/0006028].

25. A. Ritz and A. I. Vainshtein, Nucl. Phys. B **617**, 43 (2001) [arXiv:hep-th/0102121].

26. R. Portugues and P. K. Townsend, Phys. Lett. B **530**, 227 (2002) [arXiv:hep-th/0112077].

27. N. S. Manton, Phys. Lett. B **154**, 397 (1985) [Erratum-ibid. **157B**, 475 (1985)].

28. G. W. Gibbons and N. S. Manton, Phys. Lett. B **356**, 32 (1995) [arXiv:hep-th/9506052].

29. K. Lee, E. J. Weinberg and P. Yi, Phys. Rev. **D54**, 1633 (1996) [hep-th/9602167].

30. C. Fraser and T. J. Hollowood, Phys. Lett. **B402**, 106 (1997) [hep-th/9704011].
31. K. Lee and P. Yi, Phys. Rev. **D58**, 066005 (1998) [hep-th/9804174].
32. D. Tong, Phys. Lett. **B460**, 295 (1999) [hep-th/9902005].
33. D. Bak, C. Lee, K. Lee and P. Yi, Phys. Rev. **D61**, 025001 (2000) [hep-th/9906119].
34. D. Bak, K. Lee and P. Yi, Phys. Rev. **D61**, 045003 (2000) [hep-th/9907090].
35. B. Kol and M. Kroyter, *On the spatial structure of monopoles*, hep-th/0002118.
36. P. C. Argyres and K. Narayan, JHEP **0103**, 047 (2001) [arXiv:hep-th/0101114].
37. For a recent discussion, see e.g. M. M. Nieto, Phys. Lett. A **293**, 10 (2002) [arXiv:hep-th/0106233].
38. J. P. Gauntlett, N. Kim, J. Park and P. Yi, Phys. Rev. **D61**, 125012 (2000) [hep-th/9912082].
39. M. Stern and P. Yi, Phys. Rev. **D62**, 125006 (2000) [hep-th/0005275].
40. J. P. Gauntlett, C. Kim, K. Lee and P. Yi, Phys. Rev. D **63**, 065020 (2001) [hep-th/0008031].
41. S. Cecotti, P. Fendley, K. A. Intriligator and C. Vafa, Nucl. Phys. B **386**, 405 (1992) [arXiv:hep-th/9204102].
42. E. Witten, Nucl. Phys. B **188**, 513 (1981).
43. E. Witten, Phys. Lett. **B86**, 283 (1979).
44. C. Montonen and D. Olive, Phys. Lett. **B72**, 117 (1977).
45. A. Sen, Phys. Lett. **B329**, 217 (1994) [hep-th/9402032].
46. F. Ferrari, Nucl. Phys. B **501**, 53 (1997) [hep-th/9702166].
47. A. Ritz and A. Vainshtein, in preparation.
48. O. Bergman, JHEP**9905**, 004 (1999) [hep-th/9811064].
49. H. J. de Vega and F. A. Schaposnik, Phys. Rev. D **14**, 1100 (1976).
50. A. Marshakov and A. Yung, arXiv:hep-th/0202172; see also A. Yung, these proceedings.

NON-ABELIAN MONOPOLES, VORTICES AND CONFINEMENT

KENICHI KONISHI

Dipartimento di Fisica "E. Fermi" – Università di Pisa
Istituto Nazionale di Fisica Nucleare – Sezione di Pisa
Via Buonarroti, 2, Ed. C, 56127 Pisa, Italy
konishi@df.unipi.it

Three closely related issues will be discussed. Magnetic quarks having non-Abelian charges have been found recently to appear as the dominant infrared degrees of freedom in some vacua of softly broken $\mathcal{N} = 2$ supersymmetric QCD with $SU(n_c)$ gauge group. Their condensation upon $\mathcal{N} = 1$ perturbation causes confinement and dynamical symmetry breaking. We argue that these magnetic quarks can be naturally related to the semiclassical non-Abelian monopoles of the type first discussed by Goddard, Nuyts, Olive and E. Weinberg. We discuss also general properties of non-Abelian vortices and discuss their relevance to the confinement in QCD. Finally, calculation by Douglas and Shenker of the tension ratios for vortices of different N-alities in the softly broken $\mathcal{N} = 2$ supersymmetric $SU(N)$ Yang-Mills theory, is carried to the second order in the adjoint multiplet mass. A correction to the ratios violating the sine formula is found, showing that the latter is not a universal quantity.

1. Confining vacua of softly broken $\mathcal{N} = 2$ supersymmetric QCD

Recently detailed properties of confining vacua have been studied in a class of softly broken $\mathcal{N} = 2$ supersymmetric gauge theories. Confining vacua in $SU(n_c)$, $USp(2n_c)$ or $SO(n_c)$ gauge theories with softly broken $\mathcal{N} = 2$ supersymmetry, with various number of flavors $n_f < 2n_c$, $2n_c + 2$, $n_c - 2$, respectively, have been found [1,2] to fall into (roughly speaking) the following three types (see Table 1 for the phases in $SU(n_c)$ theories):

In some of the vacua (the $r = 0$ or $r = 1$ vacua of $SU(n_c)$ theories; also confining vacua of all flavorless cases [3,4,5]), the gauge group of the low-energy dual theory is the maximal Abelian subgroup $U(1)^R$, where R is the rank of the original gauge group; confinement is described by 't Hooft-Mandelstam mechanism[6].

Table 1. Phases of $SU(n_c)$ gauge theory with n_f flavors, taken from [1]. $\tilde{n}_c \equiv n_f - n_c$.

label (r)	Deg.Freed.	Eff. Gauge Group	Phase	Global Symmetry
0	monopoles	$U(1)^{n_c-1}$	Conf.	$U(n_f)$
1	monopoles	$U(1)^{n_c-1}$	Conf.	$U(n_f-1) \times U(1)$
$< [\frac{n_f}{2}]$	dual quarks	$SU(r) \times U(1)^{n_c-r}$	Conf.	$U(n_f-r) \times U(r)$
$n_f/2$	rel. nonloc.	-	SCFT	$U(\frac{n_f}{2}) \times U(\frac{n_f}{2})$
$\tilde{n}_c$	dual quarks	$SU(\tilde{n}_c) \times U(1)^{n_c-\tilde{n}_c}$	Free Mag	$U(n_f)$

In the general r vacua $(2 \le r < \frac{n_f}{2})$ of the $SU(n_c)$ theory, the effective low-energy theory is a non-Abelian $SU(r) \times U(1)^{n-r}$ gauge theory; massless magnetic monopoles in the fundamental representation of dual $SU(r)$ gauge group appear as the low-energy degrees of freedom. Their condensation, together that of Abelian monopoles of the $U(1)^{n-r-1}$ factors, describes the confinement as a generalized dual Meissner effect. The vacua in the same universality classes appear in $USp(2n_c)$ and $SO(n_c)$ theories with nonzero bare quark masses;

In the $r = \frac{n_f}{2}$ vacua of $SU(n_c)$ theory, as well as in *all* of confining vacua of $USp(2n_c)$ and $SO(n_c)$ theories with massless flavor [a], the low-energy degrees of freedom involve relatively non-local objects: the low-energy theory is a deformed superconformal theory, i.e., near an infrared fixed-point.

2. Non-Abelian Monoples

We argue first that the "dual quarks" appearing in the r-vacua of the softly broken $\mathcal{N} = 2$ $SU(n_c)$ theories can naturally be identified with the non-Abelian magnetic monopoles of the type first discussed by Goddard, Nuyts and Olive [7] and studied further by E. Weinberg [8]. Our argument is based on the simple observations as regards to their charges, flavor quantum numbers, and some general properties of electromagnetic duality [9].

2.1. *Charges of non-Abelian monopoles*

Consider [7] a broken gauge theory,

$$G \overset{\langle \phi \rangle \neq 0}{\Longrightarrow} H$$

---

[a]There are exceptions to this rule for small values of n_f and n_c, e.g., $USp(2) = SU(2)$ case. See the footnote 18 of [1].

Table 2. Some examples of dual
pairs of groups

$SU(N)/Z_N$	$\Leftrightarrow$	$SU(N)$
$SO(2N)$	$\Leftrightarrow$	$SO(2N)$
$SO(2N+1)$	$\Leftrightarrow$	$USp(2N)$

where the unbroken group H is in general non-Abelian. In order to have a finite mass, the scalar fields must behave asymptotically as

$$\mathcal{D}\phi \overset{r\to\infty}{\longrightarrow} 0 \quad\Rightarrow\quad \phi \sim U \cdot \langle\phi\rangle \cdot U^{-1}, \quad A_i^a \sim U \cdot \partial_i U^\dagger \to \epsilon_{aij}\frac{r_j}{r^3}G(r), \quad (2.1)$$

with $\mathcal{D}G = 0$, representing nontrivial elements of $\Pi_2(G/H) = \Pi_1(H)$. The function $G(r)$ can be chosen as

$$G(r) = \beta_i T_i, \qquad T_i \in \text{Cartan Subalgebra of } H. \tag{2.2}$$

Topological quantization leads to the result that the "charges" β_i take values which are weight vectors of the group $\tilde{H} = $ dual of H. The dual of a group (whose roots vectors are α's) is by definitioin has the root vectors which span dual lattice, i.e., $\tilde{\alpha} = \alpha/\alpha^2$. Examples of pairs of the duals are given in the Table 2

As an example, consider an $SU(3)$ theory broken as

$$SU(3)\overset{\langle\phi\rangle}{\longrightarrow}SU(2) \times U(1), \qquad \langle\phi\rangle = \begin{pmatrix} v & 0 & 0 \\ 0 & v & 0 \\ 0 & 0 & -2v \end{pmatrix}. \qquad (\%) \quad (2.3)$$

Take a subgroup $SU_U(2) \subset SU(3)$

$$t^4 = \frac{1}{2}\begin{pmatrix} 0 & 0 & 1 \\ 0 & 0 & 0 \\ 1 & 0 & 0 \end{pmatrix}; \quad t^5 = \frac{1}{2}\begin{pmatrix} 0 & 0 & -i \\ 0 & 0 & 0 \\ i & 0 & 0 \end{pmatrix}; \quad \frac{t^3 + \sqrt{3}t^8}{2} = \frac{1}{2}\begin{pmatrix} 1 & 0 & 0 \\ 0 & 0 & 0 \\ 0 & 0 & -1 \end{pmatrix}; \quad (2.4)$$

then

$$SU_U(2)\overset{\langle\phi\rangle}{\longrightarrow}U_U(1). \qquad (*) \tag{2.5}$$

Embedding the 't Hooft-Polyakov monopole solution $\phi(r), A(r)$ for $(*)$ [10] one gets a $SU(3)$ solution (Sol. 1) :

$$\phi = \begin{pmatrix} -\frac{1}{2}v & 0 & 0 \\ 0 & v & 0 \\ 0 & 0 & -\frac{1}{2}v \end{pmatrix} + \frac{3}{2}v\Big(t_4, t_5, \frac{t_3}{2} + \frac{\sqrt{3}t_8}{2}\Big) \cdot \hat{r}\phi(r), \tag{2.6}$$

$$\vec{A} = \Big(t_4, t_5, \frac{t_3}{2} + \frac{\sqrt{3}t_8}{2}\Big) \wedge \hat{r}A(r). \tag{2.7}$$

Together with another solution (Sol.2) with $SU_V(2) \subset SU(3)$

$$t^6 = \frac{1}{2}\begin{pmatrix} 0 & 0 & 0 \\ 0 & 0 & 1 \\ 0 & 1 & 0 \end{pmatrix}; \quad t^7 = \frac{1}{2}\begin{pmatrix} 0 & 0 & 0 \\ 0 & 0 & -i \\ 0 & i & 0 \end{pmatrix}; \quad \frac{-t^3 + \sqrt{3}t^8}{2} = \frac{1}{2}\begin{pmatrix} 0 & 0 & 0 \\ 0 & 1 & 0 \\ 0 & 0 & -1 \end{pmatrix};$$

they yield a degenerate doublet of monopoles with charges

monopoles	$\tilde{SU}(2)$	$\tilde{U}(1)$
$\tilde{q}$	$\underline{2}$	1

This construction can be generalized to cases with gauge symmetry breaking

$$SU(n)\xrightarrow{\langle\phi\rangle}SU(r)\times U^{n-r}(1), \qquad \langle\phi\rangle = \begin{pmatrix} v_1 1_{r\times r} & 0 & \cdots & 0 \\ 0 & v_2 & 0 & \cdots \\ 0 & 0 & \ddots & \cdots \\ 0 & 0 & \cdots & v_{n-r+1} \end{pmatrix}.(2.8)$$

By considering various $SU_i(2)$ subgroups ($i = 1, 2, \ldots, r$) living in $[i, r+1]$ subspace we find

(i) a degenerate r-plet of stable monopoles (q), gauge (Weyl-) transformed to each other by $SU(r) \subset SU(n)$;

(ii) Abelian monopoles (e_i), ($i = 1, 2, \ldots, n - r - 1$) of $U^{n-r}(1)$ (non degenerate).

The charges of these stable monopoles are identical to those found in the r-vacua of the softly broken $\mathcal{N} = 2$ SQCD (Table.3)! In particular, as will be shown in the next subsection these non-Abelian monopoles can acquire flavor quantum numbers through the (generalized) Jackiw-Rebbi mechanism [11].

2.2. Fermion Zero modes in non-Abelian monopole Background

We now couple fermions in the fundamental representation of the gauge group. To be concrete consider the case of a $SU(3)$ theory. The fundamental multiplet,

$$\psi_L = \psi_{L(2)} \oplus \psi_{L(0)}, \qquad \psi_R = \psi_{R(2)} \oplus \psi_{R(0)} \tag{2.9}$$

satisfies the Dirac equation $\gamma_i \mathcal{D}_i \psi = 0$. More explicitly,

$$-\vec{\sigma} \cdot \vec{p}\psi_{L(2)} - e\vec{\sigma} \cdot (\vec{t} \wedge \hat{r})A(r)\psi_{L(2)} - \frac{1}{2}v\psi_{R(2)} + 3v\vec{t} \cdot \hat{r}\psi_{R(2)}\phi(r) = 0,$$

$$-\vec{\sigma} \cdot \vec{p}\psi_{L(0)} + v\psi_{R(0)} = 0,$$

$$\vec{\sigma} \cdot \vec{p}\psi_{R(2)} + e\vec{\sigma} \cdot (\vec{t} \wedge \hat{r})A(r)\psi_{R(2)} - \frac{1}{2}v\psi_{L(2)} + \frac{3v}{2}\vec{t} \cdot \hat{r}\psi_{L(2)}\phi(r) = 0,$$

Table 3. The effective degrees of freedom and their quantum
numbers at a confining r-vacua $[2,1]$.

	$SU(r)$	$U(1)_0$	$U(1)_1$	$\ldots$	$U(1)_{n_c-r-1}$
$n_f \times q$	r	1	0	$\ldots$	0
e_1	1	0	1	$\ldots$	0
$\vdots$	$\vdots$	$\vdots$	$\vdots$	$\ddots$	$\vdots$
e_{n_c-r-1}	1	0	0	$\ldots$	1

$$\vec{\sigma} \cdot \vec{p}\psi_{R(0)} + v\psi_{L(0)} = 0. \tag{2.10}$$

Through the Yukawa coupling, the fermion acquired a mass, $m = \frac{v}{2}$. Generalizing the Jackiw-Rebbi analysis to the massive fermions, it can be shown that a normalizable zero mode exists if $3v > v$ which is obviously satisfied. Each fermion gets one zero mode; quantum mechanically, the monopoles become flavor multiplets.

An analogous construction in the case of the breaking $SU(n_c) \rightarrow SU(r) \times U(1)^{n_c-r}$, the above condition is replaced by

$$\left| \frac{v_0 - v_{r+1}}{2} \right| > \left| \frac{v_0 + v_{r+1}}{2} \right|. \tag{2.11}$$

Note that for the breaking $SU(n) \rightarrow SU(n-1) \times U(1)$ such a condition is always satisfied; otherwise, only the monopoles with VEVS satisfying the above condition will give rise to fermion zero modes.

This mechanism "explains" the low-energy degrees of freedom in the "r" vacua of softly broken $N = 2$ SQCD, with $G = SU(n_c)$, with n_f quarks:

2.3. Duality

It is also significant that, in the softly broken $\mathcal{N} = 2$ $SU(n_c)$ theory, the r vacua with a magnetic $SU(r)$ gauge group occur only for $r \leq \frac{n_f}{2}$. This is a manifestation of the fact that the quantum behavior of non-Abelian monopoles depends crucially on the massless matter fermion degrees of freedom in the fundamental theory. Indeed, the magnetic $SU(r) \times U(1)^{n_c-r}$ theory with these matter multiplets is infrared-free (i.e., non asymptotic free). This is the correct behavior as it should be dual to the original asymptotic free $SU(n_c)$ gauge theory. Note that the gauge coupling constant evolution, which appears as due to the perturbative loops of magnetic monopoles, is actually the result of, and equivalent to, the infinite sum of instanton contributions in the original $SU(n_c)$ theory.

This is perfectly analogous to the observation [12] about how the old paradox related to the Dirac quantization condition and renormalization group [13] :

$$g_e(\mu) \cdot g_m(\mu) = 2\pi n, \qquad \forall \mu, \tag{2.12}$$

is solved within the $SU(2)$ Seiberg-Witten theory.

Note that this also explains why in the pure $\mathcal{N} = 2$ $SU(n_c)$ theory or on a generic point of the Coulomb branch of the $\mathcal{N} = 2$ SQCD, the low-energy effective theory is an Abelian gauge theory [3]-[5]. Massless fermion flavors are needed in order for non-Abelian monopoles to get dressed, via a generalized Jackiw-Rebbi mechanism discussed above with a non trivial $SU(n_f)$ flavor quantum numbers and, as a result, to render the dual gauge interactions infrared-free. When this is not possible, non-Abelian monopoles are strongly coupled and do not manifest themselves as identifiable low-energy degrees of freedom.

In this respect, it is very interesting that the boundary case $r = \frac{n_f}{2}$ also occurs (confining vacua of type (iii) discussed in Introduction) within the class of supersymmetric theories considered in [1]. In these vacua, non-Abelian monopoles and dyons are strongly coupled, but still describe the low-energy dynamics, albeit via non-local effective interactions.

Non-Abelian monopoles are actually quite elusive objects. Though their presence may be detected in a semi-classical approximation, their true nature depends on the long distance physics. If the "unbroken" gauge group is dynamically broken further in the infrared such multiplets of states simply represent an approximately degenerate set of magnetic monopoles. Only if there is no further dynamical breaking do the non-Abelian monopoles transforming as nontrivial multiplets of the unbroken, dual gauge group, appear in the theory.

There are strong indications that this occurs in the r-vacua (with an effective $SU(r) \times U(1)^{n_c-r}$ gauge symmetry) of the softly broken $\mathcal{N} = 2$, $SU(n_c)$ supersymmetric QCD [1]. If our idea is correct, this is perhaps the first physical system known in which Goddard-Nuyts-Olive-Weinberg monopoles manifest themselves as infrared degrees of freedom, playing an essential dynamical role. For more about the subtle nature of nonAbelian monopoles, see [9].

3. Non-Abelian Vortices

A closely related issue is that of non-Abelian vortices [14]-[19]. If confinement is to be described as a sort of non-Abelian dual Meissner effect, the magnetic monopoles of the type discussed above condense and break the dual gauge group. As a result, the system develops vortex configurations which serve as confining strings.

3.1. *General Characterization*

This time we consider a gauge theory in which the gauge group G is spontaneously broken by the Higgs mechanism as

$$G \Longrightarrow C \tag{3.1}$$

with C a *discrete* center of the group. The general properties of the vortex, which represents a nontrivial elements of the fundamental group,

$$\Pi_1(G/C) = C, \tag{3.2}$$

are independent of the detailed form of the scalar potential or of the number of the Higgs fields present. Asymptotic form of the fields are:

$$A_i \sim \frac{i}{g} U(\phi) \partial_i U^\dagger(\phi); \quad \phi_A \sim U \phi_A^{(0)} U^\dagger, \quad U(\phi) = \exp i \sum_j^r \beta_j T_j \phi$$

where T_i's can be taken in the Cartan subalgebra of G: then

$$A_\phi \sim \frac{1}{gr} \sum_j^r \beta_j T_j$$

The vortex flux

$$\oint dx_i A_i = \frac{2\pi}{g} \sum_j^r \beta_j T_j,$$

is characterized by the "charges" β. The quantization condition

$$U(2\pi) \in C. \tag{3.3}$$

leads to the result that β_j's are weight vectors of $\tilde{G}$ (dual of G). β_j's are actually defined modulo Weyl transformations β_i's:

$$\beta' = \beta - \frac{2\alpha(\beta \cdot \alpha)}{(\alpha \cdot \alpha)},$$

where α is a root vector of G.

3.2. $SU(N)/Z_N$

The simplest system with non-Abelian vortices is $SO(3) = SU(2)/Z_2$ broken to Z_2. It has

- unique Z_2 vortex (the source charge additive *mod 2*);

- " flux "

$$\int_S F_{ij}^3 \, d\sigma_{ij} = \oint dx_i A_i^3 = \frac{2\pi n}{g}.$$

which is conserved but not gauge invariant. $n = 2$ "vortex" can be gauge-transformed away.

A more interesting system is $\mathbf{SU(3)/Z_3}$, i.e., $SU(3)$ theories with all fields in adjoint representation. The Cartan subalgebra can be taken to be

$$T_3 = \frac{1}{2\sqrt{3}} \begin{pmatrix} 1 & 0 & 0 \\ 0 & -1 & 0 \\ 0 & 0 & 0 \end{pmatrix} ; \qquad T_8 = \frac{1}{6} \begin{pmatrix} 1 & 0 & 0 \\ 0 & 1 & 0 \\ 0 & 0 & -2 \end{pmatrix} .$$

The quantization condition $U(2\pi) \in Z_3$ leads to the equations

$$\frac{\beta_3}{2\sqrt{3}} + \frac{\beta_8}{6} = -\frac{n_1}{3}, \qquad -\frac{\beta_3}{2\sqrt{3}} + \frac{\beta_8}{6} = -\frac{n_2}{3}, \qquad -\frac{1}{3}\beta_8 = -\frac{n_3}{3}, \qquad (3.4)$$

to be solved with the condition, $\sum_i n_i = 0$. The simplest N-ality (triality) one $(n_i = [1 \bmod 3])$ solutions are:

$$\beta = (-\sqrt{3}, 1), \ (\sqrt{3}, 1), \ \text{or} \ (0, -2) \ = 2N\mathbf{w} \qquad (3.5)$$

$\mathbf{w}$ = weight vector of $\mathbf{3}$. Thus the sources of the minimum vortex carry the quantum number of the quarks. The dual of the theory we are studying, $SU(3)/Z_3$, is indeed $SU(3)$!

By adding four of $(35) \to \mathbf{6^*}$ (⊞), etc., and one could construct an infinite number of triality-one solutions. However only the vortex with the lowest tension is stable.

N-ality (triality)-two solutions are found by adding vectorially the minimum solutions above. The source of these vortices correspond to the irreducible representations has charges

$$\beta = (-\sqrt{3}, -1), \ (\sqrt{3}, -1), \ \text{or} \ (0, 2),$$

= weight vectors of $\underline{3}^*$ (⊟), or

$$\beta = (-2\sqrt{3}, 2), \ (2\sqrt{3}, 2), \ (0, 2). \ (-\sqrt{3}, -1), \ (\sqrt{3}, -1), \ \text{or} \ (0, -4),$$

= $\underline{6}$ (⊏⊐). Quantum mechanically, however, the vortex with the higher tension (probably $\underline{6}$) decays through the gauge boson pair productions (Fig. 1). Somewhat similar problem of decay of metastable vortices was recently discussed by Shifman and Yung [20].

Since $\underline{3}^*$ vortex and $\underline{3}$ vortex are equivalent, there is actually a unique stable vortex with minimum $\mathbb{Z}_3$ charge in the $SU(3)$ gauge theory.

The discussion can be generalized natually to $\mathbf{SU(N)/Z_N}$ theory in Higgs phase. One finds N degenerate solutions of N-ality one

$$\beta_j = 2N\mathbf{w}_j, \qquad j = 1, 2, 3, \ldots, N$$

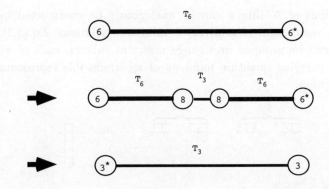

Figure 1.

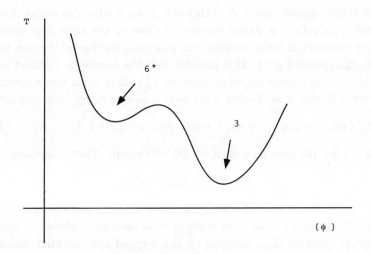

Figure 2.

$$\mathbf{w}_j = \text{the weight vectors of the } \underline{N}. \tag{3.6}$$

There are also solutions representing vortices of higher N-alities. At the N-ality two, for instance, the solutions for β have the form,

$$2N(\mathbf{w}_i + \mathbf{w}_j), \qquad i, j = 1, 2, \ldots, N. \tag{3.7}$$

They fall into two gauge inequivalent sets of vortices: their sources would carry the quantum numbers of the two irreducible representations,

$$\square\square, \quad \begin{array}{c}\square\\\square\end{array}, \tag{3.8}$$

symmetric and antisymmetric in color, respectively.

Solutions of N-ality k can be analogously be constructed by taking as β the vector sum of arbitrary k minimum solutions, Eq.(3.2). These vortices can be grouped into gauge invariant subsets, each of which has a source carrying quantum numbers of an irreducible representations of $SU(N)$ group,

$$
\overbrace{\boxed{\text{□□□}}\ldots\boxed{\text{□□}}}^{k}, \quad \overbrace{\boxed{\text{□}}\ldots\boxed{\text{□□}}}^{k-1}, \quad \ldots, \quad \boxed{\text{□}}, \tag{3.9}
$$

all having k boxes.

The vortices of N-ality, $1 \le k \le N - 1$, cannot be unwound by a gauge transformation. Nevertheless, this does not mean that each of the vortices (39) is stable against decay. A vortex of a given N-ality can decay through the pair production of gauge bosons into one of the same $\mathbb{Z}_N$ quantum number but with a lower tension, via processes similar to the one in the $SU(3)$ example of Fig. 1. It is possible that the tension is smallest in the case of the antisymmetric representation $\binom{N}{k}$. If it is so, the solution for the vortex charge β at N-ality k is truely a unique gauge-invariant set

$$
2N \{ \mathbf{w}_{i_1} + \mathbf{w}_{i_2} + \ldots + \mathbf{w}_{i_k} \quad mod \quad \alpha \}, \qquad i_m = 1, 2, \ldots, N, \tag{3.10}
$$

where α's are the root vectors of the $SU(N)$ group. These represent

$$
\Pi_1(SU(N)/Z_N) = Z_N.
$$

Which of these, apart from the smallest, N-ality one vortex, is stable against decay into a bundle of vortices with smaller N-alities, is again a dynamical question (i.e., depends on the form of the potential, values of coupling constants, quantum corrections, etc.). One would expect no universal formula for the relative tensions among vortices of different N-alities, on the general ground. However, there are some intriguing suggestions [21] that the ratios among the vortex tensions for different $\mathbb{Z}_N$ charges, found originally in the pure $\mathcal{N} = 2$ supersymmetric Yang-Mills theory (broken softly to $\mathcal{N} = 1$) [22],

$$
\frac{T_k}{T_\ell} = \frac{\sin \frac{\pi k}{N}}{\sin \frac{\pi \ell}{N}}, \tag{3.11}
$$

might be universal. The results from lattice calculations with $SU(5)$ and $SU(6)$ Yang-Mills theories [23,24] are consistent with the sine formula. More recent results on these ratios [25,26] however seem to indicate the non-universality of these ratios.

The absence of vortices of N-ality, N, can be understood since the charges corresponding to an irreducible representation with N boxes in the Young tableau, can always be screened by those of the dynamical fields (adjoint representation): the vortex is broken by copious production of massless gluons of the dual $SU(N)$ theory.

In an analogous fashon, one finds that sources of vortices in $\mathbf{USp(2N)}$ theory in Higgs phase carry the weights of the 2^N dimensional spinor representation of the dual group, $SO(2N+1)$; sources of vortices in $\mathbf{SO(2N+1)}$ theory in Higgs phase carry the weights of the $2N$ dimensional fundamental representation of $USp(2N)$, etc. For more details, see [27].

3.3. Remarks

(i) If confinement in $SU(N)$ theory is a dual Meissner effect with Olive-Montonen duality, $SU(N) \leftrightarrow SU(N)/Z_N$, then the universal $q - \bar{q}$ meson Regge trajectory will be naturally explained, in contrast to the case when the dual theory is $U(1)^{N-1}$;

(ii) Sources of the non-Abelian vortices have charge additive only *mod N*. Non-Abelian vortices are non BPS: linearized approximation is not valid in general;

(iii) Explicit construction of non-Abelian vortices [14]-[19] has been studied by using simple models for the adjoint scalar potential. However a systematic study of non-Abelian vortices, hence of non-Abelian superconductors, are still lacking.

(iv) What is the relation between vortex formation and XSB?

(v) Can we compute the ratios of vortex tensions for different N-alities (in the $SU(N)$ case)?

This last point brings us to our third issue, related to Eq.(311).

4. Non-Universal Corrections in the Tension Ratios in softly broken $\mathcal{N} = 2$ $SU(N)$ Yang-Mills

Derivation of formula such as Eq.(311) in the standard, continuous $SU(N)$ gauge theories still defies us. The first field-theoretic result on this issue was obtained by Douglas and Shenker [22], in the $\mathcal{N} = 2$ supersymmetric $SU(N)$ pure Yang Mills theory, with supersymmetry softly broken to $\mathcal{N} = 1$ by a small adjoint scalar multiplet mass m. They found Eq.(311) for the ratios of the tensions of abelian (Abrikosov-Nielsen-Olesen) [28] vortices corresponding to different $U(1)$ factors of the low-energy effective (magnetic) $U(1)^{N-1}$ theory.

The n-th color component of the quark has charges

$$\delta_{n,k} - \delta_{n,k+1}, \qquad (k = 1, 2, \ldots, N - 1;\ n = 1, 2, \ldots, N) \qquad (4.1)$$

with respect to the various electric $U_k(1)$ gauge groups. The source of the k-th ANO string thus corresponds to the N-ality k multiquark state, $|k\rangle = |q_1 q_2, \ldots q_k\rangle$, allowing a re-interpretation of Eq.(311) as referring to the ratio of the tension for different N-ality confining strings [29].

However, physics of the softly broken $\mathcal{N} = 2$ $SU(N)$ pure Yang-Mills theory is quite different from what is expected in QCD. Dynamical $SU(N) \to U(1)^{N-1}$ breaking introduces multiple of meson Regge trajectories with different slopes at low masses [29,30], a feature which is neither seen in Nature nor expected in QCD. For instance, another N-ality k state $|k\rangle' = |q_2 q_3, \ldots q_{k+1}\rangle$ acts as source of the $U_{k+1}(1)$ vortex and as the sink of the $U_2(1)$ vortex, which together bind $|k\rangle'$- anti $|k\rangle'$ states with a tension different from T_k. The Douglas-Shenker prediction is, so to speak, a good prediction for a wrong theory! Only in the limit of $\mathcal{N} = 1$ does one expect to find one stable vortex for each N-ality, corresponding to the conserved Z_N charges [29].

Within the softly broken $\mathcal{N} = 2$ $SU(N)$ theory, the two regimes can be in principle smoothly interpolated by varying the adjoint mass m from zero to infinity, adjusting appropriately Λ. At small m one has a good local description of the low-energy effective dual, magnetic $U(1)^{N-1}$ theory. The transition towards large m regime involves both perturbative and non-perturbative effects. Perturbatively, there are higher corrections due to the $\mathcal{N} = 1$ perturbation, $m \operatorname{Tr} \Phi^2$. Nonperturbatively - in the dual theory - there are productions of massive gauge bosons of the broken $SU(N)/U(1)^{N-1}$ generators, which mix different $U(1)^{N-1}$ vortices and eventually lead to the unique stable vortex with a given $\mathcal{N}$-ality.

Below is the result on the perturbative corrections to the tension ratios Eq.(311), due to the next-to-lowest contributions in m. We shall find a small non-universal correction to the sine formula Eq.(311). Our point is not that such a result is of interest in itself as a physical prediction but that it gives a strong indication for the non-universality of this formula, even though it could be an approximately a good one.

The problem of the next-to-lowest contributions in m has been already analized in $SU(2)$ theory, by Vainshtein and Yung [30] and by Hou [31], although in that case there is only one $U(1)$ factor. When only up to the order A_D term in the expansion

$$m \langle \operatorname{Tr} \Phi^2 \rangle = m\, U(A_D) = m\, \Lambda^2 \left(1 - \frac{2i A_D}{\Lambda} - \frac{1}{4} \frac{A_D^2}{\Lambda^2} + \ldots \right) \qquad (4.2)$$

is kept, the effective low energy theory turns out to be an $\mathcal{N} = 2$ SQED, A_D being an $\mathcal{N} = 2$ analogue of the Fayet-Iliopoulos term. As a result, the vortex remains BPS-saturated, and its tension is proportional to the monopole charge [30,31]. When the A_D^2 term is taken into account, the vortex ceases to be BPS-saturated: the correction to the vortex tension can be calculated perturbatively, giving rise to the results that the vacuum behaves as a type I superconductor.

Our aim here is to generalize these analyses to $SU(N)$ theory. In fact, Douglas-Shenker result Eq.(311) in $SU(N)$ theory was obtained in the BPS approximation, by keeping only the linear terms in a_{Di} in the expansion

$$U(a_{Di}) = U_0 + U_{0k}\, a_{Dk} + \frac{U_{0mn}}{2}\, a_{Dm}\, a_{Dn} + \ldots, \qquad U_{0k} = -4\,i\,\Lambda\sin\frac{\pi k}{N}. \quad (4.3)$$

The coefficients U_{0k} were computed by Douglas-Shenker [22]. Our first task is then to compute the coefficients of the second term U_{0mn}. In principle it is a straightforward matter, as one must simply invert the Seiberg-Witten formula:[b]

$$a_{Dm} = \oint_{\alpha_m} \lambda, \qquad a_m = \oint_{\beta_m} \lambda, \qquad \lambda = \frac{1}{2\pi i}\frac{x}{y}\frac{\partial P(x)}{\partial x}dx, \quad (4.4)$$

which is explicitly known, to second order. The only trouble is that a_{Dm} and a_m $(m = 1, 2, \ldots, N-1)$ are given simply in terms of N dependent vacuum parameters ϕ_i, $\sum_{i=1}^{N}\phi_i = 0$. By denoting the formal derivatives with respect to ϕ_i as $\frac{\delta}{\delta\phi_i}$, one finds

$$\sum_{i=1}^{N} \frac{\delta a_{Dm}}{\delta\phi_i}\frac{\partial\phi_i}{\partial a_{Dn}} = \delta_{mn}, \qquad \sum_{m=1}^{N-1} \frac{\partial\phi_i}{\partial a_{Dm}}\frac{\delta a_{Dm}}{\delta\phi_j} = \delta_{ij} - \frac{1}{N}, \quad (4.5)$$

which follow easily by using the constraint, $\sum_{i=1}^{N}\phi_i = 0$. In terms of $B_{mi} \equiv -i\frac{\delta a_{Dm}}{\delta\phi_i}$, $A_{mi} \equiv -i\frac{\delta a_m}{\delta\phi_i}$ which are explicitly given at the N confining vacua in [22], one then finds

$$\frac{\partial\phi_i}{\partial a_{Dm}} = -iB_{mi}; \qquad \sum_{i=1}^{N} B_{mi}B_{ni} = \delta_{mn}; \qquad \sum_{m=1}^{N-1} B_{mi}B_{mj} = \delta_{ij} - \frac{1}{N}. \quad (4.6)$$

The explicit values of B_{mi} are (see [22]):

$$B_{mi} = \frac{1}{N}\frac{\sin[\hat{\theta}_m]}{\cos[\theta_i] - \cos[\hat{\theta}_m]}; \qquad \hat{\theta}_n = \frac{\pi n}{N}; \qquad \theta_n = \frac{\pi(n-1/2)}{N}. \quad (4.7)$$

---

[b]We follow the notation of [22], with $y^2 = P(x)^2 - \Lambda^2$; $P(x) = \frac{1}{2}\prod_{i=1}^{N}(x - \phi_i)$

The definition of $u(a_{Di})$ is the following:

$$u(a_{Di}) = \sum_i \phi_i^2. \tag{4.8}$$

Then the desired coefficients can be found by the following expression, computed at $a_{Di} = 0$:

$$U_{0mn} = \frac{\partial^2 u}{\partial a_{Dm}\partial a_{Dn}} = 2\sum_k \frac{\partial \phi_k}{\partial a_{Dm}}\frac{\partial \phi_k}{\partial a_{Dn}} + 2\phi_k\frac{\partial^2 \phi_k}{\partial a_{Dm}\partial a_{Dn}}. \tag{4.9}$$

The first part of Eq.(49) becomes:

$$2\sum_k \frac{\partial \phi_k}{\partial a_{Dm}}\frac{\partial \phi_k}{\partial a_{Dn}} = -2\sum_k B_{km}B_{kn} =$$
$$-2\sum_{k,s} \frac{2}{N}\sin\left[\frac{\pi ms}{N}\right]\sin\left[\frac{\pi nk}{N}\right]\delta_{ks} = -2\delta_{mn}. \tag{4.10}$$

The evaluation of the second term is a little tricky [26]. The result is however simple:

$$2\sum_k \phi_k\frac{\partial^2 \phi_k}{\partial a_{Dm}\partial a_{Dn}} = \left(2 - \frac{1}{N}\right)\delta_{mn}, \tag{4.11}$$

thus

$$U_{0mn} = (-\frac{1}{N})\,\delta_{mn}. \tag{4.12}$$

We now use this result to calculate the corrections to the tension ratios (311) found in the lowest order. The effective Lagrangean near one of the N confining $\mathcal{N} = 1$ vacua is

$$\mathcal{L} = \sum_{i=1}^{N-1} Im\left[\frac{i}{e_{Di}^2}\left(\int d^4\theta A_{Di}A_{Di}^+ + \int d^2\theta(W_{Di})^2\right)\right] +$$
$$+ Re\left[\int d^4\theta(M_i^+ e^{V_{Di}}M_i + \tilde{M}_i^+ e^{-V_{Di}}\tilde{M}_i)\right] +$$
$$+ 2Re\left[\sqrt{2}\int d^2\theta A_{Di}M_i\tilde{M}_i + m\,U[A_{Di}]\right]. \tag{4.13}$$

The coupling constant e_{Di}^2 is formally vanishing, as

$$\frac{4\pi}{e_{Dk}^2} \simeq \frac{1}{2\pi}\ln\frac{\Lambda \sin(\hat{\theta}_k)}{a_{Dk}N}$$

where $\hat{\theta}_m \equiv \frac{\pi n}{N}$ and $a_{Dk} = 0$ at the minimum. Physically, the monopole loop integrals are in fact cut off by masses caused by the $\mathcal{N} = 1$ perturbation. The monopole becomes massive when $m \neq 0$, and $\sqrt{2}a_{Dk}$ should be replaced by the physical monopole mass $(m\Lambda \sin(\widehat{\theta}_k))^{1/2}$ which acts as the infrared cutoff for the coupling constant evolution. Thus

$$e_{Dm}^2 \simeq \frac{16\pi^2}{\ln(\frac{\Lambda \sin(\widehat{\theta}_m)}{m\,N^2})}. \tag{4.14}$$

As U_{0mn} is found to be diagonal, the description of the ANO vortices [28,14] in terms of effective magnetic Abelian theory description continues to be valid for each $U(1)$ factor. In the linear approximation $U(A_D) = m\Lambda^2 + \mu A_D$, where $\mu \equiv |4\,m\,\Lambda \sin \frac{\pi k}{N}|$ for the k-th $U(1)$ theory, the theory can be (for the static configurations) effectively reduced to an $\mathcal{N} = 4$ theory in $2{+}1$ dimensions. In this way, Bogomolny's equations for the BPS vortex can be easily found from the condition that the vacuum to be supersymmetric:

$$F_{12} = \sqrt{2}\,(\sqrt{2}M^+\tilde{M}^+ - \mu) \qquad (D_1 + iD_2)M = 0 \tag{4.15}$$

$$M = \tilde{M}^+, \qquad A_D = 0. \tag{4.16}$$

The solutions of these equations are similar to the one considered by Nielsen and Olesen:

$$M = \left(\frac{\mu}{\sqrt{2}}\right)^{1/2} e^{in\phi} f[re\sqrt{\mu}], \qquad A_\phi = -2n\frac{g(re\sqrt{\mu})}{r} \tag{4.17}$$

where

$$f' = \frac{f}{r}(1 - 2g))n \qquad g' = \frac{1}{2n}r(1 - f^2) \tag{4.18}$$

with boundary conditions $f(0) = g(0) = 0$, $f(r \to \infty) = 1$, $g(r \to \infty) = +1/2$). The tension turns out to be independent of the coupling constant: for the minimum vortex

$$T = \sqrt{2}\pi\mu = 4\sqrt{2}\pi\,|m\,\Lambda|\sin\frac{\pi k}{N}. \tag{4.19}$$

That the absolute value of m appears in Eq.(419) as it should, and also in Eq.(422) below, is not obvious. This can actually be shown by an appropriate redefinition of the field variables, used in [12], which renders all equations real.

When the second order term in $U(A_D) = \mu A_D + \frac{1}{2}\eta A_D^2$, $\eta \equiv U_{kk}$, is taken into account, the vortex ceases to be BPS saturated. The corrections to the vortex tension due to η can be taking into account by perturbation theory, following [31]. To first order, the equation for $A_{Dk} = A_D$ is

$$\nabla^2 A_D = -2e^4\eta\,(\mu - \sqrt{2}M\tilde{M}) + 2e^2 A_D(MM^+ + \tilde{M}\tilde{M}^+) \tag{4.20}$$

where unperturbed expressions from Eq.(417) can be used for M, $\bar{M}$. The vortex tension becomes simply

$$T = \int d^2x \left[(-\sqrt{2}\mu F_{12}) - 2e^2\eta A_D(\mu - \sqrt{2}M^+\bar{M}^+) \right] \tag{4.21}$$

where the second term represents the correction. By restoring the k dependence, we finally get for the tension of the k-th vortex,

$$T_k = 4\sqrt{2}\,\pi\,|m|\,\Lambda \sin\left(\frac{\pi k}{N}\right) - C\frac{16\pi^2|m|^2}{N^2 \ln\frac{\Lambda \sin(k\pi/N)}{|m|\,N^2}}, \tag{4.22}$$

where $C = 2\sqrt{2}\pi(0.68) = 6.04$. The correction term has a negative sign, independently of the phase of the adjoint mass. Note that the relation $T_k = T_{N-k}$ continues to hold. Eq.(422) is valid for $m \ll \Lambda$. Qualitative feature of this correction is shown in Fig.3, for $N = 6$.

In the above consideration, we have taken into account exactly the m^2 corrections in the F-term of the effective low-energy action. On the other hand, the corrections to the D-terms are subtler. Indeed, based on the physical consideration, a_D in the argument of the logarithm in the effective low energy coupling constant was replaced by the monopole mass, of $O(\sqrt{m\Lambda})$. This amounts to the m insertion to all orders in the loops. Such a resummation is necessitated by the infrared divergences and represents a standard procedure. Another well-known example is the chiral perturbation theory in which quark masses appear logarithmically, e.g., in the expansion of the quark condensate. This explains the non-analytic dependence on m as well as on $\frac{1}{N}$ [32].

Also, there are corrections due to nondiagonal elements in the coupling constant matrix τ_{ij}, which mix the different $U(1)$ factors [33], neglected in Eq.(413). These nondiagonal elements are suppressed by $O(\frac{1}{\log \Lambda/m})$ relatively to the diagonal ones, apparently of the same order of suppression as the correction calculated above. However, these nondiagonal elements gives rise to corrections to the tension of one order higher, $O(\frac{1}{\log^2 \Lambda/m})$, hence is negligible to the order considered.

We thus find a non-universal correction to the Douglas-Shenker formula, Eq.(311). In the process of transition towards fully non-Abelian superconductivity at large m nonperturbative effects such as the W boson productions are probably essential. Nonetheless, the presence of a calculable deviation from the sine formula is qualitatively significant and shows that such a ratio is not a universal quantity.

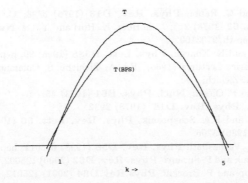

Figure 3.

Acknowledgment

The author thanks Misha Shifman and Misha Voloshin for providing us with a stimulating atmosphere and occasions for fruitful discussions.

References

1. G. Carlino, K. Konishi and H. Murayama, **JHEP 0002** (2000) 004, hep-th/0001036; **Nucl.Phys. B590** (2000) 37, hep-th/0005076; K. Konishi, Proceedings of Continuous Advances in QCD, Minneapolis, Minnesota, May 2000, hep-th/0006086; G. Carlino, K. Konishi, Prem Kumar and H. Murayama, **Nucl.Phys. B608** (2001) 51, hep-th/0104064.

2. P. C. Argyres, M. R. Plesser and N. Seiberg, **Nucl. Phys. B471** (1996) 159, hep-th/9603042; P.C. Argyres, M.R. Plesser, and A.D. Shapere, **Nucl. Phys. B483** (1997) 172, hep-th/9608129.

3. N. Seiberg and E. Witten, **Nucl.Phys. B426** (1994) 19; Erratum *ibid.* **Nucl.Phys. B430** (1994) 485, hep-th/9407087.

4. N. Seiberg and E. Witten, **Nucl. Phys. B431** (1994) 484, hep-th/9408099.

5. P. C. Argyres and A. F. Faraggi, **Phys. Rev. Lett 74** (1995) 3931, hep-th/9411047; A. Klemm, W. Lerche, S. Theisen and S. Yankielowicz, **Phys. Lett. B344** (1995) 169, hep-th/9411048; **Int. J. Mod. Phys. A11** (1996) 1929, hep-th/9505150; A. Hanany and Y. Oz, **Nucl. Phys. B452** (1995) 283, hep-th/9505075; P. C. Argyres, M. R. Plesser and A. D. Shapere, **Phys. Rev. Lett. 75** (1995) 1699, hep-th/9505100; P. C. Argyres and A. D. Shapere, **Nucl. Phys. B461** (1996) 437, hep-th/9509175; A. Hanany, **Nucl.Phys. B466** (1996) 85, hep-th/9509176.

6. G. 't Hooft, **Nucl. Phys. B190** (1981) 455. S. Mandelstam, **Phys. Lett. 53B** (1975) 476; **Phys. Rep. 23C** (1976) 245.

7. P. Goddard, J. Nuyts and D. Olive, **Nucl. Phys. B125** (1977) 1.

8. E. Weinberg, **Nucl. Phys. B167** (1980) 500; **Nucl. Phys. B203** (1982) 445.

9. S. Bolognesi and K. Konishi, hep-th/0207161 (2002).

10. G. 't Hooft, **Nucl. Phys. B79** (1974) 276; A.M. Polyakov, **JETP Lett. 20** (1974) 194.

11. R. Jackiw and C. Rebbi, **Phys. Rev. D13** (1976) 3398; C. Callias, **Comm. Math, Phys. 62** (1978) 213; J. de Boer, K. Hori and Y. Oz, **Nucl. Phys. B500** (1997) 163, hep-th/9703100.
12. M. Di Pierro and K. Konishi, **Phys. Lett. B388** (1996) 90, hep-th/9605178.
13. B. Zumino, Erice Lectures (1966), Ed. A. Zichichi; S. Coleman, Erice Lectures (1977), Ed. A. Zichichi.
14. H. Nielsen and P. Olesen, **Nucl. Phys. B61** (1973) 45.
15. H. J. de Vega, **Phys. Rev. D18** (1978) 2932.
16. H.J. de Vega and F.A. Schaposnik, **Phys. Rev. Lett. 56** (1986) 2564; **Phys. Rev. D34** (1986) 3206.
17. J. Heo and T. Vachaspati, **Phys. Rev. D58** (1998) 065011, hep-ph/9801455.
18. F.A. Schaposnik and P. Suranyi, **Phys. Rev. D62** (2000) 125002, hep-th/0005109.
19. M.A.C. Kneipp and P. Brockill, **Phys.Rev.D64** (2001) 125012, hep-th/0104171.
20. M. Shifman and A. Yung, hep-th/0205025 (2002).
21. A. Hanany, M. Strassler and A. Zaffaroni, **Nucl.Phys. B513** (1998) 87, hep-th/9707244.
22. M.R. Douglas and S.H. Shenker, **Nucl. Phys. B447** (1995) 271, hep-th/9503163.
23. B. Lucini and M. Teper, **Phys.Lett. B501** (2001) 128, hep-lat/0012025; **Phys.Rev.D64** (2001) 105019, hep-lat/0107007.
24. L. Del Debbio, H. Panagopoulos, P. Rossi and E. Vicari, **Phys.Rev. D65** (2002) 021501, hep-th/0106185; **JHEP 0201** (2002) 009, hep-th/0111090
25. C. P. Herzog and I. R. Klebanov, **Phys.Lett. B526** (2002) 388, hep-th/0111078.
26. R. Auzzi and K. Konishi, hep-th/0205172, NJP (2002), to appear.
27. K. Konishi and L. Spanu, hep-th/0106075, IJMPA (2002), to appear.
28. A.A. Abrikosov, **JETP 5** (1957) 1174.
29. M. Strassler, **Progr. Theor. Phys. Suppl. 131** (1998) 439, hep-lat/9803009.
30. A. Yung, hep-th/0005088, 3rd Moscow School of Physics and 28th ITEP Winter School of Physics, Moscow, 2000; A. Vainshtein and A. Yung, **Nucl.Phys.B614**, 3,2001, hep-th/0012250.
31. X. r. Hou, **Phys. Rev. D 63** (2001) 045015, hep-th/0005119.
32. F. Ferrari, **Nucl.Phys. B612** (2001) 151, hep-th/0106192
33. J.D. Edelstein, W.G. Fuertes, J. Mas and J. M. Guilarte, **Phys.Rev. D62** (2000) 065008, hep-th/0001184.

CASIMIR EFFECTS:
FROM GROUNDED PLATES TO THE STANDARD MODEL

ROBERT L. JAFFE

Center for Theoretical Physics
Laboratory for Nuclear Science and Department of Physics
Massachusetts Institute of Technology
Cambridge, MA 02139
E-mail: jaffe@mit.edu

I describe the framework my collaborators and I have developed for the study of one-loop quantum corrections to extended field configurations in renormalizable quantum field theories. I then present a few of the applications of these methods to problems ranging from quantum stabilization of solitons to the "classic" Casimir effect.

1. Introduction

This talk, describing some new methods and new results on an old subject in quantum field theory,[a] is offered as a birthday present to Arkady Vainshtein. Arkady and I first crossed paths (a dangerous thing to do!) when he and Ted Shuryak in Novosibirsk and Mark Soldate and I in Cambridge found ourselves working on the same problem: What are the power law corrections to deep inelastic scattering? We independently discovered the same methods, applied them to different, complementary problems, and exchanged some animated correspondence in those days before email. Amusingly, Einan Gardi will describe some recent applications of those methods here later in the week. Arkady and I first met in person in the mid-1980's in the famous Room 10 at ITEP in Moscow, where just arrived from Novosibirsk, he filled the air (an understatement) with his intensity and excitement. We've enjoyed many happy exchanges – always animated, always productive – in the years that followed. It's a pleasure to be able to celebrate this happy occasion with Nelya and Arkady and so many friends.

---

[a]The work described here is the product of a large collaboration including Eddie Farhi, Noah Graham, Peter Haagensen, Vishesh Khemani, Markus Quandt, Marco Scandurra, and Herbert Weigel. Much of this talk has been drawn from a review by Graham, Weigel, and me[1].

In this talk, I describe a program to develop reliable, accurate, and efficient techniques for a variety of calculations in renormalizable quantum field theories in the presence of background fields. These background field configurations need not be solutions of the classical equations of motion. Our calculations are exact to one loop, allowing us to proceed where perturbation theory or the derivative expansion would not be valid. For example, in a model with no classical soliton we can demonstrate the existence of a nontopological soliton stabilized at one-loop order by quantum fluctuations. We renormalize divergences in the conventional way: by combining counterterms with low-order Feynman diagrams and satisfying renormalization conditions in a fixed scheme. In this way we are certain that the theory is being held fixed as the background field is varied. Our methods are also efficient and practical for numerical computation: the quantities entering the numerical calculation are cutoff independent and do not involve differences of large numbers. The numerical calculations themselves are highly convergent.

We were originally drawn into this subject by our wish to understand what becomes of a fermion in the Standard Model when its coupling to the Higgs becomes large. There are strong reasons to believe that it does not decouple, and some indications that its character changes from a point-like object to a soliton-like object stabilized by fermion quantum fluctuations. We are still working on this hard problem (in three dimensions with both scalar and gauge background fields), however we have learned a lot along the way.

Our methods are limited to one loop and, except in special cases, to static field configurations. We also require the background field configuration to have enough symmetry that the associated scattering problem admits a partial wave expansion. The one-loop approximation includes all quantum effects at order $\hbar$. It is a good approximation for strong external fields or when the number of particles circulating in the loop becomes large. Even when it cannot be rigorously justified, the one-loop approximation can provide insight into novel structures in the same way that classical solutions to quantum field theories have done in the past. We can address a wide variety of problems, including

(1) The stabilization of solitons by quantum effects in theories that do not have classical soliton solutions.

(2) The direct calculation of induced charges, both fractional and integer, carried by background field configurations.

(3) The analysis of the divergences and physical significance of calcu-

lations of vacuum energies in the presence of boundaries – i.e., the traditional "Casimir effect".

(4) The computation of quantum fluctuations in strong external fields.

(5) Quantum contributions to the properties of branes and domain walls.

Here, time permits only the briefest introduction to our methods and a couple of examples of applications. In Section 2 I describe our method in general terms and illustrate the method with the case of a charged boson field in a bosonic background in three spatial dimensions. In Section 3 we show how the methods of dimensional regularization can be adapted to renormalize our calculations. We work in n space dimensions and show that the leading terms in the Born expansion, which diverge for integer n, can be unambiguously identified with Feynman diagrams. This approach resolves several longstanding ambiguities in Casimir calculations. In Sections 4–6 we describe three applications: In Section 4 I consider a chiral model in one space dimension and show that quantum effects of a heavy fermion can stabilize a soliton that is not present in the classical theory. We then compute corrections to the energy and central charge in $1 + 1$ dimensional supersymmetric models, again resolving longstanding questions – in this case whether the BPS bound remains saturated in the presence of quantum fluctuations. Finally, in Section 6, I give a brief overview of very recent work on the "classic" Casimir problem: the energy of quantum fields subject to boundary conditions.

2. Overview

For simplicity consider a fluctuating boson or fermion field of mass m in a static, spherically symmetric background potential $\chi(r)$ in three dimensions. Since we encounter divergences, we imagine that we have analytically continued to values of the space dimension n where the integrals are convergent. Later we provide the rigorous justification for this procedure.

We take the interaction Lagrangian $\mathcal{L}_I = g\bar{\psi}\chi\psi$ for fermions ($g\psi^\dagger\chi\psi$ for bosons) where ψ is the fluctuating field. We want to compute the one-loop "effective energy," the effective action per unit time. It is given either by the sum of all one-loop diagrams with all insertions of the background $\chi(r)$,

$$\Delta E_{\text{bare}}[\chi] \; = \; \bigcirc \; + \; \bigcirc \; + \; \bigtriangleup \; + \; \cdots \tag{1}$$

or by the "Casimir sum" of the shifts in the zero-point energies of all the small oscillation modes in the background χ,

$$\Delta E_{\text{bare}}[\chi] = \pm\frac{1}{2}\sum_j |\epsilon_j| - |\epsilon_j^0| \tag{2}$$

for bosons $(+)$ and fermions $(-)$ respectively. Both of these representations are divergent and require renormalization. We start from the second expression and work in the continuum. We rewrite the Casimir sum as a sum over bound states plus an integral over scattering states, weighted by the density of states $\rho(k)$. We subtract from the integral the contribution of the trivial background, which is given by the free density of states $\rho^0(k)$. Thus we have

$$\Delta E_{\text{bare}}[\chi] = \pm\left(\frac{1}{2}\sum_j |\omega_j| + \frac{1}{2}\int_0^\infty \omega(k)\left(\rho(k) - \rho^0(k)\right) dk\right) \tag{3}$$

where ω_j denotes the energy of the jth bound state, and $\omega(k) = \sqrt{k^2 + m^2}$.

The density of states is related to the S-matrix and the phase shifts by

$$\rho(k) - \rho^0(k) = \frac{1}{2\pi i}\frac{d}{dk}\text{Tr}\ln S(k) = \sum_\ell D^\ell \frac{1}{\pi}\frac{d\delta_\ell(k)}{dk} \tag{4}$$

where ℓ labels the basis of partial waves in which S is diagonal. D^ℓ is the degeneracy factor. For example, $D^\ell = 2\ell+1$ for a boson in three dimensions. It is convenient to use Levinson's theorem to express the contribution of the bound states to eq. (3) in terms of their binding energy. Levinson's theorem relates the number of bound states to the difference of the phase shift at $k = 0$ and ∞,

$$n_\ell^{\text{bound}} = \frac{1}{\pi}(\delta_\ell(0) - \delta_\ell(\infty)) = -\int_0^\infty dk\,\frac{d\delta_\ell(k)}{dk}. \tag{5}$$

Subtracting mn_ℓ^{bound} from the sum over bound states in eq. (3) and using eqs. (5) and (4), we obtain

$$\Delta E_{\text{bare}}[\chi] = \pm\left(\frac{1}{2}\sum_{j,\ell} D^\ell(|\omega_{j,\ell}| - m) + \int_0^\infty \frac{dk}{2\pi}(\omega(k) - m)\sum_\ell D^\ell \frac{d\delta_\ell(k)}{dk}\right) \tag{6}$$

where the sum over partial waves is to be performed before the k integration. While the phase shifts and bound state energies are finite, $\Delta E_{\text{bare}}[\chi]$ is divergent because the k-integral and ℓ-sum both diverge in the ultraviolet. To better understand the origin and character of the divergences, we go back to the diagrammatic representation of the vacuum energy, eq. (1). Since we are working with a renormalizable theory, only the first few diagrams are divergent, and these divergences can be canceled by a finite number of counterterms. The series of diagrams gives an expansion of the effective energy in powers of the background field $\chi(r)$. Likewise, the phase shift calculation can be expanded in powers of $\chi(r)$ using the Born series,

$$\delta_\ell^N(k) = \sum_{i=1}^{N} \delta_\ell^{(i)}(k) \tag{7}$$

where $\delta_\ell^{(i)}(k)$ is the contribution to the phase shift at order i in the potential $\chi(r)$. In general, the Born expansion is a poor approximation at small k, especially if the potential has bound states, when it typically does not converge. What is important for us, however, is that the contributions to ΔE_{bare} from successive terms in the Born series correspond exactly to the contributions from successive Feynman diagrams. That is, the ith term in the Born series generates a contribution to the vacuum energy which is exactly equal to the contribution of the Feynman diagram with i external insertions of χ.

This correspondence is not trivial in light of divergences. We have verified the identification for the lowest-order diagram by direct comparison in n space dimensions where both are finite. At this order the Born and Feynman contributions to $\Delta E_{\text{bare}}[\chi]$ are precisely equal as analytic functions of n as we will show in Section 3. We have also performed various numerical checks to verify the identification in higher orders.

We then define the subtracted phase shift

$$\overline{\delta}_\ell^N(k) = \delta_\ell(k) - \delta_\ell^N(k) \tag{8}$$

where we take N to be the number of divergent diagrams in the expansion of eq. (1). The effect of the Born subtraction is illustrated in Fig. 1. Note that the subtracted phase shift is large at small k, so the Born approximation is very different from the true phase shift in this region. However the Born approximation becomes good at large k, so that the subtracted phase shift vanishes quickly as $k \to \infty$.

Having subtracted the potentially divergent contributions to $\Delta E_{\text{bare}}[\chi]$ via the Born expansion, we add back in exactly the same quantities as Feynman diagrams, $\sum_{i=1}^{N} \Gamma_{\text{FD}}^{(i)}[\chi]$. We combine the contributions of the diagrams with those from the counterterms, $\Delta E_{\text{CT}}[\chi]$, and apply standard perturbative renormalization conditions. We have thus removed the divergences from the computationally difficult part of the calculation and re-expressed them as Feynman diagrams, where the regularization and renormalization have been carried out with conventional methods. This approach to renormalization in strong external fields was first introduced by Schwinger[2] in

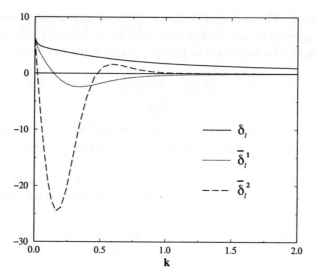

Figure 1.   Typical phase shift in three dimensions, before and after subtracting the Born approximation.

his work on QED in strong fields. Combining the renormalized Feynman diagrams,

$$\overline{\Gamma}^N_{FD}[\chi] = \sum_{i=1}^{N} \Gamma^{(i)}_{FD}[\chi] + \Delta E_{CT}[\chi] \tag{9}$$

with the subtracted phase shift calculation, we obtain the complete, renormalized, one-loop effective energy,

$$\Delta E[\chi] = \pm \frac{1}{2} \sum_{\ell} D^{\ell} \left(\sum_{j} (|\,\omega_{j,\ell}| - m) + \int_{0}^{\infty} \frac{dk}{\pi} (\omega(k) - m) \frac{d}{dk} \overline{\delta}^{N}_{\ell}(k) \right)$$
$$+ \overline{\Gamma}^{N}_{FD}[\chi] \tag{10}$$

where the two pieces are now separately finite. Since the k integral is now convergent, we are free to interchange it with the sum over partial waves or to integrate by parts. This expression is suitable for numerical computation, since it does not contain differences of large numbers. The massless limit is also smooth, except for the case of one spatial dimension, where we expect incurable infrared divergences[3].

Eq. (10) summarizes our approach. The first line is unfamiliar to particle physicists. It sums all orders in the background field χ. Though it must be computed numerically, it is finite, unambiguous, and regulator independent. We have developed efficient methods to compute phase shifts and the Born approximation[1,4]. All the potential divergences are isolated in certain low-order Feynman diagrams, where they are cancelled by counterterms in the time honored fashion of renormalizable quantum field theory.

3. Dimensional Regularization

The identification of terms in the Born series with Feynman diagrams is crucial. No arbitrariness can be tolerated in the renormalization process: if the manipulations of formally divergent quantities introduce finite ambiguities, our method is useless. There have been controversies for many years concerning how to renormalize Casimir calculations. However, we are studying a renormalizable quantum field theories. So we know that the effective energy can be calculated unambiguously. In this section, I apply the methods of dimensional regularization to scattering from a central potential and prove that the lowest-order term in the Born series is equal to the lowest-order Feynman diagram as an analytic function of n, the number of space dimensions. Since this is the most divergent diagram – quadratically divergent for $n = 3$ – we are confident that the same method will regulate all other divergences in the effective energy unambiguously.

For simplicity, I will consider the fluctuations of a single real boson, $\phi(x)$, coupled to a static, spherically symmetric background, $\chi(r)$, by $\mathcal{L}_I = |\phi(x)|^2 \chi(r)$. The generalization to fermions is discussed in Ref.[6]. For $n = 1$, $\chi(r)$ reduces to a symmetric potential with even and odd parity channels. For $n \neq 1$ the S-matrix is diagonal in the basis of the irreducible tensor representations of $SO(n)$. These are the traceless symmetric tensors of rank ℓ, where $\ell = 0, 1, 2, \ldots$. We choose a value of n between 0 and 1, where all the integrals and sums that appear in the Casimir energy converge. It is not difficult to formulate scattering theory in partial waves in n dimensions, to define phase shifts, $\delta_{n\ell}(k)$, Born approximations, $\delta_{n\ell}^{(i)}(k)$, and the density of states, $\sum_\ell \frac{1}{\pi} D_n^\ell \, d\delta_{n\ell}/dk$, where D_n^ℓ is the degeneracy of the $SO(n)$ representation labelled by ℓ.

The first Born approximation to the phase shift is

$$\delta_{n\ell}^{(1)}(k) = -\frac{\pi}{2} \int_0^\infty [J_{\frac{n}{2}+\ell-1}(kr)]^2 \chi(r) r \, dr \qquad (11)$$

and its contribution to the Casimir energy is

$$\Delta E_n^{(1)}[\chi] = \int_0^\infty \frac{dk}{2\pi} (\omega(k) - m) \sum_{\ell=0}^\infty D_\ell^n \frac{d\delta_{n\ell}^{(1)}(k)}{dk} . \qquad (12)$$

Using the Bessel function identity

$$\sum_{\ell=0}^{\infty} \frac{(2q + 2\ell)\Gamma(2q + \ell)}{\Gamma(\ell + 1)} J_{q+\ell}(z)^2 = \frac{\Gamma(2q + 1)}{\Gamma(q + 1)^2} \left(\frac{z}{2}\right)^{2q} \tag{13}$$

with $q = \frac{n}{2} - 1$, we sum over ℓ in eq. (12) and obtain

$$\Delta E_n^{(1)}[\chi] = \frac{\langle\chi\rangle(2 - n)}{(4\pi)^{\frac{n}{2}}\Gamma\left(\frac{n}{2}\right)} \int_0^{\infty} (\omega(k) - m)k^{n-3}\, dk = \frac{\langle\chi\rangle m^{n-1}}{(4\pi)^{\frac{n+1}{2}}}\Gamma\left(\frac{1 - n}{2}\right)(14)$$

which converges for $0 < n < 1$. Here $\langle\chi\rangle$ is the n-dimensional spatial average of $\chi(r)$,

$$\langle\chi\rangle = \int \chi(x)\, d^n x = \frac{2\pi^{\frac{n}{2}}}{\Gamma\left(\frac{n}{2}\right)} \int_0^{\infty} \chi(r)r^{n-1}dr. \tag{15}$$

The tadpole diagram is easily computed using dimensional regularization, and the result agrees precisely with eq. (14). Thus we can be certain that our method of subtracting the first Born approximation and adding back the corresponding Feynman diagram is correct.

4. Chiral Model in One Dimension

As a first application of our method, I show how a quantum soliton can appear in a theory with a heavy fermion. We consider a one-dimensional chiral model in which the fermion gets its mass from its coupling to a scalar condensate. It is easy to find a spatially varying scalar background which has a tightly bound fermion level. If the classical energy of the background field plus the energy of the tightly bound fermion is less than the free fermion mass m, this configuration would appear to be a stable soliton, since it is unable to decay into free fermions. However, the energy of the lowest fermion level enters at the same order in $\hbar$ as the full one-loop fermion effective energy, since the latter simply corresponds to the shift of the zero-point energies, eq. (2), of all the fermion modes. The question of stability can therefore only be addressed by computing the full one-loop effective energy. Here I summarize our analysis of this system and show that it supports stable solitons. More details of this calculation can be found in Ref. [10].

4.1. The model

We consider a chiral model in one dimension with a symmetry-breaking scalar potential. We couple a two-component real boson field $\vec{\phi} = (\phi_1, \phi_2)$

chirally to a fermion Ψ

$$\mathcal{L} = \frac{1}{2} \partial_\mu \vec{\phi} \cdot \partial^\mu \vec{\phi} - V(\vec{\phi}) + \bar{\Psi} \left\{ i\partial\!\!\!/ - G\left(\phi_1 + i\gamma_5 \phi_2\right) \right\} \Psi \qquad (16)$$

where the potential for the boson field is given by

$$V(\vec{\phi}) = \frac{\lambda}{8} \left[\vec{\phi} \cdot \vec{\phi} - v^2 + \frac{2\alpha v^2}{\lambda} \right]^2 - \alpha v^3 \left(\phi_1 - v\right) + \text{const.} \qquad (17)$$

$V(\vec{\phi})$ has its minimum at $\vec{\phi} = (v, 0)$. Terms proportional to α break the chiral symmetry explicitly. If α we set to zero, the chiral symmetry appears to break spontaneously, but quantum fluctuations in one dimension restore the symmetry[3]. For large enough α, the classical vacuum $\vec{\phi} = (v, 0)$ is stable against quantum corrections and $m = Gv$ is the fermion mass. The coefficient c in the counterterm Lagrangian

$$\mathcal{L}_{\text{CT}} = c \left(\vec{\phi} \cdot \vec{\phi} - v^2 \right) \qquad (18)$$

is fixed by the condition that the quantum corrections do not change the VEV of $\vec{\phi}$. This model has no stable soliton solutions at the classical level.

We are interested in the mass of the lightest state carrying unit fermion number. If its mass is less than m, this state is a stable soliton. We neglect boson loops, so that the effective energy is given by the sum of the classical and the fermion loop contributions, $E_{\text{tot}}[\vec{\phi}] = E_{\text{cl}}[\vec{\phi}] + E_{\text{f}}[\vec{\phi}]$. This approximation is exact in the limit where the number of independent fermion species becomes large. The fermion contribution to the effective energy is $E_{\text{f}} = E_{\text{Cas}} + E_{\text{val}}$ where E_{Cas} is the sum over zero-point energies, calculated with the methods we have developed. E_{val} is the energy required for the soliton to have unit charge. Using the methods of the previous section, we can calculate the fermion number of the background field. If a level has crossed zero, then the background field will already carry the required fermion number and $E_{\text{val}} = 0$. If the background field has zero charge, we must explicitly fill the most tightly bound level, giving $E_{\text{val}} = \epsilon_0$, where ϵ_0 is the energy of that level.

The scattering theory formalism must be extended to handle fermions. We choose backgrounds which preserve parity symmetry so the phase shifts can be labelled by the parity, $\delta_\pm$. Also we must sum contributions from particles and antiparticles, or in the context of the single particle Dirac equation, from positive and negative energies. So we define

$$\delta_F(k) = \delta_+\left(\omega(k)\right) + \delta_+\left(-\omega(k)\right) + \delta_-\left(\omega(k)\right) + \delta_-\left(-\omega(k)\right). \qquad (19)$$

Renormalization is particularly simple in this model. The first and second Born approximations corresponding to the Feynman diagrams with one and two insertions of $[\vec{\phi} - (v, 0)]$ diverge. However, the divergences are related by chiral symmetry. Both are canceled by a counterterm proportional to $\vec{\phi}^2 - v^2$. It suffices to subtract the first Born approximation to $\delta(k)$ and the part of the second related to it by chiral symmetry,

$$\delta^{(1)}(k) = \frac{2G^2}{k} \int_0^\infty dx \left(v^2 - \vec{\phi}^2(x) \right) . \tag{20}$$

The condition that the VEV of $\vec{\phi}$ does not get renormalized requires that the counterterm exactly cancel the Feynman diagrams that are added back in compensation for the Born subtractions. Thus we have

$$E_{\mathrm{Cas}}[\vec{\phi}] = -\frac{1}{2} \sum_j (|\omega_j| - m) - \int_0^\infty \frac{dk}{2\pi} \left(\omega(k) - m \right) \frac{d}{dk} \left(\delta_{\mathrm{F}}(k) - \delta^{(1)}(k) \right) .$$

4.2. Numerical studies

We consider variational *ansätze* for the background field. As $x \to \pm\infty$, $\vec{\phi}$ must go to its vacuum value, $(v, 0)$. We find that energetically favored configurations execute a loop in the (ϕ_1, ϕ_2) with radius $R > v$ so that they enclose the origin. A simple *ansatz* with these properties is

$$\phi_1 + i\phi_2 = v\{1 - R + R \exp\left[i\pi \left(1 + \tanh(Gvx/w)\right)\right]\} \tag{21}$$

with the width (w) and amplitude (R) as variational parameters. For particular model parameters G, α, λ and v, we compute $\mathcal{B} = E_{\mathrm{tot}}/m - 1$ as a function of the variational parameters w and R. We show the resulting binding energy surface in Figure 2 for one set of model parameters. The contour $\mathcal{B} = 0$ separates the region in which the effective energy of background configuration is less than m from the region in which it is larger than m. The maximal binding is indicated by a star. In Figure 3 we present the profiles ϕ_1 and ϕ_2 corresponding to this variational minimum as functions of the dimensionless coordinate $\xi = xm$. This background field configuration does not carry fermion number in this case, so the most strongly bound level must explicitly be occupied. The total charge density is shown in Figure 3. It receives contributions from the polarized fermion vacuum and from the explicitly occupied valence level, given by $\psi_0^\dagger(x)\psi_0(x)$ where $\psi_0(x)$ is the bound state wavefunction of the valence level.

Figure 4 shows the result of repeating the binding energy calculation for various sets of model parameters. When $\mathcal{B}$ is negative, the configuration is a fermion with lower energy than a fermion propagating in the trivial background. Since the true minimum of the energy will have even lower energy, we know that a soliton exists.

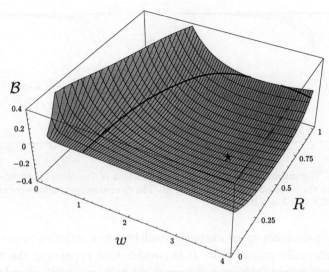

Figure 2. $\mathcal{B}$ as a function of the *ansatz* parameters for the class of *model* parameters characterized by the relations $\alpha = 0.5G^2$, $\tilde{\lambda} = G^2$, and $v = 0.375$. A solid curve marks the contour $\mathcal{B} = 0$. The star indicates the minimum at $w = 2.808$ and $R = 0.586$.

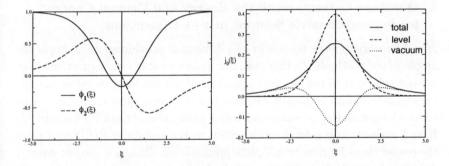

Figure 3. ϕ_1, ϕ_2, and the fermion number density j_0 at the variational minimum. The left panel shows $\phi_1(\xi)$ and $\phi_2(\xi)$, and the right panel shows the charge density $j_0(\xi)$, which gets contributions from both the polarized fermion vacuum and the filled valence level. The model parameters are as in Figure 2.

We have extended this analysis to a chiral Yukawa model with $SU(2)$ symmetry in three dimensions. The analysis is more complicated: rotational symmetry is replaced by grand spin, the sum of rotations in spatial $SU(2)$ and isospin $SU(2)$, and diagrams up to fourth order in the external field are divergent. Nevertheless, the program can still be carried out[11]. However, we do not find evidence for a bound fermionic soliton in this theory. In general, binding is weaker than in one dimension and it occurs in

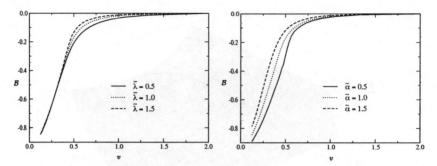

Figure 4. The maximal binding energy as a function of the model parameters as obtained from the *ansatz* eq. (21) in units of m. The dimensionless parameters are defined by $\tilde{\alpha} = \alpha/G^2$ and $\tilde{\lambda} = \lambda/G^2$.

regions of parameter space where the model or the restriction to one fermion loop is internally inconsistent. It is possible that expanding the model to also include gauge fields may change this result, and work is underway to consider this possibility.

5. Quantum Corrections to the Energy and Central Charge for Supersymmetric Solitons in 1+1 Dimensions

$N = 1$ supersymmetric models in 1+1 dimension provide another simple example of our methods. In this case, our ability to study configurations that are not solutions to the classical equations of motion (and therefore not supersymmetric) and to handle renormalization unambiguously, allows us to resolve long-standing questions regarding saturation of the BPS bound[12,13]. Here I present only a brief introduction to the results of Ref.[14] and refer the reader there for a more complete presentation. This is a subject dear to Arkady's heart. In fact, Arkady and the Mishas reached the same (correct!) conclusions using quite different methods.[15]

We consider the Lagrangian

$$\mathcal{L} = \frac{m^2}{2\lambda} \left((\partial_\mu \phi)(\partial^\mu \phi) - U(\phi)^2 + i\bar{\Psi}\partial\!\!\!/\Psi - U'(\phi)\bar{\Psi}\Psi \right) \qquad (22)$$

where ϕ is a real scalar, Ψ is a Majorana fermion, and $U(\phi) = W'(\phi)$ where $W(\phi)$ is the superpotential. If $U(\phi)^2$ is of the symmetry breaking form with equal minima at $\phi = \pm 1$, then a soliton is a solution to

$$\frac{d\phi_0(x)}{dx} = -U(\phi_0(x)) \qquad (23)$$

where $\phi_0 \to \pm 1$ as $x \to \pm\infty$. An antisoliton is obtained by sending x to $-x$.

The boson and fermion small oscillation modes are given by

$$\left(-\frac{d^2}{dx^2} + U'(\phi_0)^2 + U(\phi_0)U''(\phi_0) \right) \eta_k(x) = \omega^2 \eta_k(x) \qquad (24)$$

$$\gamma^0 \left(-i\gamma^1 \frac{d}{dx} + U'(\phi_0) \right) \psi_k(x) = \omega \psi_k(x). \qquad (25)$$

Defining

$$V(x) = U'(\phi_0)^2 + U(\phi_0)U''(\phi_0) - m^2$$
$$\tilde{V}(x) = U'(\phi_0)^2 - U(\phi_0)U''(\phi_0) - m^2 \qquad (26)$$

and squaring the Dirac equation, we obtain

$$\left(-\frac{d^2}{dx^2} + V(x) \right) \eta_k(x) = \omega^2 \eta_k(x) \qquad (27)$$

$$\begin{pmatrix} -\dfrac{d^2}{dx^2} + V(x) & 0 \\ 0 & -\dfrac{d^2}{dx^2} + \tilde{V}(x) \end{pmatrix} \psi_k(x) = k^2 \psi_k(x). \qquad (28)$$

It is easy to show that the bound state spectra of the effective scalar potentials $V(x)$ and $\tilde{V}(x)$ will always coincide, except possibly for zero modes.

5.1. Supersymmetric Spectrum

To be specific, I will consider the special case of $U(\phi) = \frac{m}{2}(\phi^2 - 1)$, where the soliton is the standard "kink," $\phi_0(x) = \tanh mx/2$. Our ability to consider configurations that are not solutions to the classical equations of motion allows us to study a sequence of background fields, $\phi_0(x, x_0)$, which interpolate between the trivial background at $x_0 = 0$ and a widely separated kink-antikink pair as $x_0 \to \infty$,

$$\phi_0(x, x_0) = \tanh \frac{m}{2}(x + x_0) - \tanh \frac{m}{2}(x - x_0) - 1. \qquad (29)$$

This procedure enables us to avoid potential ambiguities regarding the choice of boundary conditions when ϕ_0 tends toward different vacua as $x \to \pm\infty$ or at boundaries introduced to discretize the problem[12]. To stabilize an arbitrary background, $\phi_0(x, x_0)$, we must insert a source term, $J(x) = \frac{d^2\phi_0}{dx^2} - \frac{1}{2}m^2(\phi_0^3 - \phi_0)$ into the SUSY lagrangian, eq. (22). With this choice, $\phi_0(x, x_0)$ is a stationary point of the action (though not necessarily a global minimum). The source breaks supersymmetry except when $x_0 = 0$ and as $x_0 \to \infty$, but it allows us to track the properties of the system continuously from the trivial case ($x_0 = 0$) to the case of interest ($x_0 \to \infty$), both of which are supersymmetric. Technical issues associated with the source, including restoration of translation invariance and the appearance of modes with imaginary frequencies, are discussed in Ref.[16]. They do not complicate the picture presented here. To understand the subtleties of the renormalized energy calculation, it is instructive to compare the bosonic and fermionic spectra as functions of separation x_0 for kink-antikink pair. For large separation, the boson and fermion modes match, as required by supersymmetry, except we have two boson zero modes but only one fermion zero mode. Figure 5 illustrates this discrepancy.

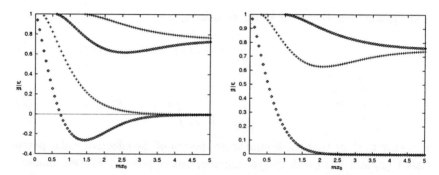

Figure 5. Bosonic (left) and fermionic (right) bound state spectra for kink/antikink background with separation $2x_0$. We display ω_B^2 for the bosonic modes and ω_F for the fermionic modes as functions of x_0. In limit of infinite separation, we have supersymmetry, so the modes match except for the zero modes.

Note that for each state in the spectrum, there is another equivalent state with the opposite sign of the energy. Since we are considering a real scalar and a Majorana fermion, in both cases we only consider one of these two states.

The zero modes matter even though they do not contribute directly to the vacuum polarization energy because they are related to the continuum through Levinson's theorem. Since the number of bound states is different for bosons and fermions, by Levinson's theorem the phase shifts at $k = 0$ must differ. Since the phase shifts are continuous, they must also differ as functions of k. Indeed, for the widely separated soliton/antisoliton pair, we find

$$\delta_B(k) - \delta_F(k) = 2\arctan\frac{m}{k} \tag{30}$$

and so the renormalized one-loop quantum correction to the energy as $x_0 \to \infty$ is

$$\Delta E = \frac{1}{2} \sum_j \left(\omega_j^B - m \right) - \frac{1}{2} \sum_j \left(\omega_j^F - m \right)$$

$$+ \int_0^\infty \frac{dk}{2\pi} \left(\sqrt{k^2 + m^2} - m \right) \frac{d}{dk} \left(\delta_B(k) - \delta_F(k) - \delta^{(1)}(k) \right)$$

$$= -\frac{m}{2} + \int_0^\infty \frac{dk}{2\pi} \left(\sqrt{k^2 + m^2} - m \right) \frac{d}{dk} \left(2 \arctan \frac{m}{k} - \frac{2m}{k} \right)$$

$$= -\frac{m}{\pi} \tag{31}$$

where we have fixed the coefficient of the counterterm

$$\mathcal{L}_{\text{ct}} = -CU''(\phi)U(\phi) - CU'''(\phi)\bar{\Psi}\Psi \tag{32}$$

by requiring that the tadpole graph vanish, with no further finite renormalizations.

We assign half this energy shift to the soliton and half to the antisoliton, so the result for the soliton is $\Delta E = -\frac{m}{2\pi}$. This assignment is supported by a careful consideration of the zero modes: for a single soliton, we find one bosonic zero mode, and "one-half" of a fermionic zero mode. Just as the bosonic zero mode reflects the breaking of translation invariance, the fermionic mode reflects the breaking of one of the two supersymmetry generators. This mode is the Majorana fermion analog of the Jackiw-Rebbi mode[8]. For large x_0, only one mode appears near $\omega = 0$ in the spectrum of positive energy solutions the Dirac equation. When we reduce to the case of a single soliton, still keeping only the positive energy states, it is weighted by one half. Indeed one can verify that the residue of the pole in the fermionic Green's function at $\kappa^2 = -m^2$ is half the usual result for a bound state. Our result has since been confirmed using the generalized effective action approach[17].

5.2. BPS bound

In the supersymmetric system, there are additional restrictions on the quantum Hamiltonian H. The central charge

$$Z = \frac{m^2}{\lambda} \int dx \, U(\phi) \frac{d\phi}{dx} \tag{33}$$

obeys the BPS bound

$$\langle H \rangle \geq |\langle Z \rangle| . \tag{34}$$

Classically, the bound is saturated,

$$E_{\rm cl} = \frac{m^2}{2\lambda} \int dx \left[\left(\frac{d\phi_0}{dx}\right)^2 + U^2(\phi_0) \right] = -\frac{m^2}{\lambda} \int dx\, U(\phi_0)\frac{d\phi_0}{dx} = -Z_{\rm cl} \quad (35)$$

and the contribution from the counterterm is equal and opposite as well,

$$\Delta E_{\rm ct} = C \int U''(\phi_0)U(\phi_0)\,dx = -C \int U''(\phi_0)\phi_0'\,dx = -\Delta Z_{\rm ct}\,. \quad (36)$$

The negative quantum correction to the energy we found in the previous section would appear to lead to a violation of the bound, which cannot be correct. The resolution is that a similar analysis, taking careful account of renormalization by using the scattering data, yields a compensating correction to the central charge. Expanding the field ϕ around the classical solution $\phi_0(x)$ gives

$$\Delta Z = \langle Z\rangle_\phi - Z_{\rm cl}$$
$$= \Delta Z_{\rm ct} + \frac{m^2}{\lambda} \int \left\langle U'\eta\eta' - \frac{1}{2}UU''\eta^2 \right\rangle_{\phi_0} dx \quad (37)$$
$$= \Delta Z_{\rm ct} + \frac{m^2}{2\lambda} \int \left\langle \left(\left(\frac{d}{dx}+U'\right)\eta\right)^2 - (\eta')^2 - \eta^2(U')^2 - UU''\eta^2 \right\rangle_{\phi_0} dx$$

where $\phi(x) = \phi_0(x) + \eta(x)$ and

$$\eta(x) = \sqrt{\frac{\lambda}{m^2}}\left(\int \frac{dk}{\sqrt{4\pi\omega_k}} \left(a_k\eta_k(x)e^{-i\omega_k t} + a_k^\dagger\eta_k^*(x)e^{i\omega_k t} \right) + \eta_{\omega=0}(x)a_{\omega=0} \right)$$

where $\omega_k = \sqrt{k^2 + m^2}$, the creation and annihilation operators obey the usual commutation relations, and we have explicitly separated the contribution of the zero mode. The other bound states are understood to give discrete contributions to the integral.

We can compute this expectation value and connect it to our scattering theory formalism using the relationship between the wavefunction and the density of states,

$$\rho(k) - \rho_0(k) = \frac{1}{\pi} \int dx \left(|\eta_k(x)|^2 - 1 \right) \quad (38)$$

yielding as a result

$$\Delta Z = \frac{1}{4}\sum_j (|\tilde\omega_j| - m) - \frac{1}{4}\sum_j (|\omega_j| - m) \quad (39)$$
$$+ \int_0^\infty \frac{dk}{4\pi}\left(\sqrt{k^2 + m^2} \right)\frac{d}{dk}\left(\tilde\delta(k) - \delta(k) + 2\delta^{(1)}(k) \right)$$
$$= \frac{m}{4} - \int_0^\infty \frac{dk}{2\pi}\left(\sqrt{k^2 + m^2} - m \right)\frac{d}{dk}\left(\arctan\frac{m}{k} - \frac{m}{k} \right) = \frac{m}{2\pi}$$

in terms of the phase shifts $\delta(k)$ and $\tilde{\delta}(k)$ and the bound states energies ω_j and $\tilde{\omega}_j$ in the potentials V and $\tilde{V}$ respectively. Comparison with eq. (31) shows that the correction to central charge for a single soliton or antisoliton satisfies $|\Delta E| = |\Delta Z|$, so the BPS bound remains saturated. This result was subsequently confirmed and extended to higher loops by Shifman, Vainshtein, and Voloshin using SUSY methods[15].

6. The "Classic" Casimir Problem

The vacuum energy of fluctuating quantum fields that are subject to boundary conditions has been studied intensely over the half-century since Casimir predicted a force between grounded metal plates[18,19,20]. The plates change the zero-point energies of fluctuating fields and thereby give rise to forces between the rigid bodies or stresses on isolated surfaces. The Casimir force between grounded metal plates has now been measured quite accurately and agrees with his prediction[21,22,23].

Casimir forces arise from interactions between the fluctuating fields and matter. Nevertheless, it is traditional to study idealized "Casimir problems" where the physical interactions are replaced *ab initio* by boundary conditions. The methods I have described to study the effective energy in the presence of background fields provide us with a way to understand under what circumstances this replacement is justified. The application of our methods to the Casimir problem is summarized in Refs.[24,25]

A real material cannot constrain modes of the field with wavelengths much smaller than the typical length scale of its interactions. In contrast, a boundary condition constrains all modes. The sum over zero point energies is highly divergent in the ultraviolet and these divergences depend on the boundary conditions. Subtraction of the vacuum energy in the absence of boundaries only removes the worst divergence (quartic in three space dimensions).

Renormalization in quantum field theory provides the only known way to regulate, discuss, and eventually remove these divergences. Our strategy, then, will be to embed the Casimir problem in a conventional quantum field theory by replacing the boundary condition by a coupling to a smooth static background field, calculating the energy in this background, and then stuying a limit in which the background field enforces the boundary condition. After renormalization, if any quantity is still infinite in the presence of the boundary condition, it will depend in detail on the properties of the material that provides the physical ultraviolet cutoff and will not exist in the idealized Casimir problem. Similar issues were discussed in the context of dispersive media in Ref.[26].

It is relatively easy to write down a QFT describing the interaction of the fluctuating field ϕ with a static background field $\sigma(\mathbf{x})$ and to choose a limit involving the shape of $\sigma(\mathbf{x})$ and the coupling strength between ϕ and σ that produces the desired boundary conditions on specified surfaces. We have developed the formalism required to compute the resulting vacuum energy in Ref. [25]. Here I focus on Dirichlet boundary conditions on a scalar field. Our methods can be generalized to the physically interesting case of conducting boundary conditions on a gauge field.

Ideally, we seek a Casimir energy that depends only on the boundary conditions and not on any other features of $\sigma(\mathbf{x})$. Therefore we do not specify any action for σ except for the standard counterterms induced by the ϕ-σ interaction. The coefficients of the counterterms are fixed by renormalization conditions applied to perturbative Green's functions. There is no ambiguity and, aside from the choice of renormalization conditions, no further freedom. Moreover, the renormalization conditions are independent of the particular choice of background $\sigma(\mathbf{x})$, so it makes sense to compare results for different geometries. The Yukawa theory with coupling g in three space dimensions gives a textbook example: The $g\bar{\psi}\sigma\psi$ coupling generates divergences in low-order Feynman diagrams proportional to σ^2, σ^4 and $(\partial\sigma)^2$ and therefore requires one to introduce a mass, a quartic self-coupling, and a kinetic term for σ. This is the only context in which one can study the fluctuations of a fermion coupled to a scalar background in three dimensions.

We couple a real scalar field ϕ to a scalar background $\sigma(\mathbf{x})$ with coupling λ, $\mathcal{L}_{\text{int}}(\phi,\sigma) = \frac{1}{2}\lambda\,\sigma(\mathbf{x})\,\phi^2(\mathbf{x},t)$. In the limit where $\sigma(\mathbf{x})$ becomes a delta function on some surface $\mathcal{S}$ ($\delta_\mathcal{S}(\mathbf{x})$) and where $\lambda \to \infty$, it is easy to verify that all modes of ϕ must vanish on $\mathcal{S}$. Call this the *Dirichlet limit*. It consists of the *singular limit*, where $\sigma(\mathbf{x})$ gets concentrated on $\mathcal{S}$, followed by the *strong coupling limit*, $\lambda \to \infty$. In general, we find that the divergence of the vacuum energy in the Dirichlet limit cannot be renormalized. Generally, even the singular limit does not lead to a finite Casimir energy except in one dimension, where the singular limit exists but the Casimir energy diverges as $\lambda\ln\lambda$ in the strong coupling limit.

Apparently the Casimir energy of a scalar field that vanishes on a surface in any dimension is infinite. However, all is not lost. The unrenormalizable divergences are localized on $\mathcal{S}$, so quantities that do not probe $\mathcal{S}$ are well defined. For example, it is relatively easy to show that the vacuum *energy density* away from $\mathcal{S}$ is well defined in the singular limit, even though the energy density on $\mathcal{S}$ diverges[27]. In the examples we have studied explicitly, the vacuum energy density even remains finite in the boundary condition limit as well. We expect that this is true in general. The forces between rigid bodies are also finite in the Dirichlet limit. But any quantity whose definition requires a deformation or change in area of $\mathcal{S}$ will pick up an infinite contribution from the surface energy density and therefore diverge. For example, we will see explicitly that the vacuum contribution to the *stress* on a circular Dirichlet shell in two dimensions is infinite, in contradiction to the claim of Ref. [28].

7. Conclusions

We have presented a general procedure that is applicable to a variety of problems in quantum field theory. It gives a concrete prescription for handling field theory divergences in a concrete way. Ref.[29] extends this approach to systems with simple time dependence and Ref. [30] applies this formalism to nonzero temperature. Work is underway to apply these techniques to Higgs/gauge solitons in the Standard Model of the weak interactions.

Acknowledgments

This work has been done in collaboration with E. Farhi, N. Graham, P. Haagensen, V. Khemani, M. Quandt, P. Sundberg, and H. Weigel. I thank Arkady for valuable discussions that helped shape the material in Section 5. This work is supported in part by the U.S. Department of Energy (D.O.E.) under cooperative research agreement #DF-FC02-94ER40818.

References

1. N. Graham, R. L. Jaffe and H. Weigel, Int. J. Mod. Phys. A **17**, 846 (2002) [arXiv:hep-th/0201148].
2. J. Schwinger, Phys. Rev. **94** (1954) 1362; J. Baacke, Z. Phys. **C53** (1992) 402.
3. S. Coleman, Commun. Math. Phys. **31** (1973) 259.
4. E. Farhi, N. Graham, P. Haagensen and R. L. Jaffe, Phys. Lett. **B427** (1998) 334.
5. F. Calegero, *Variable Phase Approach to Potential Scattering*, Acad. Press, New York and London, 1967.
6. E. Farhi, N. Graham, R.L. Jaffe, and H. Weigel, Nucl. Phys. **B595** (2001) 536.
7. R. Blankenbecler and D. Boyanovsky, Phys. Rev. **D31** (1985) 2089; R. Blankenbecler and D. Boyanovsky, Phys. Rev. **D31** (1985) 3234.
8. R. Jackiw and C. Rebbi, Phys. Rev. **D13**, 3398 (1976).

9. G. Dunne and K. Rao, Phys. Rev. **D64** (2001) 025003.

10. E. Farhi, N. Graham, R.L. Jaffe, and H. Weigel, Phys. Lett. **B475** (2000) 335, Nucl. Phys. **B585** (2000) 443,

11. E. Farhi, N. Graham, R.L. Jaffe, and H. Weigel, "Searching for Quantum Solitons in a 3+1 Dimensional Chiral Yukawa Model", arXiv:hep-th/0112217.

12. A. d'Adda and P. di Vecchia, Phys. Lett. **73B** (1978) 162, Phys. Lett. **76B** (1978) 298; R. Horsley, Nucl. Phys. **B151** (1979) 399; S. Rouhani, Nucl. Phys. **B182** (1981) 462.; H. Yamagishi, Phys. Lett. **147B** (1984) 425; A. K. Chatterjee and P. Majumdar, Phys. Rev. **D30** (1984) 844, Phys. Lett. **159B** (1985) 37; A. Uchiyama, Nucl. Phys. **B244** (1984) 57, Progr. Theor. Phys. **75** (1986) 1214, Nucl. Phys. **B278** (1986) 121; A. Rebhan and P. van Nieuwenhuizen, Nucl. Phys. **B508** (1997) 449.

13. H. Nastase, M. Stephanov, P. van Nieuwenhuizen and A. Rebhan, Nucl. Phys. **B542** (1999) 471.

14. N. Graham and R. L. Jaffe, Phys. Lett. **B435** (1998) 145, Nucl. Phys. **B544** (1999) 432, Nucl. Phys. **B549** (1999) 516.

15. M. Shifman, A. Vainshtein and M. Voloshin, Phys. Rev. **D59** (1999) 045016.

16. S. V. Bashinsky, Phys. Rev. **D61**, 105003 (2000)

17. G. Dunne, Phys. Lett. **B467** (1999) 238.

18. H. B. G. Casimir, Kon. Ned. Akad. Wetensch. Proc. **51**, 793 (1948).

19. V.M. Mostepanenko and N.N. Trunov, *The Casimir Effect and its Application*, Clarendon Press, Oxford (1997), K. A. Milton, *The Casimir Effect: Physical Manifestations Of Zero-Point Energy*, River Edge, USA: World Scientific (2001).

20. M. Bordag, U. Mohideen and V. M. Mostepanenko, Phys. Rept. **353**, 1 (2001) [arXiv:quant-ph/0106045].

21. S. K. Lamoreaux, Phys. Rev. Lett. **78**, 5 (1997).

22. U. Mohideen and A. Roy, Phys. Rev. Lett. **81**, 4549 (1998) [arXiv:physics/9805038].

23. G. Bressi, G. Carugno, R. Onofrio, and G. Ruoso, Phys. Rev. Lett. **88**, 041804 (2002)[arXiv:quant-ph/0203002]

24. N. Graham, R. L. Jaffe, V. Khemani, M. Quandt, M. Scandurra and H. Weigel, [arXiv:hep-th/0207205].

25. N. Graham R.L. Jaffe, V. Khemani, M. Quandt, M. Scandurra, and H. Weigel, arXiv:hep-th/0207120.

26. G. Barton, J. Phys. A **34**, 4083 (2001).

27. The energy density away from S is interesting in its own right; since it can be negative, it may induce interesting effects in general relativity: F. J. Tipler, Phys. Rev. Lett. **37**, 879 (1976), S. W. Hawking, Phys. Rev. D **46**, 603 (1992), L. H. Ford and T. A. Roman, Phys. Rev. D **51**, 4277 (1995), K. D. Olum, Phys. Rev. Lett. **81**, 3567 (1998), [arXiv:gr-qc/9805003]. K. D. Olum and N. Graham, [arXiv:gr-qc/0205134].

28. K. A. Milton and Y. J. Ng, Phys. Rev. D **46**, 842 (1992), S. Leseduarte and A. Romeo, Ann. Phys. **250**, 448 (1996) [arXiv:hep-th/9605022].

29. N. Graham, Phys. Lett. **B513** (2001) 112.

30. N. Graham, "Exact One-Loop Thermal Free Energies of Solitons," arXiv:hep-th/0112148.

THE SEIBERG-WITTEN MAP FOR NONCOMMUTATIVE GAUGE THEORIES

B. L. CERCHIAI, A. PASQUA, B. ZUMINO

Department of Physics, University of California
and
Theoretical Physics Group
Lawrence Berkeley National Laboratory
Berkeley, California 94720, USA
E-mail: BLCerchiai@lbl.gov, pasqua@socrates.berkeley.edu, zumino@thsrv.lbl.gov

The Seiberg-Witten map for noncommutative Yang-Mills theories is studied and methods for its explicit construction are discussed which are valid for any gauge group. In particular the use of the evolution equation is described in some detail and its relation to the cohomological approach is elucidated. Cohomological methods which are applicable to gauge theories requiring the Batalin-Vilkoviskii antifield formalism are briefly mentioned. Also, the analogy of the Weyl-Moyal star product with the star product of open bosonic string field theory and possible ramifications of this analogy are briefly mentioned.

1. Introduction

Noncommutative field theories have recently received much attention. Seiberg and Witten[1] have argued that certain noncommutative gauge theories are equivalent to commutative ones and in particular that there exists a map from a commutative gauge field to a noncommutative one, which is compatible with the gauge structure of each. This map has become known as the Seiberg-Witten (SW) map.

In two recent papers[2,3] we have discussed a cohomological method for constructing explicitly this map. Here we describe a slightly modified procedure based on the idea that the structure equations of the gauge group of the noncommutative theory are a deformation of those of the gauge group of the commutative theory. We will consider gauge theories on the noncommutative space defined by

$$[x^i \overset{\star}{,} x^j] = i\theta^{ij} , \tag{1}$$

where θ is a constant Poisson tensor. The "$\star$" operation is the associative

Weyl-Moyal product

$$f \star g = f\, e^{\frac{i}{2}\theta^{ij}\overleftarrow{\partial_i}\overrightarrow{\partial_j}}\, g \,. \tag{2}$$

We believe that our methods are much more general, and can in fact be used even when θ is not constant, but in this paper we shall make use of the fact that the x^i derivative ∂_i of functions satisfies the Leibniz rule with respect to the star product

$$\partial_i(f \star g) = (\partial_i f) \star g + f \star (\partial_i g), \tag{3}$$

just as it does with respect to the ordinary product. This simple relation requires θ to be constant.

2. Structure Equations

The structure equations of a gauge group can be expressed in terms of a ghost field $\lambda(x)$ and the gauge potential $a_i(x)$ by giving the action of the BRST operator s

$$s\lambda = i\lambda \cdot \lambda, \tag{4}$$

$$sa_i = \partial_i \lambda - ia_i \cdot \lambda + i\lambda \cdot a_i \,. \tag{5}$$

Here λ and a_i are valued in a Lie algebra and can be represented by matrices, the matrix elements of the ghost field being anticommuting functions of x. In a representation the product would imply matrix multiplication. The operator s is an odd superderivation of ghost number one

$$s(f \cdot g) = (sf) \cdot g \pm f \cdot sg, \tag{6}$$

$$s^2 = 0, \tag{7}$$

which commutes with the derivatives

$$s\partial_i = \partial_i s. \tag{8}$$

As usual, the signs in (6) depend on the parity of f. Our task is to deform the above structure equations into

$$s\Lambda = i\Lambda \star \Lambda, \tag{9}$$

$$sA_i = \partial_i \Lambda - i[A_i \overset{\star}{,} \Lambda], \tag{10}$$

where $A_i = A_i(a, \partial a, \partial^2 a, \ldots)$ is an even local functional of a_i, of ghost number zero, and $\Lambda = \Lambda(\lambda, \partial\lambda, \ldots, a, \partial a, \ldots)$ is an odd local functional of a_i and λ, of ghost number one (like λ). We take s to be undeformed and to satisfy (7), (8) and

$$s(f \star g) = sf \star g \pm f \star sg. \tag{11}$$

The solution consists in finding explicit expressions for the functionals A_i and Λ. This can be done as expansions in θ

$$\Lambda = \Lambda^{(0)} + \Lambda^{(1)} + \ldots, \quad \Lambda^{(0)} = \lambda, \tag{12}$$

$$A_i = A_i^{(0)} + A_i^{(1)} + \ldots, \quad A_i^{(0)} = a_i. \tag{13}$$

The first order terms were given already in[1]

$$\Lambda^{(1)} = \frac{1}{4}\theta^{kl}\{\partial_k\lambda, a_l\}, \tag{14}$$

$$A_i^{(1)} = -\frac{1}{4}\theta^{kl}\{a_k, \partial_l a_i + f_{li}\}, \tag{15}$$

where

$$f_{li} = \partial_l a_i - \partial_i a_l - i[a_l, a_i] \tag{16}$$

is the commutative field strength, and expressions for $\Lambda^{(2)}$ and $A_i^{(2)}$ are known,[2,4,6] see also below.

A systematic way to obtain the expansion in θ was described in[2,3] and the consistency of the procedure was demonstrated in.[7] Each order in θ is manifestly local.

One sees already from (14) and (15) that Λ and A_i cannot be Lie algebra valued in general, and we follow[5,6] by allowing them to be in the enveloping algebra of the Lie algebra of λ and a_i. A representation of this Lie algebra lifts naturally to a representation of its enveloping algebra.

3. Evolution Equations

There is an alternative approach for the study of the SW map, which is based on a differential equation.[1] Let us introduce a "time" parameter t in front of θ, in such a way that $\theta^{ij} \to t\,\theta^{ij}$, $\Lambda \to \Lambda(t)$ and $A_i \to A_i(t)$, while keeping s independent of t. Notice that Λ and A_i acquire a t-dependence through θ. Differentiating the structure equations (9) and (10) with respect to t, we obtain[a]

$$s\,\dot{\Lambda} = i\,\dot{\Lambda} \star \Lambda + i\Lambda \star \dot{\Lambda} + i\Lambda \overset{\star}{\ast} \Lambda, \tag{17}$$

$$s\,\dot{A_i} = \dot{A_i} \star \Lambda + \Lambda \star \dot{A_i} + D_i\,\dot{\Lambda} - iA_i \overset{\star}{\ast} \Lambda + i\Lambda \overset{\star}{\ast} A_i, \tag{18}$$

where

$$D_i = \partial_i - i[A_i \overset{\star}{,} \cdot] \tag{19}$$

---

[a] As customary the dot denotes differentiation with respect to t.

is the covariant derivative at time t. The star product itself depends on the evolution parameter t, and therefore it has also to be differentiated

$$\star = e^{\frac{1}{2}it\overleftarrow{\partial_i}\theta^{ij}\overrightarrow{\partial_j}}, \qquad \overset{\bullet}{\star} = e^{\frac{1}{2}it\overleftarrow{\partial_i}\theta^{ij}\overrightarrow{\partial_j}}\,\frac{i}{2}\,\overleftarrow{\partial_k}\,\theta^{kl}\,\overrightarrow{\partial_l}\,. \tag{20}$$

Explicitly this yields

$$f \overset{\bullet}{\star} g = i\frac{\theta^{kl}}{2}\,\partial_k f \star \partial_l g. \tag{21}$$

Notice that for simplicity we have restricted ourselves to a linear path in θ-space, i.e. we are considering a linear one-parameter family of deformations of θ. In principle it would be possible to consider an arbitrary variation with respect to θ corresponding to an arbitrary path in θ-space, like e.g. in.[4]

The structure of the right hand side of (17) and (18) leads in a natural way to the definition of a new operator[b] at time t:

$$\Delta_t = \begin{cases} s - i\{\Lambda \overset{\star}{,} \cdot\} & \text{on odd quantities,} \\[2mm] s - i[\Lambda \overset{\star}{,} \cdot] & \text{on even quantities.} \end{cases} \tag{22}$$

It has the following properties

$$\Delta_t A_i = \partial_i \Lambda, \quad \Delta_t^2 = 0, \quad [\Delta_t, D_i] = 0, \tag{23}$$

$$\Delta_t(f_1 \star f_2) = (\Delta_t f_1) \star f_2 \pm f_1 \star (\Delta_t f_2), \tag{24}$$

i.e. Δ_t is nilpotent, it commutes with the covariant derivative at time t and it satisfies a super-Leibniz rule. This is a consequence of the fact that

$$s^2 = 0, \qquad s\partial_i = \partial_i s, \tag{25}$$

and of the associativity of the star product. Therefore, Δ_t can be interpreted as a coboundary operator in a suitably defined cohomology.

Using the operators Δ_t and D_i the equations (17) and (18) can be rewritten as

$$\Delta_t \overset{\bullet}{\Lambda} = -\frac{1}{2}\theta^{kl}\partial_k\Lambda \star \partial_l\Lambda = -\frac{1}{2}\theta^{kl}B_k \star B_l, \tag{26}$$

$$\Delta_t \overset{\bullet}{A_i} = D_i \overset{\bullet}{\Lambda} + \frac{1}{2}\theta^{kl}\{\partial_k A_i \overset{\star}{,} \partial_l\Lambda\} = D_i \overset{\bullet}{\Lambda} + \frac{1}{2}\theta^{kl}\{\partial_k A_i \overset{\star}{,} B_l\}. \tag{27}$$

Here we have introduced the notation

$$B_i = \partial_i\Lambda, \tag{28}$$

---

[b] Δ_t is a simple generalization of the operator Δ introduced in,[2] which now should be called Δ_0. Also, what was called $\hat{\Delta}$ in[3] should now be called Δ_1.

which is useful because only derivatives of Λ enter in the right hand side of (26) and (27), but never Λ itself. The action of Δ_t in terms of these new variables A_k and B_k takes a particularly simple form

$$\Delta_t A_k = B_k, \qquad \Delta_t B_k = 0. \tag{29}$$

With this action the consistency condition that Δ_t applied to the right hand side of equation (26) vanishes is verified. For (27) we find that Δ_t on the right hand side gives $\frac{1}{2}\theta^{kl}[\Delta_t F_{ki} \stackrel{\star}{,} B_l]$. We will comment on this later in section 4.

The differential evolution equations which provide a solution to the equations (26), (27) are given by[1]

$$\dot{\Lambda} = \frac{1}{4}\theta^{ij}\{\partial_i\Lambda \stackrel{\star}{,} A_j\}, \tag{30}$$

$$\dot{A}_i = -\frac{1}{4}\theta^{kl}\{A_k \stackrel{\star}{,} \partial_l A_i + F_{li}\}, \tag{31}$$

where

$$F_{li} = \partial_l A_i - \partial_i A_l - i[A_l \stackrel{\star}{,} A_i] \tag{32}$$

is the noncommutative field strength. This can be easily checked by substituting these expressions in (26) and (27).

4. The Homotopy Operator

There is a way of computing the expressions (30) for $\dot{\Lambda}$ and (31) for $\dot{A}_i$ through a suitably defined homotopy operator K_t. Clearly, it is not possible to invert Δ_t, because it is nilpotent, but if we construct an operator such that

$$K_t\Delta_t + \Delta_t K_t = 1, \tag{33}$$

then an equation of the form

$$\Delta_t f = m, \tag{34}$$

with

$$\Delta_t m = 0, \tag{35}$$

has a solution of the type

$$f = K_t m, \tag{36}$$

because

$$m = \Delta_t K_t m + K_t \Delta_t m = \Delta_t K_t m. \tag{37}$$

The solution (36) is not unique: $K_t m + \Delta_t h$, with some appropriate h, is also a solution, since $\Delta_t^2 = 0$. This is the same method we applied for $t = 0$ in,[2,3] which closely follows the ideas developed in[8] to study anomalies in chiral gauge theories.

Let us construct such a homotopy operator K_t explicitly. We start by defining a linear operator $\tilde{K}_t$ such that

$$\tilde{K}_t B_k = A_k, \qquad \tilde{K}_t A_k = 0. \tag{38}$$

On both A_k and B_k it satisfies

$$\tilde{K}_t \Delta_t + \Delta_t \tilde{K}_t = 1. \tag{39}$$

Further, we require that it is a super-derivation

$$\tilde{K}_t(f_1 \star f_2) = (\tilde{K}_t f_1) \star f_2 \pm f_1 \star (\tilde{K}_t f_2) \tag{40}$$

and that it commutes with D_i and anticommutes with s

$$[\tilde{K}_t, D_i] = 0, \qquad \{\tilde{K}_t, s\} = 0. \tag{41}$$

Notice that due to (39) $\tilde{K}_t$ has to be odd and it decreases the ghost number by one. Moreover, it is nilpotent on A_i, B_i

$$\tilde{K}_t^2 = 0. \tag{42}$$

On monomials of higher order in A_k and B_k, the homotopy operator K_t cannot satisfy the Leibniz rule. If d is the total order of such a monomial m, then the action of K_t on it has to be defined as

$$K_t m = d^{-1} \tilde{K}_t m. \tag{43}$$

It is extended to general polynomials by linearity. Then K_t satisfies (33) and from (42) it follows that

$$K_t^2 = 0. \tag{44}$$

Now, we can use K_t to recover the solutions (30), (31) of the equations (26), (27). For Λ this is straightforward. We apply K_t to the right hand side of (26) and we get

$$\dot{\Lambda} = K_t \left(-\frac{1}{2}\theta^{kl} B_k \star B_l \right) = -\frac{1}{4}\theta^{kl} \left(\tilde{K}_t B_k \star B_l - B_k \star \tilde{K}_t B_l \right) \tag{45}$$

$$= -\frac{1}{4}\theta^{kl} \left(A_k \star B_l - B_k \star A_l \right) = \frac{1}{4}\theta^{kl} \{B_k \star A_l\},$$

which coincides with (30).

For the gauge potential, however, there is a complication. If we apply Δ_t to the right hand side of (27) we obtain

$$\Delta_t \left(D_i \dot\Lambda + \frac{1}{2}\theta^{kl}\{\partial_k A_i \stackrel{\star}{,} B_l\} \right) = \frac{1}{2}\theta^{kl}[\Delta_t F_{ki} \stackrel{\star}{,} B_l], \qquad (46)$$

where $\Delta_t F_{ki} = D_k B_i - D_i B_k + i[B_k, A_i] + i[A_k, B_i]$. This expression vanishes only if we impose the condition that

$$\Delta_t F_{ki} = 0. \qquad (47)$$

This property is true if we explicitly use the definition (28) of $B_i = \partial_i \Lambda$, but it has to be set as an additional constraint in the algebra generated by the A_i, the B_i and their derivatives. In other words, such an algebra is not free. The homotopy operator K_t can be defined only on B. In order to solve this problem, we can add to the right hand side of (27) a term which is zero by the constraint (47), but which makes the Δ_t of it vanish algebraically. For this purpose we can choose e.g. $\frac{1}{2}\theta^{kl}\{\Delta_t F_{ki} \stackrel{\star}{,} A_l\}$ and consider the expression

$$U_i \equiv D_i \dot\Lambda + \frac{1}{2}\theta^{kl}(\{\partial_k A_i \stackrel{\star}{,} B_l\} + \{\Delta_t F_{ki} \stackrel{\star}{,} A_l\}). \qquad (48)$$

Then

$$\Delta_t U_i = 0 \qquad (49)$$

algebraically and we can apply the homotopy operator to U_i and obtain (31). This is the same procedure we have proposed in[2] and[3] to treat the analogous difficulty.

5. Solutions to higher order in θ

Observe that we can recover the first order in the θ expansion as

$$\Lambda^{(1)} = \dot\Lambda(t)\,|_{t=0} = \frac{1}{4}\theta^{ij}\{\partial_i \lambda, a_j\}, \qquad (50)$$

$$A_i^{(1)} = \dot A_i(t)\,|_{t=0} = -\frac{1}{4}\theta^{kl}\{a_k, \partial_l a_i + f_{li}\}, \qquad (51)$$

which yields the well-known solution found by Seiberg and Witten.[1] More in general, once we have the expressions (30) and (31) to first order, the evolution equations provide a useful method for computing the terms of higher order in θ by just noticing that

$$\Lambda^{(n)} = \frac{1}{n!}\frac{\partial^n \Lambda(t)}{\partial t^n}\,|_{t=0}, \qquad A_i^{(n)} = \frac{1}{n!}\frac{\partial^n A_i(t)}{\partial t^n}\,|_{t=0}. \qquad (52)$$

Therefore, by simply differentiating with respect to t, we can compute $\Lambda^{(n)}$ and $A_i^{(n)}$. This is an alternative and easier technique than applying the homotopy operator order by order as suggested in.[2,3]

In particular to second order we get

$$\ddot{\Lambda} = \frac{1}{4}\theta^{kl}\left(\left\{\partial_k\dot{\Lambda}\,\overset{\star}{,}\,A_l\right\} + \left\{\partial_k\Lambda\,\overset{\star}{,}\,\dot{A}_l\right\} + \partial_k\Lambda\overset{\star}{,}A_l + A_l\overset{\star}{,}\partial_k\Lambda\right) \quad (53)$$

$$\ddot{A}_i = -\frac{1}{4}\theta^{kl}\left(\left\{\dot{A}_k\,\overset{\star}{,}\,\partial_l A_i + F_{li}\right\} + \left\{A_k\,\overset{\star}{,}\,\partial_l\dot{A}_i + \dot{F}_{li}\right\}\right. \quad (54)$$
$$\left. + A_k\overset{\star}{,}(\partial_l A_i + F_{li}) + (\partial_l A_i + F_{li})\overset{\star}{,}A_k\right).$$

Notice that the equation for $\frac{\partial^n\Lambda}{\partial t^n}$ contains $\frac{\partial^{n-1}A_i}{\partial t^{n-1}}$, while the equation for $\frac{\partial^n A_i}{\partial t^n}$ depends only on $\frac{\partial^k A_i}{\partial t^k}$, $k = 0, \ldots, n-1$. This means that the equations for A are independent from those for Λ. We need to compute A first and only afterwards we can substitute it in the expression for Λ. If we use the homotopy operator, exactly the opposite happens, we need the expression for $\Lambda^{(n)}$ first in order to obtain $A_i^{(n)}$. If we insert the expressions (30) for $\dot{\Lambda}$ and (31) for $\dot{A}_l$ we obtain

$$\ddot{\Lambda} = \frac{1}{16}\theta^{ij}\theta^{kl}\left(\left\{\{\partial_i\partial_k\Lambda\,\overset{\star}{,}\,A_j\} + \{\partial_i\Lambda\,\overset{\star}{,}\,\partial_k A_j\}\,A_l\right\}\right.$$
$$\left. - \{\partial_i\Lambda\,\overset{\star}{,}\,\{A_k\,\overset{\star}{,}\,\partial_l A_j + F_{lj}\}\}\right. \quad (55)$$
$$\left. + 2i\,[\partial_i\partial_k\Lambda\,\overset{\star}{,}\,\partial_j A_l]\right).$$

6. Ambiguities

The solution (55) has to be compared to other known solutions of the SW map at the second order, like e.g.[4] or[6]. Before doing that, let us remark that the solutions of (26) and (27) are not unique. This has been commented on by a number of authors.[2,3,4,6,9,14]

If we start with the structure equations (9), (10)

$$s\Lambda = i\Lambda\star\Lambda,$$
$$sA_i = \partial_i\Lambda - i[A_i\,\overset{\star}{,}\,\Lambda],$$

and consider a change in θ by an amount $\delta\theta$, then we see that

$$\Delta\,\delta\Lambda = -\frac{1}{2}\delta\theta^{kl}\partial_k\Lambda\star\partial_l\Lambda\,, \quad (56)$$

where the star product and the fields are at $t = 1$ and where Δ is the same as Δ_t for $t = 1$. Therefore, given a solution $(\delta\Lambda)_0$ of this equation,

$$\delta\Lambda = (\delta\Lambda)_0 + \Delta H \quad (57)$$

is also a solution, because of the nilpotency of Δ.

Similarly, for the gauge potential a change in θ induces a change in A_i determined by

$$\Delta \, \delta A_i = D_i \delta \Lambda + \frac{1}{2} \delta \theta^{kl} \, \{\partial_k A_i \stackrel{*}{,} \partial_l \Lambda\}. \tag{58}$$

Therefore, given a solution $(\delta A_i)_0$ corresponding to $(\delta \Lambda)_0$, the solution corresponding to $(\delta \Lambda)_0 + \Delta H$ is

$$\delta A_i = (\delta A_i)_0 + D_i H + S_i, \tag{59}$$

where S_i is an arbitrary local functional of ghost number 0 satisfying

$$\Delta S_i = 0. \tag{60}$$

This is a consequence of the fact that Δ commutes with the covariant derivative: $D_i \Delta = \Delta D_i$.

The ambiguities determined by H are of the form of a gauge transformation.

Due to the definition of Δ the condition (60) means that S_i transforms covariantly

$$s S_i = i[\Lambda \stackrel{*}{,} S_i]. \tag{61}$$

This covariant ambiguity is of a different type from the gauge ambiguity. It can be interpreted as a field dependent redefinition of the gauge potential.

The ambiguities of gauge type are an infinitesimal version of the Stora invariance of the structure equations (9), (10)[3]

$$\Lambda \to G^{-1} \star \Lambda \star G + i\, G^{-1} \star sG,$$
$$A_i \to G^{-1} \star A_i \star G + i\, G^{-1} \star \partial_i G, \tag{62}$$

where G is an arbitrary local functional of ghost number 0.

If we compare the solution to second order given in (55) for $t = 0$

$$\Lambda^{(2)} = \frac{1}{2} \, \ddot{\Lambda}|_{t=0} \tag{63}$$

with the solution $\Lambda'^{(2)}$ found in[4] we see that

$$\Lambda^{(2)} - \Lambda'^{(2)} = \frac{1}{64} \theta^{kl} \theta^{mn} \Delta_0 \left(\{\{D_m a_k + D_k a_m - f_{km}, a_n\}, a_l\} \right. \tag{64}$$
$$\left. - [[a_k, a_m], f_{nl}]\right),$$

which is an ambiguity of the gauge type.

7. Actions

Until this point, we have discussed the deformation of gauge structures and their representations in terms of Yang-Mills fields, without any reference to the dynamics of the fields themselves. To specify the dynamics, we must construct actions that are invariant under the deformed gauge transformations

$$
\begin{aligned}
sA_i &= \partial_i\Lambda - i[A_i \overset{\star}{,} \Lambda], \\
sF_{ij} &= -i[F_{ij} \overset{\star}{,} \Lambda].
\end{aligned}
\tag{65}
$$

The procedure is analogous to the construction of commutative Yang-Mills theory. One arrives at the expression

$$
S^{YM}[A] = -\frac{1}{4}\int d^4x\, Tr\, F_{ij} \star F^{ij},
\tag{66}
$$

where F_{ij} is the noncommutative field strength given by (32), and the trace is the ordinary matrix trace in the appropriate representation. The proof of the invariance of (66) under (65) is based on the properties

$$
\begin{aligned}
\int dx\, f \star g &= \int dx\, fg = \int dx\, g \star f, \\
\int dx\, Tr\, M \star N &= \int dx\, Tr\, N \star M,
\end{aligned}
\tag{67}
$$

the latter of which is valid for any pair of matrix valued functions, when surface terms are ignored. Hence, the integral of the trace is invariant under any cyclic permutation of its factors, also in the presence of the star product. Since the fields A and F are generally valued in the enveloping algebra, we have to use the Seiberg-Witten map in order to make sense of (66) as a theory with a finite number of degrees of freedom, namely those of a_i. To first order in the deformation parameter θ, we find

$$
S^{YM} = -\frac{1}{4}\int d^4x\, Tr\, f_{ij}f^{ij} + \frac{1}{16}\theta^{kl}\int d^4x\, Tr\, f_{kl}f_{ij}f^{ij} -
$$
$$
-\frac{1}{2}\theta^{kl}\int d^4x\, Tr\, f_{ik}f_{jl}f^{kl} + O(\theta^2),
\tag{68}
$$

where f_{ij} is the commutative field strength given by (16).

We would like to remark that at the level of free fields there is no difference between commutative and noncommutative theories, because the properties (67) guarantee that the star product disappears from any quadratic action. It is only when interaction terms are present that the commutative and the noncommutative theory and in fact different. However, interaction terms are always present in the action (66), even if the gauge group is U(1), because of the star commutator term in the expression (32) for F.

In addition to the pure Yang-Mills theory, one can construct a noncommutative version of any action with a gauge invariance, simply by replacing each ordinary product of functions with a star product, leaving the matrix multiplication and the trace unchanged, and finally expanding each noncommutative field by means of the Seiberg-Witten map associated with the deformed gauge structure. In particular, Yang-Mills theories with matter fields in various representations have been considered by several authors.

Another gauge invariant action that can be constructed in terms of gauge potentials only is the Chern-Simons action in three dimensions. Its deformed counterpart is obtained as described above and is

$$S_t^{CS}[A] = \frac{1}{4\pi} \int d^3x \, \epsilon^{klm} Tr(A_k \star \partial_l A_m - \frac{2}{3} i A_k \star A_l \star A_m), \qquad (69)$$

where the subscript t refers to the parameter of the evolution equation described in section 3.

If one were to expand (69) by means of the Seiberg-Witten map, one would find that it is in fact identical to the undeformed action.[10] In other words

$$S_1^{CS}[A] = S_0^{CS}[a]. \qquad (70)$$

This can be proven to hold at all orders in the deformation parameter, by showing instead

$$\frac{d}{dt} S_t^{CS}[A] = 0, \, \forall t. \qquad (71)$$

The total t-derivative is computed using

$$f \overset{\bullet}{\star} g = \frac{i}{2} \theta^{kl} \partial_k f \star \partial_l g \qquad (72)$$

and the evolution equation for A

$$\dot{A}_k = -\frac{1}{4} \theta^{rs} \{A_r \overset{\star}{,} \partial_s A_k + F_{sk}\}. \qquad (73)$$

In this context, it is worth noting that the WZW model in two dimensions shares the same property, namely the identity of the actions for the commutative and the noncommutative version,[11] and that the WZW model in two dimensions is related to the Chern-Simons action in three.

8. Concluding remarks

In this paper we have limited ourselves to gauge theories of the Yang-Mills type and have based our analysis on the structure equations (4) and (5) (which involve BRST transformations) and their deformation. This formulation is sufficient for Yang-Mills theories, but for gauge theories with reducible gauge transformations, such as theories with gauge potentials

418

which are differential forms of degree higher than one, it is appropriate to use the antifield formalism of Batalin and Vilkoviskii (BV). The deformation of the gauge structure should then be studied by defining generalized Seiberg-Witten maps in the context of the BV formalism.[12,13,14] The use of the master equation couples intimately the gauge transformations and the dynamics, i.e. the action functional.

The existence of the SW map, together with the understanding of its ambiguities, can be interpreted as a kind of "rigidity" of the structure of the gauge group, analogous to the rigidity of semisimple Lie algebras under smooth deformations of the structure constants: the structure constants can be brought back to their original values by performing a linear transformation on the generators. In the case of gauge groups the deformed structure equations can be reduced to the undeformed equations by expressing the deformed fields (e.g. A_i and Λ) as local functionals of the undeformed fields (a_i and λ). Strictly speaking, we have discussed only infinitesimal gauge transformations in a context in which only the space-time coordinates are deformed. Thus, we have ignored all questions for which the topology of the gauge group may be relevant when the gauge fields are quantized.[15]

As explained in the introduction, throughout this paper we have considered the case of θ^{ij} independent of x. Techniques of deformation quantization are available for an x-dependent Poisson tensor (see, e.g.,[16,17] and references therein, where general coordinate transformations for quantized coordinates are also studied). It would be interesting to extend to that case the results described in the previous sections.

Recently, several authors have pointed out the analogy of the Weyl-Moyal star product with the associative, noncommutative star product which enters in the formulation of Witten's bosonic open string field theory.[18,19,20,21,22,23] It would be interesting if methods of deformation quantization developed in the context of the Seiberg-Witten map, would turn out to be useful in string field theory.

Acknowledgements

We are very grateful to S. Schraml and A. Weinstein for many fruitful discussions. This paper is a written version of a talk given by B. Zumino at the Workshop on "Continuous Advances in QCD", Minneapolis, Minnesota, May 17-23, 2002. This work was supported in part by the Director, Office of Science, Office of High Energy and Nuclear Physics, Division of High Energy Physics of the U.S. Department of Energy under Contract DE-AC03-76SF00098 and in part by the National Science Foundation under grant PHY-0098840. B.L.C. is supported by the DFG (Deutsche Forschungsgemeinschaft) under grant CE 50/1-2.

References

1. N. Seiberg, E. Witten, *String theory and noncommutative geometry*, JHEP **9909**:032 (1999), hep-th/9908142
2. D. Brace, B. L. Cerchiai, A. Pasqua, U. Varadarajan, B. Zumino, *A Cohomological Approach to the Non-Abelian Seiberg-Witten Map*, JHEP **06**:047 (2001), hep-th/0105192
3. D. Brace, B. L. Cerchiai, B. Zumino, *Nonabelian Gauge Theories on Noncommutative Spaces*, to appear in the proceedings of '2001: A Spacetime Odissey', Michigan Center for Theoretical Physics, preprint UCB-PTH-01/29, LBNL-48673, hep-th/0107225
4. S. Goto, H. Hata, *Noncommutative Monopole at the second order in* Θ, Phys. Rev. **D 62**:085 (2000), hep-th/0005101
5. B. Jurčo, S. Schraml, P. Schupp, J. Wess, *Enveloping algebra valued gauge transformations for non-abelian gauge groups on non-commutative spaces*, Eur. Phys. J. **C17**:521 (2000), hep-th/0006246
6. B. Jurčo, L. Möller, S. Schraml, P. Schupp, J. Wess, *Construction of non-Abelian gauge theories on noncommutative spaces*, Eur. Phys. J. **C21**:383 (2001), hep-th/0104153
7. M. Picariello, A. Quadri, S. P. Sorella, *Chern-Simons in the Seiberg-Witten map for non-commutative Abelian gauge theories in 4D*, JHEP **0201**:045 (2002), hep-th/0110101
8. B. Zumino, *Chiral anomalies and differential geometry*, in 'Relativity, Groups and Topology II', Les Houches 1983, B. S. DeWitt, R. Stora (eds.), reprinted in S. B. Treiman, R. Jackiw, B. Zumino, E. Witten, 'Current algebra and anomalies', World Scientific, Singapore (1985)
9. T. Asakawa, I. Kishimoto, *Comments on Gauge Equivalence in Noncommutative Geometry*, JHEP **9911**:024 (1999), hep-th/9909139
10. N. Grandi, G. A. Silva, *Chern-Simons action in noncommutative space*, Phys.Lett. **B507**:345 (2001), hep-th/0010113
11. E. F. Moreno, F. A. Schaposnik. *Wess-Zumino-Witten and fermion models in noncommutative space*, Nucl.Phys. **B596**:439 (2001), hep-th/0008118
12. G. Barnich, M. Grigorev, M. Henneaux, *Seiberg-Witten maps from the point of view of consistent deformations of gauge theories*, JHEP **0110**:004 (2001), hep-th/0106188
13. G. Barnich, F. Brandt, M. Grigoriev, *Seiberg-Witten maps in the context of the antifield formalism*, preprint ULB-TH/01-18, hep-th/0201139
14. G. Barnich, F. Brandt, M. Grigoriev, *Seiberg-Witten maps and noncommutative Yang-Mills theories for arbitrary gauge groups*, preprint ULB-TH/02-14, MPI-MIS-44/2002, hep-th/0206003
15. A. P. Polychronakos, *Seiberg-Witten map and topology*, preprint RU-02-4-B, hep-th/0206013
16. A. S. Cattaneo, G. Felder, *A Path Integral Approach to the Kontsevich Quantization*, Commun.Math.Phys. **212**:591 (2000), math-qa/9902090
17. A. S. Cattaneo, G. Felder, *On the Globalization of Kontsevich's Star Product*

and the Perturbative Poisson Sigma Model, hep-th/0111028

18. E. Witten, *Noncommutative Geometry and String Field Theory*, Nucl. Phys. **B268**:253 (1986)

19. I. Bars, *Map of Witten's ⋆ to Moyal's ⋆*, Phys. Lett. **B517**:436 (2001), hep-th/0106157

20. I. Bars, Y. Matsuo,*Associativity Anomaly in String Field Theory*, preprint CITUSC/02-004, UT-983, hep-th/0202030

21. I. Bars, Y. Matsuo, *Computing in String Field Theory Using the Moyal Star Product*, preprint CITUSC/02-013, UT-02-08, hep-th/0204260

22. M. R. Douglas, H. Liu, G. Moore, B. Zwiebach, *Open String Star as a Continuous Moyal Product*, JHEP **0204**:022 (2002), hep-th/0202087

23. D.M. Belov, *Diagonal Representation of Open String Star and Moyal Product*, preprint hep-th/0204164

RENORMALIZATION PROPERTIES OF SOFTLY BROKEN SUSY GAUGE THEORIES

DMITRI KAZAKOV

Biological Laboratory of Theoretical Physics, Joint Institute for Nuclear Research, Dubna, Russia
and
Institute for Theoretical and Experimental Physics, Moscow, Russia

In the present review we show that renormalizations in a softly broken SUSY gauge theory are not independent but directly follow from those of an unbroken or rigid theory. This is a consequence of a treatment of a softly broken theory as a rigid one in external spurion superfield. This enables one to get the singular part of effective action in a broken theory from a rigid one by a simple modification of the couplings. Substituting the modified couplings into renormalization constants, RG equations, solutions to these equations, approximate solutions, fixed points, etc., one can get corresponding relations for the soft terms by a simple Taylor expansion over the Grassmannian variables. Some examples including the MSSM in low and high $\tan \beta$ regime, SUSY GUTs and the N=2 Seiberg-Witten model are considered.

1. Introduction

In a series of papers [1-5] we have shown that renormalizations in a softly broken SUSY theory follow from those of an unbroken one in a straightforward way. This is in agreement with the other approaches [6,7,8] and is inspired by the original observation of Ref.[9]. In what follows we give a review of our approach. It does not explicitly use the power of holomorphicity advocated by some authors [7,8], but ends up with the simple and straightforward algorithm which is easy to apply.

The main idea is that a softly broken supersymmetric gauge theory can be considered as a rigid SUSY theory imbedded into external space-time independent superfield η, so that all couplings and masses become external superfields $S(\eta, \bar\eta)$. Then, the following crucial statement is valid [1]

The statement: *In external spurion field η the UV singular part of the effective action depends on the couplings $S(\eta, \bar\eta)$, but does not depend on their derivatives:*

$$S^{eff}_{Sing}(g) \Rightarrow S^{eff}_{Sing}(S, D^2\!\!\!\!/\,S, \bar{D}^2\!\!\!\!/\,S, D^2\bar{D}^2\!\!\!\!/\,S), \tag{1}$$

where D and $\bar{D}$ are the supercovariant derivatives, and as a result has the same form in unbroken and broken cases.

With replacement of the couplings by external fields one can calculate the effective action $S_{Sing}^{eff}(g)$ assuming that the external field is a constant, i.e. in a rigid theory. This approach to a softly broken supersymmetric theory allows us to use remarkable mathematical properties of $N = 1$ SUSY field theories such holomorphicity which leads to the non-renormalization theorems, cancellation of quadratic divergences, etc.

The renormalization procedure in a softly broken SUSY gauge theory can be performed in the following way:

One takes the renormalization constants of a rigid theory, calculated in some massless scheme, substitutes instead of the rigid couplings (gauge and Yukawa) their modified expressions, which depend on a Grassmannian variable, and expand over this variable. This gives renormalization constants for the soft terms. Differentiating them with respect to a scale one can find corresponding renormalization group equations.

Thus, the soft-term renormalizations are not independent but can be calculated from the known renormalizations of a rigid theory with the help of the differential operators. Explicit form of these operators has been found in a general case and in some particular models like SUSY GUTs or the MSSM [1,3]. The same expressions have been obtained also in a somewhat different approach in Ref. [6,7,10].

In fact as it has been shown in [3] this procedure works at all stages. One can make the above mentioned substitution on the level of the renormalization constants, RG equations, solutions to these equations, approximate solutions, fixed points, finiteness conditions, etc. Expanding then over a Grassmannian variable one obtains corresponding expressions for the soft terms. This way one can get new solutions of the RG equations and explore their asymptotics, or approximate solutions, or find their stability properties, starting from the known expressions for a rigid theory.

Throughout the paper we assume the existence of some gauge and SUSY invariant regularization. Though it is some problem by itself, in principle it is solvable [11]. Provided the rigid theory is well defined, we consider the modifications which appear due to the presence of soft SUSY breaking terms. To be more precise, when discussing one, two and three loop calculations of the renormalization constants we have in mind dimensional reduction and the minimal subtraction scheme. Though dimensional reduction is not self-consistent in general [12], it is safe to use it in low orders and all the actual calculations are performed in the framework of dimensional reduction [13,14,15,16]. Nevertheless, our main formulae have general validity provided the invariant procedure exists.

Below we give some examples: the general SUSY gauge theory in higher loops, the MSSM in low $\tan \beta$ regime where analytical solutions to the one-loop RG equations are known exactly and in high $\tan \beta$ regime where analytical solutions are known in iterative or approximate form. We discuss some particular solutions like the fixed point ones and examine their properties. The method allows one to get the same type of solutions for the soft SUSY breaking terms. The other examples are the finite N=1 SUSY GUTs and the N=2 Seiberg-Witten model where exact (nonperturbative) solution is known. Here one can extend finiteness and the S-W solution to the soft terms as well.

2. Soft SUSY Breaking and the Spurion Superfields

Consider an arbitrary $N = 1$ SUSY gauge theory with unbroken SUSY within the superfield formalism. The Lagrangian of a rigid theory is given by

$$\mathcal{L}_{rigid} = \int d^2\theta \, \frac{1}{4g^2} \text{Tr} W^\alpha W_\alpha + \int d^2\bar{\theta} \, \frac{1}{4g^2} \text{Tr} \bar{W}_{\dot\alpha} \bar{W}^{\dot\alpha} \tag{2}$$

$$+ \int d^2\theta d^2\bar{\theta} \, \bar{\Phi}^i (e^V)^j_i \Phi_j + \int d^2\theta \, W + \int d^2\bar{\theta} \, \bar{W},$$

where

$$W_\alpha = -\frac{1}{4} \bar{D}^2 e^{-V} D_\alpha e^V, \quad \bar{W}_{\dot\alpha} = -\frac{1}{4} D^2 e^{-V} \bar{D}_{\dot\alpha} e^V,$$

are the gauge field strength tensors and the superpotential $\mathcal{W}$ has the form

$$W = \frac{1}{6} y^{ijk} \Phi_i \Phi_j \Phi_k + \frac{1}{2} M^{ij} \Phi_i \Phi_j. \tag{3}$$

To fix the gauge, the usual gauge-fixing term can be introduced. It is useful to choose it in the form

$$\mathcal{L}_{gauge-fixing} = -\frac{1}{16} \int d^2\theta d^2\bar{\theta} \text{Tr} \left(\bar{f}f + f\bar{f}\right) \tag{4}$$

where the gauge fixing condition is taken as

$$f = \bar{D}^2 \frac{V}{\sqrt{\xi g^2}}, \quad \bar{f} = D^2 \frac{V}{\sqrt{\xi g^2}}. \tag{5}$$

Here ξ is the usual gauge-fixing parameter. Then, the corresponding ghost term is [17]

$$\mathcal{L}_{ghost} = i \int d^2\theta \, \frac{1}{4} \text{Tr} \, b \, \delta_c f - i \int d^2\bar{\theta} \, \frac{1}{4} \text{Tr} \, \bar{b} \, \delta_{\bar{c}} \bar{f}, \tag{6}$$

where c and b are the Faddeev–Popov ghost and antighost chiral superfields, respectively, and δ_c is the gauge transformation with the replacement of gauge superfield parameters $\Lambda(\bar{\Lambda})$ by chiral (antichiral) ghost fields $c(\bar{c})$.

For our choice of the gauge-fixing condition, the gauge transformation of f looks like

$$\delta_\Lambda f = \bar{D}^2 \delta_\Lambda \frac{V}{\sqrt{\xi g^2}} = i\bar{D}^2 \frac{1}{\sqrt{\xi g^2}} \mathcal{L}_{V/2}[\Lambda + \bar{\Lambda} + \coth(\mathcal{L}_{V/2})(\Lambda - \bar{\Lambda})], \quad (7)$$

where $\mathcal{L}_X Y \equiv [X, Y]$. Equation (6) then takes the form

$$\mathcal{L}_{ghost} = -\int d^2\theta \, \frac{1}{4} \text{Tr} \, b \bar{D}^2 \frac{1}{\sqrt{\xi g^2}} \mathcal{L}_{V/2}[c + \bar{c} + \coth(\mathcal{L}_{V/2})(c - \bar{c})] + \quad h.c.$$

$$= \int d^2\theta d^2\bar{\theta} \, \text{Tr} \left(\frac{b + \bar{b}}{\sqrt{\xi g^2}} \right) \mathcal{L}_{V/2}[c + \bar{c} + \coth(\mathcal{L}_{V/2})(c - \bar{c})] \quad (8)$$

$$= \int d^4\theta \, \text{Tr} \left(\frac{b + \bar{b}}{\sqrt{\xi g^2}} \right) \left((c - \bar{c}) + \frac{1}{2}\big[V, (c + \bar{c})\big] + \frac{1}{12}\Big[V, \big[V, (c - \bar{c})\big]\Big] + ... \right).$$

The resulting Lagrangian together with the gauge-fixing and the ghost terms are invariant under the BRST transformations. For a rigid theory in our normalization of the fields, they have the form [17]

$$\delta V = \epsilon \mathcal{L}_{V/2}[c + \bar{c} + \coth(\mathcal{L}_{V/2})(c - \bar{c})],$$

$$\delta c^a = -\frac{i}{2}\epsilon f^{abc}c^b c^c, \qquad \delta \bar{c}^a = -\frac{i}{2}\epsilon f^{abc}\bar{c}^b \bar{c}^c,$$

$$\delta b^a = \frac{1}{8}\epsilon \bar{D}^2 \bar{f}^a, \qquad \delta \bar{b}^a = \frac{1}{8}\epsilon D^2 f^a. \quad (9)$$

Breaking of supersymmetry is the problem by itself. We do not discuss here the origin of SUSY breaking but rather concentrate on the consequences of it. Usually one considers the so-called "soft" breaking of SUSY, which means that the breaking terms do not spoil renormalizability of the theory and, in particular, the cancellation of quadratic divergences and are represented by the operators of dimension less than 4 [18]. Hence, to perform the SUSY breaking, that satisfies the requirement of "softness", one can introduce a gaugino mass term as well as cubic and quadratic interactions of scalar superpartners of the matter fields

$$-\mathcal{L}_{soft-br} = \left[\frac{M}{2}\lambda\lambda + \frac{1}{6}A^{ijk}\phi_i\phi_j\phi_k + \frac{1}{2}B^{ij}\phi_i\phi_j + h.c. \right] + (m^2)^i_j \phi^*_i \phi^j, \quad (10)$$

where λ is the gaugino field, and ϕ_i is the lowest component of the chiral matter superfield.

This is not the most general form of the soft terms. In principle, one can add the terms like $\bar{\psi}\psi$, $\phi^*\phi\phi$, etc. [19]. However, the conventional choice

(10) is sufficient for the goal of SUSY breaking and in what follows we stick to it.

Remarkably, one can rewrite the Lagrangian (10) in terms of N=1 superfields introducing the external spurion superfields [18] $\eta = \theta^2$ and $\bar{\eta} = \bar{\theta}^2$, where θ and $\bar{\theta}$ are the Grassmannian parameters, as [9]

$$
\mathcal{L}_{soft} = \int d^2\theta \, \frac{1}{4g^2}(1 - 2M\theta^2)\mathrm{Tr}W^\alpha W_\alpha + \int d^2\bar{\theta} \, \frac{1}{4g^2}(1 - 2\bar{M}\bar{\theta}^2)\mathrm{Tr}\bar{W}^{\dot{\alpha}}\bar{W}_{\dot{\alpha}}
$$
$$
+ \int d^2\theta d^2\bar{\theta} \;\; \bar{\Phi}^i(\delta_i^k - (m^2)_i^k\eta\bar{\eta})(e^V)_k^j\Phi_j \tag{11}
$$
$$
+ \int d^2\theta \left[\frac{1}{6}(y^{ijk} - A^{ijk}\eta)\Phi_i\Phi_j\Phi_k + \frac{1}{2}(M^{ij} - B^{ij}\eta)\Phi_i\Phi_j \right] + h.c.
$$

Thus, one can interpret the soft terms as the modification of the couplings of a rigid theory. The couplings become external superfields depending on Grassmannian parameters θ and $\bar{\theta}$. To get the explicit expression for the modified couplings, consider eqs.(11). The first two terms give [1]

$$
\frac{1}{g^2} \to \frac{1}{\tilde{g}^2} = \frac{1 - M\theta^2 - \bar{M}\bar{\theta}^2}{g^2}. \tag{12}
$$

Since the gauge field strength tensors W_α ($\bar{W}_\alpha$) are chiral (antichiral) superfields, they enter into the chiral (antichiral) integrands in eq.(11), respectively. Correspondingly, the $M\theta^2$ term of eq.(12) contributes to the chiral integral, while the $\bar{M}\bar{\theta}^2$ term contributes to the antichiral one. There is no $\theta^2\bar{\theta}^2$ term in eq.(12), since it is neither chiral, no antichiral and gives no contribution to eq.(11).

We depart here from the holomorphicity arguments [20]. Alternatively one should consider holomorphic and antiholomorphic gauge couplings and a separate non-chiral superfield to take care of the mixed term [7,8]. This is where different approaches diverge. It does not lead, however, to any practical difference in applications within the PT.

Modifying the gauge coupling in the gauge part of the Lagrangian, one has to do the same in the gauge-fixing (5) and ghost (8) parts in order to preserve the BRST invariance. Here one has the integral over the whole superspace rather than the chiral one. This means that if one adds to eq.(12) a term proportional to $\theta^2\bar{\theta}^2$, it gives a nonzero contribution. Moreover, even if this term is not added, it reappears as a result of renormalization. We suggest the following modification of eq.(12)

$$
\frac{1}{g^2} \to \frac{1}{\tilde{g}^2} = \frac{1 - M\theta^2 - \bar{M}\bar{\theta}^2 - \Delta\theta^2\bar{\theta}^2}{g^2}, \tag{13}
$$

which gives the final expression for the soft gauge coupling [4]

$$\tilde{g}^2 = g^2 \left(1 + M\theta^2 + \bar{M}\bar{\theta}^2 + 2M\bar{M}\theta^2\bar{\theta}^2 + \Delta\theta^2\bar{\theta}^2\right). \tag{14}$$

It will be clear below that it is self-consistent to put $\Delta = 0$ in the lowest order of perturbation theory, but it appears in higher orders due to renormalizations.

One has to take into account, however, that, since the gauge-fixing parameter ξ may be considered as an additional coupling, it also becomes an external superfield and has to be modified. The soft expression can be written as

$$\tilde{\xi} = \xi \left(1 + x\theta^2 + \bar{x}\bar{\theta}^2 + (x\bar{x} + z)\theta^2\bar{\theta}^2\right), \tag{15}$$

where parameters x and z can be obtained by solving the corresponding RG equation (see Appendix B).

Having this in mind, we perform the following modification of the gauge fixing condition (5) first used in [21]

$$f \to \bar{D}^2 \frac{V}{\sqrt{\tilde{\xi}\tilde{g}^2}}, \quad \bar{f} \to D^2 \frac{V}{\sqrt{\tilde{\xi}\tilde{g}^2}}. \tag{16}$$

Then, the gauge-fixing term (4) becomes

$$\mathcal{L}_{gauge-fixing} = -\frac{1}{8} \int d^2\theta d^2\bar{\theta} \, \text{Tr} \left(\bar{D}^2 \frac{V}{\sqrt{\tilde{\xi}\tilde{g}^2}} D^2 \frac{V}{\sqrt{\tilde{\xi}\tilde{g}^2}}\right). \tag{17}$$

This leads to the corresponding modification of the associated ghost term (6)

$$\mathcal{L}_{ghost} = \int d^2\theta d^2\bar{\theta} \, \text{Tr} \, \frac{1}{\sqrt{\tilde{\xi}\tilde{g}^2}} \, (b + \bar{b}) \, \mathcal{L}_{V/2}[c + \bar{c} + \coth(\mathcal{L}_{V/2})(c - \bar{c})]. \tag{18}$$

To understand the meaning of the Δ term, consider the quadratic part of the ghost Lagrangian (18)

$$\mathcal{L}_{ghost}^{(2)} = \int d^4\theta \text{Tr} \frac{1}{\sqrt{\tilde{\xi}\tilde{g}^2}} \left(1 - \frac{1}{2}M\xi\theta^2 - \frac{1}{2}\bar{M}\xi\bar{\theta}^2 - \frac{1}{2}\Delta\xi\theta^2\bar{\theta}^2\right) (b + \bar{b}) \, (c - \bar{c})$$

$$= \int d^2\theta d^2\bar{\theta} \, \text{Tr} \, \frac{1}{\sqrt{\tilde{\xi}\tilde{g}^2}} \left(1 - \frac{1}{2}\Delta\xi\theta^2\bar{\theta}^2\right) (b + \bar{b}) \, (c - \bar{c}) \tag{19}$$

$$- \frac{1}{2} \int d^2\theta \, \text{Tr} \, \frac{1}{\sqrt{\tilde{\xi}\tilde{g}^2}} M\xi bc + \frac{1}{2} \int d^2\bar{\theta} \, \text{Tr} \, \frac{1}{\sqrt{\tilde{\xi}\tilde{g}^2}} \bar{M}\xi\bar{b}\bar{c},$$

where we have used the explicit form of $\tilde{\xi}$ given in Appendix B.

If one compares this expression with the usual Lagrangian for the matter fields (11), one finds an obvious identification of the second line with the soft scalar mass term and the third line with the mass term in a superpotential. Thus, Δ plays the role of a soft mass providing the splitting in the ghost supermultiplet.

The other place where the Δ-term appears is the gauge-fixing term (17). Here it manifests itself as a soft mass of the auxiliary gauge field, one of the scalar components of the gauge superfield V.

To see this, consider the gauge-fixing term (17) in more detail. Expanding the vector superfield $V(x, \theta, \bar{\theta})$ in components

$$V(x,\theta,\bar{\theta}) = \mathbb{C}(x) + i\theta\chi(x) - i\bar{\theta}\bar{\chi}(x) + \frac{i}{2}\theta\theta N(x) - \frac{i}{2}\bar{\theta}\bar{\theta}\bar{N}(x) - \theta\sigma^\mu\bar{\theta}v_\mu(x)$$

$$+i\theta\theta\bar{\theta}[\bar{\lambda}(x) + \frac{i}{2}\bar{\sigma}^\mu\partial_\mu\chi(x)] - i\bar{\theta}\bar{\theta}\theta[\lambda + \frac{i}{2}\sigma^\mu\partial_\mu\bar{\chi}(x)] + \frac{\theta\theta\bar{\theta}\bar{\theta}}{2}[D(x) - \frac{1}{2}\Box\mathbb{C}(x)].$$

and substituting it into eq.(17) one finds

$$\mathcal{L}_{gauge-fixing} = \frac{1}{2\xi g^2}\left[-(D - \Box\mathbb{C} - \Delta\xi\mathbb{C} + \frac{i}{2}M\xi\bar{N} - \frac{i}{2}\bar{M}\xi N)^2 - (\partial^\mu v_\mu)^2\right.$$

$$+ (\bar{N} - i\bar{M}\xi\mathbb{C})\Box(N + iM\xi\mathbb{C}) - i(\lambda + \frac{1}{2}\bar{M}\xi\chi)\sigma^\mu\partial_\mu(\bar{\lambda} + \frac{1}{2}M\xi\bar{\chi})$$

$$\left. - (\lambda + \frac{1}{2}\bar{M}\xi\chi)\Box\chi - (\bar{\lambda} + \frac{1}{2}M\xi\bar{\chi})\Box\bar{\chi} - i\Box\chi\sigma^\mu\partial_\mu\bar{\chi}\right]. \tag{20}$$

One can see from eq.(20) that the parameter M, besides being the gaugino soft mass, plays the role of a mass of the auxiliary field χ, while Δ is the soft mass of the auxiliary fields N and $\mathbb{C}$. All these fields are unphysical degrees of freedom of the gauge superfield. They are absent in the Wess-Zumino gauge, however, when the gauge-fixing condition is chosen in supersymmetric form (4), this gauge is no longer possible, and the auxiliary fields χ, N, and $\mathbb{C}$ survive. Thus, the extra Δ term is associated with unphysical, ghost, degrees of freedom, just like in the component approach, one has the mass of unphysical ϵ-scalars [22]. When we go down with energy, all massive fields decouple, and we get the usual nonsupersymmetric Yang-Mills theory.

The Δ-term is renormalized and obeys its own RG equation which can be obtained from the corresponding expression for the gauge coupling via Grassmannian expansion. In due course of renormalization, this term is mixing with the soft masses of scalar superpartners and gives an additional term in RG equations for the latter.

In component formalism one has a similar term. In [10,23], the dimensional reduction (DRED) regularization is used. In this case, one is bounded to introduce the so-called ϵ-scalars to compensate the lack of bosonic degrees of freedom in 4-2ϵ dimensions. These ϵ-scalars in due course of renormalization acquire a soft mass that enters into the RG equations for soft masses of physical scalar particles. This problem has been discussed in [24]. If one gets rid of the ϵ-scalar mass by changing the renormalization scheme, DRED $\to$ DRED', there appears an additional term in RG equations for the soft scalar masses [25,26] called X [10] which usually coincides with our Δ.

Besides the modification of the gauge coupling the soft terms (10) imply the modification of the Yukawa ones and the mass term

$$y^{ijk} \to \tilde{y}^{ijk} = y^{ijk} - A^{ijk}\eta, \quad M^{ij} \to \tilde{M}^{ij} = M^{ij} - B^{ij}\eta, \quad + \; h.c.$$

There is, however, also the soft mass term $(m^2)^k_i$. To take care of it we absorb the multiplier $(\delta^k_i - (m^2)^k_i \eta\bar{\eta})$ into the redefinition of the fields Φ and $\bar{\Phi}$. This results in the additional modification of the Yukawa couplings and the mass term

$$y^{ijk} \to \tilde{y}^{ijk} = y^{ijk} - A^{ijk}\eta + \frac{1}{2}(y^{njk}(m^2)^i_n + y^{ink}(m^2)^j_n + y^{ijn}(m^2)^k_n)\eta\bar{\eta},$$

$$M^{ij} \to \tilde{M}^{ij} = M^{ij} - B^{ij}\eta + \frac{1}{2}(M^{nk}(m^2)^i_n + M^{in}(m^2)^k_n)\eta\bar{\eta}. \tag{21}$$

This completes our set of substitutions.

At the end of this section, we would like to comment on the BRST invariance in a softly broken SUSY theory. The BRST transformations (9) due to our choice of normalization of the gauge and ghost fields do not depend on the gauge coupling. Hence, in a softly broken theory they remain unchanged. One can easily check that, despite the substitution $g^2 \to \tilde{g}^2$ and $\xi \to \tilde{\xi}$, the softly broken SUSY theory remains BRST invariant [21].

3. Renormalizations in a Softly Broken SUSY Theory

The modifications of the couplings introduced above are valid not only for the classical Lagrangian but also for the quantum one. As follows from the analysis of the Feynman diagrams in superspace [27] the modification of the Feynman rules due to the soft terms does not influence the UV divergent part of the effective action [1]. This justifies the statement made above concerning the UV singular part of the effective action. The following theorem is valid:

The theorem *Let a rigid theory be renormalized via introduction of the renormalization constants Z_i, defined within some minimal subtraction massless scheme. Then, a softly broken theory is renormalized via introduction of the renormalization superfields $\bar{Z}_i$ which are related to Z_i by the* *coupling constant redefinition*

$$\tilde{Z}_i(g^2, y, \bar{y}) = Z_i(\tilde{g}^2, \tilde{y}, \tilde{\bar{y}}), \tag{22}$$

where the redefined couplings are

$$\tilde{g}_i^2 = g_i^2(1 + M_i\eta + \bar{M}_i\bar{\eta} + (2M_i\bar{M}_i + \Delta_i)\eta\bar{\eta}), \tag{23}$$

$$\tilde{y}^{ijk} = y^{ijk} - A^{ijk}\eta + \frac{1}{2}(y^{njk}(m^2)_n^i + y^{ink}(m^2)_n^j + y^{ijn}(m^2)_n^k)\eta\bar{\eta}, \tag{24}$$

$$\tilde{\bar{y}}_{ijk} = \bar{y}_{ijk} - \bar{A}_{ijk}\bar{\eta} + \frac{1}{2}(y_{njk}(m^2)_i^n + y_{ink}(m^2)_j^n + y_{ijn}(m^2)_k^n)\eta\bar{\eta},$$

Eqs.(22-24) lead to a finite renormalized softly broken SUSY theory. However, in practice it is more convenient to consider not the renormalization constants Z_i but the RG equations directly. Differentiating the renormalization constants $\tilde{Z}_i$ with respect to a scale one gets the RG functions for the soft terms of a broken theory in terms of unbroken one. The resulting soft term β functions are summarized below.

Summary of the Soft Term Renormalizations

The Rigid Terms	The Soft Terms
$\beta_{\alpha_i} = \alpha_i\gamma_{\alpha_i}$	$\beta_{M_{Ai}} = D_1\gamma_{\alpha i}$
$\beta_M^{ij} = \frac{1}{2}(M^{il}\gamma_l^j + M^{lj}\gamma_l^i)$	$\beta_B^{ij} = \frac{1}{2}(B^{il}\gamma_l^j + B^{lj}\gamma_l^i) - (M^{il}D_1\gamma_l^j + M^{lj}D_1\gamma_l^i)$
$\beta_y^{ijk} = \frac{1}{2}(y^{ijl}\gamma_l^k + \text{perm's})$	$\beta_A^{ijk} = \frac{1}{2}(A^{ijl}\gamma_l^k + \text{perm's}) - (y^{ijl}D_1\gamma_l^k + \text{perm's})$
$\Uparrow$	$(\beta_{m^2})_j^i = D_2\gamma_j^i$
chiral anomalous dim.	$\beta_{\Sigma_{\alpha_i}} = D_2\gamma_{\alpha_i}$

$$D_1 = M_{A_i}\alpha_i\frac{\partial}{\partial\alpha_i} - A^{ijk}\frac{\partial}{\partial y^{ijk}}, \qquad \bar{D}_1 = M_{A_i}\alpha_i\frac{\partial}{\partial\alpha_i} - A_{ijk}\frac{\partial}{\partial y_{ijk}}$$

$$D_2 = \bar{D}_1 D_1 + \Sigma_{\alpha_i}\alpha_i\frac{\partial}{\partial\alpha_i} + \frac{1}{2}(m^2)_n^a\left(y^{nbc}\frac{\partial}{\partial y^{abc}} + y^{bnc}\frac{\partial}{\partial y^{bac}} + y^{bcn}\frac{\partial}{\partial y^{bca}}\right.$$

$$\left. + y_{abc}\frac{\partial}{\partial y_{nbc}} + y_{bac}\frac{\partial}{\partial y_{bnc}} + y_{bca}\frac{\partial}{\partial y_{bcn}}\right),$$

$$\Sigma_{\alpha_i} = M_{A_i}\bar{M}_{A_i} + \Delta_i$$

Later on we consider some examples of the application of these formulas.

4. Grassmannian Taylor Expansion

We demonstrate now how the RG equations for the soft terms appear via Grassmannian Taylor expansion from those for the rigid couplings.

In what follows we would like to simplify the notations and consider numerical rather than tensorial couplings. When group structure and field content of the model are fixed, one has a set of gauge $\{g_i\}$ and Yukawa $\{y_k\}$ couplings. It is useful to consider the following rigid parameters $\alpha_i \equiv g_i^2/16\pi^2$, $Y_k \equiv y_k^2/16\pi^2$. Then eqs.(23-24) look like

$$\tilde{\alpha}_i = \alpha_i(1 + M_i\eta + \bar{M}_i\bar{\eta} + (M_i\bar{M}_i + \Sigma_i)\eta\bar{\eta}), \tag{25}$$
$$\tilde{Y}_k = Y_k(1 + A_k\eta + \bar{A}_k\bar{\eta} + (A_k\bar{A}_k + \Sigma_k)\eta\bar{\eta}),$$

where to standardize the notations we have redefined parameter A: $A \to Ay$ in a usual way and have changed the sign of A to match it with the gauge soft terms. Here Σ_k stands for a sum of m^2 soft terms, one for each leg in the Yukawa vertex and $\Sigma_i = M_i\bar{M}_i + \Delta_i$.

Now the RG equation for a rigid theory can be written in a universal form

$$\dot{a}_i = a_i\gamma_i(a), \qquad a_i = \{\alpha_i, Y_k\}, \tag{26}$$

where $\gamma_i(a)$ stands for a sum of corresponding anomalous dimensions. In the same notation the soft terms (25) take the form

$$\tilde{a}_i = a_i(1 + m_i\eta + \bar{m}_i\bar{\eta} + S_i\eta\bar{\eta}), \tag{27}$$

where $m_i = \{M_i, A_k\}$ and $S_i = \{M_i\bar{M}_i + \Sigma_i, A_k\bar{A}_k + \Sigma_k\}$.

Substituting eq.(27) into eq.(26) and expanding over η and $\bar{\eta}$ one can get the RG equations for the soft terms

$$\dot{\tilde{a}}_i = \tilde{a}_i\gamma_i(\tilde{a}), \tag{28}$$

Consider first the F-terms. Expanding over η one has

$$\dot{m}_i = \gamma_i(\tilde{a})|_F = \sum_j a_j\frac{\partial\gamma_i}{\partial a_j}m_j \equiv D_1\gamma_i. \tag{29}$$

This is just the RG equation for the soft terms M_i and A_k [6,1]. Proceeding the same way for the D-terms and substituting $S_i = m_i\bar{m}_i + \Sigma_i$ one has the RG equation for the mass terms

$$\dot{\Sigma}_i = \gamma_i(\tilde{a})|_D = \sum_j a_j\frac{\partial\gamma_i}{\partial a_j}(m_jm_j + \Sigma_j) + \sum_{j,k} a_ja_k\frac{\partial^2\gamma_i}{\partial a_j\partial a_k}m_jm_k \equiv D_2\gamma_i. \tag{30}$$

One can also obtain the RG equation for the individual soft masses out of field renormalization. Consider for this purpose the chiral Green function in a rigid theory. It obeys the following RG relation

$$<\Phi_i\bar{\Phi}_i> \,=\, <\Phi_i\bar{\Phi}_i>_0 e^{\int_0^t \gamma_i(a(t'))dt'}. \tag{31}$$

Making the substitution

$$< \Phi_i \bar{\Phi}_i > \; \rightarrow \; < \Phi_i \bar{\Phi}_i > (1 + m_i^2 \eta \bar{\eta}), \quad a \rightarrow \bar{a},$$

and expanding over $\eta \bar{\eta}$ (since it stands under the full Grassmann integral only D-term contributes) one has

$$m_i^2 = m_{i0}^2 + \int_0^t dt' \; \gamma_i(\bar{a}(t'))|_D. \tag{32}$$

Differentiating this relation with respect to t leads to the RG equation for the soft mass

$$\dot{m}^2{}_i = D_2 \gamma_i(a). \tag{33}$$

5. Illustration

Consider, as an illustration of the above formulas, the simplest case of a pure gauge theory [3]. In a rigid theory the coupling is renormalized as

$$\alpha^{Bare} = Z_\alpha \alpha, \quad \alpha \equiv g^2/16\pi^2. \tag{34}$$

Making the substitution $\alpha \rightarrow \bar{\alpha}$ one gets $\bar{\alpha}^{Bare} = \tilde{Z}_\alpha \bar{\alpha}$ or (up to linear terms in η)

$$\alpha^{Bare}(1 + M_A^{Bare}\eta) = \alpha(1 + M_A \eta) Z_\alpha(\alpha(1 + M_A \eta)). \tag{35}$$

After expansion over η this leads to equations

$$\alpha^{Bare} = \alpha Z_\alpha(\alpha),$$
$$M_A^{Bare}\alpha^{Bare} = M_A \alpha Z_\alpha(\alpha) + \alpha D_1 Z_\alpha,$$

where $D_1 = M_A \alpha \frac{\partial}{\partial \alpha}$ is the differential operator extracting linear terms over η. As a result, we get the bare mass

$$M_A^{Bare} = M_A + D_1 \ln Z_\alpha. \tag{36}$$

Differentiating eq.(36) with respect to a scale, one gets

$$\beta_\alpha = \alpha \gamma_\alpha, \quad \beta_{M_A} = D_1 \gamma_\alpha, \tag{37}$$

where γ_α is the gauge field anomalous dimension $\gamma_\alpha = -d \log Z_\alpha / d \log \mu^2$.

In fact, one does not need eq.(37) to get the RG equation form the gaugino mass and can get the same formulas (37) starting directly from the RG equation for α as shown above. One can go even further and consider a solution to the RG equation in a rigid theory.

Indeed, let us take a solution to the RG equation for the coupling written in quadratures

$$\int_{\alpha_0}^{\alpha} \frac{d\alpha'}{\beta(\alpha')} = \ln \frac{Q^2}{\mu^2}. \tag{38}$$

Performing the replacement of the coupling one gets

$$\int_{\alpha_0(1+M_{A0}\eta+...)}^{\alpha(1+M_A\eta+...)} \frac{d\alpha'}{\beta(\alpha')} = \ln \frac{Q^2}{\mu^2} \tag{39}$$

which after expansion over η leads to the solution for the soft mass term

$$\frac{\alpha M_A}{\beta(\alpha)} = \frac{\alpha M_{A0}}{\beta(\alpha_0)} \quad \Rightarrow \quad M_A = c_1 \frac{\beta(\alpha)}{\alpha} = c_1 \gamma(\alpha), \tag{40}$$

where α is taken from eq.(38). Thus, the solution for the gaugino mass term directly follows from the one for the rigid coupling. Eq.(40) is the first example of RG invariants first found in [28] on different grounds. Following our approach one can construct the other ones [29]. For example, one can continue the expansion up to D-terms in eq.(39), which gives a solution for the Δ term

$$\Delta = c_2 \gamma(\alpha) - c_1 \alpha \gamma'(\alpha)\gamma(\alpha). \tag{41}$$

6. Examples

6.1. General gauge theory

In the one-loop order the rigid β functions are (for simplicity, we consider the case of a single gauge coupling)

$$\beta_\alpha = \alpha \gamma_\alpha, \quad \gamma_\alpha^{(1)} = \alpha Q, \quad Q = T(R) - 3C(G), \tag{42}$$

$$\beta_y^{ijk} = \frac{1}{2}(y^{ijl}\gamma_l^k + perm's), \quad \gamma_j^{i\,(1)} = \frac{1}{2}y^{ikl}y_{jkl} - 2\alpha C(R)_j^i,$$

where $T(R), C(G)$ and $C(R)$ are the Casimir operators of the gauge group defined by

$$T(R)\delta_{AB} = Tr(R_A R_B), \quad C(G)\delta_{AB} = f_{ACD}f_{BCD}, \quad C(R)_j^i = (R_A R_A)_j^i.$$

Applying our algorithm, this leads to the following soft β functions:

$$\beta_{M_A}^{(1)} = \alpha M_A Q, \tag{43}$$

$$\beta_B^{ij\,(1)} = \frac{1}{2}B^{il}(\frac{1}{2}y^{jkm}y_{lkm} - 2\alpha C(R)_l^j) \tag{44}$$

$$+ M^{il}(\frac{1}{2}A^{jkm}y_{lkm} + 2\alpha M_A C(R)_l^j) + (i \leftrightarrow j),$$

$$\beta_A^{ijk\,(1)} = \frac{1}{2}A^{ijl}(\frac{1}{2}y^{kmn}y_{lmn} - 2\alpha C(R)_l^k) \tag{45}$$

$$+ y^{ijl}(\frac{1}{2}A^{kmn}y_{lmn} + 2\alpha M_A C(R)_l^k) + (i \leftrightarrow j, k),$$

$$[\beta_{m^2}]_j^{i\,(1)} = \frac{1}{2}A^{ikl}A_{jkl} - 4\alpha M_A^2 C(R)_j^i \tag{46}$$

$$+ \frac{1}{4}y^{nkl}(m^2)_n^i y_{jkl} + \frac{1}{4}y^{ikl}(m^2)_j^n y_{nkl} + \frac{4}{4}y^{isl}(m^2)_s^k y_{jkl}.$$

We used here the fact that in the given order the solution for Σ_α is $\Sigma_\alpha = M_A \bar{M}_A$.

In two loops the rigid anomalous dimensions are

$$\gamma_\alpha^{(2)} = 2\alpha^2 C(G)Q - \frac{2\alpha}{r}C(R)_j^i(\frac{1}{2}y^{jkl}y_{ikl} - 2\alpha C(R)_i^j), \quad r = dimG = \delta_{AA},$$

$$\gamma_j^{i\,(2)} = -(y^{imp}y_{jmn} + 2\alpha C(R)_j^p\delta_n^i)(\frac{1}{2}y^{nkl}y_{pkl} - 2\alpha C(R)_p^n) + 2\alpha^2 Q C(R)_j^i.$$

In this case, again the solution for the ghost mass Δ_α can be found analytically [4] and coincides with the mass of ϵ-scalars [30]

$$\Sigma_\alpha^{(2)} = \Delta_\alpha^{(2)} = -2\alpha[\frac{1}{r}(m^2)_j^i C(R)_i^j - M_A^2 C(G)]. \tag{47}$$

Then, the soft renormalizations are as follows:

$$\beta_{M_A}^{(2)} = 4\alpha^2 M_A C(G)Q - \frac{2\alpha M_A}{r}C(R)_j^i(\frac{1}{2}y^{jkl}y_{ikl} - 2\alpha C(R)_i^j)$$

$$+ \frac{2\alpha}{r}C(R)_j^i(\frac{1}{2}A^{jkl}y_{ikl} + 2\alpha M_A C(R)_i^j), \tag{48}$$

$$\beta_B^{ij\,(2)} = -\frac{1}{2}B^{il}(y^{jkp}y_{lkn} + 2\alpha C(R)_l^p\delta_n^j)(\frac{1}{2}y^{nst}y_{pst} - 2\alpha C(R)_p^n)$$

$$- M^{il}(A^{jkp}y_{lkn} - 2\alpha M_A C(R)_l^p\delta_n^j)(\frac{1}{2}y^{nst}y_{pst} - 2\alpha C(R)_p^n)$$

$$- M^{il}(y^{jkp}y_{lkn} + 2\alpha C(R)_l^p\delta_n^j)(\frac{1}{2}A^{nst}y_{pst} + 2\alpha M_A C(R)_p^n)$$

$$+ B^{il}\alpha^2 Q C(R)_l^j - 4M^{il}\alpha^2 Q C(R)_l^j M_A + (i \leftrightarrow j), \tag{49}$$

$$\beta_A^{ijk\,(2)} = -\frac{1}{2}A^{ijl}(y^{kmp}y_{lmn} + 2\alpha C(R)_l^p\delta_n^k)(\frac{1}{2}y^{nst}y_{pst} - 2\alpha C(R)_p^n)$$

$$+ A^{ijl}\alpha^2 Q C(R)_l^k - 4y^{ijl}\alpha^2 Q C(R)_l^j M_A$$

$$- y^{ijl}(A^{kmp}y_{lmn} - 2\alpha M_A C(R)_l^p\delta_n^k)(\frac{1}{2}y^{nst}y_{pst} - 2\alpha C(R)_p^n)$$

$$- y^{ijl}(y^{kmp}y_{lmn} + 2\alpha C(R)_l^p\delta_n^k)(\frac{1}{2}A^{nst}y_{pst} + 2\alpha M_A C(R)_p^n)$$

$$+ (i \leftrightarrow j) + (i \leftrightarrow k), \tag{50}$$

$$[\beta_{m^2}]_j^{i\,(2)} = -(A^{ikp}A_{jkn} + \frac{1}{2}(m^2)_i^i y^{lkp}y_{jkn} + \frac{1}{2}y^{ikp}y_{lkn}(m^2)_j^l)$$

$$+\frac{2}{2}y^{ilp}(m^2)^s_l y_{jsn} + \frac{1}{2}y^{iks}(m^2)^p_s y_{jkn} + \frac{1}{2}y_{ikp}(m^2)^s_n y_{jks}$$

$$+4\alpha M^2_A C(R)^p_j \delta^i_n)(\frac{1}{2}y^{nst}y_{pst} - 2\alpha C(R)^n_p)$$

$$-(y^{ikp}y_{jkn} + 2\alpha C(R)^p_j \delta^i_n)(\frac{1}{2}A^{nst}A_{pst} + \frac{1}{4}(m^2)^k_l y^{lst}y_{pst}$$

$$+\frac{1}{4}y^{nst}y_{lst}(m^2)^l_p + \frac{4}{4}y^{nlt}(m^2)^s_l y_{pst} - 2\alpha M^2_A C(R)^n_p) \qquad (51)$$

$$-(A^{ikp}y_{jkn} - 2\alpha M_A C(R)^p_j \delta^i_n)(\frac{1}{2}y^{nst}A_{pst} + 2\alpha M_A C(R)^n_p)$$

$$-(y^{ikp}A_{jkn} - 2\alpha M_A C(R)^p_j \delta^i_n)(\frac{1}{2}A^{nst}y_{pst} + 2\alpha M_A C(R)^n_p)$$

$$+12\alpha^2 M^2_A Q C(R)^i_j + 4\alpha^2 C(R)^i_j[\frac{1}{r}(m^2)^k_i C(R)^l_k - M^2 C(G)],$$

where the last term is an extra contribution due to nonzero $\Delta^{(2)}_i$ in (47).

To demonstrate the power of the proposed algorithm, we calculate the three loop gaugino mass renormalization out of a gauge β function. One has in three loops [15]

$$\gamma^{(3)}_\alpha = \alpha^3 C(G)Q[4C(G) - Q] - \frac{6}{r}\alpha^3 Q C(R)^i_j C(R)^j_i$$

$$+ \frac{3}{r}\alpha^2(y^{ikl}y_{jkl} - 4\alpha C(R)^i_j)C(R)^j_s C(R)^s_i - \frac{2}{r}\alpha^2 C(G)(y^{ikl}y_{jkl}$$

$$- 4\alpha C(R)^i_j)C(R)^j_i + \frac{3}{2r}\alpha y^{ikm}y_{jkn}(y^{nst}y_{mst} - 4\alpha C(R)^n_m)C(R)^j_i$$

$$+ \frac{1}{4r}\alpha(y^{ikl}y_{jkl} - 4\alpha C(R)^i_j)(y^{jst}y_{pst} - 4\alpha C(R)^j_p)C(R)^p_i. \qquad (52)$$

The corresponding gaugino mass renormalization is

$$\beta^{(3)}_{M_A} = 3\alpha^3 C(G)Q[4C(G) - Q]M_A - \frac{18}{r}\alpha^3 Q C(R)^i_j C(R)^j_i M_A$$

$$+ \frac{6}{r}\alpha^2(y^{ikl}y_{jkl} - 4\alpha C(R)^i_j)C(R)^j_s C(R)^s_i M_A - \frac{3}{r}\alpha^2(A^{ikl}y_{jkl}$$

$$+ 4\alpha C(R)^i_j M_A)C(R)^j_s C(R)^s_i - \frac{4}{r}\alpha^2 C(G)(y^{ikl}y_{jkl} - 4\alpha C(R)^i_j)C(R)^j_i M_A$$

$$+ \frac{2}{r}\alpha^2 C(G)(A^{ikl}y_{jkl} + 4\alpha C(R)^i_j M_A)C(R)^j_i + \frac{3}{2r}\alpha y^{ikm}y_{jkn}(y^{nst}y_{mst}$$

$$- 4\alpha C(R)^n_m)C(R)^j_i M_A - \frac{3}{2r}\alpha A^{ikm}y_{jkn}(y^{nst}y_{mst} - 4\alpha C(R)^n_m)C(R)^j_i$$

$$- \frac{3}{2r}\alpha y^{ikm}y_{jkn}(A^{nst}y_{mst} + 4\alpha C(R)^n_m M_A)C(R)^j_i$$

$$+ \frac{1}{4r}\alpha(y^{ikl}y_{jkl} - 4\alpha C(R)^i_j)(y^{jst}y_{pst} - 4\alpha C(R)^j_p)C(R)^p_i M_A$$

$$- \frac{1}{4r}\alpha(A^{ikl}y_{jkl} + 4\alpha C(R)^i_j M_A)(y^{jst}y_{pst} - 4\alpha C(R)^j_p)C(R)^p_i$$

$$- \frac{1}{4r}\alpha(y^{ikl}y_{jkl} - 4\alpha C(R)^i_j)(A^{jst}y_{pst} + 4\alpha C(R)^j_p M_A)C(R)^p_i. \tag{53}$$

To argue that a solution for Δ_i exists in all orders of PT, one can consider the so-called NSVZ-scheme [31] where the anomalous dimension γ_α is known to all orders of PT

$$\gamma_\alpha^{NSVZ} = \alpha\frac{Q - 2r^{-1}\mathrm{Tr}[\gamma C(R)]}{1 - 2C(G)\alpha}. \tag{54}$$

Then the solution for Δ_α is

$$\Delta_\alpha^{NSVZ} = -2\alpha\frac{r^{-1}\mathrm{Tr}[m^2 C(R)] - M_A^2 C(G)}{1 - 2C(G)\alpha}. \tag{55}$$

and coincides with the for ϵ-scalar mass [30,32].

6.2. The MSSM in low $\tan\beta$ regime

Consider the MSSM in low $\tan\beta$ regime. One has three gauge and one Yukawa coupling. The one-loop RG equations are [33]

$$\dot{\alpha}_i = -b_i\alpha_i^2, \quad b_i = (\frac{33}{5}, 1, -3), \quad i = 1, 2, 3, \tag{56}$$

$$\dot{Y}_t = Y_t(\frac{16}{3}\alpha_3 + 3\alpha_2 + \frac{13}{15}\alpha_1 - 6Y_t), \tag{57}$$

with the initial conditions: $\alpha_i(0) = \alpha_0$, $Y_t(0) = Y_0$ and $t = \ln(M_X^2/Q^2)$. Their solutions are given by [33]

$$\alpha_i(t) = \frac{\alpha_0}{1 + b_i\alpha_0 t}, \quad Y_t(t) = \frac{Y_0 E(t)}{1 + 6Y_0 F(t)}, \tag{58}$$

where

$$E(t) = \prod_i(1 + b_i\alpha_0 t)^{c_i/b_i}, \quad c_i = (\frac{13}{15}, 3, \frac{16}{3}), \quad F(t) = \int_0^t E(t')dt'.$$

To get the solutions for the soft terms it is enough to perform the substitution $\alpha \to \bar{\alpha}$ and $Y \to \tilde{Y}$ and expand over η and $\bar{\eta}$. Expanding the gauge coupling in (58) up to η one has (hereafter we assume $M_{i0} = M_0$)

$$\alpha_i M_i = \frac{\alpha_0 M_0}{1 + b_i\alpha_0 t} - \frac{\alpha_0 b_i\alpha_0 M_0 t}{(1 + b_i\alpha_0 t)^2} = \frac{\alpha_0}{1 + b_i\alpha_0 t} \cdot \frac{M_0}{1 + b_i\alpha_0 t},$$

or

$$M_i(t) = \frac{M_0}{1 + b_i\alpha_0 t}. \tag{59}$$

Performing the same expansion for the Yukawa coupling and using the relations

$$\left.\frac{d\tilde{E}}{d\eta}\right|_\eta = M_0 t \frac{dE}{dt}, \quad \left.\frac{d\tilde{F}}{d\eta}\right|_\eta = M_0(tE - F),$$

one finds a well known expression [33]

$$A_t(t) = \frac{A_0}{1 + 6Y_0 F} + M_0 \left(\frac{t}{E}\frac{dE}{dt} - \frac{6Y_0}{1 + 6Y_0 F}(tE - F) \right). \tag{60}$$

To get the solution for the term $\Sigma_t = \tilde{m}_t^2 + \tilde{m}_Q^2 + m_{H_2}^2$ one has to make expansion over η and $\bar{\eta}$. This can be done with the help of the following relations

$$\left.\frac{d^2\tilde{E}}{d\eta d\bar{\eta}}\right|_{\eta,\bar{\eta}} = M_0^2 \frac{d}{dt}\left(t^2 \frac{dE}{dt} \right), \quad \left.\frac{d^2\tilde{F}}{d\eta d\bar{\eta}}\right|_{\eta,\bar{\eta}} = M_0^2 t^2 \frac{dE}{dt},$$

and leads to [3]

$$\Sigma_t(t) = \frac{\Sigma_0 - A_0^2}{1 + 6Y_0 F} + \frac{(A_0 - M_0 6 Y_0(tE - F))^2}{(1 + 6Y_0 F)^2} \tag{61}$$

$$+ M_0^2 \left[\frac{d}{dt}\left(\frac{t^2}{E}\frac{dE}{dt} \right) - \frac{6Y_0}{1 + 6Y_0 F} t^2 \frac{dE}{dt} \right].$$

With analytic solutions (60,61) one can analyze asymptotics and, in particular, find the so-called infrared quasi fixed points [34] which correspond to $Y_0 \to \infty$

$$Y_t^{FP} = \frac{E}{6F}, \tag{62}$$

$$A_t^{FP} = M_0 \left(\frac{t}{E}\frac{dE}{dt} - \frac{tE - F}{F} \right), \tag{63}$$

$$\Sigma_t^{FP} = M_0^2 \left[\left(\frac{tE - F}{F} \right)^2 + \frac{d}{dt}\left(\frac{t^2}{E}\frac{dE}{dt} \right) - \frac{t^2}{F}\frac{dE}{dt} \right]. \tag{64}$$

However, the advantage of the Grassmannian expansion procedure is that one can perform it for fixed points as well. Thus the FP solutions (63,64) can be directly obtained from a fixed point for the rigid Yukawa coupling (62) by Grassmannian expansion. This explains, in particular, why fixed point solutions for the soft couplings exist if they exist for the rigid ones and with the same stability properties [35].

6.3. *The MSSM in high* $\tan\beta$ *regime*

Consider the MSSM in high $\tan\beta$ regime. One has three gauge and three Yukawa couplings. The one-loop RG equations are [33]

$$\dot{\alpha}_i = -b_i\alpha_i^2, \quad \dot{Y}_k = Y_k(\sum_i c_{ki}\alpha_i - \sum_l a_{kl}Y_l), \tag{65}$$

where $i = 1,2,3;\ k = t,b,\tau,\ \cdot \equiv d/dt,\ t = \log M_{GUT}^2/Q^2$ and

$$b_i = \{33/5, 1, -3\}, \quad a_{tl} = \{6,1,0\}, \quad a_{bl} = \{1,6,1\}, \quad a_{\tau l} = \{0,3,4\},$$
$$c_{ti} = \{13/15, 3, 16/3\}, \quad c_{bi} = \{7/15, 3, 16/3\}, \quad c_{\tau i} = \{9/5, 3, 0\}.$$

Despite a simple form of these equations, there is no explicit analytic solution similar to (58). One has either the iterative solution [36] or the approximate one [37]. In both the cases the Grassmannian expansion over η leads to the corresponding solutions for the soft terms.

Consider first the iterative solution. It can be written as [36]

$$\alpha_i = \frac{\alpha_i^0}{1 + b_i\alpha_i^0 t}, \quad Y_k = \frac{Y_k^0 u_k}{1 + a_{kk}Y_k^0 \int_0^t u_k}, \tag{66}$$

where the functions $\{u_k\}$ obey the integral system of equations

$$u_t = \frac{E_t}{(1 + 6Y_b^0 \int_0^t u_b)^{1/6}}, \quad u_\tau = \frac{E_\tau}{(1 + 6Y_b^0 \int_0^t u_b)^{1/2}},$$
$$u_b = \frac{E_b}{(1 + 6Y_t^0 \int_0^t u_t)^{1/6}(1 + 4Y_\tau^0 \int_0^t u_\tau)^{1/4}}, \tag{67}$$

and the functions E_k are given by $E_k = \prod_{i=1}^3 (1 + b_i\alpha_i^0 t)^{c_{ki}/b_i}$.

Let us stress that eqs.(66) give the exact solution to eqs.(65), while the u_k's in eqs.(67), although solved formally in terms of the E_k's and Y_k^0's as continued integrated fractions, should in practice be solved iteratively.

To get the solutions for the soft terms it is enough to perform substitution $\alpha_i \to \tilde{\alpha}_i$ and $Y_k \to \tilde{Y}_k$ and expand over η and $\bar{\eta}$. One has [38]:

$$M_i = \frac{M_i^0}{1 + b_i\alpha_i^0 t}, \quad A_k = -e_k + \frac{A_k^0/Y_k^0 + a_{kk}\int u_k e_k}{1/Y_k^0 + a_{kk}\int u_k},$$
$$\Sigma_k = \xi_k + A_k^2 + 2e_k A_k - \frac{(A_k^0)^2/Y_k^0 - \Sigma_k^0/Y_k^0 + a_{kk}\int u_k\xi_k}{1/Y_k^0 + a_{kk}\int u_k}, \tag{68}$$

where the new functions e_k and ξ_k have been introduced which obey the iteration equations. For illustration we present below the corresponding expressions for e_t and ξ_t

$$e_t = \frac{1}{E_t}\frac{d\bar{E}_t}{d\eta} + \frac{A_b^0 \int u_b - \int u_b e_b}{1/Y_b^0 + 6\int u_b},$$

$$\xi_t = \frac{1}{E_t}\frac{d^2\bar{E}_t}{d\eta d\bar{\eta}} + 2\frac{1}{E_t}\frac{d\bar{E}_t}{d\eta}\frac{A_b^0\int u_b - \int u_b e_b}{1/Y_b^0 + 6\int u_b} + 7\frac{\left(A_b^0\int u_b - \int u_b e_b\right)^2}{\left(1/Y_b^0 + 6\int u_b\right)^2}$$
$$- \frac{(\Sigma_b^0 + (A_b^0)^2)\int u_b - 2A_b^0\int u_b e_b + \int u_b \xi_b}{1/Y_b^0 + 6\int u_b}, \tag{69}$$

where the variations of $\bar{E}_k$ should be taken at $\eta = \bar{\eta} = 0$ and are given by

$$\frac{1}{E_k}\frac{d\bar{E}_k}{d\eta}\bigg|_{\eta,\bar{\eta}=0} = t\sum_{i=1}^{3} c_{ki}\alpha_i M_i^0,$$

$$\frac{1}{E_k}\frac{d^2\bar{E}_k}{d\eta d\bar{\eta}}\bigg|_{\eta,\bar{\eta}=0} = t^2\left(\sum_{i=1}^{3} c_{ki}\alpha_i M_i^0\right)^2 + 2t\sum_{i=1}^{3} c_{ki}\alpha_i (M_i^0)^2 - t^2\sum_{i=1}^{3} c_{ki}b_i\alpha_i^2 (M_i^0)^2.$$

When solving eqs.(67) and (69) in the n-th iteration one has to substitute in the r.h.s. the $(n-1)$-th iterative solution for all the corresponding functions.

The same procedure works for the approximate solutions. Once one gets an approximate solution for the Yukawa couplings, one immediately has those for the soft terms as well [37].

We consider as an illustration the approximate solution. It can be taken in the form [37]

$$Y_t^{app}(t) = \frac{Y_{t0}E_t(t)}{[1 + \frac{7}{2}(Y_{t0}F_t(t) + Y_{b0}F_b(t))]^{2/7}[1 + 7Y_{t0}F_t(t)]^{5/7}}$$

$$Y_b^{app}(t) = \frac{Y_{b0}E_b(t)}{[1 + \frac{7}{2}(Y_{t0}F_t(t) + Y_{b0}F_b(t))]^{2/7}[1 + 7Y_{t0}F_t(t)]^{2/7}}, \tag{70}$$
$$\times \frac{1}{[1 + \frac{7}{3}(3Y_{b0}F_b(t) + Y_{\tau0}F_\tau)]^{3/7}},$$

$$Y_\tau^{app}(t) = \frac{Y_{\tau0}E_\tau(t)}{[1 + \frac{21}{4}Y_{\tau0}F_\tau]^{4/7}[1 + \frac{7}{3}(3Y_{b0}F_b(t) + Y_{\tau0}F_\tau)]^{3/7}}.$$

To demonstrate the accuracy of the approximate solution (70) and the efficiency of the Grassmannian expansion, we present in Fig.1 the comparison of numerical and approximate solutions for the Yukawa couplings of a rigid theory as well as the soft terms.

One can notice perfect agreement of numerical and analytical curves. Shown also are the fixed point behaviour, again for the Yukawa couplings and for the soft terms obtained via the expansion procedure for the approximate solutions (70). The numerical curves approach the analytically calculated FP's in the infrared region.

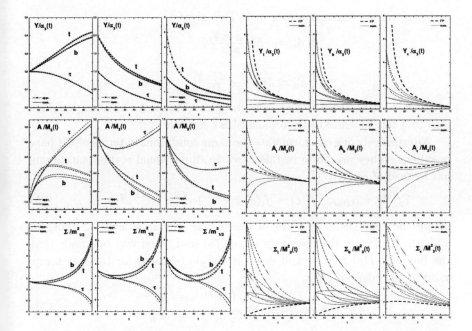

Figure 1. Comparison of numerical and approximate solutions. Dotted lines correspond to the analytical approximate solutions, solid lines to the numerical solution. Shown also are the infra-red quasi fixed points obtained via Grassmannian expansion and the numerical curves approaching them in the IR limit.

6.4. Totally all loop finite $N=1$ SUSY gauge theories

Another example of application of the same procedure is the so-called finite field theories in the framework of SUSY GUTs. These are the theories where all the UV divergences cancel and hence all the β functions vanish. This can be achieved in a rigid theory if the following two conditions are satisfied [39,40]:

• The group representations are chosen in a way to obey the sum rule

$$\sum T(R) = 3C_2(G) \tag{71}$$

• The Yukawa couplings are the functions of the gauge one

$$Y_i = Y_i(\alpha), \ Y_i(\alpha) = c_1^i \alpha + c_2^i \alpha^2 + \dots \tag{72}$$

where the coefficients c_n^i are calculated algebraically in the n-th order of PT.

To achieve the complete finiteness of the model including the soft terms, one has to modify the finiteness condition (72) as

$$\tilde{Y}_i = Y_i(\tilde{\alpha}) \tag{73}$$

and perform the expansion over $\eta, \bar{\eta}$. This gives [2]

$$
\begin{cases}
A_i = -M_A \dfrac{d \ln Y_i}{d \ln \alpha}, \\[2mm]
\Sigma_i = M_A^2 \dfrac{d}{d\alpha} \alpha \dfrac{d \ln Y_i}{d \ln \alpha},
\end{cases}
\tag{74}
$$

where $Y(\alpha)$ is assumed to be known from a rigid theory. These expressions lead to a totally finite softly broken SUSY field theory!

Alternatively one can formulate the same conditions in terms of the bare couplings. They are finite in this case. In dimensional regularization one has instead of eq.(72)

$$
Y_i^{Bare} = \alpha_{Bare} \cdot f_i(\varepsilon), \quad f_i(\varepsilon) = c_i^{(1)} + c_i^{(2)} \varepsilon + c_i^{(3)} \varepsilon^2 + \dots
\tag{75}
$$

where the coefficients $c_i^{(n)}$ are in one-to-one correspondence to those in eq.(72). Replacing the couplings in eq.(75) in a usual way one finds that the function $f(\varepsilon)$ cancels and one has simple relations for the soft terms valid in all orders of PT [2]

$$
\tilde{Y}_i^{Bare} = \tilde{\alpha}_{Bare} \cdot f_i(\varepsilon) \;\Rightarrow\;
\begin{cases}
A_i^{Bare} = -M_A^{Bare}, \\[2mm]
\Sigma_i^{Bare} = (M_A^{Bare})^2.
\end{cases}
\tag{76}
$$

These relations for the bare quantities provide the vanishing of the β functions for the soft terms.

6.5. $N=2$ SUSY Seiberg-Witten Theory

Consider now the $N=2$ supersymmetric gauge theory. The Lagrangian written in terms of $N=2$ superfields is [41]:

$$
\mathcal{L} = \frac{1}{4\pi} \mathcal{I}m Tr \int d^2\theta_1 d^2\theta_2 \frac{1}{2} \tau \Psi^2,
\tag{77}
$$

where $N=2$ chiral superfield $\Psi(y, \theta_1, \theta_2)$ is defined by constraints $\bar{D}_{\dot{\alpha}} \Psi = 0$ and $\bar{\tilde{D}}_{\dot{\alpha}} \Psi = 0$ and

$$
\tau = i\frac{4\pi}{g^2} + \frac{\theta^{\,topological}}{2\pi}.
\tag{78}
$$

The expansion of Ψ in terms of θ_2 can be written as

$$
\Psi(y, \theta_1, \theta_2) = \Psi^{(1)}(y, \theta_1) + \sqrt{2}\theta_2^\alpha \Psi_\alpha^{(2)}(y, \theta_1) + \theta_2^\alpha \theta_{2\alpha} \Psi^{(3)}(y, \theta_1),
$$

where $y^\mu = x^\mu + i\theta_1 \sigma^\mu \bar{\theta}_1 + i\theta_2 \sigma^\mu \bar{\theta}_2$ and $\Psi^{(k)}(y, \theta_1)$ are $N=1$ chiral superfields.

The soft breaking of $N=2$ SUSY down to $N=0$ can be achieved by shifting the imaginary part of τ:

$$
\mathcal{I}m\tau \to \mathcal{I}m\bar{\tau} = \mathcal{I}m\tau(1 + M_1\theta_1^2 + M_2\theta_2^2 + M_3\theta_1^2\theta_2^2)
\tag{79}
$$

This leads to

$$\Delta\mathcal{L} = \left[-\frac{M_1}{4}\lambda\lambda - \frac{M_2}{4}\psi\psi - (\frac{M_1 M_2}{4} - \frac{M_3}{4})\phi\phi + h.c. \right] - (\frac{M_1^2}{4} + \frac{M_2^2}{4})\bar{\phi}\phi,$$

where the fields λ are the gauginos, ψ and ϕ are the spinor and scalar matter fields, respectively.

Now one can use the power of duality in N=2 SUSY theory and take the Seiberg-Witten solution [42]

$$\tau = \frac{da_D}{du} \Big/ \frac{da}{du}, \tag{80}$$

where

$$a_D(u) = \frac{i}{2}(u-1)F(\frac{1}{2},\frac{1}{2},2;\frac{1-u}{2}), \quad a(u) = \sqrt{2(1+u)}F(-\frac{1}{2},\frac{1}{2},1;\frac{2}{1+u}).$$

In perturbative domain when $u \sim Q^2/\Lambda^2 \to \infty$, $a = \sqrt{2u}$, $a_D = \frac{i}{\pi}a(2\ln a + 1)$ one reproduces the well known one-loop result

$$\frac{4\pi}{g^2} = \frac{1}{\pi}[\ln\frac{Q^2}{\Lambda^2} + 3]. \tag{81}$$

Assuming that renormalizations in N=2 SUSY theory follow the properties of those in N=1, one can try to apply the same expansion procedure for a non-perturbative solution. Substituting eq.(79) into (80) with

$$u \to \tilde{u} = u(1 + M_1^0\theta_1^2 + M_2^0\theta_2^2 + M_3^0\theta_1^2\theta_2^2)$$

and expanding over θ_1^2 and θ_2^2, one gets an analog of S-W solution for the mass terms [43]:

$$M_1 = M_1^0 \frac{\mathcal{I}m\left[u\left(\frac{a_D''}{a_D'} - \frac{a''}{a'}\right)\tau\right]}{\mathcal{I}m\,\tau}, \quad M_2 = M_2^0 \frac{\mathcal{I}m\left[u\left(\frac{a_D''}{a_D'} - \frac{a''}{a'}\right)\tau\right]}{\mathcal{I}m\,\tau},$$

$$M_3 = \frac{\mathcal{I}m\left[M_3^0 u\left(\frac{a_D''}{a_D'} - \frac{a''}{a'}\right)\tau + M_1^0 M_2^0 u^2\left(\frac{a_D'''}{a_D'} - \frac{a'''}{a'} + 2\frac{a_D''}{a_D'}\left(\frac{a_D''}{a_D'} - \frac{a''}{a'}\right)\right)\tau\right]}{\mathcal{I}m\,\tau}$$

In perturbative regime one has

$$M_1 = \frac{M_1^0}{\ln Q^2/\Lambda^2 + 3}, \quad M_2 = \frac{M_2^0}{\ln Q^2/\Lambda^2 + 3}, \quad M_3 = \frac{M_3^0 - M_1^0 M_2^0}{\ln Q^2/\Lambda^2 + 3}.$$

7. Renormalization of the Fayet-Iliopoulos Term

We gave above a complete set of the rules needed for writing down the RG equations for the soft SUSY breaking terms in an arbitrary non-Abelian N=1 SUSY gauge theory. However, in the Abelian case, there exists an additional gauge invariant term, the so-called Fayet-Iliopoulos or the D-term [44]

$$\mathcal{L}_{F.I.} = \xi D = \int d^4\theta \xi V, \qquad (82)$$

which requires special consideration. In Ref. [45], it has been shown that in the unbroken theory this term is not renormalized provided the sum of hypercharges and their cubes equals zero. These requirements guarantee the absence of chiral and gravity anomalies and are usually satisfied in realistic models.

In case of a softly broken Abelian SUSY gauge theory, the F-I term happens to be renormalized even if anomalies are cancelled. The RG equation for ξ depends not only on itself, but on the other soft breaking parameters (the soft mass of chiral scalars m^2, the soft triple coupling A^{ijk} and the gaugino masses M_i). Recently, the renormalization of ξ has been performed up to three loops [46] using the component approach and/or superfields with softly broken Feynman rules. Following our main idea that renormalizations of a softly broken SUSY theory are completely defined by a rigid one, we argue that the renormalization of the F-I term, in full analogy with all the other soft terms renormalizations, is completely defined in a rigid or an unbroken theory. However, contrary to the other soft renormalizations, there is no simple differential operator that acts on the renormalization functions of a rigid theory and allows one to get the renormalization of the F-I term. One needs an analysis of the superfield diagrams and some additional diagram calculations in components.

The addition of the F-I term leads to the modification of the Lagrangian in components. The relevant part of the Lagrangian is

$$\mathcal{L} = \frac{1}{2g^2}D^2 + \xi D + D\bar{\phi}^j \mathcal{Y}^i_j \phi_i - \bar{\phi}^j (m^2)^i_j \phi_i + \dots \qquad (83)$$

where $\mathcal{Y}^i_j$ is the hypercharge matrix of the chiral supermultiplet, and $(m^2)^i_j$ is a soft scalar mass. After eliminating the auxiliary field D this becomes

$$\mathcal{L} = -\bar{\phi}^j (m^2)^i_j \phi_i - \frac{1}{2}g^2(\bar{\phi}^j \mathcal{Y}^i_j \phi_i)^2 + \dots, \qquad (84)$$

where

$$(m^2)^i_j = (m^2)^i_j + g^2 \xi \mathcal{Y}^i_j. \qquad (85)$$

From eqs.(84) and (85) it follows that the F-I term gives an additional contribution to the renormalization of the soft scalar mass $(m^2)^i_j$

$$[\beta_{m^2}]^i_j = [\beta_{m^2}]^i_j + \beta_{g^2}\xi\mathcal{Y}^i_j + g^2\beta_\xi(m^2,...)\mathcal{Y}^i_j = [\beta_{m^2}]^i_j + g^2\beta_\xi(m^2,...)\mathcal{Y}^i_j. \quad (86)$$

The last equality follows from the fact that eq.(84) does not contain ξ explicitly and, hence, ξ should be dropped from all the expressions.

There are four different types of contributions to the renormalization of the F-I term in a softly broken theory: those proportional to $(m^2)^i_j$, $M\bar{M}$, $A^{ijk}\bar{A}_{lmn}$ and $M\bar{A}_{lmn}$ ($\bar{M}A^{ijk}$).

We have found that all the information about the renormalizations of the soft SUSY breaking terms is contained in a rigid, unbroken theory. To calculate the renormalization of an additional Fayet-Iliopoulos term, one needs an analysis of superfield diagrams. To find the contribution proportional to the soft scalar mass $(m^2)^i_j$ (the square of gaugino mass $M\bar{M}$), one needs to take the self-energy diagrams for the vector superfield and replace one of the external vertices with the hypercharge $\mathcal{Y}^i_j$ by $(m^2)^i_j$ ($M\bar{M}\delta^i_j$). In this case, there is no need to do any calculations except in superfields.

The other contributions (proportional to $A\bar{A}$ and $M\bar{A}$) can be found from the analysis of the matter superfield propagator diagrams in a rigid theory and the corresponding component diagrams in a softly broken theory extracting from the latter the contribution of the tadpole graphs. In this case, one needs to calculate additionally some component diagrams the number of which is essentially reduced compared to a direct component calculation [5].

8. Conclusion

Summarizing, we would like to stress once again that is very useful to consider a spontaneously broken theory in terms of a rigid one in an external field. In case when one is able to absorb the external field into the redefinition of parameters of the original theory and perform the renormalizations for an arbitrary field, one can reproduce renormalization properties of a spontaneously broken theory from a rigid one. The Grassmannian expansion in softly broken SUSY theories happens to be a very efficient and powerful method which can be applied in various cases where the renormalization procedure in concerned. It demonstrates once again that softly broken SUSY theories are contained in rigid ones and inherit their renormalization properties.

The following statements are valid:

- All the renormalizations are defined in a rigid theory. There are no independent renormalizations in a softly broken theory.
- RG flow in a softly broken theory follows that in a rigid theory.
- This statement is true for RG equations, solutions to these equations, particular (fixed point) solutions, approximate solutions, etc.
- Renormalization of the F-I term needs a special treatment but can be also deduced from unbroken theory.

Appendix A. Three-loop renormalizations in the MSSM

In this section, we present explicit formulae for rigid and soft term renormalizations in the MSSM in the three-loop approximation in the case when we retain only α_3 and top Yukawa coupling Y_t.

The rigid renormalizations are [16]

$$\beta_{\alpha_3} = -3\alpha_3^2 + \alpha_3^2(14\alpha_3 - 4Y_t) + \alpha_3^2[\frac{347}{3}\alpha_3^2 - \frac{104}{3}\alpha_3 Y_t + 30Y_t^2], \quad (A.1)$$

$$\gamma_t = (2Y_t - \frac{8}{3}\alpha_3) - (8Y_t^2 + \frac{8}{9}\alpha_3^2) + [(30 + 12\zeta_3)Y_t^3 \quad (A.2)$$
$$+ (\frac{16}{3} + 96\zeta_3)Y_t^2\alpha_3 - (\frac{64}{3} + \frac{544}{3}\zeta_3)Y_t\alpha_3^2 + (\frac{2720}{27} + 320\zeta_3)\alpha_3^3],$$

$$\gamma_b = -\frac{8}{3}\alpha_3 - \frac{8}{9}\alpha_3^2 + [-\frac{80}{3}Y_t\alpha_3^2 + (\frac{2720}{27} + 320\zeta_3)\alpha_3^3], \quad (A.3)$$

$$\gamma_Q = (Y_t - \frac{8}{3}\alpha_3) - (5Y_t^2 + \frac{8}{9}\alpha_3^2) + [(15 + 6\zeta_3)Y_t^3 \quad (A.4)$$
$$+ (\frac{40}{3} + 48\zeta_3)Y_t^2\alpha_3 - (\frac{72}{3} + \frac{272}{3}\zeta_3)Y_t\alpha_3^2 + (\frac{2720}{27} + 320\zeta_3)\alpha_3^3],$$

$$\gamma_{H_2} = (3Y_t) - (9Y_t^2 - 16Y_t\alpha_3) \quad (A.5)$$
$$+ [(57 + 18\zeta_3)Y_t^3 + (72 - 144\zeta_3)Y_t^2\alpha_3 - (\frac{160}{3} + 16\zeta_3)Y_t\alpha_3^2],$$

$$\beta_{Y_t} = Y_t\left\{(6Y_t - \frac{16}{3}\alpha_3) - (22Y_t^2 - 16Y_t\alpha_3 + \frac{16}{9}\alpha_3^2) + [(102 + 36\zeta_3)Y_t^3\right.$$
$$\left. + \frac{272}{3}Y_t^2\alpha_3 - (\frac{296}{3} + 288\zeta_3)Y_t\alpha_3^2 + (\frac{5440}{27} + 640\zeta_3)\alpha_3^3]\right\}, \quad (A.6)$$

$$\beta_{\mu^2} = \mu^2\left\{3Y_t - (9Y_t^2 - 16Y_t\alpha_3)\right. \quad (A.7)$$
$$\left. + [(57 + 18\zeta_3)Y_t^3 + (72 - 144\zeta_3)Y_t^2\alpha_3 - (\frac{160}{3} + 16\zeta_3)Y_t\alpha_3^2]\right\},$$

Using the explicit form of anomalous dimensions calculated up to some order, one can reproduce the desired RG equations for the soft terms. In case of squark and slepton masses, they contain the contributions from unphysical masses Σ_{α_i}. To eliminate them, one has to solve the equation for Σ_{α_i}. In the case of the MSSM up to three loops, the solutions are [4]

$$\Sigma_{\alpha_1} = M_1^2 - \alpha_1\sigma_1 - \frac{199}{25}\alpha_1^2 M_1^2 - \frac{27}{5}\alpha_1\alpha_2 M_2^2 - \frac{88}{5}\alpha_1\alpha_3 M_3^2$$
$$+ \frac{13}{5}\alpha_1 Y_t(\Sigma_t + A_t^2) + \frac{7}{5}\alpha_1 Y_b(\Sigma_b + A_b^2) + \frac{9}{5}\alpha_1 Y_\tau(\Sigma_\tau + A_\tau^2), \qquad \text{(A.8)}$$

$$\Sigma_{\alpha_2} = M_2^2 - \alpha_2(\sigma_2 - 4M_2^2) - \alpha_2^2(4\sigma_2 + 9M_2^2) - \frac{9}{5}\alpha_2\alpha_1 M_1^2 - 24\alpha_2\alpha_3 M_3^2$$
$$+ 3\alpha_2 Y_t(\Sigma_t + A_t^2) + 3\alpha_2 Y_b(\Sigma_b + A_b^2) + \alpha_2 Y_\tau(\Sigma_\tau + A_\tau^2), \qquad \text{(A.9)}$$

$$\Sigma_{\alpha_3} = M_3^2 - \alpha_3(\sigma_3 - 6M_3^2) - \alpha_3^2(6\sigma_3 - 22M_3^2) - \frac{11}{5}\alpha_3\alpha_1 M_1^2 - 9\alpha_3\alpha_2 M_2^2$$
$$+ 2\alpha_3 Y_t(\Sigma_t + A_t^2) + 2\alpha_3 Y_b(\Sigma_b + A_b^2), \qquad \text{(A.10)}$$

where we have used the combinations [25]

$$\sigma_1 = \frac{1}{5}\left[3(m_{H_1}^2 + m_{H_2}^2) + 3(\tilde{m}_Q^2 + 3\tilde{m}_L^2 + 8\tilde{m}_U^2 + 2\tilde{m}_D^2 + 6\tilde{m}_E^2)\right],$$
$$\sigma_2 = m_{H_1}^2 + m_{H_2}^2 + 3(3\tilde{m}_Q^2 + \tilde{m}_L^2), \qquad \text{(A.11)}$$
$$\sigma_3 = 3(2\tilde{m}_Q^2 + \tilde{m}_U^2 + \tilde{m}_D^2),$$
$$\Sigma_t = \tilde{m}_t^2 + \tilde{m}_Q^2 + m_{H_2}^2, \quad \Sigma_b = \tilde{m}_b^2 + \tilde{m}_Q^2 + m_{H_1}^2, \quad \Sigma_\tau = \tilde{m}_\tau^2 + \tilde{m}_L^2 + m_{H_1}^2.$$

The corresponding soft term renormalizations read

$$\beta_{M_3} = -3\alpha_3 M_3 + 28\alpha_3^2 M_3 - 4Y_t\alpha_3(M_3 - A_t)$$
$$+ 347\alpha_3^3 M_3 - \frac{104}{3}\alpha_3^2 Y_t(2M_3 - A_t) + 30\alpha_3 Y_t^2(M_3 - 2A_t),$$

$$\beta_{A_t} = (6Y_t A_t + \frac{16}{3}\alpha_3 M_3) - [44Y_t^2 A_t - 16Y_t\alpha_3(A_t - M_3) - \frac{32}{9}\alpha_3^2 M_3]$$
$$+ [(306 + 108\zeta_3)Y_t^3 A_t + \frac{272}{3}Y_t^2\alpha_3(2A_t - M_3)$$
$$- (\frac{296}{3} + 288\zeta_3)Y_t\alpha_3^2(A_t - 2M_3) - 3(\frac{5440}{27} + 640\zeta_3)\alpha_3^3 M_3],$$

$$\beta_B = 3Y_t A_t - [18Y_t^2 A_t - 16Y_t\alpha_3(A_t - M_3)] + [(171 + 54\zeta_3)Y_t^3 A_t$$
$$+ (72 - 144\zeta_3)Y_t^2\alpha_3(2A_t - M_3) - (\frac{160}{3} + 16\zeta_3)Y_t\alpha_3^2(A_t - 2M_3)],$$

$$\beta_{\tilde{m}_t^2} = 2Y_t(\Sigma_t + A_t^2) - \frac{16}{3}\alpha_3 M_3^2 - 16Y_t^2(\Sigma_t + 2A_t^2) - \frac{64}{3}\alpha_3^2 M_3^2 + \frac{8}{3}\alpha_3^2\sigma_3$$
$$+ 3(30 + 12\zeta_3)Y_t^3(\Sigma_t + 3A_t^2) + (\frac{16}{3} + 96\zeta_3)Y_t^2\alpha_3[(2A_t - M_3)^2$$
$$+ 2\Sigma_t + M_3^2] - (\frac{64}{3} + \frac{544}{3}\zeta_3)Y_t\alpha_3^2[(A_t - 2M_3)^2 + \Sigma_t + 2M_3^2]$$
$$- \frac{16}{3}Y_t\alpha_3^2(\Sigma_t + A_t^2) + 4(\frac{2564}{9} + 960\zeta_3)\alpha_3^3 M_3^2 + \frac{160}{9}\alpha_3^3\sigma_3,$$

$$\beta_{\tilde{m}_b^2} = -\frac{16}{3}\alpha_3 M_3^2 - \frac{64}{3}\alpha_3^2 M_3^2 + \frac{8}{3}\alpha_3^2\sigma_3 - \frac{80}{3}Y_t\alpha_3^2[(A_t - 2M_3)^2 + \Sigma_t$$

$$+ 2M_3^2] - \frac{16}{3}Y_t\alpha_3^2(\Sigma_t + A_t^2) + 4(\frac{2564}{9} + 960\zeta_3)\alpha_3^3 M_3^2 + \frac{160}{9}\alpha_3^3\sigma_3,$$

$$\beta_{\tilde{m}_Q^2} = Y_t(\Sigma_t + A_t^2) - \frac{16}{3}\alpha_3 M_3^2 - 10Y_t^2(\Sigma_t + 2A_t^2) - \frac{64}{3}\alpha_3^2 M_3^2 + \frac{8}{3}\alpha_3^2\sigma_3$$

$$+ 3(15 + 6\zeta_3)Y_t^3(\Sigma_t + 3A_t^2) + (\frac{40}{3} + 48\zeta_3)Y_t^2\alpha_3[(2A_t - M_3)^2$$

$$+ 2\Sigma_t + M_3^2] - (\frac{72}{3} + \frac{272}{3}\zeta_3)Y_t\alpha_3^2[(A_t - 2M_3)^2 + \Sigma_t + 2M_3^2]$$

$$- \frac{16}{3}Y_t\alpha_3^2(\Sigma_t + A_t^2) + 4(\frac{2564}{9} + 960\zeta_3)\alpha_3^3 M_3^2 + \frac{160}{9}\alpha_3^3\sigma_3,$$

$$\beta_{m_{H_2}^2} = 3Y_t(\Sigma_t + A_t^2) - 18Y_t^2(\Sigma_t + 2A_t^2) + 16Y_t\alpha_3[(A_t - M_3)^2 + \Sigma_t + M_3^2]$$

$$+ 3(57 + 18\zeta_3)Y_t^3(\Sigma_t + 3A_t^2) + (72 - 144\zeta_3)Y_t^2\alpha_3[(2A_t - M_3)^2$$

$$+ 2\Sigma_t + M_3^2] - (\frac{160}{3} + 16\zeta_3)Y_t\alpha_3^2[(A_t - 2M_3)^2 + \Sigma_t + 2M_3^2]$$

$$- 16Y_t\alpha_3^2(\sigma_3 - 6M_3^2).$$

Appendix B

The RG equation for the parameter ξ in a rigid theory is

$$\dot{\xi} = -\gamma_V\xi, \tag{B.1}$$

where γ_V is the anomalous dimension of the gauge superfield. To find the soft terms $x, \bar{x}$ and z, one should solve the modified equation

$$\dot{\tilde{\xi}} = -\gamma_V(\tilde{\alpha}, \tilde{y}, \tilde{\xi})\tilde{\xi}. \tag{B.2}$$

In one-loop order $\gamma_V = (b_1 + b_2\xi)\alpha$, where $b_1 + b_2 = Q$, and the solutions are

$$x = -(M + x_0)\frac{b_1 + b_2\xi}{Q}, \qquad \bar{x} = -(\bar{M} + \bar{x}_0)\frac{b_1 + b_2\xi}{Q},$$

$$z = -(\Sigma_\alpha + z_0)\frac{b_1 + b_2\xi}{Q} + \frac{b_2\xi}{Q}(M + x_0)(\bar{M} + \bar{x}_0)\frac{b_1 + b_2\xi}{Q}, \tag{B.3}$$

where $x_0, \bar{x}_0$, and z_0 are arbitrary constants. In the Abelian case when $b_1 = Q$, $b_2 = 0$, the solutions are simplified and can be chosen as

$$x = -M(1 - \xi), \quad \bar{x} = -\bar{M}(1 - \xi), \quad z = -\Sigma_\alpha(1 - \xi) - M\bar{M}\xi(1 - \xi).$$

Together with the expression for $\tilde{\alpha}$ (14) it gives eq.(19) above.

Acknowledgments

I would like to thank my colleagues L.Avdeev, I.Kondrashuk, S.Codoban, G.Moultaka and V.Velizhanin in collaboration with whom these results have been obtained. I am grateful to the organizers of the conference "Continuous Advances in QCD-02" and especially to M.Shifman for their invitation to participate in the conference and to present this report. Many thanks are to M.Shifman, A.Vainshtein, M.Strassler, I.Jack, T.Jones and R.Rattazzi for useful discussions.

Financial support from RFBR grants # 02-02-16889 and # 00-15-96691 is kindly acknowledged.

References

1. L.A.Avdeev, D.I.Kazakov and I.N.Kondrashuk, *Nucl.Phys.* **B510** (1998) 289.
2. D. I. Kazakov, *Phys.Lett.* **B421** (1998) 211.
3. D. I. Kazakov, *Phys.Lett.* **B449** (1999) 201.
4. D. I. Kazakov and V. N. Velizhanin, *Phys.Lett.* **B485** (2000) 393.
5. D. I. Kazakov and V. N. Velizhanin, *Phys.Rev.* **D65** (2002) 085041.
6. I. Jack and D.R.T. Jones, *Phys.Lett.* **B415** (1997) 383.
7. G. F. Giudice and R. Rattazzi, *Nucl.Phys.* **B511** (1998) 25.
8. A. Nelson and M. Strassler, *JHEP* **0207** (2002) 021.
9. Y. Yamada, *Phys.Rev.* **D50** (1994) 3537.
10. I. Jack, D.R.T. Jones and A. Pickering, *Phys.Lett.* **B426** (1998) 73.
11. A. A. Slavnov and K. V. Stepanyantz, hep-th/0208006; V. K. Krivoshchekov, *Theor.Math.Phys.* **36** (1978) 291.
12. L. V. Avdeev, G. A. Chochia and A. A. Vladimirov, *Phys.Lett.* **B105** (1981) 272.
13. M. Vaughn and S. Martin, *Phys. Lett.* **B318** (1993) 331.
14. I. Jack and D.R.T. Jones, *Phys. Lett.* **B333** (1994) 372.
15. I. Jack, D.R.T. Jones and C.G. North, *Phys.Lett.* **B386** (1996) 138.
16. P.M. Ferreira, I. Jack and D.R.T. Jones, *Phys.Lett.* **B387** (1996) 80.
17. S. Ferrara and O. Piguet, *Nucl.Phys.* **B93** (1975) 261.
18. L. Girardello and M.T. Grisaru, *Nucl.Phys.* **B194** (1982) 65.
19. I. Jack and D. R. T. Jones, *Phys. Lett.* **B457** (1999) 101.
20. M. A. Shifman, A. I. Vainshtein and V. I. Zakharov, *Yad.Fiz.* **43** (1986) 1596; M. A. Shifman and A. I. Vainshtein *Nucl.Phys.* **B359** (1991) 571.
21. I. N. Kondrashuk, *J.Phys.* **A33** (2000) 6399.
22. W. Siegel, *Phys.Lett.* **B84** (1979) 193; R. van Damme and G. 't Hooft, *Phys. Lett.* **B150** (1985) 133.
23. I. Jack and D. R. T. Jones, *Phys. Lett.* **B415** (1997) 383.
24. I. Jack, D. R. T. Jones, S. P. Martin, M. T. Voughn and Y. Yamada, *Phys.Rev.* **D50** (1994) 5481.
25. S. P. Martin and M. T. Voughn, *Phys.Rev.* **D50** (1994) 2282.
26. I. Jack and D. R. T. Jones, *Phys. Lett.* **B333** (1994) 372.
27. J.A. Helayël-Neto, *Phys.Lett.* **135B** (1984) 78; F. Feruglio, J.A. Helayël-Neto and F. Legovini, *Nucl. Phys.* **B249** (1985) 533; M. Scholl, *Z. Phys.* **C28** (1985) 545.
28. J. Hisano and M. A. Shifman, *Phys.Rev.* **D56** (1997) 5475.
29. T. Kobayashi and K. Yoshioka, *Phys. Lett.* **B486** (2000)223.

448

30. I. Jack, D. R. T. Jones and A. Pickering, *Phys.Lett.* **B432** (1998) 114.
31. V. A. Novikov, M. A. Shifman, A. I. Vainshtein and V. I. Zakharov, *Nucl. Phys.* **B229** (1983) 381.
32. T. Kobayashi, J. Kubo and G. Zoupanos, *Phys.Lett.* **B427** (1998) 291.
33. L.E. Ibáñez, C. López and C. Muñoz, *Nucl.Phys.* **B256** (1985) 218.
34. C. T. Hill, *Phys.Rev.* **D24** (1981) 691; C. T. Hill, C.N. Leung and S. Rao, *Nucl. Phys.* **B262** (1985) 517.
35. I. Jack and D. R. T. Jones, *Phys. Lett.* **B443** (1998) 177.
36. G. Auberson and G. Moultaka, *Eur.Phys.J.* **C12** (2000) 331.
37. S. Codoban and D. I. Kazakov, *Eur.Phys.J* **C13** (2000) 671.
38. D. I. Kazakov and G. Moultaka, *Nucl.Phys.* **B577** (2000) 121.
39. P. C. West, *Phys.Lett.* **137B** (1984) 371; A. J. Parkes and P. C. West, *Phys.Lett.* **138B** (1984) 99; D.R.T. Jones and L. Mezinchescu, *Phys.Lett.* **138B** (1984) 293, S. Hamidi, J. Patera and J. H. Schwarz, *Phys.Lett.* **141B** (1984) 349.
40. A. V. Ermushev, D. I. Kazakov and O. V. Tarasov, *Nucl.Phys.* **B281** (1987) 72; D.R.T. Jones, *Nucl.Phys.* **B277** (1986) 153; C. Lucchesi, O. Piguet and K. Sibold, *Phys.Lett.* **201B** (1988) 241; D. I. Kazakov, *Mod.Phys.Lett.* **A9** (1987) 663.
41. see e.g. L. Alvarez-Gaume and S.F. Hassan, *Fortsch.Phys.* **45** (1997) 159.
42. N. Seiberg and E. Witten, *Nucl.Phys.* **B426** (1994) 19.
43. D. I. Kazakov and V. N. Velizhanin, *Part & Nucl.* **31** (2000) 240. Proc. of Bogoliubov Conf., Dubna, 1999.
44. P. Fayet and J. Iliopoulos, *Phys.Lett.* **B51** (1974) 461.
45. W. Fischler, H. P. Nilles, J. Polchinski, S. Raby and L. Susskind, *Phys.Rev.Lett.* **47** (1981) 757.
46. I. Jack, D. R. T. Jones and S. Parsons, *Phys.Rev.* **D62** (2000) 125022.

SUPERSYMMETRIC β FUNCTION WITH ULTRAVIOLET CHOP-OFF

A. V. SMILGA

SUBATECH, Université de Nantes
4 rue Alfred Kastler, BP 20722, Nantes 44307, France
E-mail: smilga@subatech.in2p3.fr

We notice that the renormalization of the effective charge in a 4–dimensional (supersymmetric) gauge theory is determined by the same graphs and is rigidly connected to the renormalization of the metric on the moduli space of the classical
vacua of the corresponding reduced quantum mechanical system. Supersymmetry provides constraints for possible modifications of the metric, and this might
give an alternative simple proof of nonrenormalization theorems for the original
4-dimensional theory.

1. Introduction

Consider 4–dimensional supersymmetric gauge theory placed in a small
spatial torus T^3 of size L. We assume that $g^2(L) \ll 1$ and perturbation
theory makes sense. For unitary and symplectic gauge groups G, the only
classical vacua of this theory are given by constant gauge potentials A_k, $k =
1, 2, 3$, lying in the Cartan subalgebra of the group[a] G. The low–energy
dynamics of the model is determined by the effective Hamiltonian describing
motion over the vacuum moduli space. Due to supersymmetry, the energy
of a classical vacuum configuration stays zero after loop corrections are
taken into account – no potential is generated. [b] However, supersymmetry
usually allows the existence of a nontrivial metric on the moduli space, in
which case such a metric is generated after loop corrections are taken into
account.

The loop corrections to the effective Hamiltonian were calculated first in
Ref.[3] in the simplest case of $\mathcal{N} = 1$ supersymmetric QED with two chiral
matter multiplets of opposite charges. In this case, the moduli space is

---

[a]It is not the case for higher orthogonal and exceptional groups [1], but these complications
are beyond the scope of the present talk.
[b]This is true only for nonchiral theories, which are only considered here. In the theories
with chiral matter content, the situation is more complicated [2].

represented by the constant gauge potentials A_k and their superpartners. Note that, for a field theory on T^3, the moduli space is compact, $0 \le A_k \le 2\pi/L$.

The original calculation was carried out in the Hamiltonian framework. The effective Hamiltonian is expressed in terms of A_k, $P_k = -i\partial/\partial A_k$, and the zero Fourier mode of the photino field ψ_α, $\alpha = 1, 2$. It has the form

$$\frac{1}{e^2}H^{\text{eff}} = \frac{1}{2}f(\mathbf{A})P_k^2 f(\mathbf{A}) -$$

$$-\epsilon_{jkp}\,\bar\psi\sigma_j\psi f(\mathbf{A})\partial_p f(\mathbf{A})P_k - \frac{1}{2}f(\mathbf{A})\partial_k^2 f(\mathbf{A})(\bar\psi\psi)^2 \,, \tag{1}$$

where σ_i are the Pauli matrices and

$$f(\mathbf{A}) = 1 - \frac{e^2}{4}\sum_{\mathbf{n}}\frac{1}{|\mathbf{A}L - 2\pi\mathbf{n}|^3} + \cdots \tag{2}$$

(we have rescaled $A \to A/e$ compared with the normalization of Ref. [3]). The dots stand for possible higher–loop corrections. The expression (2) is written for the theory where the charged fields are massless.

Note that the sum in the right side of Eq. (2) diverges logarithmically at large $|\mathbf{n}|$. This is none other than the effective charge renormalization

$$e^2(L) = e_0^2\left[1 - \frac{e_0^2}{4\pi^2}\ln(\Lambda_{UV}L) + \cdots\right] . \tag{3}$$

In the massive case and if the box is large enough, $\ln(\Lambda L)$ is substituted by $\ln(\Lambda/m)$. On the other hand, if we are dealing with dimensionally reduced SQED, where all Fourier harmonics with $\mathbf{n} \ne 0$ are ignored, we obtain

$$f(\mathbf{A}) = 1 - \frac{e^2}{4|L\mathbf{A}|^3} . \tag{4}$$

It is obvious that the coefficients in Eq. (4) and in Eq. (3) are related[c]. The knowledge of β–function allows one to determine the modification of the metric on the moduli space, and this is how the effective Hamiltonian for $\mathcal{N} = 1$ non-Abelian theories was evaluated in recent Ref. [5]. The inverse is also true, however, and this is one of the main emphasize of the present talk. We note that the β–function of field theories can be conveniently calculated via modification of the metric in the quantum mechanical limit where all

---

[c]The procedure for getting the aforementioned relation is similar to the T-duality transformation on D-branes in string theory (see in particular Ref. [4]). In fact, $N = 2$ SQED with doublet of hypermultiplets (plus free uncharged hypermultiplets) reduced to one dimension can be interpreted as the theory of a D0-brane in the vicinity of a D4-brane. The latter system was extensively studied [4]. In this note we concentrate, however, on the $N = 1, d = 4$ systems, which were not considered so far in the string theory framework.

nonzero Fourier harmonics are ignored. Ideologically, this is the ultraviolet cut-off procedure brought to its extreme. One can call it ultraviolet *chop-off*.

As was mentioned, the result (1) was first obtained in the Hamiltonian framework using a systematic Born–Oppenheimer expansion for H^{eff}. There is, however, a simpler way to derive the same result: to evaluate the term $\propto \dot{\mathbf{A}}\dot{\mathbf{A}}$ in the effective *Lagrangian* in a slowly varying bosonic background $\mathbf{A}(t), \psi_\alpha = 0$. Other structures in the Lagrangian can be restored using supersymmetry.

The plan of the paper is the following. In Sect. 2 we present one–loop calculations of the effective Lagrangian. We use the background field method and demonstrate that the result is given by exactly the same graphs as the graphs determining the 4–dimensional β–function. Like in 4–dimensional case, the contribution of scalar determinant cancels out in supersymmetric case, and we are left with the graphs describing fermion and gauge boson magnetic interactions. Next, we go beyond one loop and suggest in Sect. 3 a new proof of the non-renormalization theorems for four-dimensional $\mathcal{N} = 2$ and $\mathcal{N} = 4$ SYM theories. The proof would be rigourous if the relation between the $(0+1)$ and $(3+1)$ effective Lagrangians could be explicitly seen beyond one loop. In Sect. 4 we present the results of the explicit two–loop calculations for $\mathcal{L}_{\text{eff}}^{0+1}$ and discuss them.

2. Ultraviolet chop-off and β function

Let us consider for definiteness $\mathcal{N} = 1$ four–dimensional $SU(2)$ SYM theory

$$\mathcal{L} = \frac{1}{g^2} \, \text{Tr} \left(-\frac{1}{2} F_{\mu\nu}^2 + i \bar{\lambda} \not{D} \lambda \right) , \tag{5}$$

where λ are Majorana fermions in the adjoint representation of $SU(2)$ and $D_\mu = \partial_\mu - i [A_\mu, \cdot]$. Put the system in a small spatial box and impose the periodic boundary conditions. We would like to calculate quantum corrections to the effective action in the abelian background field $A_\mu = C_\mu t^3$, $C_\mu = (0, C_i)$. We assume that C_i varies slowly with time, but does not depend on spatial coordinates[d]. The background fermionic fields (superpartners of C_μ) are taken to be zero at this stage.

---

[d]The setup of the problem is basically the same as in Refs.[6],[7]. There are two differences: *(i)* We are considering the $\mathcal{N} = 1$ rather than $\mathcal{N} = 4$ theory and the corrections are not going to vanish. *(ii)* The authors of Ref. [6] did their calculation bearing in mind the geometric picture of scattered D0 branes and their background was slightly more sophisticated than ours.

The calculation can be conveniently done using background gauge method [8]. We decompose the field in the classical (abelian) background and quantum fluctuations,

$$A_\mu \to C_\mu t^3 + \mathcal{A}_\mu , \tag{6}$$

and add to the Lagrangian the gauge-fixing term

$$-\frac{1}{2g^2}(D_\mu^{\mathrm{cl}} \mathcal{A}_\mu)^2 , \tag{7}$$

where $D_\mu^{\mathrm{cl}} = \partial_\mu - i \left[A_\mu^{\mathrm{cl}}, \cdot\right]$. In what follows we use the notation $A_\mu \equiv A_\mu^{\mathrm{cl}} = C_\mu t^3$. The coefficient chosen in Eq. (7) defines the "Feynman background gauge", which is simpler and more convenient than others. Adding (7) to the first term in Eq. (5) and integrating by parts, we obtain for the gauge–field–dependent part of the Lagrangian

$$\mathcal{L}_A = -\frac{1}{2g^2} \operatorname{Tr} \left(F_{\mu\nu}^2\right) + \frac{1}{g^2} \operatorname{Tr} \left\{ \mathcal{A}_\mu \left(D^2 g_{\mu\nu} \mathcal{A}_\nu - 2i \left[F_{\mu\nu}, \mathcal{A}_\nu\right]\right)\right\} + \dots , \tag{8}$$

where the dots stand for the terms of higher order in $\mathcal{A}_\mu$. The ghost part of the Lagrangian is

$$\mathcal{L}_{ghost} = -2 \operatorname{Tr} \left(\bar{c} D^2 c\right) + \text{higher order terms} \tag{9}$$

Now we can integrate over the quantum fields $\mathcal{A}_\mu$, c, and also over the fermions using the relation

$$(i\slashed{D})^2 = -D^2 + \frac{i}{2}\sigma_{\mu\nu}F_{\mu\nu} ,$$

$\sigma_{\mu\nu} = \frac{1}{2}[\gamma_\mu, \gamma_\nu]$. We obtain the effective action as follows:

$$S_{\mathrm{eff}} = -\frac{1}{2g^2} \int_{T^3 \times R} \operatorname{Tr}(F_{\mu\nu}^2) +$$

$$\log \left(\frac{\det^{\frac{1}{4}} \left(-D^2 I + \frac{i}{2}\sigma_{\mu\nu} \left[F_{\mu\nu}, \cdot\right]\right) \det \left(-D^2\right)}{\det^{\frac{1}{2}} \left(-D^2 g_{\mu\nu} + 2i \left[F_{\mu\nu}, \cdot\right]\right)} \right) . \tag{10}$$

We see that the fermion and gauge field determinants involve, besides the term $-D^2$ which is present also in the scalar determinant, the term $\propto F_{\mu\nu}$ describing the magnetic moment interactions. An important observation is that, were these magnetic interactions absent, the contributions of the ghosts, fermions, and gauge bosons would just cancel and the effective action would not acquire any corrections. This feature is common for all supersymmetric gauge theories ($\mathcal{N} = 1$, $\mathcal{N} = 2$, $\mathcal{N} = 4$; non-Abelian and Abelian). This fact is related to another known fact that, when supersymmetric β function is calculated in the *instanton* background, only the contribution of the zero modes survives [9].

For nonsupersymmetric theories, also nonzero instanton modes provide a nonvanishing constribution in the β function. On the other hand, the contributions due to det $(-D^2)$ in the effective action do not vanish in the nonsupersymmetric case.

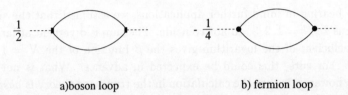

$$\frac{1}{2} \qquad \qquad \qquad - \frac{1}{4}$$

a)boson loop   b) fermion loop

Figure 1. One-loop renormalization of the kinetic term in SYM. Internal lines are Green's functions of the operator $(-D^2)$. The vertices involve the spin operator $J_{\alpha\beta}$ and are different for the fermion and gauge boson loop.

To find the magnetic contributions, we have to calculate the graphs drawn in Fig. 1. The vertices there are proportional to $\epsilon^{abc} F_{\alpha\beta} J_{\alpha\beta}$ ($J_{\alpha\beta}$ being the spin operator in the corresponding representation) and the lines are Green's functions of the operator $-D^2$. Only the color components 1 and 2 circulate in the loops. They acquire the mass $|\mathbf{C}|$ in the Abelian background $C_i t^3$. One can be convinced that gauge boson loop involves the factor -4 compared to the fermion one [the factor $\frac{1}{2} : -\frac{1}{4} = -2$ is displayed in Eq. (10) and Fig. 1 and another factor 2 comes from spin; see Eq. (16.128) in Peskin's book].

Let us calculate, say, the fermion loop. If all higher Fourier modes are "chopped off", $-D^2 \to -\partial_0^2 - \mathbf{C}^2$ and the corresponding contribution to the effective Lagrangian is

$$\frac{1}{4} \cdot \frac{1}{2} \cdot 2 \cdot \dot{C}_j \dot{C}_k \, \mathrm{Tr}\{\sigma_{0j}\sigma_{0k}\} \int_{-\infty}^{\infty} \frac{d\omega}{2\pi} \frac{1}{(\omega^2 + \mathbf{C}^2)^2} = \frac{\dot{\mathbf{C}}^2}{4|\mathbf{C}|^3} . \qquad (11)$$

(the factor $\frac{1}{4}$ is the power of the determinant in Eq. (10), $\frac{1}{2}$ comes from the expansion of the logarithm and 2 is the color factor.) Adding the gauge boson contribution and also the free bosonic term, we obtain

$$\frac{g^2}{L^3} \mathcal{L}_{\text{bos}}^{\text{eff}} = \frac{\dot{\mathbf{C}}^2}{2f^2(\mathbf{C})} \qquad (12)$$

with

$$f(\mathbf{C}) = 1 + \frac{3g^2}{4L^3|\mathbf{C}|^3} + \cdots . \qquad (13)$$

If higher Fourier modes are taken into account, we obtain in the exact analogy with Eq. (2)

$$f(\mathbf{C}) = 1 + \frac{3g^2}{4} \sum_{n_k} \frac{1}{[\sum_k (C_k L_k - 2\pi n_k)^2]^{3/2}} + \dots \,, \qquad (14)$$

where, bearing in mind further applications, we assumed that the sizes of the torus L_k, $k = 1, 2, 3$, do not coincide. The sum is divergent at large n_k. The coefficient of the logarithm gives the β function of the $\mathcal{N} = 1$ SYM theory. For sure, this could be expected in advance. What is not quite trivial, however, is that the calculation in the truncated theory is *absolutely* parallel to the well-known calculation in 4 dimensions [8]: in four dimensions the corrections to the effective action are also given by the graphs in Fig. 1, and the gauge boson and the fermion contributions in the β function have the respective coefficients[e] $4 : -1$. We will see soon that this is specific for supersymmetric theories. In nonsupersymmetric case, the β function can also be calculated with the chop-off technique, but the relevant graphs are different.

The bosonic effective action (12) can be supersymmetrized using the superfield technique developped in Ref. [10]. The explicit expression in components was written in Ref. [5]:

$$\frac{g^2}{L_1 L_2 L_3} \mathcal{L} = \frac{1}{2 f^2} \dot{C}^j \dot{C}^j + \frac{i}{2f^2} \left(\bar{\Psi} \dot{\Psi} - \dot{\bar{\Psi}} \Psi \right) - \frac{\partial_i f}{f^3} \epsilon^{ijk} \dot{C}_j \bar{\Psi} \sigma_k \Psi +$$
$$+ \frac{D^2}{2 f^2} + \frac{D \partial_i f}{f^3} \bar{\Psi} \sigma_i \Psi - \frac{1}{8} \partial^2 \left(\frac{1}{f^2} \right) \left(\bar{\Psi} \right)^2 (\Psi)^2 \qquad (15)$$

where $\Psi = \psi f$, ψ and $\bar{\psi}$ being the canonically conjugated variables of Eq. (1); D is the auxilary field. The action corresponding to the Lagrangian (15) is invariant under the transformations

$$\delta_\epsilon C_k = \bar{\epsilon} \sigma_k \Psi + \bar{\Psi} \sigma_k \epsilon \,,$$
$$\delta_\epsilon \Psi_\alpha = -i (\sigma_k \epsilon)_\alpha \dot{C}_k + \epsilon_\alpha D \,,$$
$$\delta_\epsilon D = i \left(\dot{\bar{\Psi}} \epsilon - \bar{\epsilon} \dot{\Psi} \right) \,. \qquad (16)$$

Let us briefly discuss nonsupersymmetric theories. At one–loop level, our chop-off method works (with some modifications [11]) also in the non-supersymmetric case. However, the very notion of the effective moduli-

---

[e] A remarkable fact is that one obtains the same ratio calculating the effective action in the instanton background field: the correct coefficient 6 in the β function is obtained as 8 - 2, where "8" is the number of bosonic zero modes and "2" is a half of the number of fermionic zero modes in the instanton background.

space Lagrangian is meaningless there: the potential is always generated which locks up the would-be vacuum valley [12].

3. Nonrenormalisation theorems

In the dimensionally reduced $\mathcal{N} = 4$ SYM theory (alias, maximally supersymmetric quantum mechanics, alias matrix model, alias the system of $D0$-branes), the corrections to the metric on the moduli space are absent. There are D-brane arguments in favor of this conclusion [13], it was confirmed by explicit calculation [6], and finaly *proven* using simple symmetry arguments [14]. To make the paper self-contained, we present here a somewhat refined version of these arguments.

In the maximally supersymmetric $SU(2)$ theory the effective Lagrangian is written in terms of a 9–dimensional vector C_k and a *real* 16–component $SO(9)$ spinor λ_α. The Lagrangian must be invariant with respect to the supersymmetry transformations

$$\delta_\epsilon C_k = -i\,\epsilon\gamma_k\lambda \,,$$
$$\delta_\epsilon \lambda_\alpha = \left(\gamma_k \dot{C}_k \epsilon\right)_\alpha + [M(c)\epsilon]_\alpha \,, \tag{17}$$

where ϵ is a real Grassmann spinor and γ_k are the 9–dimensional γ matrices, $\gamma_j\gamma_k + \gamma_k\gamma_j = 2\delta_{jk}$. They are all real and symmetric. The transformations (17) represent an analog of (16) with the auxiliary field expressed out.

The commutator of two SUSY transformations with parameters ϵ_1 and ϵ_2 should amount to a time translation. A trivial calculation gives

$$[\delta_1, \delta_2]C_k = -2i\,\epsilon_2\epsilon_1\dot{C}_k - i\,\epsilon_2\{\gamma_k M + M^T\gamma_k\}\epsilon_1 \,, \tag{18}$$

and we conclude that

$$\gamma_k M + M^T\gamma_k = 0 \,. \tag{19}$$

As was noticed in Ref.[14], this implies that $M = 0$. Let us prove it. On the first step, note that any M satisfying (19) commutes with all generators $J_{kj} = \frac{1}{4}[\gamma_k, \gamma_j]$ of $Spin(9)$. This means that, for any set λ belonging to the spinor representation of $Spin(9)$, the set $M\lambda$ also forms a spinor representation. Hence $M = \xi R$, where ξ is a real number and $R \in Spin(9)$. But R commutes with all generators of $Spin(9)$ and should belong to the center of $Spin(9)$, i.e. $M = \pm\xi I$. Then (19) tells us that $\xi = 0$.

When proving this, we used implicitly the fact that the real spinor representation of $SO(9)$ is *irreducible*. If it could be decomposed in a direct sum of two other representations, we could choose M as $\mathrm{diag}(\xi_1 z_1, \ \xi_2 z_2)$, with z_1 and z_2 belonging to the center of the group in the corresponding subspaces. Such M would not be necessarily proportional to I. This discussion is not purely academic. Actually, for $Spin(3)$ and $Spin(5)$, where real spinor representations are reducible, nontrivial matrices satisfying (19) exist.

The only structure not involving higher derivatives[f] and invariant with respect to the transformations (17) with $M = 0$ is

$$\frac{1}{2} \left[\dot{C}_k^2 + i\lambda\dot{\lambda} \right] .$$

Nontrivial corrections to the metric are not allowed. Bearing in mind the discussion in the previous section and assuming that the multiloop corrections to the effective chop-off and field theory Lagrangia are related to each other, as they are at the 1–loop level (see, however, next section for discussion of this issue), this simultaneously proves that the β function in $\mathcal{N} = 4$ SYM theory vanishes exactly in all loops.

In the $\mathcal{N} = 2$ case, the corrections to the metric survive, but the presence of 4 different complex supercharges dictates that the function $f^{-2}(\mathbf{C})$ ($\mathbf{C}$ is now a 5–dimensional vector. In four dimensions this corresponds to the gauge potential and a complex scalar.) is not arbitrary, but should be a harmonic function, $\Delta^{(5)} f^{-2}(\mathbf{C}) = 0$ [16,17]. The $O(5)$ invariance, which is manifest in the chop-off quantum–mechanical limit, tells us then that the only allowed form of the effective Lagrangian is

$$\mathcal{L}^{\text{eff}} = \frac{\dot{\mathbf{C}}^2}{2} \left(1 + \frac{\text{const}}{|\mathbf{C}|^3} \right) , \tag{20}$$

i.e. all the corrections beyond one loop vanish. But this means also that multiloop corrections to the β function vanish in this case.

4. Two–loop calculations

At the 2–loop level, we have *several* graphs contributing to the effective Lagrangian both in the $(0+1)$–dimensional and in the $(0+3)$–dimensional case. They are depicted in Fig. 2. We refer the reader to [18] for details and present here only the final result for the effective Lagrangian in $\mathcal{N} = 1$ supersymmetric QED. Again, it has the form (15). Its bosonic part is

$$\mathcal{L}_{\text{eff}}^{(0+1)} = \frac{\dot{\mathbf{A}}^2}{2} \left(1 + \frac{1}{2|\mathbf{A}|^3} - \frac{3}{4|\mathbf{A}|^6} + \dots \right) . \tag{21}$$

The correction $-3/(4|\mathbf{A}|^6)$ comes from the two–loop graphs in Fig. 2.

---

[f]Higher derivative terms, in particular the term $\propto (\dot{C}_k \dot{C}_k)^2$ and its superpartners are allowed. See [6,15] for detailed discussion.

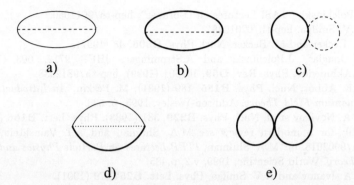

Figure 2. Graphs contributing to $\mathcal{L}_{\text{eff}}^{(2)}$. Thin solid lines describe fermions, bold lines – scalars, dashed lines - photon and dotted line - photino.

The same calculation can be done for $\mathcal{N} = 2$ QED. In this case, the correction $\propto 1/(|\mathbf{A}|^6)$ vanishes in a perfect agreement with the general theorem of the previous section.

Unfortunately, this does not look similar to

$$\mathcal{L}_{\text{eff}}^{(3+1)} = -\frac{F_{\mu\nu}^2}{4e^2(\mu)} = -\frac{F_{\mu\nu}^2}{4}\left[\frac{1}{e_0^2} + \frac{1}{4\pi^2}\ln\frac{\Lambda}{\mu} + \frac{e_0^2}{16\pi^4}\ln\frac{\Lambda}{\mu} + \dots\right] \quad (22)$$

(see e.g. [19]; μ is assumed to be much larger than the mass of charged particles). In addition, there is no obvious correspondence between the individual cotributions in $\mathcal{L}_{\text{eff}}^{(0+1)}$ and $\mathcal{L}_{\text{eff}}^{(3+1)}$. To be precise, the diagrams determining the corrections in these two cases are identical, but the results of their evaluation are not similar.

In spite of the fact that we did not establish a direct relationship between two–loop contributions to $\mathcal{L}^{\text{eff}}$ in different dimensions, we still believe that the correspondence between higher order corrections will eventually be unravelled. We do *not* believe that the similarity between nonrenormalisation theorems in 4-dimensional and in reduced theory is purely accidental. Further studies of this question are very much welcome.

References

1. E. Witten, J. High Energy Phys. **9802**, 006 (1998) ; A. Keurentjes, A. Rosly, and A.V. Smilga, Phys. Rev. **58**, 081701 (1998) ; V.G. Kac and A.V. Smilga, hep-th/9902029, published in *The Many Faces of the Superworld* (World Scientific, 2000), ed. M.A. Shifman; A. Keurentjes, J. High Energy Phys. **9905**, 001, 014 (1999).
2. A.V. Smilga, JETP **64**, 8 (1986); B.Yu. Blok, A.V, Smilga, Nucl. Phys. **B287**, 589 (1987).
3. A.V. Smilga, Nucl. Phys. **B291**, 241 (1987).

458

4. J.Polchinski, "TASI Lectures on D-branes", hep-th/9611050.
5. A.V. Smilga, hep-th/0201048.
6. K. Becker and M. Becker, Nucl. Phys. **B506**, 48 (1997).
7. M.Douglas, J.Polchinski and A.Strominger, JHEP 9712, 003 (1997); E.Akhmedov, Phys. Rev. **D59**, 101901 (1999), hep-th/9812038.
8. L.F. Abbot, Nucl. Phys. **B185**, 189 (1981); M. Peskin, *An Introduction in Quantum Field Theory*, Addison-Wesley, 1995, p. 533.
9. V.A. Novikov et al, Nucl. Phys. **B229**, 381 (1983), Phys. Lett. **B166** (1986) 329; for a modern review see M.A. Shifman and A.I. Vainshtein, hep-th/9902018, in: M.A. Shifman, *ITEP Lectures on Particle Physics and Field Theory*, World Scientific, 1999, v.2, p.485.
10. E.A. Ivanov and A.V. Smilga, Phys. Lett. **B257**, 79 (1991).
11. E.T. Akhmedov and A.V. Smilga, hep-th/0202027.
12. M. Lüscher, Nucl. Phys. **B219**, 233 (1983).
13. M.R. Douglas et al, Nucl. Phys. **B485**, 85 (1997).
14. S. Paban, S. Sethi, and M. Stern, Nucl. Phys. **B534**, 137 (1998).
15. K. Becker et al, Phys. Rev. **D56**, 3174 (1997); Y. Okawa and T. Yoneya, Nucl. Phys. **B538**, 67 (1999); R. Helling et al, Nucl. Phys. **B559**, 184 (1999).
16. D.-E. Diaconescu and R. Entin, Phys. Rev. **D56**, 8045 (1997).
17. B. Zupnik, Theor. Math. Phys. **120**, 365 (1999).
18. A. Smilga, hep-th/0205044.
19. A.I. Vainshtein and M.A. Shifman, Sov. J. Nucl. Phys. **44** (1986) 321.

NONPERTURBATIVE SOLUTION OF SUPERSYMMETRIC GAUGE THEORIES

J. R. HILLER

Department of Physics
University of Minnesota-Duluth
Duluth MN 55812 USA
E-mail: jhiller@d.umn.edu

Recent work on the numerical solution of supersymmetric gauge theories is described. The method used is SDLCQ (supersymmetric discrete light-cone quantization). An application to $N = 1$ supersymmetric Yang–Mills theory in 2+1 dimensions at large N_c is summarized. The addition of a Chern–Simons term is also discussed.

1. Introduction

Although much has been learned about supersymmetric gauge theories by analytic methods, numerical methods can yield much more of the nonperturbative structure. In particular, the method known as supersymmetric discrete light-cone quantization (SDLCQ),[1,2] an extension of ordinary DLCQ,[3,4] has been quite successful in the analysis of (1+1)-dimensional supersymmetric theories. This work has recently been extended to 2+1 dimensions[5,6,7,8] with consideration of $N = 1$ supersymmetric Yang–Mills (SYM) theory, including a Chern–Simons (CS) term.[9] The mass spectrum, Fock-state wave functions, and a stress-energy correlator are all computed. The CS term provides an effective mass that reduces the tendency of SYM to produce stringy states with many constituents. This work was done at large-N_c, but the method is also applicable to finite N_c.

As the name SDLCQ implies, light-cone coordinates[10] are used. They are defined by spacetime coordinates

$$x^{\pm} = (t \pm z)/\sqrt{2}, \quad \mathbf{x}_{\perp} = (x, y) \tag{1}$$

and momentum components

$$p^{\pm} = (E \pm p_z)/\sqrt{2}, \quad \mathbf{p}_{\perp} = (p_x, p_y). \tag{2}$$

459

The dot product of two such four-vectors then becomes

$$p \cdot x = p^+ x^- + p^- x^+ - \mathbf{p}_\perp \cdot \mathbf{x}_\perp. \tag{3}$$

The x^+ direction is treated as the direction of time evolution, which makes the conjugate variable p^- the light-cone energy. The light-cone three-momentum is $\underline{p} \equiv (p^+, \mathbf{p}_\perp)$. In a frame where the net transverse momentum $\mathbf{P}_\perp$ is zero, the mass eigenvalue problem becomes

$$2P^+ P^- |P\rangle = M^2 |P\rangle, \tag{4}$$

where $|P\rangle$ is also an eigenstate of three-momentum $\underline{P}$. The SDLCQ method provides a means to solve this eigenvalue problem with supersymmetry preserved exactly at any level of the approximation.

One of the advantages of light-cone coordinates is that there exists a well-defined Fock-state expansion for each mass eigenstate. There are no disconnected vacuum pieces, because the longitudinal momentum of each constituent, virtual or real, must be positive. SDLCQ uses a Fock-state expansion for $|P\rangle$ to obtain a matrix eigenvalue problem for the Fock-state wave functions at discrete values of the momentum. The matrix is then diagonalized by appropriate means. For large matrices the Lanczos diagonalization technique[11] has been used, as discussed in Ref. [6].

The discretization is accomplished by restricting the fields to periodic boundary conditions in a light-cone box[3,4] defined by $-L_\parallel < x^- < L_\parallel$ and $0 < x, y < L_\perp$. This leads to a discrete momentum grid

$$p^+ \to \frac{\pi}{L_\parallel} n, \quad \mathbf{p}_\perp \to \left(\frac{2\pi}{L_\perp} n_x, \frac{2\pi}{L_\perp} n_y \right). \tag{5}$$

The product $P^+ P^-$ is independent of $L_\parallel$, and the limit $L_\parallel \to \infty$ is exchanged for a limit in terms of an integer K, called the harmonic resolution,[3] defined by

$$K \equiv \frac{L_\parallel}{\pi} P^+. \tag{6}$$

Longitudinal momentum fractions $x = p^+/P^+$ then reduce to n/K. The number of particles in a Fock state is limited to K, because negative longitudinal momentum is not allowed and the individual integers n must sum to K.[a] The eigenvalue equation (4) becomes a coupled set of integral equations for the Fock-state wave functions in which the integrals are approximated by discrete sums over the momentum grid

$$\int dp^+ \int d^2 p_\perp f(p^+, \mathbf{p}_\perp) \simeq \frac{\pi}{L_\parallel} \left(\frac{2\pi}{L_\perp} \right)^2 \sum_{n, n_x, n_y} f(nP^+/K, 2\mathbf{n}_\perp \pi/L_\perp). \tag{7}$$

---

[a] Zero modes are ignored.

The harmonic resolution provides a natural cutoff for n. The transverse sums must be truncated explicitly, which is done by limiting n_x and n_y to range from $-T$ to T. The integer T can be viewed as a transverse cutoff or, at fixed dimensionful cutoff $\Lambda_\perp \equiv 2\pi T/L_\perp$, as the transverse resolution.

The distinction between DLCQ and SDLCQ lies in the choice of operator for discretization. In ordinary DLCQ one discretizes the Hamiltonian, P^-; in SDLCQ, one discretizes the supercharge Q^- and constructs P^- from the superalgebra relation

$$\{Q^-, Q^-\} = 2\sqrt{2}P^-, \tag{8}$$

which guarantees that the discrete eigenvalue problem preserves supersymmetry.[1,2] The P^- of ordinary DLCQ differs from the supersymmetric P^- by terms which disappear in the large-K limit but which break the supersymmetry at finite K.

The remainder of this paper is organized as follows. In Sec. 2, (2+1)-dimensional SYM theory is reviewed and numerical results discussed for the spectrum and for a correlator of the stress-energy tensor. Section 3 extends the study of the spectrum to include the CS term, in both a dimensionally reduced theory and the full (2+1)-dimensional case. A brief summary is given in Sec. 4.

2. SYM$_{2+1}$ theory

2.1. Formulation

The action for $N = 1$ SYM theory in 2+1 dimensions is

$$S = \int dx^+ dx^- dx_\perp \operatorname{tr}\left(-\frac{1}{4}F^{\mu\nu}F_{\mu\nu} + i\bar{\Psi}\gamma^\mu D_\mu\Psi\right), \tag{9}$$

with

$$F_{\mu\nu} = \partial_\mu A_\nu - \partial_\nu A_\mu + ig[A_\mu, A_\nu], \qquad D_\mu = \partial_\mu + ig[A_\mu, \quad]. \tag{10}$$

The fermion field is separated into chiral projections

$$\psi = \frac{1+\gamma^5}{2^{1/4}}\Psi, \qquad \chi = \frac{1-\gamma^5}{2^{1/4}}\Psi, \tag{11}$$

only one of which is dynamical. In light-cone gauge, $A^+ = 0$, with the transverse component of the gauge field $A_\perp$ written as ϕ, the action becomes

$$S = \int dx^+ dx^- dx_\perp \operatorname{tr}\left[\frac{1}{2}(\partial_- A^-)^2 + D_+\phi\partial_-\phi + i\psi D_+\psi + \right.$$

$$\left. +i\chi\partial_-\chi + \frac{i}{\sqrt{2}}\psi D_\perp\phi + \frac{i}{\sqrt{2}}\phi D_\perp\psi\right]. \tag{12}$$

The non-dynamical fields A^- and χ satisfy constraint equations

$$A^- = \frac{g}{\partial_-^2}\left(i[\phi, \partial_-\phi] + 2\psi\psi\right), \quad \chi = -\frac{1}{\sqrt{2}\partial_-}D_\perp\psi, \tag{13}$$

by which they can be eliminated from the action. The dynamical fields are expanded in terms of creation operators

$$\phi_{ij}(0, x^-, x_\perp) = \frac{1}{\sqrt{2\pi L_\perp}} \sum_{n^\perp=-\infty}^{\infty} \int_0^\infty \frac{dk^+}{\sqrt{2k^+}} \left[a_{ij}(k^+, n^\perp)e^{-ik^+x^- - i\frac{2\pi n^\perp}{L_\perp}x_\perp} \right.$$
$$\left. + a_{ji}^\dagger(k^+, n^\perp)e^{ik^+x^- + i\frac{2\pi n^\perp}{L_\perp}x_\perp} \right], \tag{14}$$

$$\psi_{ij}(0, x^-, x_\perp) = \frac{1}{2\sqrt{\pi L_\perp}} \sum_{n^\perp=-\infty}^{\infty} \int_0^\infty dk^+ \left[b_{ij}(k^+, n^\perp)e^{-ik^+x^- - i\frac{2\pi n^\perp}{L_\perp}x_\perp} \right.$$
$$\left. + b_{ji}^\dagger(k^+, n^\perp)e^{ik^+x^- + i\frac{2\pi n^\perp}{L_\perp}x_\perp} \right], \tag{15}$$

where in (2+1) dimensions $n^\perp$ is the only transverse momentum index.

The chiral components of the supercharge are

$$Q_{\text{SYM}}^+ = 2^{1/4} \int dx^- dx_\perp \operatorname{tr}\left[\phi\partial_-\psi - \psi\partial_-\phi\right], \tag{16}$$

$$Q_{\text{SYM}}^- = 2^{3/4} \int dx^- dx_\perp \operatorname{tr}\left[\partial_\perp\phi\psi + g\left(i[\phi, \partial_-\phi] + 2\psi\psi\right)\frac{1}{\partial_-}\psi\right]. \tag{17}$$

They satisfy the supersymmetry algebra

$$\{Q^+, Q^+\} = 2\sqrt{2}P^+, \quad \{Q^-, Q^-\} = 2\sqrt{2}P^-, \quad \{Q^+, Q^-\} = -4P_\perp. \tag{18}$$

This theory has the additional symmetries of transverse parity, P: $a_{ij}(k, n^\perp) \to -a_{ij}(k, -n^\perp)$, $b_{ij}(k, n^\perp) \to b_{ij}(k, -n^\perp)$ and Kutasov's transposition[12] S: $a_{ij}(k, n^\perp) \to -a_{ji}(k, n^\perp)$, $b_{ij}(k, n^\perp) \to -b_{ji}(k, n^\perp)$. These allow the matrix representation to be block diagonalized by an appropriate choice of basis. Eigenstates are labeled by the quantum numbers ± 1 associated with P and S.

2.2. Spectrum and wave functions

The main results for the spectrum of the SYM_{2+1} theory are given in Figs. 1, 11, and 12 of Ref. [6]. They show that the masses squared can be classified according to three main forms of behavior: $1/L_\perp^2$, $g^2 N_c \Lambda_\perp$, and $\Lambda_\perp^2$. In particular, the spectrum as a function of g separates into two bands, one of approximately constant $M^2 L_\perp^2$ and the other growing rapidly.

For states in the lower band, the average number of constituents increases rapidly with g. At $g \simeq 1.5\sqrt{4\pi^3/N_cL_\perp}$ the DLCQ limit of K constituents is saturated. Thus in SYM theory the low-mass states are dominated by Fock states with many constituents, in close correspondence with string theory. However, as a practical matter, the saturation means that the SDLCQ approximation breaks down, and numerical studies in this band must be limited to smaller couplings.

Within the coupling limitation, extrapolations to infinite resolution are easily done for low-mass states. One first considers M^2 as a function of $1/T$ for a sequence of fixed K values. The extrapolations to $T = \infty$ then yield M^2 as a function of $1/K$ alone, to extrapolate to $K = \infty$. The different representatives of continuum eigenstates are disentangled by studying their properties, such as average constituent content and momentum. Of course, the different P and S symmetry sectors are explicitly separated at the start. Typical extrapolations are illustrated in Ref. [6] with plots in Figs. 4-8 and results in Tables II and III.

For the spectrum as a whole there is a curious behavior with respect to the average number of fermion constituents $\langle n_F \rangle$. Calculations for transverse resolution $T = 1$ and longitudinal resolutions $K = 5$ and 6, and for many different coupling strengths, show a gap between $\langle n_F \rangle = 4$ and 6, where no state is found.

Wave functions are also obtained in the diagonalization process. In the analysis of the spectrum they were used to compute various average quantities that helped identify states computed at different resolutions. More of the form of the wave function is revealed in the structure function

$$
g_A(n, n^\perp) = \sum_{q=2}^{K} \sum_{n_1,\ldots,n_q=1}^{K-q} \sum_{n_1^\perp,\ldots,n_q^\perp=-T}^{T} \delta\left(\sum_{i=1}^{q} n_i - K\right) \delta\left(\sum_{j=1}^{q} n_j^\perp\right)
$$

$$
\times \sum_{l=1}^{q} \delta_n^{n_l} \delta_{n^\perp}^{n_l^\perp} \delta_{A_l}^{A} |\psi(n_1, n_1^\perp; \ldots; n_q, n_q^\perp)|^2 , \tag{19}
$$

where A and A_l represent the statistics (bosonic or fermionic) of the probed type and the l-th constituent, respectively, and ψ is a Fock-state wave function. In the lower band the shapes are typically simple and are found to confirm the identification of states at different resolutions. In the upper band, there are complicated shapes with multiple bumps in transverse momentum, such as in Figs. 13 and 14 of Ref. [6].

2.3. *Stress-energy correlator*

Consider the following correlator of the stress-energy component T^{++}:

$$F(x^+, x^-, x^\perp) \equiv \langle 0|T^{++}(x^+, x^-, x^\perp)T^{++}(0,0,0)|0\rangle\,, \qquad (20)$$

at strong coupling,[7] as an example of what one might compare with a supergravity approximation to string theory for small curvature.[13] In the discrete approximation, F can be written as

$$F(x^+, x^-, 0) = \sum_{n,m,s,t} \left(\frac{\pi}{2L_\parallel^2 L_\perp}\right)^2 \qquad (21)$$

$$\times \langle 0|\frac{L_\parallel}{\pi}T(n,m)e^{-iP_{op}^- x^+ - iP^+ x^-}\frac{L_\parallel}{\pi}T(s,t)|0\rangle\,,$$

where

$$\frac{L_\parallel}{\pi}T^{++}(n,m)|0\rangle = \frac{\sqrt{nm}}{2}\text{tr}\left[a_{ij}^\dagger(n,n_\perp)a_{ji}^\dagger(m,m_\perp)\right]|0\rangle \qquad (22)$$

$$+ \frac{(n-m)}{4}\text{tr}\left[b_{ij}^\dagger(n,n_\perp)b_{ji}^\dagger(m,m_\perp)\right]|0\rangle\,.$$

Insertion of a complete set of bound states $|\alpha\rangle$ with light-cone energies $P_\alpha^- = (M_\alpha^2 + P_\perp^2)/P^+$ at resolution K (and therefore $P^+ = \pi K/L_\parallel$) and with total transverse momentum $P_\perp = 2T\pi/L_\perp$ yields

$$\frac{1}{\sqrt{-i}}\left(\frac{x^-}{x^+}\right)^2 F(x^+, x^-, 0) = \sum_\alpha \frac{1}{2(2\pi)^{5/2}}\frac{M_\alpha^{9/2}}{\sqrt{r}}K_{9/2}(M_\alpha r)\frac{|\langle u|\alpha\rangle|^2}{L_\perp K^3|N_u|^2}\,, \qquad (23)$$

with $r^2 = x^+ x^-$, $x_\perp = 0$, and

$$|u\rangle = N_u \frac{L_\parallel}{\pi}\sum_{n,m}\delta_{n+m,K}\delta_{n_\perp+m_\perp,N_\perp}T(n,m)|0\rangle\,. \qquad (24)$$

Here N_u is a normalization factor such that $\langle u|u\rangle = 1$. The sum over the full set of eigenvalues can be avoided by a Lanczos-based technique.[14]

For free particles, $(x^+/x^-)^2 F$ has a $1/r^6$ behavior.[7] In the interacting case, this behavior should be recovered for small r, where the bound states behave as free particles. Because this behavior depends on having a mass spectrum that extends to infinity, the finite resolution of the numerical calculation yields only $1/r^5$; however, the $1/r^6$ behavior is recovered in a careful limiting process.

The behavior for large r is determined by the massless states. Because this theory has zero central charge, there are exactly massless Bogomol'nyi–Prasad–Sommerfeld (BPS) states at any coupling. However, their wave functions remain sensitive to the coupling, and, at a particular (resolution dependent) value of g, the correlator is exactly zero in the large-r limit. The associated 'critical' value of g increases in proportion to the square root of the transverse resolution T.[7]

3. SYM-CS theory

The following CS term can be added to the Lagrangian:

$$L_{\text{CS}} = \frac{\kappa}{2}\epsilon^{\mu\nu\lambda}\left(A_\mu\partial_\nu A_\lambda + \frac{2i}{3}gA_\mu A_\nu A_\lambda\right) + \kappa\bar{\Psi}\Psi. \tag{25}$$

This induces an additional term in the supercharge

$$\kappa Q_{\text{CS}}^- \equiv -2^{3/4}\kappa\int dx^-\,\partial_-\phi\frac{1}{\partial_-}\psi, \tag{26}$$

which generates in P^- terms proportional to κ^2 that act like a constituent mass squared. The presence of an effective mass reduces the tendency for low-mass states to be composed of high-multiplicity Fock states. This creates a theory in which the eigenstates are more likely to be QCD-like, *i.e.* valence dominated, and improves the applicability of the SDLCQ approximation to a greater range of couplings.[15,16,8]

The dominance of the valence state is most prominent in the dimensionally reduced theory.[15] This (1+1)-dimensional theory is obtained by requiring the fields to be constant in the transverse direction and replacing $\partial_\perp$ by zero in the full supercharge $Q_{\text{SYM}}^- + \kappa Q_{\text{CS}}^-$. The SYM contribution is then proportional to g. Figure 3 of Ref. [15] illustrates the dramatic reduction in the average number of constituents as the ratio κ/g is increased. Also, there are states for which the mass is nearly independent of g at fixed κ. These are identified as approximate BPS states and are the reflection of the massless BPS states in the underlying SYM theory.[16] This behavior can be seen in Fig. 1 of Ref. [16].

Similar anomalously light states appear in the full (2+1)-dimensional theory.[8] The bulk of the spectrum is driven to large M^2 values as g is increased, but one or more states remain at low values. The presence of the transverse degree of freedom makes this more difficult to disentangle numerically, because the matrices are larger and because one must consider the transverse resolution limit. Also, the eigenstates are less valence-dominated at stronger YM coupling, which makes the SDLCQ approximation less useful. However, structure functions have been extracted at intermediate coupling to show that the approximate BPS states have a distinctly flat dependence in longitudinal momentum. Figure 3b of Ref. [8] gives an example of this behavior.

4. Summary

This work shows that one can compute spectra, wave functions, and matrix elements nonperturbatively in supersymmetric theories. The introduction of a Chern–Simons term brings an effective constituent mass which has the effect of reducing the tendency of SYM theory to form stringy, low-mass states with many constituents. Instead, the lowest-mass states tend to be dominated by their valence Fock state. The massless BPS states of SYM theory survive in SYM-CS theory as states with masses nearly independent of the YM coupling.

A number of interesting issues remain to be explored. They include theories in 3+1 dimensions, matter in the fundamental representation,[17] and supersymmetry breaking. All of these are important for making contact with QCD.

Acknowledgments

This talk was based on work done in collaboration with S.S. Pinsky and U. Trittmann and supported in part by the U.S. Department of Energy and by grants of computing time from the Minnesota Supercomputing Institute.

References

1. Y. Matsumura, N. Sakai, and T. Sakai, *Phys. Rev.* **D52**, 2446 (1995).
2. O. Lunin and S. Pinsky, *AIP Conf. Proc.* **494**, 140 (1999), hep-th/9910222.
3. H.-C. Pauli and S.J. Brodsky, *Phys. Rev.* **D32**, 1993 (1985); **D32**, 2001 (1985).
4. S.J. Brodsky, H.-C. Pauli, and S.S. Pinsky, *Phys. Rep.* **301**, 299 (1997), hep-ph/9705477.
5. P. Haney, J.R. Hiller, O. Lunin, S. Pinsky, and U. Trittmann, *Phys. Rev.* **D62**, 075002 (2000), hep-th/9911243.
6. J.R. Hiller, S. Pinsky, and U. Trittmann, *Phys. Rev.* **D64**, 105027 (2001), hep-th/0106193.
7. J.R. Hiller, S. Pinsky, and U. Trittmann, *Phys. Rev.* **D63**, 105017 (2001), hep-th/0101120.
8. J.R. Hiller, S.S. Pinsky, and U. Trittmann, to appear in *Phys. Lett.* **B**, hep-th/0206197.
9. G.V. Dunne, "Aspects of Chern–Simons Theory," Lectures at the 1998 Les Houches NATO Advanced Studies Institute, Session LXIX, *Topological Aspects of Low Dimensional Systems*, edited by A. Comtet *et al.*, pp. 177-263, (Springer–Verlag, Berlin, 2000), hep-th/9902115.
10. P.A.M. Dirac, *Rev. Mod. Phys.* **21**, 392 (1949).
11. C. Lanczos, *J. Res. Nat. Bur. Stand.* **45**, 255 (1950); J. Cullum and R.A. Willoughby, *Lanczos Algorithms for Large Symmetric Eigenvalue Computations* (Birkhauser, Boston, 1985), Vol. I and II.
12. D. Kutasov, *Nucl. Phys.* **B414**, 33 (1994).

467

13. J. Maldacena, *Adv. Theor. Math. Phys.* **2**, 231 (1998).
14. J.R. Hiller, O. Lunin, S. Pinsky, and U. Trittmann, *Phys. Lett.* **B482**, 409 (2000), hep-th/0003249.
15. J.R. Hiller, S.S. Pinsky, and U. Trittmann, *Phys. Rev.* **D65**, 085046 (2002), hep-th/0112151.
16. J.R. Hiller, S.S. Pinsky, and U. Trittmann, submitted for publication, hep-th/0203162.
17. Some work on fundamental matter has been done in 1+1 dimensions; see O. Lunin and S. Pinsky, *Phys. Rev.* **D63**, 045019 (2001).

TESTING ADS/CFT CORRESPONDENCE WITH WILSON LOOPS

K. ZAREMBO

Department of Theoretical Physics
Uppsala University
Box 803, SE-751 08 Uppsala, Sweden
E-mail: Konstantin.Zarembo@teorfys.uu.se

ITEP, B. Cheremushkinskaya 25, 117259 Moscow, Russia

Wilson loops in the context of AdS/CFT correspondence are reviewed with emphasis on Wilson loops protected or partially protected by supersymmetry.

1. Introduction

It is probably not an exaggeration to say that the most remarkable aspect of supersymmetry is possibility to obtain exact results without doing exact calculations. Supersymmetry leads to cancellations of quantum corrections which sometimes are strong enough to protect certain quantities from being renormalized. The protected quantities then depend only on a small and controllable part of otherwise complicated dynamics. One of the best known results of this kind is the exact β-function in a generic supersymmetric Yang-Mills (SYM) theory [1]:

$$\beta(\alpha) = -\frac{\alpha^2}{2\pi} \frac{3T(G) - \sum_I T(R_I)(1 - \gamma_I)}{1 - \frac{\alpha}{2\pi} T(G)}, \qquad (1)$$

where $T(G)$ and $T(R_I)$ are the usual group factors of gauge and matter superfields and γ_I are anomalous dimensions of the latter.

Using this exact result, it is easy to show that the β function of maximally supersymmetric $N = 4$ SYM theory is equal to zero. In the $N = 1$ language, the field content of $N = 4$ SYM consists of gauge supermultiplet and three matter supermultiplets with zero anomalous dimensions. All fields are in the adjoint representation, so $T(R_I) = T(G)$. According to (1), the β function vanishes, and $N = 4$ SYM is a completely finite, superconformal field theory [2,3,4]. Another remarkable property of this theory is it's holographic duality to string theory in the background which is the direct product of Anti-de-Sitter space and the five-sphere: $AdS_5 \times S^5$ [5,6,7,8]. The finiteness and the superconformal invariance play an important role in the AdS/CFT correspondence.

If AdS radius is rescaled to unity, the line element in $AdS_5 \times S^5$ becomes dimensionless, and the metric written with coordinates (x^μ, z^i), $\mu = 1 \ldots 4$, $i = 4 \ldots 9$ takes the form:

$$ds^2 = \frac{dx^\mu dx^\mu + dy^i dy^i}{y^2}. \tag{2}$$

The unit 6-vector z^i/z parameterizes S^5 and x^μ, z are the coordinates of AdS^5. The AdS^5 and S^5 have equal radii of curvature. The boundary of the space is at $z = 0$ and the AdS horizon is at $z = \infty$. The string tension in the AdS units is also dimensionless and, according to the AdS/CFT dictionary, is related to the Yang-Mills coupling as follows:

$$T = \sqrt{g_{YM}^2 N}/2\pi \tag{3}$$

The sigma model coupling is the inverse of that. The string coupling g_s and the Yang-Mills coupling g_{YM} are related by

$$g_s = 4\pi g_{YM}^2. \tag{4}$$

With these identifications, the string theory and the SYM theory are conjectured to be exactly equivalent. This equivalence is a remarkable and extremely non-trivial fact. Weaker and computationally more useful versions of the AdS/CFT duality are obtained by taking limits (table 1). The 't Hooft large-N limit of the gauge theory [9] coincides with the classical string theory ($g_s \to 0$) on the $AdS^5 \times S^5$ background. This establishes a long-suspected equivalence between planar diagrams of the large-N limit and free strings. Even though strings do not interact in the large-N limit, their quantization in the AdS background remains an open problem and only semiclassical limit of large tension is well understood. Since tension is proportional to the square root of the 't Hooft coupling, strings become semiclassical in the strong-coupling limit of SYM theory, which is opposite to the perturbative weak-coupling regime. There is a lot of indications that observables on SYM theory smoothly interpolate between these two regimes, but the direct comparison of field theory with strings is possible only in few cases.

I will review Wilson loops in the context of AdS/CFT correspondence with most emphasis on the cases when supersymmetry allows to get exact results in field theory, which can then be directly compared to predictions of string theory at strong coupling.

Table 1. Different limits of the AdS/CFT correspondence.

$N = 4$ SYM	String theory in AdS$^5 \times$ S^5	
Yang-Mills coupling: g_{YM}	String coupling: g_s	
Number of colors: N	String tension: T	
Level 1: Exact equivalence		
$g_s = g_{YM}^2/4\pi, \quad T = \sqrt{g_{YM}^2 N/2\pi}$		
Level 2: Equivalence in the 't Hooft limit		
$N \to \infty, \quad \lambda = g_{YM}^2 N$-fixed	$g_s \to 0, \quad T$-fixed	
(planar limit)	(non-interacting strings)	
Level 3: Equivalence at strong coupling		
$N \to \infty, \quad \lambda \gg 1$	$g_s \to 0, \quad T \gg 1$	
	(classical supergravity)	

2. Supersymmetric Wilson loops

The members of $N = 4$ multiplet are gauge fields A_μ, six scalars Φ_i ($i = 4 \ldots 9$) and four Majorana fermions Ψ^A. It is convenient to put fermions into a single Majorana-Weyl spinor of $Spin(9,1)$[a]. The Euclidean action then takes the following form:

$$S = \frac{1}{g^2} \int d^4x \, \mathrm{tr} \left\{ \frac{1}{2} F_{\mu\nu}^2 + (D_\mu \Phi_i)^2 - \frac{1}{2} [\Phi_i, \Phi_j]^2 + \bar\Psi \Gamma^\mu D_\mu \Psi \right.$$
$$\left. + i\bar\Psi \Gamma^i [\Phi_i, \Psi] \right\}, \tag{5}$$

where $\Gamma^M = (\Gamma^\mu, \Gamma^i)$ are ten-dimensional Dirac matrices. The action is invariant under supersymmetry transformations, whose action on the bosonic fields is

$$\delta_\epsilon A_\mu = \bar\Psi \Gamma^\mu \epsilon,$$
$$\delta_\epsilon \Phi_i = \bar\Psi \Gamma^i \epsilon, \tag{6}$$

where the parameter of transformation ϵ is a ten-dimensional Majorana-Weyl spinor.

The Wilson loop operator which is dual to the string in Anti-de-Sitter space [10] is a hybrid of the usual non-Abelian phase factor and the scalar loop of [11]:

$$W(C, \theta) = \frac{1}{N} \, \mathrm{tr} \, \mathrm{P} \exp \int ds \, \left(iA_\mu(x)\dot{x}^\mu + \Phi_i(x)\theta^i |\dot{x}| \right). \tag{7}$$

---

[a] Working with Minkowski spinors in Euclidean space is somewhat awkward. Since most of the results will not depend on the signature of the metric, the Dirac matrices will be assumed to anti-commute on the Euclidean metric in what follows.

Here, $x^\mu(s)$ parameterizes contour C in $\mathbb{R}^4$ and θ^i is a unit six-vector: $\theta^i \theta^i = 1$, which also forms a closed contour in S^5.

The condition for the Wilson loop to preserve a part of the supersymmetry is

$$\left(i\Gamma^\mu \dot{x}^\mu + \Gamma^i \theta^i |\dot{x}|\right) \epsilon = 0. \tag{8}$$

Since the linear combination of Dirac matrices $i\Gamma^\mu \dot{x}^\mu + \Gamma^i \theta^i |\dot{x}|$ squares to zero, equation (8) has eight independent solutions for any given s. In general, these solutions will depend on s, so an arbitrary Wilson loop is only locally supersymmetric. Though local supersymmetry is not a symmetry of the action, it has important implications. For instance, it insures that Wilson loop correlators for smooth contours do not suffer from UV divergences and need not be renormalized. The requirement that ϵ is s-independent is a constraint on $x^\mu(s)$ and $\theta^i(s)$. The number of linearly independent ϵ's that satisfy eq. (8) determines the number of conserved supercharges.

The simplest supersymmetric operator is the Wilson line with constant $\dot{x}^\mu$ and θ^i. It preserves 1/2 of the supersymmetry [12]. Wilson loops with smaller degree of supersymmetry can be constructed by relating the space-time and the S^5 contours [13]:

$$\theta^i = M^i_\mu \frac{\dot{x}^\mu}{|\dot{x}|}, \tag{9}$$

where M^i_μ is a projector on a four-plane in six-dimensional space:

$$M^i_\mu M^i_\nu = \delta_{\mu\nu}. \tag{10}$$

A particular form of M^i_μ is not important because of $SO(4) \times SO(6)$ global symmetry of $N = 4$ SYM. With this choice of θ^i, the Wilson loop operator becomes

$$W_s(C) = \frac{1}{N} \operatorname{tr} \mathrm{P} \exp \oint_C dx^\mu \left(i A_\mu + M^i_\mu \Phi_i\right). \tag{11}$$

The condition of supersymmetry for this operator, unlike (8), is independent of the position on the contour and always has at least one solution. Generally, Wilson loop (11) preserves 1/16 of $N = 4$ supersymmetry, but if contour C lies in a three-dimensional hyperplane (if $\dot{x}^0$ is identically zero, for instance), the supersymmetry is enhanced. Three-dimensional loops commute with two supercharges. Planar loops preserve yet larger amount of supersymmetry: if contour C lies in a two-dimensional plane, the corresponding Wilson loop is 1/4 BPS.

3. Exact results

If the amount of supersymmetry is sufficiently large, supersymmetric Wilson loops do not receive quantum corrections. 1/2 BPS states are always protected from renormalization, which means that an expectation value of the straight Wilson line is equal to one. It turns out that 1/4 BPS planar Wilson loops also do not renormalize. Wilson loops with less supersymmetry seem to get quantum corrections. Actually, loop corrections to all supersymmetric Wilson loops cancel to rather high order, but string theory computation at strong coupling indicates that non-planar Wilson non-trivially depend on the 't Hooft coupling [13]. The amount of supersymmetry and non-renormalization properties of various Wilson loops are summarized in table 2.

Table 2. The amount of supersymmetry and (non)-renormalization of supersymmetric Wilson loops.

Dimensionality	Amount of supersymmetry	Non-renormalization
4D	1/16	–
3D	1/8	–
2D	1/4	$\checkmark$
1D	1/2	$\checkmark$

Conformal invariance of $N = 4$ SYM implies that expectation values of Wilson loops for contours related by a conformal transformation should be the same. This statement is correct, unless one of the contours goes to infinity. Then, the conformal transformation leads to an 'anomaly'. The straight line and the circle is an example of contours, which are related by conformal transformation, but whose expectation values are different. There expectation value of the line is exactly equal to one, but the circle is renormalized by quantum corrections. The origin of this 'conformal anomaly' is explained in [14,15]. The circular Wilson loop is invariant under a linear combination of supersymmetry and superconformal charges [16], which leads to cancellation of some, but not all quantum corrections. It was argued [14] that all anomalous contributions to the vev of circular Wilson loop are captured by diagrams without internal vertices. These diagrams can be explicitly resummed. The details of the calculations can be found elsewhere [17,18]. The result (in the planar limit) is

$$\langle W(\text{circle}) \rangle = \frac{2}{\sqrt{\lambda}} I_1 \left(\sqrt{\lambda} \right). \tag{12}$$

In accord with expectations from string theory, the square root of the 't Hooft coupling explicitly appears here, though the expectation value is a power series in λ.

If the size of the contour (the circle in our case) is small compared to some other scale in the problem, Wilson loop can be expanded in local operators. The coefficients of the operator product expansion can be extracted from the normalized two-point correlation function:

$$\frac{\langle W(C) O^A(L) \rangle_c}{\langle W(C) \rangle} = C_A \frac{R^{\Delta_A}}{L^{2\Delta_A}} + \dots \tag{13}$$

where R is the radius of the loop and $L \gg R$. The dimensionless coefficient C_A is the weight with which a conformal primary $O^A(0)$ enters the OPE of the Wilson loop $W(C)$. The omitted terms correspond to descendants and are of higher order in R/L.

The OPE coefficients of the circular Wilson loop for chiral primary operators (CPOs) can again be computed in the free-field approximation. This particularly interesting set of operators are traces of the scalar fields,

$$O_k^I = \frac{(8\pi^2)^{k/2}}{\sqrt{k}\lambda^{k/2}} C_{i_1 \dots i_k}^I \, \mathrm{tr}\, \Phi^{i_1} \dots \Phi^{i_k}, \tag{14}$$

where $C_{i_1 \dots i_k}^I$ are totally symmetric traceless tensors which are normalized as

$$C_{i_1 \dots i_k}^I C_{i_1 \dots i_k}^J = \delta^{IJ}. \tag{15}$$

Here, we are following the convention of refs. [19,20]. The sum of the planar diagrams without internal vertices for OPE coefficients can be expressed in terms of modified Bessel functions [21]:

$$\frac{\langle W(\text{circle}) O_k^I \rangle}{\langle W(\text{circle}) \rangle} = 2^{k/2-1} \sqrt{k\lambda} \frac{I_k\left(\sqrt{\lambda}\right)}{I_1\left(\sqrt{\lambda}\right)} \frac{R^k}{L^{2k}} Y^I(\theta). \tag{16}$$

We expect that this expression is exact in the large N limit. The symmetry factor $Y^I(\theta)$ is the spherical function associated with each CPO:

$$Y^I(\theta) = C_{i_1 \dots i_k}^I \theta^{i_1} \dots \theta^{i_k}. \tag{17}$$

The expectation values (12) and (16) were obtained by resummation of perturbative series, but can be analytically continued to strong coupling, where they can be compared to AdS/CFT predictions of string theory.

4. Wilson loops in string theory

The expectation value of the Wilson loop is the string partition function 10,22:

$$\langle W(C,\theta) \rangle = \int_{\text{reg}} DX^M D\vartheta^\alpha Dh_{ab}$$

$$\times \exp\left(-\frac{\sqrt{\lambda}}{4\pi} \int_D d^2\sigma \sqrt{h} h^{ab} G_{MN} \partial_a X^M \partial_b X^N + \text{ferm.} \right) (18)$$

The sigma-model metric is defined by the line element (2). The string world sheet extends to the boundary of AdS space, where it terminates on the contour C. In other words, the sigma-model path integral is supplemented by the boundary conditions:

$$X^\mu|_{\partial D} = x^\mu(s), \qquad Z|_{\partial D} = \varepsilon, \qquad \Theta^i|_{\partial D} = \theta^i(s). \qquad (19)$$

Θ^i parameterizes a position of the world sheet in S^5. The regularization parameter ε cuts off divergences which arise because of the $1/z^2$ singularity of the metric at the boundary. At the end, ε should be sent to zero. If the string partition function is appropriately defined, the divergences appear only in the intermediate calculations and eventually cancel. The correct definition of the partition function involves the Legendre transform in Z 12. Explicit implementation of the Legendre transform is somewhat cumbersome, but, fortunately, in the semiclassical approximation, the Legendre transform amounts in dropping $1/\varepsilon$ divergences whenever they appear.

It is not known how to solve the $AdS_5 \times S^5$ sigma model exactly. The only simplification occurs at large 't Hooft coupling, when the sigma model becomes weakly coupled, and the partition functions can be computed in the saddle-point approximation. Minima of the string action correspond to minimal surfaces in $AdS_5 \times S^5$, whose boundary is the contour C. The action at the saddle point is the area of the minimal surface:

$$A(C) = \int d^2\sigma \sqrt{\det_{ab} G_{MN} \partial_a X_{\text{cl}}^M \partial_b X_{\text{cl}}^N}, \qquad (20)$$

which should be regularized by subtraction of the boundary divergence. The α' expansion of the sigma model, obtained by expanding around the classical solution and integrating out fluctuations, yields $1/\sqrt{\lambda}$ expansion of the Wilson loop expectation value. There are additional complications if the classical solution depends on parameters (has moduli). The moduli give rise to zero mode fluctuations which should be separated by introducing collective coordinates. Each moduli integration is accompanied by a factor of $\alpha'^{1/2}$, which, in the present case, should be identified with $\lambda^{1/4}$. The gauge fixing of the world sheet diffeomorphism invariance also produces a non-trivial factor, because the usual conformal gauge leaves residual three-parametric gauge freedom in the disk partition function [23]. The residual gauge symmetries give a factor of $\lambda^{-1/4}$ each [14]. Consequently, the semi-classical partition function for the Wilson loop vev has the following general form:

$$\langle W(C,\theta)\rangle = \text{const} \cdot \lambda^{(N_{z.m.}-3)/4} \exp\left(-\frac{\sqrt{\lambda}}{2\pi}A(C)\right) + \ldots, \qquad (21)$$

where $N_{z.m.}$ is the number of zero modes, or the number of moduli in the classical solution X_{cl}^M. An overall constant comes from the integration over non-zero-mode fluctuations. Corrections to the leading semiclassical approximation are power series in $1/\sqrt{\lambda}$.

Let us first consider $1/4$ supersymmetric Wilson loops. Their expectation values should not receive quantum corrections and, consequently, should be equal to one. That implies that corresponding minimal surfaces should have zero area. The area indeed vanishes for anti-parallel lines [10] and for supersymmetric circular loop [13], because of the cancellations between the S^5 and the AdS_5 contributions. It is likely that the area vanishes for all planar supersymmetric contours, but general proof is lacking. It is relatively easy to see that the zero mode prefactor turns to one for $1/4$ supersymmetric loops, because minimal surfaces then depend on exactly three parameters and $N_{z.m.} = 3$. Indeed, the S^5 part of $1/4$ BPS loops is a projection of the tangent vector of the space-time contour. For planar loops, this projection is always an equatorial circle of S^5. Evidently, minimal surface which ends on the equatorial circle covers some two-dimensional hemi-sphere. We can parameterize S^5 as $(\cos\psi\cos\varphi, \cos\psi\sin\varphi, \mathbf{n}\sin\psi)$, where $\mathbf{n}$ is a four-dimensional unit vector. Then, the minimal surface will extend in the ψ and φ directions, while we can choose $\mathbf{n}$ so that it is constant along the minimal surface. Different $\mathbf{n}$'s will correspond to different minimal surfaces, so the minimum of the string action is degenerate. The moduli space of different minimal surfaces is a three-sphere, consequently, the number of zero modes is three, as predicted by supersymmetry.

Let us now explore the strong coupling limit of the expectation value of the circular loop (12):

$$\langle W(\text{circle})\rangle = \sqrt{\frac{2}{\pi}}\,\lambda^{-3/4}\,e^{\sqrt{\lambda}} \qquad (\lambda \to \infty). \qquad (22)$$

The string partition function has exactly the same form. Computing the area of minimal surface associated with the circle, one finds -2π [12,20], in complete agreement with the exponent in (22)! Also the zero-mode factor $\lambda^{-3/4}$ comes out right, since the moduli space shrinks to zero size for a minimal surface which sits at one point in S^5. The exact expression (12) smoothly interpolates between perturbative series in λ and the strong coupling regime, where the natural expansion parameter is $1/\sqrt{\lambda}$. This latter expansion is equivalent to the α' expansion of the world-sheet sigma model.

The method to compute OPE coefficients of Wilson loops in AdS/CFT correspondence was proposed in [20]. For the two-point correlators of the circular loop with CPOs, the authors of [20] got

$$\frac{\langle W(C) O_k^I \rangle}{\langle W(C) \rangle} = 2^{k/2-1} \sqrt{k\lambda} \, \frac{R^k}{L^{2k}} Y^I(\theta) \quad (\lambda \to \infty). \tag{23}$$

If we take the strong-coupling limit of the field-theory correlation function (16), we arrive at the same expression. Again, the resummed perturbative expression smoothly interpolates between the weak and the strong coupling regimes.

5. Discussion

In some cases, Wilson loop expectation values are protected by supersymmetry from receiving quantum corrections. While at weak coupling that leads to cancellations between different diagrams, absence of corrections at strong coupling has non-trivial implications for minimal surfaces in $AdS_5 \times S^5$. Explicit calculation of the area and zero-mode counting indeed agree with non-renormalization of 1/4 BPS Wilson loops. Quantum corrections to superconformally invariant but not supersymmetric Wilson loops do not cancel, but seem to be under control, which allows to trace the dependence on the 't Hooft coupling from the perturbative regime to the strong coupling asymptotics described by string theory.

The strict strong coupling limit of AdS/CFT correspondence, usually associated with AdS supergravity, is by now well understood, but not much is known about genuine string theory in $AdS_5 \times S^5$. In fact, Wilson loops allow to go beyond supergravity and to probe stringy degrees of freedom even in the strict strong coupling limit. In that case, going beyond semiclassical approximation also proves very difficult. One way to learn something about strings in AdS (which reduces to some sort of semiclassical approximation) is to consider states with large quantum numbers, R-charge or spin [24,25]. It would be interesting to study Wilson loops in this latter context. For instance, OPE coefficients of a Wilson loop for operators of large R-charge may have some semiclassical string description.

Acknowledgments

I am grateful to Gordon Semenoff for collaboration on most of the work reported here. The work was supported by STINT grant IG 2001-062 and by Royal Swedish Academy of Sciences.

References

1. V. A. Novikov, M. A. Shifman, A. I. Vainshtein and V. I. Zakharov, Nucl. Phys. B **229**, 381 (1983).
2. S. Mandelstam, Nucl. Phys. B **213**, 149 (1983).
3. L. Brink, O. Lindgren and B. E. Nilsson, Phys. Lett. B **123**, 323 (1983).
4. P. S. Howe, K. S. Stelle and P. C. West, Phys. Lett. B **124**, 55 (1983).
5. J. Maldacena, Adv. Theor. Math. Phys. **2**, 231 (1998) [Int. J. Theor. Phys. **38**, 1113 (1998)] [hep-th/9711200].
6. S. S. Gubser, I. R. Klebanov and A. M. Polyakov, Phys. Lett. B **428** (1998) 105 [arXiv:hep-th/9802109].
7. E. Witten, Adv. Theor. Math. Phys. **2**, 253 (1998) [hep-th/9802150].
8. O. Aharony, S. S. Gubser, J. Maldacena, H. Ooguri and Y. Oz, Phys. Rept. **323** (2000) 183 [arXiv:hep-th/9905111].
9. G. 't Hooft, Nucl. Phys. B **72**, 461 (1974).
10. J. Maldacena, Phys. Rev. Lett. **80**, 4859 (1998) [arXiv:hep-th/9803002].
11. Y. M. Makeenko, Phys. Lett. B **212** (1988) 221.
12. N. Drukker, D. J. Gross and H. Ooguri, Phys. Rev. D **60**, 125006 (1999) [arXiv:hep-th/9904191].
13. K. Zarembo, hep-th/0205160.
14. N. Drukker and D. J. Gross, J. Math. Phys. **42** (2001) 2896 [arXiv:hep-th/0010274].
15. G. W. Semenoff and K. Zarembo, Nucl. Phys. Proc. Suppl. **108** (2002) 106 [short version of arXiv:hep-th/0202156].
16. M. Bianchi, M. B. Green and S. Kovacs, JHEP **0204** (2002) 040 [arXiv:hep-th/0202003].
17. J. K. Erickson, G. W. Semenoff and K. Zarembo, Nucl. Phys. B **582**, 155 (2000) [arXiv:hep-th/0003055].
18. K. Zarembo, "Wilson Loops in $N = 4$ SYM Theory", in: *Continuous Advances in QCD 2000*, ed. M.B. Voloshin (World Scientific, 2001) p. 42.
19. S. Lee, S. Minwalla, M. Rangamani and N. Seiberg, Adv. Theor. Math. Phys. **2**, 697 (1998) [hep-th/9806074].
20. D. Berenstein, R. Corrado, W. Fischler and J. Maldacena, Phys. Rev. D **59**, 105023 (1999) [hep-th/9809188].
21. G. W. Semenoff and K. Zarembo, Nucl. Phys. B **616** (2001) 34 [arXiv:hep-th/0106015].
22. S. J. Rey and J. Yee, Eur. Phys. J. C **22** (2001) 379 [arXiv:hep-th/9803001].
23. O. Alvarez, Nucl. Phys. B **216** (1983) 125.
24. D. Berenstein, J. M. Maldacena and H. Nastase, JHEP **0204**, 013 (2002) [arXiv:hep-th/0202021].
25. S. S. Gubser, I. R. Klebanov and A. M. Polyakov, arXiv:hep-th/0204051.

PERTURBATIVE – NONPERTURBATIVE CONNECTION IN QUANTUM MECHANICS AND FIELD THEORY

GERALD V. DUNNE

Department of Physics, University of Connecticut, Storrs CT 06269, USA

On the occasion of this ArkadyFest, celebrating Arkady Vainshtein's 60th birthday, I review some selected aspects of the connection between perturbative and nonperturbative physics, a subject to which Arkady has made many important contributions. I first review this connection in quantum mechanics, which was the subject of Arkady's very first paper. Then I discuss this issue in relation to effective actions in field theory, which also touches on Arkady's work on operator product expansions. Finally, I conclude with a discussion of a special quantum mechanical system, a quasi-exactly solvable model with energy-reflection duality, which exhibits an explicit duality between the perturbative and nonperturbative sectors, without invoking supersymmetry.

1. Divergence of perturbation theory

> "The majority of nontrivial theories are seemingly unstable at some phase of the coupling constant, which leads to the asymptotic nature of the perturbative series."
> A. Vainshtein, 1964 [1]

In this talk I review some aspects of the historical development of the connection between perturbative and nonperturbative physics. It is particularly appropriate to look back on this subject on the occasion of Arkady Vainshtein's 60th birthday, because this has been a central theme of many of Arkady's great contributions to theoretical physics. In fact, in his very first physics paper [1], now almost 40 years ago, Arkady made a fundamental contribution to this subject. This paper was published as a Novosibirsk report and so has not been widely circulated, especially in the West. For this ArkadyFest, Misha Shifman has made an English translation of this paper, and both the original Russian and the translation are reprinted in these Proceedings.

The physical realization of the possibility of the divergence of perturbation theory is usually traced back to a profound and influential paper by Dyson [2], in which he argued that QED perturbation theory should be divergent. Dyson's argument goes like this: a physical quantity in QED, computed using the standard rules of renormalized QED perturbation theory, is expressed as a perturbative series in powers of the fine structure constant, $\alpha = \frac{e^2}{4\pi}$:

$$F(e^2) = c_0 + c_2 e^2 + c_4 e^4 + \ldots \tag{1}$$

Now, suppose that this perturbative expression is convergent. This means that in some small disc-like neighborhood of the origin, $F(e^2)$ has a well-defined convergent approximation. In particular, this means that within this region, $F(-e^2)$ also has a well-defined convergent expansion. Dyson then argued on physical grounds that this cannot be the case, because if $e^2 < 0$ the vacuum will be unstable. This, he argued, is because with $e^2 < 0$ like charges attract and it will be energetically favorable for the vacuum to produce $e^+ e^-$ pairs which coalesce into like-charge blobs, a runaway process that leads to an unstable state:

"Thus every physical state is unstable against the spontaneous creation of large numbers of particles. Further, a system once in a pathological state will not remain steady; there will be a rapid creation of more and more particles, an explosive disintegration of the vacuum by spontaneous polarization."
F. J. Dyson, 1952 [2]

The standard QED perturbation theory formalism breaks down in such an unstable vacuum, which Dyson argued means that $F(-e^2)$ cannot be well-defined, and so the original perturbative expansion (1) cannot have been convergent.

Dyson's argument captures beautifully an essential piece of physics, namely the deep connection between instability and the divergence of perturbation theory. The argument is not mathematically rigorous, and does not *prove* one way or another whether QED perturbation theory is convergent or divergent, or analytic or nonanalytic. However, it is nevertheless very suggestive, and has motivated many subsequent studies in both quantum mechanics and quantum field theory.

At roughly the same time, C. A. Hurst [3] and W. Thirring [4] (see also A. Petermann [5]) showed by explicit computation that perturbation theory diverges in scalar ϕ^3 theory. Both Hurst and Thirring found lower bounds on the contribution of Feynman graphs at a given order of perturbation theory, and showed that these lower bounds were themselves factorially divergent. Hurst used the parametric representation of an irreducible, renormalized and finite ϕ^3 Feynman graph, to show that the magnitude of this graph was bounded below:

$$|I| \geq \frac{n^{-n+3/2-E/2} \, e^{2n-2} \, \pi^{-(n-E+5)/2}}{(2\pi)^F \, 2^{n+1/2-3E/2} \, 3^{4n-7/2} \, \gamma^{(n+E-4)/2}} \, \lambda^n \qquad (2)$$

Here n is the loop order, λ is the cubic coupling constant, E is the number of external lines, $F = \frac{1}{2}(3n - E)$ is the number of internal lines, and γ is a constant depending on the external momenta. This lower bound is found by clever rearrangements of the parametric representation, together with the identity

$$\prod_{i=1}^{F} \left(\frac{1}{p_i^2 + \kappa^2} \right) \geq \frac{F^F}{\left(\sum_{i=1}^{F} p_i^2 + F \kappa^2 \right)^F} \qquad (3)$$

The second important piece of the argument is to show that there are no sign cancellations which would prevent this lower bound from a typical graph from being used to obtain a lower bound on the total contribution at a given order. This requires some technical caveats – for example, for a two-point function one requires $p^2 < m^2$. The final piece of Hurst's argument is the fact that the number of distinct Feynman diagrams at n^{th} loop order grows like $(\frac{n}{2})! \, n!$.

Together, the lower bound (2), the nonalternation of the sign, and the rapid growth of the number of graphs, lead to a lower bound for the total contribution at n-loop order (with E external lines):

$$\sum I \geq C^n \, n^{n/2+5/2-E/2} \, \lambda^n \qquad (4)$$

Here C is a finite constant, independent of n. Therefore, Hurst concluded that in $\lambda\phi^3$ theory, perturbation theory diverges for any coupling λ. He also suggests that a similar argument should hold for $\lambda\phi^4$ theory, and comments:

> "If it be granted that the perturbation expansion does not lead to a convergent series in the coupling constant for all theories which can be renormalized, at least, then a reconciliation is needed between this and the excellent agreement found in electrodynamics between experimental results and low-order calculations. It is suggested that this agreement is due to the fact that the S-matrix expansion is to be interpreted as an asymptotic expansion in the fine-structure constant ..."
> C. A. Hurst, 1952 [3]

Thirring's argument [4] was similar in spirit, although he concentrated on the ϕ^3 self-energy diagram. Thirring found a set of graphs that were simple enough that their contribution could be estimated and bounded below, while plentiful enough that they made a divergent contribution to the perturbative series. He noted that the proof relied essentially on the fact that certain terms always had the same sign, and traced this fact to the hermiticity of the interaction. He found the following (weaker) lower bound, valid for $p^2 < m^2$:

$$\Delta(p^2) \geq \sum_n C(p^2) \left(\frac{\lambda e}{4\pi m 3^{5/2}} \right)^n \frac{(\frac{n}{2} - 2)!}{n^2} \tag{5}$$

Thirring concluded that there was no convergence for any λ. His final conclusion was rather pessimistic:

> "To sum up, one can say that the chances for quantized fields to become a mathematically consistent theory are rather slender."
> W. Thirring, 1953 [4]

These results of Dyson, Hurst and Thirring, provide the backdrop for Arkady's first paper [1], "Decaying systems and divergence of perturbation theory", written as a young student beginning his PhD at Novosibirsk. I encourage the reader to read Arkady's paper – it is simple but deep. I paraphrase the argument here. The main contribution of his paper was to provide a quantitative statement of the relation between the divergence of perturbation theory and the unstable nature of the ground state in ϕ^3 theory.

Motivated by the earlier results for ϕ^3 theory (in 4 dimensions), Arkady had the clever idea to consider ϕ^3 theory in $0 + 1$ dimensions, which is just quantum mechanics. Here it is natural to consider the hamiltonian

$$H = \frac{1}{2}\dot{\phi}^2 + \frac{1}{2}m^2\phi^2 - \lambda\phi^3 \tag{6}$$

and the ground state $|\psi\rangle$ such that $H|\psi\rangle = E|\psi\rangle$. To make connection

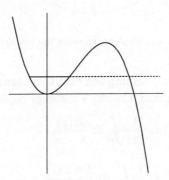

Figure 1. Unstable ground state for $V(\phi) = \frac{1}{2}\phi^2 - \lambda\phi^3$

with the field theory results, note that the two-point function

$$i\,G(t) = \frac{\langle \psi | T\,\phi(t)\,\phi(0)\,|\psi\rangle}{\langle \psi | \psi\rangle} \tag{7}$$

is related to the energy E as

$$i\,G(t=0) = \frac{1}{m}\frac{dE}{dm} = \frac{1}{m^2}\left(E - 5\lambda^2\frac{dE}{d\lambda^2}\right) \tag{8}$$

where in the last step we have used the fact that, by dimensional reasoning, the energy E can be expressed as $E = m\,f(\frac{\lambda^2}{m^5})$. Thus, if the perturbative expression for the two-point function diverges, the expression for the ground state energy, $E = E(\lambda^2)$, should also diverge. One subtlety here is that the state $|\psi\rangle$ is clearly unstable. Arkady showed in an appendix [1] how to deal with this, by considering the adiabatic evolution of a stable state into an unstable state. In particular this suggests that the expression for $E = E(\lambda^2)$ must have a cut along the positive λ^2 axis, as shown in Fig. 2, with an associated jump in the imaginary part of E across this cut.

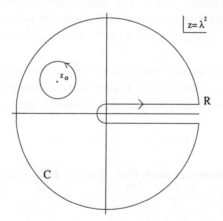

Figure 2. The complex $z = \lambda^2$ plane, showing the cut along the positive z-axis.

Under the (important) assumption that there are no other cuts or poles in the complex $z = \lambda^2$ plane, Cauchy's theorem implies that :

$$E(z_0) = \frac{1}{2\pi i}\oint_C dz\,\frac{E(z)}{z - z_0}$$

$$= \frac{1}{\pi}\int_0^R dz\,\frac{Im\,E(z)}{z - z_0}$$

$$= \sum_{n=0}^{\infty} z_0^n \left(\frac{1}{\pi} \int_0^R dz \, \frac{Im \, E(z)}{z^{n+1}} \right) \tag{9}$$

Thus, the perturbative expansion coefficients are explicitly related to the moments of the imaginary part of the energy along the cut. Furthermore, it is clear from (9) that at large n (i.e., at large order in perturbation theory), the dominant contribution comes from the behavior of $Im E(z)$ as $z \to 0$. This observation is very important, because the $z = \lambda^2 \to 0$ limit is a semiclassical limit (note that the barrier height goes like $\frac{1}{\lambda^2}$, and the barrier width like $\frac{1}{\lambda}$). Hence, in this limit the imaginary part of the energy may be estimated using semiclassical techniques, such as WKB. Simple scaling shows that the leading WKB approximation for the imaginary part of the energy has the form

$$Im \, E(z) \sim \frac{a}{\sqrt{z}} e^{-b/z} \quad , \quad z \to 0 \tag{10}$$

where a and $b > 0$ are (calculable) constants. Note, of course, that this expression is nonperturbative in $z = \lambda^2$.

Now consider the perturbative expansion for the lowest energy

$$E(\lambda^2) = \frac{1}{2} + \sum_{n=1} c_n \lambda^{2n} \tag{11}$$

Inserting the WKB estimate (10) into the dispersion relation (9), we see that at large order the perturbation theory coefficients should behave as

$$c_n \sim \frac{a}{\pi} \int_0^{\infty} dz \, \frac{e^{-b/z}}{z^{n+3/2}} = \frac{a}{\pi} \frac{\Gamma(n + \frac{1}{2})}{b^{n+1/2}} \tag{12}$$

So, this argument suggests that the perturbative expansion (11) for $E(\lambda^2)$ should be a divergent nonalternating series. Indeed, it is straightforward to do this perturbative calculation to very high orders, and to do the WKB calculation precisely, and one finds excellent agreement [6]. If the hamiltonian is rescaled as $H = p^2 + \frac{1}{4}x^2 - \lambda x^3$ (this scaling makes the expansion coefficients integers), then the leading growth rate for large n is

$$c_n \sim -\frac{(60)^{n+1/2}}{(2\pi)^{3/2}} \, \Gamma(n + \frac{1}{2}) \left[1 - \frac{169}{60(2n-1)} + O\left(\frac{1}{n^2}\right) \right] \tag{13}$$

which agrees with Arkady's form (12), and fits beautifully the growth rate of the actual expansion coefficients [6]. Indeed, the factorial growth of the perturbative coefficients kicks in rather early, as illustrated in Fig. 3.

The most important physics lessons from Arkady's paper [1] are :

(i) the divergence of perturbation theory is related to the possible instability of the theory, at some phase of the coupling.

(ii) there is a precise *quantitative* relation (9) between the large-order divergence of the perturbative coefficients and nonperturbative physics.

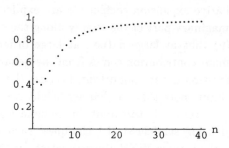

Figure 3.   The ratio of the exact perturbative coefficients c_n in (11) to the leading WKB expression from (13), as a function of the order n.

In more modern language, this divergence associated with instability and tunneling is a divergence due to instantons. This idea has become a cornerstone of quantum field theory [7]. However, since this time, it has been found that in quantum field theory (as distinct from quantum mechanics) there is yet another source of divergence in perturbation theory – this divergence is due to "renormalons", which arise essentially because of the running of the coupling constant, and can be related to special classes of diagrams [8]. For recent developments in this important subject, see the talks by M. Beneke, I. Balitsky and E. Gardi in these Proceedings.

The connection between the large-order behavior of quantum mechanical perturbation theory and WKB methods was developed independently, and was probed in great depth, by Bender and Wu, who studied the quartic anharmonic oscillator [9]. Bender and Wu developed recursion techniques for efficiently generating very high orders of perturbation theory, and compared these results with higher orders of the WKB approximation. Their WKB analysis is a *tour de force*, and the agreement with the large order perturbative coefficients is spectacular.

The dependence of the ground state energy on the coupling λ in the $\lambda\phi^4$ case is different from the $\lambda\phi^3$ case. As is clear from Fig. 4, in the $\lambda\phi^4$ case the instability arises when the coupling λ changes sign from positive to negative. Thus, the cut is expected along the *negative* real axis in the complex $z = \lambda$ plane, which led Bender and Wu to the following dispersion relation:

$$c_n = \frac{1}{2\pi i} \int_{-\infty}^{0} \frac{dz}{z^n} \lim_{\epsilon \to 0} [F(z + i\epsilon) - F(z - i\epsilon)] \tag{14}$$

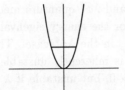 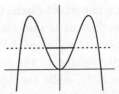

Figure 4. Stable ground state for the potential $V = x^2 + \lambda x^4$, on the left, and unstable ground state for the potential $V = x^2 - \lambda x^4$, on the right.

where $F(\lambda) = \frac{1}{\lambda}(E(\lambda) - \frac{1}{2})$ has one subtraction. Here, Bender and Wu made use of some rigorous results [10] concerning the analyticity behavior of $E(\lambda)$ in the complex λ plane : (i) $|E(\lambda)| \sim |\lambda|^{1/3}$ for large $|\lambda|$, (ii) $E(\lambda)$ is analytic in the cut λ-plane, with the cut along the negative real axis, and (iii) the expansion $E(\lambda) \sim \frac{1}{2} + \sum_{n=1}^{\infty} c_n \lambda^n$ is asymptotic.

The dispersion relation (14) relates the perturbative expansion coefficients for $F(\lambda)$ to the discontinuity of $F(\lambda)$ across the cut. Bender and Wu used high orders of WKB to compute this nonperturbative imaginary part, thereby providing an extremely precise connection between the large orders of perturbation theory and semiclassical tunneling processes. They found that the expansion coefficients are alternating and factorially divergent

$$c_n = (-1)^{n+1} \frac{3^n \sqrt{6} \, \Gamma(n + \frac{1}{2})}{\pi^{3/2}} \left[1 - \frac{95}{72n} + O\left(\frac{1}{n^2}\right) \right] \tag{15}$$

Bender and Wu extended these results to the general anharmonic oscillator:

$$\left(-\frac{d^2}{dx^2} + \frac{1}{4}x^2 + \frac{\lambda}{2^N} x^{2N} - E(\lambda) \right) \psi = 0 \tag{16}$$

Then the kth energy level has an asymptotic series expansion

$$E^{k,N}(\lambda) \sim k + \frac{1}{2} + \sum_{n=1}^{\infty} c_n^{k,N} \lambda^n \tag{17}$$

where the perturbative expansion coefficients $c_n^{k,N}$ are related to the lifetime of the kth unstable level when the coupling λ is negative:

$$c_n^{k,N} = \frac{(-1)^{n+1}(N-1)2^k \Gamma(nN - n + k + \frac{1}{2})}{\pi^{3/2} k! 2^n} \left(\frac{\Gamma(\frac{2N}{N-1})}{\Gamma^2(\frac{N}{N-1})} \right)^{nN-n+\frac{1}{2}} [1 + \ldots] \tag{18}$$

Once again, these growth estimates are derived from WKB and fit the actual perturbative coefficients with great precision [9].

An important distinction between the $\lambda\phi^3$ and $\lambda\phi^4$ quantum mechanical oscillators is that the perturbative series for the energy eigenvalue is *nonalternating* in the $\lambda\phi^3$ case, and *alternating* in the $\lambda\phi^4$ case. This is directly related to the fact that the $\lambda\phi^3$ case is inherently unstable (for any real λ), while the $\lambda\phi^4$ case is stable if $\lambda > 0$, but unstable if $\lambda < 0$. Even though the quartic anharmonic oscillator with $\lambda > 0$ is completely stable, the perturbative expression for the ground state energy is divergent, because the theory with $\lambda < 0$ is unstable.

It is interesting to note that there has been much recent interest in certain nonhermitean hamiltonians, whose spectra appear to be completely real, despite the nonhermicity [11]. For example, there is very strong numerical evidence [12,13] that the hamiltonian

$$H = p^2 + \frac{1}{4}x^2 + i\epsilon x^3 \tag{19}$$

has a completely real spectrum. For the massless case, $H = p^2 + i\epsilon x^3$, the reality of the spectrum has in fact been proved rigorously [14]. These results are nicely consistent with Arkady's analysis of the $\lambda\phi^3$ quantum mechanical model, since an imaginary coupling corresponds to $z = \lambda^2$ being real and negative. In this region it was assumed that $E(z)$ is analytic (recall from Fig. 2 that the cut is along the *positive* z axis), and the perturbative series is divergent but alternating, and can be analyzed using various standard (Padé and Borel) techniques [12,13], yielding excellent agreement with numerical integration results.

A useful mathematical technique for dealing with divergent series is Borel summation [15,16,17,10]. This method is best illustrated by the paradigmatic case of the factorially divergent series. The series

$$f(g) \sim \sum_{n=0}^{\infty} (-1)^n n! \, g^n \tag{20}$$

is clearly divergent, and is alternating if $g > 0$. Using the standard integral representation, $n! = \int_0^\infty ds \, e^{-s} s^n$, and formally interchanging the summation and integration, we can write

$$f(g) \sim \frac{1}{g} \int_0^\infty ds \, \frac{e^{-s/g}}{1+s} \tag{21}$$

This integral representation is *defined* to be the Borel sum of the divergent series in (20). The advantage of the integral is that it is convergent for all $g > 0$. To be more precise, all this actually shows is that the integral in (21) has the same asymptotic series expansion as the divergent series in (20). In order for this identification between the series and the Borel integral to be unique, various further conditions must be satisfied [15,16,17]. In some quantum mechanical examples it is possible to study these conditions rigorously [10], but unfortunately this is usually impossible in realistic quantum field theories. This means we are often confined to "experimental mathematics" when applying Borel techniques to perturbation theory in QFT. Nevertheless, I prefer the attitude of Heaviside:

> "The series is divergent; therefore we may be
> able to do something with it"
> O. Heaviside, 1850 – 1925

to the (older) attitude of Abel:

> "Divergent series are the invention of the devil, and it is shameful to
> base on them any demonstration whatsoever"
> N. H. Abel, 1828

Continuing the paradigm in (20), when $g < 0$ the series (20) becomes nonalternating. Then the same formal steps lead to the following representation:

$$f(-g) \sim \frac{1}{g} \int_0^\infty ds \, \frac{e^{-s/g}}{1-s} \tag{22}$$

Clearly there is a problem here, as there is a pole on the contour of integration, and so an ambiguity enters in the way one treats this pole. This means that the nonalternating factorially divergent series in (20), with $g < 0$, is *not* Borel summable. For example, the principal parts prescription leads to the following imaginary part for the nonalternating series :

$$Im f(-g) = \frac{\pi}{g} \exp\left[-\frac{1}{g}\right] \tag{23}$$

This imaginary contribution is nonperturbative in the expansion parameter g. It is not seen at any finite order in perturbation theory. However, the imaginary part is inherently ambiguous in the absence of further physical information beyond the series expansion (20) itself.

Despite this ambiguity, it should be clear that the Borel approach provides a natural formalism with which to analyze the problem of the divergence of perturbation theory. It captures the essence of the connection with nonperturbative tunneling, and associates such nonperturbative effects with the unstable cases of the $\lambda\phi^3$ oscillator and the $\lambda\phi^4$ oscillator with $\lambda < 0$, for which the perturbative series is indeed nonalternating and factorially divergent.

Similar formal expressions exist for the Borel sum of $f(g) \sim \sum c_n g^n$, if the c_n are not simply factorial as in (20), but have the general form:

$$c_n \sim \beta^n \, \Gamma(\gamma n + \delta) \tag{24}$$

where β, $\gamma > 0$ and δ are constants. Then the Borel sum approximation is

$$f(g) \sim \frac{1}{\gamma} \int_0^\infty \frac{ds}{s} \left(\frac{1}{1+s} \right) \left(\frac{s}{\beta g} \right)^{\delta/\gamma} \exp\left[- \left(\frac{s}{\beta g} \right)^{1/\gamma} \right] \tag{25}$$

For $g < 0$ the nonalternating series has an imaginary part:

$$Im \, f(-g) \sim \frac{\pi}{\gamma} \left(\frac{1}{\beta g} \right)^{\delta/\gamma} \exp\left[- \left(\frac{1}{\beta g} \right)^{1/\gamma} \right] \tag{26}$$

In the next section we will use these relations in an explicit example.

Much more could be said about the divergence of perturbation theory, both in quantum mechanics and field theory. Lipatov [18] generalized the instanton technique to scalar field theory, showing that large orders of perturbation theory may be described by pseudoclassical solutions of the classical field equations, together with quantum fluctuations. This approach built on instanton results of Langer [19] in his classic study of metastability. Perturbation theory for systems with degenerate minima was investigated in [20], and a new twist on the perturbative–nonperturbative connection for this degenerate case is discussed in section 3 of this talk. For further references, I refer the interested reader to the review [21] of Le Guillou and Zinn-Justin as an excellent source.

2. Effective actions, OPEs, and divergent series

As mentioned previously, it is extremely difficult, even in quantum mechanics, to prove truly rigorous results concerning the divergence of perturbation theory [10]; in quantum field theory we are even more restricted when it comes to rigor. However, the study of effective actions is an example in QFT where some rigorous results are possible. This also makes connection with the subject of operator product expansions, which is another subject to which Arkady has made seminal contributions.

For this talk, I consider the QED effective action, which encodes nonlinear interactions due to quantum vacuum polarization effects, such as light-by-light scattering. The effective action is defined via the determinant that is obtained when the electron fields are integrated out of the QED functional integral:

$$S[A] = -\frac{i}{2} \log \det \left(\slashed{D}^2 + m^2 \right) \tag{27}$$

where $\not{D} = \gamma^\mu D_\mu = \gamma^\mu(\partial_\mu - ieA_\mu)$ is the Dirac operator in the classical gauge field background A_μ. The effective action has a natural perturbative expansion in terms of the electromagnetic coupling e, as represented in Fig. 5. Indeed, by charge conjugation invariance (Furry's theorem), the

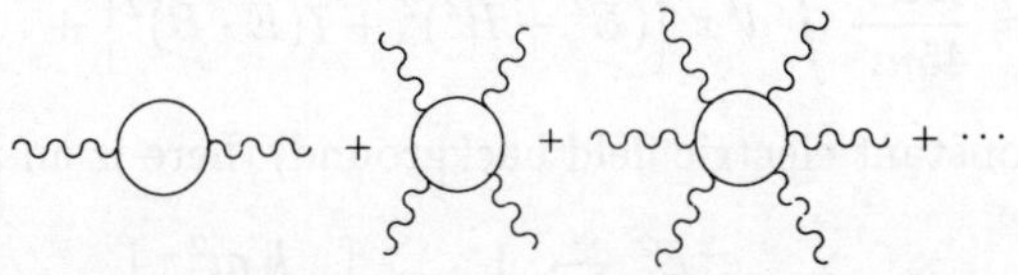

Figure 5. Perturbative expansion of the one-loop effective action

expansion involves only even numbers of external photon lines, which means that the perturbative series is actually a series in the fine structure constant $\alpha = \frac{e^2}{4\pi}$. Another natural expansion is the "effective field theory" expansion (or "large mass" expansion):

$$S[A] = m^4 \sum_n c_n \frac{O^{(n)}}{m^n} \tag{28}$$

Here $O^{(n)}$ represents gauge invariant and Lorentz invariant terms constructed from the field strength $F_{\mu\nu}$, and having mass dimension n. For example, at mass dimension 8, we can have $(F_{\mu\nu}F^{\mu\nu})^2$ or $(F_{\mu\nu}\tilde{F}^{\mu\nu})^2$, while at mass dimension 10 we could have a term $(\partial_\mu F_{\nu\rho}\partial^\mu F^{\nu\rho})(F_{\alpha\beta}F^{\alpha\beta})$. As shown by Arkady and his collaborators [22,23], such an expansion is related to the operator product expansion (OPE), such as that for

$$\Pi_{\mu\nu} = (q_\mu q_\nu - q^2 g_{\mu\nu}) \sum_n c_n(Q^2) \langle O_n \rangle \tag{29}$$

In the special case where the classical background has constant field strength $F_{\mu\nu}$, the perturbative and large mass expansions coincide. This case of a constant background field was solved by Euler and Heisenberg [24] (see also Weisskopf [25] and Schwinger [26]), who obtained an exact nonperturbative expression for the effective action:

$$S[A] = \frac{1}{8\pi^2} \int_0^\infty \frac{ds}{s^3} e^{-im^2 s} \left\{ (es)^2 |\mathcal{G}| \cot \left[es \left(\sqrt{\mathcal{F}^2 + \mathcal{G}^2} + \mathcal{F} \right)^{\frac{1}{2}} \right] \right.$$

$$\left. \times \coth \left[es \left(\sqrt{\mathcal{F}^2 + \mathcal{G}^2} - \mathcal{F} \right)^{\frac{1}{2}} \right] - 1 + \frac{2}{3}(es)^2 \mathcal{F} \right\} \tag{30}$$

Here $\mathcal{F} = \frac{1}{4}F_{\mu\nu}F^{\mu\nu}$, and $\mathcal{G} = \frac{1}{4}F_{\mu\nu}\tilde{F}^{\mu\nu}$, are the two Lorentz invariant combinations. The -1 term in the integrand corresponds to the zero-field subtraction of $S[0]$, while the last term, $\frac{2}{3}(es)^2 \mathcal{F}$, corresponds to charge renormalization [26]. There are several important physical consequences of this result [24,25,26]. First, expanding to leading order in the fields, we find the famous light-by-light term:

$$S = \frac{2\alpha^2}{45m^4} \int d^4x \left[(\vec{E}^2 - \vec{B}^2)^2 + 7(\vec{E} \cdot \vec{B})^2 \right] + \dots \tag{31}$$

Second, for a constant electric field background, there is an imaginary part

$$Im\, S = \frac{e^2 E^2}{8\pi^3} \sum_{k=1}^{\infty} \frac{1}{k^2} \exp\left[-\frac{k\, m^2 \pi}{eE} \right] \tag{32}$$

which gives the pair production rate due to vacuum polarization.

The Euler-Heisenberg result (30) provides an excellent example of the application of Borel summation techniques, as we now review. Consider first of all the case of a uniform magnetic field background of strength B. Then the full perturbative expansion of the Euler-Heisenberg result is

$$S = -\frac{e^2 B^2}{2\pi^2} \sum_{n=0}^{\infty} \frac{\mathcal{B}_{2n+4}}{(2n+4)(2n+3)(2n+2)} \left(\frac{2eB}{m^2} \right)^{2n+2} \tag{33}$$

Viewed as a low energy effective action, the "low energy" condition here is simply that the characteristic energy scale for electrons in the magnetic background, $\hbar \frac{eB}{mc}$, is much smaller than the electron rest mass scale mc^2. The expansion coefficients in the series (33) involve the Bernoulli numbers $\mathcal{B}_{2n}$, which alternate in sign and grow factorially in magnitude [27]. Thus, the series (33) is an alternating divergent series. In fact, the expansion coefficients are:

$$\begin{aligned} c_n &= \frac{2^{2n} \mathcal{B}_{2n+4}}{(2n+4)(2n+3)(2n+2)} \\ &= (-1)^{n+1} \frac{\Gamma(2n+2)}{8} \left[\frac{1}{\pi^{2n+4}} + \frac{1}{(2\pi)^{2n+4}} + \frac{1}{(3\pi)^{2n+4}} + \dots \right] \end{aligned} \tag{34}$$

If we keep just the leading term in (34), then the expansion coefficients are of the form in (24), so that the Borel prescription (25) yields the leading Borel approximation for S as

$$S_{\text{Borel}} \sim \frac{e^2 B^2}{4\pi^6} \int_0^\infty ds\, \frac{s}{1 + s^2/\pi^2} \exp\left[-\frac{m^2 s}{eB} \right] \tag{35}$$

In Fig. 6 this leading Borel approximation is compared with successive terms from the perturbative expansion (33). Clearly the Borel representation is far superior for larger values of the expansion parameter $\frac{eB}{m^2}$.

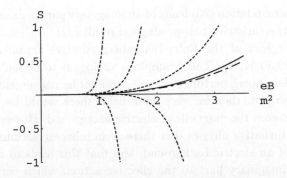

Figure 6. Comparison of the leading Borel expression (35) [short-long-dash curve], as a function of $\frac{eB}{m^2}$, with the exact expression (36) [solid curve], and successive partial sums from the series (33) [short-dash curves]. The leading Borel expression is much better than the series expressions for $\frac{eB}{m^2} \geq 1$.

Actually, in this Euler-Heisenberg case, we can do even better since we are in the unusual situation of having the *exact* expression (34) for the perturbative coefficients, to all orders. Furthermore, the subleading terms in (34) are also of the form (24), for which we can once again use the Borel prescription (25). Including these subleading terms, we find

$$S = -\frac{e^2 B^2}{8\pi^2} \int_0^\infty \frac{ds}{s^2} \left(\coth s - \frac{1}{s} - \frac{s}{3} \right) \exp\left[-\frac{m^2 s}{eB} \right] \tag{36}$$

where we have used the trigonometric expansion [27]:

$$\sum_{k=1}^\infty \frac{-2s^3}{k^2\pi^2(s^2 + k^2\pi^2)} = \coth s - \frac{1}{s} - \frac{s}{3} \tag{37}$$

It is interesting to note that the correct renormalization subtractions appear here. The integral representation (36) is precisely the Euler-Heisenberg expression (30), specialized to the case of a purely magnetic field. Thus, the "proper-time" integral representation (36) is the Borel sum of the divergent perturbation series (33). Conversely, the divergent series (33) is the asymptotic expansion of the nonperturbative Euler-Heisenberg result (30).

Now consider the case of a purely *electric* constant background, of strength E. Perturbatively, the only difference from the constant magnetic case is that we replace $B^2 \rightarrow -E^2$. This is because the only Lorentz invariant combination is $(E^2 - B^2)$ (clearly, $\vec{E} \cdot \vec{B} = 0$ if one or other of E or B is zero). Thus, the alternating series in (33) becomes nonalternating, without changing the magnitude of the expansion coefficients. Applying the

Borel dispersion relation (26) leads to an imaginary part in exact agreement with the nonperturbative pair-production result (32).

The divergence of the Euler-Heisenberg effective action is very well known [28,29,30,31,32]. This example is analogous to Dyson's argument. The Euler-Heisenberg perturbative series *cannot* be convergent, because if it were convergent, then for very weak fields there would be no essential difference between the magnetic or electric background. However, we know from nonperturbative physics that there *is* an inherent vacuum instability in the case of an electric background, and that this leads to an exponentially small imaginary part to the effective action, which corresponds to pair production due to vacuum polarization.

The phenomenon of pair production from vacuum is fascinating, but is very difficult to observe because the exponential suppression factor is exceedingly small for realistic electric field strengths. The critical electric field at which the exponent becomes of order 1 is $E_c = \frac{m^2 c^3}{e\hbar} \sim 10^{16} \, V/cm$, which is still several orders of magnitude beyond the peak fields obtained in the most intense lasers. See [33] for a recent discussion of the prospects for observing vacuum pair production using an X-ray free-electron laser.

Even if one can produce such an intense background electric field, it is clear that it will not be a uniform field. Thus, it is important to ask how the Euler-Heisenberg analysis is modified when the background field is inhomogeneous. This is a very difficult problem in general. The most powerful approach is through semiclassical WKB techniques [34,35]. However, here I turn this question around and ask what this issue can tell us about the *series* expansion of the effective action when the background is inhomogeneous [36]. If the field strength is inhomogeneous, then the large mass expansion (28) of the effective action involves many more terms at a given order, since we can now include terms involving derivatives of $F_{\mu\nu}$. In fact [37], the number of terms appears to grow factorially fast: 1, 2, 7, 36, ... But there are two obvious problems with quantifying this divergence. First, the expansion is not a true series, for the simple reason that at successive orders many completely new types of terms appear. Second, to learn anything nonperturbative one needs to go to very high orders in the derivative expansion, which is extremely difficult.

Fortunately, there is a special case in which both these problems are circumvented at once [36]. Consider the particular inhomogeneous magnetic field pointing in the z-direction, but varying in the x-direction as:

$$\vec{B}(x) = \vec{B} \, \text{sech}^2 \, (x/\lambda) \tag{38}$$

where λ is an arbitrary inhomogeneity scale. In such a background the effective action can be computed in closed-form using nonperturbative techniques [38]. Also, since the inhomogeneity of the background is encoded solely in the dependence on the scale λ, the large mass expansion (28) can be expressed as a true series. That is, at a given order in the derivative expansion, all terms with the same number d of derivatives combine to give a single contribution $\propto \frac{1}{(m\lambda)^d}$. Moreover, since this case is exactly soluble, we have access to *all orders* of the derivative expansion. Indeed, for the inhomogeneous magnetic background (38), the inverse mass expansion (28) becomes a double sum

$$S = -\frac{m^4}{8\pi^{3/2}} \sum_{j=0}^{\infty} \frac{1}{(m\lambda)^{2j}} \sum_{k=2}^{\infty} \left(\frac{2eB}{m^2} \right)^{2k} \frac{\Gamma(2k+j)\Gamma(2k+j-2)\mathcal{B}_{2k+2j}}{j!(2k)!\Gamma(2k+j+\frac{1}{2})} \quad (39)$$

with the two expansion parameters being the derivative expansion parameter, $\frac{1}{m\lambda}$, and the perturbative expansion parameter, $\frac{eB}{m^2}$. Note that the expansion coefficients are known to all orders, and are relatively simple numbers, just involving the Bernoulli numbers and factorial factors. It has been checked in [38] that the first few terms of this derivative expansion are in agreement with explicit field theoretic calculations, specialized to this particular background.

Given the explicit series representation in (39), we can check that the series is divergent, but Borel summable. This can be done in several ways. One can either fix the order k of the perturbative expansion in (39) and show that the remaining sum is Borel summable, or one can fix the order j of the derivative expansion in (39) and show that the remaining sum is Borel summable. Or, one can sum explicitly the k sum, for each j, as an integral of a hypergeometric function, and show that for various values of $\frac{eB}{m^2}$, the remaining derivative expansion is divergent but Borel summable. These arguments do not prove rigorously that the double series is Borel summable, but give a strong numerical indication that this is the case. It is interesting to note that an analogous double-sum structure also appears in the renormalon-OPE analysis in the talk by E. Gardi in these Proceedings.

It is also possible to compute the closed-form effective action for a time dependent electric field pointing in the z-direction:

$$\vec{E}(t) = \vec{E} \operatorname{sech}^2 (t/\tau) \quad (40)$$

where τ characterizes the temporal inhomogeneity scale. A short-cut to the answer is to note that we can simply make the replacements, $B^2 \to -E^2$, and $\lambda^2 \to -\tau^2$, in the magnetic case result (39). In particular this has the consequence that the alternating divergent series of the magnetic case becomes a non-alternating divergent series, just as was found in the

Euler-Heisenberg constant-field case. Fixing the order j of the derivative expansion, the expansion coefficients behave for large k (with j fixed) as

$$c_k^{(j)} = \frac{(-1)^{j+k}\Gamma(2k+j)\Gamma(2k+j+2)\mathcal{B}_{2k+2j+2}}{\Gamma(2k+3)\Gamma(2k+j+\frac{5}{2})} \sim 2\frac{\Gamma(2k+3j-\frac{1}{2})}{(2\pi)^{2j+2k+2}} \quad (41)$$

Note that these coefficients are non-alternating and grow factorially with $2k$, as in the form of (24). Applying the Borel dispersion formula (26) gives

$$ImS^{(j)} \sim \frac{m^4}{8\pi^3}\left(\frac{eE}{m^2}\right)^{5/2} \exp\left[-\frac{m^2\pi}{eE}\right]\frac{1}{j!}\left(\frac{m^4\pi}{4\tau^2e^3E^3}\right)^j \quad (42)$$

Remarkably, this form can be resummed in j, yielding a leading exponential

$$ImS \sim \frac{m^4}{8\pi^3}\left(\frac{eE}{m^2}\right)^{5/2} \exp\left[-\frac{m^2\pi}{eE}\left\{1-\frac{1}{4}\left(\frac{m}{eE\tau}\right)^2\right\}\right] \quad (43)$$

We recognize the first term in the exponent as the leading exponent in the constant field result (32). Thus, the second term may be viewed as the leading *exponential* correction to the constant-field answer (32). This is exactly what we set out to find, and we see that it arose through the divergence of the derivative expansion. I stress that this exponential correction is not accessible from low orders of the derivative expansion. This gives a Dyson-like argument that the derivative expansion *must* be divergent, since if it were not divergent, there would be no essential difference between the electric and magnetic cases, and there would be no correction to the exponent of the imaginary part of the effective action. However, we know, for example from WKB, that there *is* such a correction, and so the derivative expansion must be divergent.

In fact, the situation is even more interesting than the result (43) suggests. We could instead have considered doing the Borel resummation for the j summations, at each fixed k. Then for large j, the coefficients go as

$$c_j^{(k)} = (-1)^{j+k}\frac{\Gamma(j+2k)\Gamma(j+2k-2)\mathcal{B}_{2k+2j}}{\Gamma(j+1)\Gamma(j+2k+\frac{1}{2})} \sim 2^{\frac{9}{2}-2k}\frac{\Gamma(2j+4k-\frac{5}{2})}{(2\pi)^{2j+2k}} \quad (44)$$

which once again are non-alternating and factorially growing. Applying the Borel dispersion formula (26) gives

$$ImS^{(k)} \sim \frac{m^{3/2}}{4\pi^3\tau^{5/2}}\frac{(2\pi eE\tau^2)^{2k}}{(2k)!}e^{-2\pi m\tau} \quad (45)$$

Once again, this leading form can be resummed, yielding

$$ImS \sim \frac{m^{3/2}}{8\pi^3\tau^{5/2}} \exp\left[-2\pi m\tau\left(1-\frac{eE\tau}{m}\right)\right] \quad (46)$$

Note that this leading exponential form of the imaginary part is different from that obtained in (43), and moreover, it is different from the constant-field case (32). The resolution of this puzzle is that there are two competing leading exponential behaviours buried in the double sum (39), and the question of which one dominates depends crucially on the relative magnitudes of the two expansion parameters, the derivative expansion parameter, $\frac{1}{m\tau}$, and the perturbative expansion parameter, $\frac{eE}{m^2}$. Another important quantity is their *ratio*, since this sets the scale of the corresponding gauge field:

$$\frac{A(t)}{m} = \frac{eE\tau \tanh(t/\tau)}{m} \sim \frac{eE\tau}{m} = \frac{eE/m^2}{1/(m\tau)} \tag{47}$$

Thus, it is natural to define a "nonperturbative" regime, in which $\frac{eE\tau}{m} \gg 1$. Then $m\tau \gg \frac{m^2}{eE}$, so that the dominant exponential factor is $\exp[-\frac{m^2}{eE}] \gg \exp[-2\pi m\tau]$. In this regime, the leading imaginary contribution to the effective action is given by the expression (43), and we note that it is indeed nonperturbative in form, and the correction in the exponent is in terms of the small parameter $\frac{m}{eE\tau} \ll 1$. On the other hand, in the "perturbative" regime, where $\frac{eE\tau}{m} \ll 1$, this means that $m\tau \ll \frac{m^2}{eE}$, so that the dominant exponential factor is $\exp[-2\pi m\tau] \gg \exp[-\frac{m^2}{eE}]$. In this regime, the leading imaginary contribution to the effective action is given by the expression (46), and is in fact perturbative in nature, despite its exponential form.

These results are completely consistent with the WKB approach developed by Brézin and Itzykson [34] and Popov [35]. For a time-dependent electric background $E(t) = \dot{A}_z(t)$, in the z-direction, the WKB expression for the imaginary part of the effective action is:

$$ImS \sim \int d^3k \, \exp[-\pi\Omega] \tag{48}$$

where $\Omega = \frac{2i}{\pi} \int_{\text{tp}} \sqrt{m^2 + k_\perp^2 + (k_z - eA_z(t))^2}$. Applying this WKB analysis to the (exactly soluble) case $E(t) = E \operatorname{sech}^2(t/\tau)$, one obtains precisely the leading results (43) or (46), depending on whether we are in the "nonperturbative" $\frac{eE\tau}{m} \gg 1$, or "perturbative" $\frac{eE\tau}{m} \ll 1$ regime. This serves as a useful cross-check of the somewhat formal Borel analysis.

3. Perturbative – nonperturbative duality in QES systems

In this last section I discuss some recent results [39] concerning a new type of perturbative – nonperturbative connection that has been found in certain special quantum mechanical systems. We benefited from discussing these systems with Arkady, and I hope he enjoys the results!

Quasi-exactly solvable (QES) systems are those for which some finite portion of the energy spectrum can be found exactly using algebraic means [40]. A positive integer parameter J characterizes the 'size' of this exact portion of the spectrum. Two simple examples are : $V = x^6 - (4J - 1)x^2$, and $V = \sinh^2 x - (2J - 1)\cosh x$. For a QES system it is possible to define a quadractic form, $H = \sum_{a,b} c_{ab} J_a J_b + \sum_a d_a J_a$, in terms of $sl(2)$ generators of spin J, such that the eigenvalues of the algebraic matrix H are the QES eigenvalues of the original system. It is interesting that algebraic hamiltonians of this form are widely used in the study of tunneling phenomena in single-molecule magnets [41].

In [42], the large J limit of QES systems was identified as a semiclassical limit useful for studying the *top* of the quasi-exact spectrum. It was found that remarkable factorizations reduce the semiclassical calculation to simple integrals, leading to a straightforward asymptotic series representation for the *highest* QES energy eigenvalue. The notion of energy-reflection (ER) symmetry was introduced and analyzed in [43]: for certain QES systems the QES portion of the spectrum is symmetric under the energy reflection $E \leftrightarrow -E$. This means that for a system with ER symmetry, there is a precise connection between the top of the QES spectrum and the bottom of the spectrum. Coupled with the semiclassical large J limit, the ER symmetry therefore relates semiclassical (nonperturbative) methods with perturbative methods [43]. In this section I discuss a class of *periodic* QES potentials for which the ER symmetry is in fact the fixed point (self-dual point) of a more general duality transformation. The duality between weak coupling and semiclassical expansions applies not just to the asymptotic series for the locations of the bands and gaps, but also to the exponentially small widths of bands and gaps.

Consider the quasi-exactly solvable (QES) Lamé equation [44]:

$$\left\{ -\frac{d^2}{d\phi^2} + J(J + 1)\,\nu\,\mathrm{sn}^2(\phi|\nu) - \frac{1}{2}J(J + 1) \right\}\Psi(\phi) = E\,\Psi(\phi). \quad (49)$$

Here $\mathrm{sn}(\phi|\nu)$ is the doubly-periodic Jacobi elliptic function [44,27], the coordinate $\phi \in R^1$, and E denotes the energy eigenvalue. The real elliptic parameter ν lies in the range $0 \leq \nu \leq 1$. The potential in (49) has period $2K(\nu)$, where $K(\nu)$ is the elliptic quarter period. The parameter ν controls the period of the potential, as well as its strength: see Fig. 7. As $\nu \to 1$, the period $2K(\nu)$ diverges logarithmically, $2K(\nu) \sim \ln(\frac{16}{1-\nu})$, while as $\nu \to 0$, the period tends to a nonzero constant: $2K(\nu) \to \pi$. In the Lamé equation (49), the parameter J is a positive integer (for non-integer J, the problem is not QES). This parameter J controls the depth of the wells of the potential;

the constant subtraction $-\frac{1}{2}J(J+1)$ will become clear below.

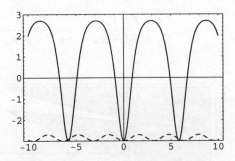

Figure 7. The potential energy in (49) as a function of ϕ, for $J = 2$. The solid curve has elliptic parameter $\nu = 0.95$, for which the period is $2K(0.95) \approx 5.82$. The dashed curve has $\nu = 0.05$, for which the period is $2K(0.05) \approx 3.18$. Note how different the two potentials are; and yet, their spectra are related by the duality transformation (53).

It is a classic result that the Lamé equation (49) has bounded solutions $\Psi(\phi)$ with an energy spectrum consisting of exactly J bands, plus a continuum band [44]. It is the simplest example of a "finite-gap" model, there being just a finite number, J, of "gaps" in the spectrum. This should be contrasted with the fact that a generic periodic potential has an *infinite* sequence of gaps in its spectrum [45]. We label the band edge energies by E_l, with $l = 1, 2, \ldots, (2J+1)$. Thus, the energy regions, $E_{2l-1} \leq E \leq E_{2l}$, and $E \geq E_{2l+1}$, are the allowed bands, while the regions, $E_{2l} < E < E_{2l+1}$, and $E < E_1$, are the gaps.

Another important classic result [46,47] concerning the Lamé model (49) is that the band edge energies E_l, for $l = 1, \ldots, 2J+1$, are simply the eigenvalues of the finite dimensional $(2J+1) \times (2J+1)$ matrix

$$H = J_x^2 + \nu J_y^2 - \frac{1}{2}J(J+1)\,\mathbf{I} \tag{50}$$

where J_x and J_y are $su(2)$ generators in a spin J representation and $\mathbf{I}$ is the unit matrix. Thus the Lamé band edge spectrum is *algebraic*, requiring only the finding of the eigenvalues of the finite dimensional matrix H in (50). For example, for $J = 1$ and $J = 2$, the eigenvalues of H are:

$$J = 1 \quad : \quad E_1 = -1 + \nu\,,$$
$$E_2 = 0\,,$$
$$E_3 = \nu\,; \tag{51}$$
$$J = 2 \quad : \quad E_1 = -1 + 2\nu - 2\sqrt{1 - \nu + \nu^2}\,,$$

$$E_2 = -2 + \nu\,,$$
$$E_3 = -2 + 4\nu\,,$$
$$E_4 = 1 + \nu\,,$$
$$E_5 = -1 + 2\nu + 2\sqrt{1 - \nu + \nu^2}\,. \tag{52}$$

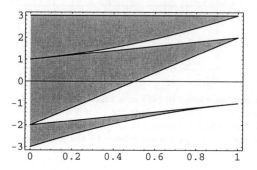

Figure 8.   The energy bands (52) for the $J = 2$ Lamé system (49), as a function of the elliptic parameter ν. The shaded areas on this plot are the bands, while the unshaded areas are the gaps. The top band actually continues up to $E \to \infty$.

The spectrum of the Lamé system (49) has a special duality:

$$E[\nu] = -E[1 - \nu] \tag{53}$$

That is, the spectrum of the Lamé system (49), with elliptic parameter ν, is the energy reflection of the spectrum of the Lamé system with the dual elliptic parameter $1 - \nu$. In particular, for the band edge energies, E_l, which are the eigenvalues of the finite dimensional matrix H in (50), this means

$$E_l[\nu] = -E_{2J+2-l}[1 - \nu] \qquad , \qquad l = 1, 2, \ldots, 2J + 1 \tag{54}$$

This duality can be seen directly in the eigenvalues of the $J = 1$ and $J = 2$ examples in (51) and (52). The proof of the duality (54) for the band edge energies is a trivial consequence of the algebraic realization (50), since

$$J_x^2 + \nu J_y^2 - \frac{1}{2}J(J+1)\,\mathbf{I} = -\left[\left(J_z^2 + (1-\nu)J_y^2\right) - \frac{1}{2}J(J+1)\,\mathbf{I}\right] \tag{55}$$

Noting that $[J_z^2 + (1-\nu)J_y^2]$ has the same eigenvalues as $[J_x^2 + (1-\nu)J_y^2]$, the duality result (54) follows. It is instructive to see this duality in graphical form, in Fig. 8, which shows the band spectra as a function of ν. The transformation $\nu \to 1-\nu$, with the energy reflection $E \to -E$, interchanges the shaded regions (bands) with the unshaded regions (gaps). The fixed point, $\nu = \frac{1}{2}$, is the "self-dual" point, where the system maps onto itself; here the energy spectrum has an exact energy reflection (ER) symmetry.

In fact, the duality relation (53) applies to the entire spectrum, not just the band edges (54). This is a consequence of Jacobi's imaginary transformation [27]. Making the coordinate transformation

$$\phi' = i\,(\phi - K - i\,K')\,, \tag{56}$$

the Lamé equation (49) transforms into

$$\left\{ -\frac{d^2}{d\phi'^2} + J(J+1)\,(1-\nu)\,\mathrm{sn}^2(\phi'|1-\nu) - \frac{1}{2}J(J+1) \right\} \Psi(\phi') = -E\,\Psi(\phi') \tag{57}$$

So solutions of (49) are mapped to solutions of the dual equation (57), with $\nu \to 1 - \nu$, and with a sign reflected energy eigenvalue: $E \to -E$.

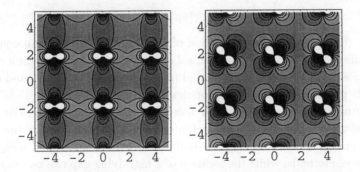

Figure 9. Contour plots in the ϕ plane of the real (left) and imaginary (right) parts of the potential function $sn^2(\phi|\nu)$, for $\nu = \frac{1}{2}$. Note that the potential is real along the real ϕ axis, and along the line $Re(\phi) = K(\frac{1}{2}) \approx 1.85$ (as well as the periodic displacements). This corresponds to the rotation (56).

To see why bands and gaps are interchanged under our duality transformation (56), recall that the independent solutions of the original Lamé equation (49) can be written as products of theta functions [44], and under the change of variables (56), these theta functions map into the same theta functions, but with dual elliptic parameter. However, they map from bounded to unbounded solutions (and vice versa), because of the "i" factor appearing in (56). Thus, the bands and gaps become interchanged.

As an aside, I mention that the Lamé system plays a distinguished role in the theory of $su(2)$ BPS monopoles. For example, Ward has shown [47

] that the Lamé equation factorizes, using the Nahm equations (which are fundamental to the construction of monopole solutions)

$$\frac{dT_j}{ds} = \mp \frac{i}{2} \epsilon_{jkl} [T_k, T_l] \tag{58}$$

To see this, define the quaternionic operators

$$\Delta_\pm = \mathbf{1}_{2n} \frac{d}{ds} \pm T_j(s) \otimes \sigma_j \tag{59}$$

where $T_j(s)$, for $j = 1, 2, 3$, take values in an n-dimensional represenation of the Lie algebra $su(2)$. Then

$$\Delta_\pm \Delta_\mp = \left(\mathbf{1}_n \frac{d^2}{ds^2} - T_j T_j \right) \otimes \mathbf{1}_2 + \left(\mp \frac{dT_j}{ds} - \frac{i}{2} \epsilon_{jkl} [T_k, T_l] \right) \otimes \sigma_j \tag{60}$$

and this combination is real if T_j satisfy the Nahm equations (58). Furthermore, using the solution $T_1 = -\sqrt{\nu}\,\mathrm{sn}(s)\,t_1$, $T_2 = i\sqrt{\nu}\,\mathrm{cn}(s)\,t_2$, $T_3 = i\,\mathrm{dn}(s)\,t_3$, where t_j are $su(2)$ generators, it follows from properties of the elliptic functions that

$$T_j T_j = (t_1^2 + t_2^2 + t_3^2)\,\nu\,\mathrm{sn}^2(s) - (\nu\,t_2^2 + t_3^2) \tag{61}$$

Thus, (60) provides a factorization of the Lamé operator. Subsequently, Sutcliffe showed [48] that the spectral curve for the Nahm data for a charge $n = 2j + 1$ $su(2)$ monopole is related to a j-gap Lamé operator, and corresponds physically to $2j + 1$ monopoles aligned along an axis.

Returning to the perturbative–nonperturbative duality (53), I first discuss how this operates for the *locations* of the bands and gaps in the Lamé spectrum. The location of a low-lying band can be calculated using perturbative methods, while the location of a high-lying gap can be calculated using semiclassical methods. The exact duality (53) between the top and bottom of the spectrum provides an explicit mapping between these two sectors. Defining $\kappa = \sqrt{J(J+1)}$, we see that $1/\kappa$ is the weak coupling constant of the perturbative expansion. Simultaneously, $1/\kappa$ plays the role of $\hbar$ in the quasiclassical expansion.

In the limit $J \to \infty$, the width of the lowest band becomes very narrow, so it makes sense to estimate the "location" of the band. In fact, the width shrinks exponentially fast, so we can estimate the location of the band to within exponential accuracy using elementary perturbation theory. A straightforward calculation [39] shows that the lowest energy level is

$$E_0 = -\frac{1}{2}\kappa^2 \left[1 - \frac{2\sqrt{\nu}}{\kappa} + \frac{\nu + 1}{2\kappa^2} + \frac{(1 - 4\nu + \nu^2)}{8\sqrt{\nu}\kappa^3} + O\left(\frac{1}{\kappa^4}\right) \right]. \tag{62}$$

We now consider the semiclassical evaluation of the location of the highest gap, in the limit $J \to \infty$. First, note that for a given ν, as $J \to \infty$ the highest gap lies *above* the top of the potential. Thus, the turning points lie off the real ϕ axis. For a periodic potential the gap edges occur when the discriminant [45] takes values ± 1. By WKB, the discriminant is

$$\Delta(E) = \cos\left(\frac{1}{\hbar} \sum_{n=0}^{\infty} \hbar^n S_n(P)\right), \tag{63}$$

where P is the period, and $S_n(x)$ are the standard WKB functions which can be generated to any order by a simple recursion formula [16]. The Jth gap occurs when the argument of the cosine in the discriminant (63) is $J\pi$:

$$\kappa \sum_{n=0}^{\infty} \frac{1}{\kappa^n} S_n(2K) = J\pi = \pi\kappa\left(1 - \frac{1}{2\kappa} + \frac{1}{8\kappa^2} - \frac{1}{128\kappa^4} + \ldots\right) \tag{64}$$

where $2K$ is the period of the Lamé potential, and where on the right-hand side we have expressed J in terms of the effective semiclassical expansion parameter $1/\kappa$. This relation (64) can be used to find an expansion for the energy of the Jth gap by expanding

$$E = \frac{1}{2}\kappa^2 + \sum_{\ell=1}^{\infty} \frac{\varepsilon_\ell}{\kappa^{\ell-2}}. \tag{65}$$

The expansion coefficients ε_ℓ are fixed by identifying terms on both sides of the expansion in (64). A straightforward calculation [39] leads to

$$E = \frac{1}{2}\kappa^2 - \kappa\sqrt{1-\nu} + \frac{2-\nu}{4} + \frac{(-2+2\nu+\nu^2)}{16\kappa\sqrt{1-\nu}} + \ldots. \tag{66}$$

Comparing with the perturbative expansion (62) we see that the semiclassical expansion (66) is indeed the dual of the perturbative expansion (62), under the duality transformation $\nu \to 1-\nu$ and $E \to -E$.

This illustrates the perturbative – nonperturbative duality for the *locations* of bands and gaps. However, it is even more interesting to consider this duality for the *widths* of bands and gaps, because the calculations of widths are sensitive to exponentially small contributions which are neglected in the calculations of the locations.

The width of a low-lying band can be computed in a number of ways. First, since the band edge energies are given by the eigenvalues of the finite dimensional matrix H in (50), the most direct way to evaluate the width of the lowest-lying band is to take the difference of the two smallest eigenvalues of H. This leads [49] to the exact leading behavior, in the limit $\nu \to 1$, of the width of the lowest band, for any J:

$$\Delta E_{\text{band}}^{\text{algebraic}} = \frac{8J\,\Gamma(J+1/2)}{4^J\sqrt{\pi}\,\Gamma(J)}(1-\nu)^J\left(1 + \frac{J-1}{2}(1-\nu) + \ldots\right) \tag{67}$$

This clearly shows the exponentially narrow character of the lowest band.

In the instanton approximation, tunneling is suppressed because the barrier height is much greater than the ground state energy of any given isolated "atomic" well. The instanton calculation for the Lamé potential can be done in closed form [49], leading in the large J and $\nu \to 1$ limit to

$$\Delta E_{\text{band}}^{\text{instanton}} \sim \frac{8J^{3/2}}{\sqrt{\pi}\,4^J}\,(1-\nu)^J\left[1 + \frac{J-1}{2}(1-\nu) + \ldots\right], \qquad (68)$$

which agrees perfectly with the large J limit of the exact algebraic result (67). Thus, this example gives an *analytic* confirmation that the instanton approximation gives the correct leading large J behavior of the width of the lowest band, as $\nu \to 1$.

Having computed the width of the lowest band by several different techniques, both exact and nonperturbative, we now turn to a perturbative evaluation of the width of the highest gap. First, taking the difference of the two largest eigenvalues of the finite dimensional matrix H in (50), it is straightforward to show that as $\nu \to 0$, for any J, this difference gives

$$\Delta E_{\text{gap}}^{\text{algebraic}} = \frac{8J\,\Gamma(J+1/2)}{4^J\sqrt{\pi}\,\Gamma(J)}\,\nu^J\left(1 + \frac{J-1}{2}\nu + \ldots\right), \qquad (69)$$

which is the same as the algebraic expression (67) for the width of the lowest band, with the duality replacement $\nu \to 1-\nu$. But it is more interesting to try to find this result from perturbation theory. From (69) we see that the width of the highest gap is of Jth order in perturbation theory. So, to compare with the semi-classical (large J) results for the width of the lowest band, we see that we will have to be able to go to very high orders in perturbation theory. This provides a novel, and very direct, illustration of the connection between nonperturbative physics and high orders of perturbation theory.

It is generally very difficult to go to high orders in perturbation theory, even in quantum mechanics. For the Lamé system (49) we can exploit the algebraic relation to the finite-dimensional spectral problem (50). However, since H in (50) is a $(2J+1) \times (2J+1)$ matrix, the large J limit is still non-trivial. Nevertheless, the high degree of symmetry in the Lamé system means that the perturbative calculation can be done to arbitrarily high order [39]. The result is that the splitting between the two highest eigenvalues arises at the Jth order in perturbation theory, and is given by

$$\Delta E_{\text{gap}}^{\text{pert.theory}} = \frac{8}{4^{2J}}\frac{(2J)!}{[(J-1)!]^2}\,\nu^J = \frac{8J\,\Gamma(J+1/2)}{4^J\sqrt{\pi}\,\Gamma(J)}\,\nu^J \qquad (70)$$

This is in complete agreement with (69), and by duality agrees also with the nonperturbative results for the width (67) of the lowest band.

4. Conclusions

To conclude, there are many examples in physics where there are divergences in perturbation theory which can be associated with a potential instability of the system, thereby providing an explicit bridge between the nonperturbative and perturbative regimes. While this is not the only source of divergence, it is an important one which involves much fascinating physics. Arkady has made many important advances in this subject. On this occasion it is especially appropriate to give him the last word:

> "The majority of nontrivial theories are seemingly unstable at some phase of the coupling constant, which leads to the asymptotic nature of the perturbative series." A. Vainshtein, 1964 [1]

References

1. A.I. Vainshtein, "Decaying systems and divergence of perturbation theory", Novosibirsk Report, December 1964, reprinted in Russian, with an English translation by M. Shifman, in these Proceedings of QCD2002/ArkadyFest.
2. F.J. Dyson, "Divergence of Perturbation Theory in Quantum Electrodynamics", Phys. Rev. **85**, 631 (1952).
3. C.A. Hurst, "An example of a divergent perturbation expansion in field theory", Proc. Camb. Phil. Soc. **48**, 625 (1952); "The enumeration of graphs in the Feynman-Dyson technique", Proc. Roy. Soc. A **214**, 44 (1952).
4. W. Thirring, "On the divergence of perturbation theory for quantized fields", Helv. Phys. Acta **26**, 33 (1953).
5. A. Petermann, "Divergence of perturbation expansion", Phys. Rev. **89**, 1160 (1953).
6. R. Yaris et al, "Resonance calculations for arbitrary potentials", Phys. Rev. A **18**, 1816 (1978); J.E. Drummond, "The anharmonic oscillator: perturbation series for cubic and quartic distortion", J. Phys. A **14**, 1651 (1981).
7. V.A. Novikov, M.A. Shifman, A.I. Vainshtein and V.I. Zakharov, "ABC of instantons", in M. Shifman, *ITEP Lectures in Particle Physics and Field Theory*, (World Scientific, Singapore, 1999), Vol. 1, p 201.
8. G. 't Hooft, "Can We Make Sense Out Of 'Quantum Chromodynamics'?," in *The whys of subnuclear physics*, ed. A Zichichi, (Plenum, NY, 1978); reprinted in G. 't Hooft, *Under the spell of the gauge principle*, (World Scientific, Singapore, 1994); B. Lautrup, "On High Order Estimates In QED," Phys. Lett. B **69**, 109 (1977).
9. C.M. Bender and T.T. Wu, "Anharmonic oscillator", Phys. Rev. **184**, 1231 (1969); "Large-order behavior of perturbation theory", Phys. Rev. Lett. **27**, 461 (1971); "Anharmonic oscillator II. A study of perturbation theory in large order", Phys. Rev. **D7**, 1620 (1973).
10. B. Simon, "Coupling constant analyticity for the anharmonic oscillator", Ann. Phys. **58**, 76 (1970); J.J. Loeffel, A. Martin, B. Simon and A.S. Wightman, "Padé Approximants And The Anharmonic Oscillator," Phys. Lett. B **30**, 656 (1969); J.J. Loeffel and A. Martin, CERN preprint TH-1167 (1970).

504

11. C.M. Bender, S. Boettcher and P. Meisinger, "PT-Symmetric Quantum Mechanics," J. Math. Phys. **40**, 2201 (1999), [arXiv:quant-ph/9809072].

12. C.M. Bender and G.V. Dunne, "Large-order Perturbation Theory for a Non-Hermitian PT-symmetric Hamiltonian," J. Math. Phys. **40**, 4616 (1999), [arXiv:quant-ph/9812039].

13. C.M. Bender and E.J. Weniger, "Numerical evidence that the perturbation expansion for a non-hermitean PT-symmetric hamiltonian is Stieltjes", J. Math. Phys. **42**, 2167 (2001), [arXiv:math-ph/0010007].

14. P. Dorey, C. Dunning and R. Tateo, "Spectral equivalences from Bethe ansatz equations," J. Phys. A **34**, 5679 (2001), [arXiv:hep-th/0103051].

15. G. Hardy, *Divergent Series* (Oxford Univ. Press, 1949).

16. C.M. Bender and S.A. Orszag, *Advanced Mathematical Methods for Scientists and Engineers* (McGraw-Hill, New York, 1978).

17. J. Zinn-Justin, *Quantum Field Theory and Critical Phenomena*, (Clarendon Press, Oxford, 1996).

18. L.N. Lipatov, "Divergence of the perturbation-theory series and the quasi-classical theory", Sov. Phys. JETP **45**, 216 (1977).

19. J.S. Langer, "Theory of the condensation point", Ann. Phys. **41**, 108 (1967).

20. E. Brézin, G. Parisi and J. Zinn-Justin, "Perturbation theory at large orders for a potential with degenerate minima", Phys. Rev. D **16**, 408 (1977); E.B. Bogomolnyi, "Calculation of instanton–anti-instanton contributions in quantum mechanics", Phys. Lett. B **91**, 431 (1980).

21. J.C. Le Guillou and J. Zinn-Justin (Eds.), *Large-Order Behaviour of Perturbation Theory*, (North Holland, Amsterdam, 1990).

22. V.A. Novikov, M.A. Shifman, A.I. Vainshtein and V.I. Zakharov, "Wilson's Operator Expansion: Can It Fail?," Nucl. Phys. B **249**, 445 (1985) [Yad. Fiz. **41**, 1063 (1985)]; "Operator Expansion In Quantum Chromodynamics Beyond Perturbation Theory," Nucl. Phys. B **174**, 378 (1980).

23. V.A. Novikov, M.A. Shifman, A.I. Vainshtein and V.I. Zakharov, "Calculations In External Fields In Quantum Chromodynamics: Technical Review", Fortsch. Phys. **32**, 585 (1985).

24. W. Heisenberg and H. Euler, "Folgerungen aus der Diracschen Theorie des Positrons", Z. Phys. **98**, 714 (1936).

25. V. Weisskopf, "Uber die Elektrodynamik des Vakuums auf Grund der Quantentheorie des Elektrons", Kong. Dans. Vid. Selsk. Math-fys. Medd. XIV No. 6 (1936), reprinted in *Quantum Electrodynamics*, J. Schwinger (Ed.) (Dover, New York, 1958).

26. J. Schwinger, "On Gauge Invariance and Vacuum Polarization", Phys. Rev. **82**, 664 (1951).

27. M. Abramowitz and I. Stegun, *Handbook of Mathematical Functions*, (Dover, 1990).

28. B. L. Ioffe, "On the divergence of perturbation series in quantum electrodynamics", Dokl. Akad. Nauk SSSR **94**, 437 (1954) [in Russian].

29. V. Ogievetsky, "On a possible interpretation of perturbation series in quantum field theory", Dokl. Akad. Nauk SSSR **109**, 919 (1956) [in Russian].

30. S. Chadha and P. Olesen, "On Borel Singularities In Quantum Field Theory," Phys. Lett. B **72**, 87 (1977).

31. V.S. Popov, V.L. Eletsky and A.V. Turbiner, "Higher Order Theory Of Perturbations And Summing Series In Quantum Mechanics And Field Theory,", Sov. Phys. JETP **47**, 232 (1978).

32. A. Zhitnitsky, "Is an Effective Lagrangian a Convergent Series?", Phys. Rev. D **54**, 5148 (1996).

33. A. Ringwald, "Pair production from vacuum at the focus of an X-ray free electron laser," Phys. Lett. B **510**, 107 (2001), [arXiv:hep-ph/0103185]; R. Alkofer, M.B. Hecht, C.D. Roberts, S.M. Schmidt and D.V. Vinnik, "Pair Creation and an X-ray Free Electron Laser," Phys. Rev. Lett. **87**, 193902 (2001) [arXiv:nucl-th/0108046].

34. E. Brézin and C. Itzykson, "Pair production in Vacuum by an Alternating Field", Phys. Rev. D **2**, 1191 (1970); E. Brézin, Thèse (Paris, 1970).

35. V.S. Popov, "Pair production in a variable external field (quasiclassical approximation)", Sov. Phys. JETP **34**, 709 (1972).

36. G.V. Dunne and T.M. Hall, "Borel summation of the derivative expansion and effective actions," Phys. Rev. D **60**, 065002 (1999), [arXiv:hep-th/9902064].

37. D. Fliegner, P. Haberl, M.G. Schmidt and C. Schubert, "The higher derivative expansion of the effective action by the string inspired method. II," Annals Phys. **264**, 51 (1998), [arXiv:hep-th/9707189].

38. G.V. Dunne and T.M. Hall, "An Exact QED_{3+1} Effective Action", Phys. Lett. B **419**, 322 (1998) [arXiv:hep-th/9710062]; D.Cangemi, E.D'Hoker and G.V. Dunne, "Effective energy for QED in (2+1)-dimensions with semilocalized magnetic fields: A Solvable model," Phys. Rev. D **52**, 3163 (1995) [arXiv:hep-th/9506085].

39. G.V. Dunne and M.A. Shifman, "Duality and Self-Duality (Energy Reflection Symmetry) of Quasi-Exactly Solvable Periodic Potentials,", Ann. Phys. **299**, 143 (2002), [arXiv:hep-th/0204224].

40. For a review see : M.A. Shifman, *ITEP Lectures in Particle Physics and Field Theory*, (World Scientific, Singapore, 1999), Vol. 2, p. 775.

41. E.M. Chudnovsky and J. Tejada, *Macroscopic quantum tunneling of the magnetic moment*, (Cambridge University Press, 1998).

42. C.M. Bender, G.V. Dunne and M. Moshe, "Semiclassical Analysis of Quasi-Exact Solvability," Phys. Rev. A **55**, 2625 (1997), [arXiv:hep-th/9609193].

43. M.A. Shifman and A.V. Turbiner, "Energy Reflection Symmetry of Lie-Algebraic Problems: Where the Quasiclassical and Weak Coupling Expansions Meet," Phys. Rev. A **59**, 1791 (1999), [arXiv:hep-th/9806006].

44. E. Whittaker and G. Watson, *Modern Analysis*, (Cambridge, 1927).

45. W. Magnus and S. Winkler, *Hill's Equation*, (Wiley, New York, 1966).

46. Y. Alhassid, F. Gürsey and F. Iachello, "Potential scattering, transfer matrix and group theory", Phys. Rev. Lett. **50** (1983) 873.

47. R.S. Ward, "The Nahm equations, finite-gap potentials and Lamé functions", J. Phys. A **20** (1987) 2679.

48. P.M. Sutcliffe, "Symmetric Monopoles And Finite-Gap Lamé Potentials," J. Phys. A **29**, 5187 (1996).

49. G.V. Dunne and K. Rao, "Lamé instantons," JHEP **0001**, 019 (2000), [arXiv:hep-th/9906113].

SECTION 5.

COSMOLOGY AND AXIONS

AXIONS: PAST, PRESENT, AND FUTURE

MARK SREDNICKI

Department of Physics
University of California
Santa Barbara, CA 93106, USA

I give a pedagogical and historical introduction to axion physics, and briefly review the present status of axions in our understanding of particle physics and cosmology.

1. The Old and New U(1) Problems

Consider quantum chromodynamics (QCD) with color group SU(3) and three flavors of massless quarks. The lagrangian is invariant under an additional global flavor group, $U(3)_L \times U(3)_R$. Although we will use it anyway, this "left–right" terminology is actually somewhat misleading; in four dimensions, a right-handed fermion field is the hermitian conjugate of a left-handed fermion field, and so we could (and will) adopt the convention that all fundamental fermion fields are left-handed (with, of course, right-handed hermitian conjugates). In this case, there would be three left-handed fields transforming as **3**'s of $SU(3)_{color}$, and three left-handed fields transforming as $\bar{\bf 3}$'s of $SU(3)_{color}$; thus the flavor group would be better named $U(3)_3 \times U(3)_{\bar{3}}$. [We note in passing that if quarks were in a real or pseudo-real representation of the color group, such as **3** of $O(3)_{color}$ or **2** of $SU(2)_{color}$, the lagrangian flavor group would be $U(6)$, since in this case there is no distinction between left-handed quarks and left-handed antiquarks.] For now we ignore the effects of the quantum anomaly, which will of course play a key role later.

QCD exhibits dynamical breaking of the flavor symmetry; this can be understood as the formation of a quark condensate,

$$\langle 0 | \chi_{\alpha i a} \tilde{\chi}^{\beta}_{\bar{\jmath} b} | 0 \rangle = -\tfrac{1}{6} \Lambda^3 \, \delta_\alpha{}^\beta \, \delta_{i\bar{\jmath}} \, \varepsilon_{ab} \; . \tag{1}$$

Here $\chi_{\alpha i a}$ is a left-handed fermion field transforming as a **3** of $SU(3)_{color}$ with color index $\alpha = 1, 2, 3$, "left" flavor index $i = 1, 2, 3$, and left-handed spinor index $a = 1, 2$; $\tilde{\chi}^{\beta}_{\bar{\jmath} b}$ is a left-handed fermion field transforming as a $\bar{\bf 3}$ of $SU(3)_{color}$ with color index $\beta = 1, 2, 3$, "right" flavor index $\bar{\jmath} = 1, 2, 3$, and left-handed spinor index $b = 1, 2$; ε_{ab} is the antisymmetric invariant symbol of SU(2); and Λ is a parameter with dimensions of mass. Under a U(3)$_L$ transformation, $\tilde{\chi}$ is unchanged, and $\chi_{\alpha ia} \to L_i{}^j \chi_{\alpha ja}$, where L is a unitary matrix; under a U(3)$_R$ transformation, χ is unchanged, and $\tilde{\chi}^\beta{}_{\bar{\imath} b} \to (R^*)_{\bar{\imath}}{}^{\bar{\jmath}} \tilde{\chi}^\beta{}_{\bar{\jmath} b}$, where R^* is an independent unitary matrix (the complex conjugation is a notational convention). The condensate is unchanged only by transformations in the "vector" subgroup U(3)$_V$ specified by $R = L$. Thus, U(3)$_L \times$ U(3)$_R$ is spontaneously broken down to U(3)$_V$. [I leave it as an exercise for the reader to show that the unbroken subgroup would be O(6) if quarks were **3**'s of O(3)$_{\text{color}}$, and Sp(6) if quarks were **2**'s of SU(2)$_{\text{color}}$.] The nine broken generators lead us to expect nine Goldstone bosons; these can be thought of as long wavelength excitations of the condensate,

$$\langle 0|\chi_{\alpha ia}(x)\tilde{\chi}^\beta{}_{b\bar{\jmath}}(x)|0\rangle = -\tfrac{1}{6}\Lambda^3\,\delta_\alpha{}^\beta\,\varepsilon_{ab}\,U(x)_{i\bar{\jmath}}\,, \tag{2}$$

where U is a unitary matrix field. Under a U(3)$_L \times$ U(3)$_R$ transformation,

$$U(x)_{i\bar{\jmath}} \to L_i{}^k\,(R^*)_{\bar{\jmath}}{}^{\bar{\ell}}\,U(x)_{k\bar{\ell}}\,, \tag{3}$$

or equivalently $U \to LUR^\dagger$. We can write an effective lagrangian for U; it must of course be U(3)$_L \times$ U(3)$_R$ invariant. There are no allowed terms with no derivatives, and two with two derivatives:

$$L = -\tfrac{1}{4}f_\pi^2\,\text{Tr}\,\partial^\mu U^\dagger \partial_\mu U - \tfrac{1}{18}(f_1^2 - \tfrac{3}{2}f_\pi^2)\partial^\mu(\det U^\dagger)\partial_\mu(\det U) + \ldots\,, \tag{4}$$

where f_π and f_1 are parameters with dimensions of mass. If we write

$$U(x) = \exp\left[-i\sum_{a=1}^{8}\lambda^a \pi^a(x)/f_\pi - i\pi^9(x)/f_1\right]\,, \tag{5}$$

where the Gell-Mann λ matrices are hermitian and normalized via $\text{Tr}\,\lambda^a\lambda^b = 2\delta^{ab}$, then the hermitian Goldstone fields $\pi^a(x)$ ($a = 1,\ldots,9$) have canonical kinetic terms, $L = -\tfrac{1}{2}\partial^\mu\pi^a\partial_\mu\pi^a$. The parameter f_π is called the *pion decay constant*, and it has the measured value of 92.4 MeV. (It is measured via the $\pi \to \mu\nu$ decay rate; see Ref. [1] for details.)

In the real world, the three light quarks have small masses:

$$L_{\text{mass}} = -M^{i\bar{\jmath}}\,\varepsilon^{ab}\,\chi_{\alpha ia}\tilde{\chi}^\alpha{}_{b\bar{\jmath}} + \text{h.c.}\,, \tag{6}$$

where M is in general a complex matrix with no particular symmetries; $M^\dagger M$ has eigenvalues m_u^2, m_d^2, and m_s^2, where $m_{u,d,s}$ are the three light-quark masses. M can be made diagonal with positive real entries m_u, m_d, and m_s via a U(3)$_L \times$ U(3)$_R$ transformation that leaves the rest of the lagrangian unchanged. In terms of the effective lagrangian, it is easy to see from eqs. (1) and (6) that

$$L_{\text{mass}} = \Lambda^3\,\text{Tr}\,(MU + \text{h.c.})\,. \tag{7}$$

Expanding in the Goldstone fields, we get

$$L_{\text{mass}} = -\tfrac{1}{4}\Lambda^3 \, \text{Tr}\,[M(\lambda^a \lambda^b + \lambda^b \lambda^a)]\pi^a \pi^b / f_\pi^2 \,, \tag{8}$$

where $\lambda^9 \equiv (f_\pi/f_1)I$. The key conclusion we want to obtain is reached most easily by taking the exact isospin limit $m_u = m_d \equiv m$; then the eigenvalues of the mass-squared matrix in eq. (8) are $m_{1,2,3}^2 = 4\Lambda^3 m/f_\pi^2$ (these are the three pions) and $m_{4,5,6,7}^2 = 2\Lambda^3(m + m_s)/f_\pi^2$ (these are the four kaons). The π^8 and π^9 fields have a mass-squared matrix that must be diagonalized; taking $m \ll m_s$, the eigenvalues are $2\Lambda^3 m_s(\tfrac{4}{3}f_\pi^{-2} + f_1^{-2})$ and $12\Lambda^3 m/(f_\pi^2 + \tfrac{4}{3}f_1^2)$. This implies that there is a fourth pion-like particle with mass less than $\sqrt{3}\,m_\pi$, where $m_\pi = m_{1,2,3} = 135\,\text{MeV}$ is the neutral pion mass. There is, in nature, no such particle. This discrepancy between theory and experiment is the "U(1) problem"[2] or (nowadays) the "old U(1) problem".

The old U(1) problem is solved by the axial anomaly[3]: while the lagrangian is invariant under $\text{U}(3)_L \times \text{U}(3)_R$, the measure over the fermion fields in the functional integral is not[4]; the measure transforms nontrivially under the "axial" $\text{U}(1)_A$ subgroup $L = R^\dagger = e^{i\alpha}I$. Furthermore, the presence of instanton solutions[5] of the euclidean field equations of QCD allow us to see explicitly the physical effects of the anomaly in semiclassical calculations[6]. Thus, when we formulate an effective lagrangian for the low-energy fields, we have no reason to expect it to be invariant under this subgroup. This allows us to add terms involving $\det U$ (and its derivatives) to the lagrangian[7]:

$$L_{\text{anom}} = f_\pi^4 \sum_{n=1}^{\infty} c_n \, (\det U)^n + \text{h.c.} + \ldots \, . \tag{9}$$

In the absence of quark masses, the functional integral of QCD is CP invariant. In the effective theory, CP exchanges U and $U^\dagger$; CP invariance then implies that the c_n coefficients in eq. (9) must all be real. The main effect of L_{anom} is to give the π^9 field a large mass, $m_9^2 = 18Cf_\pi^4/f_1^2$, where $C = \sum_{n=1}^{\infty} c_n n^2$; the π^9 field is then essentially removed from the effective theory, and the mass of the π^8 field (physically, the η meson) is given by $\tfrac{8}{3}\Lambda^3 m_s/f_\pi^2$.

A shadow of the π^9 field remains, however; in the presence of L_{anom}, we cannot remove the overall phase of the quark mass matrix M via a $\text{U}(3)_L \times \text{U}(3)_R$ transformation. If $\det M = m_u m_d m_s e^{i\theta}$, and if we then make a $\text{U}(3)_L \times \text{U}(3)_R$ transformation on U that brings M to the form $\text{diag}(m_u, m_d, m_s)$, the c_n's change: $c_n \to c_n e^{-in\theta}$. This would result in CP violating effects in hadron interactions; the most visible of these would be an electric dipole moment for the neutron[8]. Experimental limits on the electric dipole moment of the neutron imply $\theta < 10^{-9}$. This, then, is the "strong CP problem", or the "new U(1) problem": why is θ so small?

One possibility is that $\det M = 0$ because $m_u = 0$; then there is no phase to remove from the quark mass matrix. It turns out that this possibility cannot be ruled out solely by experimental evidence (since higher-order corrections in the d and s quark masses can mimic a nonzero u quark mass[9]), but it does require some severe theoretical contortions (such as the invalidity of the large-N_{color} expansion to all orders); see Ref. [10] for a discussion.

2. The Peccei–Quinn Solution

Let us choose to work in a field basis in which the c_n's are real and the quark-mass matrix M is diagonal, but contains the unremovable phase θ: $M = \mathrm{diag}(m_u e^{i\theta}, m_d, m_s)$. Now the potential energy is minimized at nonzero values of π^3, π^8, and π^9. For small θ, this minimum energy works out to be

$$V_{\min}(\theta) = \frac{m_u m_d m_s}{m_u m_d + m_d m_s + m_s m_u} \Lambda^3 \theta^2 . \tag{10}$$

This suggests a possible resolution of the strong-CP problem. If θ was some sort of dynamical variable, then it would want to relax to zero, in order to minimize the energy. This is the essential physics of the solution proposed by Peccei and Quinn[11]. In their model, θ is effectively replaced by $a(x)/f_a$, where $a(x)$ is a new scalar field, the *axion*, and f_a is a new parameter, the *axion decay constant*. The observed value of θ is now zero, since that is what minimizes the energy.

Of course, this new field provides us with a new particle as well. Its mass is given by

$$\begin{aligned}
m_a^2 &= \frac{1}{f_a^2} \frac{d^2 V_{\min}(\theta)}{d\theta^2}\bigg|_{\theta=0} \\
&= \frac{2 m_u m_d m_s}{m_u m_d + m_d m_s + m_s m_u} \frac{\Lambda^3}{f_a^2} \\
&= \left[\frac{m_u m_d m_s/(m_u + m_d)}{m_u m_d + m_d m_s + m_s m_u} \right] \frac{f_\pi^2}{f_a^2} m_\pi^2 ,
\end{aligned} \tag{11}$$

where we have used $m_\pi^2 = 2(m_u + m_d)\Lambda^3/f_\pi^2$ to get the last line. The factor in brackets has the numerical value[10] 0.225 ± 0.005. We see that if $f_a \gg f_\pi$, then the axion is very light.

Peccei and Quinn did not present their solution in this language. They were concerned with constructing a renormalizable extension of the Standard Model of quarks and leptons that did not suffer from the strong CP problem. In their model, they used two Higgs fields, instead of one: one to give mass to the up quarks, and the other to give mass to the down quarks (and, somewhat incidentally, the charged leptons). The axion then appears as the relative phase of these two fields, and it is massless (classically) if the lagrangian is invariant under an extra U(1) symmetry, now known as *Peccei-Quinn symmetry*, that is spontaneously broken by the Higgs vacuum expectation values. In working out the physics, Peccei and Quinn assumed that the anomaly would give a large mass to the axion; the fact that it has a small mass was noticed by Weinberg[12] and Wilczek[13]. In the Peccei-Quinn model, the axion decay constant is given by $f_a = \frac{1}{6}\sin(2\beta)v_{EW}$, where $v_{EW} = (\sqrt{2}G_F)^{-1/2} = 246\,\mathrm{GeV}$ is the electroweak scale, and $\tan\beta$ is the ratio of the two Higgs-field VEVs. The couplings of the axion to quarks, leptons, and photons are all proportional to $1/f_a$, and hence of typical weak-interaction strength. Thus the axion should be roughly as visible as a neutrino. It soon became apparent that there is no such particle in nature.

3. Invisible Axions

From our presentation in the last section, it should be clear that there is no fundamental principle that relates f_a to v_{EW}; this is simply a consequence of the specific model explored by Peccei and Quinn. Making f_a larger would make the axion lighter, but also more weakly coupled to ordinary matter of all kinds, and hence harder to produce or detect.

The first successful theoretical proposal to detach f_a from v_{EW} came from Kim[14]. His model is just the Standard Model (with the usual one Higgs doublet), plus an additional sector consisting of a Higgs field that is an SU(3)×SU(2)×U(1) singlet, with a Yukawa coupling to a new quark field that is an SU(2) singlet and carries an essentially arbitrary hypercharge. (Thus, the electric charge of this quark is equal to its hypercharge.) The Peccei-Quinn symmetry of this model involves only these new fields. The singlet Higgs field gets a vacuum expectation value v, breaking the PQ symmetry. Since this Higgs VEV does not break any gauge symmetries, it can be arbitrarily large. If $v \gg v_{EW}$, then the phase of this field becomes the axion, and $f_a \simeq v$. The new quark gets a mass given by v times the value of the Yukawa coupling. Kim's paper was submitted to *Physical Review Letters* in February 1979 and published in July 1979.

Next came the proposal of Zhitnitskiĭ[15]; his model consisted of the Peccei-Quinn model (with its two Higgs doublets), plus an extra singlet Higgs field with (as in Kim's model) a large VEV v. This Higgs singlet is coupled to the two doublets in such a way as to preserve the Peccei-Quinn symmetry of the original lagrangian. The large singlet VEV spontaneously breaks the PQ symmetry, and (as in Kim's model) this results in the phase of this field becoming the axion, with $f_a \simeq v$. Zhitnitskiǐ's paper was submitted to *Yadernaya Fizika* in May 1979 and published in February 1980. The English translation (by Jonathan Rosner) appeared in the *Soviet Journal of Nuclear Physics* about a year later.

Neither of these papers attracted much attention at the time. I will now play amateur sociologist and attempt to guess why.

The title of Kim's paper, "Weak-Interaction Singlet and Strong CP Invariance," gives little clue as to its content. The abstract, however, is clear and succinct: "Strong CP invariance is *automatically* preserved by a spontaneously broken chiral $U(1)_A$ symmetry. A weak-interaction singlet heavy quark Q, a new scalar meson σ^0, and *a very light axion* are predicted. Phenomenological implications are included." In the introduction, Kim takes a different tack, beginning with a classification scheme for theories of CP violation (his categories are *hard, dynamical,* and *spontaneous*), and remarking that "a *simple* theory should belong to one of these classifications for *completeness* of the gauge theory of weak and electromagnetic interactions." He then presents his model. However, Kim does not emphasize the arbitrariness of f_a; when he needs a numerical value, he uses $f_a = 100\,\text{TeV}$, and he takes $100\,\text{GeV}$ for the mass of the new quark. He considers the new quark (and the heavy radial excitation of the new Higgs, σ^0) to be important aspects of the model's phenomenology. He concludes: "What are the possible experiments to prove the present scheme? Probably high-precision experiments of the axion search will do. But the easier verification of the weak-interaction singlets Q and σ^0 in pp or $\bar{p}p$ machines (e^+e^- annihilation machine also if Q is charged) will shed light on the whole idea of the spontaneously broken chiral $U(1)_A$ invariance and the multiple vacuum structure of QCD." A new quark at the electroweak scale with nonstandard electroweak interactions that could not be fit into grand unification schemes must have seemed to many at the time to be an unattractive proposition.

Zhitnitskiǐ's paper, titled "On the possible suppression of axion-hadron interactions," is clear, succinct, and correct throughout. His abstract reads, "A possible mechanism for strong suppression of the axion-fermion interaction is considered. Two models in which this mechanism is realized are described in detail." The emphasis is on the axion coupling to quarks; the couplings to leptons and photons (which are also suppressed) are not mentioned. Also, the axion mass is mentioned only in the paper's last sentence: "Here the mass of the axion is $m_a \sim 100\,\delta\,\text{keV}$," where $\delta \simeq v_{\text{EW}}/v$. In addition to the model discussed above, Zhitnitskiǐalso presents a model based on the gauge group $SU(3) \times SU(2)_L \times SU(2)_R \times U(1)$; in this model, no new

Higgs fields are needed other than the ones already needed to break $SU(2)_R$. The emphasis on a left–right symmetric model probably contributed to this paper's initially limited impact; these models were by then waning in popularity.

Next came the paper entitled "Can confinement ensure natural CP invariance of the strong interactions?" by Shifman, Vainshtein, and Zakharov (SVZ)[16], which was submitted to *Nuclear Physics B* in July 1979 and published in April 1980. SVZ point out that the usual argument for the physical effect of the θ parameter rests on the properties of long-range instanton field configurations, and it is not immediately obvious how to treat these in the presence of color confinement. Could it be that color confinement renders θ unobservable after all? Their answer is *no*; my presentation in Sect. 1 of low-energy physics with $\theta \neq 0$ essentially follows their argument. They show, in particular, that nonzero θ results in a nonzero amplitude for $\eta \to \pi\pi$ decay. So the effect of θ is really there, and the strong CP problem is really a problem. This analysis makes up the bulk of the paper.

So, what to do? SVZ begin the last section of their paper by stating that, "In fact, the problem of the θ-term cannot be solved by QCD alone since it is intimately related to the origin of the quark masses and the mass generation is considered to be a prerogative of the weak interactions." They then note than even the phase of an arbitrarily heavy quark would contribute to θ (which is just the phase of the determinant of the quark mass matrix), and that this would even apply to the mass of a Pauli-Villars regulator field! "Since θ is directly observable (see above)," they write, "it seems to be a unique example of how a laboratory experiment can shed light on regulator properties. Alternatively, it is very suspicious."

Their suspicions aroused, they then proceed to construct a model identical to Kim's. They conclude their paper with this paragraph: "If v becomes arbitrarily large, the axion interaction with normal hadrons vanishes. (The same is true for its mass.) The new quark also becomes unobservable. This theoretical phantom still restores the natural P and T invariance of QCD. Although the model discussed is evidently a toy one, one might hope that something of the kind happens in unified theories."

Here, then, is a clear presentation of the key point, by well known and respected physicists, published in the most widely read particle-physics journal of the time. Still, the initial impact was minimal. Why?

One clue can be found in this paragraph, near the top of the paper's second page: "The overwhelming reaction to this observation is that the example of the one-instanton solution does prove CP violation in the presence of the θ-term. Our final result complies with the conclusion on CP violation so that most readers can, justifiably, lose interest in the paper at this point." Presumably, most readers did just that!

The fourth discovery of the arbitrariness of f_a came in the paper "A simple solution to the strong CP problem with a harmless axion," by Dine, Fischler, and myself[17], submitted to *Physics Letters B* in May 1981 and published in August 1981. We had been working on (very messy) models of supersymmetric technicolor, and had found an axion with a large decay constant in one of them. We began wondering whether this phenomenon could be duplicated in a simple Higgs model, and quickly constructed one (which turned out to be identical to Zhitnitskiĭ's). We were excited at first, then wary. This was too easy! If it really worked, surely others would have thought of it before.

So we began trying to search the literature (without, of course, a computer database). We found Kim's paper pretty quickly. We were disappointed, but luckily our specific model was different: we didn't have an extra quark, only extra Higgs fields. We felt this would be considered more palatable. Also, we could fit our model into the GUT paradigm of the day. Our abstract reflects these points: "We describe a simple generalization of the Peccei-Quinn mechanism which eliminates the strong CP problem at the cost of a very light, very weakly coupled axion. The mechanism requires no new fermions and is easily implemented in grand unified theories." As the paper was being typed (by a *secretary*, on a *typewriter*), we continued to poke around in the library, because we were still worried. Eventually we came across the SVZ paper. We were relieved to find that their model was the same as Kim's. In our text, we talked about "Kim's model," and obviously we needed to change this to "the model of Kim and Shifman, Vainshtein, and Zakharov". Too late! Our paper was typed, and re-typing was out of the question. We were at least able to add "see also SVZ" at the end of the reference to Kim. It came out looking as though we thought the SVZ paper was somehow inferior to the Kim paper. I don't know if anyone ever noticed this, but I would like to take this occasion of Arkady's sixtieth birthday to apologize to him and his collaborators!

Much to our surprise, our paper received a great deal of attention. Wise, Georgi, and Glashow[18] constructed an explicit SU(5) model, and made the important point that solving the strong CP problem in this way did not require any extra fine-tuning in the Higgs sector. They also coined the phrase "invisible axion," which has stuck. Nilles and Raby constructed a similar supersymmetric model[19]. Many more papers followed. The two original models became known as the KSVZ and DFS models. Zhitnitskiĭ's paper was still generally unknown and unreferenced.

In 1986, I had a (wrong, as it turned out) idea about the electric dipole moment of the neutron, and consulted Shalabin's 1983 review[20]. In it was a reference to Zhitnitskiĭ, right where I expected a reference to DFS. Wondering what this paper could be, I looked it up, began reading, and felt a flush of embarrassment wash over me as I realized that (a) it had a model identical to ours and (b) it was written and published long before ours. I began to reference it (and to ask others to reference it in axion papers that I refereed). I don't know if this was the catalyst or not, but eventually the DFS model became the DFSZ model.

4. Axion Astrophysics and Cosmology

Detailed discussions of the axion couplings to hadrons, electrons, and photons can be found in Ref. [21]. All are proportional to $1/f_a$, and the strongest limits on them come from astrophysical processes: axion emission from various objects (red giants, stars in globular clusters, supernova 1987A, etc.) must be slow enough to avoid changing the physics. The results are model dependent, but roughly require $f_a > 6 \times 10^9$ GeV, or $m_a < 0.01$ eV. For a review, see Ref. [22].

The axion can play a cosmological role[23] for larger values of f_a. At high temperatures, the instanton effects which give rise to the axion mass go away. In the very early universe, then, the value of θ is undetermined, and presumably varies slowly from place to place. Inflation would pick out a particular value in our horizon volume. If subsequent reheating is to a temperature below f_a, this value is frozen in place until much later, when the universe cools and the axion mass appears. The axion field then begins coherent oscillations around its minimum. If the initial value of θ is $O(1)$, then the energy stored in these oscillations would overclose the universe if f_a is larger than about 10^{12} GeV. If f_a is near this value, this energy (which is in the form of axions with near zero momentum) could be the cold dark matter.

Things are more complicated if there is no inflation, or reheating after inflation to a temperature above f_a. In this case a network of cosmic-string defects forms in the axion field[24], and the evolution of this network is a complicated numerical problem. The axions from string decay contribute at least as much dark matter as the axions from the initial misalignment, and possibly much more; see Ref. [22] for more details.

The possibility that axions form the cold dark matter leads to the exciting possibility that we might be able to detect them. The axion-photon interaction lagrangian is

$$L_{a\gamma\gamma} = \frac{\alpha}{2\pi f_a}(C - C')a\mathbf{E}\cdot\mathbf{B} ,$$

(12)

where C and C' are numerical constants. This interaction allows an axion to convert to a photon in a magnetic field, and Sikivie proposed using this effect to search for dark-matter axions[25]. The value of C is computed in terms of the Peccei-Quinn charges of the fermions, and C' arises from axion-pion and axion-eta mixing; it is model independent, and given by

$$C' = \frac{2}{3}\frac{m_u m_d + 4m_d m_s + m_s m_u}{m_u m_d + m_d m_s + m_s m_u}$$
$$= 1.93 \pm 0.04 .$$

(13)

Unfortunately, C is typically positive, so there is a cancelation. In grand-unifiable models (that is, models where the fermions come in SU(5) multiplets with the same PQ charge; whether the extra heavy particles implied by unification exist or not is irrelevant) such as the DFSZ model,

$$C_{\text{DFSZ}} = \frac{8}{3} .$$

(14)

In KSVZ models with a single heavy quark of charge Q,

$$C_{\text{KSVZ}} = 6Q^2 .$$

(15)

If axions are the dark matter, $m_a \sim 1$ to $100\,\mu\text{eV}$ is the most interesting mass range ($1\,\mu\text{eV} = 10^{-6}\,\text{eV}$). Experiments to search for them are currently underway, and one[26] has ruled KSVZ axions (with $Q = 0$) in the mass range $2.8 \pm 0.5\,\mu\text{eV}$ as a significant component of the local dark matter. The experiments are continuing; for a review, see Ref. [27].

5. The Future of Axion Physics

Over twenty years have passed since the invention of the invisible axion, and we still do not know whether or not this is the correct solution to the strong CP problem. It is vitally important that the current searches continue until they have covered at least the most plausible mass and coupling ranges; it would be a great shame if such important physics surrounded us, and we left it undiscovered. Of course, even if these searches do not find dark matter axions, this means only that the dark matter is not axions, and not that axions do not exist. There are many cosmological scenarios for this possibility.

Also, it should be noted that axions arise rather naturally in superstring models, and it may be that any ground state of string/M theory that resembles the Standard Model always includes an axion with, say, $f_a \sim (v_{EW} M_{Planck})^{1/2} \sim 10^{11}$ GeV. However at present we seem a long way from being to reach this sort of conclusion.

The road ahead for axion physics is thus likely to be a hard one, both theoretically and experimentally. But the reward for a successful traversal will make the journey worthwhile.

I would like to thank the organizing committee for Arkadyfest—Keith Olive, Mikhail Shifman, and Mikhail Voloshin—for the opportunity to present this talk, and Arkady Vainshtein for providing the occasion and the inspirational physics. I also acknowledge the NSF for financial support.

References

1. H. Georgi, *Weak Interactions and Modern Particle Theory* (Benjamin, 1984).
2. S. Weinberg, "The U(1) problem," *Phys. Rev.* **D11**, 3583 (1975).
3. S. Adler, "Axial-vector vertex in spinor electrodynamics," *Phys. Rev.* **177**, 2426 (1969); J. S. Bell and R. Jackiw, "A PCAC puzzle: $\pi^0 \to \gamma\gamma$ in the sigma -model," *Nuovo Cimento* **A60**, 47 (1969); W. Bardeen, "Anomalous Ward identities in spinor field theories," *Phys. Rev.* **184**, 1848 (1969);
4. K. Fujikawa, "Path-integral measure for gauge-invariant fermion theories," *Phys. Rev. Lett.* **42**, 1195 (1979).
5. A. A. Belavin, A. M. POolyakov, A. S. Schwartz, and Yu. S. Tyuptin, "Pseudoparticle solutions of the Yang-Mills equations," *Phys. Lett.* **B59**, 85 (1975);
6. G 't Hooft, "Symmetry breaking through Bell-Jackiw anomalies," *Phys. Rev. Lett.* **37**, 8 (1976), and "Computation of the quantum effects due to a four-dimensional pseudoparticle," *Phys. Rev.* **D14**, 3432 (1976); R. Jackiw and C. Rebbi, "Vacuum periodicity in a Yang-Mills quantum theory," *Phys. Rev. Lett.* **37**, 172 (1976); C. G. Callan, R. F. Dashen, and D. J. Gross, "The structure of the gauge theory vacuum," *Phys. Lett.* **B63**, 334 (1976).
7. E. Witten, "Large N chiral dynamics," *Ann. Phys.* **128**, 363 (1980).
8. V. Baluni, "CP-nonconserving effects in quantum chromodynamics," *Phys. Rev.* **D19**, 2227 (1979); R. J. Crewther, P. Di Vecchia, G. Veneziano, and E. Witten, "Chiral estimate of the electric dipole moment of the neutron in quantum chromodynamics," *Phys. Lett.* **B88**, 123 (1979).
9. D. B. Kaplan and Manohar, "Current-mass ratios of the light quarks," *Phys. Rev. Lett.* **56**, 2004 (1986).
10. H. Leutwyler, "Bounds on the light quark masses," hep-ph/9601234, *Phys. Lett.* **B374**, 163 (1996), and "The ratios of the light quark masses," hep-ph/9602366, *Phys. Lett.* **B378**, 313 (1996).
11. R. Peccei and H. R. Quinn, "CP conservation in the presence of pseudoparticles," *Phys. Rev. Lett.* **38**, 1440 (1977), and "Constraints imposed by CP conservation in the presence of pseudoparticles," *Phys. Rev.* **D16**, 1791 (1977).
12. S. Weinberg, "A new light boson?", *Phys. Rev. Lett.* **40**, 223 (1978).
13. F. Wilczek, "Problem of strong P and T invariance in the presence of instantons," *Phys. Rev. Lett.* **40**, 279 (1978).

14. J. E. Kim, "Weak interaction singlet and strong CP invariance," *Phys. Rev. Lett.* **43**, 103 (1979).

15. A. P. Zhitnitskiĭ, "On the possible suppression of axion-hadron interactions," *Yad. Fiz.* **31**, 497 (1980) [*Sov. J. Nucl. Phys.* **31**, 260 (1980)].

16. M. A. Shifman, A. I. Vainshtein, and V. I. Zakharov, "Can confinement ensure natural CP invariance of strong interactions?", *Nucl. Phys.* **B166**, 493 (1980).

17. M. Dine, W. Fischler, and M. Srednicki, "A simple solution to the strong CP problem with a harmless axion." *Phys. Lett.* **B104**, 199 (1981).

18. M. Wise, H. Georgi, and S. L. Glashow, "SU(5) and the invisible axion," *Phys. Rev. Lett.* **47**, 402 (1981).

19. H. P. Nilles and S. Raby, "Supersymmetry and the strong CP problem," *Nucl. Phys.* **B198**, 102 (1982).

20. E. P. Shalabin, "Electric dipole moment of the neutron in gauge theory," *Ups. Fiz. Nauk.* **139**, 561 (1983) [*Sov. Phys. Usp.* **26**, 297 (1983)].

21. D. B. Kaplan, "Opening the axion window," *Nucl. Phys.* **B260**, 215 (1985); M. Srednicki, "Axion couplings to matter: CP-conserving parts," *Nucl. Phys.* **B260**, 689 (1985); H. Georgi, D. B. Kaplan, and L. Randall, "Manifesting the invisible axion at low energies," *Phys. Lett.* **B169**, 73 (1986).

22. G. G. Rafelt, "Axions and other very light bosons: astrophysical constraints," in "Review of particle properties," K. Hagiwara et al., *Phys. Rev.* **D66**, 010001 (2002).

23. J. Preskill, M. B. Wise, and F. Wilczek, "Cosmology of the invisible axion," *Phys. Lett.* **B120**, 127 (1983); L. F. Abbott and P. Sikivie,, "A cosmological bound on the invisible axion," *Phys. Lett.* **B120**, 133 (1983); M. Dine and W. Fischler, "The not so harmless axion," *Phys. Lett.* **B120**, 137 (1983).

24. R. L. Davis, "Cosmic axions from cosmic strings," *Phys. Lett.* **B180**, 225 (1986).

25. P. Sikivie, "Experimental tests of the 'invisible' axion," *Phys. Rev. Lett.* **51**, 1415 (1983).

26. S. J. Asztalos et al, "Experimental constraints on the axion dark matter halo density," *Astrophys. J.* **571**, L27 (2002).

27. C. Hagmann, K. van Bibber, and L. J. Rosenberg, "Axions and other very light bosons: experimental limits," in "Review of particle properties," op. cit.

QCD VACUUM AND AXIONS: WHAT'S HAPPENING?

GREGORY GABADADZE AND M. SHIFMAN

Theoretical Physics Institute, University of Minnesota, Minneapolis, MN 55455

A deeper understanding of the vacuum structure in QCD invites one to rethink certain aspects of the axion physics. The recent advances are mostly due to developments in supersymmetric gauge theories and the brane theory, in which QCD can be embedded. They include, but are not limited to, the studies of metastable vacua in multicolor gluodynamics, and the domain walls. We briefly review basics of the axion physics and then present a modern perspective on a rich interplay between the QCD vacuum structure and axion physics.

1. Introduction

Almost 25 years elapsed since the axion was introduced in particle physics [1,2] as a possible solution of the strong CP problem. Since then, it became a text-book and encyclopedia subject. For instance, Oxford University's *"A Dictionary of Science"* defines axion as "a hypothetical elementary particle postulated to explain why there is no observed CP violation (see CP invariance) in the strong interaction (see fundamental interactions). Axions have not been detected experimentally, although it has been possible to put limits on their mass and other properties from the effects that they would have on some astrophysical phenomena (e.g. the cooling of stars). It has also been suggested that they may account for some or all of the missing mass in the universe."

While at the early stages the axion physics was considered predominantly in the context of quantum chromodynamics, at present the center of gravity of the axion studies shifted in astrophysics. It was realized rather early that the axion was a viable dark matter candidate. The research on this aspect of the axion physics quickly picked up and never subsided. Extensive investigations were and are being carried out in the astrophysical community. At the same time, after the rapid advances in the 1980's, the QCD practitioners seemingly lost interest in this subject. The reason is obvious: the progress in understanding the QCD vacuum structure was painfully slow. The prevailing impression was that "nothing happened here," so there was no motivation for revisiting QCD-related aspects of the axion physics.

In this review we will try to argue that "something interesting happened here." A substantial progress has been achieved in the recent years mainly due to insights in QCD obtained from supersymmetry and the brane theory. The existence of a multitude of (quasi) stable vacua at large N_c and "abnormally" thin domain walls with "abnormally" heavy excitations are just a few topics of interest that should be mentioned in this context. A deeper understanding of the QCD vacuum structure requires a reassessment of a number of issues of direct relevance to axions. After a brief summary of basics of the axion physics we review these new developments.

2. The strong CP problem

2.1. *The θ term*

After the discovery of asymptotic freedom in QCD in 1973[3,4] for a short while it was believed that QCD possesses the same natural conservation laws as its more primitive predecessor, QED. The discovery that P and T conservation in QCD is not natural came as a shocking surprise. This fact was realized with the advent of instantons[5] which demonstrated that the so-called θ term

$$\Delta \mathcal{L}_\theta \equiv \frac{\theta}{32\,\pi^2}\, G^a_{\mu\nu}\, \tilde{G}^a_{\mu\nu}\,, \tag{1}$$

does not necessarily vanish. Here the dual field strength is defined as

$$\tilde{G}^a_{\mu\nu} \equiv (1/2)\, \varepsilon_{\mu\nu\alpha\beta}\, G^a_{\alpha\beta}\,.$$

(The indices are assumed to be contracted via the flat space metric). The operator $G\tilde{G}$ has dimension four, it can and should be added to the QCD Lagrangian. The θ term (1) violates P and T invariance (and hence, it violates CP since CPT is preserved). Note that the analogous term $\Delta \mathcal{L} \sim F\tilde{F}$ in QED has no impact on the theory whatsoever. What is the difference?

The θ term can be rewritten as follows

$$\Delta \mathcal{L}_\theta = \theta\, \partial_\mu\, K_\mu\,, \tag{2}$$

where K_μ is the Chern-Simons current defined as

$$K_\mu = \frac{1}{16\,\pi^2}\, \varepsilon_{\mu\nu\alpha\beta} \left(A^a_\nu\, \partial_\alpha\, A^a_\beta + \frac{1}{3}\, f_{abc} A^a_\nu\, A^b_\alpha\, A^c_\beta \right)\,. \tag{3}$$

Being a total derivative, the θ term does not affect the equations of motion. At a naive level, one can discard in the action the integrals over full derivatives. This was a rationale behind the original belief that QCD naturally conserves P and T.

The instantons revealed the fact that the vacuum structure in QCD is more complicated than that in QED. In particular, the field configurations with the instanton boundary conditions give rise to a nonvanishing

$$(\Delta S_\theta)_{\text{one inst}} = \int d^4 x \, (\Delta \mathcal{L}_\theta)_{\text{one inst}} = \theta \,. \tag{4}$$

The integral over the full derivative does not vanish.[a] Therefore, CP-violating effects may be present in strong interactions.

We pause here to make an explanatory remark regarding Eq. (4). A key notion is the topological charge $\mathcal{V}$ of a gauge field configuration,

$$\mathcal{V} \equiv \int d^4 x \, \partial_\mu K_\mu = \int d^3 x \, K_0(x, t)|_{t=-\infty}^{t=+\infty}$$

$$\equiv \mathcal{K}(t = +\infty) - \mathcal{K}(t = -\infty) \,, \tag{5}$$

where $\mathcal{K}$ is usually referred to as the Pontryagin number. The topological charge is zero for any perturbative gauge fields — such fields are said to have trivial topology. The instanton field configuration has a nontrivial topology. In the $A_0 = 0$ gauge it interpolates between $A_m(x, t \to -\infty) = 0$, $m =, 1, 2, 3$, and

$$A_m(x, t \to +\infty) = U^+ \partial_m U \,, \tag{6}$$

where the matrix U is Poyakov's hedgehog,

$$U(\vec{x}) = \exp\left(-\frac{i\pi \vec{x} \vec{\sigma}}{\sqrt{\vec{x}^2 + \rho^2}}\right). \tag{7}$$

The Pontryagin number for U reduces to

$$\mathcal{K} = \frac{1}{24\pi^2} \int d^3 x \, \varepsilon_{ijk} \, \text{Tr} \left(U^+ \partial^i U\right) \left(U^+ \partial^j U\right) \left(U^+ \partial^k U\right) = 1, \tag{8}$$

implying that the instanton topological charge is unity.

2.2. Superselection rule and θ sectors

As we pointed out in the previous section, the value of the parameter θ a priori can be arbitrary. The theories with different values of θ describe different worlds which do not "communicate" with each other. In other words, the worlds with different θ belong to distinct superselection sectors.[6,7]

---

[a]We jumped here from the Minkowski to the Euclidean formulation of the theory. In passing from Minkowski to Euclidean, the θ term (1) acquires an "i" factor, so does the integration measure in the action. Since this is a text-book topic, we will pass freely from Minkowski to Euclidean and back making no explicit statements as to which space any given formula belongs to.

To see that this is the case and to further elucidate the role of θ let us consider pure gluodynamics in the Hamilton gauge $A_0 = 0$. In this gauge the Lagrangian does not depend on A_0 and the Gauss' law (which in other gauges could have been obtained by varying the action with respect to A_0) is imposed as a constraint on physical states

$$D^i \, G_{i0} \, |\text{Phys}\rangle \; = \; 0 \, . \tag{9}$$

This gauge fixing does not eliminate, however, the gauge freedom completely. Purely spatial gauge transformations (independent of the time variable) are still allowed. The generator of these residual gauge transformations can be written as

$$\mathcal{G}(\alpha) \; \equiv \; \exp \left(i \int d^3 x \, \text{Tr} \, D^i \, G_{i0}(\vec{x}) \, \alpha(\vec{x}) \right) , \tag{10}$$

where $\alpha = \alpha^a \, t^a$ and the trace Tr runs over the color indices. The generator $\mathcal{G}(\alpha)$ acts on the spatial components of the gauge fields,

$$\mathcal{G}^+ \, A_k \, \mathcal{G} \; = \; U^+ \, (A_k + i \, \partial_k) \, U \, , \tag{11}$$

where $U \equiv \exp(i\alpha(x))$. Furthermore, it is straightforward to show that the operator $\mathcal{G}$ does not commute with $\mathcal{V}$ if the corresponding gauge transformations give rise to a nonzero right-hand side in Eq. (8) (the latter are called *large gauge transformations*),

$$[\mathcal{G} \, \mathcal{V}] \neq 0 \, . \tag{12}$$

Therefore, an eigenstate of $\mathcal{V}$ cannot be a physical state. Instead, the physical state is defined as a superposition of the eigenstates of $\mathcal{V}$

$$|\theta\rangle \; = \; \sum_{n=-\infty}^{+\infty} e^{i \theta n} \, |n\rangle \, , \tag{13}$$

where $\mathcal{V}|n\rangle = n \, |n\rangle$.

In other words, in the infinitely-dimensional space of fields there is one direction parametrized by the variable $\mathcal{K} = \int K_0 d^3 x$ which forms a closed circle. The wave function (as a function of $\mathcal{K}$) is the Bloch superposition (13). The parameter θ is nothing but a "quasimomentum."[8,6,7] In QCD it is called the vacuum angle.

In this formulation the θ angle enters as an arbitrary phase in (13) and varies in the interval from 0 to 2π. Physics must be 2π periodic in θ. At $\theta \neq 0$ or π one can expect P and T noninvariant effects.

It is straightforward to show that for any *gauge invariant* operator $\hat{O}$

$$\langle \theta' | \, \hat{O} \, |\theta\rangle \; \sim \; \delta_{\theta' \, \theta} \, . \tag{14}$$

Therefore, no gauge invariant operator can transform a state of one θ world into a state of another θ world. The different θ worlds are disconnected from each other.

2.3. *Constraints on θ*

As was mentioned, nonvanishing θ leads to CP violating observables in QCD. (The point $\theta = \pi$ will be discussed separately). It is known that strong interactions conserve CP. Hence, a natural question arises as to what are the experimental constraints on the value of θ.

In a full theory, with quarks, there is an additional contribution to the CP odd part of the Lagrangian. It comes from the imaginary phases of the quark mass matrix $\mathcal{M}$. These phases can be rotated away from the mass matrix by chiral transformations of the quark fields. However, because of the axial anomaly,[9] which manifests itself as a noninvariance of the Feynman integral measure under the chiral transformations at the quantum level,[10] the quark mass matrix phases appear in front of the $G\tilde{G}$ term in the QCD Lagrangian. Therefore, the actual parameter that sets the magnitude of CP violation in QCD is

$$\theta + \arg\left(\det \mathcal{M}\right) . \tag{15}$$

In what follows, for simplicity of notation, we will denote this parameter by the same letter θ, implying that the part $\arg\left(\det \mathcal{M}\right)$ is included by default.

Perhaps, the most pronounced effect of the θ term is generation of a nonzero neutron electric dipole moment (nEDM). The latter is parametrized by the following effective Lagrangian

$$\mathcal{L}_{nEDM} = \frac{d_n^\gamma}{2} \, \bar{n} \, i \, \gamma_5 \, \sigma_{\mu\nu} \, n \, F^{\mu\nu} , \tag{16}$$

where the photon field strength is $F_{\mu\nu} = \partial_\mu \mathcal{A}_\nu - \partial_\nu \mathcal{A}_\mu$, and n stands for the neutron. Moreover, $\sigma_{\mu\nu}$ is the antisymmetric product of two Dirac's gamma matrices, $\sigma_{\mu\nu} \equiv [\gamma_\mu \gamma_\nu]/2i$.

In the presence of the θ term, nEDM can be found from the following matrix element:

$$\langle n(p_f) \, \gamma(k) \, | e \, J_\mu^{\text{em}} \, \mathcal{A}^\mu \cdot i \int d^4x \, \Delta \, \mathcal{L}_\theta \, | n(p_i) \rangle$$

$$= d_n^\gamma \, \bar{n}(p_f) \, \gamma_5 \, \sigma_{\mu\nu} \, n(p_i) \, k^\mu \epsilon^\nu(k) , \tag{17}$$

where J_μ^{em} is the quark electromagnetic current. The momentum carried by the photon, $k_\mu = (p_f)_\mu - (p_i)_\mu$, equals to the difference between the final and initial momenta of the neutron while $\epsilon^\nu(k)$ denotes the photon polarization four-vector.

The matrix element on the left-hand side of (17) is a highly nonperturbative object. Its calculation in QCD is nontrivial. Nevertheless, there are a number of different methods by which nEDM was estimated. We will list them below. The bag model calculation was performed in Ref. 11. The result is

$$d_n^\gamma|_{\text{Bag}} \simeq \theta\, 2.7 \cdot 10^{-16}\, e \cdot \text{cm}\,. \tag{18}$$

Shortly after Ref. 11, the chiral logarithms (CL) method was used [12] leading to the estimate

$$d_n^\gamma|_{\text{CL}} \simeq \theta\, 5.2 \cdot 10^{-16}\, e \cdot \text{cm}\,. \tag{19}$$

The method of chiral perturbation theory (ChPT) was further advanced in Ref. 13 with the following result

$$d_n^\gamma|_{\text{ChPT}} \simeq \theta\, 3.3 \cdot 10^{-16}\, e \cdot \text{cm}\,. \tag{20}$$

Finally, the most recent paper on QCD sum rule (SR) calculations [14] of d_n^γ gives [15]

$$d_n^\gamma|_{\text{SR}} \simeq \theta\, 1.2 \cdot 10^{-16}\, e \cdot \text{cm}\,. \tag{21}$$

All results above have a considerable uncertainty, of at least 50%, which reflects a variety of uncertainties inherent to nonperturbative QCD calculations. Even though the results scatter by a factor of several units, it is beyond any doubt that $d_n^\gamma|_{\text{theor}} \sim \theta\, 10^{-16}\, e \cdot \text{cm}$.

This number should be compared with the most recent experimental result for nEDM presented in Ref. 16

$$|d_n^\gamma|_{\text{exp}} < 6.3 \cdot 10^{-26}\, e \cdot \text{cm}\,. \tag{22}$$

One gets a very strong constraint on the value of θ,

$$|\theta| \lesssim 10^{-9}\,. \tag{23}$$

We see that if the value of θ is nonvanishing it has to be unnaturally small. There is no *a priori* reason why two terms in (15), the bare theta parameter and $\arg(\det \mathcal{M})$, should cancel each other with such an extraordinary accuracy, of one part in 10^9 or better. A dynamical mechanism is needed to explain the unnatural smallness of the θ term.

Before proceeding further, let us mention that other CP odd effects are induced by the θ term too. They impose less stringent bounds on θ, however. For instance, a nonzero θ gives rise to a nonvanishing amplitude of a CP violating decay $\eta \to \pi^+ \pi^-$ for which [17,13]

$$\mathrm{Br}(\eta \to \pi^+ \pi^-) \simeq \theta^2 \, 1.8 \cdot 10^2 \, .$$

The experimental limit for this decay is $\mathrm{Br}_{\exp}(\eta \to \pi^+ \pi^-) < 1.5 \cdot 10^{-3}$. This yields a constraint $|\theta| < 3 \cdot 10^{-3}$, much weaker than (23).

2.4. *Can QCD solve the strong* CP *problem itself ?*

Thus, theorists' task is to try to find a mechanism which would make CP conservation in strong interactions natural. Two alternative approaches are logically possible. One can invoke physics beyond QCD (this approach will be discussed in the bulk of this review) or one can try a minimalistic standpoint and ask whether QCD could solve the strong CP problem itself, with no new physics.

An obvious solution of the latter type exists: were one of the quarks massless, e.g., $m_u = 0$, then all θ effects would be unobservable. In this case there is a global $U_A(1)$ symmetry of the chiral rotations of the u quark field, $u_R \to \exp(i\beta) \, u_R$, $u_L \to \exp(-i\beta) \, u_L$, which can eliminate θ altogether. However, $m_u = 0$ does not go through on phenomenological grounds,[18] and at present this scenario may be safely discarded.

A more intricate solution could exist if confinement itself were to ensure the effects of the θ term to be screened. As far as we know, this question was first raised by A. Polyakov shortly after the discovery of the strong CP problem. His argumentation was as follows. The θ term in the action is the integral over the full derivative, which can be operative only if there are long-range components of the gauge fields. In the quasiclassical approximation such components are certainly present, as is evident from instanton calculations. However, this approximation misses the most salient feature of QCD, color confinement, which might eliminate long-range interactions (i.e. "screen" color) and make the integral over the full derivative vanish. That's exactly what happens in the (1+2)-dimensional Polyakov model of color confinement.[19]

This issue was studied in Ref. 17, with the negative conclusion. The problem is that there are two effects in QCD which depend on the above "screening": the heaviness of the η' mass (the so-called U(1) problem) and CP conservation/violation. Even though we do not know precisely how exactly the confinement mechanism works in QCD, we know for a fact that η' is split from the octet of the Goldstone bosons. This knowledge (plus some reasonable arguments regarding the value of chiral and/or $1/N_c$ corrections) is sufficient to show that confinement in QCD does not eliminate the θ dependence and thus does not solve the strong CP problem. We briefly outline this line of reasoning below.

The flavor singlet meson, the η', is significantly heavier than the flavor octet Goldtones, $m_{\eta'} \approx 958\,\text{MeV}$. As was shown by Weinberg in the pre-QCD era, were the η' a Goldstone, its mass would be constrained by $m_{\eta'} < \sqrt{3} m_\pi$. This suggests that, unlike the octet of the Goldstone bosons, the η' is not massless in the chiral limit (unless chiral expansion is invalid). This is the only fact we will need.

An extra contribution to the η' mass comes from nonperturbative effects due to the axial anomaly, as was exemplified [20] by the same instantons.

To quantify the effect on the theoretical side, let us introduce the correlator of the topological charge densities (in the literature it is referred to as the topological susceptibility)

$$\mathcal{X} = -i \int d^4x \, \langle 0|T\, Q(x)\, Q(0)\, |0\rangle \,, \tag{24}$$

where

$$Q \equiv \frac{1}{32\,\pi^2} G_{\mu\nu}^a \, \tilde{G}_{\mu\nu}^a \,. \tag{25}$$

According to the low-energy theorem derived in Refs. 21,22 in the leading approximation of the $1/N_c$ expansion, the quantity $\mathcal{X}$ is saturated by the η' contribution implying the following formula for the η' mass:

$$m_{\eta'}^2 = \frac{6\,\mathcal{X}}{f_\pi^2} + \mathcal{O}\,(m_q) + \mathcal{O}\left(\frac{1}{N_c^2}\right), \tag{26}$$

where the topological susceptibility $\mathcal{X}$ on the right-hand side is evaluated in pure gluodynamics, the Yang-Mills theory with no light quarks. In order for the η' mass to be nonzero in the chiral limit, $\mathcal{X}$ should be nonzero in pure gluodynamics.

A substantial amount of theoretical evidence is accumulated in the last 20 years showing that the topological susceptibility in pure Yang-Mills does not vanish. The lattice [23] and the QCD sum rule studies [24] yield $\mathcal{X} \simeq (180\,\text{MeV})^4 \neq 0$. This successfully takes care of the $U(1)$ problem.

Having nonzero topological susceptibility in pure gluodynamics means, *per se*, that there is a sensitivity to the parameter θ in this theory. Indeed, $\mathcal{X}$ is nothing but the second derivative of the vacuum energy with respect to θ taken at $\theta = 0$

$$\mathcal{X} = -\frac{\partial^2}{\partial \theta^2} \left(\frac{\ln Z_\theta}{V}\right)\Bigg|_{\theta=0} \,. \tag{27}$$

In the theory with the light quarks included, the topological susceptibility can be calculated by applying chiral perturbation theory (see e.g.

Ref. 17). As expected on general grounds, in this case $\mathcal{X} \propto m_q$ provided the η' is split from the octet of the Goldstones; otherwise the η' contribution cancels the $O(m_q)$ term in the topological susceptibility. However, in the theory with the light quarks it is much more instructive to calculate directly CP odd decay rates, for instance the rate of $\eta \to \pi^+ \pi^-$. This amplitude is forbidden by CP. In the same way as with the topological susceptibility, the chiral low-energy theorem yields[17] a nonvanishing amplitude $O(m_q)$ provided the η' is split from the octet of the Goldstones.

Therefore, since QCD does solve the U(1) problem — it does split the η' from the octet Goldstones — it cannot solve the strong CP problem[17] without help from outside.[b]

3. In search of a solution beyond QCD

3.1. *Peccei-Quinn mechanism*

The first dynamical mechanism solving the strong CP problem was proposed by Peccei and Quinn.[27] The main observation of Ref. 27 is as follows: if there is a U(1) axial symmetry in the theory

$$q_L \to e^{i\alpha} q_L, \qquad q_R \to e^{-i\alpha} q_R, \tag{28}$$

then the θ term can be removed from the Lagrangian, much like in the case of one massless quark discussed in the previous section. Below this

---

[b]The situation seems to be quite clear in this respect, nevertheless, an attempts to develop models of confinement that would solve the strong CP problem are not abandoned. Let us comment on a proposal of Ref. 25 where two distinct "topological susceptibilities" are defined: the local and global ones. Let V_c denote the volume at which the confinement effects take place, and V be the total volume of space time $V_c \ll V \to \infty$. According to Ref. 25, the local topological susceptibility is $\mathcal{X}_{\text{loc}} = \int_{V_c} d^4x \langle 0|T Q(x) Q(0)|0\rangle$, while the global one is $\mathcal{X}_{\text{glob}} = \int_V d^4x \langle 0|T Q(x) Q(0)|0\rangle$. The author claims that the solution of the U(1) problem requires that $\mathcal{X}_{\text{loc}} \neq 0$, while the solution of the strong CP implies $\mathcal{X}_{\text{glob}} = 0$, and both conditions may be dynamically compatible, so that both strong CP and U(1) problems could be solved simultaneously. The underlying dynamics outlined in Ref. 25 is a special interaction between instantons which "screens" them.

There are a number of objections to this suggestions. First, it is the global topological susceptibility that enters in the Witten-Veneziano relation and determines the η' mass modulo $1/N_c$ corrections. Even if we forget about theoretical calculations of this quantity demonstrating that it does not vanish, we know that the η' is split from the Goldstone octet because the Weinberg relation $m_{\eta'} < \sqrt{3}m_\pi$ is grossly violated. (Of course, the validity of the chiral expansion is assumed; otherwise the Weinberg relation is meaningless.) If the η' is split, there is no way out:[17] θ-induced effects are observable.

In terms of the model suggested in Ref. 25 this means that if one were able to complete the calculations at the level of physical observables, one would find that the η' is not split from the octet of the Goldstones or, more likely, that the required interaction between instantons is not sustainable. The latter variant was advocated in Ref. 26.

symmetry will be referred to as $U(1)_{PQ}$.

To see whether this symmetry is present in the Standard Model with one Higgs doublet let us consider the Yukawa sector and restrict ourselves to the first generation quarks (consideration in the general case is quite similar),

$$\lambda_u \bar{Q}_L \phi u_R + \lambda_d \bar{Q}_L \phi_c d_R + \text{H.c.} + V(\phi^+ \phi), \qquad (29)$$

where Q_L is the left-handed $SU(2)_W$ quark doublet, $(\phi_c)_i = \epsilon_{ij} \phi^{*j}$ is the charge conjugate Higgs field, and V denotes the Higgs potential. Although the first term in this expression is invariant under the transformations

$$q_L \rightarrow e^{i\alpha} q_L, \quad q_R \rightarrow e^{-i\alpha} q_R, \quad \phi \rightarrow e^{2i\alpha} \phi, \qquad (30)$$

there is a second term which is not invariant under (30) since ϕ_c transforms as conjugate to ϕ. Therefore, the one Higgs doublet SM is *not* invariant under $U(1)_{PQ}$ and the strong CP cannot be solved.

However, as was pointed out in Ref. 27, the required $U(1)_{PQ}$ symmetry is present in SM with *two* Higgs doublets — let us call them ϕ and χ. In this case the Yukawa sector for the first generation quarks reads

$$\lambda_u \bar{Q}_L \phi u_R + \lambda_d \bar{Q}_L \chi^* d_R + \text{H.c.} + V(\phi^+ \phi, \chi^+ \chi, (\phi^+ \chi)(\chi^+ \phi)) \quad (31)$$

It is invariant under

$$q_L \rightarrow e^{i\alpha} q_L, \quad q_R \rightarrow e^{-i\alpha} q_R, \quad \phi \rightarrow e^{2i\alpha} \phi, \quad \chi \rightarrow e^{-2i\alpha} \chi. \qquad (32)$$

This fact can be used to solve the strong CP problem.[27] The symmetry (32) is explicitly broken by the axial anomaly. As a result, instantons induce an effective potential for the θ term. The potential can be calculated in a certain approximation.[27] The crucial model-independent fact is that the resulting potential is minimized at a zero value of the CP violating phase.[27]

3.2. Weinberg-Wilczek axion

When the Higgs fields develop vacuum expectation values (VEV's) the local electroweak symmetry group is spontaneously broken. This gives masses to the intermediate $W^{\pm}$ and Z vector bosons. Simultaneously, the global $U(1)_{PQ}$ is spontaneously broken too. Spontaneous breaking of the *global* symmetry leads to the emergence of a massless Goldstone boson, the axion in the present case.[1,2] In SM with two Higgs doublets the axion is given by the following superposition

$$a \equiv \frac{1}{v} (v_\phi \text{Im}\phi_0 - v_\chi \text{Im}\chi_0), \qquad (33)$$

where ϕ_0 and χ_0 denote the neutral components of the Higgs doublets. Moreover, $v \equiv \sqrt{v_\phi^2 + v_\chi^2} \simeq 250\,\text{GeV}$, and v_ϕ and v_χ are the vacuum expectation values of ϕ and χ, respectively. In this approximation the axion is massless. However, as we mentioned above, nonperturbative QCD effects (such as instantons) give rise to a potential for the axion. Hence, the axion acquires a nonzero mass which can be estimated as follows:[1,2]

$$m_a \simeq \frac{f_\pi m_\pi}{v} \simeq 100\,\text{KeV} . \tag{34}$$

Moreover, the axion decay constant is $1/v$. Therefore, we see that the Weinberg-Wilczek (WW) axion mass and decay constant are tied to the electroweak symmetry breaking scale v. This turns out to be too much of a constraint, and, as we will discuss in Sec. 3.5, the WW axion is excluded on the basis of existing experimental data.

3.3. KSVZ axion

If the scale of PQ symmetry breaking is much higher than the electroweak scale, then according to (34), the axion is much lighter and its decay constant is much smaller. Such an "invisible" axion would not be in conflict with experimental data.

A scenario with the harmless axion was first proposed in Refs. 28 and 17 (the KSVZ axion). In the latter paper it was called *phantom axion*. Needless to say that to untie the axion from the electroweak scale one has to decouple the corresponding scalar fields from the known quarks and couple them to hypothetical (very) heavy fermion fields carrying color.

In more detail, one introduces a complex scalar field Φ coupled to a hypothetical electroweak singlet, a quark field Q in the fundamental representation of color $SU(3)$,

$$\Delta \mathcal{L} = \Phi \bar{Q}_R Q_L + \text{H.c.} . \tag{35}$$

The modulus of Φ is assumed to develop a large vacuum expectation value $f/\sqrt{2}$, while the argument of Φ becomes the axion field a, modulo normalization,

$$a(x) = f\alpha(x) , \quad \alpha(x) \equiv \text{Arg}\Phi(x) , \quad f \gg \Lambda . \tag{36}$$

Then the low-energy coupling of the axion to the gluon field is

$$\Delta \mathcal{L} = \frac{1}{f}\, a\, \frac{1}{32\pi^2}\, G_{\mu\nu}^a \tilde{G}_{\mu\nu}^a , \tag{37}$$

so that the QCD Lagrangian depends on the combination $\theta + \alpha(x)$.

In general, one could introduce more than one fundamental field Q, or introduce them in a higher representation of the color group. Then, the axion-gluon coupling (37) acquires an integer multiplier N,

$$\Delta \mathcal{L}' = \frac{1}{f} \, a \, N \, \frac{1}{32\pi^2} \, G^a_{\mu\nu} \tilde{G}^a_{\mu\nu} \,. \tag{38}$$

This factor N (not to be confused with the number of colors N_c nor with $\mathcal{N}$ of extended supersymmetry) is sometimes referred to as the axion index. The minimal axion corresponds to $N = 1$. In the general case the QCD Lagrangian depends on the combination $\theta + N\alpha(x)$. As previously, non-perturbative QCD effects generate a potential for $\theta + N\alpha(x)$. The latter is minimizes at the value $\theta + N\alpha_{\mathrm{vac}} = 0$, i.e., the strong CP problem is automatically solved.

3.4. ZDFS axion

An alternative way to introduce an "invisible" axion was proposed in Refs. 29 and 30, (the ZDFS axion). In this proposal one maintains the PQ symmetry of the two doublet SM but separates the scales of the PQ and electroweak breaking.[29,30] To this end the SM Lagrangian is extended – a scalar SM singlet field Σ is added,

$$\lambda_u \, \bar{Q}_L \, \phi \, u_R + \lambda_d \, \bar{Q}_L \, \chi^* \, d_R + \text{H.c.} +$$

$$V(\phi^+ \phi, \ \chi^+ \chi, \ (\phi^+ \chi)(\chi^+ \phi), \ \Sigma^+ \Sigma, \ (\phi^+ \chi) \Sigma^2) \,. \tag{39}$$

One notes that this expression is invariant under the following axial transformations

$$q_L \to e^{i\alpha} \, q_L \,, \quad q_R \to e^{-i\alpha} \, q_R \,, \quad \phi \to e^{2i\alpha} \, \phi \,, \quad \chi \to e^{-2i\alpha} \, \chi \,, \quad \Sigma \to e^{2i\alpha} \, \Sigma \tag{40}$$

Upon spontaneous breaking of this symmetry the Goldstone particle, an axion, emerges as a superposition

$$a \equiv \frac{1}{V} \, (v_\phi \mathrm{Im}\phi_0 - v_\chi \mathrm{Im}\chi_0 + v_\Sigma \, \mathrm{Im}\Sigma) \,, \tag{41}$$

where $V \equiv \sqrt{v_\phi^2 + v_\chi^2 + v_\Sigma^2}$, and v_ϕ, v_χ and v_Σ are the vacuum expectation values of ϕ, χ and Σ, respectively. The vacuum expectation value of Σ does not have to be related to the electroweak symmetry breaking scale. In fact, it can be as large as the GUT scale. If so, the axion is light and its decay constant is tiny. We will discuss experimental bounds on these quantities in the next section.

3.5. *Constraints on the axion mass*

As we discussed in the previous sections, the PQ symmetry is explicitly broken by the axial anomaly. Therefore, the axion is a *pseudo* Goldstone boson. Nonperturbative QCD effects induce the axion mass. For further discussions it is convenient to parametrize the axion mass as follows:

$$m_a \simeq 0.6\,\text{eV}\,\frac{10^7\,\text{GeV}}{f}, \qquad (42)$$

where f is the axion decay constant determined by the PQ breaking scale.

In general, while discussing phenomenological constrains on the axion mass, one should distinguish between the KSVZ and ZDFS cases. The axion couplings to matter are different in these two scenarios. In particular, the KSVZ axion has no tree-level couplings to the standard model quarks and leptons. However, the aim of the present section is to summarize briefly an order of magnitude constraints on the axion mass. For this goal the effects which distinguish between the KSVZ and ZDFS axions will not be important (for detailed studies see Ref. 31 and citations therein).

The quantity $1/f$ sets the strength of the axion coupling. Light axions can be produced in stars and a part of the energy of a star can be carried away by those axions. Stars can loose energy due to the production of light axions in the following possible processes

(i) Nucleon-nucleon bremsstrahlung: $N + N \to N + N + a$;

(ii) The Primakoff process: $\gamma \leftrightarrow a$ conversion in the electromagnetic field of a nucleus;

(iii) Photoproduction on an electron: $\gamma + e^- \to e^- + a$;

(iv) Electron bremsstrahlung on a nucleus: $e^- + (A, Z) \to e^- + a + (A, Z)$;

(v) Photon fusion: $\gamma + \gamma \to a$;

Unless $1/f$ is really small, the emission of axions that are produced in the above reactions would lead to unacceptable energy loss by the star. This leads to the following lower bound on the axion decay constant:

$$f \gtrsim 10^9\,\text{GeV}.$$

It is remarkable that cosmology puts [32,33,34] an *upper* bound on f. The latter comes about as follows. If f is too large then the axion coupling $1/f$ is very small. As a result, during the course of cosmological evolution of the universe axions decouple early and begin to oscillate coherently. There are two major mechanisms by which the energy density stored in these oscillations can be dissipated – the Hubble expansion of the universe and the particle production by axions. However, if $f \gtrsim 10^{12}\,\text{GeV}$, neither of these mechanisms are effective (the axion coupling is too small). As a result, at some point of the evolution the axion energy density exceeds the critical energy density and over-closes the universe. In order for this not to happen one should impose the constraint $f \lesssim 10^{12}\,\text{GeV}$.

Summarizing, we obtain the following order of magnitude bounds on f and m_a:

$$10^9\,\text{GeV} \lesssim f \lesssim 10^{12}\,\text{GeV}, \quad 10^{-6}\,\text{eV} \lesssim m_a \lesssim 10^{-3}\,\text{eV}. \qquad (43)$$

For further details see, e.g., Ref. 31.

4. The vacuum structure in large N_c gluodynamics

The early studies [35] of the chiral Ward identities in QCD revealed that the vacuum energy density depends on the vacuum angle θ through the ratio θ/N_f, where N_f is the number of quarks with mass $m_q \ll \Lambda$. Shortly after it was shown in Refs. 36 and 37 that this structure occurs naturally, provided that there exist N_f states in the theory such that one of them is the true vacuum, while others are local extrema; all are intertwined in the process of "the θ evolution." Namely, in passage from $\theta = 0$ to $\theta = 2\pi$, from $\theta = 2\pi$ to $\theta = 4\pi$, and so on, the roles of the above states interchange: one of the local extrema becomes the global minimum and *vice versa*. This would imply, with necessity, that at $\theta = k\pi$ (where k is an odd integer) there are two degenerate vacuum states. Such a group of intertwined states will be referred to as the "vacuum family." The crossover at $\theta = \pi$, 3π, etc. is called the Dashen phenomenon. [38]

This picture was confirmed by a detailed examination of effective chiral Lagrangians [36,37,39,40] (for a recent update see Ref. 41). For two and three light quarks with equal masses it was found that the vacuum family consists of two or three states respectively; one of them is a global minimum of the potential, while others are local extrema.[c] At $\theta = \pi$ the levels intersect. Thus, Crewther's dependence [35] on θ/N_f emerges.

On the other hand, the examination of the effective chiral Lagrangian with the realistic values of the quark masses, $m_d/m_u \sim 1.8$, $m_s/m_d \sim 20$, yields [36,37,41] a drastically different picture – the vacuum family disappears (shrinks to one state); the crossover phenomenon at $\theta = \pi$ is gone as well.

This issue remained in a dormant state for some time. Recently arguments were given that the "quasivacua" (i.e. local minima of the energy

---

[c] We stress that the states from the vacuum family need not necessarily lie at the minima of the energy functional. As was shown by Smilga,[41] at certain values of θ some may be maxima. Those which intersect at $\theta = k\pi$ (k odd) are certainly the minima at least in the vicinity of $\theta = k\pi$.

functional), which together with the true vacuum form a vacuum family, is an indispensable feature of gluodynamics. The first argument in favor of this picture derives [42] from supersymmetric gluodynamics, with supersymmetry softly broken by a gluino mass term. The same conclusion was reached in Ref. 43 based on a D-brane construction in the limit of large N_c. One can see that in both approaches the number of states in the vacuum family scales as N_c. In fact, in Ref. 43 the expression for the theta dependence of the vacuum energy in the large N_c pure Yang-Mills (YM) model was derived from a D-brane construction. It has [43] the following form: [d]

$$\mathcal{E}(\theta) = C \, \min_k \, (\theta + 2\pi k)^2 \, + \mathcal{O}\left(\frac{1}{N_c}\right), \qquad (44)$$

where C is some constant independent of N_c and k stands for an integer number. This expression has a number of interesting features which might seem a bit puzzling from the field theory point of view. Indeed, in the large N_c limit there are N_c^2 degrees of freedom in gluodynamics, thus, naively, one would expect that the vacuum energy density in this theory scales as $\sim N_c^2$. However, the leading term in Eq. (44) scales as ~ 1. As a natural explanation, one could conjecture that there should be a colorless massless excitation which saturates the expression for the vacuum energy density (44). However, pure gluodynamics generates a mass gap and there are no physical massless excitations in the model. Thus, the origin of Eq. (44) seems to be a conundrum. We will discuss and resolve this puzzle in the next section.

Note, that an additional argument in favor of the vacuum family may be found in a cusp structure which develops once one sums up [44] sub-leading in $1/N_c$ terms in the effective η' Lagrangian. At large $N_c = \infty$ the states from the vacuum family are stable, and so are the domain walls interpolating between them. [43,45]

When $N_c < \infty$ the degeneracy and the vacuum stability is gone, strictly speaking. It is natural to ask what happens if one switches on the axion field. This generically leads to the formation of the axion domain walls. The axion domain wall [46] presents an excellent set-up for studying the properties of the QCD vacuum under the θ evolution. Indeed, inside the axion wall, the axion field (which, in fact, coincides with an effective θ) changes slowly from zero to 2π. The characteristic length scale, determined by the inverse axion mass m_a^{-1}, is huge in the units of QCD, Λ^{-1}. Therefore, by visualizing a set of spatial slices parallel to the axion wall, separated by distances $\gg \Lambda^{-1}$,

---

[d]It has been conjectured long time ago in Ref. 36.

one obtains a chain of QCD laboratories with distinct values of θ_{eff} slowly varying from one slice to another. In the middle of the wall $\theta_{\text{eff}} = \pi$.

Intuitively, it seems clear that in the middle of the axion wall, the effective value of $\theta_{\text{eff}} = \pi$. Thus, in the central part of the wall the hadronic sector is effectively in the regime with two degenerate vacua, which entails a *stable* gluonic wall as a core of the axion wall. In fact, we deal here with an axion wall "sandwich." Its core is the so-called D-wall, see Ref. 47.

Below we will discuss this idea more thoroughly. We also address the question whether this phenomenon persists in the theory with light quarks, i.e., in real QCD. Certainly, in the limit $N_c = \infty$ the presence of quarks is unimportant, and the axion wall will continue to contain the D-wall core. As we lower the number of colors, however, below some critical number it is inevitable that the regime must change, the gluonic core must disappear as a result of the absence of the crossover. The parameter governing the change of the regimes is Λ/N_c as compared to the quark mass m_q. At $m_q \ll \Lambda/N_c$, even if one forces the axion field to form a wall, effectively it is screened by a dynamical phase whose origin can be traced to the η', so that in the central part of the axion wall the hadronic sector does not develop two degenerate vacua. The D-walls cannot be accessed in this case via the axion wall.

The issue of hadronic components of the axion wall in the context of a potential with cusps [44] were discussed in Refs. 48,49,50. However, the gluonic component of the axion walls was not studied. The η' component in the axion walls was discussed in Refs. 51,48.

4.1. *Arguments from supersymmetric gluodynamics*

First we will summarize arguments in favor of the existence of a nontrivial vacuum family in pure gluodynamics.

The first indication that the crossover phenomenon may exist in gluodynamics comes [42] from supersymmetric Yang-Mills theory, with supersymmetry being broken by a gluino mass term. The same conclusion was reached in Ref. [43] based on a D-brane construction in the limit of large N_c. In both approaches the number of states in the vacuum family is N_c.

The Lagrangian of softly broken supersymmetric gluodynamics is

$$L = \frac{1}{g^2} \left\{ -\frac{1}{4} G_{\mu\nu}^a G_{\mu\nu}^a + i\, \bar{\lambda}_{\dot{\alpha}}^a D^{\dot{\alpha}\alpha} \lambda_{\alpha}^a - (m\lambda_{\alpha}^a \lambda^{a\alpha} + \text{H.c.}) \right\}$$
$$+ \theta\, \frac{1}{32\pi^2}\, G_{\mu\nu}^a \tilde{G}_{\mu\nu}^a \,, \tag{45}$$

where m is the gluino mass which is assumed to be small, $m \ll \Lambda$ (here we rescaled the gluon and gluino fields so that $1/g^2$ appears as a common multiplier in the Lagrangian).

There are N_c distinct chirally asymmetric vacua, which (in the $m = 0$ limit) are labeled by

$$\langle\lambda^2\rangle_\ell = N_c\Lambda^3 \exp\left(i\,\frac{\theta + 2\pi\ell}{N_c}\right), \quad \ell = 0, 1, ..., N_c - 1. \tag{46}$$

At $m = 0$ there are stable domain walls interpolating between them.[52] Setting $m \neq 0$ we eliminate the vacuum degeneracy. To first order in m the vacuum energy density in this theory is

$$\mathcal{E} = \frac{m}{g^2}\langle\lambda^2\rangle + \text{H.c.} = -m\,N_c^2\Lambda^3 \cos\frac{\theta + 2\pi\ell}{N_c}. \tag{47}$$

Degeneracy of the vacua is gone. As a result, all the metastable vacua will decay very quickly. Domain walls between them, will be moving toward infinity because of the finite energy gradient between two adjacent vacua. Eventually one ends up with a single true vacuum state in the whole space.

For each given value of θ the ground state energy is given by

$$\mathcal{E}(\theta) = \min_\ell\left\{-m\,N_c^2\,\Lambda^3 \cos\frac{\theta + 2\pi\ell}{N_c}\right\}. \tag{48}$$

At $\theta = \pi$, 3π, ..., we observe the vacuum degeneracy and the crossover phenomenon. If there is no phase transition in m, this structure will survive, qualitatively, even at large m when the gluinos disappear from the spectrum, and we recover pure gluodynamics.

Based on a D-brane construction Witten showed[43] that in pure $SU(N_c)$ (non-supersymmetric) gluodynamics in the limit $N_c \to \infty$ a vacuum family does exist:[e] the theory has an infinite group of states (one is the true vacuum, others are non-degenerate metastable "vacua") which are intertwined as θ changes by $2\pi \times$ (integer), with a crossover at $\theta = \pi\times$(odd integer). The energy density of the k-th state from the family is

$$\mathcal{E}_k(\theta) = N_c^2\,\Lambda^4\,F\left(\frac{\theta + 2\pi k}{N_c}\right), \tag{49}$$

---

[e]This was shown in Ref. 43 assuming that there is no phase transition in a certain parameter of the corresponding D-brane construction. In terms of gauge theory, this assumption amounts of saying that there is no phase transition as one interpolates to the strong coupling constant regime. Thus, the arguments of Ref. 43 have the same disadvantage as those of SUSY gluodynamics where one had to assume the absence of the phase transition in the gluino mass.

where F is some 2π-periodic function, and the truly stable vacuum for each θ is obtained by minimizing $\mathcal{E}_k$ with respect to k,

$$\mathcal{E}(\theta) = N_c^2 \, \Lambda^4 \min_k F\left(\frac{\theta + 2\pi k}{N_c}\right) , \tag{50}$$

much in the same way as in Eq. (48).

At very large N_c Eq. (50) takes the form

$$\mathcal{E}(\theta) = \Lambda^4 \min_k (\theta + 2\pi k)^2 + \mathcal{O}\left(\frac{1}{N_c}\right) . \tag{51}$$

The energy density $\mathcal{E}(\theta)$ has its absolute minimum at $\theta = 0$. At $N_c = \infty$ the "vacua" belonging to the vacuum family are stable but non-degenerate. To see that the lifetime of the metastable "vacuum" goes to infinity in the large N_c limit one can consider the domain walls which separate these vacua.[45,53] These walls are seen as wrapped D-branes in the construction of Ref. 43, and they indeed resemble many properties of the QCD D-branes on which a QCD string could end. We refer to them as D-walls because of their striking similarity to D2-branes. The consideration of D-walls has been carried out[45] and leads to the conclusion that the lifetimes of the quasivacua go to infinity as $\exp(\text{const } N_c^4)$.

Moreover, it was argued[54,47] that the width of these wall scales as $1/N_c$ both, in SUSY and pure gluodynamics. To reconcile this observation with the fact that masses of the glueball mesons scale as N_c^0, we argued[47] that there should exist heavy (glue) states with masses $\propto N_c$ out of which the walls are built. The D-brane analysis[55], effective Lagrangian arguments and analysis of the wall junctions,[56] support this interpretation. These heavy states resemble properties of D0-branes. The analogy is striking; as D0-branes make D2-branes from the standpoint of the M(atrix) theory,[57] so these QCD "zero-branes" make QCD D2-branes (domain walls).[f] The distinct vacua from the vacuum family differ from each other by a restructuring of these heavy degrees of freedom. They are essentially decoupled from the glueballs in the large N_c limit.

Now we switch on the axion

$$\Delta \mathcal{L} = \frac{1}{2} f^2 (\partial_\mu \alpha)(\partial^\mu \alpha) + \frac{\alpha}{32\pi^2} \, G_{\mu\nu}^a \tilde{G}_{\mu\nu}^a , \tag{52}$$

with the purpose of studying the axion walls. The potential energy $\mathcal{E}(\theta)$ in Eq. (50) or (51) is replaced by $\mathcal{E}(\theta + \alpha)$.

Since the hadronic sector exhibits a nontrivial vacuum family and the crossover[g] at $\theta = \pi, 3\pi$, etc., strictly speaking, it is impossible to integrate

---

[f] See also closely related discussions in Ref. 58.
[g] For nonminimal axions, with $N \geq 2$, the crossover occurs at $\alpha = k\pi/N$.

out completely the hadronic degrees of freedom in studying the axion walls. If we want to resolve the cusp, near the cusp we have to deal with the axion field *plus* those hadronic degrees of freedom which restructure. In the middle of the wall, at $\alpha = \pi$, it is mandatory to jump from one hadronic vacuum to another – only then the energy of the overall field configuration will be minimized and the wall be stable. Thus, in gluodynamics the axion wall acquires a D-wall core by necessity.

One can still integrate out the heavy degrees of freedom everywhere except a narrow strip (of a hadronic size) near the middle of the wall. Assume for simplicity that there are two states in the hadronic family. Then the low-energy effective Lagrangian for the axion field takes the form (83). The domain wall profile will also exhibit a cusp in the second derivative. The wall solution takes the form:

$$\alpha(z) = \begin{cases} 8 \arctan\left(e^{m_a z} \tan\frac{\pi}{8}\right), & \text{at } z < 0 \\ -2\pi + 8 \arctan\left(e^{m_a z} \tan\frac{3\pi}{8}\right), & \text{at } z > 0, \end{cases} \tag{53}$$

at $N_c = 2$ (the wall center is at $z = 0$).

Examining this cusp with an appropriately high resolution one would observe that it is smoothed on the hadronic scale, where the hadronic component of the axion wall "sandwich" would become visible. The cusp carries a finite contribution to the wall tension which cannot be calculated in the low-energy approximation but can be readily estimated, $T_{\text{core}} \sim \Lambda^3 N_c$. In subsequent sections we will examine this core manifestly in a toy solvable model.

4.2. *Arguments from D-brane construction*

The theta dependent vacuum energy (44) is related to the correlator measuring the vacuum fluctuations of the topological charge. The question which arises here is whether this can be seen from the original string theory computation.[43] We are going to discuss below how the string theory calculation suggests that the vacuum energy (44) should indeed be related to the vacuum fluctuations of the topological charge. In fact, we argue that this is related to the instantons carrying $D0$-brane charge in the Type IIA fourbrane construction of the four-dimensional YM model.

In general, a great deal of information can be learned on nonperturbative phenomena in four-dimensional gauge theories by obtaining these models as a low-energy realization of certain D-brane configurations,[59] and/or using a duality of large N superconformal gauge theories and string theory compactified on certain spaces (see Refs. 60 and 61,62). This duality, being a powerful technique, has also been generalized for the case of non-supersymmetric models.[63] This was applied to study various dynamical issues in large N_c pure Yang-Mills theory.[64,65,66,67,68,69]

To begin with let us recall how the theta dependent vacuum energy appears in the brane construction of the four-dimensional YM model.[43] One starts with Type IIA superstring theory on $\mathcal{M} \equiv \mathbf{R}^4 \times \mathbf{S}^1 \times \mathbf{R}^5$, with N_c coincident $D4$-branes.[63] The $D4$-brane worldvolume is assumed to be $\mathbf{R}^4 \times \mathbf{S}^1$ and the fermion boundary conditions on $\mathbf{S}^1$ are chosen in such a way that the low-energy theory on the worldvolume is pure non-supersymmetric $U(N_c)$ YM theory.[63] In the dual description, the large N_c limit of the $SU(N_c)$ part of this theory can be studied by string theory on a certain background.[60,61,62,63] It was shown in Ref. 43 that the theta dependent vacuum energy (44) arises in the dual string description due to the $U(1)$ gauge field B_M, $M = 1,..,5$. To find out what this corresponds to in the original gauge theory language recall that this $U(1)$ field is nothing but the Ramond-Ramond (RR) one-form of Type IIA theory. Furthermore, once the gauge theory is realized in the Type IIA fourbrane construction, the Wess-Zumino-Witten (WZW) term present in the worldvolume effective action defines the correspondence between the gauge theory operators on one side and the string theory Ramond-Ramond fields on the other side. In the case at hand the worldvolume WZW term looks as follows:

$$S_{\text{WZW}} = \frac{1}{8\pi^2} \int_\Omega B \wedge \text{Tr}\, G \wedge G, \qquad (54)$$

where Ω denotes the worldvolume of a wrapped fourbrane, $\Omega \equiv \mathbf{R}^4 \times \mathbf{S}^1$. In accordance with the general principles of the large N_c AdS/CFT correspondence[60,61,62] the classical action for the RR one-form on the string theory side defines the YM correlation functions of the composite operator $G\tilde{G}$ (since this is the operator which couples to the corresponding RR field in (54)). Thus, it is not surprising that the theta dependent vacuum energy which is defined by the RR one-form in the string theory calculation is related to the nonzero value of the topological susceptibility in the gauge theory studies. The physical reason for this correspondence, as we have mentioned above, is the special property of the gauge theory instantons in the fourbrane construction. Indeed, in accordance with (54) the RR one-form couples to the topological charge density $G\tilde{G}$, on the other hand the RR one-form couples by definition to $D0$-branes. Thus, the gauge theory instantons in this case carry zerobrane charge. This is the physical reason for the correspondence discussed above.

4.3. *Derivation of vacuum energy in QCD*

The aim of this section is to derive Eq. (44) in pure YM model and, in particular, to identify the degrees of freedom which are responsible for the theta dependent vacuum energy density.

In the quasi-classical approach the theta dependence can be calculated using instantons.[5] In a simplest approximation of non-interacting instantons the theta angle enters the Euclidean space partition function in the following form:

$$\exp\left(-\frac{8\pi^2}{g^2} \pm i\theta\right) \equiv \exp\left(-N_c\frac{8\pi^2}{\lambda} \pm i\theta\right), \tag{55}$$

where g stands for the strong coupling constant. λ denotes the 't Hooft's coupling $\lambda \equiv N_c \, g^2$ which is kept fixed in the large N_c limit. The expression above vanishes in the large N_c limit, so does the theta dependence in (55). However, this conclusion cannot be extrapolated to the infrared region of the model. The limitations of the expression (55) prevent one to do so. Indeed, the quasi-classical approximation is valid in the limit of small coupling constant (see, for instance, discussions in Ref. 70). Once quantum corrections are taken into account the coupling constant g^2 in (55) becomes a scale dependent quantity. In fact, it will depend upon an instanton size $g^2 = g^2(\rho)$. For small size instantons the running coupling is small and the quasi-classical approximation in (55) holds. However, for large size instantons, i.e. large couplings, it is not even clear whether the notion of a single instanton is a legitimate approximation. The overlap between instantons can be big in this case and some more complicated field configurations should be relevant for the description of physical phenomena.[71] In any event, the expression (55) is no longer reliable in the strong coupling limit. Thus, the conclusion that the theta dependence goes away in the large N_c limit cannot be justified. One way to study the infrared region is to look for some appropriate composite colorless excitations for which the notion of an asymptotic state can be used. We will start by searching for these excitations in pure Yang-Mills theory. To proceed, let us recall that the topological susceptibility, $\mathcal{X}$, is a nonzero number in pure gluodynamics (we rewrite it in the following form)

$$\mathcal{X} = -i \int \partial^\mu \, \partial^\nu \langle 0|T \, K_\mu(x)K_\nu(0)|0\rangle \, d^4x \neq 0 \,. \tag{56}$$

Here, K_μ as before denotes the Chern-Simons current. As we discussed earlier, the value of $\mathcal{X}$ in large N_c pure YM theory determines the η' meson mass in full QCD with massless quarks via the Witten-Veneziano formula[21,22]

$$m_{\eta'}^2 f_{\eta'}^2 \propto \mathcal{X},$$

with $f_{\eta'}$ being the η' meson decay constant.[h]

In what follows it will prove convenient to introduce a new variable by rewriting the expression for the topological charge density Q in terms of a four-index (four-form) tensor field $H^{\mu\nu\alpha\beta}$:

$$Q = \frac{\varepsilon_{\mu\nu\alpha\beta} H^{\mu\nu\alpha\beta}}{4!}, \tag{57}$$

where the four-form field $H^{\mu\nu\alpha\beta}$ is the field strength for the three-form potential $C_{\mu\nu\alpha}$:

$$H_{\mu\nu\alpha\beta} = \partial_\mu C_{\nu\alpha\beta} - \partial_\nu C_{\mu\alpha\beta} - \partial_\alpha C_{\nu\mu\beta} - \partial_\beta C_{\nu\alpha\mu}. \tag{58}$$

The $C_{\mu\nu\alpha}$ field is defined as a composite operator of the gluon fields A_μ^a:

$$C_{\mu\nu\alpha} = \frac{1}{16\pi^2} (A_\mu^a \overline{\partial}_\nu A_\alpha^a - A_\nu^a \overline{\partial}_\mu A_\alpha^a - A_\alpha^a \overline{\partial}_\nu A_\mu^a + 2f_{abc} A_\mu^a A_\nu^b A_\alpha^c). \tag{59}$$

Here, f_{abc} denote the structure constants of the corresponding $SU(N_c)$ gauge group. The right-left derivative in this expression is defined as $A\overline{\partial}B \equiv A(\partial B) - (\partial A)B$. Notice, that the $C_{\nu\alpha\beta}$ field is not a gauge invariant quantity; if the gauge transformation parameter is Λ^a, the three-form field transforms as $C_{\nu\alpha\beta} \to C_{\nu\alpha\beta} + \partial_\nu \Lambda_{\alpha\beta} - \partial_\alpha \Lambda_{\nu\beta} - \partial_\beta \Lambda_{\alpha\nu}$, where $\Lambda_{\alpha\beta} \propto A_\alpha^a \partial_\beta \Lambda^a - A_\beta^a \partial_\alpha \Lambda^a$. However, one can check that the expression for the field strength $H_{\mu\nu\alpha\beta}$ is gauge invariant.

It has been known for some time[72] that the $C_{\nu\alpha\beta}$ field propagates long-range correlations if the topological susceptibility is nonzero in the theory. The easiest way to see this is to turn to the notion of the Kogut-Susskind pole.[73] Let us consider the correlator of the vacuum topological susceptibility at a nonzero momentum. In this case $\mathcal{X}$ is defined as the zero momentum limit of the correlator of two Chern-Simons currents multiplied by two momenta:

$$\mathcal{X} = -i \lim_{q \to 0} q^\mu q^\nu \int e^{iqx} \langle 0|T K_\mu(x) K_\nu(0)|0\rangle d^4x. \tag{60}$$

Since this expression is nonzero, it must be that the correlator of two Chern-Simons currents develops a pole as the momentum vanishes, the Kogut-Susskind pole.[73]

---

[h]Below, unless stated otherwise, we will not distinguish between $\mathcal{X}$ and its large N_c limit. The constant contact term in the definition of $\mathcal{X}$ will also be omitted for simplicity.

Given that the correlator of two Chern-Simons currents has a pole and that the Chern-Simons current and the three-form $C_{\nu\alpha\beta}$ field are related, one concludes that the $C_{\nu\alpha\beta}$ field also has a nonzero Coulomb propagator.[72] Thus, the $C_{\nu\alpha\beta}$ field behaves as a massless collective excitation propagating a long-range interaction.[72] These properties, in the large N_c limit, can be summarized in the following effective action for the $C_{\nu\alpha\beta}$ field:

$$S_{\text{eff}} = -\frac{1}{2 \cdot 4! \, \mathcal{X}} \int H_{\mu\nu\alpha\beta}^2 \, d^4x - \frac{\theta}{3!} \int_{\partial\Gamma} C_{\nu\alpha\beta} \, dx^\nu \wedge dx^\alpha \wedge dx^\beta + \dots \quad (61)$$

where the dots denote higher-dimensional operators. The first term in this expression yields the correct Coulomb propagator for the three-form $C_{\nu\alpha\beta}$ field. The second term is just the usual CP odd θ term of the initial YM action written as a surface integral at spatial infinity $\partial\Gamma$. Notice that higher dimensional terms are not explicitly written in this expression. There might be two types of higher dimensional contributions in (61). First of all, there are terms with higher powers of derivatives of the fields. These terms are suppressed by momenta of the "massless" three-form field and do not contribute to the zero momentum vacuum energy of the system. In addition, there might be higher dimensional terms with no additional derivatives. In the next section we will present some of these contributions and show that they are suppressed by higher powers of $1/N_c$.

In what follows we are going to study the large N_c effective action given in Eq. (61).[i] In particular, we will calculate the ground state energy of the system in the large N_c limit using the effective action (61). In fact, we will derive Eq. (44).

Before we turn to this calculation let us mention that Maxwell's equations for a free four-form field-strength $H_{\mu\nu\alpha\beta}$ yield only a constant solution in $(3 + 1)$-dimensional space-time.[76] The reason is as follows. A four-form potential has only one independent degree of freedom in four-dimensional space-time, let us call it Σ. Then, the four Maxwell's equations written in terms of the Σ field require that this field is independent of the all four space-time coordinates, thus the solution can only be a space-time constant. As a result, the free $H_{\mu\nu\alpha\beta}$ field propagates no dynamical degrees of freedom in $(3 + 1)$-dimensions. However, this field can be responsible for a positive vacuum energy density in various models of Quantum Field

---

[i]The action (61) is not an effective action in the Wilsonian sense. It is rather related to the generating functional of one-particle-irreducible diagrams of the composite field in the large N_c limit. The effective action in Eq. (61) is not to be quantized and loop diagrams of that action are not to be taken into account in calculating higher order Green's functions. The analogous effective action for the CP even part of the theory was constructed in Refs. 74,75.

Theory (see Ref. 77). Thus, studying classical equations of motion for the $H_{\mu\nu\alpha\beta}$ field one can determine the value of the ground state energy given by configurations of $H_{\mu\nu\alpha\beta}$. We are going to solve explicitly the classical equations of motion for the effective action (61). Then, the energy density associated with those solutions will be calculated.

Let us start with the equations of motion. Taking the variation of the action (61) with respect to the $C_{\nu\alpha\beta}(z)$ field one gets

$$\partial^{\mu} H_{\mu\nu\alpha\beta}(z) = \theta \; \mathcal{X} \int_{\partial\Gamma} \delta^{(4)}(z - x) \; dx_{\nu} \wedge dx_{\alpha} \wedge dx_{\beta}. \tag{62}$$

This equation can be solved exactly in four-dimensional space-time.[76] The solution is the sum of a particular solution of the inhomogeneous equation and a general solution of the corresponding homogeneous equation:

$$H_{\mu\nu\alpha\beta}(z) = \theta \; \mathcal{X} \int \delta^{(4)}(z - x) \; dx_{\mu} \wedge dx_{\nu} \wedge dx_{\alpha} \wedge dx_{\beta} + b \; \varepsilon_{\mu\nu\alpha\beta}. \tag{63}$$

The integration constant b, if nonzero, induces an additional CP violation beyond the existed θ angle. However, periodicity of the θ angle with respect to shifts by $2\pi \times$ (integer) allows for some nonzero b proportional to $2\pi\mathbf{Z}$. As a result, the general solution to the equation of motion reads as follows:

$$H_{\mu\nu\alpha\beta} = -(\theta + 2\pi k) \; \mathcal{X} \; \varepsilon_{\mu\nu\alpha\beta} . \tag{64}$$

Thus, the different vacua are labeled by the integer k and the order parameter for these vacua in the large N_c limit can be written as:

$$\langle G\tilde{G} \rangle_k = (\theta + 2\pi k) \; \mathcal{X}. \tag{65}$$

As a next step let us compute the vacuum energy associated with the solution given in Eq. (64). The density of the energy-momentum tensor for the action (61) takes the form

$$\Theta_{\mu\nu} = -\frac{1}{3! \; \mathcal{X}} \left(H_{\mu\alpha\beta\tau} H_{\nu}^{\;\alpha\beta\tau} - \frac{1}{8} g_{\mu\nu} H_{\rho\alpha\beta\tau}^2 \right). \tag{66}$$

Using the expression (64) one calculates the corresponding energy density[j] $\mathcal{E}_k$,

$$\mathcal{E}_k(\theta) = \frac{1}{2}(\theta + 2\pi k)^2 \; \mathcal{X}. \tag{67}$$

---

[j]Note that the total YM energy density should contain some negative constant related to the nonzero value of the gluon condensate.[14] This constant is subtracted from the expression for the energy discussed in this work. The energy density (67) is normalized as $\mathcal{E}_0(\theta = 0) = 0$, and for $k = 0$ was discussed in Ref. 78.

Since the $H_{\mu\nu\alpha\beta}$ field does not propagate dynamical degrees of freedom the expression above is the total energy density of the system given by the action (61).[k]

Before we go further let us stop here to discuss some of the consequences of Eq. (67). First of all, let us notice that the result (67), as well as Eq. (44), is only valid in the limit of infinite N_c. Below we will calculate subleading order corrections to Eqs. (44,67) and argue that these expressions can also be used as a good approximation for large but finite N_c. The constant C emerging in (44) is related to the topological susceptibility as follows:

$$\mathcal{X}|_{N_c \to \infty} = 2C.$$

Thus, the vacuum energy (44,67) is defined by vacuum fluctuations of the topological charge measured by $\mathcal{X}$.

The crucial feature of (67) is that it defines an infinite number of vacua. The true vacuum is obtained by minimizing (67) with respect to k as in (44):

$$\mathcal{E}_0(\theta) = \frac{1}{2} \, \mathcal{X} \, \min_k \, (\theta + 2\pi k)^2 \, .$$

This expression is periodic[43] with respect to shifts of θ by $2\pi\mathbf{Z}$ and is also a smooth function of θ except for $\theta = \pi$ (see also discussions below). Thus, there is an infinite number of the false vacua in the theory.[43] The fate of these states will be discussed in Sec. 5.

Let us now consider full QCD with three quark flavors. We are going to write down a low-energy effective Lagrangian for this case and then gradually decouple quarks by taking the quark masses to infinity. The resulting effective Lagrangian should be giving the energy density for pure Yang-Mills theory.

In the large N_c expansion the effective Lagrangian of QCD with three flavors takes the form:[79,39,40]

$$\mathcal{L}(U, U^*, Q) = \mathcal{L}_0(U, U^*) + \frac{1}{2} \, i \, Q(x) \, \mathrm{Tr} \left(\ln U - \ln U^* \right) +$$
$$\frac{1}{2 \, \mathcal{X}} \, Q^2(x) + \theta \, Q(x) + \frac{B}{2\sqrt{2}} \, \mathrm{Tr} \, (MU + M^*U^*) + \dots \, , \qquad (68)$$

---

[k]One might wonder whether the same result is obtained if one treats θ not as a constant multiplying Q in the Lagrangian, but as the phase that the states acquire under a topologically non-trivial gauge transformations. In this case the arbitrary integration constant in Eq. (63) has to be chosen in such a way which would guarantee a proper θ dependence of the VEV of the topological charge density. This would leave the results of our discussion without modifications.

where U denotes the flavor group matrix of pseudoscalar mesons, $\mathcal{L}_0$ denotes the part of the Lagrangian which contains the meson fields only,[79,39,40] B is some constant related to the quark condensate, and M stands for the meson mass matrix (for recent discussions of the effective chiral Lagrangian approach see Ref. 80). Higher order terms in (68) are suppressed by quadratic and higher powers of $1/N_c$. In order to study vacuum properties, we concentrate on the low-momentum approximation. The Lagrangian presented above can be used to solve the $U(1)$ problem.[21,22] Indeed, the field Q enters the Lagrangian in a quadratic approximation and can be integrated out. As a result, the flavor singlet meson, the η', gets an additional contribution into its mass term. This leads to the Witten-Veneziano relation and the solution of the $U(1)$ problem without instantons.[21,22] In the present case we would like to follow an opposite way. Namely, we are going to make quarks very heavy and integrate them out keeping the field Q in the Lagrangian. In the limit $m_q \to \infty$ one finds that $M \to \infty$. Thus, the low-energy effective Lagrangian which is left after the mesons are integrated out will take the form:

$$\mathcal{L}_{\text{eff}}(Q) = \frac{1}{2\,\chi}\, Q^2(x) + \theta\, Q(x) + \mathcal{O}\left(\frac{\Lambda_{\text{QCD}}^2}{M^2},\ \frac{\partial^2 Q^2}{\Lambda_{\text{QCD}}^{10}},\ \frac{1}{N_c^2}\right). \quad (69)$$

Rewriting the field Q in terms of the "massless" tensor $C_{\alpha\beta\gamma}$ as in the previous section, one finds that the expression (69) is nothing but the Lagrangian presented in (61). Thus, the higher order terms neglected in (61) which could contribute to the vacuum energy at zero momenta would correspond to higher corrections in $1/N_c$. In fact, the subleading corrections to the effective Lagrangian (68) can also be found.[81] These terms are proportional (with the corresponding dimensionful coefficients) to the following expressions:

$$\frac{\text{const.}}{N_c^2}\, Q^2\, \text{Tr}\,(\partial_\mu U\, \partial_\mu U^*), \qquad \frac{\text{const.}}{N_c^2}\, Q^4\,. \quad (70)$$

The terms listed above are suppressed in the effective Lagrangian by the factor $1/N_c^2$. As a next step, we can include the terms (70) into the full effective Lagrangian and then integrate the heavy meson fields out. The net result of this procedure is that the terms proportional to Q^4 appear in the effective Lagrangian for pure YM theory. This, in its turn, modifies the equation of motion for the single component of $H_{\mu\nu\alpha\beta}$ considered in the previous section. Performing the calculation of the vacuum energy in the same manner as discussed above we find the following result for the energy density:

$$\mathcal{E}_k(\theta) = \frac{1}{2}\, \mathcal{X}\, \left(\theta + 2\pi k\right)^2 + \frac{\text{const.}}{N_c^2}\, \mathcal{X}\, \left(\theta + 2\pi k\right)^4 + \mathcal{O}\!\left(\frac{1}{N_c^3}\right). \quad (71)$$

In this expression a constant emerges as a result of integration of the equation of motion.[1] Notice that the topological susceptibility in the expression above is also defined in the corresponding order in the large N_c expansion: $\mathcal{X} = 2C + \mathcal{X}_1/N_c + \mathcal{X}_2/N_c^2$. Thus, the expressions (67,71) could in principle give a reasonable approximation for big enough but otherwise *finite* N_c. The true vacuum energy density, $\mathcal{E}_0(\theta)$, can be obtained by minimizing the expression (71) with respect to k as in (44). Then, $\mathcal{E}_0(\theta)$ satisfies the relation $\partial_\theta^2 \mathcal{E}_0(\theta)\,|_{\theta=0} = \mathcal{X}$, no matter what is the value of the constant in (71).

5. Dynamics of false vacua

In this section we will discuss the dynamics of the false vacua present in the theory. In accordance with (67,71) there are an infinite number of vacua for any given value of the theta angle. Clearly, not all of these are degenerate. As we discussed, the true vacuum state is defined by minimizing the expressions (67,71) with respect to k. All the other states are false vacua with greater values of the energy density. There is a potential barrier that separates a given false vacuum state from the true one. Thus, a false vacuum can in general decay[83] into the true state through the process of bubble nucleation.[m] The decay rates for these vacua were evaluated in Ref. 45. In this section we analyze the fate of the false vacua for different realizations of the initial conditions in which the system is placed. For the sake of simplicity we will be discussing transitions between the vacuum states labeled by k' and k for different values of these integers. The first two cases considered in this section were studied in Refs. 43 and 45, the remainder of the section follows Ref. 53.

5.1. *False vacua with* $k' \sim 1$

In this subsection we consider the system which in its initial state exists in a false vacuum with k' of order ~ 1. Let us start with the case when N_c is

---

[1]The numerical value of this constant was recently calculated on lattices.[82]

[m]This decay can go through the Euclidean "bounce" solution.[84] Though the existence of the bounce for this case is not easy to understand within the field theory context, nevertheless, one could be motivated by the brane construction where this object appears as a sixbrane bubble wrapped on a certain space.[43]

a large but finite number so that the formula (67) (or (71)) is still a good approximation. Since there exists the true vacuum state with less energy, the false vacuum can "decay" into the true one via the bubble nucleation process. That is to say, there is a finite probability to form a bubble with the true vacuum state inside. The shell of the bubble is a domain wall which separates the false state from the true one. The dynamical question we discuss here is whether it is favorable energetically to create and expand such a bubble. Let us study the energy balance for the case at hand. While creating the shell of the bubble one looses the amount of energy equal to the surface area of the bubble multiplied by the tension of the shell. On the other hand, the true vacuum state is created inside the bubble, thus, one gains the amount of energy equal to the difference between the energies of the false and true states. The energy balance between these two effects defines whether the bubble can be formed, and, whether the whole false vacuum can transform into the true one by expanding this bubble to infinity. Let us start with the volume energy density. The amount of the energy density which is gained[n] by creating the bubble is

$$\Delta \mathcal{E} = \frac{1}{2} \, \mathcal{X} \, \left[(\theta + 2\pi)^2 - \theta^2 \right] = 2\pi \, \mathcal{X} \, (\theta + \pi).$$

Thus, $\Delta \mathcal{E}$ scales as ~ 1 in the large N_c limit as long as the volume of the bubble is finite. Let us now turn to the surface energy which is lost. This energy is defined as:

$$E_s = T_D \times (\text{surface area}) . \tag{72}$$

The tension of the wall between the adjacent vacua, T_D, should scale as $T_D \sim N_c$ in the large N_c limit. Hence, the surface energy will also scale as $\sim N_c$. Thus, the process of creation of a finite volume bubble in the large N_c limit is not energetically favorable. Indeed, the amount of energy which is lost while creating the shell is bigger than the amount which is gained. In terms of the false vacuum decay width this means that the width of this process is suppressed[45] in the large N_c limit,

$$\frac{\Gamma}{\text{Volume}} \propto \exp\left(- a N_c^4 \right), \tag{73}$$

where a stands for a positive constant.[45] Thus, one concludes that in the limit $N_c \to \infty$ the false vacua with $k' \sim 1$ are stable.[43,45]

---

[n]In this subsection we assume that $\theta \neq \pm\pi$. The case $\theta = \pi$ will be considered below.

5.2. *False vacua with* $k' \sim N_c$

Here we study the fate of the false vacua with $k' \sim N_c$. We discuss a possibility that these vacua can decay into a state k with $k' - k \sim N_c$ and $k' + k \sim N_c$. As in the previous subsection, we are going to study the energy balance for the bubble nucleation process. The amount of the volume energy density which is gained by creating such a bubble in the large N_c limit scales as follows:

$$\Delta \mathcal{E} \propto \mathcal{X} \, N_c^2.$$

Thus, the volume energy which is gained increases as $\sim N_c^2$. Let us now turn to the surface energy which is lost while nucleating a bubble. This is defined as $E_s' = T_D' \times$ (surface area), where T_D' denotes the tension of the domain wall interpolating between the vacua labeled by k' and k. Since $k' - k \sim N_c$ these vacua are not neighboring ones. Thus, in general, there is no reason to expect that the tension of the wall interpolating between these vacua scales as $\sim N_c$. T_D' might scale as $\sim N_c^2$ at most (as the energy of a generic configuration in a model with N_c^2 degrees of freedom). However, even in the case when $T_D' \sim N_c^2$ the volume energy which is gained is at least of the same order as the surface energy which is lost. Hence, it is energetically favorable to increase the radius of such a bubble (since the volume energy scales as a cubic power of the radius while the surface energy scales only as a quadratic power of the effective size). Thus, the bubble nucleation process will not be suppressed and the false vacua with $k' \sim N_c$ will eventually decay into the true ground state. Note, that the state $k' = N_c$ can as well decay into the neighboring vacuum $k = N_c - 1$ which subsequently is allowed to turn into the ground state.

5.3. *Parallel domain walls*

In this subsection we consider the special case when all the vacua are present simultaneously in the initial state of the model. This can be achieved, for instance, by placing in space an infinite number of parallel domain walls separating different vacua from each other. It is rather convenient to picture these walls as parallel planes. Each vacuum state is sandwiched between the corresponding two domain walls (two planes) separating this state from the neighboring vacua. Each domain is labeled by k and in accordance with (67,71) is characterized by the corresponding value of the vacuum energy. Furthermore, the order parameter $\langle G\tilde{G} \rangle$ takes different values in these vacua in accordance with (65). Let us turn to the true vacuum state. For simplicity we assume that this state is given by $k = 0$

(which corresponds to $|\theta|$ being less than π). The corresponding vacuum energy is the lowest one. Consider the two states which are adjacent to the true vacuum. These states have the energy density bigger than that of the true vacuum. Thus, there is a constant pressure acting on the domain walls separating the true vacuum from the adjacent false ones. This pressure will tend to expand the domain of the true vacuum. In fact, for large but finite N_c, the pressure will indeed expand the spatial region of the true vacuum by moving apart the centers of the domain walls sandwiching this state. The very same effect will be happening between any two adjacent vacua. Indeed, let us calculate the jump of the energy density between the two vacua labeled by k' and k:

$$\Delta \mathcal{E}_{k'k} = 2\pi \; \mathcal{X} \; (k' - k) \left(\theta + \pi(k' + k) \right) . \tag{74}$$

As far as N_c is large but finite, the walls will start to accelerate. Farther the wall is located from the true vacuum (i.e. larger the sum $k' + k$), bigger the initial acceleration of the wall is going to be; i.e., the walls will start to move apart from each other with the following initial acceleration:

$$a_{k'k} \propto \Lambda_{\mathrm{YM}} \; \frac{(k' - k) \, [\theta + \pi(k' + k)]}{N_c} . \tag{75}$$

For finite N_c all the walls will be moving to spatial infinity and the whole space will eventually be filled with the true vacuum state. On the other hand, when $N_c \to \infty$ the picture is a bit different. There are a number of interesting cases to consider:

First of all let us set $k' - k = 1$ and k' , $k \sim 1$. Then, in the limit $N_c \to \infty$ the acceleration $a_{k'k} \to 0$. Thus, the neighboring walls stand still if they had no initial velocity. The physical reason of this behavior is as follows. Although there is a constant pressure of order ~ 1 acting on the wall, nevertheless, the wall cannot be moved because the mass per unit surface area of the wall tends to infinity in the limit $N_c \to \infty$.

The second interesting case would be when the constant pressure produced by the energy jump between some neighboring vacua is of order $\sim N_c$. In this case it will be possible to accelerate these walls up to the speed of light and send them to spatial infinity. Indeed, if $k' - k = 1$ but $k' + k \sim N_c$, then the wall between these two vacua starts moving with a non-vanishing acceleration which scales as follows:

$$a_{k'k} \propto \Lambda_{\mathrm{YM}} \frac{\pi(k' + k)}{N_c} \sim \mathcal{O}(1). \tag{76}$$

Thus, these walls will eventually be approaching spatial infinity with a speed of light even in the limit of infinite N_c.

In addition to the effects emphasized above there might also be decays of the false vacua happening in each particular domain. As we discussed in the previous subsections, for large but finite N_c all the false vacua will be nucleating bubbles with energetically favorable phases inside and expanding these bubbles to infinity. Thus, for large but finite N_c, there are two effects which eliminate the false vacua: The moving walls are sweeping these states to infinity, and, in addition, these vacua are decaying via bubble nucleation processes.

What happens for an infinite N_c? As we learned above there are an infinite number of domains which will stay stable in that limit and the corresponding false vacua would not decay because of the exponential suppression. Thus, there are an infinite number of inequivalent spatial regions which are separated by domain walls. Consider one of the regions sandwiched between two domain walls. The three-form field $C_{\mu\nu\alpha}$ will couple to the walls and the large N_c effective action for this case will look as follows:

$$\tilde{S} = S_{\text{eff}} + \sum_{i=k,\ k+1} \mu_i \int_{\mathcal{W}_i} C_{\mu\nu\alpha}\, dx^\mu \wedge dx^\nu \wedge dx^\alpha, \qquad (77)$$

where S_{eff} is defined in (61), μ_i stands for the coupling of the three-form potential to a corresponding domain wall; $\mathcal{W}_i$ denotes the worldvolume of the wall. In this case the domain wall can be regarded as a source of the corresponding three-form potential. This is reminiscent to what happens in the large N_c supersymmetric YM model.[85]

5.4. Domain walls at $\theta = \pi$

If $\theta = \pi$, the initial classical Lagrangian is CP invariant. Indeed, under CP transformations $\theta = \pi$ goes into $-\pi$. Since π and $-\pi$ angles are equivalent, CP is a symmetry of the Lagrangian. However, in accordance with (65), this symmetry is spontaneously broken by the vacuum of the theory. Thus, one finds the following two degenerate true vacua:

$$\mathcal{E}_{k=0} = \mathcal{E}_{k=-1} = \frac{1}{2}\, \mathcal{X}\, \pi^2 \,. \qquad (78)$$

These two vacua are labeled by the order parameter (65). In the $k = 0$ state $\langle G\tilde{G}\rangle = \pi\mathcal{X}$ and in the $k = -1$ state $\langle G\tilde{G}\rangle = -\pi\mathcal{X}$. As a result of the spontaneous breaking of a discrete symmetry there should be a domain wall separating these two vacua. Let us consider the case discussed in the previous section. Namely, let us choose the initial condition of the system as a state where all the possible vacua are simultaneously realized in space. That is, there are an infinite number of domain walls (parallel planes) dividing space into an infinite number of domains with different values of the vacuum energy density labeled by k. As we mentioned above, only two of these domains have equal minimal energy density given in (78). The domain wall separating these two vacua, as we will see below, is somewhat special. In accordance with the discussions in the previous subsection, for large but finite N_c, all the walls merging with the false vacua will tend to rush to spatial infinity. The final stable state of the model can be described as a space separated into two parts by a single domain wall. To the left (right) of the wall one discovers the phase with $k = -1$ with the corresponding order parameter $\langle G\tilde{G} \rangle = -\pi \mathcal{X}$, and, to the right (left) of the wall one finds the state with $k = 0$ and $\langle G\tilde{G} \rangle = \pi \mathcal{X}$. Vacuum energies of these two states are degenerate.

In the case of infinite N_c the picture is slightly different. As elucidated in the previous subsection, there will be an infinite number of stable vacua. The domain wall separating the two true vacua can be regarded in this case as the fixed plane under $\mathbf{Z}_2$ transformations of the coordinate transverse to the plane. The three-form field $C_{\mu\nu\alpha}$ will be able to couple to this wall in a manner discussed in the previous subsection.

6. Axions and vacuum structure in gluodynamics

6.1. *Two scenarios*

The invisible axion is very light. Integrating out all other degrees of freedom and studying the low-energy axion effective Lagrangian must be a good approximation. The axion effective potential in QCD can be of two distinct types.

Assuming that for all values of θ the QCD vacuum is unique one arrives at the axion effective Lagrangian of the form

$$\mathcal{L}_a = f^2 \left[\frac{1}{2}(\partial_\mu \alpha)^2 + m_a^2 \left(\cos(\alpha + \theta) - 1 \right) \right] . \tag{79}$$

The axion potential does not have to be (and generically is not) a pure cosine; it may have higher harmonics. In the general case it is a smooth periodic function of $\alpha + \theta$, with the period 2π. For illustration we presented the potential as a pure cosine. This does not change the overall picture in the qualitative aspect.

As we will see below, a smooth effective potential of the type (79) emerges even if the (hadronic) vacuum family is non-trivial, but the transition between the distinct hadronic vacua does not occur inside the axion wall. This is the case with very light quarks, $m_q \ll \Lambda/N_c$. In the opposite limit, one arrives at the axion potential with cusps, considered below.

In the theory (79) one finds the axion walls interpolating between the vacuum state at $\alpha = -\theta$ and the same vacuum state at $\alpha = -\theta + 2\pi$,

$$\alpha(z) + \theta = 4 \arctan\left(e^{m_a z}\right), \tag{80}$$

where the wall is assumed to lie in the xy plane, so that the wall profile depends only on z. This is the most primitive "2π wall."

The tension of this wall is obviously of the order of

$$T_1 \sim f^2 m_a. \tag{81}$$

Taking into account that $f^2 m_a^2 \sim \mathcal{X}$ where $\mathcal{X}$ is the topological susceptibility of the QCD vacuum, we get

$$T_1 \sim \mathcal{X}/m_a. \tag{82}$$

The inverse proportionality to m_a is due to the fact that the transverse size of the axion wall is very large.

Let us now discuss the axion effective potential of the second type. In this case the potential has cusps, as is the case in pure gluodynamics, where the axion effective Lagrangian is of the form

$$\mathcal{L}_a = \frac{f^2}{2}(\partial_\mu \alpha)^2 + \min_\ell \left\{ N_c^2 \Lambda^4 \cos \frac{\alpha + 2\pi\ell}{N_c} \right\}, \tag{83}$$

(see more detailed discussions below). Here the θ angle was included in the definition of the axion field. The axion wall interpolates between $\alpha = 0$ and $\alpha = 2\pi$.

What is the origin of this cusp? The cusps reflect a restructuring in the hadronic sector. When one (adiabatically) interpolates in α from 0 to 2π a gluonic order parameter, for instance $\langle G\tilde{G} \rangle$, necessarily experiences a restructuring in the middle of the wall corresponding to the restructuring of heavy gluonic degrees of freedom. In other words, one jumps from the hadronic vacuum which initially (at $\alpha = 0$) had $\langle G\tilde{G} \rangle = 0$ into the vacuum in which initially $\langle G\tilde{G} \rangle \neq 0$. Upon arrival to $\alpha = 2\pi$, we find $\langle G\tilde{G} \rangle = 0$ again. This implies that the central part of such an axion wall is dominated by a gluonic wall. Thus, the cusp at $\alpha = \pi$ generically indicates the formation of a hadronic core, the D-wall[47] in the case at hand.

Returning to the question of the tension we note that

$$\chi \sim \Lambda^4 N_c^0, \quad m_a \sim \Lambda^2 N_c^0 f^{-1} \text{ in pure gluodynamics,}$$

$$\chi \sim \Lambda^3 N_c m_q, \quad m_a \sim \Lambda^{3/2} m_q^{1/2} N_c^{1/2} f^{-1} \text{ in QCD with light quarks} \tag{84}$$

which implies, in turn,

$$T_1 \sim \begin{cases} f\Lambda^2 N_c^0 \text{ in pure gluodynamics} \\ f\Lambda^{3/2} m_q^{1/2} N_c^{1/2} \text{ in QCD with light quarks}. \end{cases} \tag{85}$$

Here m_q is the light quark mass.

The presence of the large parameter f in T_1 makes the axion halo the dominant contributor to the wall tension. The contribution of the hadronic component contains only hadronic parameters, although it may have a stronger dependence on N_c. Examining the cusp with an appropriately high resolution one would observe that it is smoothed on the hadronic scale, where the hadronic component of the axion wall "sandwich" would become visible. The cusp carries a finite contribution to the wall tension which cannot be calculated in the low-energy approximation.[86] To this end one needs to consider the hadronic core explicitly. The tension of the core $T_{\text{core}} \sim \Lambda^3 N_c$, while the tension of the axion halo $T_{\text{halo}} \sim f\Lambda^2$ (in pure gluodynamics).

We pause here to make a comment on the literature. The consideration of the axion walls in conjunction with hadrons dates back to the work of Huang and Sikivie, see Ref. 46. This work treats the Weinberg-Wilczek $N = 2$ axion in QCD with two light flavors, which is replaced by a chiral Lagrangian for the pions, to the leading order (quadratic in derivatives and linear in the light quark masses). It is well-known[36,37,41] that in this theory the crossover phenomenon takes place at $m_u = m_d$. In the realistic situation, $(m_d - m_u)/(m_d + m_u) \sim 0.3$ considered in Ref. 46, there is no crossover. The pions can be integrated over, leaving one with an effective Lagrangian for the axion of the type (79) (with $\alpha \to 2\alpha$). The potential is not pure cosine, higher harmonics occur too. The axion halo exhausts the wall, there is no hadronic core in this case.

At the same time, Huang and Sikivie (see Ref. 46) found an explicit solution for the "π^0" component of the wall. In fact, this is an illusion. The Huang-Sikivie (HS) solution refers to the *bare* π^0 field. To find the physical π^0 field one must diagonalize the mass matrix at every given value of α (the bare $f\alpha$ is the physical axion field up to small corrections $\sim f_\pi^2/f^2$ where f_π stands for the pion decay constant). Once this is done, one observes that the physical pion field, which is a combination of the bare pion and $f\alpha$, is not excited in the HS solution. The equation (2.16) in the HS paper is exactly the condition of vanishing of the physical pion in the wall profile. This explains why the wall thickness in the HS work is of order m_a^{-1}, with no traces of the m_π^{-1} component. The crossover of the hadronic vacua at $\alpha = \pi/2$ (remember, this is $N = 2$ model) could be recovered in the Huang-

Sikivie analysis at $m_u = m_d$. However, the chiral pion Lagrangian predicts in this case the vanishing of the pion mass in the middle of the wall, for accidental reasons. This is explained in detail by A. Smilga, Ref. 41.

6.2. *An Illustrative model*

To find the axion walls with D-wall core one has to solve QCD, which is way beyond our possibilities. Our task is more modest. We would like to obtain a qualitative description of the axion wall sandwich which, with luck, can become semi-quantitative. To this end we want to develop toy models. An obvious requirement to any toy model is that it must qualitatively reproduce the basic features of the vacuum structure which we expect in QCD. In SUSY gluodynamics it was possible to write down a toy model with a Z_{N_c} symmetry [87] which "integrates in" the heavy degrees of freedom and allows one to investigate the BPS domain walls in the large N_c limit [54] (see also Ref. 88). We will suggest a similar model in (nonsupersymmetric) QCD, then switch on axions, and study the axion domain walls in a semi-realistic setting. In this model we will be able to find exact solutions for D-walls and axion walls.

Here we suggest a simple toy model which has a proper vacuum structure. In what follows an appropriate (complex) glue order parameter is denoted by Φ. The modulus and phase of this field describe respectively the 0^{++} and 0^{-+} channels of the theory.

Our toy model Lagrangian is

$$\mathcal{L} = N_c^2 (\partial_\mu \Phi)^* (\partial_\mu \Phi) - V(\Phi, \Phi^*), \qquad V = V_0 + V_1,$$

$$V_0 = N_c^2 A^2 \left| 1 - \Phi^{N_c} e^{-i\theta} \right|^2,$$

$$V_1 = \left\{ -\frac{\mathcal{X} N_c^2}{2} \Phi \left[1 + \frac{1}{N_c} (1 - \Phi^{N_c} e^{-i\theta}) \right] + \frac{\mathcal{X} N_c^2}{2} \right\} + \text{H.c.}. \quad (86)$$

Here A is a numerical constant of order one, and $\mathcal{X}$ is the vacuum topological susceptibility in pure gluodynamics (note that $\mathcal{X}$ is independent of N_c). The scale parameter Λ is set to unity.

This model has the vacuum family composed of N_c states. Indeed, the minima of the energy are determined from the equations

$$\left. \frac{\partial V}{\partial \Phi} \right|_{\text{vac}} = \left. \frac{\partial V}{\partial \Phi^*} \right|_{\text{vac}} = 0, \quad (87)$$

which have the following solutions (we put temporarily $\Lambda = 1$):

$$\Phi_{\ell\text{vac}} = \exp\left(i \frac{\theta + 2\pi\ell}{N_c} \right), \quad \ell = 0, 1, ..., N_c - 1. \quad (88)$$

In the ℓ-th minimum V_0 vanishes, while V_1 produces a non-vanishing vacuum energy density,

$$\mathcal{E}_\ell = \mathcal{X} N_c^2 \left\{ 1 - \cos\left(\frac{\theta + 2\pi\ell}{N_c}\right) \right\}. \tag{89}$$

For each given θ the genuine vacuum is found by minimization,

$$\mathcal{E}(\theta) = N_c^2 \mathcal{X} \min_\ell \left\{ 1 - \cos\left(\frac{\theta + 2\pi\ell}{N_c}\right) \right\}. \tag{90}$$

The remaining $N_c - 1$ minima are quasivacua. Once the heavy field Φ is integrated out, the vacuum energy is given by the expression (90); it has cusps at $\theta = \pi, 3\pi$ and so on. Needless to say that the potential (86) has no cusps.

We will first consider the model (86) without the axion field, at $\theta = 0$, in the limit $N_c \to \infty$. In this limit the false vacua from the vacuum family are stable.

The classical equation of motion defining the wall is

$$N_c^2 \, \Phi^{*\prime\prime} = \frac{\partial V}{\partial \Phi}, \tag{91}$$

where primes denote differentiation with respect to z (we look for a solution which depends on the z coordinate only).

This is a differential equation of the second order. It is possible, however, to reduce it to a first order equation. Indeed, Eq. (91) has an obvious "integral of motion" ("energy"),

$$N_c^2 \, \Phi^{*\prime} \Phi^\prime - V = \text{Const} = 0, \tag{92}$$

where the second equality follows from the boundary conditions. In the large N_c limit one can parametrize the field Φ as follows ($\rho \sim 1$):

$$\Phi \equiv 1 + \frac{\rho}{N_c}. \tag{93}$$

Taking the square root of Eq. (92), substituting Eq. (93) and neglecting the terms of the subleading order in $1/N_c$ we arrive at

$$\bar\rho^\prime = iAN_c \, (1 - \exp\rho). \tag{94}$$

The phase on the right-hand side can be chosen arbitrarily. The choice in Eq. (94) is made in such a way as to make it compatible with the boundary conditions for the wall interpolating between $\Phi_{\text{vac}} = 1$ and $\Phi_{\text{vac}} = \exp(2\pi i/N_c)$. This is precisely the expression [54,88] that defines the domain walls in SUSY gluodynamics. It is not surprising that the same equation determines the D-walls in non-SUSY gluodynamics – the fermion-induced effects are not important for D-walls in the large N_c limit.

The solution of this equation was obtained in Ref. 88. In the parametrization $\rho = \sigma + i\tau$ the solution takes the form:

$$\cos\tau = (\sigma + 1)\exp(-\sigma),$$

$$\int_{\sigma(0)}^{\sigma(z)} [\exp(2t) - (1+t)^2]^{-1/2}\, dt = -AN_c|z|. \tag{95}$$

The real part of ρ is a bell-shaped function with an extremum at zero; it vanishes at $\pm\infty$. The imaginary part of ρ, on the other hand, changes its value from 0 to 2π. This determines a D-wall in the large N_c gluodynamics. The width of the wall scales as $1/N_c$.

The solution presented above is exactly the same as in SUSY gluodynamics. This is not surprising since the *ansatz* (93) implies that V_1 does not affect the solution – its impact is subleading in $1/N_c$, while V_0 is exactly the same as in the SUSY-gluodynamics-inspired model of Ref. 54. Moreover, for the same reason the domain wall junctions emerging in this model will be exactly the same as in the SUSY-gluodynamics-inspired model.[47] Inclusion of V_1 in the subleading order makes the wall to decay.

Inclusion of the $N = 1$ axion field amounts to the replacement

$$\theta \to \theta + \alpha$$

in Eq. (86), plus the axion kinetic term

$$\mathcal{L}_{\text{kin}} = \left(\frac{f^2 + 2\Phi^*\Phi}{2}\right)(\partial_\mu\alpha)^2 + iN_c(\partial_\mu\alpha)(\Phi^*\partial_\mu\Phi - \Phi\partial_\mu\Phi^*). \tag{96}$$

The occurrence of the mixing between α and the phase of Φ is necessary, as is readily seen from the softly broken SUSY gluodynamics. (To get the potential of the type (86) in this model, one must eliminate the $G\tilde{G}$ term by a chiral rotation. Then $m \to m\exp((\theta + \alpha)/N_c)$ and, additionally one gets $\partial_\mu\alpha\times$ [the gluino axial current].) The term $2\Phi^*\Phi$ in the brackets has to be included to reproduce the correct mass for the axion after the physical heavy state is integrated out. The presence of this term signals that QCD dynamics generates not only the potential for the axion but also modifies its kinetic term. On the other hand, since $\Phi^*\Phi \leq \Lambda^2$ and, moreover, $\Lambda \ll f$, this term can be neglected for all practical purposes.

We are interested in the configuration with α interpolating between 0 and 2π. The phase of Φ will first adiabatically follow α/N_c, then at $\alpha \approx \pi$, when the phase of Φ is close to π/N_c, it will very quickly jump by $-2\pi/N_c$, and then it will continue to grow as α/N_c, so that when α reaches 2π the phase of Φ returns to zero. This jump is continuous, although it occurs at a scale much shorter than m_a^{-1}. This imitates the D-wall core of the axion wall. One cannot avoid forming this core, since otherwise the interpolation would not connect degenerate states – on one side of the wall we would have (hadronic) vacuum, on the other side an excited state.

In the large N_c limit one can be somewhat more quantitative. Indeed, in this approximation the model admits the exact solutions. The gluonic core of the wall has the same form as before, Eq. (95), but the phase τ is now substituted by the superposition $\tau - (\alpha + \theta)$ since the axion field is mixed with the phase of the Φ field.

This very narrow core is surrounded by a diffused axion halo. The axion field is described in this halo by the solution to the Lagrangian (83). This takes the form:

$$\theta + \alpha(z) = -2\pi + 4N_c \arctan\left(e^{m_a z} \tan\frac{\pi}{4N_c}\right), \quad z < 0,$$

$$\theta + \alpha(z) = -4N_c \arctan\left(e^{-m_a z} \tan\frac{\pi}{4N_c}\right), \quad z > 0. \quad (97)$$

Thus, we find explicitly the stable axion wall with a D-wall core. Note that this is a usual "2π" wall as it separates two identical hadronic vacua. As we discussed in the introduction, this wall can decay quantum mechanically. However, its lifetime is infinite for all practical purposes. Moreover, this wall is harmless cosmologically. It will be produced bounded by global axion strings in the early universe. Bounded walls shrink very quickly by decaying into axions and hadrons.

7. Axions and vacuum structure in QCD with light quarks

So far we discussed pure gluodynamics with the axion. Our final goal is to study QCD with $N_f = 3$. There are two, physically distinct regimes to be considered in this case. In real QCD

$$m_u, m_d \ll m_s \sim \frac{\Lambda}{N_c}, \quad m_u, m_d, m_s \ll \Lambda. \quad (98)$$

In this regime the consideration of the chiral Lagrangians,[36,37,41] does not exhibit the vacuum family. We will comment on why the light quarks screen the vacuum family of the glue sector, so that the axion domain wall provides no access to it. In the limit (98) the effects due to the D-walls will be marginal.

On the other hand, in the genuinely large N_c limit

$$\frac{\Lambda}{N_c} \ll m_u, m_d \ll m_s \ll \Lambda, \quad (99)$$

physics is rather similar to that of pure gluodynamics. The light quarks are too heavy to screen the vacuum family of the glue sector

In what follows we study the axion walls and their hadronic components in the limits (98) and (99), separately.

7.1. *One light quark*

To warm up, let us start from the theory with one light quark. In the limit of large N_c this introduces a light meson, "η'". An appropriate effective Lagrangian can be obtained by combining the vacuum energy density of gluodynamics with what remains from the chiral Lagrangian at $N_f = 1$,

$$\mathcal{L} = \frac{F^2}{2}(\partial_\mu \beta)^2 - V(\beta),$$

$$V = -m_q \Lambda^3 N_c \cos \beta + \min_\ell \left\{ -N_c^2 \Lambda^4 \cos \frac{\beta + \theta + 2\pi\ell}{N_c} \right\}. \quad (100)$$

Here β is the phase of $U \sim \bar{q}_L q_R$, while $F^2 \sim \Lambda^2 N_c$ is the "η'" coupling constant squared. The product $F\beta$ is the "η'" field. The first term in V corresponds to the quark mass term, $\mathcal{M}U + $ h.c. At $N_c = \infty$ the second term in V becomes $(\beta+\theta)^2$. It corresponds to $(i\ln \det U + \theta)^2$ in Eq. (11) in Ref. 36. The subleading in $1/N_c$ terms sum up into a 2π periodic function of the cosine type, with the cusps. It is unimportant that we used cosine in Eq. (100). Any 2π periodic function of this type would lead to the same conclusions. The second term in Eq. (100) differs from the vacuum energy density in gluodynamics by the replacement $\theta \to \beta + \theta$.

If $m_q \ll \Lambda/N_c$, the first term in V is a small perturbation; therefore, in the vacuum, $\beta + \theta = 2\pi k$, and, hence, the θ dependence of the vacuum energy is

$$\mathcal{E}_{\text{vac}}(\theta) = -m_q \Lambda^3 N_c \cos \theta. \quad (101)$$

It is smooth, 2π periodic and proportional to m_q as it should be on general grounds in the theory with one light quark.

The condition $m_q \ll \Lambda/N_c$ precludes us from sending $N_c \to \infty$. The would be "2π" wall in the variable β is expected to be unstable. This is due to the fact that at $N_c \sim 3$ the absolute value of the quark condensate $\bar{\psi}\psi$ is not "harder" than the phase of the condensate β, and the barrier preventing the creation of holes in the "2π" wall is practically absent.

If one closes one's eyes on this instability one can estimate that the tension of the "η'" wall is proportional to $\Lambda^3 N_c^{1/2}$, with a small correction $m_q \Lambda^2 N_c^{3/2}$ from the quark mass term. The tension of the D wall core is, as previously, $\Lambda^3 N_c$.

In the opposite limit

$$m_q \gg \frac{\Lambda}{N_c}, \qquad \text{but } m_q \text{ still} \ll \Lambda, \tag{102}$$

the situation is trickier. Now the first term in V is dominant, while the second is a small perturbation. There are N_c distinct vacua in the theory,

$$\beta_\ell = -\frac{2\Lambda}{m_q N_c}(\theta + 2\pi\ell). \tag{103}$$

Then the θ dependence of the vacuum energy density is

$$\mathcal{E}_{\text{vac}}(\theta) = \Lambda^4 \min_\ell (\theta + 2\pi\ell)^2, \tag{104}$$

this is similar to that in the theory without light quarks (i.e., the same as in gluodynamics). The "η'" wall is stable at $N_c \to \infty$, with a a D-wall core in its center. The η' wall is a "2π" wall.

¿From this standpoint the quark with the mass (102) is already heavy, although the "η'" is still light on the scale of Λ,

$$M_{\eta'} \sim m_q^{1/2} \Lambda^{1/2} \ll \Lambda.$$

So far the axion was switched off. What changes if one includes it in the theory?

The Lagrangian now becomes

$$\mathcal{L} = \frac{F^2}{2}(\partial_\mu \beta)^2 + \frac{f^2}{2}(\partial_\mu \alpha)^2 - V(\beta, \alpha),$$

$$V = -m_q \Lambda^3 N_c \cos\beta + \min_\ell \left\{ -N_c^2 \Lambda^4 \cos\frac{\beta + \alpha + 2\pi\ell}{N_c} \right\}, \tag{105}$$

where the θ angle is absorbed in the definition of the axion field.

The bare "η'" mixes with the bare axion. It is easy to see that in the limit $m_q \ll \Lambda/N_c$ the physical "η'" is proportional to $\beta + \alpha$, rather than to β. Therefore, even if we force the axion wall to develop, (i.e. α to evolve from 0 to 2π) the "η'" wall need not develop. It is energetically expedient to have $\beta + \alpha = 0$. Thus, the effect of the axion field on the hadronic sector is totally screened by a dynamical phase β coming from the quark condensate. In other words, the axion wall with the lowest tension corresponds to the frozen physical "η'",

$$\beta + \alpha = 0.$$

There is no hadronic core. The tension of this wall is determined from the term $\propto m_q \Lambda^3 N_c$.

(If one wishes, one could add the "η'" wall to the axion wall. Then the "η'" wall, with the D-wall core will appear in the middle of the axion wall, but they are basically unrelated. This will be a secondary phenomenon, and the D wall core will be, in fact, the core of the "η'" wall rather than the axion wall.)

If the quark mass is such that (102) applies, then the axion field α cannot be screened, since we cannot freeze $\beta + \alpha$ everywhere in the axion wall profile at zero – at $m_q \gg \Lambda/N_c$, β is proportional to the physical "η'" and is much heavier than the axion field. Thus, in this case the axion wall will be described by the Lagrangian (83) and will have a D-wall core. One may also add, on top of it, the "η'" wall. This will cost $m_q^{1/2}\Lambda^{5/2}N_c$ in the wall tension – still much less than $\Lambda^3 N_c$ of the D-wall core of the axion wall.

The limit (102) is unrealistic. Moreover, in this limit the D walls taken in isolation, without the axion walls, are stable by themselves, although they interpolate between nondegenerate states.[45]

7.2. Three Light Quarks

Let us turn the case of three light flavors. The physical picture is quite similar to that of the one-flavor case, see Sec. 7.1.

We assume the mass matrix $\mathcal{M}$ in the meson Lagrangian to be diagonal. Therefore, we will looking for a diagonal $U(3)$ meson matrix which minimizes the potential,

$$U = \text{diag}\left(e^{i\phi_1}, \; e^{i\phi_2}, \; e^{i\phi_3}\right) . \tag{106}$$

The potential takes the form

$$V = -\sum_i m_i \Lambda^3 N_c \cos\phi_i + \min_\ell \left\{ -N_c^2 \Lambda^4 \cos \frac{\sum_i \phi_i + \theta + 2\pi\ell}{N_c} \right\} \tag{107}$$

As before, we will consider two limiting cases, (98) and (99).

Let us switch off the axion field first. In the limit of genuinely light quarks, Eq. (98), when the second term in the potential (107) is dominant, the solutions for ϕ's were found in Refs. 36,37. They satisfy to the relation $\phi_3 \simeq 0$ and $\phi_1 + \phi_2 = -\theta$. The corresponding expression for the vacuum energy density is

$$\mathcal{E}_{\text{vac}}(\theta) = -N_c \Lambda^3 \sqrt{m_u^2 + m_d^2 + 2m_u\, m_d \cos\theta} . \tag{108}$$

As in Sec. 7.1, we deal here with a smooth single-valued function of θ. The inclusion of the axion replaces $\theta \to \theta + \alpha \to \alpha$. The physical η' field

562

is given by the sum $\sum_i \phi_i + \alpha$. It is energetically favorable to freeze this state. Thus, the situation is identical to that in the one-flavor case: even if the axion wall is forced to develop, the physical η' wall (which is now the $\sum_i \phi_i + \alpha$ wall) does not have to occur. Effectively, the vacuum angle is screened, and there is no D-wall core in the axion wall.

If, nonetheless, the η' wall is formed due to some cosmological initial conditions, it will have a D-wall core (albeit the η' wall is unstable in the limit at hand and cannot be considered in the static approximation). The would-be η' wall is independent of the axion wall; its effect on the axion wall formation is rather irrelevant.

In addition to this, a "2π" wall could develop for nonsinglet mesons at certain values of the quark masses. There is nothing new we could add to this issue which is decoupled from the issue of the vacuum family in the glue sector and D-walls.

We now pass to the opposite limit (99), when the first term in the potential (107) is dominant. As in Sec. 7.1, there are N_c distinct vacua with the energy given by (104). It is straightforward to show that the potential for the axion in this case is of the form (83), with the cusps which signal the presence of the D-wall core. This is similar to what happens in gluodynamics. One cannot avoid having an η' wall in the middle of the axion wall, which entails a D-wall too. The D-walls separate the degenerate vacua. Since they "live" in the middle of the axion wall, they are perfectly stable.

(In addition, there can be "2π" walls in either of ϕ's or their linear combinations. However, these latter are unstable and do not appear in the physical spectrum of the theory.)

Acknowledgments

The authors would like to thank S. Dimopoulos, A. Smilga, L. Susskind and E. Zhitnitsky for usefull discussions. The work is supported by DOE grant DE-FG02-94ER408.

References

1. S. Weinberg, Phys. Rev. Lett. **40**, 223 (1978).
2. F. Wilczek, Phys. Rev. Lett. **40**, 279 (1978).
3. D. J. Gross and F. Wilczek, Phys. Rev. Lett. **30** (1973) 1343.
4. H. D. Politzer, Phys. Rev. Lett. **30** (1973) 1346.

5. A. A. Belavin, A. M. Polyakov, A. S. Shvarts and Y. S. Tyupkin, Phys. Lett. B **59** (1975) 85.

6. R. Jackiw and C. Rebbi, Phys. Rev. Lett. **37**, 172 (1976).

7. C. G. Callan, R. F. Dashen and D. J. Gross, Phys. Lett. B **63**, 334 (1976).

8. V.N. Gribov, 1975, unpublished.

9. S. L. Adler, Phys. Rev. **177**, 2426 (1969).
   J. S. Bell and R. Jackiw, Nuovo Cim. A **60**, 47 (1969).

10. K. Fujikawa, Phys. Rev. Lett. **42**, 1195 (1979).

11. V. Baluni, Phys. Rev. D **19**, 2227 (1979).

12. R. J. Crewther, P. Di Vecchia, G. Veneziano and E. Witten, Phys. Lett. B **88**, 123 (1979) [Erratum-ibid. B **91**, 487 (1980)].

13. A. Pich and E. de Rafael, Nucl. Phys. B **367**, 313 (1991).

14. M.A. Shifman, A.I. Vainshtein, V.I. Zakharov, Nucl. Phys. **B147** (1979) 385.

15. M. Pospelov and A. Ritz, Phys. Rev. Lett. **83**, 2526 (1999) [hep-ph/9904483].

16. P. G. Harris *et al.*, Phys. Rev. Lett. **82** (1999) 904.

17. M. A. Shifman, A. I. Vainshtein and V. I. Zakharov, Nucl. Phys. B **166**, 493 (1980).

18. H. Leutwyler, Nucl. Phys. Proc. Suppl. **94**, 108 (2001) [hep-ph/0011049].

19. A. M. Polyakov, Nucl. Phys. B **120**, 429 (1977).

20. G. 't Hooft, Phys. Rev. Lett. **37**, 8 (1976).

21. E. Witten, Nucl. Phys. **B156** (1979) 269.

22. G. Veneziano, Nucl. Phys. B **159**, 213 (1979).

23. B. Alles, M. D'Elia, A. Di Giacomo, Nucl. Phys. **B494** (1997) 281; hep-lat/9605013.

24. G. Grunberg, Phys. Rev **D30** (1984) 1570.

25. S. Samuel, Mod. Phys. Lett. A **7**, 2007 (1992).

26. N. J. Dowrick and N. A. McDougall, Nucl. Phys. B **399**, 426 (1993).

27. R. D. Peccei and H. R. Quinn, Phys. Rev. Lett. **38**, 1440 (1977). R. D. Peccei and H. R. Quinn, Phys. Rev. D **16**, 1791 (1977).

28. J. E. Kim, Phys. Rev. Lett. **43**, 103 (1979).

29. A. R. Zhitnitsky, Sov. J. Nucl. Phys. **31** (1980) 260 [Yad. Fiz. **31** (1980) 497].

30. M. Dine, W. Fischler and M. Srednicki, Phys. Lett. B **104**, 199 (1981).

31. G. G. Raffelt, Ann. Rev. Nucl. Part. Sci. **49**, 163 (1999) [hep-ph/9903472].

32. J. Preskill, M. B. Wise and F. Wilczek, Phys. Lett. B **120**, 127 (1983).

33. L. F. Abbott and P. Sikivie, Phys. Lett. B **120**, 133 (1983).

34. M. Dine and W. Fischler, Phys. Lett. B **120**, 137 (1983).

35. R. J. Crewther, Phys. Lett. **B70**, 349 (1977); Acta Phys. Austriaca Suppl. **19**, 47 (1978).

36. E. Witten, Ann. Phys. **128**, 363 (1980).

37. P. Di Vecchia and G. Veneziano, Nucl. Phys. **B171**, 253 (1980).

38. R. Dashen, Phys. Rev. **D3**, 1879 (1971).

39. C. Rosenzweig, J. Schechter and C. G. Trahern, Phys. Rev. **D21**, 3388 (1980).

40. R. Arnowitt, Pran Nath, Phys. Rev. **D23** (1981) 473; Nucl. Phys. **B209** (1982) 234, 251;

41. M. Creutz, Phys. Rev. **D52**, 2951 (1995); [hep-th/9505112];
    A. V. Smilga, Phys. Rev. **D59**, 114021 (1999) [hep-ph/9805214];
    M. H. Tytgat, Phys. Rev. D **61**, 114009 (2000) [hep-ph/9909532].

42. M. Shifman, Prog. Part. Nucl. Phys. **39**, 1 (1997); [hep-th/9704114]; N. Evans, S. D. Hsu and M. Schwetz, Phys. Lett. **B404**, 77 (1997); [hep-th/9703197].

43. E. Witten, Phys. Rev. Lett. **81**, 2862 (1998); [hep-th/9807109].

44. I. Halperin and A. Zhitnitsky, Phys. Rev. **D58**, 054016 (1998) [hep-ph/9711398]; Mod. Phys. Lett. **A13**, 1955 (1998) [hep-ph/9707286]; I. Halperin and A. Zhitnitsky, Nucl. Phys. **B539**, 166 (1999) [hep-th/9802095]; I. Halperin and A. Zhitnitsky, Phys. Rev. Lett. **81**, 4071 (1998) [hep-ph/9803301]; A. R. Zhitnitsky, Nucl. Phys. Proc. Suppl. **73** (1999) 647.

45. M. Shifman, Phys. Rev. **D59**, 021501 (1999) [hep-th/9809184].

46. P. Sikivie, Phys. Rev. Lett. **48**, 1156 (1982);
P. Sikivie, *Axions In Cosmology*, Report UF-TP-83-6, Based on lectures given at 21st Schladming Winter School, Schladming, Austria, Feb 26 - Mar 6, 1982; M. C. Huang and P. Sikivie, Phys. Rev. **D32**, 1560 (1985); S. Chang, C. Hagmann and P. Sikivie, Phys. Rev. **D59**, 023505 (1999) [hep-ph/9807374].

47. G. Gabadadze and M. Shifman, Phys. Rev. **D61**, 075014 (2000); [hep-th/9910050].

48. T. Fugleberg, I. Halperin and A. Zhitnitsky, Phys. Rev. **D59**, 074023 (1999); [hep-ph/9808469].

49. I. Halperin and A. Zhitnitsky, Phys. Lett. **B440**, 77 (1998) [hep-ph/9807335].

50. M. M. Forbes and A. R. Zhitnitsky, Phys. Rev. Lett. **85**, 5268 (2000) [hep-ph/0004051].

51. N. Evans, S. D. Hsu, A. Nyffeler and M. Schwetz, Nucl. Phys. **B494**, 200 (1997) [hep-ph/9608490].

52. G. Dvali and M. Shifman, Phys. Lett. **B396**, 64 (1997); [hep-th/9612128], (E) **B407**, 4521 (1997).

53. G. Gabadadze, Nucl. Phys. **B552**, 194, (1999); [hep-th/9902191].

54. G. Dvali, G. Gabadadze, Z. Kakushadze, Nucl. Phys. **B562**, 158 (1999); [hep-th/9901032].

55. J. Polchinski and M. J. Strassler, hep-th/0003136 ; A. R. Frey, JHEP **0012**, 020 (2000) [hep-th/0007125].

56. A. Gorsky and M. Shifman, Phys. Rev. **D61**, 085001 (2000); [hep-th/9909015].

57. T. Banks, W. Fischler, S. H. Shenker and L. Susskind, Phys. Rev. **D55**, 5112 (1997) [hep-th/9610043].

58. G. Gabadadze and Z. Kakushadze, Mod. Phys. Lett. **A15**, 293 (2000) [hep-th/9905198]; Mod. Phys. Lett. **A14**, 2151 (1999) [hep-th/9908039].

59. For a review, see:
A. Giveon and D. Kutasov, Rev. Mod. Phys. **71**, 983 (1999) [hep-th/9802067].

60. J. M. Maldacena, Adv. Theor. Math. Phys. **2**, 231 (1998) [Int. J. Theor. Phys. **38**, 1113 (1999)] [hep-th/9711200].

61. S. S. Gubser, I. R. Klebanov and A. M. Polyakov, Phys. Lett. B **428**, 105 (1998) [hep-th/9802109].

62. E. Witten, Adv. Theor. Math. Phys. **2**, 253 (1998) [hep-th/9802150].

63. E. Witten, Adv. Theor. Math. Phys. **2**, 505 (1998) [hep-th/9803131].

64. D. J. Gross and H. Ooguri, Phys. Rev. D **58**, 106002 (1998) [hep-th/9805129].

65. C. Csaki, H. Ooguri, Y. Oz and J. Terning, JHEP **9901**, 017 (1999) [hep-

th/9806021];

R. de Mello Koch, A. Jevicki, M. Mihailescu and J. P. Nunes, Phys. Rev. D **58**, 105009 (1998) [hep-th/9806125];

M. Zyskin, Phys. Lett. B **439**, 373 (1998) [hep-th/9806128].

66. H. Ooguri, H. Robins and J. Tannenhauser, Phys. Lett. B **437**, 77 (1998) [hep-th/9806171].

67. I. R. Klebanov, "From threebranes to large N gauge theories," hep-th/9901018.

68. I. R. Klebanov and A. A. Tseytlin, Nucl. Phys. B **546**, 155 (1999) [hep-th/9811035].

69. J. A. Minahan, JHEP **9901**, 020 (1999) [hep-th/9811156].

70. S. Coleman, "Aspects of Symmetry", Cambridge University Press, 1985.

71. E.V. Shuryak, Nucl.Phys. **B203** (1982) 93;
    D.I. Diakonov, V.Y. Petrov, Nucl.Phys. **B245** (1984) 259;
    T. DeGrand, A. Hazenfratz, T.G. Kovacs, Nucl. Phys. **B505** (1997) 417; hep-lat/9705009.

72. M. Lüscher, Phys. Lett. **78B** (1978) 465.

73. J. Kogut, L. Susskind, Phys. Rev. **D11** (1975) 3594.

74. J. Schechter, Phys. Rev. **D21** (1980) 3393.

75. A.A. Migdal, M.A. Shifman, Phys. Lett. **114B** (1982) 445.

76. A. Aurilia, Phys. Lett. **81B** (1979) 203.

77. A. Aurilia, H. Nicolai, P.K. Townsend, Nucl. Phys. **B176**(1980)509.

78. G. Gabadadze, Phys. Rev. **D58** (1998) 094015; hep-ph/9710402.

79. P. Di Vecchia, G. Veneziano, Nucl. Phys. **B171** (1980) 253.

80. A. Pich, Rept. Prog. Phys. **58** (1995) 563; hep-ph/9502366.

81. P. Di Vecchia, F. Nicodemi, R. Pettorino, G. Veneziano, Nucl. Phys. **B181** (1981) 318.

82. L. Del Debbio, H. Panagopoulos and E. Vicari, *Theta dependence of SU(N) gauge theories*, hep-th/0204125.

83. M. B. Voloshin, I. Yu. Kobzarev, L. B. Okun', Yad. Fiz. **20** (1974) 1229 (Sov. J. Nucl. Phys. **20** (1975) 644);
    S. Coleman, Phys. Rev **D15** (1977) 2929; Erratum-ibid. **D16** (1977) 1248.

84. S. Coleman, Phys. Rev **D15** (1977) 2929; (E) **D16** (1977) 1248;
    C. Callan, S. Coleman, Phys. Rev. **D16** (1977) 1762.

85. G. R. Dvali, G. Gabadadze and Z. Kakushadze, Nucl. Phys. B **562**, 158 (1999) [hep-th/9901032].

86. I. I. Kogan, A. Kovner and M. Shifman, Phys. Rev. **D57**, 5195 (1998); [hep-th/9712046].

87. G. Gabadadze, Nucl. Phys. **B544**, 650 (1999); [hep-th/9808005].

88. G. Dvali and Z. Kakushadze, Nucl. Phys. **B537**, 297 (1999); [hep-th/9807140].

DILUTING COSMOLOGICAL CONSTANT VIA LARGE DISTANCE MODIFICATION OF GRAVITY

GIA DVALI

Department of Physics, New York University, New York, NY 10003

GREGORY GABADADZE AND M. SHIFMAN

Theoretical Physics Institute, University of Minnesota, Minneapolis, MN 55455

We review a solution (Ref. 1) of the cosmological constant problem in a brane-world model with infinite-volume extra dimensions. The solution is based on a nonlinear generally covariant theory of a metastable graviton that leads to a large-distance modification of gravity. ¿From the extra-dimensional standpoint the problem is solved due to the fact that the four-dimensional vacuum energy curves mostly the extra space. The four-dimensional curvature is small, being inversely proportional to a positive power of the vacuum energy. The effects of infinite-volume extra dimensions are seen by a brane-world observer as nonlocal operators. ¿From the four-dimensional perspective the problem is solved because the zero-mode graviton is extremely weakly coupled to localized four-dimensional sources. The observable gravity is mediated not by zero mode but, instead, by a metastable graviton with a lifetime of the order of the present-day Hubble scale. Therefore, laws of gravity are modified in the infrared above the Hubble scale. Large wave-length sources, such as the vacuum energy, feel only the zero-mode interaction and, as a result, curve space very mildly. Shorter wave-length sources interact predominantly via exchange of the metastable graviton. Because of this, all standard properties of early cosmology, including inflation, are intact.

1. Introduction

The cosmological constant problem can be cast into two questions. First, there is an old question:

(i) Why is the vacuum energy determined by a momentum scale much smaller than any reasonable cut-off scale in effective field theory of particle interactions?

This is sometimes referred to as the "old" cosmological constant problem. In view of the present astrophysical observations[2] one should also find an answer to the second question:

(ii) How come that the vacuum energy and the matter energy are comparable today? Do we live in a special epoch?

This is the so-called "cosmic coincidence" problem. To our knowledge the only existing framework, which can address both questions simultaneously, is the anthropic approach.[3,4] Below we will concentrate on a dynamical solution of the "old" cosmological constant problem suggested in Ref. 1 which is based on a nonlinear and generally covariant model of metastable graviton proposed in Refs. 5,6. We have nothing to say about the "cosmic coincidence" problem.

Before reviewing the solution of Ref. 1 let us formulate the question properly. As it stands, the question (i) is ill-posed for our purposes. Let us first recall why the vacuum energy in the universe is normally assumed to be small.

In general relativity the cosmological expansion of the universe is described by a standard metric

$$ds^2 = dt^2 - a^2(t)\, d\vec{x}^2 \,. \tag{1}$$

Here $a(t)$ is the scale factor and t is time in the co-moving coordinate system (we assume for definiteness that three-dimensional curvature is zero). In the presence of the vacuum energy density $\mathcal{E}$, the scale factor $a(t)$ obeys the Friedmann equation

$$H^2 \equiv \left(\frac{\dot{a}}{a}\right)^2 = \frac{\mathcal{E}}{3\, M_{\mathrm{Pl}}^2} \,, \tag{2}$$

where $\dot{a} \equiv d\,a/dt$ (for simplicity we set $\mathcal{E} > 0$).

We have no direct experimental way to measure $\mathcal{E}$. Instead, we measure space-time curvature by cosmological observations, and then determine $\mathcal{E}$ through Eq. (2). Thus, claiming that $\mathcal{E}$ should be small we implicitly assume that Eq. (2) is valid for arbitrarily large length scales. This assumption does not need to be true.

Many approaches to the cosmological constant problem were designed to dynamically cancel the right-hand side of Eq. (2). Within four-dimensional local field theories with a finite number of degrees of freedom dynamical cancellation is ruled out by Weinberg's no-go theorem.[7] We take an alternative route. Following Ref. 1, we accept that the vacuum energy in our world can be large, but we question the validity of Eq. (2) in the extreme low-energy approximation.

We use the framework of Refs. 5,6 where gravity in general, and the Friedmann equation, in particular, is modified for wave lengths larger than a certain critical value. The cosmological constant problem is then remedied in the following way: Due to large-distance modification of gravity the energy density $\mathcal{E} \gtrsim (1\ \mathrm{TeV})^4$ does not curve the space as it would in the conventional Einstein gravity. Therefore, the observed space-time curvature is small, despite the fact that $\mathcal{E}$ is huge (as it comes out naturally). This is the most crucial point of the approach of Ref. 1 — the point where we depart from the previous investigations.

Before we come to details, let us briefly discuss the Weinberg no-go theorem [7] adapted to the present case (for details see Ref. 1). The theorem states that $\mathcal{E}$ cannot be canceled (without fine-tuning) in any effective four-dimensional theory that satisfies the following conditions:

(a) General covariance;

(b) Conventional four-dimensional gravity is mediated by a *massless* graviton;

(c) Theory contains a finite number of fields below the cut-off scale;

(d) Theory contains no negative norm states.

Since we do not try to cancel $\mathcal{E}$ but rather intend to suppress the space-time curvature induced by it, one could have argued that the theorem is inapplicable to begin with. However, this argument is incomplete and unsatisfactory. The point is that in any theory obeying conditions (a-d) Eq. (2) is valid. Therefore, in any such theory small H^2 would require small $\mathcal{E}$.

We conclude that a successful solution must violate at least one condition in (a-d). The solution of Ref. 1 does violate (b) and (c). In particular, the zero-mode graviton, that mediates gravity in the far-infrared, has a coupling which is many orders of magnitude smaller than the Newton coupling. That is why large value of $\mathcal{E}$ does not induce huge curvature.

The mode that mediates gravity between shorter wave-length sources, such as most of the sources in the observable part of the universe, is coupled to the sources with the usual Newton constant. This gives rise to conventional interactions for observable sources in the universe.

Note that the division in two different modes is purely for convenience. In general one could just drop the mode expansion language altogether and state that the strength of gravity depends on the wave length of the source. Although in the present case condition (c) is violated, in other possible realizations of the present idea this may not be necessary. At the same time, violation of condition (b) seems inevitable in any model that solves the cosmological constant problem through large-distance modification of gravity.

In four-dimensional general relativity gravitational interactions are mediated (at least at distances $r \gtrsim 0.1$ mm) by a massless spin-2 particle, the graviton $h_{\mu\nu}$. General covariance (and the absence of ghosts and tachyons) requires the universality of the graviton coupling to matter

$$h_{\mu\nu} T^{\mu\nu} . \tag{3}$$

General covariance also fixes uniquely the low-energy effective action to be the Einstein-Hilbert action

$$S = \frac{M_{\mathrm{Pl}}^2}{2} \int d^4x \sqrt{g} \, (R - 2\Lambda) , \tag{4}$$

where the cosmological constant $\Lambda = -\mathcal{E}/M_{\mathrm{Pl}}^2$ is included. The universal coupling to all sorts of energy, including the vacuum energy, is the reason for the emergence of the cosmological constant problem.

If the measured four-dimensional gravity were not mediated by an exactly massless state, the universality of coupling could be avoided. Thus, one may hope that in such theories very large wave-length sources (such as the vacuum energy) may effectively decouple from four-dimensional gravity, eliminating the cosmological constant problem. This is what happens in theories with infinite-volume extra dimensions. This phenomenon can be understood from the point of view of a four-dimensional brane observer as a modification of the equation of motion for the graviton. In conventional linearized general relativity the graviton obeys the following free field equation:

$$\nabla^2 h_{\mu\nu} = 0, \tag{5}$$

while in the present case [8,1] (see also Sec. 4) the graviton obeys a modified linearized equation

$$\left\{ 1 + \frac{M_*^{N-2}}{c_1(N) \, M_*^{N-2} + c_2(N) \, (\nabla^2)^{\frac{N-2}{2}}} \frac{1}{r_c^2 \nabla^2} \right\} \nabla^2 h_{\mu\nu} = 0, \tag{6}$$

where M_* is the scale of higher dimensional theory, $c_{1,2}$ are some constants, and N denotes the number of extra dimensions. $r_c = M_{\mathrm{Pl}}/M_*^2$ is the critical crossover distance beyond which gravity is modified, as described by Eq. (6). The effective strength of gravity is set by the operator in the braces, which tends to unity for the short wave length ($\ll r_c$) gravitons. As a result, Eq. (6) becomes indistinguishable from Eq. (5). Thus, in this region, the theory reproduces predictions of ordinary gravity with the standard G_N coupling. However, in the deep-infrared region, where the momenta are smaller than r_c^{-1}, new terms in (6) dominate over the conventional term (5). Hence, infrared physics is modified. It is this modification that serves as a loophole in the no-go theorem for solving the cosmological constant problem. It is remarkable that such a modification takes place in a manifestly generally covariant theory.

2. A brief overview of the model

The effective low-energy action in these theories takes the form [5,6]

$$S = M_*^{2+N} \int d^4x \, d^N\rho \sqrt{G} \mathcal{R} + \int d^4x \sqrt{\bar{g}} \left(\mathcal{E} + M_{\text{ind}}^2 R + \mathcal{L}_{\text{SM}} \right). \quad (7)$$

Here G_{AB} stands for a $(4+N)$-dimensional metric $(A, B = 0, 1, 2, ..., 3+N)$, while ρ are "perpendicular" coordinates. For simplicity we do not consider brane fluctuations. Thus, the induced metric on the brane is given by

$$\bar{g}_{\mu\nu}(x) \equiv G_{\mu\nu}(x, \rho_n = 0), \qquad (n = 4, ..., 3+N). \quad (8)$$

The first term in (7) is the bulk Einstein-Hilbert action for $(4 + N)$-dimensional gravity, with the fundamental scale M_*. The expression in (7) has to be understood as an effective low-energy Lagrangian valid for graviton momenta smaller than M_*. We imply that, in addition, there are an infinite number of gauge-invariant high-dimensional bulk operators suppressed by inverse powers of M_*.

The $M_{\text{ind}}^2 R$ term in (7) is the four-dimensional Einstein-Hilbert (EH) term of the induced metric. This term plays a crucial role. It ensures that the laws of four-dimensional gravity are reproduced at observable distances on the brane despite the fact that there is no localized zero-mode graviton. Its coefficient M_{ind} is another parameter of the model. $\mathcal{E}$ denotes the brane tension. Thus, the low-energy action, as it stands in (7), is governed by three parameters M_*, M_{ind} and $\mathcal{E}$. Both parameters M_* and M_{ind} are perturbatively-stable under quantum corrections. The parameter $\mathcal{E}$ plays the role of the vacuum energy.

For $\mathcal{E} = 0$ the gravitational dynamics on the brane is as follows. The infinite extra space notwithstanding, a brane observer measures four-dimensional gravitational interaction up to a certain critical distance r_c. The potential between two static sources on the brane scales as

$$V(r) \propto -\frac{1}{M_{\text{ind}}^2 \, r}, \quad (9)$$

for distances in the interval

$$M_*^{-1} \lesssim r \lesssim r_c, \quad (10)$$

where

$$r_c \sim \begin{cases} M_{\text{ind}}^2/M_*^3 & \text{for} \quad N = 1 \\ M_{\text{ind}}/M_*^2 & \text{for} \quad N > 1 \end{cases}. \quad (11)$$

However, at distances smaller than M_*^{-1} and larger than r_c gravity changes. An important requirement [9] is that gravity must be soft above the scale

M_*. The expression (9) fixes $M_{\text{ind}}^2 = M_{\text{Pl}}^2/2$. In order for the late-time cosmology to be standard we require that $r_c \sim H_0^{-1} \sim 10^{28}$ cm. This restricts the value of the bulk gravity scale to lie in the ball-park

$$10^{-3} \text{ eV} \lesssim M_* \lesssim 100 \text{ MeV} \quad \text{for} \quad N = 1,$$
$$M_* \sim 10^{-3} \text{ eV} \quad \text{for} \quad N > 1. \tag{12}$$

Inclusion of non-vanishing $\mathcal{E} \gg M_*^4$ triggers inflation on the brane. The inflation rate is rather peculiar and is given by [1]

$$H^2 \sim M_*^2 \left(\frac{M_*^4}{\mathcal{E}} \right)^{\frac{2}{N-2}}. \tag{13}$$

(This formula is not applicable to the $N = 2$ case; see Ref. 10 for a discussion of induced gravity in this case.) Therefore, a large vacuum energy density $\mathcal{E}$ causes a small rate of inflation for $N > 2$.

The following question immediately come to one's mind: How can such a behavior be understood from the standpoint of a four-dimensional observer on the brane? We will answer this question below using the linearized analysis of Ref. 5.

The effect can be best understood in terms of the four-dimensional mode expansion. From this perspective a high-dimensional graviton represents a continuum of four-dimensional states and can be expanded in these states. Below we will be interested only in spin-2 components for which the KK decomposition can schematically be written as follows:

$$h_{\mu\nu}(x, \rho_n) = \int d^N m \, \epsilon_{\mu\nu}^m(x) \, \sigma_m(\rho_n), \tag{14}$$

where $\epsilon_{\mu\nu}^m(x)$ are four-dimensional spin-2 fields of mass m and $\sigma_m(\rho_n)$ are their wave-function profiles in extra dimensions. The strength of individual mode coupling to a brane source is given by the value of the wave function at the position of the brane, that is $\sigma_m(0)$. Four-dimensional gravity on the brane is mediated by exchange of all the above modes. Each of these modes gives rise to a Yukawa type gravitational potential. The net result is

$$V(r) \propto \frac{1}{M_*^{2+N}} \int_0^\infty dm \, m^{N-1} |\sigma_m(0)|^2 \frac{e^{-rm}}{r}. \tag{15}$$

It is a crucial property of the model (7) that four-dimensional gravity on the brane is recovered for $r \ll r_c$ due to the fact that modes with $m > 1/r_c$ have suppressed wave-functions [11,12] and, therefore, the above integral is effectively cut-off at the upper limit at $m \sim 1/r_c$. Most easily this can be seen from the propagator analysis. Gravitational potential (15) on the brane is mediated by an "effective" 4D graviton which can be defined as

$$h_{\mu\nu}(x,0) = \int d^N m \, \epsilon_{\mu\nu}^m(x) \, \sigma_m(0). \tag{16}$$

The Green's function for this state can be defined in a usual way. Using (16) and orthogonality of the $\epsilon_{\mu\nu}^m(x)$ states we obtain

$$\mathcal{G}(x-x',0)_{\mu\nu,\gamma\delta} = \langle h_{\mu\nu}(x,0) \, h_{\gamma\delta}(x',0) \rangle = \int d^N m |\sigma_m(0)|^2 \langle \epsilon_{\mu\nu}^m(x) \epsilon_{\gamma\delta}^m(x') \rangle. \tag{17}$$

¿From now on we will suppress the tensor structure, which is not essential for this discussion. Passing to the Euclidean momentum space we get the following expression for the scalar part of the propagator

$$G(p,0) = \int dm \, m^{N-1} \frac{|\sigma_m(0)|^2}{m^2 + p^2}. \tag{18}$$

This is the spectral representation for the Green's function

$$G(p,0) = \int ds \, \frac{\rho(s)}{s + p^2}, \tag{19}$$

with $s \equiv m^2$ and

$$\rho(s) = \frac{1}{2} s^{\frac{N-2}{2}} |\sigma_{\sqrt{s}}(0)|^2. \tag{20}$$

Therefore, the spectral representation can be simply understood as the Kaluza-Klein mode expansion (14). Then, the wave-function suppression of the heavy modes can be read off from Eqs. (18) and (19) by using the explicit form [6,8] of the propagator $G(p)$,

$$G(p,0) = \frac{1}{M_{\text{Pl}}^2 p^2 + M_*^{2+N} D^{-1}(p,0)}, \tag{21}$$

where $D^{-1}(p,0)$ is the inverse Green's function of the bulk theory with no brane. For the purposes of the present discussion it is enough to note that at large momenta $p \gg r_c^{-1}$ the above propagator behaves as [5,6]

$$G(p,0) \simeq \frac{1}{M_{\text{Pl}}^2 p^2}, \tag{22}$$

which is the propagator of a massless four-dimensional graviton with the coupling $1/M_{\text{Pl}}^2$. Substituting (22) in the left-hand side of (19) we find that the function $\rho(s)$ must be suppressed at $s \gg r_c^{-2}$. If so, the relation (20) implies that the wave functions of the heavy modes must be vanishingly small as well.

For the $N = 1$ case both the propagator[5] and the wave-function profiles can be evaluated analytically,[12]

$$G(p, 0) = \frac{1}{M_{\text{Pl}}^2 p^2 + 2M_*^3 p}, \qquad (23)$$

and

$$|\sigma_m(0)|^2 = \frac{4}{4 + m^2 M_{\text{Pl}}^4 / M_*^6}. \qquad (24)$$

This shows that all the modes which are heavier than $r_c^{-1} = M_*^3 / M_{\text{Pl}}^2$ are suppressed on the brane. Substituting (24) into (15) we derive the usual Newton potential (9) at distances $r \ll r_c$.

One can interpret the above Green's function as describing a metastable state that decays into the bulk states with the lifetime $\tau_c \sim r_c$. A remarkable fact is that the existence of such a metastable state is perfectly compatible with the exact four-dimensional general covariance.

3. Dilution of the cosmological constant

It is instructive to rederive the above-mentioned results with extra dimensions being compactified at very large cosmological distances. For nonvanishing $\mathcal{E}$ the compactification is not trivial. If we were to set the compactification radius L smaller than the gravitational radius of the brane [13,14]

$$\rho_g \sim M_*^{-1} \left(\frac{\mathcal{E}}{M_*^4} \right)^{\frac{1}{N-2}}, \qquad (25)$$

this would distort the brane gravitational background very strongly. Therefore, L has to be at least somewhat larger than ρ_g. For realistic values of $\mathcal{E}$ and M_* this leads[1] to an estimate $L \gtrsim H_0^{-1} \sim 10^{28}$ cm. In what follows for simplicity we will not distinguish the values of L, ρ_g and H_0^{-1}, in spite of the fact that there should be an order of magnitude difference between these scales ($L \gtrsim \rho_g \gtrsim H_0^{-1}$). We simply set all these values in the ball-park of H_0^{-1}. Let us now address the question: what does a 4D observer see on the brane?

Because the space is compactified, there is a mass gap in the KK modes. Start from the zero-mode massless graviton. This mode couples universally to the brane matter and vacuum energy. However, because the compactification scale is huge $L \sim \rho_g \sim H_0^{-1}$ the coupling of the zero mode G_{zm} is tiny. Indeed,

$$G_{\text{zm}} \sim \frac{1}{M_*^{2+N} L^N + M_{\text{Pl}}^2}. \qquad (26)$$

The first term in the denominator is due to the compactness of the extra space and the second term is due to the induced EH term in (7). In particular, $M_*^{2+N} L^N \gg M_{\mathrm{Pl}}^2$ and, therefore,

$$G_{\mathrm{zm}} \ll G_N. \tag{27}$$

Besides the zero mode, there is a tower of massive KK modes, with masses quantized in the units of H_0. *A priori* each of these modes is important for interactions at observable distances $\ll L$, for instance in the solar system. However, due to the presence of the induced EH term on the brane (7) the wave functions of these modes are suppressed on the 4D world volume. The net result due to the KK modes can now be summarized as a single metastable graviton with the lifetime $\sim L \sim H_0^{-1}$. The two descriptions — one in terms of the KK modes, and the other one in terms of the metastable graviton — are complimentary to each other.

The coupling of this metastable graviton to matter is determined by the coefficient in front of the induced EH term in (7). Since the latter is set to be M_{Pl}, the metastable mode couples to matter with the Newton coupling G_N. Therefore, at observable distances, that are somewhat smaller than 10^{28} cm, the laws of 4D gravity are enforced by a metastable graviton which at these scales can be treated as a (almost) stable spin-2 state interacting with G_N. For these scales the presence of the zero mode is irrelevant because of its tiny coupling G_{zm}. This provides conventional gravity and cosmology for any time scale up until today.

Let us now turn to distances somewhat larger that 10^{28} cm. There, the metastable mode does not mediate interactions (since its lifetime is less than H_0^{-1}). Instead, one should think in terms of the massive KK modes. Since the mass of the lightest KK mode is $\sim H_0$, the interactions due to the KK modes are exponentially suppressed at distances larger than H_0^{-1}. Therefore, one is left with the zero mode interactions only. Any source which has characteristic wave lengths larger than H_0^{-1}, (i.e., the characteristic momenta smaller than H_0) will feel gravity only due to the zero mode. Because the coupling of the zero mode is tiny (26), the curvature R produced by this source will also be small

$$R \sim G_{\mathrm{zm}} \mathcal{E} \sim \frac{\mathcal{E}}{M_*^{2+N} L^N} \ll \frac{\mathcal{E}}{M_{\mathrm{Pl}}^2}. \tag{28}$$

Since $R \sim H^2$ and $L \sim \rho_g$ where ρ_g is defined in (25) we find from (28)

$$H^2 \sim M_*^2 \left(\frac{M_*^4}{\mathcal{E}} \right)^{\frac{2}{N-2}}. \tag{29}$$

This coincides with (13). Therefore, the small inflation rate (small curvature) is due to the fact that the zero mode is very weakly coupled to vacuum energy. Summarizing, compactification of extra dimensions with the ultra-large radius (25) produces essentially the same physical picture as in the model with infinite extra dimensions.

4. Four-dimensional picture on the brane

In this section we return to the uncompactified case. Four-dimensional gravitational interactions are mediated by a gapless infinite tower of Kaluza-Klein modes the wave functions of which are suppressed on the brane. Alternatively, these interactions can be thought of to be mediated by a *single metastable* four-dimensional graviton. The expression for the two-point Green's function on the brane[5] leads to the equation for this four-dimensional "effective" graviton,[15]

$$\left(\nabla^2 + m_c \sqrt{\nabla^2}\right) h_{\mu\nu} = 8\pi G_N \left(T_{\mu\nu} - \frac{1}{3}\eta_{\mu\nu} T\right). \tag{30}$$

Here $T \equiv T^\nu_\nu$ and $m_c \equiv r_c^{-1} \sim M_*^3/M_{\text{Pl}}^2$. This refers to one extra dimension,[a] $N = 1$. As before, M_* denotes an ultraviolet scale at which gravity breaks down as an effective field theory. On phenomenological grounds, this can be any scale in the interval $10^{-3}\,\text{eV} \lesssim M_* \lesssim M_{\text{Pl}}$. As we discussed before, in the five-dimensional model we have $10^{-3}\,\text{eV} \lesssim M_* \lesssim 100$ MeV, so that $m_c \lesssim H_0 \sim 10^{-42}$ GeV is less than the Hubble scale H_0.

We would like to provide a nonlinear completion of Eq. (30). Let us start with the right-hand side (r.h.s.) of the equation. It is known that in the massive (metastable) graviton theory the tensorial structure of the graviton propagator is affected by *nonlinear* corrections.[17,18] The expression which takes these nonlinear effects into account can be parametrized as follows:

$$\text{r.h.s.} = 8\pi G_N \left\{T_{\mu\nu} - \mathcal{K}\left(\frac{m_c}{\mu}\right)\eta_{\mu\nu} T\right\}, \tag{31}$$

where $\mathcal{K}(m_c/\mu)$ is a function of the physical scale μ set by the source T and of the distance from the source.[17,18] For large (unobservable) distances it gives rise to $\mathcal{K}(m_c/\mu) \simeq 1/3$, i.e., the result in (30). However, at measurable distances (e.g. solar system) we have $\mathcal{K}(m_c/\mu) \simeq 1/2$ due to nonlinear effects. Hence, the discontinuity[19,20,21] in the mass parameter m_c is in fact absent in nonlinear theory of massive gravity[17] and, in particular, in the

---

[a]Two branches of the square root in (30) lead to the standard and self-inflationary cosmological solutions.[16]

model of metastable graviton,[5] as was shown in Refs. 18,22,23,24 (see also Ref. 25). Therefore, for the distances of practical interest

$$\text{r.h.s.} = 8\pi G_N \left(T_{\mu\nu} - \frac{1}{2}\eta_{\mu\nu}T \right), \tag{32}$$

which is a correct tensorial structure of the conventional Einstein equation.

The next step is to perform a nonlinear completion of the left-hand side (l.h.s.) of Eq. (31). The latter procedure is somewhat arbitrary from the point of view of pure 4D theory, however it is uniquely fixed by the higher dimensional theory.[24] Here we are interested in qualitative features which are independent of the form of the completion. To this end one can use the substitution

$$h_{\mu\nu} \to \frac{1}{\nabla^2} R_{\mu\nu}, \tag{33}$$

where $R_{\mu\nu}$ denotes the Ricci tensor. Furthermore, performing simple algebra we find the following nonlinear completion of Eq. (31):

$$\mathcal{G}_{\mu\nu} + \frac{m_c}{\sqrt{\nabla^2}}\mathcal{G}_{\mu\nu} = 8\pi G_N T_{\mu\nu}, \tag{34}$$

where $\mathcal{G}_{\mu\nu} \equiv R_{\mu\nu} - (1/2)g_{\mu\nu}R$. This should be compared with the conventional Einstein equation

$$\mathcal{G}_{\mu\nu} = 8\pi G_N T_{\mu\nu}. \tag{35}$$

Note that the second term on the l.h.s in Eq. (34) does not appear in the Einstein equation. In fact, this is the very same term that dominates in the infrared region. The Bianchi identity imposes a new constraint on possible gravitational backgrounds $m_c \nabla^\mu (\nabla^2)^{-1/2}\mathcal{G}_{\mu\nu} = 0$ (in the linearized approximation the latter reduces to a gauge fixing condition). One can also note that from the standpoint of four-dimensional theory a violation of unitarity takes place. This corresponds to the fact that the metastable graviton can decay into KK modes. It is clear that the full five-dimensional unitarity is preserved while it can be violated in any given four-dimensional subspace.

As we have already mentioned, the procedure of the nonlinear completion used above is not unique. From the 4D point of view there is no guiding principle for this completion. In general, an infinite number of nonlinear terms with arbitrary coefficients can be added to the r.h.s. of (33). This would lead to an infinite number of new terms on the l.h.s. of Eq. (34). However, what is critical is that there is a *unique* set of these terms which complete the 4D theory to a higher dimensional theory that we started from. The purpose of the exercise performed above is to demonstrate that the presence of these nonlocal terms leads to modification of gravity in the infrared.

Similar arguments apply to higher co-dimensions. In a model with $(4 + N)$ dimensions[6] the equation analogous to (31) takes the form[1]

$$\left\{ \nabla^2 + \frac{M_*^{N+2}/M_{\rm Pl}^2}{c_1(N)\, M_*^{N-2} + c_2(N)\, (\nabla^2)^{\frac{N-2}{2}}} \right\} h_{\mu\nu}$$

$$= 8\,\pi\, G_N \left\{ T_{\mu\nu} - \mathcal{K} \left(\frac{m_c}{\mu} \right) \eta_{\mu\nu}\, T \right\}, \tag{36}$$

where $c_{1,2}(N)$ are some constants that depend[11,8] on N (we do not discuss here the $N = 2$ case for which there are logarithmic functions of ∇^2 on the l.h.s. in (36)). Using the same method as above we propose a nonlinear completion of the equation which looks as follows:

$$\mathcal{G}_{\mu\nu} + \left\{ \frac{M_*^{N+2}/M_{\rm Pl}^2}{c_1(N)\, M_*^{N-2} + c_2(N)\, (\nabla^2)^{\frac{N-2}{2}}} \right\} \frac{1}{\nabla^2}\, \mathcal{G}_{\mu\nu} = 8\,\pi\, G_N\, T_{\mu\nu}. \tag{37}$$

One observes the same pattern: the second term on the l.h.s. dominates in the infrared. Hence, infrared gravity is modified. This is a necessary condition for the present approach to solve the cosmological constant problem. However, it is not sufficient, generally speaking. As we argued in Ref. 1, and in the previous sections, only $N > 2$ models can solve the cosmological constant problem.

5. Killing cosmological constant does not kill inflation

The present scenario solves the cosmological constant problem due to modification of 4D gravity at very low energies. Due to this modification a strictly constant vacuum energy curves four-dimensional space extremely weakly. As we have seen, the contribution to the effective Hubble expansion rate from the vacuum energy $\mathcal{E}$ is set by an inverse power of $\mathcal{E}$, see Eq. (13). A question to be discussed in this section is whether the dilution of cosmological constant kills inflation. Naively this seems inevitable. According to the inflationary scenario, our 4D universe underwent the period of an exponentially fast expansion (for a review see Ref. 26). In the standard case this is achieved by introducing a slow-rolling scalar field, the inflaton. During the slow-roll epoch the potential energy dominates and mimics an approximately constant vacuum energy. This leads to an accelerated growth of the scale factor.

However, as we have just argued, in the present case any constant vacuum energy is diluted according to Eq. (13). If so, it seems that the same should happen to any inflationary energy density. In other words, one would naively expect that standard inflation with the energy density $\mathcal{E}_{\text{inf}}$ should generate the acceleration rate $H \sim M_* \left(M_*^4/\mathcal{E}_{\text{inf}}\right)^{1/N-2}$. If such a relation indeed took place during the slow-roll period this would eliminate inflation and all the benefits one gets with it. Fortunately, this is not the case as we will now argue.

The relation (13) is only true for energy sources with wave lengths $\gg r_c \sim H_0^{-1}$. Brane sources of shorter wave lengths gravitate according to the conventional 4D laws since they are coupled to the resonance graviton. Furthermore, let us recall that $\mathcal{E}_{\text{inf}}$ is *time-dependent* during slow-roll inflation, with a typical time scale that is smaller than r_c by many orders of magnitude. For instance, consider inflation with $\mathcal{E}_{\text{inf}} \gtrsim (1 \text{ TeV})^4$. For such inflation to solve the horizon and flatness problems and generate density perturbations, $\mathcal{E}_{\text{inf}}(t)$ must be approximately constant on the time scale of the inflationary Hubble time $\sim M_P/\sqrt{\mathcal{E}_{\text{inf}}} \lesssim \text{mm}$. This is at least 30 orders of magnitude smaller than our r_c. That is why inflation in our scenario proceeds in the same manner as in the conventional setup. Below we will consider this issue in more detail.

Let us first consider the simplest nonlinearly-completed effective 4D equation (37). Although it is equivalent to the high-dimensional model (7) only in the linearized approximation, nevertheless, it captures the most essential feature — the large-distance modification of gravity. Taking the trace of the above equation we get the following relation between the curvature and stress tensor

$$\left[1 + \frac{M_*^{N-2}}{c_1 M_*^{N-2} + c_2 \nabla^{N-2}} \frac{1}{(r_c \nabla)^2}\right] R = 8\pi G_N T_\mu^\mu. \tag{38}$$

Let us consider a constant vacuum energy $\mathcal{E}$ as the source in the r.h.s. of this equation. In the conventional Einstein gravity this would induce curvature $R \sim G_N \mathcal{E}$.

However, this cannot be a solution in our case since the second term on the l.h.s. diverges for a constant curvature. This divergence is just an artifact of our simplified non-linear completion in which we keep only finite number of the ∇^{-1} operators. Careful resummation of all such terms should give rise to a small but finite curvature as it is apparent from the high-dimensional description (see Eq. (13)).

What is important, however, is the fact that the nonlocal terms dominate for constant curvature, while they are negligible for time-dependent sources with characteristic wavelengths $\ll r_c$. For instance, let us assume T_μ^μ is the energy density of a slowly-rolling inflaton field. For the standard inflation, a typical time scale of change in the energy density is by many orders of magnitude smaller than any reasonable value of r_c. For a typical slow-roll inflation with $H \sim 10^{12}$ GeV we get

$$\left[\frac{1}{(r_c \nabla)^2}\right] R \sim 10^{-108} R. \tag{39}$$

From this estimate it is clear that the nonlocal terms are negligible, and the standard relation is intact.

Were we able to calculate all nonlocal terms in 4D world volume, we would recover the series of the form

$$\left[1 + \frac{M_*^{N-2}}{c_1 M_*^{N-2} + c_2 \nabla^{N-2}} \frac{1}{(r_c \nabla)^2} + \sum_n \frac{a_n}{(r_c \nabla)^n}\right] R = 8\pi G_N T_\mu^\mu, \tag{40}$$

where the coefficients a_n are, generally speaking, functions of metric invariants that die away in the linearized flat-space limit. For the time independent curvature, such as the one induced by a constant vacuum energy, each term diverges. Hence, explicit summation must be performed in (40). In each particular case the summation should be possible, in principle. However, in practice it may be much easier to solve directly the high-dimensional equations. This, as we know, indicates that the resulting curvature is given by Eq. (13) and is tiny.

The proportionality of the curvature to the negative power of vacuum energy is a peculiar property of large wave-length sources. Nothing of the kind happens for shorter wave-length sources. For the slow-roll inflationary curvature the contribution from nonlocal terms is completely negligible. This is why we expect that inflation, as well as all "short distance" physics, obey the conventional laws. A simple dimensional analysis shows that deviations from the conventional behavior must be suppressed by powers of λ/r_c, where λ is a typical wave length (or time scale) of the system. Thus, our solution predicts measurable deviations from predictions of the Einstein gravity for sufficiently large objects. In fact this was explicitly demonstrated in Ref. 23 in the five-dimensional example where the deviation for the Jupiter orbit might be potentially observable.

580

Acknowledgments

We would like to thank S. Dimopoulos, A. Vainshtein and A. Vilenkin for useful discussions on the subject of the paper. The work of G.D. is supported in part by a David and Lucile Packard Foundation Fellowship for Science and Engineering, by Alfred P. Sloan foundation fellowship and by NSF grant PHY-0070787. G.G. and M.S. are supported by DOE grant DE-FG02-94ER408.

References

1. G. Dvali, G. Gabadadze and M. Shifman, *Diluting cosmological constant in infinite volume extra dimensions*, hep-th/0202174.
2. A. G. Riess *et al.* [Supernova Search Team Collaboration], Astron. J. **116**, 1009 (1998) [astro-ph/9805201];
   S. Perlmutter *et al.* [Supernova Cosmology Project Collaboration], Astrophys. J. **517**, 565 (1999) [astro-ph/9812133].
3. S. Weinberg, Phys. Rev. Lett. **59**, 2607 (1987).
4. J. Garriga, M. Livio, A. Vilenkin, Phys. Rev. D **61**, 023503 (2000) [astro-ph/9906210]; J. Garriga and A. Vilenkin, Phys. Rev. D **61**, 083502 (2000) [astro-ph/9908115]; Phys. Rev. D **64**, 023517 (2001); [hep-th/0011262]. For the recent review, see, A. Vilenkin, Talk given at XVIIIth IAP conference *"Nature of dark energy"*, Paris, July 1-5, 2002.
5. G. R. Dvali, G. Gabadadze and M. Porrati, Phys. Lett. B **485**, 208 (2000) [hep-th/0005016].
6. G. R. Dvali and G. Gabadadze, Phys. Rev. D **63**, 065007 (2001) [hep-th/0008054].
7. S. Weinberg, Rev. Mod. Phys. **61**, 1 (1989).
8. G. Dvali, G. Gabadadze, X. Hou and E. Sefusatti, *See-saw modification of gravity*, hep-th/0111266.
9. G. R. Dvali, G. Gabadadze, M. Kolanovic and F. Nitti, hep-th/0106058; Phys. Rev. D **64**, 084004 (2001) [hep-ph/0102216].
10. O. Corradini, A. Iglesias, Z. Kakushadze and P. Langfelder, Phys. Lett. B **521**, 96 (2001) [hep-th/0108055].
11. M. Carena, A. Delgado, J. Lykken, S. Pokorski, M. Quiros and C. E. Wagner, Nucl. Phys. B **609**, 499 (2001) [hep-ph/0102172].
12. G. R. Dvali, G. Gabadadze, M. Kolanovic and F. Nitti, Phys. Rev. D **64**, 084004 (2001) [hep-ph/0102216].
13. R. Gregory, Nucl. Phys. B **467**, 159 (1996) [hep-th/9510202].
14. C. Charmousis, R. Emparan and R. Gregory, JHEP **0105**, 026 (2001) [hep-th/0101198].
15. C. Deffayet, G. R. Dvali and G. Gabadadze, Phys. Rev. D **65**, 044023 (2002) [astro-ph/0105068].
16. C. Deffayet, Phys. Lett. B **502**, 199 (2001) [hep-th/0010186].

17. A. I. Vainshtein, Phys. Lett. **39B**, 393 (1972) .

18. C. Deffayet, G. R. Dvali, G. Gabadadze and A. I. Vainshtein, Phys. Rev. D **65**, 044026 (2002) [hep-th/0106001].

19. Y. Iwasaki, Phys. Rev. D **2** (1970) 2255.

20. H. van Dam and M. Veltman, Nucl. Phys. **B22**, 397 (1970) .

21. V. I. Zakharov, JETP Lett. **12**, 312 (1970) .

22. A. Lue, *Cosmic strings in a brane world theory with metastable gravitons*, hep-th/0111168.

23. A. Gruzinov, *On the graviton mass*, astro-ph/0112246.

24. M. Porrati, Phys. Lett. B **534**, 209 (2002) [hep-th/0203014].

25. I. I. Kogan, S. Mouslopoulos and A. Papazoglou, Phys. Lett. B **503**, 173 (2001) [hep-th/0011138];
    M. Porrati, Phys. Lett. B **498**, 92 (2001) [hep-th/0011152].

26. A.D. Linde, *Particle physics and inflationary cosmology*, Harwood Academic, Switzerland (1990).

A STABLE H DIBARYON:
DARK MATTER CANDIDATE WITHIN QCD?

GLENNYS R. FARRAR

Center for Cosmology and Particle Physics
Department of Physics, New York University, NY, NY 10003, USA
E-mail: farrar@physics.nyu.edu

Particle physics arguments suggest that the H-dibaryon – a state consisting of two u, two d, and two s quarks – may have a mass $\approx 1.5 \pm 0.2$ MeV, and that $r_H \lesssim \frac{1}{4} r_N$. Remarkably, the observed stability of nuclei and other experimental limits not exclude this scenario at present, as discussed here. If they are present in sufficient abundance, relic H's would be the cold dark matter. Tests of this scenario are discussed

1. Introduction

Despite intensive experimental and observational efforts, the identity of the dark matter particle remains a mystery. So far, all phenomenologically satisfactory proposals for dark matter have required invoking physics beyond the standard model. Yet large portions of the parameter space for popular beyond-the-standard-model particle physics candidates have been excluded, and the remaining allowed regions of parameter space are increasingly difficult to motivate.

Surprisingly, a suitable dark matter candidate may be provided by QCD. The non-perturbative forces of QCD confinement could be such that the H-dibaryon, which consists of two u, two d, and two s quarks, is the lowest energy-per-baryon state of matter. If so, the H is stable and there will be a relic population of H's. The effects responsible for confinement and hadronic mass generation may have a temperature dependence which would allow the relic H abundance to have the required value; that question will be addressed elsewhere[a]. Here we will concentrate on the issue of the possible

---

[a]Due to the exponential sensitivity of the relic density to the H mass and freezeout temperature, greater care must be given to the temperature dependence of condensates and masses, and to competing reactions, than in the simplistic estimate given in my ArkadyFest talk.

stability of the H.

If the H were stable, nuclei would be unstable. However nuclear decay rates are extremely sensitive to the short distance properties of nuclei and the size of the H. Within constraints imposed by present observations, nuclear lifetimes can easily be beyond experimental sensitivity. Other constraints on a stable H are also reviewed. Finally we discuss how this scenario can be experimentally confirmed or excluded. It is virtually indistinguishable cosmologically from conventional CDM, but may be accessible in direct searches.

2. Properties of the H

As the density of a fermionic system such as quarks in a neutron star increases, it eventually becomes energetically favorable to replace some of the light quarks ($m_u \approx 5$ MeV, $m_d \approx 10$ MeV) with strange quarks ($m_s \approx 50-100$ MeV), thereby reducing the fermi-levels of the u, d quarks at the expense of the mass of the strange quarks. Thus in systems of ultrahigh baryon number, the ground state of matter is presumably strange quark matter[1,2]. In addition to this purely "exchange force" effect which would be present even without QCD interactions, the interaction between quarks depends on their relative spin and color configuration. The spin-color configuration of a color singlet state of quarks is controlled by its overall spin and flavor state due to Fermi statistics. As initially pointed out by Jaffe in the bag model, QCD binding is particularly strong in the flavor singlet, spin-0 state of 6 quarks ($uuddss$). He argued that two Λ's (uds; $m_\Lambda = 1115$ MeV) could form a bound state he called the H-dibaryon[3], based on a perturbative bag model estimate of the hyperfine interaction. With a mass in the $1.9 - 2.2$ GeV region, the H would be strong-interaction-stable: it could not decay to $\Lambda\Lambda$, but would not be absolutely stable because it could decay to two nucleons via weak interactions. There have been many searches for such an H, and recent results which appear to rule it out[4,5]. These will be reconsidered below, after the new proposal is outlined.

There are two lines of reasoning which indicate that the H might have a low enough mass to be stable. The first was put forth some time ago and is purely empirical and phenomenological. It starts from the observation[6,7] that the properties of the problematic $\frac{1}{2}^-$ baryon resonance $\Lambda(1405)$ and its spin 3/2 partner $\Lambda(1520)$ are nicely explained if these are "hybrid" baryons – bound states of uds quarks in a flavor-singlet color-octet state with a gluon. The similarity of $\Lambda(1405)$, $\Lambda(1520)$ and glueball masses (~ 1.5 GeV) suggests, in the hybrid-baryon interpretation of the $\Lambda(1405)$ and $\Lambda(1520)$, that the color singlet bound state of two $(uds)_8$'s would also have a mass ≈ 1.5 GeV[7]. The $\Lambda(1405)$ and $\Lambda(1520)$ have eluded satisfactory explanation in other models, as explained in [8,7], so this indirect hint of a light H is worth pursuing.

The second line of reasoning that suggests the H may be stable starts from the improved understanding of the phase structure of QCD due to Wilczek and colleagues, which has given persuasive evidence for color-flavor locking[9,10]. They apply a gap-equation approach at high chemical potential, so their results are not directly applicable to the H , but they are suggestive that non-perturbative attractive forces in the color-flavor-spin singlet 6 quark state may be very strong. It has also been noted that instantons produce a strong attraction in the scalar diquark channel[11], explaining the observed quark-diquark structure of the nucleon. The H has a color-flavor-spin structure which permits a large proportion of the quark pairwise interactions to be in this highly attractive spin-0, color $\bar{3}$ channel, again suggesting the possibilty of a light H [b].

A new observation offered here is that there is a symmetry reason the H in particular could be anomalously light compared to the naive quark model. If the QCD attraction leading to color-flavor locking were sufficiently strong, and m_u were exactly zero instead of $\lesssim 5$ MeV $<< \Lambda_{QCD} \sim$ 100 MeV, then baryon number could be spontaneously violated due to formation of a $< uuddss >$ condensate and the H would be the Goldstone boson. This is not in contradiction to the Vafa-Witten theorem, which

---

[b]It was suggested in ref. [12], based on an instanton-gas estimate giving $m_H = 1718$, that stable H 's could be the messenger particles accounting for ultrahigh energy cosmic rays (UHECRs), since the GZK energy threshold increases in proportion to the mass of the messenger[13] and there is evidence for a directional association between the highest energy UHECRs distant shrouded AGN's[14]. However the proposal that the H is the UHE messenger particle can be excluded by the following argument. The H 's, being neutral, cannot themselves be accelerated via shock acceleration or other astrophysical acceleration mechanisms which are applicable to protons, and producing them via a beam-dump mechanism from protons accelerated in a powerful AGN[14] would in the absence of fine tuning imply a far larger flux of protons than messengers, since the production cross section for H 's is very small as shown here. But as shown in [15], the GZK problem is not overcome by a messenger mechanism unless the flux of protons from the source is less than a few percent of the messenger flux from the source. Furthermore, the acceleration mechanism suggested in ref. [12], that high energy H 's could be generated by supernova implosion to strange matter, would yield H 's with GeV energies rather than $\sim 10^{11}$ GeV energies as required for them to be UHECRs. Nor did ref. [12] consider the relic dark matter possibility or address the crucial problems associated with a stable H : nuclear stability, exotic isotopes, and doubly-strange hypernuclei.

excludes the spontaneous breaking of a vectorial global symmetry *in a theory with massive fermions*[16]. The VW theorem shows that the Goldstone-boson-wannabe's mass would only vanish in the limit that the lightest quark mass vanishes.

The analogy $(uds)_8 \leftrightarrow g$ suggests that the radius of the H is comparable to that of the glueball, i.e., a factor 4 - 6 smaller than a nucleon[17]. A small radius is also expected on the basis of instanton-liquid arguments, which explain why $r_\pi = 0.38$ fm and $r_N = 0.88$ fm. More generally, a small radius would naturally be associated with tightly bound state. As we will see below, a small radius is phenomenologically essential if the H is stable, and we will therefore assume henceforth that $r_H \lesssim r_N/4$. An instanton analysis of the mass and size of the H would be valuable.

Lattice QCD efforts to determine the H mass have been unable to obtain clear evidence for a bound H. The analysis above suggests that this may be due to inadequate spatial resolution. Even the best lattice calculation used ≈ 0.15 fm lattice spacing [18], whereas the lattice spacing should be much smaller than the size of the state to get a robust result.

The H has spin-0 and therefore cannot have a dipole or higher multipole moment. If it were pointlike it would have no interactions, since it is color and charge neutral. It interacts with other hadrons only if it has a non-vanishing color charge radius, thus a crude estimate of its scattering cross section with a nucleon is $\sigma_{HN} \sim (r_H/r_N)^4 \sigma_{NN}$. The value of the coefficient in this scaling relationship reflects the density of the color charge inside the H and nucleon; taking it to be ≈ 1 gives $\sigma_{HN} \approx 0.05 \ (5R_H/R_N)^4$ mb. If the column density of color charge inside the H is sufficient that it is opaque to another H, the H-H cross section would be geometric: $\sigma_{HH} \approx 4\pi R_H^2 \approx 4(5R_H/R_N)^2$ mb. Because the H wavefunction is spatially so dissimilar to that of the nucleon, conversion of pairs of baryons to an H is highly suppressed. This will be calculated below.

3. Potential obstacles to a stable H

• *Stringent bounds exclude exotic-mass isotopes.* If the H were to bind to nuclei and had a relic abundance sufficient to account for dark matter, it could be excluded because there are strong limits on the abundance of exotic isotopes of many nuclei. However in this scenario the H does not bind to nuclei. Exchange of a single pseudoscalar meson, which produces a strong and long-range attraction between nucleons, is not possible because the H is a flavor-singlet so absorption of a flavor octet meson leads to a flavor-octet dibaryon which is not a bound state. Nor does exchange of a pair of pseudoscalar mesons provide adequate attraction, since the flavor-octet dibaryon intermediate state has a small amplitude of creation and a large energy denominator. The attractive interaction between nucleons and H due to exchange of flavor singlet states like the σ and glueballs are unlikely to be strong enough to produce a bound state, given the short range of these attractive potentials, even if $g_{HH\sigma} \times g_{NN\sigma}$ or $g_{HH\sigma} \times g_{NN\sigma}$ were ≈ 1. However in fact the $H - H - \sigma$ vertex is small because of the small size of the H, and the $N - N - G$ vertex is small due to the small size of the glueball, so we conclude the H does not bind to nuclei. Quantitative results will appear in a forthcoming paper[19].

• *There is no evidence for an H in accelerator searches.* There have been many unsuccessful attempts to produce and detect the H, but they were all designed to be sensitive to m_H around 2 GeV or higher and they are generally inapplicable for an H whose mass is ~ 1.5 GeV. Even in the conventional picture of a loosely bound H it is difficult to make reliable production cross section estimates and therefore it has not been possible to interpret the absence of a signal for H production as proof the H does not exist. In the present scenario the small size of the H means its production cross section is highly suppressed, so direct H-search experiments could not be expected to have observed a signal. The most sensitive experiment to put limits on a light stable strongly interacting particle is ref. [20] which uses time-of-flight, but it is not sensitive to masses below ~ 2 GeV because of neutron background, and therefore does not constrain this scenario.

• *Stability of nuclei and neutron stars.* Conversion of two nucleons to an H is suppressed because the process must change strangeness by two units and is thus second-order in the weak interactions. In addition, the small size of the H, along with the hard core repulsion in the nucleon-nucleon wavefunction in a nucleus, produces a suppression in the amplitude for six quarks in two nucleons to convert to an H. Taking dimensional factors in the weak amplitudes to be 100 MeV, the rate at which two nucleons in a nucleus convert to an H with emission of a pion is approximately

$$\Gamma \approx (sin\theta_C \ G_F(100\text{MeV})^2)^4 \alpha_s O^2 \cdots, \tag{1}$$

where O is the N-N-H spatial wavefunction overlap and the ellipsis includes factors of order 1 and the flavor-color wavefunction overlap. This leads to a lifetime for nuclear disintegration $\tau \sim 1 \text{ yr}/O^2$. Super-Kamiokande has an exposure of $\approx 10^{31}$ nuclei-years so although a limit has not been published, a lifetime shorter than $\approx 10^{30}$ years should produce an observable rate of

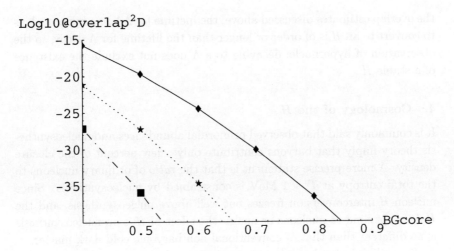

Figure 1.

appearance of pions with energy ~ 350 MeV.

The spatial overlap of the quark wavefunctions can be estimated using the Isgur-Karl oscillator model for quarks in the baryons, convoluting with the nuclear two-body wavefunction for nucleons to be in a relative s-wave, and making the ansatz that the wavefunction for the quarks in the H is of the Isgur-Karl form but with the characteristic length scale decreased by the factor r_H/r_N. As expected, the result is extremely sensitive to the short distance nuclear wavefunction and the size of the H. Nuclear wavefunctions are not constrained by experiment for distances $\lesssim 0.7$ fm[21], so for illustration a Bethe-Goldstone wavefunction with hard core radius $r_c = 0.4-0.7$ fm is used. Fig. 1 shows the square of the spatial wavefunction overlap amplitude. The flavor wavefunction overlap of 6 quarks in two nucleons with those in an H is not included in Fig. 1. It implies a substantial further suppression in O^2, so the total wavefunction overlap can easily be small enough to preclude a detectable signal in Super-K.

• *Existence of doubly-strange hypernuclei.* Recently, evidence has been presented for the production of hypernuclei containing two Λ's[4,5]. If two Λ's in a nucleus combine to produce an H before decaying via the weak interaction ($\tau_\Lambda = 3\ 10^{-10}$ sec), the observed signatures for Λ decay would not be seen. The lifetime for H production is about $10^{-23}/O^2$ sec. Given the overlap estimates discussed above, the lifetime for $\Lambda\Lambda$ in a hypernucleus to convert to an H is of order or longer than the lifetime for Λ decay, so the observation of hypernuclei decaying to a Λ does not exclude the existence of a stable H.

4. Cosmology of the H

It is commonly said that observed primordial abundances and nucleosynthesis theory imply that baryons contribute only a few percent of the closure density. A more precise statement is that the *ratio* of ordinary nucleons to the total entropy at $T \sim 1$ MeV is constrained by nucleosynthesis. Since nucleon-H interconversion freezes out well above nucleosynthesis, and the H does not bind to nucleons or nuclei, the effect of H's on nucleosynthesis is no different than that of conventional non-baryonic cold dark matter.

Spergel and Steinhardt[22] estimated that a dark matter self-interaction satisfying $\sigma_{XX}/m_X \approx 0.1 - 1$ barn/GeV would account for apparent discrepancies between the predictions of collisionless cold dark matter and observations at small scales, without affecting the excellent predictions of CDM for the dark matter distributions at large scales. Although the H is strongly interacting in the technical sense that its interactions derive from QCD, $\sigma_{HH} \ll 0.1$ mb is too small for it to be called SIDM in the Spergel-Steinhardt sense. Nor do most experimental constraints on a SIDM particle apply to the H, as we shall see below.

5. Testing the H Dark Matter Hypothesis

If relic H's provide the observed dark matter density, their number density in our galactic neighborhood is the dark matter mass density divided by the H mass, or ≈ 0.2 cm^{-3}. The solar system is moving at ~ 300 km/sec with respect to the galaxy, so the H flux on the Earth's atmosphere is ~ 50 $m^{-2}sr^{-1}s^{-1}$. The column depth of the atmosphere is ~ 1000 g/cm^2, which is about 0.1 interaction lengths for an $H-N$ cross section of 0.03 mb. The typical kinetic energy of the H in the Earth rest frame is $\sim$ keV, but the fractional energy transfer is small if the detector is not made of light nuclei. Data from the XQC balloon experiment is nearly sensitive enough to find dark matter if the H is responsible. Ref. [23] used XQC data to set a rough limit of $\sigma_{HN} \gtrsim 0.03$ mb – tantalizingly close to the naive estimate above. If a method of discriminating the H flux from background can be devised, a dark matter search could be done at the Earth's surface. Furthermore, the sensitivity of balloon experiments can probably be increased sufficiently to probe smaller $H - N$ cross sections.

6. Discussion

To summarize, it has been proposed here that the H dibaryon may be the ground state of baryonic matter. We have shown that this is consistent with particle physics constraints, if the H is sufficiently compact. A stable H would explain the observed properties of dark matter, if its relic abundance is adequate. A lattice gauge theory calculation of the H mass and wavefunction, at very much finer lattice spacing, could dispose of the possibility that the H is stable, or if a low mass H is indicated, it would motivate further experimental and theoretical work to discover it.

This research was supported in part by NSF-PHY-99-96173 and NSF-PHY-0101738. I wish to acknowledge the contributions of my graudate students Olaf Kittel and Gabrijela Zaharijas who have worked on several topics related to the stable-H possibility with me. I have also benefited from discussions with many colleagues, particularly G. Baym, A. Bondar, G. Efstathiou, D. Eisenstein, R. Jaffe, T. Kajita, M. May, M. Ramsey-Musolf, M. Rees, E. Shuryak, D. Spergel, P. Steinhardt, P. Vogel, F. Wilczek, and M. Zaldarriaga.

It is a special pleasure to present the conjecture of a stable H at the ArkadyFest, because Arkady and his colleagues have lead the way toward developing a physical understanding of the perplexing mass hierarchies that non-perturbative QCD dynamics produce. In the case of the pion and rho masses, and other properties of hadrons, observations forced us to recognize and accept the challenge posed by QCD. Hopefully, QCD theory is now well enough developed to confirm or dispose of the conjecture that the H is stable, yet experience up to now suggests the definitive answer may only come experimentally.

References

1. Edward Witten. Cosmic separation of phases. *Phys. Rev.*, D30:272–285, 1984.
2. Edward Farhi and R. L. Jaffe. Strange matter. *Phys. Rev.*, D30:2379, 1984.
3. R. Jaffe. Perhaps a stable dihyperon... *Phys. Rev. Lett.*, 38:195, 1977.
4. J. K. Ahn et al. *Phys. Rev. Lett.*, 87:132504, 2001.
5. H. Takahashi et al. *Phys. Rev. Lett.*, 87:212502, 2001.
6. Glennys R. Farrar. Detecting gluino-containing hadrons. *Phys. Rev. Lett.*, 76:4111–4114, 1996.
7. Olaf Kittel and Glennys R. Farrar. Masses of flavor singlet hybrid baryons. 2000.

8. R. H. Dalitz. Note on the $\lambda(1405)$. *Eur. Phys. J*, C15:748, 2000.
9. Mark G. Alford, Krishna Rajagopal, and Frank Wilczek. Color-flavor locking and chiral symmetry breaking in high density QCD. *Nucl. Phys.*, B537:443–458, 1999.
10. Thomas Schafer and Frank Wilczek. Superconductivity from perturbative one-gluon exchange in high density quark matter. *Phys. Rev.*, D60:114033, 1999.
11. R. Rapp, Thomas Schafer, Edward V. Shuryak, and M. Velkovsky. Diquark bose condensates in high density matter and instantons. *Phys. Rev. Lett.*, 81:53–56, 1998.
12. N. I. Kochelev. Ultra-high energy cosmic rays and stable h-dibaryon. *JETP Lett.*, 70:491–494, 1999.
13. Daniel J. H. Chung, Glennys R. Farrar, and Edward W. Kolb. Are ultrahigh energy cosmic rays signals of supersymmetry? *Phys. Rev.*, D57:4606–4613, 1998.
14. Glennys R. Farrar and Peter L. Biermann. Correlation between compact radio quasars and ultra-high energy cosmic rays. *Phys. Rev. Lett.*, 81:3579–3582, 1998.
15. Glennys R. Farrar and Tsvi Piran. Deducing the source of ultrahigh energy cosmic rays. 2000.
16. Cumrun Vafa and Edward Witten. Parity conservation in qcd. *Phys. Rev. Lett.*, 53:535, 1984.
17. Thomas Schafer and Edward V. Shuryak. Glueballs and instantons. *Phys. Rev. Lett.*, 75:1707–1710, 1995.
18. A. Pochinsky, J. W. Negele, and B. Scarlet. Lattice study of the h dibaryon. *Nucl. Phys. Proc. Suppl.*, 73:255–257, 1999.
19. G. R. Farrar and G. Zaharijas. in preparation.
20. H. Richard Gustafson, Cyril A. Ayre, Lawrence W. Jones, Michael J. Longo, and P. V. Ramana Murthy. Search for new massive longlived neutral particles. *Phys. Rev. Lett.*, 37:474, 1976.
21. Akmal and V. Pandharipande. *Phys. Rev.*, C56:2261, 1997.
22. David N. Spergel and Paul J. Steinhardt. Observational evidence for self-interacting cold dark matter. *Phys. Rev. Lett.*, 84:3760–3763, 2000.
23. Benjamin D. Wandelt et al. Self-interacting dark matter. 2000.

SECTION 6.
ARKADYFEST

ARKADY IN SIBERIA

EDWARD SHURYAK

Department of Physics and Astronomy, State University of New York,
Stony Brook, NY 11794, USA

These are random recollections on the "middle years" of Arkady's life in science, from the late 1960's to late 1980's. One cannot write about Arkady in Siberia without first describing a general atmosphere in the Budker Institute of Nuclear Physics and Akademgorodok, Novosibirsk, in the 1960's and 1970's. I guess it may be interesting for our international friends and colleagues to learn about it, and to those who were there at the time, to recall Akademgorodok once again.

First, about the place. Siberia, with its area of the size of the whole US, is still poorly populated, and for a reason. Arkady's parents moved there during the World War II, fleeing from advancing German troops from their native Donetsk in Ukraine.

Although there are several large industrial cities in Siberia (e.g. Novosibirsk's population is about 1.5 million) there were no strong universities [a] or technical schools there until the end of the 1950's, when Khrushchev, with his characteristic decisiveness, endorsed construction of one of the world largest scientific centers. A large variety of research institutes covering all hard sciences are situated there, and a brand new Novosibirsk State University. In the 1960's, Akademgorodok had a population of about 30 thousand, (eventually it grew to over 100 thousand) with, perhaps, half of them involved in scientific research, in one way or another.

Arkady (and his wife Nelly, and many friends) happened to be in the first graduating class of the newly-born University. I am sure their student years were very colorful, but I cannot say anything about this because I moved to Akademgorodok later, in 1964 at age 16, spent my last high-school year in a FMS (a specialized physics and mathematics school the

---

[a] Well, there was Tomsk University, since the mid-nineteen century, but I met no physicists from this university so far.

NSU had established to attract the brightest), entered the university next year, and met Arkady for the first time in his capacity of an instructor of the Quantum Mechanics course (with Prof. S.T. Belyaev as the lecturer) in 1966.

One good thing about this University was that there were practically no funds for professors' salaries, and nearly all of them were from research Institutes. Teaching there simply fell on people who had enough energy to do it, practically for the fun of it, with salaries being rather symbolic, even by Russian standards.

Another good thing — following from the first — was that policies toward student curriculum were rather liberal. In order to demonstrate a good deal of enthusiasm aimed at jump-starting immediate work at the front-line of science, let me give my own example. In my *freshman* year I took a course by Yu.B. Rumer and A.I. Fet "Unitary Symmetries," on SU(3) and quarks.[b] Taking it *before* quantum mechanics had little sense, but *after* it, quantum mechanics looked like an enlightenment sent by God.

I attended many Arkady's talks and started communicating with him since about 1967, when I was allowed to attend seminars at the Institute of Nuclear Physics. It is now called the Budker Institute, and very rightly so: Gersh Budker indeed managed to build a world-class laboratory in this remote place, and he did so against quite visible and ever growing hostility towards him on the part of the Akademgorodok and Novosibirsk local authorities. A pioneer of electron-electron, electron-positron and proton-antiproton colliders in 1960, he pointed out already in the mid-1970's that the future of high energy physics lay in linear electron colliders.[c] Perhaps, a less known talent of Budker (which would be so cherished in this country) was his ability to invent applications of accelerator technology, produce hardware and make good deals with industry. Due to this, his institute was, to a large extent, a self-sustaining enterprise.[d]

Gersh Budker was not only *the* Director but a true intellectual center of this Institute. I am sure all of us, who were lucky to communicate with him, or just hear his talks or lectures. will never forget him. Arkady and

---

[b]Note that the timing of this course was quite remarkable since the year was 1965, just the next year after seminal Gell-Mann and Ne'eman's papers.

[c]This fact got recognition only this year in the US Long Range Plan.

[d]I recall that in one of his speeches Panofsky addressed Budker as "a director of a capitalistic Institution in a socialistic country", while he (Panofsky) referred to himself as a director in the inverse situation— he had to beg for money from the funding agencies all the time. This feature which was always handy but became crucial for the survival of the Institute in the early 1990's when the infrastructure of what remained of the Soviet Union nearly collapsed.

myself, as members of various "round tables" (something like a standing committee meeting each week at 12, for an hour or more), were seeing him for his last years, dealing with science, strategic and day-by-day issues as they were coming along. This is where we learned what physics is all about and how one should deal with it (yes, of course, with a good joke).

One of the most memorable moments of a strong interaction between Budker and Arkady happened sometime in the early 1970's. Budker got fascinated then by prospects of studying CP violation in kaon-antikaon system originating from the ϕ-meson decays[e] and started asking theorists a series of pointed questions, trying to work out an experimental program. Questions were copious and appeared in rapid sequences: basically only Arkady was able to provide answers, usually right away at the blackboard. It went on for several months at these weekly meetings, in small installments but with an increasing sophistication. The audience watched them both in amazement. It was a good lesson: neither of these two could possibly proceed by himself, and yet doing it together they worked out a beautiful program of experiments, with fine interplay of Bose statistics, interfering amplitudes and CP violation. Another lesson: as far as I know, neither Budker nor Arkady cared to write down and publish what they had done: for them, understanding was enough a reward by itself.

Let me now come to an important problem which Budker Institute had in the 1960's: its theory group lacked its own intellectual center.[f] However, as we all know, theoretical physics is transferred mostly as a kind of Olympic flame, from one leader to the next. Budker knew it, tried to seduce one or another senior theorist (such as Yakov Zeldovich) but it did not work. Professor V.M. Galitsky played this role for a while, but was gone well before my time. Galitsky was Arkady's physics adviser. He realized that Arkady was interested not in many-body theory he could teach him, but, rather, in high-energy physics. Departing from Novosibirsk, Galitsky managed to "sell" Arkady to Boris Ioffe.[g] It was a tremendous piece of luck: it gave Arkady not only Boris Ioffe as an excellent adviser, but the ability to come regularly to ITEP, sometimes for an extended time. ITEP had one of the most established theory groups in Russia, hand-picked by

---

[e] Again, this was way before anybody else thought of this possibility; even now, thirty years later, this is not yet done but is supposed to be done at Frascati. Similar physics could be studied at b factories, another project developed early in Budker's Institute.

[f] Eventually S. Belyaev, V. Baier, I. Khriplovich, B. Chirikov and others created schools of their own, but that took time.

[g] Ioffe recalled that it was a not-so-well-heard phone conversation he received in a noisy corridor of his then-"communal" apartment; Ioffe just did not managed to say "no". He did not regret this later, as far as I could tell.

Landau and Pomeranchuk, with a large number of young and active people. They became Arkady's life-long collaborators and friends.

Some ITEP theorists were allowed to go to international conferences. Upon return, they would bring news which were then reported at ITEP seminars and discussed at length. Thus, on a large number of occasions, it was Arkady (a frequent visitor to ITEP) who told us what was happening in the world. And there were plenty of things to discuss: the emergence of the Weinberg-Salam model, then QCD, then the "November Revolution" of 1974 when J/ψ was discovered. Yulik Khriplovich and Arkady Vainshtein's review in Uspekhi, [h] on gauge theories of weak interactions, was an example of how close they were to the front-line of research, and how well prepared they were for applications of all these new theories. Celebrated "penguin diagrams" by Arkady and collaborators were a prime example of this era.

Now I should describe my interactions with Arkady in the late 1970's, when the celebrated QCD sum rules appeared, focusing the thoughts of many on the mysteries of nonperturbative QCD. In 1976-1978 I completed my first set of finite-temperature QCD papers, and thought I was ready to jump into the game. However, running after the Vainshtein-Zakharov-Shifman trio was not an easy thing, even with generous explanations which I could always get from Arkady. Only one of my QCD-sum-rule-related papers got any notice, the 1981 one, [i] which was the first occurrence of the "heavy quark symmetry".

As a part of my efforts to catch up with these guys, I convinced Arkady to work with me on a project close to QCD sum rules, the so called higher-twist effects in deep inelastic scattering. We wrote two papers, [j] both quite reasonably cited in the literature now, but both having a completely negligible effect then. These papers had interesting physics points, but mostly were about derivation of some lengthy general formulae for these effects. Obviously, I was not very useful in that, frankly just a drag to Arkady, who knew how to deal with technical problems *en route*. His view however was that if there were two authors, both should independently derive *all* formulae, from the beginning to the very end, and only then compare the whole

---

[h] A.I. Vainshtein and I.B. Khriplovich, *Renormalizable Models of the Electromagnetic and Weak Interactions*, Uspekhi Fiz. Nauk, **112**, 685 (1974) [Soviet Physics - Uspekhi, **17**, 263 (1974)].

[i] E. V. Shuryak, *Hadrons Containing a Heavy Quark and QCD Sum Rules* Nucl. Phys. B **198**, 83 (1982).

[j] E. V. Shuryak and A. I. Vainshtein, *Theory Of Power Corrections To Deep Inelastic Scattering In Quantum Chromodynamics. Parts 1 and 2*, Nucl. Phys. B **199**, 451 (1982), and Nucl. Phys. B **201**, 141 (1982).

thing. After I weeded out all my mistakes in derivation of the operator expansion expressions and thought it was finally over, Arkady announced that without radiative corrections responsible for the mixing of the operators under consideration, the paper was incomplete and could not be published. It was the first time in my life I had to deal with a "perfectionist"-type theorist. It was not an easy experience. What added to frustration is that after a very influential (at least for me) paper [k] of the same trio plus Novikov entitled "Are all hadrons alike?" which appeared at the same time, we both knew that there was much more in the QCD vacuum than any operator product expansion could explain. And we both realized that, doing the same expansion but inside the nucleon, one could not possibly come to a different conclusion. The lesson I would like to deduce from this part of the story is that it almost never pays to join already existing development and simply widen its applications. It is much more instructive to think about its limitations and deep origins of these limitations.

To go back to a lighter part, Arkady was also my first teacher in downhill skiing. (In this case, as in science, he is not responsible for my bad style.) Unlike in physics, in this case he only considered it important to get a newcomer on the lifts and get him/her as high as he/she may be fooled to go. Then he would give primary instructions and, convinced that the person would know by himself/herself how to fall in a proper time and has enough common sense not to get killed, would happily disappear.

Of course, the Siberian flats around Novosibirsk were not suited for down-hill skiing, so Arkady's instruction took place at Bakuriani Winter School in Georgia.[l] The same is true for many people in the room: let as recall and thank the Bakuriani School organizers once again. I recall, once we came to the slope, on top of a large truck as usual, but that day there was a problem with the lift. All except Arkady (and Pontecorvo Jr.) disappointedly went back: these two were not intimidated at all, they climbed the mountain Kokhta (not a small one) *twice* this day, with skis and boots and other heavy stuff in their hands, and happily glided down. I mention this episode because, obviously, Arkady's attitude toward scientific problems is exactly the same.

Many good features of Arkady has been discussed: now let me come to "problems". Arkady had no students, at least during the Novosibirsk years (and perhaps, beyond). A simplistic theory of this phenomenon goes

---

[k] V.A. Novikov, M.A. Shifman, A.I. Vainshtein, and V.I. Zakharov, Nucl. Phys. **B191**, 301 (1981).

[l] I mean here Georgia in the former Soviet Union. How can we forget the famous statement from a local Georgian lift operator: "Physicist? Then pay."

as follows: Arkady is not patient enough, he solves any problem he can think of too quickly, he does not need a student to slow him down. I think the true explanation is the opposite: my observation was he has infinite time and infinite patience. Explaining something, Arkady simply cannot stop until he is convinced the other person got it, to the tiniest detail. It may go on for hours or days, and Arkady will put away any part of his own work to do so. But as one very seasoned person (in a good sportish shape then) told me, after physics conversations with Arkady he used to have strong headaches. Another one went as far as to suggest that after such conversations one always has a feeling of being hit by a passing truck.

This leads me to a final proposition. Arkady: your noble age notwithstanding, please proceed with the intensity you like. Still, please, take it somewhat easier on others...

OF A SUPERIOR BREED

VLADIMIR ZELEVINSKY

Department of Physics and Astronomy, Michigan State University,
East Lancing, MI 48824, USA

To me Arkady always seemed to be a striking and extremely impressive sample of a human breed, an evidence for the existence of a person who very naturally reached a summit of human abilities.

I got acquainted with him (and Nelly) almost 40 years ago, in the fall of 1962. I was not particularly close to him at that time. But I remember very well that Professor Viktor M. Galitsky, a wise man and excellent physicist himself, was of an extremely high opinion of Arkady, his new Siberian student. Various conversations inevitably used to end in a comparison of Arkady with several Moscow graduates who came to Siberia with Galitsky. Sure enough, this comparison was not in favor of the Muscovites, although some of them later made quite successful careers in theoretical physics.

I can honestly say (and I am sure that this is not only my opinion) that very soon we accepted and got used to the fact that Arkady was stronger, deeper, smarter, and so on, than anybody else in our circle. This was merely a fact of life. Later Arkady became really a legendary figure, omnipotent and omni-knowledgeable, capable of helping in any problem related to physics, science at large, and everyday life ...

I have never worked with Arkady directly, as a co-author. My memory keeps, however, a few interesting "snapshots." They reflect two types of Arkady's behavior in response to my rather frequent attempts to seek Arkady's advice regarding particular scientific problems (I know that other people had quite similar experiences too). I do not remember anymore which questions were raised. This is not so important, after all. What is important is that Arkady's response was always either of one type or another. Either he would immediately know the correct answer (or the problem was so stupid that the answer was trivial right from the start, from his standpoint, of course). In this case he would start his reply with something like that:

— "Of course, you very well understand yourself that... "

And in a few minutes the author of the question would be forced to confess that, certainly, the answer was absolutely clear, and that this transparency was obvious even before the question was asked ...

A little bit different (and more rare) version of the situation was that Arkady would not know the answer immediately. This could happen if the question was related to a scientific area remote from Arkady's current interests. Then one could have enjoyed the most remarkable performance: Arkady would switch on his phenomenal thinking machine , starting from scratch, frequently on blackboard. Usually the desired answer would be found very quickly. It would happen so naturally that the inquirer would usually get puzzled: why the hell he was unable to arrive at the same result by himself. The immediate punishment for weaker intellectual abilities was unavoidable: Arkady would go into all details and consequences, often far away from the original question, and continue his explanations to the point when the inquirer would become fully exhausted and unable to grasp anything ...

My memoir would not be complete without at least a few words on Arkady in everyday life. I remember, for instance, that one nice morning we woke up in an apartment of Victor Chernyak in the East-Siberian city of Irkutsk, on the shore of the famous Lake Baikal, where there was a conference. We stayed there overnight — Arkady, myself and my young sister-in-law who traveled with me to tour Lake Baikal. And, gosh, this morning was special — I am sorry to say, something happened to the sewer system in Chernyak's apartment building, and a part of sewage water gushed out to the floor from nowhere. We discovered this disaster after waking up, when the disaster had already happened. Arkady was *the only* person who did not lose his spirit, in this tragicomic situation, and organized, in a business-like manner, our damage control operation in the most efficient way, using all available improvised means.

Women see and evaluate things differently, they have another kind of vision. That's why I want to conclude my mini-essay by a passage written by my wife Vera:

"Arkady is a truly outstanding person, outstanding in all meanings of this word, including his appearance. He is immediately singled out in the crowd, everybody says that.

What makes Arkady so remarkable? First and foremost, his outstanding intelligence. This is obvious. There is something else, however. Each facet of his personality is bright: absolute selflessness, optimism without limits, almost childish ... No matter what he does — physics research, hiking, wrestling with computers — he does it with full concentration, leaving everything else aside, forgetting about the outside world, his family including. Everyone who had the pleasure of hiking, skiing or dancing with Arkady at least once will confirm this. I cannot forget a sauna festival Arkady and Nelly once arranged at their *dacha* near Novosibirsk. Lots of people came, they were so different and so sincere, as probably never before. I think, that was due to an atmosphere of a "festival of life" ... Arkady made it happen. I asked Nelly how she could cope with such pace of life.

— Sometimes I get tired, terribly tired, she answered. My housewife's side has to be sacrificed. So what? We are always surrounded by great people. We will never be alone.

By the way, the very same fall Arkady was repairing something on the roof of their *dacha* cottage, broke it and fell through. Luckily, there were no dire consequences.

Sincerity — that's Arkady's precious gift attracting to him all of us. We all remember how hard it was to survive back there. One could not survive without support of one's friends. And Arkady was very generous with his support. A long time ago we were moving from one apartment to another. Arkady immediately volunteered to be a mover. I remember him grabbing a refrigerator, putting it on his back and crawling with it to the second floor of our new apartment building. He did it alone. Then he found out that our kids had already gone to bed by the time he was done with the refrigerator. He just dropped by their bedroom, and in a second it exploded, a joyful chorus of of three happy "piglets." My "piglets" were happy. Absolute sincerity is the advantage of children. There are not so many adults who have it to that extent. This is God's gift to Arkady, perhaps even more precious than his intellect."

DEFYING ZENO'S PROCEDURE

VALENTIN SOKOLOV

Budker Institute of Nuclear Physics, University,
East Lancing, MI 48824, USA

It is somewhat strange but, in spite of my quite vivid memory of Arkady's excellent personality and many conversations and discussions I had with him, I cannot recall anything reasonably coherent. There were very few amusing or funny incidents since he was always wise and did everything so well. Still it is my feeling that, being very certain and resolute in his scientific judgments, he may look rather indecisive in everyday life and sometimes just sinks in all those "from the one hand..., but from the other...".

I remember how I heard of Arkady for the first time. Roald Sagdeev was one of the examiners at Novosibirsk University (most probably, this was an examination in classical electrodynamics). After the examination Sagdeev shared with us his impressions. He repeatedly suggested the same problem to many students and nobody was able to solve it. The problem contained a sequence of events, and the question was what will be the result after a great many steps. Roald knew a trick which allowed one to obtain the result in a rather economic way. There was only one student in class who solved the problem correctly though by a lengthy direct summation. Of course, this was Arkady. However, the most memorable was Roald's sad tone when he pensively said:

— " Well, hmm, this student solves any problem he is asked to."

Alas, I cannot recall now the problem itself...

Another episode, I am afraid, can be interesting only to a narrow circle of people. After one of our traditional tea-and-cake gatherings on the "theoretical floor" at the Budker Institute, (that's where all theorists had their offices) a good piece of cake was left over and brought to Pavel Isaev's office. I happened to be in this office at that time, and watched people entering from time to time. Everybody would bashfully cut off a half of the remaining piece, leaving another half to the next newcomer —- exactly

Zeno's procedure! I made a comment, something about how considerate people in our small theoretical community were. Almost immediately after my comment, Arkady entered the office and made the entire procedure convergent in one step — just by swallowing the whole remaining piece! And he could not understand why everybody bursted into laughter so loudly....

With Arkady everything always was quite normal. He is excellent and that is it. Isn't this strange? He is really an absolutely remarkable person. And everybody remembers his brilliance rather than some particular amusing events.

REMINISCENCES IN PASTELS *

M. SHIFMAN

Theoretical Physics Institute, University of Minnesota
Minneapolis MN 55455, USA

Glimpses of ITEP

For about twenty years, I was a member of the ITEP theory group. ITEP was more than an institute. It was our refuge where the insanity of the surrounding reality was, if not eliminated, was reduced to a bearable level. Doing physics there was something which gave a meaning to our lives, making it interesting and even happy. Our theory group was like a large family. As in any family, of course, this did not mean that everybody loved everybody else, but we knew that we had to stay together and to rely on each other, no matter what, in order to survive and to be able to continue doing physics. This was considered by our teachers to be the most important thing, and this message was always being conveyed, in more than one way, to young people joining the group. We had a wonderful feeling of stability in our small brotherhood. A feeling so rare in the western laboratories where a whirlpool of postdocs, visitors, sabbatical years come and go, there are a lot of new faces, and a lot of people whom you do not care so much about.

The rules of survival were quite strict. First, seminars – what is now known worldwide as the famous Russian-style seminars. The primary goal of the speaker was to *explain* to the audience his or her results, not merely to advertise them. And if the results were nontrivial, or questionable or just unclear points would surface in the course of the seminar, the standard two hours were not enough to wind up. Then the seminar could last for three or even four hours, until either everything was clear or complete exhaustion, whichever came first. I remember one seminar in Leningrad in 1979, when

---

*the first part of this article is an abbreviated version of the foreword to m. shifman, *itep lectures on particle physics and field theory*, (world scientific, 1999).

Gribov was still there, which started at eleven in the morning. A lunch break was announced from two to three, and then it continued from three till seven in the evening.

In ITEP we had three, sometimes more, theoretical seminars a week. The most important were a formal seminar on Mondays, and an informal coffee seminar which at first took place every Friday at 5 o'clock, when the official work day was over, but later was shifted to Thursdays, at the same time. Usually, these were by far the most exciting events of the week. The leaders and the secretaries of the seminars were supposed to find exciting topics, either by recruiting ITEP or other "domestic" authors, or, quite often, by picking up a paper or a preprint from the outside world and asking somebody to learn and report the work to the general audience. This duty was considered to be a moral obligation. The tradition dated back to the time when Pomeranchuk was the head of the theory group, and its isolation had been even more severe than during my times. As a matter of fact, in those days there were no preprints, and getting fresh issues of *Physical Review* or *Nuclear Physics* was not taken for granted at all. When I, as a student, joined the group – this was a few years after Pomeranchuk's death – I was taken, with pride, to the Pomeranchuk memorial library, his former office where a collection of his books and journals was kept. Every paper, in every issue, was marked by Chuk's hand (that's how his students and colleagues would refer to him), either with a minus or a plus sign. If there was plus, there would also be the name of one of his students who had been asked to "dig into" the paper and give a talk for everyone's benefit. This was not the end of the story, however. Before the scheduled day of the seminar, Pomeranchuk would summon the speaker-to-be to his office to give a pre-talk to him alone, so that he could judge whether the subject had been worked out with sufficient depth and that the speaker was "ripe enough" to face the general audience and their blood-thirsty questions. In my time, the secretaries of the seminars were less inclined to sacrifice themselves to that extent, but, still, it was not uncommon that pre-talks were arranged for unknown, young or inexperienced speakers.

Scientific reports of the few chosen to travel abroad for a conference or just to collaborate for a while with western physicists, were an unquestionable element of the seminar routine. The attendance of an international conference by A or B by no means was considered as a personal matter of A and B alone. Rather, these rare lucky guys were believed to be our ambassadors, and were supposed to represent the whole group. In practical terms, this meant that once you had made your way to a conference, you could be asked to present important results of other members of the group.

Moreover, you were supposed to attend as many talks as physically possible, including those which did not exactly belong to your field, make extensive notes and then, after returning home, deliver an exhaustive report of all new developments discussed, all interesting questions raised, rumors, etc.

The scientific rumors, as well as nonscientific impressions, were like an exotic dessert, usually served after nine. I remember that, after his first visit to the Netherlands, Simonov mentioned that he was very surprised to see a lot of people on the streets just smiling. He said he could not understand why they looked so relaxed. Then he added that he finally figured out why: "... because they were not concerned with building communism..." This remark almost immediately became known to "Big Brother" who was obviously watching us this evening, as usual, and it cost Simonov a few years of sudden "unexplainable allergy" to any western exposure. His "health condition", of course, would not allow him to accept any invitation to travel there. I cannot help mentioning another curious episode with Big Brother. Coffee, which we used to have during the coffee seminars, was prepared in turn, by all members of the group. Once, when it was Ioffe's turn, he brought a small bottle of cognac and added a droplet or two in every cup. I do not remember why, perhaps, it was his birthday or something like that. That was Friday evening. Very early on the next Monday morning, he was summoned to the corresponding ITEP branch office to give explanations concerning his "obviously subversive activities"!

The coffee seminars typically lasted till nine, but sometimes much later, for instance, in the stormy days of the November revolution in 1974. The few months following the discovery of J/ψ were the star days of QCD and, probably, the highest emotional peak of the ITEP theory group. Never were the mysteries of physics taken so close to our hearts as then. There was a spontaneously arranged team of enthusiasts working practically nonstop. A limit to our discussions was set only by the schedule of the Moscow metro – those who needed to catch the last train had to be leaving before 1 a.m.

The ITEP seminars were certainly one of the key elements in shaping the principles and ideals of our small community, but not the only one. The process of selecting students who could eventually grow up into particle theorists played a crucial role and was, probably, as elaborate as the process of becoming a knight of the British crown. Every year we had about 20 new students, at the level roughly corresponding to that of graduate students in American universities. They came mostly from the Moscow Institute for Physics and Technology, a small elite institution near the city, a counterpart of MIT in the States. Some students were from the Moscow Engineering and Physics Institute, and a few from the Moscow State University. They were offered (actually, obliged to take) such a spectrum of courses in special disciplines which I have never heard of anywhere else in the world: everything from radiophysics and accelerator physics; several levels of topics in quantum mechanics, including intricacies of theory of scattering; radiation theory and nuclear physics; mathematical physics (consisting of several separate parts); not less than three courses in particle phenomenology (weak, electromagnetic and strong interactions); quantum electrodynamics, numerous problem-solving sessions, etc. And yet, only those who successfully passed additional examinations, covering the famous course of theoretical physics by Landau and Lifshitz, were allowed, after showing broad erudition and ingenuity in solving all sorts of tricky problems, to join the theory group. Others were supposed to end up as experimentalists or engineers. Needless to say, the process of passing these examinations could take months, even years, and was notoriously exhausting, but there was never a lack of volunteers trying their luck. They were always seen around Ter-Martirosian and Okun who were sort of responsible for the program. It should be added that the set of values to be passed from the elders to the young generations included the idea that high energy physics is an experimental science that *must* be very closely related to phenomena taking place in nature. Only those theoretical ideas which, at the end of the day, could produce a number which could be confronted with phenomenology were cherished. Too abstract and speculative constructions, and theoretical phantoms, were not encouraged, to put it mildly. The atmosphere was strongly polarized against what is now sometimes called "theoretical theory". Even extremely bright students, who were too mathematically oriented, like, say, Vadim Knizhnik, were having problems in passing these examinations. Vadim, by the way, never made it to the end, got upset and left ITEP. Well, nothing is perfect in this world, and I do not want to make an impression that the examination routine in the ITEP theory group was without flaws.

The ITEP theory group was large – about 50 theorists – and diverse. Moreover, it was a natural center of attraction for the whole Moscow particle physics community. Living in the capital of the last world empire had its advantages. There is no question, it was the evil empire, but what was good, as it usually happens with any empire, all intellectual forces tended to cluster in the capital. So, we had a very dynamic group where virtually every direction was represented by at least several theorists, experts in the given field. If you needed to learn something new, there was an easy way to do it, much faster and more efficient than through reading journals or textbooks. You just needed to talk to the right person. Educating others, sharing your knowledge and expertise with everybody who might be interested, was another rule of survival in our isolated community. In such an environment, different discussion groups and large collaborations were naturally emerging all the time, creating a strong and positive coherent effect. The brain-storming sessions used to produce, among other results, a lot of noise, so once you were inside the old mansion occupied by the theorists, it was very easy to figure out which task force was where – just step out in the corridor and listen. And, certainly, all these sessions were open to everybody.

The isolation of the ITEP theory group had a positive side effect. Everybody, including the youngest members, could afford to work on problems not belonging to the fashion of the day, without publishing a single line for a year or two. Who cared about what we were doing there anyway? This was okay. On the other hand, it was considered indecent to publish results of dubious novelty, incomplete results (of the status report type) or just papers with too many words per given number of formulae. Producing dense papers was a norm. This style, which was probably perceived by the outside readers as a chain of riddles, is partly explained by tradition, presumably dating back to the Landau times. It was also due to specific Soviet conditions, where everything was regulated, including the maximal number of pages any given paper could have. Compressing derivations and arguments to the level considered acceptable, was an art which had its grandmasters.

It is high time for Arkady to appear on these pages. Arkady Vainshtein was especially good at inventing all sorts of tricks which allowed him to squeeze in extra formulae with very few explanatory remarks. I remember that in 1976, when we were working on the large JETP paper on penguins in weak decays,[a] we had to make 30 pages out of the original 60-page preprint version, and he managed to do that without losing any equations and even inserting a few extra ones! This left a strong impression on me.

By the way, about penguins. From time to time students ask how this word could possibly penetrate high energy physics. This is a funny story indeed. The first paper where the graphs that are now called penguins were considered in the weak decays appeared[b] in JETP Letters in 1975,

---

[a]By "we" I mean Zakharov, Vainshtein and myself. Arkady Vainshtein had a permanent position at the Budker Institute of Nuclear Physics in Novosibirsk. He commuted between Moscow and Novosibirsk for many years, and was considered, essentially, as a member of the ITEP theory group. The large penguin paper was published in Zh. Eksp. Teor. Fiz. **72** (1977) 1275 [Sov. Phys. JETP **45** (1977) 670].

[b]A. Vainshtein, V. Zakharov and M. Shifman, *Pis'ma ZhETF* **22** (1975) 123 [*JETP Lett.*

and there they did not look like penguins at all. Later on they were made
to look like penguins:

and called penguins by John Ellis. Here is his story as he recollects it
himself.

"Mary K. [Gaillard], Dimitri [Nanopoulos] and I first got interested in
what are now called penguin diagrams while we were studying CP violation
in the Standard Model in 1976... The penguin name came in 1977, as
follows.

In the spring of 1977, Mike Chanowitz, Mary K and I wrote a paper
on GUTs predicting the b quark mass before it was found. When it was
found a few weeks later, Mary K, Dimitri, Serge Rudaz and I immediately
started working on its phenomenology. That summer, there was a student
at CERN, Melissa Franklin who is now an experimentalist at Harvard. One
evening, she, I and Serge went to a pub, and she and I started a game of
darts. We made a bet that if I lost I had to put the word penguin into
my next paper. She actually left the darts game before the end, and was
replaced by Serge, who beat me. Nevertheless, I felt obligated to carry out
the conditions of the bet.

For some time, it was not clear to me how to get the word into this
b quark paper that we were writing at the time. Then, one evening, after
working at CERN, I stopped on my way back to my apartment to visit some
friends living in Meyrin where I smoked some illegal substance. Later, when
I got back to my apartment and continued working on our paper, I had a
sudden flash that the famous diagrams look like penguins. So we put the
name into our paper, and the rest, as they say, is history."

A few touches on Arkady's portrait

You can view the previous part as an extended introduction intended to
convey a flavor of the epoch. Of course, it would be better if I could write

---

22 (1975) 55].

about the Institute of Nuclear Physics in Novosibirsk, of which Arkady was a permanent member. This institution was a remarkable phenomenon in the USSR. I do not think it had parallels. Budker was running it on a unique fuel, a mixture of east and west, capitalist entrepreneurship and communist reality, the usual Russian sloppiness and equally usual creativity. I heard many incredible legends about it from Khriplovich, Eidelman, Zolotorev and others. It is a pity that neither of them volunteered to put·these stories in writing. I was in this Institute perhaps a dozen of times. Each time it was a short visit, however — from a few days to a couple of weeks — too short a time to become an insider. Writing a glorious chronicle of the Budker Institute of Nuclear Physics,[c] with all anecdotal evidence (which does deserve to be preserved for the future generations) included, is a task for other people.

As I have already mentioned, Arkady Vainshtein was considered, essentially, as a member of the ITEP theory group. He would visit two or three times a year, each time staying for a a month or more. The 1974/75 academic year was special. Arkady's daughter Tanya got sick: an awkward move during a physical exercise led to a spine injury. Out of all clinics in the USSR only one could provide necessary medical treatment. Sure enough, this was a Moscow clinic. It was very hard to get her admitted to this clinic for treatment, but Budker made it happen. He gave a one-year paid leave of absence to Arkady, and sent him to Moscow. For me it was a blessing in disguise.

It was Arkady and Valya Zakharov who got me involved, in earnest, in quantum chromodynamics. This happened in the late fall or winter of 1973, in the very beginning of my PhD work. This involvement shaped my entire career.

Arkady is a deep thinker. He is the deepest thinker of all people I am closely acquainted with. When he gets seriously interested in a certain physics problem — let us call it "problem A" — his mind sends a powerful urge to start digging. The outside world ceases to exist, the work continues almost on the 24/7 basis. A sophisticated fantasmagoric construction gradually emerges in Arkady's mind. Being left to himself, he would never return back. The problem A would lead to a set of subproblems a_1, a_2, and so on, which, in turn, would continuously evolve into a set of sub-subproblems $\alpha_{1\ell}$, $\alpha_{2\ell}$, etc. Let alone related problems B, C, D, ... The fractal nature of such an approach requires from Arkady a noncommensu-

---

[c] In the 1970's Budker was still alive, and one could hardly even dream that a time would come when the Institute, his child, would bear his name.

rate amount of time and effort. A little baroque exercise at level α whose impact on the general picture is minute, is as important to him as everything else. It may take weeks or months. Nevermind. Being left to himself, Arkady would never say: "this is *the* answer, I pause here to let other people know of what I have achieved." For him, the pleasure of finding out how things work is sufficient by itself. You may call him superperfectionist. Yes, that's the right word, extreme perfectionist.

Only strong external impulses can extract him from the deepening fractal structure of his making. The onset of the vacation season may serve as such an impulse. Another option is to distract him by suggesting a new and more challenging problem. In this latter case the attraction of the new problem must be overwhelming, to overcome the inertia of the original motion.

Upon forced return from the n-th intellectual journey, nothing can be taken for granted with Arkady. Even a solid baggage of results and insights acquired *en route* is no guarantee that the corresponding paper will ever see the light of the day. To make a decision to start writing a paper is a torture for Arkady. Even more so the process of writing. Every research project, its merits notwithstanding, has loose ends and dark corners. At the discussion stage everything is volatile, up in the air. What was a loose end today might find a perfect match tomorrow. But when you put this on paper, this is it. Every string of Arkady's superperfectionist *ego* protests. The necessity to document things before they are fully complete (and they never are) burns Arkady out. Literally.

I remember a funny story that happened in 1982. We were working on a large project entitled *Two-Dimensional Sigma Models: Modeling Nonperturbative Effects of Quantum Chromodynamics.* [d] A motivation for this project was "donated" to us by Sasha Polyakov. As usual, Sasha had a wealth of interesting calculations in his treasure trove which he did not consider to be important enough to warrant publication. In a private discussion he made a remark which turned Arkady on. At that time we were excited about the gluon condensate which we had introduced just a few years earlier. [e] Polyakov said:

"Look, guys, both $G_{\mu\nu}^2$ in Yang-Mills and $(\partial_\mu \vec{n})^2$ in the $O(3)$ sigma model are negatively defined in the Euclidean. And in general, these theories are very similar. You claim that the gluon condensate is positive. I

---

[d] V. Novikov, M. Shifman, A. Vainshtein, and V. Zakharov, Phys. Rept. **116** (1984) 103.

[e] M. Shifman, A. Vainshtein, and V. Zakharov, Nucl. Phys. **B 147** (1979) 385.

found $\langle(\partial_\mu\vec{n})^2\rangle$ in the sigma model, and I am certain that this condensate is negative. How come?"

The work on this project lasted for over a year; by 1983 the material accumulated became so vast it was hard to manage. A paper was drafted in Moscow and was sent to Arkady, who at that time was in Novosibirsk. He was supposed to read the draft, make any corrections/alterations he wanted, and then return it back.

When I say the paper was drafted I mean it. It was a hand-written manuscript. We had no access to photocopying machines. The copy sent to Novosibirsk (through a reliable person, certainly not by mail) was the only one.

In the subsequent telephone conversations Arkady seemed to deliberately avoid this topic. This went on and on. In half a year I came to Novosibirsk, and discovered the truth.

Arkady would carry the draft in his briefcase — in the morning from his home to office, where he would put it on his desk, open and look in desperation at all those disgusting logical leaps, omissions and other shortcomings which are unavoidable in the first draft, being unable to delve there, postponing the beginning of the work till the evening, when he would carry the draft in the opposite direction. Next day — the same story ... One night something happened in his garage, which required an immediate intervention. There was no electricity there and Arkady had to make an improvised torch. He fished out a few sheets of paper from his briefcase to lighten the place. In haste he did not notice that this was a good portion of the unlucky draft. When it was all over, he was just afraid to tell us of what had happened. It was Nelly who told me about the burnt manuscript when I came to Novosibirsk.

Well, they say manuscripts do not burn. It took us about a year to produce a new one. I hasten to add that a new version was much better than previous. Arkady's misadventure turned out to be a blessing in disguise.

By the way, I have just mentioned the telephone conversations. Physics issues were discussed in the telephone conversations with Arkady on a regular basis. That's how we worked together. It was not allowed to call long distance from ITEP (at least, it was not allowed to me). So, I had to call from my home phone. As a result, my phone bills exceeded any reasonable number I could afford. (What I could afford was close to zero, if not negative, anyway). The large JETP paper on penguins was done essentially in the telephone mode. After that my wife revolted. I had to limit phone physics from my home phone to one hour a week at most. Fortunately, by that time Arkady discovered that Budker's policy on long-distance calls was much more liberal than that of ITEP — Arkady could call us from his office with very mild limitations.

In retrospect, trying to summarize what was typical for our scientific and nonscientific interactions over the years, I see, first of all, endless and *very exhausting* (but very fulfilling, too) discussions of various physics issues. My collaboration with Arkady lasts for almost 30 years. He was and still is one of my teachers. I am happy that I had the opportunity to discuss with him all aspects of high energy physics an almost infinite number of times.

I see, very clearly in my memory, other episodes too. For instance, guess what was the major concern of esteemed Professor Arkady Vainshtein each time he would come to Moscow, towards the end of his visit? He always had a huge backpack with him. Real huge. And each time before returning home to Novosibirsk he used to spend two or three days hunting for food and other basic necessities (such as toothpaste, razor blades and the like), which in the 1980's could still be found, from time to time, in Moscow but were obliterated in Novosibirsk stores. I close my eyes and see him leaving, with his backpack (weighing, perhaps, 30 kilos) full of oranges, cheese, shoes for his lovely daughters and other similar exotic stuff which was not considered by communists to be vitally important for the survival of the country.

The shortage (or, better to say, almost complete absence) of everything in Novosibirsk had a positive side effect on scientific aspirations and careers of the Siberian physicists. First of all, nothing distracted young people from work. More importantly, there was a primitive but very powerful direct relation between one's promotion and one's nutrition. Basic goods were rationed and delivered to the Novosibirsk scientific community through a system of the so-called distribution centers closed to general public. One's scientific standing was in one-to-one correspondence with the access to higher-level centers. Young researchers at the pre-PhD stage were entitled to next-to-nothing. Getting PhD was a step forward. PhD holders (in Russian they are called "Candidates of Science") could get meat and other protein-rich products. Of course, the amount was very limited, which kept them aggressive in their research work. (And young people should do research aggressively, I think everybody will agree.) Here it should be explained that the academic hierarchy in Russia follows the German rather than the Anglo-American pattern. An approximate equivalent of PhD in the US is the *Candidate of Science* degree. The highest academic degree, doctoral, is analogous to the German *Habilitation*. The doctoral dissertation is usually prepared at a mature stage of the academic career; only a fraction of the *Candidate* degree holders make it to the doctoral level. Well, defending the doctoral dissertation was a major leap, opening access to a distribution center almost as good as the one for Academicians. Doctors of Science were supposed to have meat in their diet on a regular basis.

I do not really know whether this long digression belongs here. Upon reflection, I decided to keep it because it gives an idea of the environment in which Arkady lived and worked for many years.

In spite of our 30 friendly years, surprisingly, I cannot say that I know Arkady well, beyond physics. Complicated processes take place deep inside him, and one can only guess of what is going on from rare outbursts. Perhaps, I have a general idea, but details and nuances are blurred ... The only thing of which I am certain, is that Arkady is the most selfless person of all people I am closely acquainted with. (Remember, I started this section on the same note). If he sees that someone needs his help, he is always ready to help. There is no limit to his patience. If there is something he can share — be it his computer or skiing skills, or just his strong shoulders — he will always offer his assistance, generously investing his time, with no back thoughts.

BILL FINE, TPI AND ARKADY

STEVE GASIOROWICZ

Theoretical Physics Institute, University of Minnesota
Minneapolis MN 55455, USA

The Theoretical Physics Institute at the University of Minnesota is a direct result of the interest and generosity of Bill Fine. It was roughly 20 years ago that he and I became acquainted, and I discovered that Bill had a deep interest in physics, specifically High Energy physics. It was through conversations about this subject that we came to a point at which Bill indicated that he wanted to do something for the field: the idea of a theoretical physics institute was born! Bill and I tried to do some fund-raising, but the general public, or at least the part that we could approach seemed less than enthusiastic about giving money. Furthermore, the college administration at the time was also less than interested (Bill and I talked about "the instinct for the capillary"). In 1985 Minnesota hosted the 6th Workshop on Grand Unification, and on this occasion that Gloria Lubkin entered the picture. It was she who pointed out that the proposal was on too small a scale and that it was necessary to bring the top levels of the University administration into the planning. She suggested bringing in Leo Kadanoff as spokesman and potential director to give reality to the proposal. In the summer of 1986, during a festive and intensive get together in Minneapolis, Bill and Leo, with strong support by Chuck Campbell, outgoing head of the School of Physics and Astronomy, and Marvin Marshak, his successor, persuaded then-President Ken Keller of the merits of building a Theoretical Physics Institute at the University. Building on a very generous pledge by Bill Fine, the University committed itself to matching Bills gift to create two chairs (subsequently split into three) and to provide permanent funding of a magnitude to support an active, vibrant institute. The Theoretical Physics Institute (renamed the William I Fine Theoretical Physics Institute on the occasion of the 15-th anniversary of its creation) became a reality. I was appointed acting director and during 1987-89 conducted a vigorous search for director. In 1989 in a fortunate alignment of stars, several things happened: (1) *perestroika*, (2) Larry McLerran became the first director and (3) Larry with the strong support of Gloria Lubkin— an active member of the oversight committee —decided to take advantage of the unique opportunities provided by (1). Larry had been to Russia many times, and knew at any given time where to find people. The people we recruited were known to us, at least by name, although I had met Misha Voloshin in Aspen and at DESY earlier. In any case, the first recruits were Boris Shklovskii in condensed matter physics, Misha Voloshin and Arkady. Misha Shifman came a year later, as did Leonid Glazman, and and a few years later, Anatoly Larkin. The first year was quite miraculous. In addition to these people, a large number of visitors came. Since we could not pay them a regular salary —this was still the time when the Soviet government wanted a cut of the pay — the whole group lived on per diems and were housed together at 110 Grant, a comfortable highrise in the center of town. I can only describe it as a year-long summer camp. The tradition, born in periods of deprivation, that if you could get hold of some good food you had a party, carried over, and there were *always* parties. Arkady and Nelly were among the main organizers of social activities and took it as their duty to look after the guests. Arkady may have been the only person who was an experienced driver, and our aged Subaru became the vehicle that brought people to and from the airport, to and from 110 Grant. It was a time when we learned about Russian-style seminars—you bring sandwiches, a thermos and sometimes a sleeping bag. I discovered that if you asked Arkady a question, he could not only answer it, but had probably written a paper about it. Ten or so years later, everything settled into something of a routine. Most of the families settled in, their children moved to successful careers, but there is still something magical about being on the 4-th floor of the physics building. When the door is open, its like being at an opera (Mussorgsky?): you dont understand a word, but t he music is powerful and enchanting (and loud!). The creation of what some people have called Moscow (and Novosibirsk) on the Mississippi has been a wonderful adventure, and the new friendships we have made with Arkady and Nelly, and with all the other newcomers, have enriched us enormously. So thank you Bill, thank you Mr. Gorbachev and thank you Arkady!

DECAYING SYSTEMS AND DIVERGENCE OF THE SERIES OF PERTURBATION THEORY. PREPRINT (1964) BY A.I. VAINSHTEIN.

Foreword and translation by M. Shifman

Perturbation theory is the most common tool applied for calculations in quantum mechanics and, especially, field theory. In weakly coupled theories, such as quantum electrodynamics or electroweak model, calculations based on the Feynman graphs (which represent a particular order in the perturbative series) are innumerable. This approach has a solid theoretical foundation, and its remarkable success is no surprise.

There is a deep general question as to the nature of the coupling constant expansion. Half a century ago Dyson argued [1] the the series in α are asymptotic in quantum electrodynamics. The essence of his argument is as follows. Consider N charged moving particles, of one and the same charge e, assuming that $N \gg 1$. The energy E of this system can be represented as

$$E = NT + \frac{N^2}{2} e^2 V, \tag{1}$$

where T is the average kinetic energy per particle. The second term represents the Coulomb energy: V stands for the average inverse distance between the particles, $V = \langle r^{-1} \rangle > 0$. The factor $N^2/2$ represents the number of the interacting pairs (in fact, it should be $N(N-1)/2$, but this distinction is negligible at large N). For positive $\alpha \equiv e^2$ the system is stable. However, if α becomes negative, then the potential part of the energy E becomes attractive, and at sufficiently large N it will always take over the kinetic part. Thus, at $N \geq N_* = -T/(V\alpha)$ the energy E of the conglomerate becomes negative and an instability develops. This instability is due to the fact that a spontaneous pair creation becomes energetically expedient. The particles of charge e are attracted to the conglomerate; those of charge $-e$ run away to infinity. The more pairs are produced, the more negative E becomes. This phenomenon — instability — occurs irrespective of the value of α. Of course, the critical value N_* becomes exceedingly larger as $\alpha \to 0$.

Dyson concludes that physical quantities in quantum electrodynamics cannot be analytic in α, and the point $\alpha = 0$ is singular. If so, the expansion in the powers of α cannot be convergent.

Being brilliant, Dyson's argument is qualitatitve. Many years had elapsed before quantitative methods were developed allowing one to calculate the divergence of the perturbative series in high orders. A breakthrough, which paved the way to quantitative analysis, became possible when it was found that : (i) the divergence of the perturbative series at high orders, at physical values of the coupling constant, is related (via the dispersion relation in the coupling constant) to the imaginary part which develops at unphysical values of the coupling constant, when the system under consideration becomes unstable (ii) this imaginary part, in turn, is related to the barrier-penetration phenomenon and can be calculated quasiclassically at small unphysical values of the coupling constant; (iii) the rate of divergence at high orders is fully determined by the tunneling amplitude at weak coupling.

This result was first obtained in quantum mechanics and is usually credited to Bender and Wu [2] (see e.g. such authoritative source as Le Guillou and Zinn-Justin's compilation [3]). Bender and Wu's paper, a benchmark in this area of research, was written in 1972. Very few theorists know that the very same construction was worked out in 1964 in Soviet Union. In fact, this was one of the first research projects of Arkady Vainshtein, who at that time was a student at the Novosibirsk University and Novosibirsk Institute of Nuclear Physics (currently, the Budker Institute of Nuclear Physics). His paper was published in 1964 in Russian, as a Novosibirsk Institute of Nuclear Physics Report [4], which obviously hindered its recognition in the western high-energy physics community. Only experts in the Soviet community were aware of Vainshtein's construction, in particular, Lev Lipatov and Eugene Bogomolny, whose works on the divergences of the perturbative series are well-known.

Now, almost 40 years later, original Vainshtein's report became a rarity, it can hardly be found even in large libraries. I decided to correct the situation, and make it available to the high-energy physics community. On occasion of Arkady's 60th birthday I translated the paper in English. Below you will find both, the English translation and the Russian original.

1. F.J. Dyson, *Phys. Rev.* **85**, 631 (1952).

2. C.M. Bender and T.T. Wu, *Phys. Rev.* **D7**, 1620 (1973).

3 J.C. Le Guillou and J. Zinn-Justin (Eds), *Large-Order Behaviour of Perturbation Theory* (North-Holland, Amsterdam, 1990).

4. A.I. Vainshtein, *Decaying Systems and Divergence of the Series of Perturbation Theory*, Novosibirsk Institute of Nuclear Physics Report, Decmber 1964.

INSTITUTE OF NUCLEAR PHYSICS OF THE SIBERIAN BRANCH
OF THE USSR ACADEMY OF SCIENCES

Preprint

A.I. Vainshtein

DECAYING SYSTEMS AND DIVERGENCE OF THE SERIES

OF PERTURBATION THEORY

Novosibirsk — 1964

620

Abstract

One-dimensional field models are considered. If, for a certain sign of the coupling constant λ, the spectrum is continuous, the perturbative series for the propagator diverges. At large n the n-th term of the perturbative series has the form $n!\,(\alpha\lambda)^n$.

1. In Ref. 1 Dyson argued that the perturbation theory series are divergent in quantum electrodynamics. Dyson's argument was based on the observation that the world in which the square of the electric charge e^2 is negative, has no ground state and decays. Therefore, it is hard to imagine that such a situation can be described by functions analytic in e^2 at $e^2 = 0$.

Thirring investigated [2] the theory with the interaction $L_{\text{int}} = \lambda\varphi^3$ and showed that the perturbative series for the polarization operator diverges in the domain of momenta $p^2 < m^2$. At large n the terms of the perturbative series are shown to have the form

$$C(\alpha\lambda)^n\,\frac{(n-4)!}{n^2}\,,$$

where C and α are functions of p^2. The model considered by Thirring is an example of an unstable theory. One can readily show, by virtue of a direct variational method, that there is no ground state in the model of Ref. 3.

We will show that instability of the system implies a divergence of the perturbation theory series in a one-dimensional model.

2. Consider a model in which field operators φ depend only on time, the spatial coordinates are absent. The Hamiltonian and equal-time commutation relations have the form

$$H = \frac{1}{2}\,(\dot\varphi)^2 + \frac{m^2}{2}\varphi^2 + V(\varphi)\,, \qquad [\varphi(t)\,,\dot\varphi(t)] = i\,. \tag{1}$$

This is the Hamiltonian and commutation relations of the conventional quantum-mechanical anharmonic oscillator with the frequency $\omega = m$ and mass $\mu = 1$.

For definiteness let us choose the interaction in the form

$$V(\varphi) = -\lambda\varphi^3\,. \tag{2}$$

From what follows it will be clear that in fact our consideration is applicable to all decay-permitting interactions.

Repeating the proof due to Thirring [2] in the one-dimensional case, for the interaction $\lambda\varphi^3$, we will arrive at a result which is identical to that of the four-dimensional problem, namely that the series for the polarization operator diverges, the divergence being the same as in four dimensions. We will connect this divergence with the fact that the system at hand can decay.

Let us consider the causal Green function for the field φ. In the interaction representation it is defined as

$$iG(\tau) = \frac{(0|T\,\varphi(\tau)\,\varphi(0)|0)}{S_{00}} \tag{3}$$

where $\varphi(\tau)$ is the field operator in the interaction representation. It is assumed that the interaction switches on adiabatically.

If we now pass to the Heisenberg operators $\underline{\phi}(\tau)$, we will get

$$iG(\tau) = \frac{(0|S(\infty,0)[T\,\underline{\phi}(\tau)\,\underline{\phi}(0)]S(0,-\infty)|0)}{(0|S(\infty,0)\,S(0,-\infty)|0)},$$

$$\underline{\phi}(\tau) = S^+(\tau,0)\varphi(\tau)S(\tau,0). \tag{4}$$

3.   Usually the state $|\psi\rangle = S(0,-\infty)|0)$ is considered to be the physical vacuum. If the physical vacuum does exist, this is ensured by adiabatic switching on — the interaction turns on adiabatically. In the model under consideration there is no physical vacuum, the system is unstable. In such cases the mathematical vacuum passes into a corresponding quasi-level after the interaction is turned on adiabatically. The quasi-level is a state with a complex energy describing a decay.

To show this we will consider the problem of an oscillator with the frequency changing with time as

$$\omega^2\left(1 - \gamma e^{-\alpha|t|}\right).$$

If $\gamma > 1$ then at $t = 0$ the oscillator turns upside down, and the physical vacuum is absent. It turns out, that if at $t \to -\infty$ we start from the ground state of the oscillator, at $t = 0$ we arrive at a state which, in the limit $\alpha \to 0$ has the energy

$$E = -\frac{i\omega}{2}\sqrt{\gamma - 1},$$

and describes the decay. A detailed solution is given in Appendix A.

It is interesting to note that the state $(0|S(\infty,0) = \langle\psi|$ is not obtained from $S(0,\infty)|0)$ by Hermitean conjugation; this is due to the fact that the stability condition $S(\infty,-\infty)|0) = |0)$ is not satisfied. The state

$\langle\psi| = (0|S(\infty,0)$ is Hermitean conjugate to the state describing the process reverse to decay. The energy of such state is complex conjugated to that of the quasilevel. In what follows, we will call such state anti-quasilevel.

4. In Appendix B a relation between the energy of the state $|\psi\rangle$ and $G(\tau)|_{\tau=0}$ and $G(p)|_{p=0}$ is derived. Here $G(p)$ is the Fourier-transform of $G(\tau)$. Therefore, for studying the analytical properties of $G(\tau)|_{\tau=0}$ and $G(p)|_{p=0}$, as functions of the coupling constant, it is sufficient to study the analytic behavior of $E(\lambda^2)$ where $E(\lambda^2)$ is the energy of the state $|\psi\rangle = S(0,-\infty)|0\rangle$. The equation for $|\psi\rangle$ is

$$H|\psi\rangle = E|\psi\rangle, \qquad H = \frac{\dot{\varphi}^2}{2} + \frac{m^2\varphi^2}{2} - \lambda\varphi^3. \tag{5}$$

This is a conventional differential equation for anharmonic oscillator. At $\varphi \to -\infty$ the wave function $\psi(\varphi)$ falls off exponentially, while at $\varphi \to \infty$ there is only an outcoming wave. (The constant λ is assumed to be positive.)

Let us continue $\psi(\varphi)$, defined for positive λ, to complex values of λ. Then λ is complex in Eq. (5). Let us now examine the boundary conditions.

At positive λ

$$\psi(\varphi) \to \frac{C}{\sqrt{p}} \exp\left[i\int^\varphi p\,d\varphi\right], \quad \text{at } \varphi \to +\infty,$$

$$\psi(\varphi) \to \frac{C'}{\sqrt{p}} \exp\left[-i\int^\varphi p\,d\varphi\right], \quad \text{at } \varphi \to -\infty,$$

$$p = \sqrt{2E - m^2\varphi^2 + 2\lambda\varphi^3}. \tag{6}$$

These are the well-known quasiclassical asymptotics. One can assert that $\psi(\varphi)$ has the same asymptotics for all complex λ in the upper half-plane of the parameter λ. Indeed, as long as λ is in the upper half-plane, $\psi(\varphi)$ falls off exponentially at $\varphi \to \pm\infty$, and the growing exponent cannot appear.

Let us pass in the λ plane from the positive semi-axis to negative, via the upper half-plane. Then Eq. (6) implies that after the rotation $\psi(\varphi)$ falls off exponentially at $\varphi \to +\infty$, while at $\varphi \to -\infty$ the wave function $\psi(\varphi)$ represents a wave running into the well. That is to say, starting from the problem of a quasilevel, we arrived at the problem of anti-quasilevel. This means that the function $E(\lambda^2)$ has a cut along the positive semi-axis in the λ^2 plane. The imaginary part $\mathrm{Im}E(\lambda^2)$ experiences a jump on this cut.

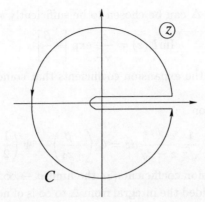

Figure 1. Integration contour in the integral (7).

Consider the integral

$$\int_C dz \, \frac{E(z)}{z - z_0} = 2\pi i \, E(z_0), \qquad z = \lambda^2, \tag{7}$$

where the contour C is indicated in Fig. 1. The radius of the circle in Fig. 1 is Δ, while $|z_0| < \Delta$. The integral over the circle is an analytic function for all z_0 inside the circle; therefore, we will not consider it since we are interested in the part of $E(z)$ nonanalytic at the origin,

$$E(z_0) = \frac{1}{\pi} \int_0^\Delta dz \, \frac{\operatorname{Im} E(z)}{z - z_0}. \tag{8}$$

The expansion of $E(z_0)$ in z_0 (at $z_0 \to 0$) immediately follows from Eq. (8),

$$E(z_0) = \sum_{n=0}^\infty z_0^n \left(\frac{1}{\pi} \int_0^\Delta dz \, \frac{\operatorname{Im} E(z)}{z^{n+1}} \right). \tag{9}$$

The integrals on the right-hand side converge since $\operatorname{Im} E(z)$ falls off exponentially at $z \to 0$.

Indeed, $\operatorname{Im} E(\lambda^2)$ is proportional (at $\lambda \to 0$) to the barrier transmission coefficient

$$D = \exp \left[-2 \int_{\varphi_1}^{\varphi_2} d\varphi \, \sqrt{m^2 \varphi^2 - 2\lambda \varphi^3 - 2E} \right].$$

where the integral is to be taken between two turning points. At $\lambda \to 0$

$$D = \frac{\alpha'}{\sqrt{\lambda^2}} \exp \left[-\frac{\beta' m^5}{\lambda^2} \right], \tag{10}$$

where α' and β' are constants.

Since the radius Δ can be chosen to be sufficiently small, substituting

$$\text{Im} E(z) = \frac{\alpha}{\sqrt{z}} \exp\left[-\frac{\beta}{z}\right]$$

in Eq. (9) one gets the expansion coefficients that coincide with the exact ones at $n \to \infty$.

Thus, the function

$$\overline{E}(z_0) = \frac{\alpha}{\pi} \int_0^\infty \frac{1}{\sqrt{z}} \frac{e^{-\beta/z}}{z - z_0} dz = C \left(-\frac{\beta}{z_0}\right)^{1/2} \Psi\left(\frac{1}{2}, \frac{1}{2}, -\frac{\beta}{z_0}\right) \tag{11}$$

has the same expansion coefficients, in the limit $n \to \infty$, as the exact $E(z)$. (The fact that we added the integral from Δ to ∞ is of no importance, since this added integral gives a function analytic at the point $z_0 = 0$. Moreover, $\Psi\left(\frac{1}{2}, \frac{1}{2}, x\right)$ is the degenerate hypergeometric function.)

The expansion of $\overline{E}$ has the form

$$\overline{E}(z) = C \sum_{n=0}^\infty \left(\frac{z}{\beta}\right)^n \Gamma\left(n + \frac{1}{2}\right). \tag{12}$$

Since $E(\lambda^2)$ is related to the Green function as follows (see Appendix B)

$$iG(\tau)|_{\tau=0} = \frac{1}{m^2}\left[E - 5\lambda^2 \frac{\partial E}{\partial \lambda^2}\right],$$

$$\tilde{G}(p)\Big|_{p=0} = -\frac{1}{m^2} + \frac{9\lambda^2}{m^8}\left[25\,(\lambda^2)^2 \frac{\partial^2 E}{\partial(\lambda^2)^2} + 35\lambda^2 \frac{\partial E}{\partial \lambda^2} - 3E\right], \tag{13}$$

we arrive at the conclusion that the perturbation theory series diverges: the expansion coefficients grow factorially at large n. Nevertheless, the series is asymptotic.

5. One can consider interactions of other types in perfectly the same way. It is clear that if a given system decays after the interaction switches on, then the imaginary part of the quasilevel energy is a quantity which is exponentially small in the limit of the vanishing coupling constant. This implies the factorial growth of the expansion coefficients in the coupling constant series. There will be no potential barrier if $m = 0$. But in this case the perturbation theory integrals diverge at the lower limit of integration.

Of particular interest is the interaction $V = -\lambda\varphi^4$. At $\lambda > 0$ this interaction corresponds to a decaying system. Performing the same consideration as for $\lambda\varphi^3$ we will obtain that at large n the terms of the perturbative series for $E(\lambda)$ grow as

$$C \left(\frac{\lambda}{\gamma}\right)^n \Gamma\left(n + \frac{1}{2}\right).$$

At $\lambda < 0$ the system is stable. However, the perturbative series is the same both for positive and negative λ. Thus, there emerges a situation of the type suggested by Dyson in quantum electrodynamics. It is interesting that all terms of the series are of the same sign in the instability domain of negative λ.

The majority of nontrivial theories are seemingly unstable at some phase of the coupling constant, which leads to asymptotic nature of the perturbative series. Equations in such theories have solutions nonanalytic in the coupling constant at the origin. It is unclear, though, to which extent these solutions are physical in the instability domain. In such theories the point $\lambda = 0$ is a branching point. Moreover, if $m \neq 0$, it presents an essential singularity.

What remains unclear is the relation between the decaying nature of the system and the perturbative series divergence in four-dimensional theories. One should note that if small spatial momenta are of importance, the theory becomes one-dimensional; however, the impact of renormalizations calls for a study. It is interesting that if an arbitrary graph with the vanishing external momenta is considered in four dimensions, one can readily get the following inequality:

$$\int \frac{1}{m^2 - p_1^2} \cdots \frac{1}{m^2 - p_i^2} \prod d^4 q$$

$$\geq \frac{1}{m^{6i}} \left[\int \frac{1}{m^2 - (p_1^0)^2} \cdots \frac{1}{m^2 - (p_i^0)^2} \prod dq^0 \right]^4 \tag{14}$$

where p_k are the internal line momenta, q are the integration momenta, while p_k^0 stand for the time-like components of p_k. Then what appears on the right-hand side of Eq. (14) is the corresponding one-dimensional diagram, and we could built a minorant for the four-dimensional theory, if it were not for the necessity of renormalizations.

I am deeply grateful to V.M. Galitsky for suggesting me this topic for research and for guidance. I would like to thank I.B. Khriplovich for valuable discussions.

Appendix A

Let us consider an oscillator with the time-dependent frequency changing as

$$\omega^2 \left(1 - \gamma e^{-\alpha |t|} \right).$$

We are interested in the time development of the state which tends to the ground state of the frequency ω oscillator at $t \to -\infty$. We demote it $\Psi_\alpha(t)$. In the interaction representation $\Psi_\alpha(t)$ satisfies the following equation:

$$i \frac{\partial \Psi_\alpha}{\partial t} = -\frac{\gamma \omega^2}{4} e^{-\alpha |t|} x^2(t) \Psi_\alpha, \tag{A.1}$$

$$x(t) = \frac{1}{\sqrt{2}} \left[a^+(t) + a^-(t) \right] = \frac{a^+ e^{i\omega t} + a^- e^{-i\omega t}}{\sqrt{2}}, \tag{A.2}$$

where

$$[a^+, a^-] = -1.$$

We look for $\Psi_\alpha(t)$ in the form

$$\Psi_\alpha(t) = K_\alpha(t) \exp \left[(a^+)^2 f_\alpha(t) \right] |0\rangle \tag{A.3}$$

where $K_\alpha(t)$ and $f_\alpha(t)$ are functions of time, while $|0\rangle$ is the ground state of the oscillator with the frequency ω, so that $a|0\rangle = 0$. Substituting Eq. (A.3) in (A.1) and performing the commutation we obtain terms with $\exp((a^+)^2 f_\alpha)|0\rangle$ and $(a^+)^2 \exp((a^+)^2 f_\alpha)|0\rangle$. Requiring the coefficients in front of these terms to vanish, we arrive at the following equations:

$$i \frac{K'_\alpha}{K_\alpha} = -\frac{\gamma \omega}{4} e^{-\alpha |t|} (1 + 2 f_\alpha), \tag{A.4}$$

$$i f'_\alpha = -\frac{\gamma \omega}{4} e^{-\alpha |t|} \left(e^{2i\omega t} + 4 f_\alpha + 4 f_\alpha^2 e^{-2i\omega t} \right), \tag{A.5}$$

with the boundary condition

$$f_\alpha(t) \to 0 \quad \text{at} \quad t \to -\infty.$$

Consider $t < 0$ and introduce a new function $y(t)$

$$f_\alpha(t) = -\frac{1}{i\gamma \omega} e^{(2i\omega - \alpha)t} \frac{y'(t)}{y(t)} - \frac{1}{i\gamma \omega} \left(-i\omega + \frac{\alpha}{2} \right) e^{(2i\omega - \alpha)t} - \frac{1}{2} e^{2i\omega t} . \tag{A.6}$$

Then we get the following equation for $y(t)$:

$$y'' + \left[\left(\omega + \frac{i\alpha}{2} \right)^2 - \gamma \omega^2 e^{\alpha t} \right] y = 0. \tag{A.7}$$

The solution of this equation is

$$y(t) = C_1 J_\nu(z) + C_2 J_{-\nu}(z), \tag{A.8}$$

where

$$\nu = \frac{2i\omega}{\alpha} - 1, \qquad z = \frac{2i\omega\sqrt{\gamma}}{\alpha} e^{\alpha t/2},$$

and $J_\nu(z)$ is the Bessel function. Using the fact that $f_\alpha(t)$ vanishes at $t \to -\infty$ we find

$$f_\alpha(t) = -\frac{1}{i\gamma\omega} e^{(2i\omega-\alpha)t} \left[-i\omega + \frac{\alpha}{2} + \frac{\lambda\omega}{2} e^{\alpha t} + \frac{d(J_\nu(z))/dt}{J_\nu(z)} \right]. \tag{A.9}$$

We are interested in the limit

$$\lim_{\alpha \to 0} \Psi_\alpha(0) = \Psi(0).$$

Using the quasiclassical asymptotics of the Bessel functions [4] we find

$$f(0) = \frac{1}{\gamma} \left[1 - \frac{\gamma}{2} + i\sqrt{\gamma-1} \right], \quad \gamma > 1,$$

$$f(0) = \frac{1}{\gamma} \left[1 - \frac{\gamma}{2} - \sqrt{1-\gamma} \right], \quad \gamma < 1. \tag{A.10}$$

At $\gamma < 1$ we arrive at the ground state of the oscillator with the frequency $\omega\sqrt{1-\gamma}$, a "physical vacuum". If $\gamma > 1$ then

$$\Psi(0) = \exp\left((a^+)^2 f(0) \right) |0\rangle$$

has the following form in the x representation:

$$\Psi(0) = \exp\left(i\omega\sqrt{\gamma-1}\, \frac{x^2}{2} \right). \tag{A.11}$$

This state describes an outflux of particles from the origin to $\pm\infty$. The energy of this state is

$$E = -\frac{i\omega}{2}\sqrt{\gamma-1}.$$

We see that $n = E/\omega$ is an adiabatic invariant for complex ω too.

If one considers the state which at $t \to \infty$ goes to the vacuum of the frequency ω oscillator, one obtains that at $t = 0$ and $\alpha \to 0$ one deals with the state of an anti-quasilevel with

$$E = \frac{i\omega}{2}\sqrt{\gamma-1}.$$

Appendix B

The Green function in the Heisenberg representation has the form

$$iG(\tau) = \frac{\langle\psi|\underline{\phi}(\tau)\underline{\phi}(0)|\psi\rangle}{\langle\psi|\psi\rangle} . \tag{B.1}$$

The connection between $G(\tau)|_{\tau=0}$ and the energy of the state $|\psi\rangle$ is known [5]. We will derive this relation for completeness, however. The equation for $|\psi\rangle$ is

$$(H - E)|\psi\rangle = 0, \qquad H = \frac{1}{2}\dot{\varphi}^2 + \frac{m^2}{2}\varphi^2 - \lambda\varphi^3 . \tag{B.2}$$

Differentiating Eq. (B.2) with respect to m^2 and multiplying by $\langle\psi|$ from the left we get

$$iG(\tau)|_{\tau=0} = \frac{\langle\psi|\underline{\phi}^2(0)|\psi\rangle}{\langle\psi|\psi\rangle} = 2\frac{\partial E}{\partial m^2} . \tag{B.3}$$

From dimensional arguments

$$E = m\,\Phi\left(\frac{\lambda^2}{m^5}\right) .$$

Therefore,

$$iG(\tau)|_{\tau=0} = \frac{1}{m^2}\left(E - 5\lambda^2\frac{\partial E}{\partial\lambda^2}\right) . \tag{B.4}$$

Let us derive now a relation between E and $G(p)|_{p=0}$ where

$$G(\tau) = \int_{-\infty}^{\infty}\frac{dp}{2\pi}\,G(p)\,e^{-ip\tau} .$$

First of all let us note that given the interaction $\lambda\varphi^3$

$$\overline{\varphi} = \frac{\langle\psi|\underline{\phi}(0)|\psi\rangle}{\langle\psi|\psi\rangle} \neq 0 .$$

Therefore, there is a constant in τ part in $G(\tau)$ having no physical meaning. In the p representation it yields $\delta(p)$. It is more correct to consider

$$i\tilde{G}(\tau) = \frac{T\,\langle\psi|(\underline{\phi}(\tau) - \overline{\varphi})\,(\underline{\phi}(0) - \overline{\varphi})|\psi\rangle}{\langle\psi|\psi\rangle} = iG(\tau) - \overline{\varphi}^2 . \tag{B.5}$$

Using the definition of $\underline{\phi}(\tau)$ in terms of the Schrödinger φ,

$$\underline{\phi}(\tau) = e^{iH\tau}\,\varphi\,e^{-iH\tau}$$

we have

$$\tilde{G}(p) = \frac{1}{\langle\psi|\psi\rangle} \left\langle \psi \left| (\varphi - \overline{\varphi}) \left[\frac{1}{p - (H - E - i\epsilon)} \right.\right.\right. \tag{B.6}$$

$$\left.\left.\left. - \frac{1}{p + (H - E - i\epsilon)} \right] (\varphi - \overline{\varphi}) \right| \psi \right\rangle ,$$

$$\tilde{G}(p)\Big|_{p=0} = -\frac{2}{\langle\psi|\psi\rangle} \left\langle \psi \left| (\varphi - \overline{\varphi}) \frac{1}{H - E} (\varphi - \overline{\varphi}) \right| \psi \right\rangle . \tag{B.7}$$

Moreover, $i\epsilon$ can be omitted since

$$\langle\psi|\varphi - \overline{\varphi}|\psi\rangle = 0 . \tag{B.8}$$

Let us introduce a term $f\varphi$ in the Hamiltonian, where f is a parameter. Then we differentiate Eq. (B.2) twice with respect to f,

$$(H - E) \frac{\partial|\psi\rangle}{\partial f} + \left(\frac{\partial H}{\partial f} - \frac{\partial E}{\partial f} \right) |\psi\rangle , \tag{B.9}$$

$$(H - E) \frac{\partial^2|\psi\rangle}{\partial f^2} + 2 \left(\frac{\partial H}{\partial f} - \frac{\partial E}{\partial f} \right) \frac{\partial|\psi\rangle}{\partial f} - \frac{\partial^2 E}{\partial f^2} |\psi\rangle . \tag{B.10}$$

Now multiplying by $\langle\psi|$ from the left we get

$$\frac{\partial E}{\partial f} = \frac{1}{\langle\psi|\psi\rangle} \left\langle \psi \left| \frac{\partial H}{\partial f} \right| \psi \right\rangle = \overline{\varphi} , \tag{B.11}$$

$$\frac{\partial^2 E}{\partial f^2} = \frac{2}{\langle\psi|\psi\rangle} \left\langle \psi \left| \frac{\partial H}{\partial f} - \frac{\partial E}{\partial f} \right| \frac{\partial\psi}{\partial f} \right\rangle = \overline{\varphi} . \tag{B.12}$$

Furthermore, Eq. (B.9) implies

$$\frac{\partial|\psi\rangle}{\partial f} = -\frac{1}{H - E} \left(\frac{\partial H}{\partial f} - \frac{\partial E}{\partial f} \right) |\psi\rangle . \tag{B.13}$$

In Eq. (B.13) one can add $|\psi\rangle$ with an arbitrary coefficient. This additional term will vanish, however, upon substitution in Eq. (B.12). Thus,

$$\frac{\partial^2 E}{\partial f^2} = -\frac{2}{\langle\psi|\psi\rangle} \left\langle \psi \left| (\varphi - \overline{\varphi}) \frac{1}{H - E} (\varphi - \overline{\varphi}) \right| \psi \right\rangle . \tag{B.14}$$

Now one can set $f = 0$, arriving at

$$\frac{\partial^2 E}{\partial f^2} = \tilde{G}(p)\Big|_{p=0} . \tag{B.15}$$

The relation (B.15) can be rewritten in terms of derivatives over λ^2. To this end we introduce a new operator η instead of φ,

$$\varphi = \eta + \varphi_0, \qquad \varphi_0 = \frac{m^2 - \sqrt{m^4 + 12\lambda f}}{6\lambda}. \tag{B.16}$$

Then the Hamiltonian does not contain terms linear in η, and one can write

$$E = E^0 + M\Phi\left(\frac{\lambda^2}{M^5}\right), \tag{B.17}$$

$$E^0 = \frac{m^2\varphi_0^2}{2} + f\varphi_0 - \lambda\varphi_0^3, \qquad M^2 = m^2 - 6\lambda\varphi_0. \tag{B.18}$$

Then Eq. (B.15) goes into

$$\tilde{G}(p)\Big|_{p=0} = -\frac{1}{m^2} + \frac{9\lambda^2}{m^8}\left[25\,(\lambda^2)^2\,\frac{\partial^2 E}{\partial(\lambda^2)^2} + 35\lambda^2\frac{\partial E}{\partial\lambda^2} - 3E\right]. \tag{B.19}$$

References

1. F.J. Dyson, *Phys. Rev.* **85**, 631 (1952).

2. W. Thirring, *Helv. Phys. Acta* **26**, 33 (1953).

3. G. Baym, *Phys. Rev.* **117**, 886 (1960).

4. A.Z. Patashinsky, V.L. Pokrovsky, and I.M. Khalatnikov, *ZhETF* **44**, 2062 (1961).

5. V.M. Galitsky and A.B. Migdal, *ZhETF* **34**, 139 (1958).

Responsible for the release — I.B. Khriplovich
Signed into print on December 2, 1964 — Glavlit MH00663
Number of copies printed — 150
Order number 052, Free of Charge

Printed in the Institute of Nuclear Physics of the Siberian Branch of the USSR Academy of Sciences

ИНСТИТУТ ЯДЕРНОЙ ФИЗИКИ СИБИРСКОГО ОТДЕЛЕНИЯ АН СССР

Препринт

А.И. Вайнштейн

РАСПАДАЮЩИЕСЯ СИСТЕМЫ И РАСХОДИМОСТЬ

РЯДА ТЕОРИИ ВОЗМУЩЕНИЙ

Новосибирск — 1964 год

Аннотация

Рассматриваются одномерные полевые модели. Если при каком-либо знаке константы связи λ спектр является непрерывным, то ряд теории возмущений для пропагатора расходится, причем общий член ряда имеет вид $n!\,(\alpha\lambda)^n$ для больших n.

1. Дайсон в работе [1] привел аргументы в пользу того, что ряды теории возмущений в квантовой электродинамике являются расходящимися. Он основывался на том, что мир, в котором квадрат заряда e^2 отрицателен, не имеет основного состояния и распадается. Поэтому трудно себе представить, что такая ситуация может описываться функциями аналитичными по e^2 в точке $e^2 = 0$.

Тирринг [2] исследовал теорию с взаимодействием $L_{\text{int}} = \lambda\varphi^3$ и показал, что ряд теории возмущений для поляризатсионного оператора расходится в области импульсов $p^2 < m^2$. При больших n члены ряда имеют вид

$$C(\alpha\lambda)^n \frac{(n-4)!}{n^2},$$

где C и α — функции p^2. Рассмотренная Тиррингом модель является примером неустойчивой теории. С помощью прямого вариационного метода легко показать, что в модели нет нижнего состояния [3].

Мы покажем, что распадность системы приводит к расходимости ряда теории возмущений в одномерной модели.

2. Рассмотрим модель, в которой полевые операторы φ зависят только от времени, то есть нет пространственных координат. Гамильтониан и одновременные перестановочные соотношения имеют вид

$$H = \frac{1}{2}\,(\dot\varphi)^2 + \frac{m^2}{2}\varphi^2 + V(\varphi)\,, \qquad [\varphi(t)\,, \dot\varphi(t)] = i\,. \tag{1}$$

Это — гамильтониан и перестановочные соотношения обычного квантово-механического нелинейного осциллятора с частотой $\omega = m$ и массой $\mu = 1$.

Взаимодействие $V(\varphi)$ возьмём для определенности в виде

$$V(\varphi) = -\lambda\varphi^3 \,. \tag{2}$$

Из дальнейшего будет видно, что рассмотрение пригодно для всех распадных взаимодействий.

Повторяя доказательство Тирринга [2] в одномерном случае для взаимодействия $\lambda\varphi^3$, мы придём к такому же, как и в четырехмерном варианте, результату, то-есть, что ряд для поляризационного оператора расходится, причем таким же образом. Мы свяжем это с распадностью системы.

Будем рассматривать причинную функцию Грина поля φ. В представлении взаимодействия она определяется как

$$iG(\tau) = \frac{(0|T\,\varphi(\tau)\,\varphi(0)|0)}{S_{00}} \tag{3}$$

где $\varphi(\tau)$ — полевой оператор в представлении взаимодействия. Усреднение идет по математическому вакууму. Предполагается адиабатическое включение.

Если мы перейдём к гейзенберговским операторам $\underline{\phi}(\tau)$, то получим

$$iG(\tau) = \frac{(0|S(\infty,0)[T\,\underline{\phi}(\tau)\,\underline{\phi}(0)]S(0,-\infty)|0)}{(0|S(\infty,0)\,S(0,-\infty)|0)} \,,$$

$$\underline{\phi}(\tau) = S^+(\tau,0)\varphi(\tau)S(\tau,0) \,. \tag{4}$$

3. Обычно состояние $|\psi\rangle = S(0,-\infty)|0)$ считают равным физическому вакууму. Если физический вакуум существует, то это обеспечивается адиабатическим включением взаимодействия. В рассматриваемой модели физического вакуума нет, система неустойчива. В таких случаях математический вакуум при адиабатическом включении взаимодействия переходит в соответствующий квазиуровень — состояние с комплексной энергией, описывающее распад. Чтобы показать это мы рассмотрели задачу об осцилляторе, у которого частота менялась со временем как

$$\omega^2\left(1 - \gamma e^{-\alpha|t|}\right) \,.$$

Если $\gamma > 1$, то при $t = 0$ осциллятор был перевернут, и физический вакуум отсутствовал. Действительно, оказалось, что если при $t \to -\infty$ мы имели основное состояние, то при $t = 0$

мы приходим к состоянию, которые в пределе $\alpha \to 0$ имеет энергию

$$E = -\frac{i\omega}{2}\sqrt{\gamma - 1}\,,$$

и описывает распад. Подробное решение дано в приложении А.

Интересно отметить, что состояние $(0|S(\infty, 0) = \langle\psi|$ не получается эрмитовым сопряжением из $S(0, \infty)|0\rangle$, что связано с невыполнением условия устойчивости $S(\infty, -\infty)|0\rangle = |0\rangle$. Состояние $\langle\psi| = (0|S(\infty, 0)$ является эрмитово сопряженным к состоянию, описывающему процесс, обратный к распаду. Энергия такого состояния комплексна сопряжена к энергии квазиуровня. Такое состояние мы в дальнейшем будем называть антиквазиуровнем.

4. В приложении В выведена связь между энергией состояния $|\psi\rangle$ и $G(\tau)|_{\tau=0}$ и $G(p)|_{p=0}$. Здесь $G(p)$ — Фурье-образ $G(\tau)$. Поэтому для изучения аналитических свойств $G(\tau)|_{\tau=0}$ и $G(p)|_{p=0}$, как функции константы связи, достаточно это сделать для $E(\lambda^2)$, где $E(\lambda^2)$ — энергия состояния $|\psi\rangle = S(0, -\infty)|0\rangle$. Уравнение для $|\psi\rangle$ таково

$$H\,|\psi\rangle = E\,|\psi\rangle\,, \qquad H = \frac{\dot{\varphi}^2}{2} + \frac{m^2\varphi^2}{2} - \lambda\varphi^3\,. \tag{5}$$

Это — обычное дифференциальное уравнение нелинейного осциллятора. При $\varphi \to -\infty$ волновая функция $\psi(\varphi)$ экспоненциально падает, а при $\varphi \to \infty$ имеется только выходящая волна. (Мы считаем $\lambda > 0$).

Продолжим $\psi(\varphi)$, определённую для положительных λ, на комплексные λ. Тогда в (5) λ комплексно. Разберёмся, что будет с граничными условиями.

При положительных λ

$$\psi(\varphi) \to \frac{C}{\sqrt{p}}\exp\left[i\int^{\varphi} p\,d\varphi\right]\,, \quad \varphi \to +\infty\,,$$

$$\psi(\varphi) \to \frac{C'}{\sqrt{p}}\exp\left[-i\int^{\varphi} p\,d\varphi\right]\,, \quad \varphi \to -\infty\,,$$

$$p = \sqrt{2E - m^2\varphi^2 + 2\lambda\varphi^3}\,. \tag{6}$$

Это — известные квазиклассические асимптотики. Можно утверждать, что $\psi(\varphi)$ имеет эти же асимптотики и для всех комплексных λ в верхней полуплоскости параметра λ. Действительно, пока λ находится в верхней полуплоскости, $\psi(\varphi)$ экспоненциально падает при $\varphi \to \pm\infty$, и растущая экспонента не может появиться.

Перейдём в плоскости λ с положительной полуоси на отрицательную через верхнюю полуплоскость. Тогда из (6) видно, что после поворота $\psi(\varphi)$ экспоненциально падает при $\varphi \to +\infty$, а при $\varphi \to -\infty$ представляет собой волну, бегущую в яму. То-есть, начав с задачи о квазиуровне, мы пришли к задаче об антиквазиуровне. Это означает, что функция $E(\lambda^2)$ имеет в плоскости λ^2 разрез по положительной полуоси. На этом разрезе терпит скачок $\mathrm{Im}E(\lambda^2)$.

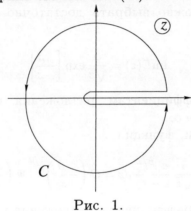

Рис. 1.

Рассмотрим интеграл

$$\int_C dz\, \frac{E(z)}{z - z_0} = 2\pi i\, E(z_0), \qquad z = \lambda^2, \tag{7}$$

где контур C показан на рис. 1. Радиус окружности на рис. 1 равен Δ, $|z_0| < \Delta$. Интеграл по окружности является аналитической функцией для всех z_0 внутри окружности, поэтому мы не будем его рассматривать, так как нас интересует неаналитическая в нуле часть $E(z)$,

$$E(z_0) = \frac{1}{\pi} \int_0^\Delta dz\, \frac{\mathrm{Im}\, E(z)}{z - z_0}. \tag{8}$$

Из (8) сразу следует разложение $E(z_0)$ в ряд при $z_0 \to 0$,

$$E(z_0) = \sum_{n=0}^\infty z_0^n\, \left(\frac{1}{\pi} \int_0^\Delta dz\, \frac{\mathrm{Im}\, E(z)}{z^{n+1}} \right). \tag{9}$$

Интегралы сходятся, так как $\operatorname{Im} E(z)$ экспоненциально убывает при $z \to 0$. Действительно, $\operatorname{Im} E(\lambda^2)$ при $\lambda \to 0$ пропорциональна коэффициенту прохождения через барьер

$$D = \exp\left[-2\int_{\varphi_1}^{\varphi_2} d\varphi\ \sqrt{m^2\varphi^2 - 2\lambda\varphi^3 - 2E}\right].$$

Интеграл берется между двумя точками поворота. При $\lambda \to 0$

$$D = \frac{\alpha'}{\sqrt{\lambda^2}}\exp\left[-\frac{\beta' m^5}{\lambda^2},\right], \tag{10}$$

где α' и β' — константы.

Так как Δ можно выбрать достаточно малым, то подстановка в (9)

$$\operatorname{Im} E(z) = \frac{\alpha}{\sqrt{z}}\exp\left[-\frac{\beta}{z}\right]$$

приведёт к коэффициентам разложения совпадающим с точными при $n \to \infty$.

Таким образом, функция

$$\overline{E}(z_0) = \frac{\alpha}{\pi}\int_0^\infty \frac{1}{\sqrt{z}}\frac{e^{-\beta/z}}{z - z_0}\,dz = C\left(-\frac{\beta}{z_0}\right)^{1/2}\Psi\left(\frac{1}{2},\frac{1}{2},-\frac{\beta}{z_0}\right) \tag{11}$$

имеет такие же коэффициенты разложения в пределе $n \to \infty$, как и точная $E(z)$. (Добавление интеграла от Δ до ∞ не имеет значения, так как этот интеграл даёт функцию аналитическую в точке $z_0 = 0$. Напомним, что $\Psi\left(\frac{1}{2},\frac{1}{2},x\right)$ — вырожденная гипергеометрическая функция).

Разложение $\overline{E}$ имеет вид

$$\overline{E}(z) = C\sum_{n=0}^\infty \left(\frac{z}{\beta}\right)^n \Gamma\left(n + \frac{1}{2}\right). \tag{12}$$

Так как $E(\lambda^2)$ связана с функцией Грина соотношениями (см. приложение B)

$$iG(\tau)\big|_{\tau=0} = \frac{1}{m^2}\left[E - 5\lambda^2\frac{\partial E}{\partial\lambda^2}\right],$$

$$\tilde{G}(p)\Big|_{p=0} = -\frac{1}{m^2} + \frac{9\lambda^2}{m^8}\left[25\,(\lambda^2)^2\,\frac{\partial^2 E}{\partial(\lambda^2)^2} + 35\lambda^2\frac{\partial E}{\partial\lambda^2} - 3E\right], \tag{13}$$

то мы приходим к выводу, что ряд теории возмущений расходится, причем коэффициенты разложения факториально растут при больших n. Тем не менее, ряд является асимптотическим.

5. Совершенно аналогичным способом можно рассмотреть другие типы взаимодействий. Понятно, что если при включении взаимодействия система распадается, то мнимая часть энергии квазиуровня есть величина экспоненциально малая при константе связи стремящейся к нулю, что приводит к факториальному росту коэффициентов ряда по степеням константы связи. Потенциального барьера не будет, если $m = 0$. Но в этом случае интегралы теории возмущений расходятся на нижнем пределе.

Представляет интерес взаимодействие $V = -\lambda\varphi^4$. При $\lambda > 0$ оно соответствует распадающейся системе. Проделав такое же как и для $\lambda\varphi^3$ рассмотрение, мы получим что при больших n члены ряда для $E(\lambda)$ ведут себя как

$$ C\left(\frac{\lambda}{\gamma}\right)^n \Gamma\left(n + \frac{1}{2}\right). $$

При $\lambda < 0$ система устойчива. Но ряд будет общим и для положительных λ и для отрицательных. Ситуация возникает такого типа, которую предполагает Дайсон для квантовой электродинамики. Интересно, что все члены ряда имеют один знак в области неустойчивости.

Повидимому, большинство нетривиальных теорий являются неустойчивыми при какой либо фазе константы связи, что приводит к асимптотичности рядов. Уравнения теорий имеют решения неаналитичные по константе связи в нуле, правда, неясно, насколько эти решения физичны в области неустойчивости. Точка $\lambda = 0$ является в таких теориях точкой ветвления и, если $m \neq 0$, существенно особой точкой.

Остаётся неясным вопрос о связи распадности с расходимостью ряда в четырехмерных теориях, хотя надо отметить, что, если играют роль малые пространственные импульсы, то теория становися одномерной, однако требует выяснения влияние перенормировок. Интересно, что если рассмотреть произвольную диаграмму четырехмерной теории с внешними импульсами равными нулю, то легко получить неравенство

$$\int \frac{1}{m^2 - p_1^2} \cdots \frac{1}{m^2 - p_i^2} \prod d^4 q$$

$$\geq \frac{1}{m^{6i}} \left[\int \frac{1}{m^2 - (p_1^0)^2} \cdots \frac{1}{m^2 - (p_i^0)^2} \prod dq^0 \right]^4, \qquad (14)$$

где p_k — импульсы внутренних линий, q — импульсы интегрирования, p_k^0 — временные компоненты. Тогда в правой части (14) стоит соответствующая одномерная диаграмма, и мы могли бы поставить миноранту для четырехмерной теории, если бы не было необходимости перенормировок.

Приношу глубокую благодарность В.М. Галицкому за предложение темы и руководство работой. Автор благодарен И.Б. Хрипловичу за ценные дискуссии.

Приложение А

Рассмотрим осциллятор у которого частота меняется со временем по закону

$$\omega^2 \left(1 - \gamma\, e^{-\alpha|t|}\right).$$

Нас интересует развитие во времени состояния, которое при $t \to -\infty$ стремится к вакууму осциллятора с частотой ω. Обозначим его через $\Psi_\alpha(t)$. В представлении взаимодействия $\Psi_\alpha(t)$ удовлетвотяет уравнению

$$i\frac{\partial \Psi_\alpha}{\partial t} = -\frac{\gamma\omega^2}{4}\, e^{-\alpha|t|}\, x^2(t)\Psi_\alpha\,, \tag{A.1}$$

$$x(t) = \frac{1}{\sqrt{2}}\left[a^+(t) + a^-(t)\right] = \frac{a^+ e^{i\omega t} + a^- e^{-i\omega t}}{\sqrt{2}}\,, \tag{A.2}$$

где

$$[a^+, a^-] = -1\,.$$

Ищем $\Psi_\alpha(t)$ в виде

$$\Psi_\alpha(t) = K_\alpha(t)\exp\left[(a^+)^2 f_\alpha(t)\right]|0\rangle \tag{A.3}$$

где $K_\alpha(t)$ и $f_\alpha(t)$ — функции времени, $|0\rangle$ — вакуум осциллятора с частотой ω, так что $a|0\rangle = 0$. Подставляя (A.3) в (A.1) и производя коммутации, мы получим члены с $\exp((a^+)^2 f_\alpha)|0\rangle$ и с $(a^+)^2\exp((a^+)^2 f_\alpha)|0\rangle$. Приравняв коэффициенты при них нулю, мы придём к уравнениям

$$i\frac{K'_\alpha}{K_\alpha} = -\frac{\gamma\omega}{4}\, e^{-\alpha|t|}\left(1 + 2f_\alpha\right), \tag{A.4}$$

$$if'_\alpha = -\frac{\gamma\omega}{4}\, e^{-\alpha|t|}\left(e^{2i\omega t} + 4f_\alpha + 4f_\alpha^2 e^{-2i\omega t}\right), \tag{A.5}$$

с граничным условием

$$f_\alpha(t) \to 0 \quad \text{at} \quad t \to -\infty\,.$$

Будем рассматривать $t < 0$ и введем новую функцию $y(t)$,

$$f_\alpha(t) = -\frac{1}{i\gamma\omega}\, e^{(2i\omega - \alpha)t}\, \frac{y'(t)}{y(t)} - \frac{1}{i\gamma\omega}\left(-i\omega + \frac{\alpha}{2}\right) e^{(2i\omega - \alpha)t} - \frac{1}{2} e^{2i\omega t}. \tag{A.6}$$

Для $y(t)$ получим уравнение

$$y'' + \left[\left(\omega + \frac{i\alpha}{2} \right)^2 - \gamma\omega^2\, e^{\alpha t} \right] y = 0\,. \tag{A.7}$$

Его решение таково

$$y(t) = C_1 J_\nu(z) + C_2 J_{-\nu}(z)\,, \tag{A.8}$$

где

$$\nu = \frac{2i\omega}{\alpha} - 1\,, \qquad z = \frac{2i\omega\sqrt{\gamma}}{\alpha}\, e^{\alpha t/2}\,,$$

и $J_\nu(z)$ — функция Бесселя.

Используя обращение $f_\alpha(t)$ в нуль при $t \to -\infty$, найдем

$$f_\alpha(t) = -\frac{1}{i\gamma\omega}\, e^{(2i\omega - \alpha)t} \left[-i\omega + \frac{\alpha}{2} + \frac{\lambda\omega}{2}\, e^{\alpha t} + \frac{d(J_\nu(z))/dt}{J_\nu(z)} \right]. \tag{A.9}$$

Нас интересует

$$\lim_{\alpha \to 0} \Psi_\alpha(0) = \Psi(0)\,.$$

Воспользовавшись квазиклассическими асимптотиками функции Бесселя [4], найдём

$$f(0) = \frac{1}{\gamma} \left[1 - \frac{\gamma}{2} + i\sqrt{\gamma - 1} \right]\,, \quad \gamma > 1\,,$$

$$f(0) = \frac{1}{\gamma} \left[1 - \frac{\gamma}{2} - \sqrt{1 - \gamma} \right]\,, \quad \gamma < 1\,. \tag{A.10}$$

При $\gamma < 1$ мы приходим к основному состоянию осциллятора с частотой $\omega\sqrt{1 - \gamma}$ — "физическому вакууму". При $\gamma > 1$

$$\Psi(0) = \exp\left((a^+)^2 f(0) \right) |0\rangle$$

в x-представлении имеет вид

$$\Psi(0) = \exp\left(i\omega\sqrt{\gamma - 1}\, \frac{x^2}{2} \right)\,. \tag{A.11}$$

Это состояние описывает разлетание частиц из области начала координат на $\pm\infty$. Его энергия равна

$$E = -\frac{i\omega}{2}\sqrt{\gamma - 1}\,.$$

Мы видим, что $n = E/\omega$ является адиабатическим инвариантом и для комплексных ω.

Если рассмотреть состояние, которое при $t \to \infty$ переходит в вакуум осциллятора с частотой ω, то получим, что при $t = 0$ и $\alpha \to 0$ мы имеем состояние антиквазиуровня с

$$E = \frac{i\omega}{2}\sqrt{\gamma - 1}\,.$$

Приложение B

Функция Грина в гейзенберговском представлении определяется как

$$iG(\tau) = \frac{\langle\psi|\underline{\phi}(\tau)\underline{\phi}(0)|\psi\rangle}{\langle\psi|\psi\rangle}\,. \qquad (\text{B.1})$$

Связь между $G(\tau)|_{\tau=0}$ и энергией состояния $|\psi\rangle$ известна [5]. Но для полноты изложения мы её выведем. Уравнение для $|\psi\rangle$ таково

$$(H - E)\,|\psi\rangle = 0\,, \qquad H = \frac{1}{2}\,\dot{\varphi}^2 + \frac{m^2}{2}\,\varphi^2 - \lambda\varphi^3\,. \qquad (\text{B.2})$$

Дифференцируя (B.2) по m^2 и умножая на $\langle\psi|$ слева, получим

$$iG(\tau)|_{\tau=0} = \frac{\langle\psi|\underline{\phi}^2(0)|\psi\rangle}{\langle\psi|\psi\rangle} = 2\,\frac{\partial E}{\partial m^2}\,. \qquad (\text{B.3})$$

Из размерных соображений

$$E = m\,\Phi\left(\frac{\lambda^2}{m^5}\right)\,.$$

Поэтому

$$iG(\tau)|_{\tau=0} = \frac{1}{m^2}\left(E - 5\lambda^2\frac{\partial E}{\partial \lambda^2}\right)\,. \qquad (\text{B.4})$$

Выведем теперь связь E и $G(p)|_{p=0}$, где

$$G(\tau) = \int_{-\infty}^{\infty}\frac{dp}{2\pi}\,G(p)\,e^{-ip\tau}\,.$$

Прежде всего отметим, что при взаимодействии $\lambda\varphi^3$

$$\overline{\varphi} = \frac{\langle\psi|\underline{\phi}(0)|\psi\rangle}{\langle\psi|\psi\rangle} \neq 0\,.$$

Поэтому в $G(\tau)$ есть постоянная по τ часть, не имеющая физического смысла, которая в p-представлении даёт $\delta(p)$. Правильнее рассматривать

$$i\tilde{G}(\tau) = \frac{T\,\langle\psi|(\underline{\phi}(\tau) - \overline{\varphi})\,(\underline{\phi}(0) - \overline{\varphi})|\psi\rangle}{\langle\psi|\psi\rangle} = iG(\tau) - \overline{\varphi}^2\,. \tag{B.5}$$

Используя определение $\underline{\phi}(\tau)$ через шредингеровское φ,

$$\underline{\phi}(\tau) = e^{iH\tau}\,\varphi\,e^{-iH\tau}$$

имеем

$$\tilde{G}(p) = \frac{1}{\langle\psi|\psi\rangle}\left\langle\psi\left|(\varphi - \overline{\varphi})\left[\frac{1}{p - (H - E - i\epsilon)}\right.\right.\right. \tag{B.6}$$

$$\left.\left.\left. - \frac{1}{p + (H - E - i\epsilon)}\right](\varphi - \overline{\varphi})\right|\psi\right\rangle,$$

$$\left.\tilde{G}(p)\right|_{p=0} = -\frac{2}{\langle\psi|\psi\rangle}\left\langle\psi\left|(\varphi - \overline{\varphi})\frac{1}{H - E}(\varphi - \overline{\varphi})\right|\psi\right\rangle. \tag{B.7}$$

заметим, что $i\epsilon$ можно опустить, так как

$$\langle\psi\,|\varphi - \overline{\varphi}|\,\psi\rangle = 0\,. \tag{B.8}$$

Введем в гамильтониан член $f\varphi$, где f — параметр. Продиффенцируем уравнение (В.2) дважды по параметру f,

$$(H - E)\frac{\partial|\psi\rangle}{\partial f} + \left(\frac{\partial H}{\partial f} - \frac{\partial E}{\partial f}\right)|\psi\rangle, \tag{B.9}$$

$$(H - E)\frac{\partial^2|\psi\rangle}{\partial f^2} + 2\left(\frac{\partial H}{\partial f} - \frac{\partial E}{\partial f}\right)\frac{\partial|\psi\rangle}{\partial f} - \frac{\partial^2 E}{\partial f^2}|\psi\rangle. \tag{B.10}$$

Умножая на $\langle\psi|$ слева, получим

$$\frac{\partial E}{\partial f} = \frac{1}{\langle\psi|\psi\rangle}\left\langle\psi\left|\frac{\partial H}{\partial f}\right|\psi\right\rangle = \overline{\varphi}, \tag{B.11}$$

$$\frac{\partial^2 E}{\partial f^2} = \frac{2}{\langle\psi|\psi\rangle}\left\langle\psi\left|\frac{\partial H}{\partial f} - \frac{\partial E}{\partial f}\right|\frac{\partial\psi}{\partial f}\right\rangle = \overline{\varphi}. \tag{B.12}$$

Из (В.9) находим $\partial|\psi\rangle/\partial f$,

$$\frac{\partial|\psi\rangle}{\partial f} = -\frac{1}{H - E}\left(\frac{\partial H}{\partial f} - \frac{\partial E}{\partial f}\right)|\psi\rangle. \tag{B.13}$$

К (В.13) можно прибавить с произвольным коэффициентом $|\psi\rangle$, но при подстановке в (В.12) эта добавка даст нуль,

$$\frac{\partial^2 E}{\partial f^2} = -\frac{2}{\langle\psi|\psi\rangle} \left\langle \psi \left| (\varphi - \overline{\varphi}) \frac{1}{H - E} (\varphi - \overline{\varphi}) \right| \psi \right\rangle . \tag{В.14}$$

Теперь f можно положить равным нулю,

$$\frac{\partial^2 E}{\partial f^2} = \tilde{G}(p)\big|_{p=0} . \tag{В.15}$$

Связь (В.15) можно записать через производные по λ^2. Для этого вводим вместо φ новый оператор η,

$$\varphi = \eta + \varphi_0 , \qquad \varphi_0 = \frac{m^2 - \sqrt{m^4 + 12\lambda f}}{6\lambda} . \tag{В.16}$$

Тогда гамильтониан не содержит линейного по η члена, и можно записать

$$E = E^0 + M\Phi\left(\frac{\lambda^2}{M^5}\right) , \tag{В.17}$$

$$E^0 = \frac{m^2\varphi_0^2}{2} + f\varphi_0 - \lambda\varphi_0^3 , \qquad M^2 = m^2 - 6\lambda\varphi_0 . \tag{В.18}$$

Тогда (В.15) перейдёт в

$$\tilde{G}(p)\big|_{p=0} = -\frac{1}{m^2} + \frac{9\lambda^2}{m^8} \left[25\,(\lambda^2)^2\,\frac{\partial^2 E}{\partial(\lambda^2)^2} + 35\lambda^2\frac{\partial E}{\partial\lambda^2} - 3E \right] . \tag{В.19}$$

Литература

1. F.J. Dyson, *Phys. Rev.* **85**, 631 (1952).

2. W. Thirring, *Helv. Phys. Acta* **26**, 33 (1953).

3. G. Baym, *Phys. Rev.* **117**, 886 (1960).

4. А.З. Паташинский, В.Л. Покровский, И.М. Халатников, ЖЭТФ, **4**, 2062 (1961).

5. В.М. Галицкий, А.Б. Мигдал, ЖЭТФ, **34** , 139 (1958).

Ответственный за выпуск — И.Б. Хриплович
Подписано к печати МН00663 2.12.1964
Формат бумаги 270 × 290, тираж 150
Заказ № 052. Бесплатно

Отпечатано на ротапринте в Институте ядерной физике СО АН СССР.